SMART ACTUATION AND SENSING SYSTEMS – RECENT ADVANCES AND FUTURE CHALLENGES

Edited by **Giovanni Berselli**, **Rocco Vertechy** and **Gabriele Vassura**

Smart Actuation and Sensing Systems – Recent Advances and Future Challenges
http://dx.doi.org/10.5772/2760
Edited by Giovanni Berselli, Rocco Vertechy and Gabriele Vassura

Contributors

A. Spaggiari, G. Scirè Mammano, E. Dragoni, Adelaide Nespoli, Carlo Alberto Biffi, Riccardo Casati, Francesca Passaretti, Ausonio Tuissi, Elena Villa, William Coral, Claudio Rossi, Julian Colorado, Daniel Lemus, Antonio Barrientos, Simone Pittaccio, Stefano Viscuso, Carmine Maletta, Franco Furgiuele, Daniele Davino, Alessandro Giustiniani, Ciro Visone, Thomas G. McKay, Benjamin M. O'Brien, Iain A. Anderson, P. Maiolino, A. Ascia, M. Maggiali, L. Natale, G. Cannata, G. Metta, Doan Ngoc Chi Nam, Ahn Kyoung Kwan, Zheng Chen, T. Um, Hilary Bart-Smith, Weihua Li, Tongfei Tian, Haiping Du, José G. Martínez, Joaquín Arias-Pardilla, Toribio F. Otero, Yusuke Hara, Shingo Maeda,Takashi Mikanohara, Hiroki Nakagawa, Satoshi Nakamaru, Shuji Hashimoto, Quoc-Hung Nguyen, Seung-Bok Choi, D. Q. Truong, Makoto Nokata, Laura Rodríguez-Arco, Ana Gómez-Ramírez, Juan D.G. Durán, Modesto T. López-López, Enrico Zenerino, Joaquim Girardello Detoni, Diego Boero, Andrea Tonoli, Marcello Chiaberge, Angelo Bonfitto, Mario Silvagni, Lester D. Suarez, Qining Wang, Jinying Zhu, Yan Huang, Kebin Yuan, Long Wang, I. Gaiser, R. Wiegand, O. Ivlev, A. Andres, H. Breitwieser, S. Schulz, G. Bretthauer, Lucia Seminara, Luigi Pinna, Marco Capurro, Maurizio Valle, Yasuhide Shindo, Fumio Narita, A. Paternoster, R. Loendersloot, A. de Boer, R. Akkerman, Ipek Basdogan, Utku Boz, Serkan Kulah, Mustafa Ugur Aridogan, Marcus Neubauer, Sebastian M. Schwarzendahl, Xu Han

Published by InTech

Janeza Trdine 9, 51000 Rijeka, Croatia

Publishing Process Manager Dragana Manestar
Technical Editor InTech Prepress, Novi Sad
Cover InTech Design team

Edition 2014

Additional hard copies can be obtained from orders@intechopen.com

Smart Actuation and Sensing Systems – Recent Advances and Future Challenges,
Edited by Giovanni Berselli, Rocco Vertechy and Gabriele Vassura
p. cm.
ISBN 978-953-51-0798-9

For EU product safety concerns:
IN TECH d.o.o., Prolaz Marije Krucifikse Kozulić 3, 51000 Rijeka, Croatia,
info@intechopen.com or visit our website at intechopen.com.

Meet the editors

Giovanni Berselli received the Laurea degree (cum laude) in mechanical engineering from the University of Modena and Reggio Emilia, Modena Italy, and the Ph.D. degree in the mechanics of machines from the University of Bologna, Bologna, Italy, in 2004 and 2009, respectively. He is currently an Assistant Professor with the University of Modena and Reggio Emilia. He was a Research Assistant with the Department of Mechanical Engineering, Monash University, Melbourne, Australia, and with Centros de Estudios y Investigaciones Tecnica de Gipuzlcoa (CEIT), Escuela Superior de Ingenieros de la Universidad de Navarra, San Sebastian, Spain. His research interests include the design and control of compliant robotic systems and smart actuators, dexterous robotic hands, and engineering methods and tools for automated systems.

Rocco Vertechy received the Laurea degree (cum laude) in mechanical engineering and the Ph.D. degree in mechanics of machines from the University of Bologna, Bologna, Italy, in 2001 and 2005, respectively. He is currently an Assistant Professor with the Perceptual Robotics Laboratory, Scuola Superiore Sant'Anna, Pisa, Italy. He was a Research Assistant with the Department of Mechanical Engineering, University of Canterbury, Christchurch, New Zealand, and a Visiting Researcher with the Robotics Locomotion Laboratory, Stanford University, Stanford, C A. He was engaged in research on the kinetostatic analysis and synthesis of parallel manipulators. His current research interests include the design and control of robotic systems which physically interact with both humans and unknown constrained environments, as well as of solid-state mechatronic devices which are safe, affordable, and disposable.

Gabriele Vassura received the Laurea degree (cum laude) in mechanical engineering from the University of Bologna, Bologna, Italy, in 1972. He is currently an Associate Professor at DIEM, Mechanical Engineering Department, University of Bologna. His teaching activity is in the field of machine design, mainly focused on mechanical design of automatic machines and robots. His main interests in research are the design of dexterous robotic hands, simulation of pneumatic actuation systems, and design methodologies for automated systems.

Contents

Preface

In the last few decades, much effort has been directed towards the development of mechatronic devices capable of interacting safely and effectively with unstructured environments and humans. On one hand, these research activities highlighted the limits of traditional sensorymotor technologies in terms of flexibility and responsiveness to ever changing scenarios. On the other hand, the fascinating world of *smart structures and materials,* which is somehow the most natural engineering answer to the challenge of adaptability, is still far from meeting strict industrial requirements such as reliability, damage-tolerance, ease-of-usage and cost-effectiveness. In particular, even if it is possible to envisage futuristic solid-state machines with unconventional morphing shapes, it would be too presumptuous to say that every *smart* devices have already transitioned from basic research to practically useful and well-engineered products.

Trivially speaking, a device might be called *smart* if it can sense and respond to the surrounding environment in a predictable and useful manner via the integration of an actuation system, a network of proprioceptive and exteroceptive sensors, and a suitable controller. Such devices, possibly powered with a minimum amount of energy, usually include one or more *smart materials* which exhibit some coupling between multiple physical domains (e.g. piezoelectric materials, shape memory alloys, magento/electro rheological fluids). There are instances where the breakthrough from proof-of-concept laboratory rigs into commercial applications has already seen the light. For example, piezo-actuators and sensors are state-of-the-art technology. In the same way, shape-memory-alloys are widely used in many biomedical applications. In other cases, such as magneto/electro-active polymer actuators and generators, the technology is rather new and its potential may not be fully exploited at the current level of knowledge.

Regardless of the aforementioned considerations, the effort towards the technical maturity of any *smart* device requires the combined action of different research fields ranging from material science, mechanical and electrical engineering, chemistry and physics. Hence, it is strongly believed that the tremendous growth of research and industrial projects concerning *smart* systems in the last 20 years has been principally due to the synergistic cooperation of universities, government institutions and industries and to the birth of under- and post-graduate courses where a multidisciplinary approach is now a *de-facto* standard.

Therefore, if a path towards the future has been traced and if interdisciplinarity is the key to success, it is surely valuable to combine researchers and scientists from different fields into a single virtual room. That is indeed the objective of the present book, which tries to summarize in an edited format and in a fairly comprehensive manner, many of the recent technical research accomplishments in the area of *Smart Actuators* and *Smart Sensors*. Current and future challenges for the optimal design, modelling, control and technological implementation of the next-generation adaptive mechatronic systems are treated with the objective to provide a reference point on the current state-of-the-art, to propose future research activities and to stimulate new ideas. As long as the authorship is taken from disparate disciplines, the book hopefully reflects the multicultural nature of the field and will allow the reader to taste and appreciate different points of view, different engineering methods and different tools that must be jointly considered when designing and realizing *smart actuation and sensing systems*.

Giovanni Berselli, Ph.D.
Interdepartmental Center INTERMECH Mo.Re.
DIEF – "Enzo Ferrari" Engineering Department
University of Modena and Reggio Emilia
Modena,
Italy

Section 1

SMA-Based Systems

Chapter 1

Optimum Mechanical Design of Binary Actuators Based on Shape Memory Alloys

A. Spaggiari, G. Scirè Mammano and E. Dragoni

Additional information is available at the end of the chapter

http://dx.doi.org/10.5772/50147

1. Introduction

Shape memory alloys are smart materials which have the ability to return to a memorized shape when heated. When an SMA is below its transformation temperature (martensitic phase), it has a low yield strength and can be deformed quite easily and behaves like a pseudoplastic solid. When the deformed material is heated above its transformation temperature there is a change in its crystal structure which causes the return to its original shape (austenitic phase). During this transformation the SMA element can generate a net force, behaving like an intrinsic actuator. The most common shape memory material is a nickel and titanium alloy called Nitinol [1]. SMAs have very good electrical and mechanical properties, high corrosion resistance and biocompatibility. When an electric current is injected in the SMA element, it can generate enough heat to cause the phase transformation due to joule effect. Thanks to their unique behaviour shape memory alloys have become a valuable industrial choice in the engineering world. Pseudo-plasticity, superelasticity, and shape memory effect [2] are increasingly used in many applications including actuators, constant-force springs, and adaptive damping systems. While the application of the superelastic effect is quite well established and understood for the manufacturing of medical devices with peculiar properties, the use of the shape memory effect for building solid state actuators is still characterized by a trial-and-error approach. Although the thermo mechanical phenomena behind the behaviour of SMAs are theoretically well known [3] - [4], there is an open challenge for engineering methods to assist the designer in exploiting these alloys for the development of industrial devices.

1.1. Design methodology review for SMA actuators

Many SMAs applications have been studied in recent years. Kuribayashi [5] proposes a rotary joint based on a bending SMA actuator, while, in [6] design and applications of SMA

actuators are presented. Microrobots can be developed using SMA as shown in [7] where there is a basic method to design the SMA spring based on a thermo-electromechanical approach. Reynaerts et al. [8] present design considerations concerning the choice of the active element to evaluate SMA actuator efficiency. Lu et al. [9] design a high strain shape memory actuator taking into account pseudoplasticity and compared its performance with traditional actuators. Due to SMA high non linearity, design curves relying on experiments are proposed in [10] to assess the SMA actuator geometry. A comprehensive review of applications of SMAs in the field of mechanical actuation has recently been published in [11]. Jansen et al [12] develop a linear actuator used as a drive module in an angular positioning mini-actuator. This architecture allows both large force and long strokes to be obtained. Strittmatter et al. [13] propose a SMA actuator for the activation of a hydraulic valve, biased by a conventional spring. Bellini et al. [14] propose a linear SMA actuator able to vary the air inflow for internal combustion engines, improving gas combustion and leading to higher efficiency. Haga et al. [15] propos a mini-actuator to be used in Braille displays. Elwaleed et al. [16] develop a SMA beam actuator able to amplify the SMA actuator strain using elastically instable beams. Among the proposed actuator there is lack of simple design instruments to provide basic information to the designer, either due to specific constraints of application or due to the high complexity of the thermomechanical material models used. In order to answer for an analytical design methodology, the author described in several technical publications a set of equations useful for linear [18] and rotary application [19]. The authors developed the design equations both for SMA actuators under a general system of external forces [20] and produced the design formulae to increase the output stroke thanks to negative stiffness compensators [21]. Moreover two peculiar systems were designed and developed: a telescopic actuator [22] and a wire on drum system [23]. The present work reviews and improves the design rules developed by the authors and set them in a coherent formal analytical framework. Design examples are provided to illustrate the step-by-step application of the design optimization procedures for realistic case studies.

1.2. Challenges and issues in SMA actuators design

The three main challenges in SMA actuator design are: obtaining a simple and reliable material model, increasing the stroke of the actuator and finding design equations to guide the engineer in dimensioning the actuator. To design SMA actuators, a material model must describe the mechanical behaviour of the alloy in two temperature ranges: below the temperature M_f, at which the austenite-martensite transformation is finished (OFF or deactivated or cold state) and above the temperature, A_f, at which the martensite-austenite transformation is completed (ON or activated or hot state). In these two conditions, inside the shape memory material there is only one stable crystalline phase and therefore the macroscopic mechanical properties are known.

The authors proposed two simple material models to describe SMA behaviour, both of them describing the mechanical behaviour of a SMA element at high temperatures (austenitic phase) as linear, characterized by the elastic modulus E_A. The first one approximates the martensitic

behaviour as a linear one [18] (Figure 1a). This behaviour is by no means obvious, thus an explanation is needed. The typical stress-strain curve of a shape memory material, shown in Figure 1a, has two characteristic paths corresponding to a low temperature (martensitic curve) and to a high temperature (austenitic curve). Both curves consist of an initial elastic portion (OA and OC), followed by a constant-stress plateau. In the practical use of shape memory alloys for making actuators, the materials remain within the linear elastic range at the higher temperature but are strained beyond the elastic limit when cooled to the lower temperature. The maximum strain, ε_{adm} is small enough to ensure the desired fatigue life [24-25] but large enough to maximize the stroke of the actuator for a given amount of material involved. Since this analysis is aimed at dimensioning the binary actuator for the extreme positions, irrespective of the intermediate state, the behaviour of the material is approximated by segments OA' (martensite) and OC (austenite) in Figure 1a. The reference elastic moduli are E_M and E_A, respectively. This is also true if the shear behaviour of the shape memory material is considered, when the maximum shear strain, γ_{adm}, replaces ε_{adm} and the shear moduli G_M and G_A replace E_M and E_A.

The second material model describes the behaviour at cold temperatures (martensitic phase) with a bilinear law [21] as shown in the stress-strain diagram of Figure 1b. The model is defined by a first leg OD with an elastic modulus E_{MA} and a second leg DE with a gradient E_{MB}. We define ε_g as the deformation of the occurrence of the change of slope. Due to the bilinear stress-strain response, the SMA elements used in the actuators (springs or wires) also have bilinear force-displacement behaviour when disabled. The elastic moduli E_A, E_{MA} and E_{MB} are replaced by the stiffness K_A, K_{MA}, K_{MB}, while the deformation ε_g is replaced by the displacement x_g. Assuming that the geometric changes related to the deformations of the springs do not influence the elastic constant value, a parameter of merit of the SMA material, s_1, can be defined as follows:

$$s_1 = \frac{E_A}{E_M} = \frac{G_A}{G_M} = \frac{k_a}{k_m} \tag{1}$$

This non dimensional group expresses for both models the shape memory capability of the alloy, the larger s_1 the better the material is. The only difference is that in case of the simplest linear model (Figure 1a) the denominator is the secant modulus, while in case of bilinear model (Figure 1b) the denominator is E_{MA} the modulus of the first linear martensitic region. In order to balance the active SMA element authors evaluates the influence of three backup elements: a constant force (Section 3.1) a conventional spring (Section 3.2) and an antagonistic SMA (Section 3.3). Moreover the authors propose two compensator systems in order to increase the stroke of the actuator. The systems are either based on a leverage (rocker-arm, Section 4.2.1) or on articulated mechanisms (double quadrilateral, Section 4.2.2), which can be considered as backup elements with negative stiffness.

2. Design procedure of binary SMA actuators

In this section a design procedure for binary SMA actuators is described and discussed. The system is made up by a generic actuator which moves the output port by means of an elastic system containing an active SMA element and a bias (backup) element.

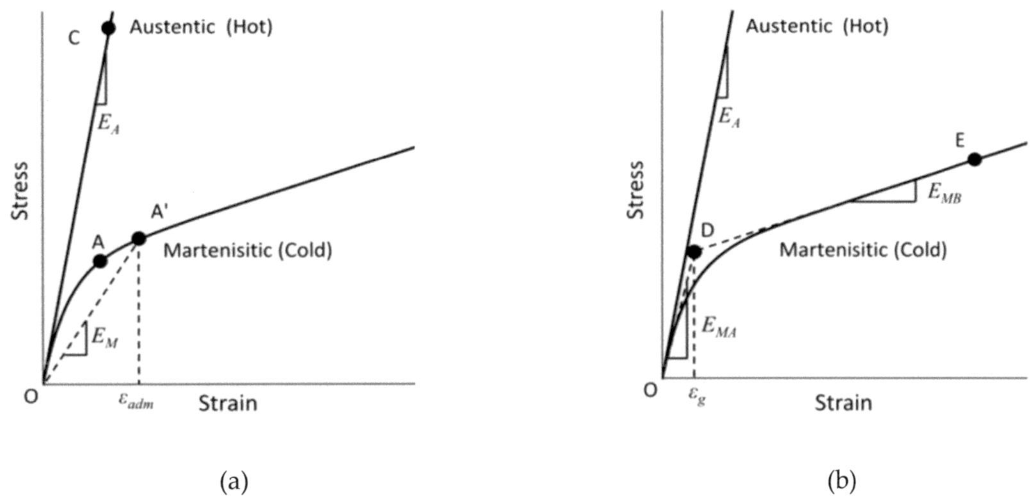

Figure 1. Linear (a) and bilinear (b) material model for martensitic phase of SMA elements

According to the particular means of applying the bias force, the three cases shown in Figure 2 are analyzed:

- a primary SMA spring biased by a constant force (Figure 2a);
- a primary SMA spring biased by a traditional spring (Figure 2b);
- two antagonist SMA springs (when one is hot, the other one is cold, Figure 2c).

Without loss of generality, each spring in Figure 2 is modeled as a traction spring exhibiting a linear force-deflection relationship. While the assumption of linearity is obvious for the traditional spring in Figure 2b, the linear behaviour for the SMA spring is an approximation needed to use the model in Figure 1a.

Each actuator presented in Figure 2 is intended to move the output port E through a total useful stroke Δx when working against an external dissipative force F_F and an external conservative force F_0. The force F_F is always opposite to the velocity of the cursor and is assumed to be constant. For example, if the cursor is subject to dry friction forces characterized by a static value F_S and a dynamic value $F_D < F_S$, the design dissipative force $F_F = \text{MAX}(F_S, F_D) = F_S$ will be adopted for the calculation.

This approach comprises every possible external constant load, as exemplified by the membrane pump shown in Figure 3a. This example represents the most general case of external constant forces the actuator has to deal with. The SMA actuator undergoes the following external loads: two generic dissipative forces F_1 (force during aspiration) and F_2 (force during pumping) and a conservative force F_u due to the gravity force on the piston. The dissipative force F_1 and the force F_u act together against the primary spring, when the piston is moving towards the primary SMA spring (inlet of the fluid). By contrast, the dissipative force F_2 acts on the cursors against the force F_u and the primary spring, when the cursor moves away from the primary SMA spring (outlet of the fluid). This force system can

be always reduced to the two forces considered in the method: a conservative one, F_0 and two symmetric dissipative ones, F_F, rearranging the forces as follows:

$$F_0 = F_u + \frac{F_1 - F_2}{2} \quad , \quad F_F = \frac{F_1 + F_2}{2} \tag{2}$$

The direction of the forces depends on the piston speed, v, as shown in Figure 3b-c.

A second non dimensional parameter, s_2, useful in the configurations with two springs, is defined as the ratio between the minimum value assumed by the stiffness of bias spring 2 and the stiffness of active spring 1 in the cold state:

$$s_2 = \frac{K_{2\min}}{K_{1SC}} \tag{3}$$

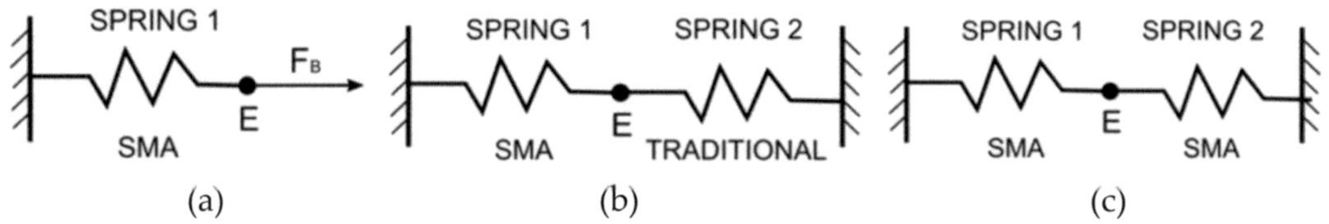

Figure 2. Three cases of the shape memory actuator biased by: a constant force (a), a traditional spring (b), a shape memory spring (c).

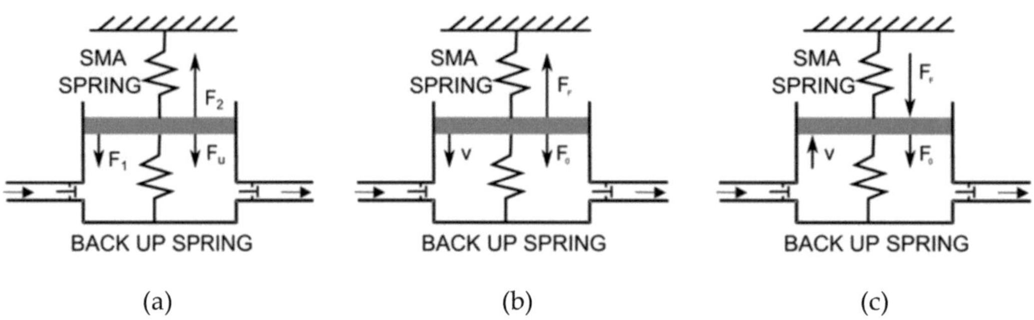

Figure 3. Example of generic conservative and dissipative forces acting on the system (a), equivalent forces when SMA is inactive (b) and when SMA is active (c)

The minimum value assumed by the stiffness of spring 2, K_{2min}, coincides with the only stiffness of spring 2, K_{2C}, if spring 2 is a traditional one, while it coincides with K_{2SC} if spring 2 is an active one. A third dimensionless parameter, s_F, is introduced in order to consider the influence of the dissipative forces in the motion of the SMA actuator. This parameter is defined as the ratio between the dissipative force F_F and the maximum force sustained by the primary spring in the cold state, calculated as the product between the cold spring stiffness, K_{1SC}, and the maximum deflection, L_C-L_{01}, in the cold state:

$$s_F = \frac{F_F}{K_{1SC} \cdot (L_C - L_{01})} \tag{4}$$

The fourth dimensionless parameter, s_0, is introduced in order to consider the influence of the conservative force in the motion of the SMA actuator. This parameter is defined as the ratio between the conservative force F_0 and the maximum force sustained by the primary spring in the cold state, calculated as the product between the cold spring stiffness, K_{1SC}, and the maximum deflection, L_C-L_{01}, in the cold state:

$$s_0 = \frac{F_0}{K_{1SC} \cdot (L_C - L_{01})} \tag{5}$$

The combination of definitions (4) and (5) gives the relationship between s_F and s_0:

$$s_0 = \frac{F_0}{F_F} \cdot s_F \tag{6}$$

2.1. SMA actuator backed up by a constant force

Figure 4a shows the actuator biased by a constant force in three characteristic positions. Figure 4b shows the relationship between the applied force and the spring deflection during the travel of the actuator between these positions.

The procedure can be retrieved from [20]. The useful stroke of the actuator is obtained as:

$$\Delta x = \frac{F_B}{K_{1SC}} \cdot \frac{(s_1 - 1 - 2s_F)}{s_1 \cdot (s_F + 1 - s_0)} \tag{7}$$

The bias force is:

$$F_B = F_F \cdot \frac{s_F + 1 - s_0}{s_F} \tag{8}$$

Thus, the stroke can be cast as:

$$\Delta x = (L_C - L_{01}) \frac{s_1 - 1 - 2s_F}{s_1} \tag{9}$$

Eq. (7) shows that meaningful strokes ($\Delta x > 0$) are only possible if parameter s_F is lower than the following critical value.

$$s_{Fcr} = \frac{s_1 - 1}{2} \tag{10}$$

Moreover, a possible way to choose the value of s_F consists on studying the free length of the active spring, L_{01}, and its wire diameter, d_1, as described below in eq. 14.

2.1.1. *Embodiment of SMA actuator*

If helical springs or wires are considered, two fundamental expressions can be written for the diameter or the wire and for the length of the springs, as a function of the stiffness, K and of the deflection of the SMA element, f.

$$d = m_d \cdot \sqrt{K \cdot f} \tag{11}$$

$$L_0 = m_l \cdot f \tag{12}$$

The constant m_d and m_f depends on the embodiment and can be calculated as shown in the examples in Section 3.1.2 and Section 3.2.2.

A generic expression of the free length, L_{01}, can be written as follows:

$$L_{01} = m_l \cdot \Delta x \cdot \frac{s_1}{s_1 - 2 \cdot s_F - 1} \tag{13}$$

In (13) the free length, L_{01}, depends on the particular embodiment of the spring, but it is always minimized if s_F is chosen as small as possible. The minimum value of s_F can be determined by fixing the maximum allowable wire diameter, d_l, of the SMA element. This maximum value can be determined, for example, using cooling time considerations, the bigger the wire, the slower the cooling, [8], [18].

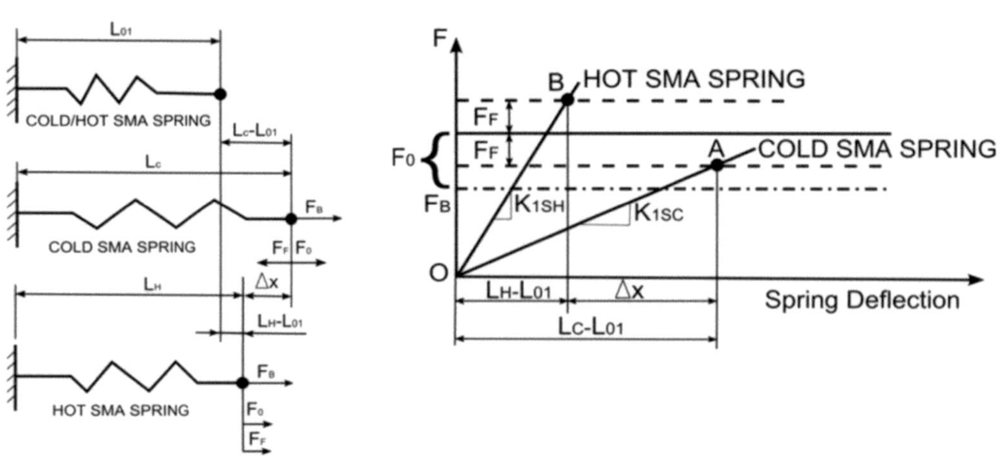

(a) (b)

Figure 4. Actuator model (a) and force-deflection diagram (b) of the shape memory actuator biased by a constant force

Combining expression (11) with (1) and (3), the following relationship can be used to determine the minimum value of s_F.

$$d_1 = m_{d1} \cdot \sqrt{\frac{F_F}{s_F}} \tag{14}$$

Where m_{d1} depends only on the embodiment of the SMA primary spring.

2.1.2. *Case study: SMA wire based actuator*

The equations (11) and (12) represent the relationships between stiffness, *K*, and free length, L_0, either for a spring or a wire.

In particular the stiffness of a SMA actuator in wire form is obtained using the coefficient:

$$m_{dw} = \sqrt{\frac{4}{\pi \cdot E_m \cdot \varepsilon_{adm}}} \tag{15}$$

where it is shown that the limit is given by the maximum strain of the SMA wire in martensitic phase. The free length is obtained from eq. (11) by considering the following coefficient for a SMA wire, which comes out simply from the definition of axial strain in a rod.

$$m_{lw} = 1/\varepsilon_{adm} \tag{16}$$

2.2. SMA actuator backed up by an elastic spring

Figure 5a shows the actuator biased by a traditional spring in three characteristic positions. Figure 5b displays the relationship between the applied force and the deflection of each spring during the thermal operation of the actuator between those positions.

The detailed procedure can be retrieved from [20]. The pre-stretch is

$$p = \left(L_C - L_{01}\right) \frac{s_2 + s_F + 1 - s_0}{s_2} \tag{17}$$

$$\Delta x = p \frac{s_2 \cdot \left(s_1 - 2 \cdot s_F - 1\right)}{\left(s_1 + s_2\right) \cdot \left(s_2 + s_F + 1 - s_0\right)} \tag{18}$$

The following alternative expression of the overall travel can be written

$$\Delta x = \left(L_C - L_{01}\right) \frac{s_1 - 1 - 2s_F}{s_1 + s_2} \tag{19}$$

Lastly, Figure 5b shows that the maximum deflection in the cold state of the bias spring amounts to p-(L_H-L_{01}). This expression can be written as:

$$p-(L_H-L_{01})=(L_C-L_{01})\cdot\frac{s_1\cdot(s_2+s_F+1)-s_2\cdot s_F-s_0\cdot(s_1+s_2)}{s_2\cdot(s_1+s_2)} \tag{20}$$

2.2.1. Embodiment of the SMA actuator

If helical springs or wires are considered, the two fundamental expressions (11) and (12) can be written for the diameter of the wire and for the length of each spring. Following [20] a generic expression of the total length can be obtained, which depends on the dimensionless parameters s_1, s_2, s_0 and s_F and on the embodiment of the springs.

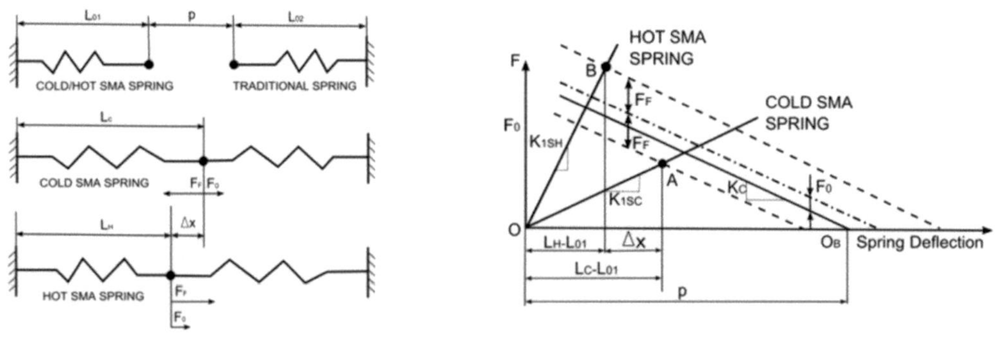

Figure 5. Actuator model (a) and force-deflection diagram (b) of the shape memory actuator biased by a traditional spring

This expression is minimized if s_F is chosen as small as possible and if a precise value of s_2^* is chosen, obtained by equating to zero the derivative of the total length of the actuator. The optimal value s_2^* can be expressed as:

$$s_2^*=\sqrt{\frac{s_1\cdot(m_{l2}+1)\cdot(s_F+1-s_0)}{(m_{l1}+1)}} \tag{21}$$

As a first attempt, it is possible to determine the m_l coefficients considering the desired embodiment of each spring and a value of C = 7, because the value of s_2^* is not greatly affected by C.

2.2.2. Case study: SMA spring based actuator

The equations (11) and (12) represent the relationships between stiffness, K, and free length, L_0, either for a spring or a wire.

In particular the wire diameter of a helical spring can be cast as follows [26]:

$$d=\sqrt{\frac{4\cdot K\cdot f\cdot K_b}{\pi\cdot G\cdot\gamma_{adm}}} \tag{22}$$

Were K is the stiffness of the spring, f is the deflection of the spring and K_b is the coefficient of Bergstrasser, given by the following relationship:

$$K_b = \frac{4 \cdot C + 2}{4 \cdot C - 3} \tag{23}$$

Thus for a spring the m_{ds} coefficient in expression (11) can be expressed as:

$$m_{ds} = 4\sqrt{\frac{(C+2)}{\pi \cdot G \cdot \gamma_{adm} \cdot (4C-3)}} \tag{24}$$

The external diameter of a helical spring can be calculated using the definition of the spring index ($C = D/d$) and the number of active coils is given by:

$$N = \frac{G \cdot d}{8 \cdot C^3 \cdot K} \tag{25}$$

The fully compressed length can be calculated using expression:

$$L_{fc} = Nd = \frac{G \cdot d^2}{8 \cdot C^3 \cdot K} \tag{26}$$

The free length is:

$$L_0 = 1.15 \cdot Nd + f = 1.15 \cdot \frac{G \cdot f \cdot d^2}{8 \cdot C^3 \cdot F} + f \tag{27}$$

Thus the coefficient to calculate the free length in eq. (12) is:

$$m_{ls} = \frac{1.15 + \pi \cdot \gamma_{adm} \cdot C^2}{\pi \cdot \gamma_{adm} \cdot C^2} \tag{28}$$

2.3. SMA actuator backed up by an antagonist SMA spring

Figure 6a shows the actuator biased by a second SMA spring in three characteristic positions. Figure 6b shows the relationship between the applied force and the deflection of each spring during the thermal operation of the actuator between these positions.

Following the procedure described in [20], the prestretch p can be written:

$$p = (L_C - L_{01}) \cdot \frac{s_1 \cdot s_2 + s_F + 1 - s_0}{s_1 \cdot s_2} \tag{29}$$

The stroke can be cast as:

$$\Delta x = p \cdot \frac{s_2 \cdot \{(s_1 - 1 - s_F) \cdot (s_1 + 1) - s_0 \cdot (s_1 - 1)\}}{(s_1 + s_2) \cdot (s_1 \cdot s_2 + s_F + 1 - s_0)} \tag{30}$$

The following expression for the optimal value s_2^* is found:

$$s_2^* = \sqrt{s_F + 1 - s_0} \tag{31}$$

The following alternative expression of the overall travel can be obtained:

$$\Delta x = \left(L_C - L_{01}\right) \cdot \frac{\left(s_1 - 1 - s_F\right)\cdot\left(s_1 + 1\right) + s_0 \cdot \left(s_1 - 1\right)}{s_1 \cdot \left(s_1 + s_2\right)} \tag{32}$$

Both Equations (30) and (32) demonstrate that meaningful strokes ($\Delta x > 0$) are only possible if parameter s_F is lower than the following critical value.

$$s_{Fcr} = \frac{F_F \cdot \left(s_1 - 1\right)^2}{F_0 \cdot \left(s_1 - 1\right) + F_F \cdot \left(s_1 + 1\right)} \tag{33}$$

Figure 6b shows that the maximum deflection of the bias spring in the cold state amounts to $p - \left(L_H - L_{01}\right)$. This expression can be written as:

$$p - \left(L_H - L_{01}\right) = \left(L_C - L_{01}\right) \cdot \frac{s_1 \cdot s_2 - s_2 \cdot s_F + s_F + 1 - s_0 \cdot \left(s_2 + 1\right)}{s_2 \cdot \left(s_1 + s_2\right)} \tag{34}$$

2.3.1. *Embodiment of SMA antagonist actuator*

If helical springs or wires are considered, the two fundamental expressions (11) and (12) can be written for the diameter of the wire and for the length of each spring. Following [20] a generic expression of the total length can be obtained, which depends on the dimensionless parameters s_1, s_2, s_0 and s_F and on the embodiment of the springs.

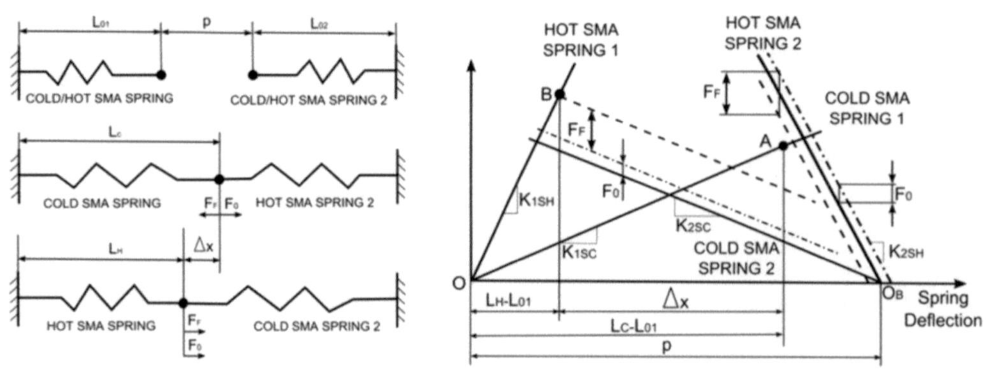

(a) (b)

Figure 6. Actuator model (a) and force-deflection diagram (b) of the agonist antagonist shape memory actuator

This expression is minimized if s_F is chosen as small as possible and if an optimal value for s_2^* is chosen, obtained equating to zero the derivative of the total length of the actuator. The optimal value s_2^* can be expressed as:

$$s_2^* = \sqrt{\frac{(m_{l2}-1)\cdot(s_F+1-s_0)}{(m_{l1}-1)}} \tag{35}$$

Expression (35) reduces to expression (31) if the same spring embodiment is considered for spring 1 and spring 2. As a first attempt, it is possible to determine the m_l coefficients considering the desired embodiment of each spring and a value of $C = 7$, because the value of s_2^* is not greatly affected by C. The minimum value of s_F can be determined by fixing the maximum allowable wire diameter, d_l, of the SMA element. This maximum value can be determined, for example, using cooling time considerations. Relationship (14) can be used to determine the minimum value of s_F.

2.4. Design procedures for a binary SMA actuator

The step by step procedures which guide the designer to apply the above design methods are described in Table 1. The first four steps are the same irrespective of the backup element used in the actuator. From step 5, the procedure is threefold.

These procedures ensure that any actuator biased by a one of the above described back up elements, with the calculated stiffness K_C and containing whatever SMA spring, with the selected material parameter s_1 and the calculated cold stiffness K_{lSC}, satisfies the design problem (useful stroke Δx, design dissipative force F_F and design conservative force F_0) when assembled with the calculated pre-stretch p.

2.4.1. *Case study: SMA based swing louver*

In this section, the complete design procedure is carried out numerically for the actuation of a swing louver exploited to direct the air flow in domestic air conditioners. An actuator made up of an SMA spring biased by an antagonist SMA spring is designed here as a possible alternative solution to conventional electric motors and linkages. The SMA actuator acts on the louver with a known arm to make the louver swing. The design parameters are: required stroke: 5mm; dissipative force: 5N, conservative force: 2N. The material considered is Nitinol, with the following properties, $\gamma_{adm} = 0.02$, to ensure a fatigue limit over 500 thousand cycles, austenitic shear modulus G_a=23000 MPa, equivalent martensitic shear modulus G_m=8000 MPa.

The procedure starts calculating the non dimensional groups from eq. (1), (10), (31) and (5):

$$s_1 = \frac{G_A}{G_M} = \frac{23000}{8000} = 2.875 \tag{36}$$

Backup: Constant Force (Section 3.1)	Backup: Traditional spring (Section 3.2)	Backup: antagonist SMA spring (Section 3.3)
—the stroke of the actuator, Δx, the design dissipative force, F_F, and the design conservative force, F_0, are accepted from design specifications; —a shape memory material for the spring and a maximum strain, ε_{adm} or γ_{adm}, are selected, after which parameter $s_1 = E_A/E_M = G_A/G_M$ is known; —The spring embodiment is chosen and the parameters ml_1, ml_2 and md_1 are calculated as shown in Section 3.1.2 or 3.2.2. Otherwise, this step is skipped; —a s_F value lower than the critical value (10) is selected, possibly using (14);		
—the value of s_0 is calculated using definition (6); —the bias force, F_B, is calculated using eq. (8); —the maximum deflection of the primary spring, L_C-L_{01}, is retrieved from (9); —the cold stiffness, K_{1SC}, of the spring is calculated from eq. (7);	—the value of s_0 is calculated using definition (6); —the lowest affordable value of parameter s_2 is chosen (to exploit the primary spring), or the optimal value (21) is adopted (to conserve overall space). If the parameters ml_1 and ml_2 cannot be determined, expression (31) can be used for a sub-optimal design; —the maximum deflection of the primary spring, L_C-L_{01}, is retrieved from (19); —the cold stiffness, K_{1SC}, of the spring is calculated from eq. (4); —the pre-stretch of the system, p, is calculated from (17); —the stiffness of the traditional spring, K_C, is calculated from eq. (3); —the maximum deflection of the traditional spring, p-(L_H- L_{01}), is calculated from (20).	—a s_F value lower than the critical value (10) is selected. The parameter is chosen as small as possible to reduce the size of the actuator, respecting the constraint of the diameter of the wire, d, given by (14); —the value of s_0 is calculated using definition (6); —the lowest affordable value of parameter s_2 is chosen (to exploit the primary spring), alternatively the optimal value (35) is adopted (to conserve overall space). If the parameters ml_1 and ml_2 cannot be determined, expression (31) can be used for a sub-optimal design; —the maximum deflection of the primary spring, L_C-L_{01}, is retrieved from (32); —the cold stiffness, K_{1SC}, is calculated from eq. (4); —the overall pre-stretch of, p, is calculated from (29); —the stiffness of the secondary SMA spring, K_{2SC}, is calculated from eq. (3); —the maximum deflection of the secondary SMA spring, p-(L_H- L_{01}), is calculated from (34).

Table 1. Step by step procedure for each Backup element considered

$$s_{Fcr} = \frac{s_1 - 1}{2} = \frac{2.875 - 1}{2} = 0.9375 \quad (37)$$

$$s_F = 0.25 \le s_{Fcr} \quad (38)$$

$$s_2 = s_2^* = \sqrt{s_1 \cdot (s_F + 1)} = \sqrt{2.875 \cdot (0.25 + 1)} = 1.896 \quad (39)$$

$$s_0 = \frac{F_0}{F_F} \cdot s_F = \frac{2}{5} \cdot 0.25 = 0.1 \; s_0 = \frac{F_0}{F_F} \cdot s_F = \frac{2}{5} \cdot 0.25 = 0.1 \quad (40)$$

the maximum deflection of the primary spring, L_C-L_{01}, is retrieved from (32);

$$(L_C - L_{01}) = \Delta x \cdot \frac{s_1 \cdot (s_1 + s_2)}{(s_1 - 1 - s_F) \cdot (s_1 + 1) + s_0 \cdot (s_1 - 1)} = \\ = 5 \cdot \frac{2.875 \cdot (2.875 + 1.896)}{(2.875 - 1 - 0.5) \cdot (2.875 + 1) + 0.1 \cdot (2.875 - 1)} = 12.43\,mm \quad (41)$$

while eq. (4) allows the SMA spring cold stiffness to be calculated

$$K_{1SC} = \frac{F_F}{s_F \cdot (L_C - L_{01})} = \frac{5}{0.25 \cdot 12.43} = 1.61\,N/mm \quad (42)$$

the overall pre-stretch of, p, is calculated from (29) and K_C, is calculated from eq. (3)

$$p = (L_C - L_{01}) \cdot \frac{s_1 \cdot s_2 + s_F + 1 - s_0}{s_1 \cdot s_2} = 12.43 \cdot \frac{2.875 \cdot 1.896 + 0.25 + 1 - 0.1}{2.875 \cdot 1.896} = 15.01\,mm \quad (43)$$

$$K_{2\min} = K_{2SC} = s_2 \cdot K_{1SC} = 1.896 \cdot 1.61 = 3.052\,N/mm \quad (44)$$

$$p - (L_H - L_{01}) = (L_C - L_{01}) \cdot \frac{s_1 \cdot s_2 - s_2 \cdot s_F + s_F + 1 - s_0 \cdot (s_2 + 1)}{s_2 \cdot (s_1 + s_2)} = \\ = 12.43 \cdot \frac{2.875 \cdot 1.896 - 1.896 \cdot 0.25 + 0.25 + 1 - 0.1 \cdot (1.896 + 1)}{1.896 \cdot (2.875 + 1.896)} = 8.16\,mm \quad (45)$$

the maximum deflection of the secondary SMA spring, p-(L_H-L_{01}), is calculated from eq. (34).

The detailed design of primary SMA spring is given by combination of eq. (11) and (24) regarding the spring wire diameter, considering a spring index C=7

$$d = 4\sqrt{\frac{(C+2) K_{1SC} (L_C - L_{01})}{\pi \cdot G \cdot \gamma_{adm} \cdot (4C - 3)}} = 4\sqrt{\frac{(7+2) \cdot 1.61 \cdot 12.43}{\pi \cdot 8000 \cdot 0.02 \cdot (4 \cdot 7 - 3)}} = 0.48\,mm \quad (46)$$

the free length of the primary spring is given by the combination of eq. (12) and (28)

$$L_0 = \frac{1.15 + \pi \cdot \gamma_{adm} \cdot C^2}{\pi \cdot \gamma_{adm} \cdot C^2} \cdot \left(L_C - L_{01}\right) = \frac{1.15 + \pi \cdot 0.02 \cdot 7^2}{\pi \cdot 0.02 \cdot 7^2} \cdot 12.43 = 17.07\,mm \tag{47}$$

The same procedure is applied to calculate the design parameter of the antagonist spring.

3. Design of compensated SMA actuators

SMA based actuators develop significant forces but are usually characterized by low strokes. The stroke is primarily limited by the maximum strain that the alloy can withstand for the expected life as shown in [24], [25]. The backup element needed to recover the stroke prevents the SMA element from recovering completely its shape is a further cause of stroke loss. Furthermore, the force delivered by shape memory actuators varies linearly with the displacement while the external load is usually constant. The design of a SMA actuator ensures that the minimum actuator force is sufficient to contrast the external load [18], providing an additional cause of stroke reduction.

This section introduces a compensation system to store available power from the SMA element in high force positions and then return this power in low force positions [19]. The same principle was successfully applied for electroactive polymer actuators by introducing compliant mechanisms [27]. The compensation system adopted has a negative elastic characteristic, generating a decreasing force as the deformation increases.

3.1. Principle of elastic compensation

The force-deflection diagram in Figure 7 shows the characteristic lines of an SMA actuator in the austenitic (hot SMA, solid circles) and martensitic (cold SMA, solid squares) states. When the SMA element is backed up by a linear spring with the linear characteristic shown with hollow triangles, the net stroke under no external force is S_{spring}. Starting from the same maximum deflection and force of the cold SMA, the stroke increases to S_{weight} when the active element is backed up by a constant force (crossed horizontal line). The improvement is consequent upon the reduced stiffness (lower contrasting forces) of the backup element which allows the SMA to recover a greater share of deformation.

By learning from this positive trend, it is easily seen that an even greater stroke (S_{comp}) is achieved if the backup element displays a negative slope (hollow circles) so that the contrast force would decrease with increasing deflection. Energetically, the compensation system accumulates energy from the SMA element in the positions where the SMA force is high (right-hand side in Figure 7) and releases that energy to the actuators in the positions where the SMA force is low (left-hand side in Figure 7).

As shown in the subsequent sections, a backup element with negative slope as in Figure 7 can be achieved by exploiting one of the many spring-assisted mono or bistable mechanisms described in the technical literature. The use of an elastic compensation system requires the introduction of hard stops to prevent the SMA elements from over-straining. In the case of a single-SMA actuator (see Figure 7), a single hard stop is needed and the behaviour in the

absence of power becomes monostable. In the case of a double-SMA actuator, two hard stops are required and the behaviour in the absence of power becomes bistable. Although the advantages of the compensation system in terms of force and stroke apply to both single-SMA and two-SMA actuators, the improvements are more pronounced for the two-SMA actuator. The general theory [21] demonstrates that a single-SMA actuator can generate a truly constant output force on only one direction of motion. By contrast, a two-SMA actuator can achieve a constant-force profile in both directions. Further advantages of the compensated architecture over regular SMA actuators are the existence of definite equilibrium positions when the power is shut off, the enforcement of precise mono (single-SMA) or binary (two-SMA) positioning, and the possibility of easy control strategies. The input data required for the design of a shape memory actuator are the value of the guaranteed minimum useful force in the two directions of activation (F_{ON}, F_{OFF} in the case of a single SMA element, F_{ON1}, F_{ON2} when there are two opposing SMA elements), the value of the stroke desired, S, and the type of alloy used for the active elements (i.e. s_l, s_m, s_g).

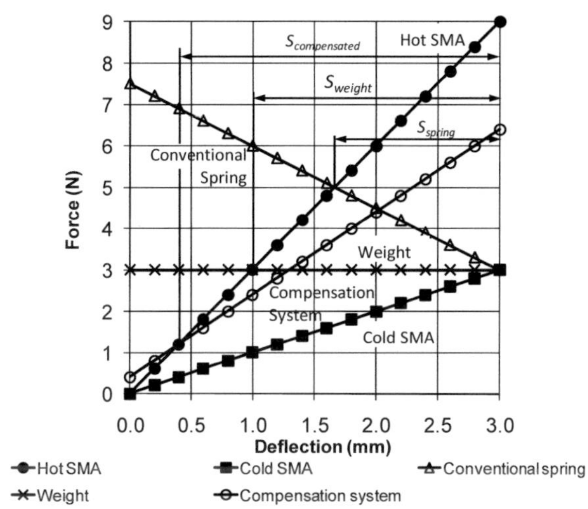

Figure 7. Force-deflection curves of a single-SMA actuator backed up by: a conventional spring, a constant force and an elastic compensation system

3.2. Material model: Definitions

In order to design compensated shape memory actuators the bilinear model for the martensitic state of SMA (Figure 1b) was used. Since ε_g (then x_g) is very small ($0.2\% < \varepsilon_g < 0.5\%$) the force-stroke characteristic of the SMA elements can be approximated as a linear trend (D'E in Figure 8), with slope K_{MB}, starting from the force F_{0m} in correspondence to zero displacement (Figure 8). The force F_{0m} is calculated as:

$$F_{0m} = (K_{MA} - K_{MB})x_g \tag{48}$$

The force F_{SMA_ON} produced by the generic shape memory element in the activated state (austenite) is given by the line OC in Figure 8 and is:

$$F_{SMA\_ON} = K_A x \tag{49}$$

In the cold state (martensite), the force F_{SMA_OFF} is given by the line D'E in Figure 8:

$$F_{SMA\_OFF} = F_{0m} + K_{MB} x \tag{50}$$

In addition to s_1 already defined in eq. (1), to facilitate the design of the actuator, it is useful to define two other dimensionless coefficients:

$$s_m = \frac{E_{MB}}{E_{MA}} = \frac{K_{MB}}{K_{MA}} \tag{51}$$

$$s_g = \frac{\varepsilon_g}{\varepsilon_{adm}} = \frac{x_g}{x_{adm}} \tag{52}$$

The parameter s_m is the ratio between the two stiffness of the alloy at cold temperatures and it is a characteristic of the shape memory material. The parameter s_g is the ratio of the deflection, x_g, at which the change in stiffness in the martensitic state is recorded and the maximum allowable deflection x_{adm}, related to the maximum deformation, ε_{adm}, admissible for the material. To limit the size of the device, it is convenient to assume the $x_{maz} = x_{adm}$ so as to completely exploit the material.

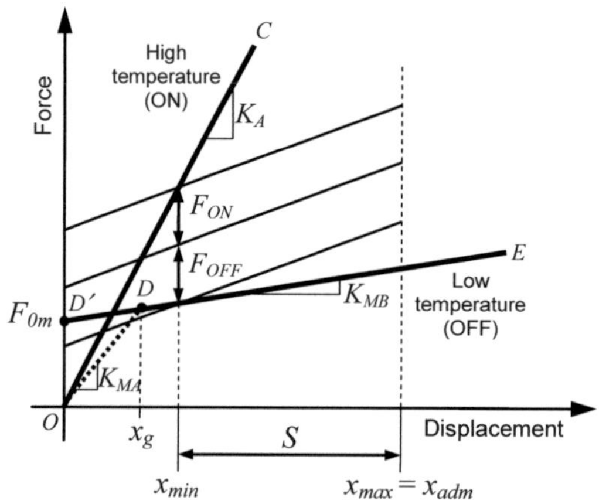

Figure 8. Force-deflection curves of a two-SMA actuator superimposed on the characteristic lines of the compensator.

3.3. Design of compensated single SMA actuators

To design compensated actuators with a single SMA element, it is convenient to introduce the ratio, γ, between the stroke and the maximum deflection of the SMA element:

$$\gamma = S / x_{adm} \tag{53}$$

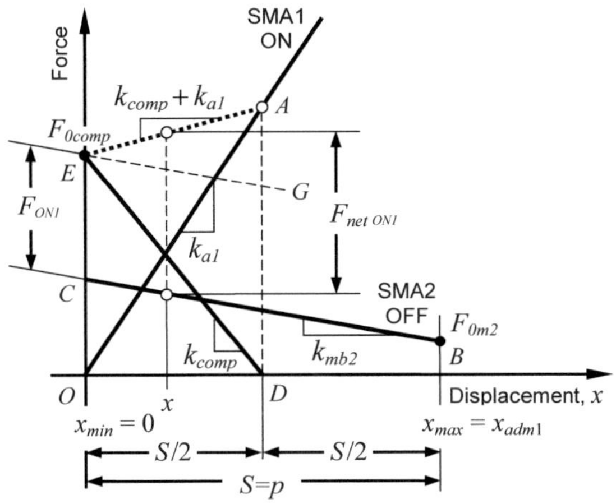

Figure 9. Force-deflection curves of a two-SMA actuator compensated by a generic negative elastic spring (line *ED*)

The axial dimension of the actuator is reduced as the ratio γ tends to one because in this case all the deformation that the material can sustain is used to produce a useful stroke. However, the value γcannot reach the unity because the minimum deflection would be zero and the SMA element would not be able to exert any useful force.

The force delivered by the shape memory element can be calculated with (49), in the austenitic state, and with the approximated expression (50), in the martensitic state. Figure 8 shows that the high-temperature curve (OC) and the low-temperature curve (A'B) diverge. Thus, the most unfavorable position in which the design force is required is where the difference between the two curves (useful force) is minimal. In this position, the sum of the design forces in both directions must equal the distance between the ON straight line and the OFF straight line. Assuming conventionally $F_{ON} > 0$ (because it agrees with the force exerted by the SMA element) and $F_{OFF} < 0$ (because it is opposed to the force produced by the SMA) we have $(F_{ON} - F_{OFF}) = F_{SMA\_ON}\ (x = x_{min}) - F_{SMA\_OFF}\ (x = x_{min})$. Recalling (48),(49),(50) and introducing the dimensionless parameters (1),(51),(52) this condition becomes:

$$F_{ON}-F_{OFF}=\frac{k_{MA}S}{\gamma}\left[\left(s_1-s_m\right)\left(1-\gamma\right)-s_g\left(1-s_m\right)\right] \tag{54}$$

The stiffness needed to fulfill the minimum force in both directions and to achieve the desired stroke is obtained by solving (54) with respect to the stiffness k_{MA}:

$$k_{MA}=\frac{\gamma\left(F_{ON}-F_{OFF}\right)}{S\left[\left(s_1-s_m\right)\left(1-\gamma\right)-s_g\left(1-s_m\right)\right]} \tag{55}$$

The next step is to set the desired characteristics of the compensation element in terms of stiffness (K_{comp}) and of force ($F_{comp}\big|_{x=x_{\min}}$), delivered at position $x = x_{min} = S\ (1-\gamma)\ /\gamma$.

Target behaviour	Value of k_{comp}
Balanced force variation in ON and OFF modes	$-(K_{MB}+K_A)/2$
Constant force in OFF mode	$-K_A$
Constant force in ON mode	$-K_{MB}$

Table 2. Suggested values for the compensation stiffness to achieve specific behaviour of the actuator

From (1) and (55) the stiffness K_A of the SMA element in the austenitic state is obtained.

Though the compensation stiffness (always negative) could be set as desired in the range $-K_{MB} \div -K_A$, Table 1 shows how to choose the value of K_{comp} to obtain favourable operating characteristics: for $K_{comp} = -K_A$, the force of the activated actuator is constant and for $K_{comp} = -K_{MB}$ the force of the deactivated actuator is constant. By setting $K_{comp} = -0.5(K_A + K_{MB})$, the deviation from uniformity of both states of actuator is minimized.

The compensator force the at position $x = x_{min}$ to obtain the desired behavior must be:

$$F_{comp}\Big|_{x=x_{\min}} = F_{ON} - K_A S\left(\frac{1-\gamma}{\gamma}\right) \tag{56}$$

The compensator always acts with a force increasing with x and opposite to the force delivered by the SMA element.

3.3.1. Case study: single SMA compensated actuator

In this section, the design procedure of a single SMA compensated actuator is carried out numerically for a generic system with the following technical specifications:

- Required stroke: 10mm
- Minimum force in the right to left displacement F_{ON}: 10N
- Minimum force in the left to right displacement F_{OFF}: -5N
- Minimization of the variation force in overall travel

The material considered is a Nitinol wire, with austenitic Young modulus E_a=75GPa, martensitic Young modulus E_{MA}=28GPa, bilinear gradient E_{MB}=5GPa, strain at the threshold of pseudo-plastic regime ε_g=0.4% maximum deformation ε_{adm}=4%.

The non dimensional groups are calculated as follows:

$$s_1 = \frac{E_A}{E_{MA}} = \frac{75}{28} = 2.68 \tag{57}$$

$$s_m = \frac{E_{MB}}{E_{MA}} = \frac{5}{28} = 0.18 \tag{58}$$

$$s_g = \frac{\varepsilon_g}{\varepsilon_{adm}} = \frac{0.004}{0.04} = 0.1 \tag{59}$$

An optimal value for the ratio γ=0.75 can be chosen as a tradeoff between axial dimension and oversizing of the active SMA elements.

From (55) it is possible to obtain the cold state stiffness of the SMA wire:

$$K_{MA} = \frac{\gamma\left(F_{ON} - F_{OFF}\right)}{S\left[\left(s_1 - s_m\right)\left(1-\gamma\right) - s_g\left(1 - s_m\right)\right]} = \frac{0.75\left[10 - \left(-5\right)\right]}{10\left[\left(2.68 - 0.18\right)\left(1-0.75\right) - 0.1\left(1-0.18\right)\right]} = \tag{60}$$
$$= 2.07\frac{N}{mm}$$

Considering (1) and (51) and the last expression the cold state stiffness K_A and the post elastic martensitic stiffness K_{MB} can be obtained:

$$K_A = s_1 K_{MA} = 2.68 \cdot 2.07 = 5.55 N / mm \tag{61}$$

$$K_{MB} = s_m K_{MA} = 0.18 \cdot 2.07 = 0.373 N / mm$$

From (53) is computed x_{adm}, the maximum elongation of the SMA wire:

$$x_{adm} = \frac{S}{\gamma} = \frac{10}{0.75} = 13.33 mm \tag{62}$$

Taking into account the maximum deformation of the material ε_{adm} the wire length l_0 is:

$$l_0 = \frac{x_{adm}}{\varepsilon_{adm}} = \frac{13.33}{0.04} = 333.3 mm \tag{63}$$

From the austenitic stiffness and the wire length the wire diameter d is immediately obtained:

$$d = \sqrt{\frac{4K_A l_0}{E_A \pi}} = \sqrt{\frac{4 \cdot 5.55 \cdot 333.3}{75000\pi}} = 0.177 mm \tag{64}$$

The compensator system preload is obtained through (56).This value is negative because in its minimum stroke position $x = x_{min}$= 3.33mm, the preload is opposed to SMA active element action:

$$F_{comp}\Big|_{x=x_{\min}} = F_{ON} - K_A S\left(\frac{1-\gamma}{\gamma}\right) = 10 - 5.55 \cdot 10\left(\frac{1-0.75}{0.75}\right) = -8.5 N \tag{65}$$

In order to minimize the force fluctuations in both travel directions the optimum stiffness of compensation system is obtained from (Table 1) and is:

$$k_{comp} = -\frac{K_{MB} + K_A}{2} = -\frac{0.373 + 5.55}{2} = -2.96\frac{N}{mm} \qquad (66)$$

3.4. Design of compensated actuators with two antagonistic SMA elements

The input data required for the design of a two antagonistic shape memory actuators are: the value of the minimum useful forces in the two directions of activation F_{ON1}, F_{ON2} the value of the stroke desired, S, and the active elements material characteristics (i.e. s_1, s_m and s_g). In the following equations, assuming that the axis of the actuator is horizontal, the forces are assumed positive when directed from right to left and negative when directed from left to right. The displacements are assumed positive from left to right.

For a compensated actuator with two antagonistic SMA element it can be shown [21] that it is always possible to set the ratio γ (53) to 1, so that the actuator stroke is equal to the maximum deflection imposed on the SMA elements (x_{min}=0).

If both active elements are constituted by the same alloy (same parameters s_1, s_m, s_g) is possible to demonstrate [21] that the optimum rigidity of the two elements will be equal ($K_{A1}=K_{A2}=K_A$ and $K_{MB1}=K_{MB2}=K_{MB}$).

The force generated by element 1 in the ON state is:

$$F_{SMA1\_ON} = K_A x \qquad (67)$$

In the OFF state, using (3) for the martensite behavior, the force in element 1 is:

$$F_{SMA1\_OFF} = F_{0m} + K_{MB} x \qquad (68)$$

For element 2, the force output depends on the difference between the prestretch p of the two SMA elements and the position of the actuator. Assuming γ=1 the prestretch p is equal to the stroke S, and the force of element 2 in the ON and OFF state is respectively:

$$F_{SMA2\_ON} = -K_A (S - x) \qquad (69)$$

$$F_{SMA2\_OFF} = -F_{0m} - K_{MB}(S - x) \qquad (70)$$

The force delivered from the compensator is:

$$F_{comp} = F_{0comp} + k_{comp} x \qquad (71)$$

where F_{0comp} indicates the force of the compensator at position $x = 0$.

Figure 9 shows the force displacement diagram of a two-SMA actuator and helps to understand the relationship between the variables involved in the equations.

Line OA in Figure 9 represents the characteristics of the austenitic (hot) SMA1 element. Segment BC represents the martensitic (cold) response of the antagonistic SMA2 element. Line DE is the characteristic of the compensation spring, with point D corresponding to the centerpoint of the total stroke S of the actuator. Line EG represents an ideally constant external load of amplitude F_{ON1}. The situation beyond point D is obtained by extrapolating linearly all the characteristic lines shown in Figure 9. Line AE corresponds to the characteristic of the SMA1 element and the compensation spring combined. At any position x, the difference between lines AE and BC gives the net output force of the actuator ($F_{net\ ON1}$) when element SMA1 is activated. When element SMA1 is disabled and SMA2 is enabled, the chart becomes similar to Figure 9, with all the lines mirrored with respect to the line AD.

The optimal performance of the actuator in Figure 9 is achieved when lines AE is parallel to line BC, so that the net output force of the actuator ($F_{net\ ON1}$) equals the external load (F_{ON1}) at any position x. Scirè and Dragoni [21] demonstrated that the optimal actuator meets the design specifications (i.e. the given output forces F_{ON1}, F_{ON2} and the net stroke S) when the following relationships hold true:

$$K_{MA} = \frac{\left(F_{ON1} - F_{ON2}\right)}{S\left[\left(s_1 - s_m\right) + 2s_g\left(s_m - 1\right)\right]} \tag{72}$$

$$k_{comp} = -\frac{\left(F_{ON1} - F_{ON2}\right)\left(s_1 + s_m\right)}{S\left[\left(s_1 - s_m\right) + 2s_g\left(s_m - 1\right)\right]} \tag{73}$$

$$F_{0comp} = \frac{s_g\left(F_{ON1} + F_{ON2}\right)\left(s_m - 1\right) + \left(F_{ON1}s_1 - F_{ON2}s_m\right)}{\left(s_1 - s_m\right) + 2s_g\left(s_m - 1\right)} \tag{74}$$

The force generated by the actuator in the case of element 1 ON and element 2 OFF is:

$$F_{net\_ON1} = K_A x + F_{0comp} + k_{comp} x - F_{0m} - K_{MB}\left(S - x\right) \tag{75}$$

while the force generated in the case of element 2 ON and element 1 OFF is:

$$F_{net\_ON2} = F_{0m} + K_{MB} x + F_{0comp} + k_{comp} x - K_A\left(S - x\right) \tag{76}$$

From (1) and (55) the stiffness K_A of the SMA element in the austenitic state is obtained.

The actuator designed by (72)-(74) provides a constant force in both directions. In the compensated actuators with two SMAs, the force of the compensator becomes zero and changes sign at mid-stroke. In the range $0 < x < S/2$ the force of the compensator has the same direction as the force exerted by element 1, while in the range $S/2 < x < S$ it has the same direction as the force generated by element 2.

3.4.1. *Case study: Double SMA compensated actuator*

This section describes the design of double SMA compensated actuator under the hypothesis of using traction Nitinol springs with the following characteristics:

- Wire diameter: 0.65mm
- Mean coil diameter: 6.8mm
- Number of active coils: 25.5
- Free length: 25mm
- Austenitic stiffness: K_A: 0.0615N/mm
- Martensitic stiffness: K_{MA}: 0.0414N/mm
- Pseudo-plastic stiffness: K_{MB}: 0.0156N/mm
- Deflection at the threshold of pseudo-plastic regime x_g: 23mm

The desired output stroke is S =75mm, with the maximum possible force in both directions.

In this case the dimensional parameter γ can be taken equal to one and consequently the maximum extension of the active elements x_{adm} is equal to the desired stroke S.

The non dimensional parameters are:

$$s_1 = \frac{K_A}{K_{MA}} = \frac{0.0615}{0.0414} = 1.486 \quad (77)$$

$$s_m = \frac{K_{MB}}{K_{MA}} = \frac{0.0156}{0.0414} = 0.377 \quad (78)$$

$$s_g = \frac{x_g}{x_{adm}} = \frac{23}{75} = 0.306 \quad (79)$$

Using the above dimensionless parameters, eq. (72) gives the maximum force differential $(F_{ON1} - F_{ON2})$ that the actuator can produce in both directions as:

$$\begin{aligned} F_{ON1} - F_{ON2} &= K_{MA} S\left[(s_1 - s_m) + 2s_g(s_m - 1)\right] = \\ &= 0.0414 \cdot 75\left[(1.486 - 0.377) + 2 \cdot 0.306(0.377 - 1)\right] = 2.26N \end{aligned} \quad (80)$$

Assuming that the actuator generates the same output force regardless of the direction of motion gives:

$$F_{ON1} = -F_{ON2} = 1.13N \quad (81)$$

From (73) the overall stiffness of the compensation system k_{comp} is calculated:

$$\begin{aligned} k_{comp} &= -\frac{(F_{ON1} - F_{ON2})(s_1 + s_m)}{S\left[(s_1 - s_m) + 2s_g(s_m - 1)\right]} = -\frac{2.26(1.486 + 0.377)}{75\left[(1.486 - 0.377) + 2 \cdot 0.306(0.377 - 1)\right]} = \\ &= -0.0771N/mm \end{aligned} \quad (82)$$

Likewise, (74) gives the force F_{0comp} that the compensator must apply to the SMA elements at the position $x = 0$:

$$F_{0comp} = \frac{s_g\left(F_{ON1} + F_{ON2}\right)\left(s_m - 1\right) + \left(F_{ON1}s_1 - F_{ON2}s_m\right)}{\left(s_1 - s_m\right) + 2s_g\left(s_m - 1\right)} = \\ = \frac{0.306\left[1.13 + \left(-1.13\right)\right]\left(0.377 - 1\right) + \left[1.13 \cdot 1.486 - \left(-1.13\right) \cdot 0.377\right]}{\left(1.486 - 0.377\right) + 2 \cdot 0.306\left(0.377 - 1\right)} = 2.893N \tag{83}$$

3.5. Design of the compensator system

This section describes the elastic constants of the shape memory elements and the maximum deflections that they will undergo, calculated from the design data (force and stroke required of the actuator) following the procedure of the previous section. Thanks to these parameters the shape memory elements (wires or springs) can be described using classical engineering formulas. The method also provides the properties needed by the elastic compensation to meet the required performance. Two elastic compensation systems are described in detail in this section: 1) a rocker-arm mechanism and 2) a double articulated quadrilateral.

3.5.1. Rocker-arm compensator

The compensation mechanism shown in Figure 10 is made up of a rocker-arm R hinged in G to the frame T. The (conventional) compensation spring S_c (with free length L_{0Trad} and spring rate k_{Trad}) connects the extremity F of the shortest side of the rocker-arm to point E of the frame. Point O (the output port of the actuator) is used to connect the primary elastic elements of the actuator (SMA1 and SMA2), which are also fixed at points P and Q to the frame. In the case of a single-SMA actuator, the active element (SMA1) is placed at the bottom of the device, element SMA 2 disappears and the contrasting element is made up by the conventional compensation spring, S_c.

The mechanism in Figure 10 has a position of unstable equilibrium, corresponding to the configuration where the axis EF of the compensation spring S_c passes through the hinge G of the rocker-arm. In this position, the force exerted by the compensator on the slider O is null. In the case of an actuator with a single active element, the compensation spring S_c is always placed to the right of hinge G in order to exert a contrasting force on the SMA element 1 needed to deform it in the cold state. For an actuator with two opposing SMA elements, the spring S_c is located in an unstable position (line EF passes through G) when point O is at the center of the stroke (S). In this way, the compensation mechanism helps active element SMA1 for the lower half of the stroke and the element SMA2 for the upper half of the stroke. If in Figure 10 the absolute value of the angle γ is small, i.e. if $l_0+x-(a+d) << c$ then the following expressions apply with good approximation:

$$F_{comp}(x) = -\frac{abF_{Trad}\beta}{c\left(a+b\right)} \tag{84}$$

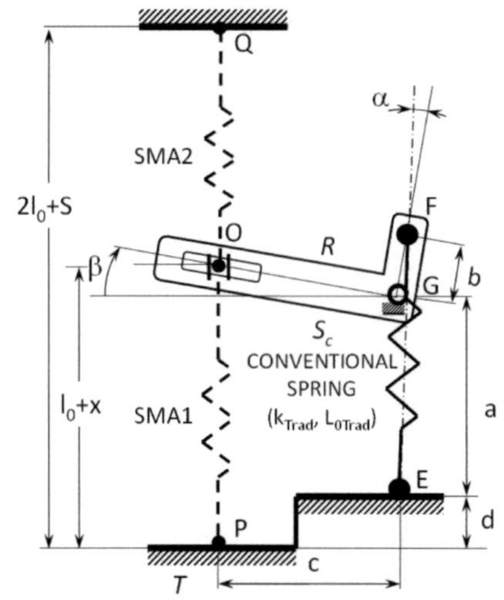

Figure 10. Oscillating rocker arm compensation mechanism

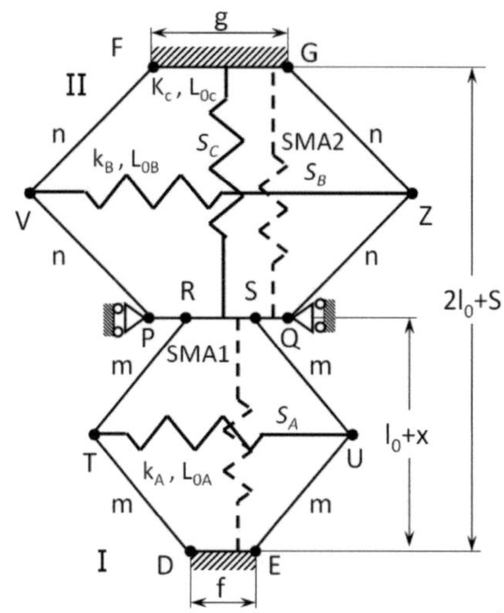

Figure 11. Double articulated quadrilateral compensation mechanism

$$k_{Comp} = \frac{\partial F_{Comp}}{\partial x} = -\frac{abF_{Trad}}{(a+b)c^2} \tag{85}$$

When lengths b,c,d have been arbitrarily decided along with the stiffness of the traditional spring k_{Trad}, from (84) (calculated for $x=x_{min}$) and (85) the expressions for a and L_{0Trad} are derived which define the correct sizing of the compensation mechanism:

$$L_{0Trad} = \frac{\left[k_{comp}\left(A-b+d-l_0\right)+F_{comp}\Big|_{x_{min}}\right]\left(c^2\,k_{comp}^2\,\gamma - B\right)}{B\,k_{comp}} \tag{86}$$

$$a = l_0 - A - d - \frac{F_{comp}\big|_{x_{\min}}}{k_{comp}} \tag{87}$$

with:

$$A = S\left(\frac{\gamma - 1}{\gamma}\right) \tag{88}$$

$$B = b k_{Trad}\, \gamma \left[k_{comp}(d - l_0 + A) + F_{comp}\big|_{x_{\min}} \right] \tag{89}$$

In Equations (86) – (89) the term $F_{comp}\big|_{x_{\min}}$ is given by (56) or by (74) respectively for the single SMA or two opposing SMAs actuator. Likewise, the stiffness k_{comp}, is obtained from Table 1 or from (73) for single SMA or for the two opposing SMAs actuator respectively.

3.5.2. *Double articulated quadrilateral compensator*

Figure 11 illustrates a second elastic compensation system based on the use of two articulated quadrilaterals (I and II), which are connected together in series. The first quadrilateral (I) is made up of four equal rods DT, EU, RT and SU, hinged together in T and U and fixed to the frame with hinges D and E. In the upper part, the rods are hinged in R and S to the bar PQ, which can translate vertically. The rods DT, EU, RT and SU have length *m*, while the horizontal segments DE and RS have length *f*. The conventional spring S_A, with stiffness k_A and free length L_{0A}, is stretched between hinges T and U.

The second quadrilateral (II) is made up of four equal rods VP, ZQ, VF and ZG, connected with internal hinges in V and Z and fixed to the frame with hinges F and G. The length of the rods VP, ZQ, VF and ZG is *n*, while the length of the horizontal segments PQ and FG is *g*.

Quadrilateral II contains two conventional springs, S_B and S_C. Spring S_B, with stiffness k_B and free length L_{0B}, is stretched horizontally between hinges V and Z. Spring S_C, with stiffness k_C and free length L_{0C}, is stretched vertically between the horizontal sides FG and PQ. Member PRSQ, which is common to the two quadrilaterals, represents the output ports of the actuator. In addition to the conventional springs, S_A, S_B and S_C, the mechanism in Figure 11 contains also the primary active elements of the actuator. Active element SMA1 is hosted by the quadrilateral I and connects the base DE to the output port PRSQ. The second primary element SMA2 (if applicable) is hosted by quadrilateral II and connects the frame FG to the output port PRSQ.

For the sake of simplicity, all the springs shown in Figure 11 (both SMA and conventional) are traction springs. However, the mechanism can work just as well with all compression springs, which can be designed with the same equations presented below.

By imposing the vertical equilibrium of the member PRSQ in Figure 11 and excluding the forces exerted by the shape memory elements, it is possible to obtain the compensation force as a function of the position *x* of the output port.

Imposing that:

$$L_{0A} = f \qquad L_{0B} = g \qquad k_B = k_C \tag{90}$$

the compensation force becomes:

$$F_{comp}(x) = -k_A(l_0 + x) + k_C L_{0C} \tag{91}$$

i.e. it depends linearly on x as desired in an ideal compensation system. The compensation stiffness implied in (91) is:

$$k_{comp} = \frac{\partial F_{comp}}{\partial x} = -k_A \tag{92}$$

From (91) the force exerted by the compensation system at $x{=}x_{min}$ is:

$$F_{comp}\Big|_{x=x_{min}} = -k_A(l_0 + x_{min}) + k_C L_{0C} \tag{93}$$

Conditions (92) and (93) complete the design of the compensation mechanism in Figure 11. The value k_{comp} is taken from Table 1, for the case of a single SMA, or alternatively from (73) for two SMA elements. Similarly, the value $F_{comp}\Big|_{x=x_{min}}$ is taken from (56) or (74), for one or two SMA elements, respectively.

By solving (93) with respect to the product $k_C L_{0C}$, you can get the correct design of spring S_C. From condition (90), we see that the springs S_B and S_C must have the same stiffness and that the free length of spring S_B must be g. It is thus convenient to set to g the free length of spring C as well, so as to have two identical springs S_B and S_C in quadrilateral II with spring rate:

$$k_C = k_B = \left(F_{comp}\Big|_{x=x_{min}} + k_A(l_0 + x_{\min}) \right) \Big/ g \tag{94}$$

4. General discussion

4.1. Performances of the uncompensated SMA actuators

The analytical framework described in Paragraph 3 considers both dissipative and conservative forces acting on the system. Closed-form relationships are developed, which are the basis of a step-by-step procedure for an optimal design of the entire actuator (both with antagonist SMA, bias elastic element or backup force). Specific formulas for dimensioning of the SMA element in the form of straight wires and helical traction springs are also presented.

The physical meaning of dimensionless design parameters s_1, s_2, s_F and s_0 deserves specific comments. Parameter s_1 is an intrinsic property of the SMA material, the higher s_1, the better the SMA behaviour. Parameter s_2 is a measure of the relative stiffness between the bias spring (be it traditional or SMA) and the primary SMA spring. Optimal values of parameter

s_2 maximize the overall travel. Low values of s_2 lead to an increase in stroke and in the actuator dimension. Parameter s_F quantifies the influence of dissipative forces on the stroke of the actuator. For low values of s_F the dissipative force is negligible in comparison with the forces expressed by the backup systems and the actuator moves freely. The upper limit of s_F is reached when the dissipative force becomes high enough to prevent the movement of the actuator. Parameter s_0, is the equivalent of s_F for the conservative force. Since the conservative force helps the backup system, high values of s_0 improve the actuator performances.

Also for the three backup systems analyzed, some specific comments are reported. Using a force as back up for the primary spring may be convenient when there is important conservative force acting on the system which helps the restoring process. Often this solution is not feasible due to dimensional constraint which prevents dead loads to be used. The binary actuator with a traditional spring backup gives the worst performances in terms of output characteristic, due to disadvantages of the elastic slope of the traditional spring. By analyzing Section 3.2 equations it is found that for a given cursor stroke, Δx the optimal value of s_2 provided by (21) minimizes the pre-stretch of the system, p, and tends to minimize the total size of the actuator. Eq.(18) shows that the pre-stretch of the system, p, acts on the stroke, Δx, as a pure gain. Figure 12a displays the normalized stroke, $\Delta x/p$, as a function of the other variables: s_0 (used as parameter of the family of surfaces), s_2 and s_F (these two variables are plotted along the axes), taking $s_1=3$ because this is a typical value for the elastic moduli ratio of commercial shape memory alloy. Eq. (18) and Figure 12a also highlight that for each combination of s_0 and s_F exists an optimal value, s_2^*, of parameter s_2 which maximizes the normalized stroke. Eq. (19) shows that for each combination of s_1 and s_F the stroke increases monotonically when parameter s_2 decreases, implying a more flexible bias spring. Both equations (18) and (19) prove that meaningful strokes ($\Delta x>0$) are only possible if parameter s_F is lower than s_{Fcr} of equation (10).

Section 3.3 shows that the binary actuator with a SMA spring backup gives the best performances in terms of output characteristic, both respect to elastic backup and bias force backup. Equation (30) shows that also in this case the pre-stretch of the system, p, acts on the stroke, Δx, as a pure gain. Figure 12b displays the normalized stroke, $\Delta x/p$, as a function of the other variables: s_0 (used as a parameter of the family of surfaces), s_2 and s_F (these two variables are plotted along the axes), taking $s_1=3$ (a typical value for elastic moduli in austenitic and martensitic phase of a commercial SMA). Equation (30) and Figure 12b also highlight that for each combination of s_1 and s_F an optimal value, s_2^*, of parameter s_2 exists that maximizes the normalized stroke. Equation (30) shows that for each combination of s_1 and s_F the stroke increases monotonically when parameter s_2 decreases, implying a more flexible SMA bias spring. Correspondingly, equation (29) shows that the pre-stretch of the system, p, tends to infinity. If (31) is used to choose s_2, the global pre-stretch, p, is minimized and the total length of the actuator tends to be reduced, but it is not minimized. To obtain a value of s_F and a value of s_2 able to minimize the total length of the actuator, expressions that depend on the embodiment of the springs are needed.

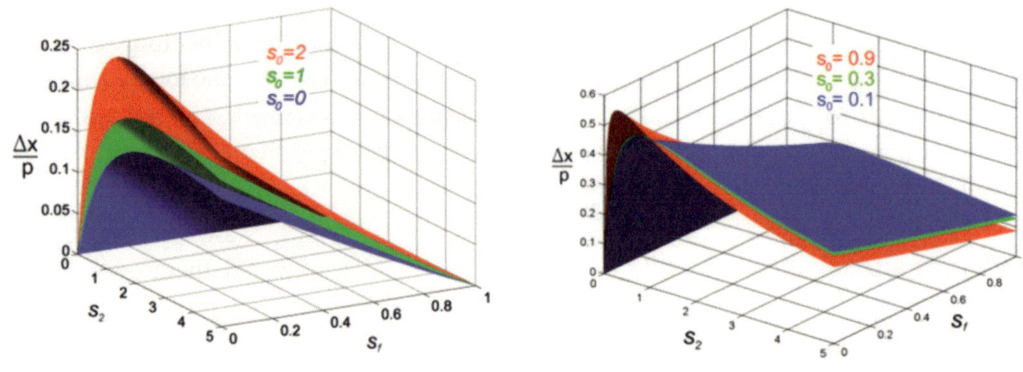

(a) (b)

Figure 12. Normalized stroke of a shape memory actuator backed up by a traditional spring (a) or by an antagonist SMA spring (b) as a function of the parameters s_2 , s_F and s_0.

4.2. Performances of the compensated SMA actuators

The performances of SMA compensated actuators are improved compared to traditional SMA actuators. A comparison between a traditional and a compensated SMA actuators, being equal the active elements and their maximum deformations, shows that the compensated system has longer strokes. Conversely, being equal the output stroke, the deformations in the compensated SMA system are lower than the traditional one. In general the compensated SMA actuator is able to obtain the same performances of a traditional system with lower maximum deformation in the material, while for a given admissible deformation the active elements of a compensated SMA are more stressed because they work against both the external load and the compensation system during the first part of the stroke. The energy stored in the compensator is released in the second part of the stroke helping in keeping the output force constant.

Experimental comparative tests on a SMA actuator with or without a rocker-arm compensator [28] demonstrated that the stroke of the compensated system is always greater regardless the external load. The stroke incremented by 2.5 times with no load up to 22 times with the maximum design force applied. Furthermore, the net stroke of the compensated actuator is not dependent on the applied force. This peculiarity simplifies both the design and the selection of the actuator for a given application. Other remarkable advantages of the compensated actuator lie in the nearly-constant value of the output force and in the fastest response compared to traditional systems, even with longer output strokes. The fastest response is due to the lower difference between internal external forces in the end positions. By contrast, the response time (the time needed to achieve an incipient displacement) is considerably longer for the compensated actuator. This behaviour is due to the high stress at which the SMA springs operate before activation, which requires a greater

degree of martensite-austenite transformation (hence a greater temperature) to start the motion. In general, the compensated actuator has no intrinsic positions of stable equilibrium so mechanical hard stops are needed to prevent dangerous over deformation in the SMA. One hard stop provides a monostable behaviour for the single SMA, the two hard stops needed for the double SMA produces a bistable behaviour. The characteristic of mono or bistability leads to great operative advantages because the actuator is able to: maintain a specific position even without power; achieve precise and repetitive positioning; facilitate control strategies of the overall system.

5. Conclusion

The optimum mechanical design of shape memory based actuators is performed by means of analytical description of the governing equations of the system, providing an easy to follow design procedure for engineers working with SMA devices. Since no complex thermo-mechanical models are involved, the closed form equations developed are based on two simple constitutive models: a linear or a bilinear behaviour for the martensitic phase and a linear relationship for the austenitic phase. Three backup elements are considered to recover the stroke: a constant force, a traditional elastic spring and antagonist SMA element, leaving the designer the opportunity to adopt those which fits the application best. The external load is as general as possible, because a system of both dissipative and conservative forces is taken into account. The actuator performances are improved considering an elastic compensation instead of a backup element. This negative stiffness element can enhance either the stroke or the force of the SMA actuator depending on the needs. The output force design equations both for the SMA actuators and for the compensator are given and the detailed description of the compensated SMA system grants an immediate comprehension of the whole system. In order to give an operative guideline for SMA actuators design, several numerical examples are provided. The proposed procedure is applied to real case studies and to a specific configuration of SMA active elements (wires or springs). The benefits of the described methodology are critically discussed and compared. In terms of mechanical performance the agonist antagonist SMA system gives the best performances both for compensated and uncompensated systems, while optimum values for the non dimensional parameters are provided in order to obtain compact and efficient SMA based actuators. The analytical equations developed in this Chapter allow a SMA based actuator to be correctly designed in order to match the specifications and give useful information to optimize size and mechanical response of the system with no need of complex numerical simulations.

Author details

A. Spaggiari*, G. Scirè Mammano and E. Dragoni
Department of Science and Methods for Engineering, University of Modena and Reggio Emilia, via Amendola, Campus S. Lazzaro Reggio Emilia, Italy

* Corresponding Author

6. References

[1] Otsuka K Wayman C M (2002) Shape memory Materials, Cambridge University Press, Cambrdige UK.

[2] Lagoudas DC, (2008) Shape Memory Alloys Encyclopedia of Aerospace Engineering John Wiley & Sons, New York.

[3] Boyd JG, Lagoudas DC (1996) A Thermodynamical Constitutive Model for Shape Memory Materials. Part I. Int. J. Plast 12(6): 805-842.

[4] Arghavania J, Auricchio F, Naghdabadia R and Sohrabpoura S (2009) A 3-D phenomenological constitutive model for shape memory alloys under multiaxial loadings Int. J. Plast. 26(7): 976-991.

[5] Kuribayashi, K. (1989) Millimeter-sized Joint Actuator using a Shape Memory Alloy, Sensor Actuator, 20(1-2): 57-64.

[6] Mertmann M, Vergani G (2008) Design and application of shape memory actuators, Eur. Phys. J. Special Top. 158:221–230.

[7] Kim B, Lee MG, Lee YP, Kim YI, Lee GH (2006) An earthworm-like micro robot using shape memory alloy actuator, Sensors and Actuators A: 125(2): 429-437,

[8] Reynaerts D. Van Brussel H (1998) Design aspects of shape memory actuators Mechatronics, 8: 635-656.

[9] Lu A, Grant D, Hayward V (1997) Design and comparison of high strain shape memory alloy actuators" Proc.IEEE Robotics and Automation 1(20): 260-265.

[10] Bergamasco M, Dario P Salsedo F (1990) Shape Memory Alloy Microactuators Sensors Actuators A 21(1-3): 253-257.

[11] Nespoli A, Besseghini S, Pittaccio S, Villa E, Viscuso S (2010) The high potential of shape memory alloys in developing miniature mechanical devices: A review on shape memory alloy mini-actuators. Sensors and Actuators A: Physical 158(1): 149-160.

[12] Jansen S, Breidert J Welp EG (2004) Positioning actuator based on shape memory wires, ACTUATOR Proc. of 9th International Conference on New Actuators.

[13] Strittmatter J, Gümpel P (2004) Shape memory actuator for hydraulic valve ACTUATOR Proc. of 9th International Conference on New Actuators.

[14] Bellini A, Colli M, Dragoni E (2009) Mechatronic Design of a Shape Memory Alloy Actuator for Automotive Tumble Flaps: A Case Study, IEEE Trans on Industrial Electronics, 56(7):2644-2656.

[15] Haga Y et al. (2005) Dynamic Braille display using SMA coil actuator and magnetic latch, Sens. Actuators A, 119: 316-322.

[16] Elwaleed AK, et al., (2008) A new method for actuating parallel manipulators, Sens. Actuators A 147:593–599.

[17] Migliavacca F. et al. (2004) Stainless and shape memory alloy coronary stents: a computational study on the interaction with the vascular wall, Biomechan Model Mechanobiol 2:205-217.

[18] Spinella I, Dragoni E (2009) Design equations for binary shape memory actuators under dissipative forces. Pro Insn Mech Engrs C: J. Mech. Eng. Sci 223(C3): 531–543.

[19] Spinella I, Scirè Mammano G, Dragoni E (2009) Conceptual design and simulation of a compact shape memory actuator for rotary motion. J. Mater. Eng. Perform. 18(5-6): 638–648.

[20] Spaggiari A, Spinella I, Dragoni E (2012) Design equations for binary shape memory actuators under arbitrary external forces, Journal of Intelligent Material Systems and Structures (In press) DOI: 10.1177/1045389X12444491

[21] Scirè Mammano G, Dragoni E (2011) Increasing stroke and output force of linear shape memory actuators by elastic compensation. Mechatronics 21(3): 570-580.

[22] Spaggiari A, Spinella I, Dragoni E (2011) Design of a Telescopic Linear Actuator Based on Hollow Shape Memory Springs, J. Mater. Eng. Perform. 20: 489-496.

[23] Scirè Mammano G, Dragoni E (2011) Modelling of Wire-on-Drum Shape Memory Actuators for Linear and Rotary Motion". J. Intell. Mater. Syst. Struct. 22(11): 1129-40

[24] Lagoudas D C, Miller D A, Rong L, Kumar P K (2009) Thermomechanical fatigue of shape memory alloys. Smart Materials and Structures 18:085021.

[25] Scirè Mammano G, Dragoni E (2012) Functional fatigue of Ni–Ti shape memory wires under various loading conditions. Int. J. fatigue (In press) http://dx.doi.org/10.1016/j.ijfatigue.2012.03.004

[26] Budynas R, Nisbett R, (2006) Shigley's Mechanical Engineering Design, McGrow Hill, New York.

[27] Berselli G, Vertechy R, Vassura G, Parenti Castelli V (2009) Design of a single-acting constant-force actuator based on dielectric elastomers. J Mech Robot 1 (3): 031007.1:031007.7.

[28] Scirè Mammano G, Dragoni E (2011) Design and Testing of an Enhanced Shape Memory Actuator Elastically Compensated by a Bi-Stable Rocker-Arm. Proc.ASME SMASIS Conference 18 – 21, Scottsdale, Arizona, USA, ISBN: 978-0-7918-4415-1.

Chapter 2

New Developments on Mini/Micro Shape Memory Actuators

Adelaide Nespoli, Carlo Alberto Biffi, Riccardo Casati, Francesca Passaretti, Ausonio Tuissi and Elena Villa

Additional information is available at the end of the chapter

http://dx.doi.org/10.5772/50473

1. Introduction

Shape Memory Alloys (SMAs) are functional materials, characterized by some attractive features, such as pseudo-elasticity (PE) and shape memory effect (SME) [1]. The latter consist in the capability of the material to recover high deformation values by heating and can be considered very suitable for actuation applications.

In the actuation field, the most common shape memory alloy is quasi-equiatomic NiTi system, commercially known as Nitinol®. Ti-rich NiTi compounds show characteristic transformation temperatures higher than the room one and they can recover high values of deformation. Moreover, these intermetallic compounds are widely used since they exhibit high thermal and mechanical cycling stability. Other NiTi-based alloys are also employed for this kind of applications. In particular, NiTiCu, with Cu substituting Ni in the 3-10at% range, or NiTiCo system are used respectively when the application requires narrow thermal hysteresis or high stiffness.

In Figure 1, the comparison between differential scanning calorimetry (DSC) data derived from NiTi wire (200 μm in diameter) in fully annealed condition and after aging (500ºC for 10 minutes) are reported. As it can be seen, the fully annealed NiTi alloy goes into a one-step transformation both during cooling and heating. In this case, shape memory effect occurs by a martensitic transformation (MT) between a low-temperature monoclinic structure, B19', called martensite, and a high-temperature body-centered cubic parent phase, B2, called austenite. When the material is aged at specific temperatures, this transition may occur in two steps in association with a trigonal phase, called R-phase [1].

During heating, the phase transformation, named inverse MT, is defined by austenitic start (As) and finish (Af) temperatures. Similarly, during cooling direct MT is defined by martensitic start (Ms) and finish (Mf) temperatures, respectively.

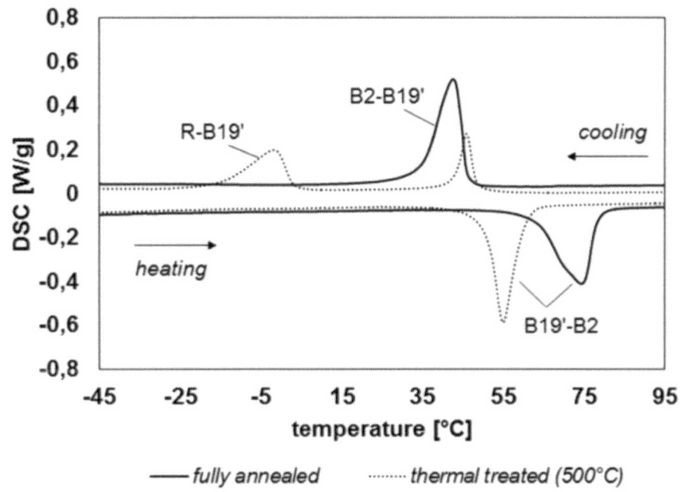

Figure 1. DSC scan of 200μm NiTi wire in the fully annealed state and after aging (500°C for 10 minutes).

Then, the deformation induced by a state of stress on a SMA element can be recovered heating the material over the austenitic characteristic temperatures: the shape recovery starts at As and finishes at Af.

The transition temperatures are strongly influenced by the applied load, since it stabilizes the martensitic structure causing an increase of all transition temperatures. This behavior of SMA is well described by the Clausius-Clapeyron law [1-5].

Furthermore, SMA actuators generally need a training to stabilize their fatigue properties in terms of transformation temperatures and strain recovery. For example, a possible procedure consists in the thermal cycling under a constant load. Figure 2 depicts the evolution of the temperature-stroke relationship during the first thermo-mechanical cycles. It is seen that the stabilization of the material behavior can be reached after few thermal loops.

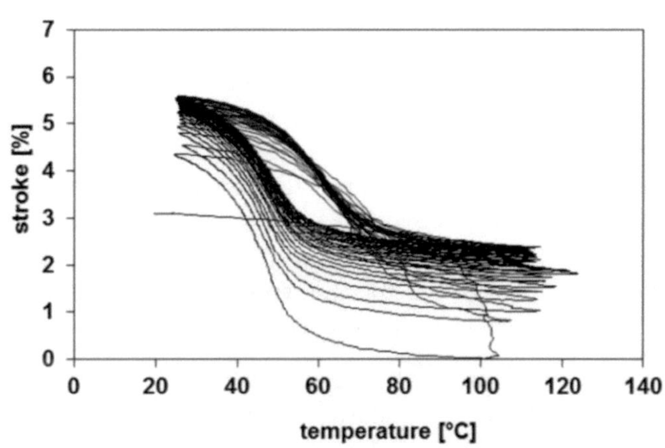

Figure 2. Example of thermo-mechanical hysteresis typical of a SMA material.

Since SMA can generate mechanical work, they are employed as actuators. The functional properties of SMA element used in actuators strictly depend on several factors, such as the chemical composition and the thermo-mechanical history[1,4,5].

Nowadays, the tendency is to produce ever smaller actuators because of the continued miniaturization of all the industrial products. In general, a mini-actuator may be composed by only a SMA element or by a mini-modular device, in which the SMA element is embedded and acts as the core and active part.

In this chapter, the fundamental operating parameters, which affect the performance of shape memory thin wires, are reviewed. Then, the influence of electrical heating conditions performed by different waveforms on functional fatigue of NiTi micro-wires is also reported.

When the SMA element is embedded into a mini-modular mechanical device, its shape could be a serious problem as spaces are very restricted. To solve this drawbacks, a new SMA conformation suitable for the mini and micro scales is presented: it consists of a planar wavy formed NiTi wire, called snake-like arrangement. Currently, this configuration of the SMA wire is principally exploited in the textile and medical domains [6-8]. n the micro scale, original works are reported by Mineta et al. [7-8], who produced different SMA snake elements by means of electrochemical etching for obtaining bending motion of active catheters. Khol et al. [9] used the micro snake geometry to activate a microgripper system. Moreover, Leester-Schadel et al. [10] adopted laser technology to produce micro SMA snake actuators and then used the batch fabrication process to obtain more articulated samples. In this chapter the main mechanical performance of this unusual geometry and its exploitation in a mini-modular mechanical device are presented.

Another topic of this work is related to a current research on fabrication and characterization of SMA micro-snakes, by means of laser technology and following polishing processes. The reason why this non-traditional production technology has been chosen for this purpose is due to its flexibility in the machining of small features, high productivity and repeatability [11-12]. Considering the direction of miniaturization of the products during the recent years [13], in the last part of the chapter the rescaling of SMA mini-snakes down to micro-scale is presented. The attention is also focused on the evaluation of the capability of laser micro-cutting industrial process in the fabrication of micro-snakes from NiTi SMA sheets [14-17]. The evaluation of the thermo-mechanical properties of the produced actuator in correspondence of the different fabrications steps is then presented and discussed in order to describe the behavior of the micro-snake-like element for actuation.

2. SMA linear actuators

The first case study is the application of a simple straight wire as base element for actuation systems [18]. The most important functional properties, which characterize shape memory wire for actuators, are the capability to recover high strain values, the characteristic transformation temperatures and the thermo-mechanical cycling stability [1]. These

characteristics of SMA can be tailored acting on chemical composition, thermal and mechanical processing, shape setting treatment and training [1, 20-25].

In recent years, the good ductility of NiTi intermetallic has been exploited for the production of few tens micron wires [21, 26, 27].

Thanks to their small dimensions, such thin wires permit to reach very short cooling time but, obviously, they can be employed just in all those applications in which high loads are not required. The fabrication of these products has led to the miniaturization of SMA components for actuators that are nowadays employed for new kind of devices such as optical image stabilizer and autofocus for small camera [26]

During working life, the SMA wire is subjected to thermo-mechanical cycling and accumulates plastic deformation, elongating irreversibly and reducing its diameter (see Figure 3). This behavior has to be considered in actuator design. Another important point to consider is the unsteadiness of the functional properties of shape memory wires during the working life. The plastic deformation induced during the thermo-mechanical cycling leads to a change of microstructure of SMA wire. Lattice defects, dislocations and nano-scaled precipitates are introduced into the matrix and they cause significant changes in transformation temperature and in the capability of the material to recover a deformation [28-30], as depicted in Figure 4.

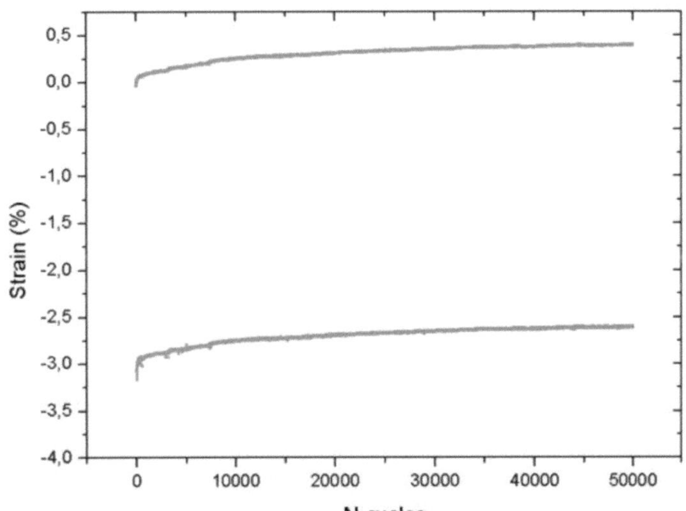

Figure 3. Thermal cycling of a 80μm wire under constant stress (200MPa). The heating was performed by electrical pulse, and stroke was limited at 3%.

In most cases, SMA wires used as actuator, are heated by means of an electrical pulse (Joule effect) and cooled by natural air convection. So, it is very easy to reach short actuation time (heating time) but the reset time (cooling time) cannot be substantially controlled. Slight improvement in the reduction of reset time is achieved positioning the shape memory wire horizontally [19], as shown in Figure 5. In this figure, it can be noticed that after heating (at

200s) the strain is maintained to two constant values for both the two configurations till time is near to 230s and 250s for horizontal and vertical position respectively. After these points, the two wires start the cooling process which results to be faster for the horizontally positioned wire than the vertical one. As an example, it can be seen that at 275s the horizontally positioned wire recovers half of its maximum strain while at the same time the vertical positioned wire has just started the cooling route.

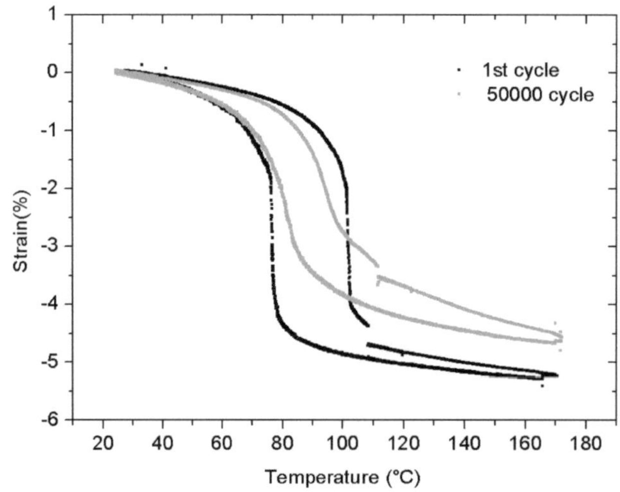

Figure 4. Thermal loop under constant stress (200MPa). The heating was performed by a thermal chamber. The test was carried out on a 80μm wire before and after 5x10[4] thermal cycles under constant stress (200MPa).

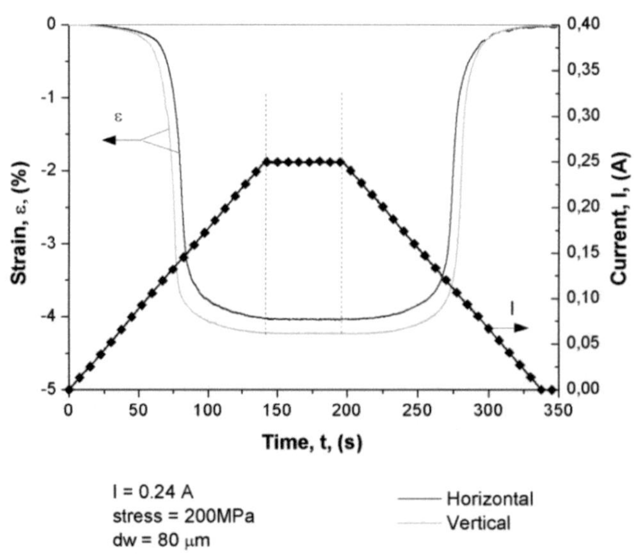

Figure 5. Comparison between the mechanical performance of horizontal and vertical configuration of 80μm SMA wire used as actuator [19].

As previously stated, the heating process of a SMA actuator can be easily obtained by Joule effect. In this case, it has to be considered that the shape of the electrical pulse used for the actuation strongly affects the functional properties of SMA [27]. As an example, Figure 6 shows the time required by the wires to recover the 3% of deformation (actuation time) when it is heated by two different current pulses. At the 1st cycle ramp and step electrical pulse employed to heat two 80μm wires were designed to have the same electrical efficiency but the actuation time is 400ms by step current pulse while it is 623ms by ramp. After $5{\cdot}10^3$ cycles the wire heated by step pulse employs 618ms, so 218ms more than the time employed at the first cycle. As opposite, the ramp pulse leads to an increase of just 22ms. After $5{\cdot}10^4$ cycles the actuation time related to the wire heated by step is even higher than the one related to wire heated by ramp. This behavior leads to a drastic decrease of the step heating method efficiency (from 0.91% to 0.55%), vice versa the efficiency of the ramp heating method does not change so much (from 0.91% to 0.75%).

Then, in order to achieve fatigue performance acceptable for the specific device, the right electrical pulse has to be chosen considering the number of cycles that the SMA actuator has been designed to work.

Recently, the effect of drawing procedure on functional fatigue of thin NiTi wires has also been investigated [21]. Basically, 80μm NiTi wires were produced through two different drawing procedures reaching the same final cold working level before shape setting. These two processes differ for the number of drawing steps carried out to reach a certain cold working level before each heat treatment. After the last thermal treatment the specimen that underwent to a less number of drawing steps shows a narrow thermal hysteresis, even after thermo-mechanical cycling (see Figure 7). It means that adopting a severe drawing procedure leads to improve the actuation performance.

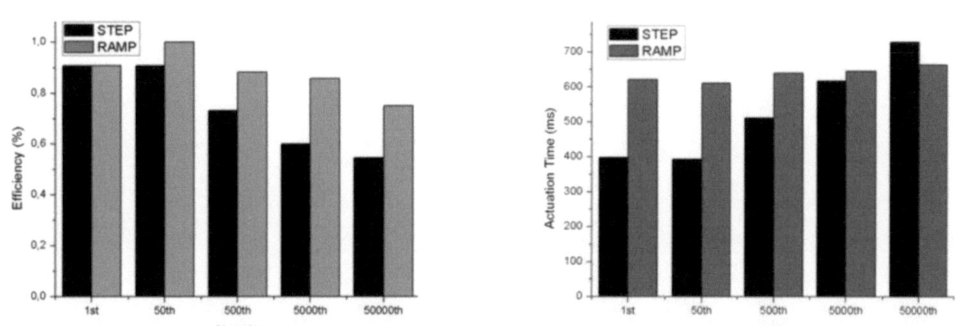

Figure 6. Actuation time tests performed by electrical heating (on the left) and by efficiencies (on the right).

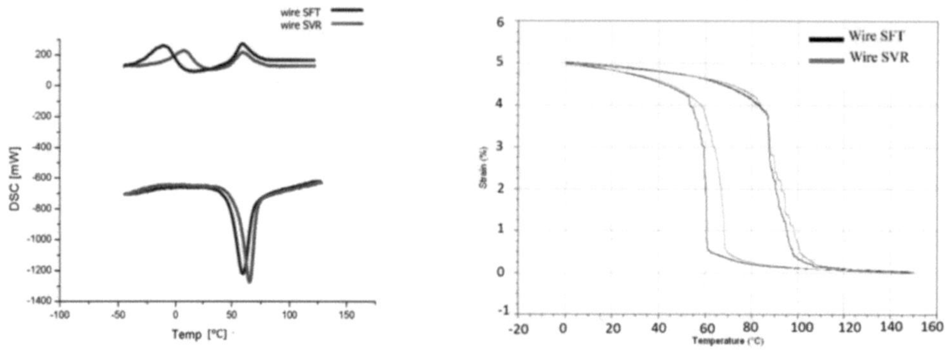

Figure 7. DSC scans, on the left, and thermal loop under constant tensile stress, on the right, of two 80μm NiTi wires obtained by a severe (SVR) and soft (SFT) drawing procedure.

3. Snake-like SMA actuators

The main focus of miniaturization is to develop devices with high mechanical performance in very limited spaces. In the SMA field, the mini-actuator may be composed by only the SMA element or by a mini-modular device in which the SMA element is embedded and acts as the core part. In the latter case, some issues typical of SMA miniaturization should be taken into a serious consideration: first of all, the fixing points of the SMA element and the electrical contacts should be adequately designed to prevent abrupt rupture, and then the use of a wire geometry generally implies the employment of additional mechanical parts which may affect the long run performance of the SMA element.

The mechanical work resulting from the SMA shape recovery is used to produce linear or rotational motion in smart actuators. To attain high strokes in limited space, designers developed several miniature devices which include SMA elements with a straight or a helical spring shape [31]. These two shapes present two opposite mechanical behaviors: the straight conformation allows high forces but smart wire arrangements should be employed in order to achieve also high strokes. As an example, Jansen et al. used two sets of seven pulleys to arrange a 1000mm long SMA wire with a diameter of 0.200mm, and attain 3% of stroke under a constant load of about 130MPa [32]. The use of additional mechanical components may affect the overall performance of the SMA element, leading to friction which could influence the final functional parameters of the device. As opposite, the helical spring shape guarantees high strokes without the use of any further mechanical part but fairly small forces can be attained especially when embedded in miniature devices.

Hence, straight and helical spring conformations represent two opposite limits in the force versus stroke plane. In order to answer both to the need of stroke and force between these two limits, new SMA geometries should be studied.

In a recent work, a new SMA conformation has been proposed [33]; it consists in a planar wavy formed NiTi wire, named snake-like arrangement. To understand the mechanical performance of this unusual conformation as respect to other shapes, three NiTi samples

with different geometry, but with identical elemental composition, thermal history and power consumption were tested by means of a standard tensile test. In this investigation the NiTi wire was formed to have the straight conformation, the helical spring and the snake-like arrangements, and austenite and martensite mechanical responses were assessed separately. As can be seen in Figure 8 the snake-like shape has a mechanical behavior between the wire and the helical spring ones.

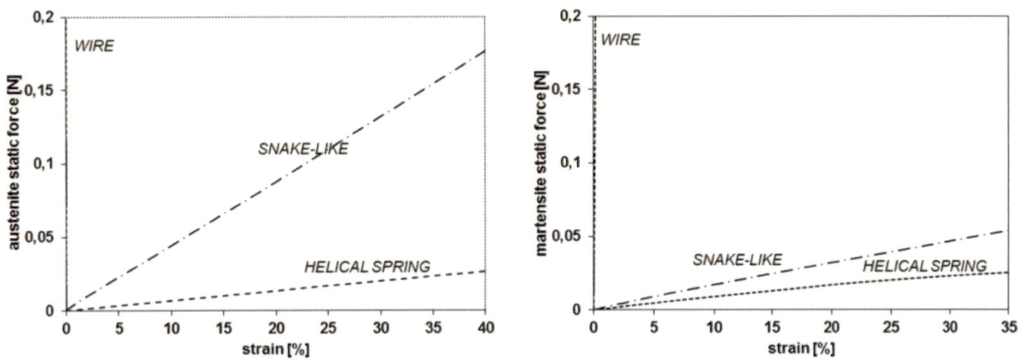

Figure 8. Comparison between austenite and martensite mechanical behavior of different NiTi wire geometries for actuators.

The snake-like conformation is characterized by four main parameters: the number of curvatures, N, the curvature radius, R, the distance between two consecutive curvatures, D, and the height, H. A snake-like specimen can be prepared by constraining a SMA wire in the snake-like arrangement by using a drilled aluminum bar and a series of nails; heating at high temperature is finally used to fix the shape. Figure 9 reports a snake-like sample prepared from a commercial NiTi wire having the diameter of 0.2mm, formed at 500°C during ten minutes and quenched in water at room temperature, with N=4, H=5.27mm±0.07mm, R=0.54mm±0.03mm and D=0.82mm±0.06mm.

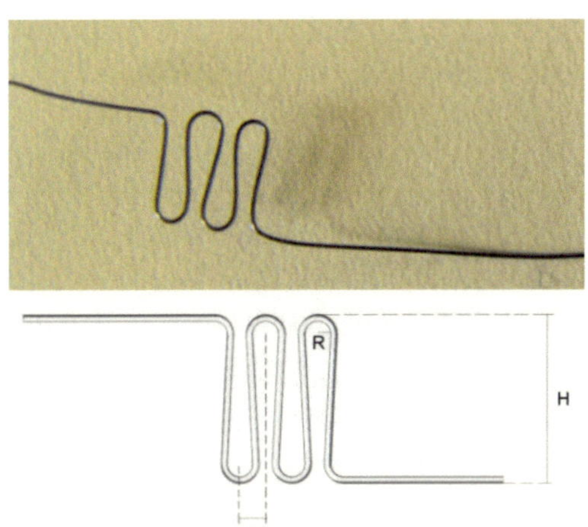

Figure 9. Snake-like NiTi wire with N=4, H=5.27mm±0.07mm, R=0.54mm±0.03mm and D=0.82mm±0.06mm.

As previously stated, during the ten minutes formation process the curved parts of the snake element are in contact with metal devices. This fact does not modify the calorimetric properties of the SMA element in all its length, as can be seen in Figure 10 where the overlapping of the DSC scans relative to the straight and the curved segments is clearly visible.

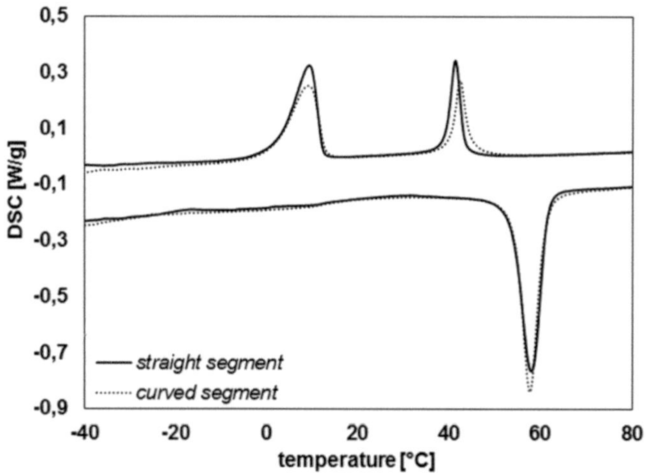

Figure 10. DSC scans of the straight and the curved part of the snake-like NiTi wire.

For small deformation values (see Figure 11), this snake-like shape shows an almost complete recovery of the deformation both for austenite and for martensite.

The relative higher output stroke performed by the snake-like NiTi wire as respect to the straight wire conformation is substantially due to the bend/unbend movement of the curvatures. In order to avoid plastic deformation and to achieve simultaneously a significant output stroke, the snake sample, formed as previously explained, should be stressed with a load near 0.1N, as forces lower and higher than 0.1N do not guarantee a proper functioning of this snake element: in fact, lower forces do not induce martensite formation, as opposite higher forces cause plastic deformations. As an example, in Figure 12 the curves derived from five thermal cycles of the snake-like NiTi sample under two constant loads, 0.1N and 0.2N is depicted. By comparing the two trends, it can be seen that in both cases the snake sample lose part of its shape recovery during cycling; this is much more visible for the specimen stressed with 0.2N.

Fatigue life was studied under the constant load of 0.1N; the snake-like sample was electrically activated with 0.6A with a maximum power consumption of 1W and a work frequency of 0.1Hz. Fracture takes place at the apex of that curvature most far from the applied load and from Figure 13 it can be observed that it occurs near the $7{,}5x10^4$ cycle. The fractured section was analyzed through SEM observation, see the inset of Figure 13; it can be seen a region with the striation of the crack propagation departing from a nucleation site placed on the inner surface of the snake and a region characterized by dimples, typical of ductile materials.

After the cycling test, the DSC scan shows a feeble shifting of the transformation temperatures towards higher values; a better definition of the rhombohedral phase during heating is also visible (see Figure 14).

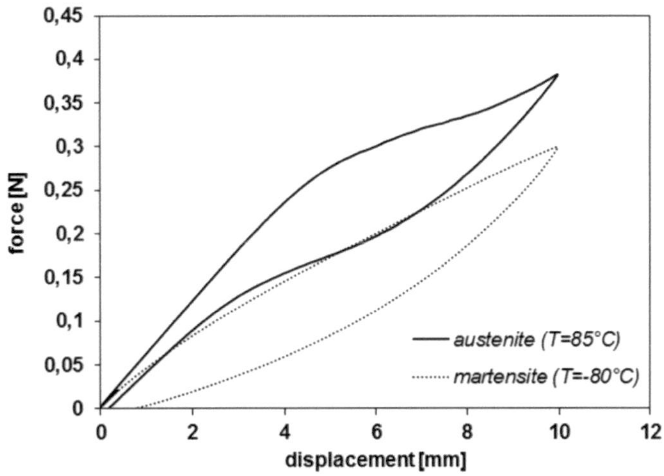

Figure 11. Force versus displacement of austenite and martensite of the snake-like NiTi wire.

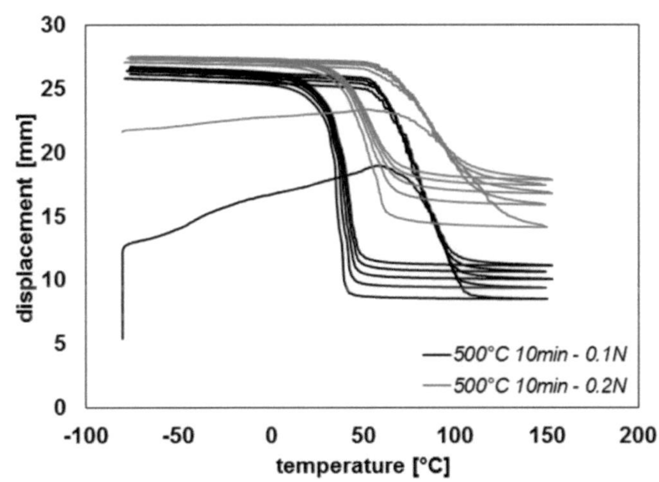

Figure 12. Hysteresis curves of the snake-like NiTi sample treated at 500°C for 10 minutes under 0.1N and 0.2N constant load.

When embedded in mini-modular device, the snake-like wire does not lose its mechanical performance. As an example, in Figure 15 a miniature rotational actuator composed by two mutual antagonist NiTi snake-like wire is reported [34]. The device exerts a stroke of about 120° under a constant torque of 0.1Nmm, as shown in Figure 16; it is activated by 0.6A with a maximum power consumption of 1W and a work frequency of 0.05Hz. Due to the particular construction, plastic deformation of the SMA element is prevented as the SMA elements never exceed 6mm of deformation.

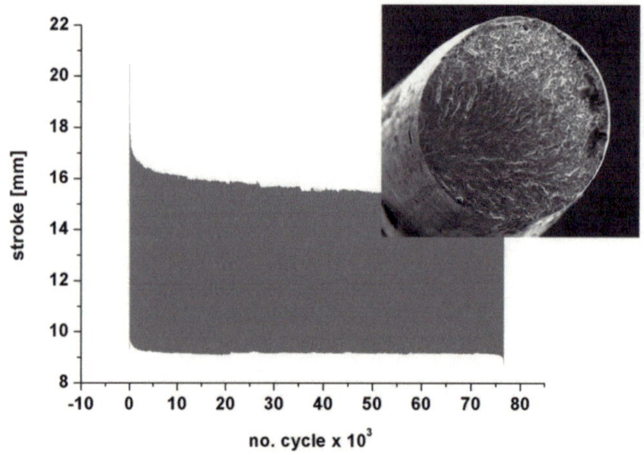

Figure 13. Snake-like NiTi sample fatigue test result under 0.1N load.

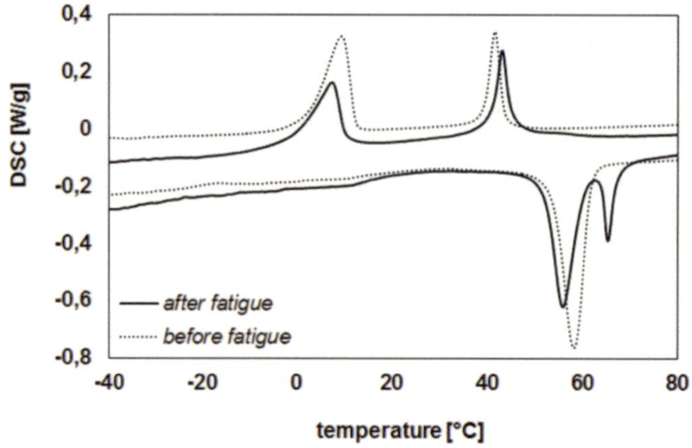

Figure 14. Snake-like NiTi sample DSC scan before and after the fatigue test

Figure 15. Mini rotational actuator activated by two antagonists snake-like NiTi wires.

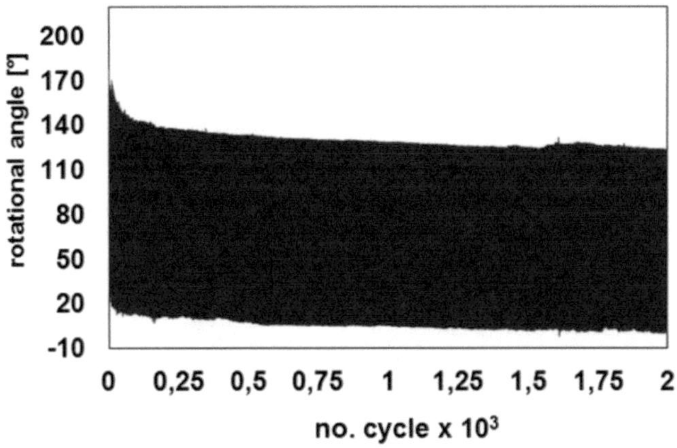

Figure 16. Fatigue test results of the mini rotational actuator under 0.1Nmm constant torque.

3.1. Micro snake-like shape memory alloy elements

In the last years the trend of miniaturization of components and products has been evidently introduced in several industrial fields, such as biomedical, electronics, aerospace and mechanics [35-36]. SMAs, as it already happens in biomedical applications [14,15,37-40], can be machined using laser machining with appreciated and high qualitative results [12]. The decision to investigate this method for the production of SMA actuators is considered an interesting topic because it shows some relevant advantages in comparison with other technologies, such as high productivity, high quality and repeatability, when applied in the industrial world.

Hence, the snake geometry can be easily scale down to the micro scale by the employment of laser technology with appreciated and high qualitative results. Because laser material processing is a thermal process, in which an heat source is used to remove a certain volume of material from the work piece, some thermal damages are evidently obtained, such as melted material, heat affected zone and oxides. To remove these defects, chemical etching and electro-chemical polishing have to be performed on the snake SMA samples [37,41]. Figure 17 depicts SEM images of the snake NiTi sample after laser, chemical etching and electro-chemical processes; top and bottom surfaces as well as snake sample section are shown. It can be seen that laser damages are mainly visible on the bottom surface and that the chemical etching remove the great part of defects. Moreover, a significant loss of mass and consequent reduction of the dimensions of the micro-snake can be evidently observed from SEM pictures due to the material removal of the post-processing. The snake NiTi sample geometry and calorimetric properties change after each process, as can be seen in Figure 18 and in Figure 19 respectively. In particular, the geometrical dimension are referred to a micro-snake obtained starting from a 120μm NiTi sheet. As concern the DSC data, it can be observed that the laser machining produces a refinement of the rhombohedral phase peak both on cooling and heating while the martensite peak barely change as respect to the unworked 120mm thick sheet, being almost spread and invisible. The laser process also yields a conspicuous diminution of the rhombohedral phase transformation temperatures

and of the martensite starting temperature while the other characteristic transformation temperatures do not have an analogous variation being almost unchanged. Chemical etching and electro-polishing steps do not visibly alter any phase transformation temperature both on cooling and on heating, even if some little change due probably to the hydrogen introduction in the chemical process, is visible [45,46].

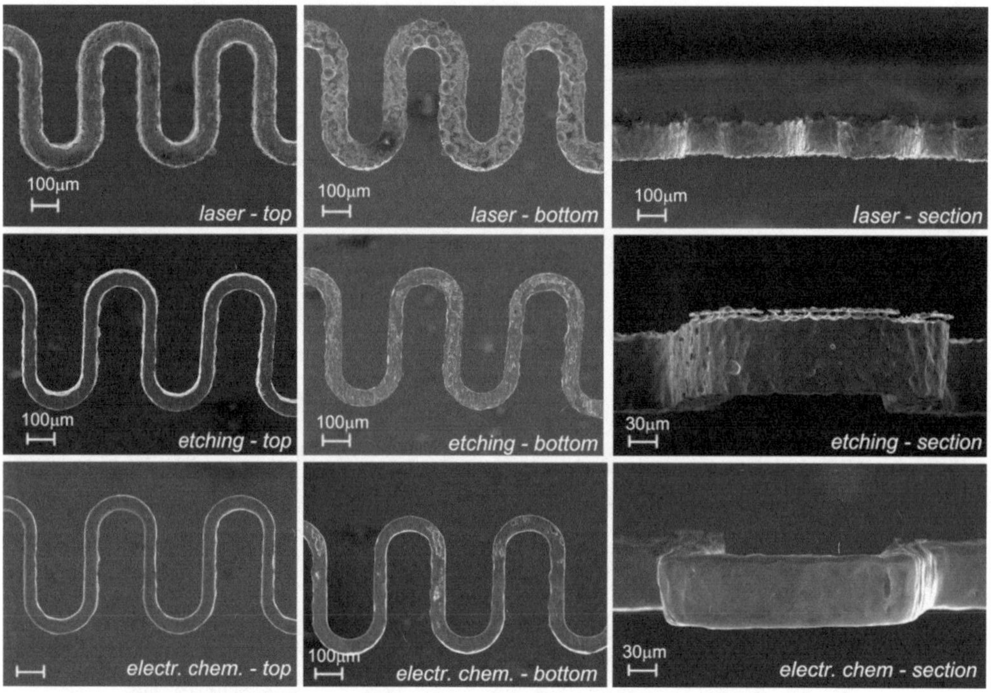

Figure 17. SEM images of the snake NiTi sample surfaces and section after laser, chemical etching and electro-polishing processes.

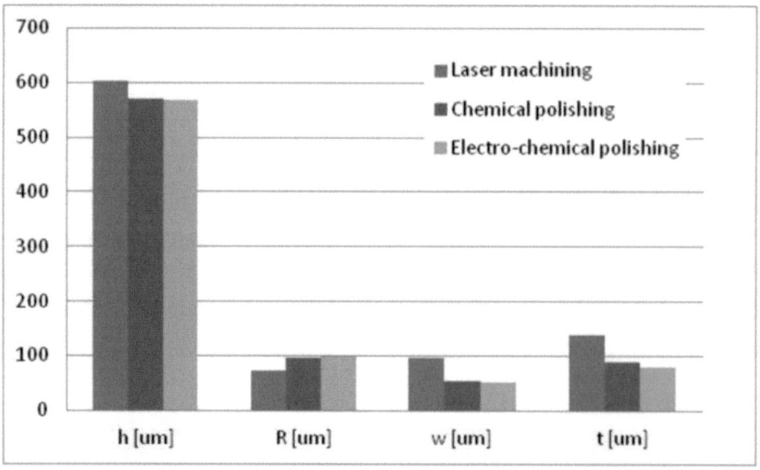

Figure 18. Variation of the geometrical parameters of the micro-snake during the fabrication process (h: high, R: radius of curvatures, w: width, t: thickness).

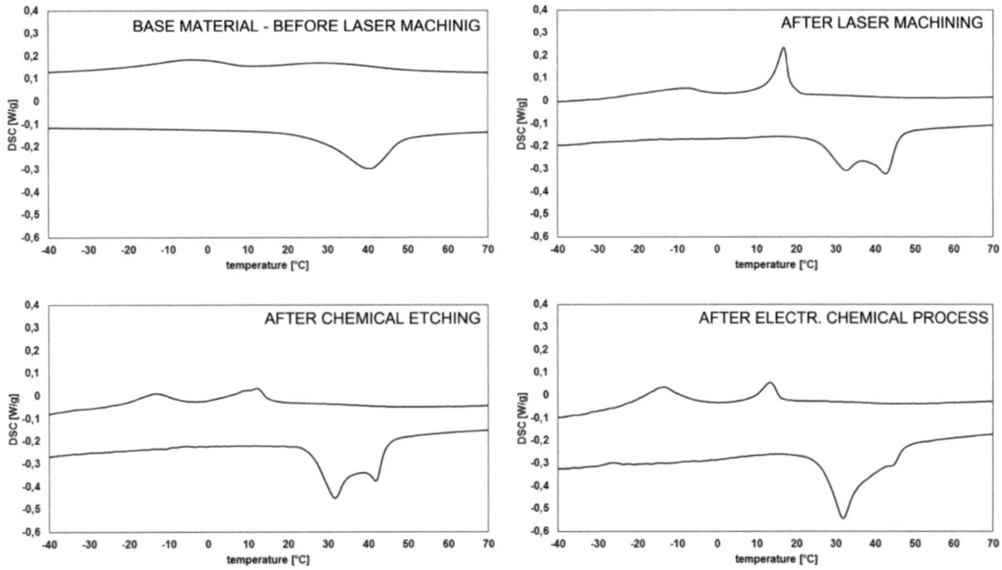

Figure 19. DSC scans before laser machining and in correspondence of each fabrication process.

The fabrication process previously described can be used to produce snake SMA sample of different length. As an example, Figure 20 depicts the SEM image of a micro-snake NiTi sample with 19 curvatures (i.e. 3.8mm of length) obtained starting from a 120μm NiTi sheet treated at 400°C for 15 minutes and quenched in water.

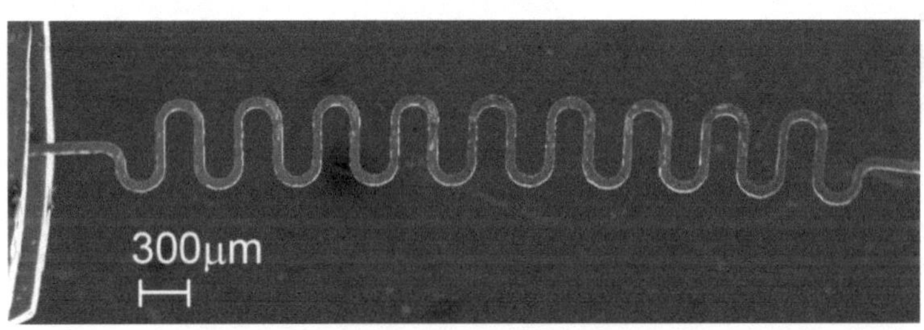

Figure 20. SEM image of a micro-snake NiTi sample obtained starting from a 120μm NiTi sheet.

Strain-recovery and thermo-mechanical cycling tests demonstrate the high performance of this micro-snake sample. As an example, under the constant load of 16mN, it reaches about 40% of recovery with a cycling stabilization within few hundreds of cycles, when heated by electrical current (130mA) with a maximum power consumption of 0.3W and a frequency of 0.15Hz.

4. Conclusions

In this work, new developments on SMA actuators in the mini/micro scale are presented: the attention was focused on straight NiTi thin wires and mini/micro snake-like NiTi samples.

Several working parameters have proved to strongly affect the functional and fatigue properties of SMA actuator, such as the electrical pulse performed to heat a wire, the

different drawing procedures carried out to produce the wire or the wire arrangement into the device.

Then, the functional parameters of a snake-like NiTi wire demonstrate that this kind of geometry may be a good alternative to the most common geometries. It can generate a shape recovery in a very large range of deformation comparing to the straight SMA wire. This point is very beneficial for practical uses since the new proposed SMA conformation just occupies small space to generate the same displacement as the straight one, and it demonstrates to maintain its general mechanical performance when embedded in a mini-modular mechanical device. Finally, the presented fabrication process adopted for the manufacturing of micro-actuators looks to be very attractive (laser process can be performed on a large family of materials with high repeatability). The thermo-mechanical behavior of the micro-snake actuator can be considered interesting in the range of low applied load values with a good cycling stability.

Author details

Adelaide Nespoli*, Carlo Alberto Biffi, Riccardo Casati,
Francesca Passaretti, Ausonio Tuissi and Elena Villa
Consiglio Nazionale delle Ricerche, Istituto per l'Energetica e le Interfasi (CNR-IENI), Lecco, Italy

Acknowledgement

Authors would like to thank Prof. Barbara Previtali, head of Sitec-Laboratory for Industrial Laser Applications of Mechanical Department of Politecnico di Milano, for her precious support on the laser processing.

5. References

[1] Kazuhiro Otsuka, Clarence M Wayman (1998) ShapeMemoryMaterials. Cambridge University Press. 284 p.
[2] Hiroyasu Funakubo (1984) Shape memory alloys. London: Gordon and Breach Science Publishers.
[3] Miller DA, Lagoudas DC (2001) Influence of cold work and heat treatment on the shape memory effect and plastic strain development of NiTi. Mater. sci. eng. A. 308:161–75.
[4] Fukuda T, Kakeshita T, Kitayama M, Saburi T (1995) Effect of aging on martensitic transformation in a shape memory Ti–40.5Ni–10Cu alloy. J. phys. IV. 5:717–22.
[5] Degeratu S, Rotaru P, ManoleaGh, Manolea HO, Rotaru A (2009) Thermal characteristics of Ni–Ti SMA (shape memory alloy) actuators. J. therm. anal. calorim. 97:695–700.
[6] Villa E, Arnaboldi S, Tuissi A, Giacomelli M, Turco E (2009) Mechanical analysis of hybrid textile composites with NiTi wires. J. mater. eng. perform. 18(5–6):517–21.

* Corresponding Author

[7] Mineta T, Mitsui T, Watanabe Y, Kobayashi S, Haga Y, Esashi M (2001) Batch fabricated flat meandering shape memory alloy actuator for active catheter. Sens. actuators A. 88:112–20.

[8] Mineta T, Mitsui T, Watanabe Y, Kobayashi S, Haga Y, Esashi M (2002) An active guide with shape memory alloy bending actuator fabricated by room temperature process. Sens.actuators A. 97–98:632–7.

[9] Kohl M, Krevet B, Just E (2002) SMA microgripper system. Sens. and actuators A. 97-98:646-652.

[10] Leester-Schadel M, Hoxhold B, Lesche C, Demming S, Buttgenbach S (2008) Micro actuators on the basis of thin SMA foils. Microsyst. technol.14:697-704.

[11] John F.Ready, Dave F. Larson (2001) LIA Handbook of Laser Materials Processing. Laser Institute of America, Magnolia Publishing, Inc. p715.

[12] Dubey AK, Yadava V (2008) Laser beam machining—A review. Int. J. mac. tools manuf. 48:609–628.

[13] Gower MC (2000) Industrial applications of laser micromachining. Opt. Express 7:56-57

[14] Li C, Nikumb S, Wong F (2006) An optimal process of femtosecond laser cutting of NiTi shape memory alloy for fabrication of miniature devices. Opt. and lasers in eng. 44:1078–1087

[15] Momma C, Knop U, Nolte S (1999) Laser cutting of slotted tube coronary stents –state of the art and future developments. Prog. in biomed. res. 39-44.

[16] Buttgenbach S, Butefish S, Leester-Schadel M, Wogersien A (2001) Shape Memory Microactuators. Microsys. technol. 7:165-170.

[17] Meijer J, Du K, Gillner A, Hoffmann D, Kovalenko VS, Masuzawa T, Ostendorf A, Poprawe R, Schulz W (2002) Laser machining by short and ultrashort pulses, state of the art and new opportunities in the age of the photons, CIRP Ann. Manufact. Technol. 51:531-550.

[18] Saikrishna CN, Ramaiah KV, Allam Prabhu S, Bhaumik SK (2009) On stability of NiTi wire during thermo-mechanical cycling. Bull. mater. sci. 32:343–352

[19] Zanotti C, Giuliani P, Arnaboldi S, Tuissi A (2010) Analysis of wire position and operating conditions of functioning of NiTi wires for shape memory actuators. Proc. SMST 2010. 688-696.

[20] Frenzel J, George EP, Dlouhy A, Somsen CH, Wagner MFX, Eggeler G (2010) Influence of Ni on martensitic phase transformations in NiTi shape memory alloys. Actamater. 58:3444–3458.

[21] Casati R, Tuissi A, Belochapkine S, Dickinson C, Tofail SAM Thin NiTi wires with reduced thermal hysteresis for shape memory actuators. doi:10.1142/S1793604 7125000991250009

[22] Miller DA, Lagoudas DC (2001) Influence of cold work and heat treatment on the shape memory effect and plastic strain development of NiTi. Mat. Sci. Eng. A. 308:161-175.

[23] Todoroki T, Tamura H (1987) Effect of heat treatment after cold working on the phase transformation in TiNi alloy. Tran. of the jpn. Inst. of met. 28:83-94.

[24] Frick CP, Ortega AM, Tyber J, Maksound AEM, Maier HJ, Liu Y, Gall K (2005) Thermal processing of polycrystalline NiTi shape memory alloys. Mat. sci. eng. A. 405:34-49

[25] Contardo L, Guenin G (1990) Training and two way memory effect in Cu-Zn-Al alloy. Acta metall. mater. 38:1267-1272.

[26] http://www.cambridgemechatronics.com/

[27] Casati R, Tuissi A (2011) Effect of current pulses on fatigue of thin NiTi wires for shape memory actuators. Proc. SMST 2011. Submitted

[28] Wagner MFX, Dey SR, Gugel H, Frenzel J, Somsen Ch, Eggeler G (2010) Effect of low-temperature precipitation on the transformation characteristics of Ni-rich NiTi shape memory alloys during thermal cycling. Intermet. 18:1172-1179.

[29] Simon T, Kroger A, Somsen C, Dlouhy A, Eggeler G (2010) On the multiplication of dislocations during martensitic transformation in NiTi shape memory alloys. Acta mater. 58:1850-1860.

[30] Ibarra A, San Juan J, Bocanegra EH, Nò ML (2007) Evolution of microstructure and thermomechanical properties during superelastic compression cycling in Cu-Al-Ni single crystal. Acta Mater. 55:4789-4798.

[31] Nespoli A, Besseghini S, Pittaccio S, Villa E, Viscuso S (2010) The high potential of shape memory alloys in developing miniature mechanical devices: A review on shape memory alloy mini-actuators. Sens. and actuators A.158:149–160

[32] Jansen S, Breidert J, Welp EG (2004) Positioning actuator based on shape memory wires. Actuators 2004. 9th International Conference on New Actuators.

[33] Nespoli A, Villa E, Besseghini S (2011) Thermo-mechanical properties of snake-like NiTi wires and their use in miniature devices. J. therm.anal calorim. DOI 10.1007/s10973-011-1324-0

[34] Nespoli A, Bassani E, Besseghini S, Villa E (2010) Rotational mini-actuator activated by two antagonist shape memory alloy wires. Phis. Procedia 10:182-188

[35] Hsu T (2002) Miniaturization – A paradigm shift in advanced manufacturing and education. Proc. IEEE/ASME

[36] Feldmann K, Franke J, Schüßler F (2010) Development of micro assembly processes for further miniaturization in electronics production. CIRP Annals –Manufact. technol. 59:1–4

[37] Kathuria YP (2005) Laser microprocessing of metallic stent for medical therapy. J. of Mater. Process. Technol. 170:545-550

[38] Meng H, Liao J, Zhou Y, Zhang Q (2009) Laser micro-processing of cardiovascular stent with fiber laser cutting system.Opt. And laser technol. 41:300– 302

[39] Previtali B, Arnaboldi S, Bassani P, Biffi CA, Lecis N, Tuissi A, Carnevale M, LoConte A (2010) Microcutting of NiTiCu alloy with pulsed fiber laser.ISBN 978-0-7918-3877-8, Order No. I844CD

[40] BiffiC A, Bassani P, Tuissi A, Carnevale M, Lecis N, LoConte A, Previtali B (2011) Flexural Vibration Suppression of Glass Fiber/CuZnAl SMA Composite. Funct. Mater. Lett., DOI 10.1142/S1793604712500142

[41] Raval A, Choubey A, Engineer C, Kothwala D (2004) Development and assessment of 316LVM cardiovascular stents. Mater.sci. and eng. A.386:331–343.

[42] Kudesia SS, Solana P, Rodden W, Hand DP, Jones J (2005) Appropriate Regimes of Laser Drilling Models Containing Melt Eject Mechanisms. Journal of Laser Applications, 14:159-164.

[43] Rao BT, Kaul R, Tiwari P, Nath AK (2005) Inert gas cutting of titanium sheet with pulsed mode CO2 laser. Opt. and Lasers in Eng. 43:1330–1348

[44] Biffi CA, Lecis N, Previtali B, Vedani M, VimercatiG (2010) Fiber laser microdrilling and its effect on material microstructure. The int. j. of adv. manuf. Technol. 51:983-994

[45] Mazzolai FM, Biscarini A, Coluzzi B, Mazzolai G, Villa E, Tuissi A (2007) Low frequency internal friction of hydrogen-free and hydrogen-doped NiTi alloys. Acta mater.55:4243-425

[46] Biscarini A, Coluzzi B, Mazzolai G, Mazzolai FM, Tuissi A (2003) Mechanical spectroscopy of H-free and H-doped NiTiCu shape memory. J. of alloy and compd.356:669-672

Chapter 3

SMA-Based Muscle-Like Actuation in Biologically Inspired Robots: A State of the Art Review

William Coral, Claudio Rossi, Julian Colorado, Daniel Lemus and Antonio Barrientos

Additional information is available at the end of the chapter

http://dx.doi.org/10.5772/50209

1. Introduction

New actuation technology in functional or "smart" materials has opened new horizons in robotics actuation systems. Materials such as piezo-electric fiber composites, electro-active polymers and shape memory alloys (SMA) are being investigated as promising alternatives to standard servomotor technology [52]. This paper focuses on the use of SMAs for building muscle-like actuators. SMAs are extremely cheap, easily available commercially and have the advantage of working at low voltages.

The use of SMA provides a very interesting alternative to the mechanisms used by conventional actuators. SMAs allow to drastically reduce the size, weight and complexity of robotic systems. In fact, their large force-weight ratio, large life cycles, negligible volume, sensing capability and noise-free operation make possible the use of this technology for building a new class of actuation devices. Nonetheless, high power consumption and low bandwidth limit this technology for certain kind of applications. This presents a challenge that must be addressed from both materials and control perspectives in order to overcome these drawbacks. Here, the latter is tackled. It has been demonstrated that suitable control strategies and proper mechanical arrangements can dramatically improve on SMA performance, mostly in terms of actuation speed and limit cycles.

Due to their limitations, SMAs have not raised the attention of the robotics technology for several years. However, recent studies have demonstrated that by (i) finding suitable niches of application, (ii) dedicated mechatronics design, and (iii) ad-hoc control strategies, SMAs can effectively be used as an alternative actuation technology in a wide spectrum of applications and robotic systems. Indeed, as it will be introduced in this chapter, careful control design that takes into account the particular characteristics of the material coupled with proper mechanic design, play a significant role for an efficient use of SMAs. Even so, it is clear that SMAs (and smart materials in general) cannot, at this stage, be thought as a universal substitute for classical servomotor technology. However, niches of applications can be found that greatly benefit from this technology. Bio-inspired artificial systems are one such niche.

Although SMAs are mostly used as actuators, they also have sensing capabilities. Despite most of the SMA physical parameters are strongly related in a nonlinear hysteresis fashion, the electrical resistance varies linearly with the strain of the alloy. Because strain is kinematically related to the motion of the actuator (either linear motion or rotational), the electrical resistance and the motion produced by the actuator are both linearly related. This linear relationship between resistance variation and motion is achieved because the martensite fraction is kinematically coupled to the motion, and the martensite fraction is what drives the resistance changes. This issue is an advantage for developing closed-loop position controllers that regulate the SMA actuation. In fact, most of the applications involving position linear control of SMAs, feedback electrical resistance measurements to estimate the motion generated by the actuator. This avoids the inclusion of external position sensors for closing the control loop.

SMAs are used in a variety of applications [46],[40],[56],[29],[27],[80]. Their special properties have aroused great expectations in various technologies and industries; it can be used to generate a movement or storing energy. In addition, its scope covers many sectors ranging from the use in deployable satellite antennas for different sensors to machinery, to materials for the construction of suspension bridges or anti-seismic devices. In general, all applications somehow depend on the effect of action-reaction of the material and the conditions under which particular application takes place, which make the SMAs a functional material.

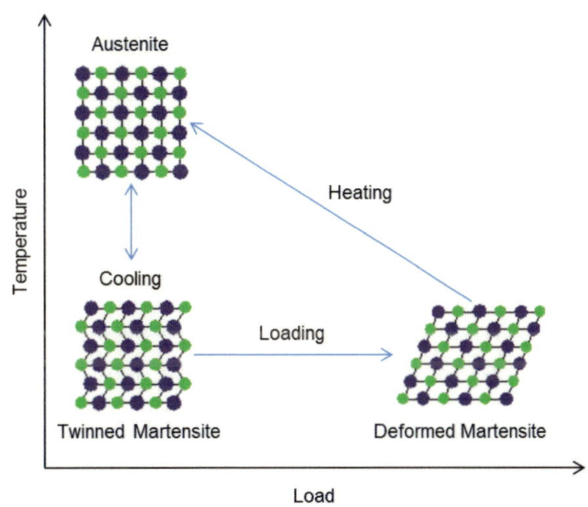

Figure 1. Microscopic viewpoint of the Shape Memory Effect

For instance, they are being used in many non-invasive surgery devices [45],[21],[62],[23],[43] and biomedicine, taking advantage of their large strains and their capability to recover the shape when the load is removed. This property allows applications in devices such as stents, tubular prosthetic devices, because it restores the ability of flow of any bodily duct affected by a narrowing.

In classical robotic systems, linear actuation systems have been proposed using SMAs. The focus of this chapter is on bio-inspired robotics. SMA-based actuators provide a suitable technology as muscle-like actuation mechanisms, which resemble the mechanics of muscles in biological systems. For this reason in the last years a number of bio-inspired robots have been

designed adopting SMA technology. In this paper, we review the main prototypes, organizing them according to the mean (water, air, ground), and on main morphological characteristics (full body actuation or appendices only).

2. Shape memory alloys background

2.1. Principle of operation

Shape Memory Alloys are metallic materials with the ability to "remember" a determined shape, even after a severe deformation produced by a thermal stimulus. In the case of metallic alloys, the shape memory effect consists on a transition that occurs between two solid phases, one of low temperature or martensitic and other of high temperature or austenitic. The material is deformed in the martensitic phase and retrieves, reversibly, its original dimensions by heating above a critical transition temperature. The terms martensite and austenite originally referred only to the steel phases, however these terms have been extended referring not only to the material but also the kind of transformation. Thereby, the martensite steel involves a change of volume and shape, while the SMA has basically a change of length.

In general, NiTi (Nickel-Titanium) SMAs are the most common alloys used. This is basically because these materials are intrinsically susceptible of use both as sensors and actuators, which makes them suitable for integration in smart structures. NiTi SMAs work based on the shape memory effect, which essentially takes place by the influence of temperature change of the material; i.e. the temperatures at which the martensitic and austenite phase transformations begin and end. Figure 1 depicts how these changes occur at the microscopic level of the material. The phase transition occurs when the material is heated or cooled. In general, there is a certain temperature range for the transition, which is mainly defined by the manufacturer.

SMAs normally exhibit one-way shape memory effect, also called memory effect in a simple manner. The alloy deforms upon heating but cooling does not change the shape unless it is stressed again. The percentage of deformation of NiTi alloys (% of strain) is about five percent, a range considerably higher if one considers that the deformation of common steel allows only an average of two percent. Currently, SMAs that exhibit two-way shape memory effect are also manufactured. In this case, the alloy expands by heating above the range of transition temperature and spontaneously contract when cooled again below this temperature [47]. To produce the double shape memory effect, the material is subjected to heat treatment, also called training. This training-phase forces the material to remember both heating and cooling states.

From the microscopic viewpoint (Figure 1), all the physical properties of the alloy vary depending on the phase, i.e. from cooling to heating and vice versa. Some of these properties refer to corrosion resistance, elasticity, damping capacity, strain, stress, electrical resistance, and temperature. Therefore, shape memory alloys behave in a thermo-mechanical way, with all these variables strongly coupled within a nonlinear hysteresis fashion.

Table 1 shows the commercial characteristics of SMAs depending on the diameter of the wires (NiTiNol®). From the table it can be noticed their high electrical power consumption. In robotics applications, power consumption is a critical issue due to the level autonomy of the robotic system is fully dependent on the capacity of the onboard batteries.

Diameter Size *inches(mm)*	Resistance	Pull Force *pounds* *(grams)*	Approximate Current for 1 Second Contraction (mA)	Cooling Time $158^{\circ}F, 70^{\circ}C$ "LT" Wire (seconds)	Cooling Time $194^{\circ}F, 90^{\circ}C$ "HT" Wire (seconds)
0.001 (0.025)	36.2 (1425)	0.02 (8.9)	45	0.18	0.15
0.0015 (0.038)	22.6 (890)	0.04 (20)	55	0.24	0.2
0.002 (0.050)	12.7 (500)	0.08 (36)	85	0.4	0.3
0.003 (0.076)	5.9 (232)	0.18 (80)	150	0.8	0.7
0.004 (0.10)	3.2 (126)	0.31 (143)	200	1.1	0.9
0.005 (0.13)	1.9 (75)	0.49 (223)	320	1.6	1.4
0.006 (0.15)	1.4 (55)	0.71 (321)	410	2	1.7
0.008 (0.20)	0.74 (29)	1.26 (570)	660	3.2	2.7
0.010 (0.25)	0.47 (18.5)	1.96 (891)	1050	5.4	4.5
0.012 (0.31)	0.31 (12.2)	2.83 (1280)	1500	8.1	6.8
0.015 (0.38)	0.21 (8.3)	4.42 (2250)	2250	10.5	8.8
0.020 (0.51)	0.11 (4.3)	7.85 (3560)	4000	16.8	14

Table 1. Characteristics of NiTinol®SMA wires [6].

2.2. Improving the performance of SMA actuators

One of the main limitations in SMA actuation speed is due to high latency that the the cooling time of the wire implies. Despite increasing the input heating power can reduce the heating time, large cooling times limit the operation frequency of the actuator. On average, NiTi wires with a diameter of $127\mu m$ typically requires an electrical current input about $320mA$ to contract in about $1s$ (nominal heating time) and relax in approximately $1.4s$ (nominal cooling time). In this case both contraction and recovery times would set a nominal actuation frequency about $0.416Hz$, quite slow for many applications requirements.

Research to overcome this limitation has been oriented towards developing cooling systems for SMAs, aimed at decreasing the nominal cooling time involved during the recovery process. In this direction, temperature control methods have been proposed in [41]. Cooling systems based on Peltier cells [18] or active cooling [67], have been commonly used. However, nowadays bio-inspired robotic systems tend to be small and light, therefore other methodologies for enhancing SMA actuation speed must be addressed. For several years different strategies have been proposed to implement rapid control in the SMA wires [12],[19],[70],[72],[11],[71],[75]. A system consisting of rapid heating of the SMA was proposed by [12] aimed at increasing the overall actuation frequency by means of overloading the operation of SMAs. The term overloading refers to increasing the amount of input heating power to be delivered to the SMA wires. In [74] experiments carried out using a two degree-of-freedom Pantograph robot actuated by an antagonistic pair of SMA wires acting

as linear actuators have shown how the nominal actuation frequency was increased from $0.416Hz$ to $1Hz$.

Overloading should be monitored in order to avoid overheating problems that may cause physical damage of the shape memory effect. In [71], further research in this direction allowed for the introduction of a force control architecture with the proper mechanisms for safe overload the operation of SMA actuators. In the prototypes described in Section 4, we have used a control architecture similar to the one described in [71], which makes use of proper mechanisms to overload the operation of SMAs. However, these mechanisms have been adapted to work within a position control scheme, avoiding the need of including external force sensors. Section 4.2 will detail on this issue.

Besides rapid heating techniques to overload SMA operation, further investigations have been also carried out to verify whether SMAs can respond to high frequencies. In [64] and [73] experiments have demonstrated that NiTi SMA wires with a diameter of $0.1mm$ can respond up to frequencies of $2KHz$. This high-frequency response corresponds to small-signal heating currents inputs with frequencies of that magnitude. These results allow for the development of small-signal high-bandwidth controllers capable of improving SMA performance, but more important, eliminating the limit cycles of operation of SMAs. In other approaches, $20 - 30Hz$ limit cycles have been observed, whereas in [19], [50] at approximately $100 - 200Hz$. In this regard, the use of high-bandwidth force sensors might be suitable for developing a SMA force feedback control system.

2.3. Modeling and control

The physical behavior of SMAs is more complex than many common materials: the stress-strain relationship is nonlinear, hysteresis is presented, large reversible strains are exhibited, and it is temperature dependent. This thermo-mechanical relationship can be described by formulating phenomenological models. Tanaka in [69] was one of the pioneers to study a stress-induced martensite phase transformation, proposing an unified one-dimensional phenomenological model that makes use of three state variables to describe this process: temperature, strain, and martensite fraction. His main contribution was to demonstrate that the rate of stress is a function of strain, temperature and martensite fraction rates. Later, Elahinia [7], [8] proposed an enhanced phenomenological model compared to other works [40], [31], [69] and also addressed the nonlinear control problem. This model was able to better describe the behavior of SMAs in cases where the temperature and stress states changed simultaneously. Their model was verified against experimental data regarding a SMA-actuated robotic arm [10].

Phenomenological models may provide some insights of SMA thermo-mechanical behavior that facilitate the development of control procedures. To control purposes, parameters' tuning is highly dependent of a modeling stage, but definitively phenomenological models are not the best choice for control design, especially if the goal is related to improving actuation speed. In this direction system identification is a promising alternative. As noted by [40], [19], [11], [71], [20], identified linear models for SMA can be developed. It has been demonstrated that the AC response of NiTi SMA wires behave as a first order low-pass filter.

Section 4 of this paper details two different approaches for modeling SMAs; one based on identifying how electrical resistance change as a function of the input current [59], and the other based on identifying how the output torque produced by an antagonistic pair of SMA actuators change as as a function of the applied power [5]. Furthermore, [5] details how to take advantage of phenomenological models for simulating overheating problems when a SMA wires are overloaded. Attempting to perform this analysis on the real SMA actuators might cause several damage to the structure. Phenomenological models are really useful for determining the upper limits of applied input heating currents.

The control methods presented in [59] and [5] have been conceived for controlling a pair of antagonist SMA actuators. The antagonistic configuration is useful for having SMA actuators where each direction of motion can be controlled. In [59], the antagonistic pair of actuators must bend the structure of the fish robot, whereas in [5] the antagonistic SMA actuators are connected to a joint for providing the rotational motion. Other approaches in [11], [30], [41], [28] have demonstrated the advantages of using an antagonistic arrangement in terms of controllability. When the active actuator is being heated while the passive (antagonistic) is cooling, hysteresis effects are reduced due to the external stress that the active actuator applies on the inactive one above the austenite finish temperature.

3. Bio-inspired robots with SMA muscle-like actuation

The use of SMAs as artificial muscles allows for more realistic bio-inspired actuation presented in nature [26]. SMA wires acting as muscle fibers can respond upon electrical signals, taking advantage of the large pull force and its excellent strength-weight tradeoff. Currently, the use of SMAs in biomimetic robotic systems [3],[13],[9] can be found in ground, water and air robots, in many sizes, including those micro-robots or microstructures [25],[78],[53],[35],[63],[82],[37],[37]. In the following sections, we describe the most representative bio-inspired robots and structures that integrate SMAs as muscle-like actuation mechanisms.

3.1. Water

Biologically inspired robots that operate in water can be found in two categories. Firstly, robots that use SMAs for actuating appendices (fins), and secondly, robots that use SMAs to actuate the robot's body. In the latter body actuation is used for undulatory motion (fish-like robots). Some animals can move by bending their body in such a way to produce a backward-propagating propulsive wave. The movement obtained by bending a continuous structure is much more natural than others where joints are presented. In Section 4.2 our bending structure prototype is presented.

3.1.1. A micro-robot fish with embedded SMA wire actuated by flexible biomimetic fin

In [78] it has been proposed a micro robot fish that uses a flexible biomimetic fin propeller with embedded SMA wires to mimic the musculature and flexible bending of squid fin. The propulsion consists of an active component (the biomimetic fin) and a passive component (the caudal fin). The biomimetic fin-based propulsion mechanism is an actuator that combines the SMA wire and an elastic substrate.

This micro-robot fish introduces a new concept in the world of biomimetic robotics due to its ability to swim noiseless. This means the robot avoid the use of any traditional components like gears, bearings and joints, only using the SMAs as actuators that produce the propulsion. The robot is able to achieve a swimming speed of $112mm/s$ when the SMAs actuate at $2.1Hz$ (contracting upon electrical heating), and a minimum turning radius of $136mm$, which makes the robot the fastest micro robot-fish compared to other prototypes that use IPMCs like actuators [14],[54],[33]. The authors have measured the robot performance based on the Strouhal number [15], [24], which typically varies between 0.25 to 0.35 for the biological counterparts. Their robot has a Strouhal number of 0.58 (at maximum swimming speed). This upper value highlights the optimal movement of the robot, however, high amounts of input power have been required to actuate the SMAs.

3.1.2. Towards a biologically inspired small-scale water jumping robot

In [63], the locomotion description of a water-jumping robot that mimic the ability of the water striders and the fishing spider to jump on the water surface. This biomimetic robot achieves a vertical jumping motion by pushing the water surface. The motion is triggered with a latch driven by the SMA actuator.

As a result of the research, quantification of Re = 260 (Reynolds number is the ratio of inertial over viscous forces), Bo = 0.0054 (Bond number is the ratio of the buoyancy to the surface tension) and We = 4.7 (Weber number is the ratio of the inertia to the surface tension) and the B_a (Baudoin number is the ratio of the body weight to the surface tension) suggest that the physics of jumping in this robot is similar to those of the fishing spider. The Bond, Weber number and Baudoin numbers are explained by [15], [24]. In terms of actuation, the SMA allows the robot to be extremely light (mass of $0.51g$), which it is essential to ensure the buoyancy on water. The maximum jumping height is $26mm$, 26% of the height reached when jumping on ground ($53.1mm$). This prototype is the first concept of jumping robot that integrates SMAs within a structure with an overall mass of $1g$. The robot requires $2W$ of power consumption in order to generate a force of $1.35mN$.

3.1.3. A micro biomimetic manta ray robot fish actuated by SMA

In [79] a manta ray robot fish actuated by SMA wires is designed. Figure 2 shows the prototype of the robot. Two pectoral fins arranged in triangular-shaped made of latex with a thickness of 0.2mm form the fin surface.

This micro manta ray was the first prototype that uses SMAs to generate thrust. This robot is capable to swim forward and turn. The sweep back angle of the pectoral fins is 20°. A maximum swimming speed of $57mm/s$ was achieved and the maximum amplitude of the motion was $40mm$. All the biomimetic fins are open-loop controlled.

3.1.4. Controlling a lamprey-based robot with an electronic nervous system

In [81] a sea Lamprey has been developed. The robot consists of a cylindrical electronics bay propelled by an undulatory body axis. SMA actuators generate propagating flexion waves in five undulatory segments of a polyurethane strip. The lamprey robot Figure 3(a) consists on a cylindrical hull that houses the electronics and battery pack. In this application, the authors use a neuronal network that allows the robot to be controlled in real time. This neuronal

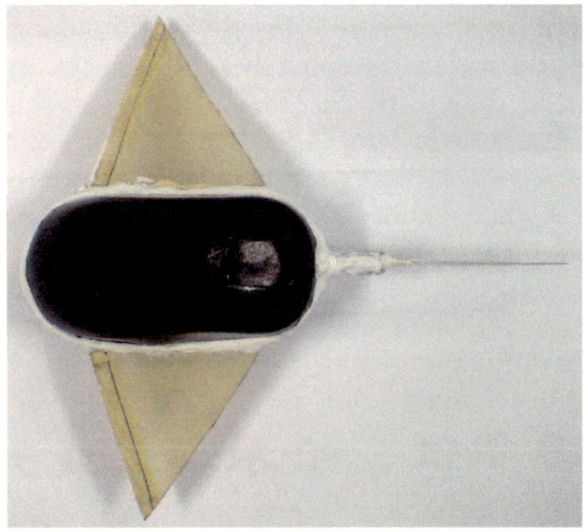

Figure 2. Micro biomimetic manta ray robot fish [79].

network also drives the control of the SMAs. The results have shown the system can reject disturbances thanks to the robustness of the nonlinear controller [55]. Each SMA wire drains $1.5A$ of electrical current when activated.

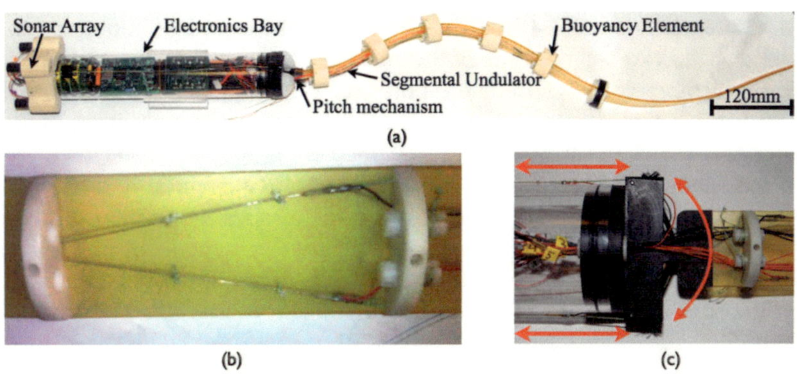

Figure 3. (a) Lamprey Robot with sonar array, (b) Lateral view of tail segment showing nitinol actuator, Teflon vertebra and tensioning nuts and (c) Lateral view of pitch mechanism [81].

3.1.5. A biomimetic robotic jellyfish (Robojelly) actuated by shape memory alloy composite actuators

The newest and more advanced aquatic robot that uses SMA actuators is a jellyfish robot designed by [68]. The hydrogen-fuel-powered robot called "Robojelly" mimics the propulsion, morphology, kinematics and physical appearance of a medusa (jellyfish); the Aurelia aurita species. The bio-inspired actuators are made of silicone, SMA wires and spring steel.

The development of Robojelly has introduced a systematic method for the design and fabrication of SMA-based actuators called BISMAC (bio-inspired shape memory alloy composite). This method allows for bending the structure of the robot by means of SMA contraction [77]. Thanks to the BISMAC SMA arrangement, this robot was capable to mimic the physics and swimming characteristics of jellyfish in terms of A. aurita's bell geometry, passive relaxation mechanism, neutral buoyancy, frequency of motion, and

deformation-to-flap motion profiles. The structure can be bended by the SMAs actuators (deformation), and then a flap motion of the bell-segment structures provide the propulsion. The Robojelly was able to produce enough thrust to propel itself and achieve a proficiency of $0.19s - 1$ which is comparable to the natural medusa at $0.25s - 1$. The robot consumes an average of 16.74W over its 14th cycle of actuation. This robot confirms the fact that most aquatic biomimetic robots use SMA wires combined with other materials to create SMA-based actuators. This characteristic shows the flexibility of the SMA to work in combination with other materials.

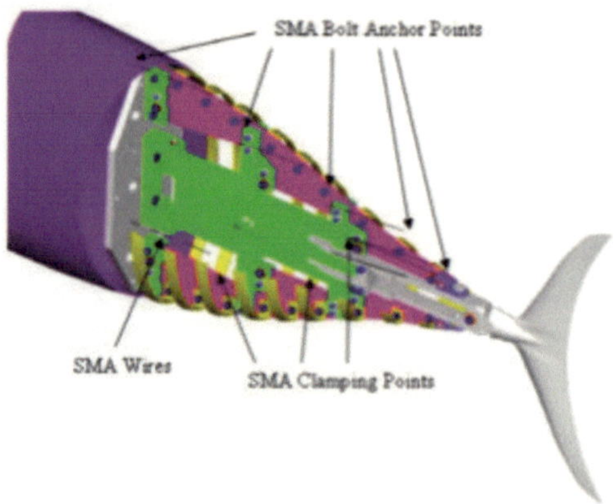

Figure 4. A CAD representation of the tail of Suleman's Tuna [65].

3.1.6. Design and testing of a biomimetic tuna using shape memory alloy induced propulsion

In [65], an SMA actuated tail inspired by a Bluefin tuna is proposed. Figure 4 shows the tail cutaway. This fish-like robot has length of $1m$ capable of the biological carangiform swimming mode. The maximum tail beat frequency was $0.5Hz$ due to the limitations of the SMA. Even at this low frequency, the power requirements were significant. The minimum and maximum power consumptions were calculated to be $292.8W$ and $333.6W$ respectively.

3.2. Air

In aerial bio-inspired robots most of the applications are appendices. Here, we can identify two main categories: insects and birds. To the best of the authors' knowledge only one robotic flying insect has been developed, apart form the jumping robot described earlier. This can be explained by the flapping frequency needed, far form the SMA's capabilities, and also by their power requirements. For these reasons insect-like flying robots mostly adopt piezo-electric actuators. In fact, the flying insect prototype described below uses SMA to fold and unfold the wings, and not for the primary flapping motion. Despite SMA actuation speed does not allow the actuation of flapping wings, it could allow for other kind of wing actuation, such as morphing-wings.

3.2.1. Recent progress in developing a beetle-mimicking flapping-wing system

In [51] a beetle-like insect robot inspired by the Allomyrina Dichotomapresents is presented. This robot features a morphing-wing airfoil capable of folding and unfolding the hind wing

using SMA wires. A single small size DC motor drives the flapping mechanism. Figure 5 shows the prototype and the unfolding of the artificial flapping/morphing wing device.

Similar folding ratio of the robot's wings has been observed in comparison with the biological counterpart, accounting for 1.7 of value. On average, wing unfolding was completed within about 3s and the wing folded in about 4s.

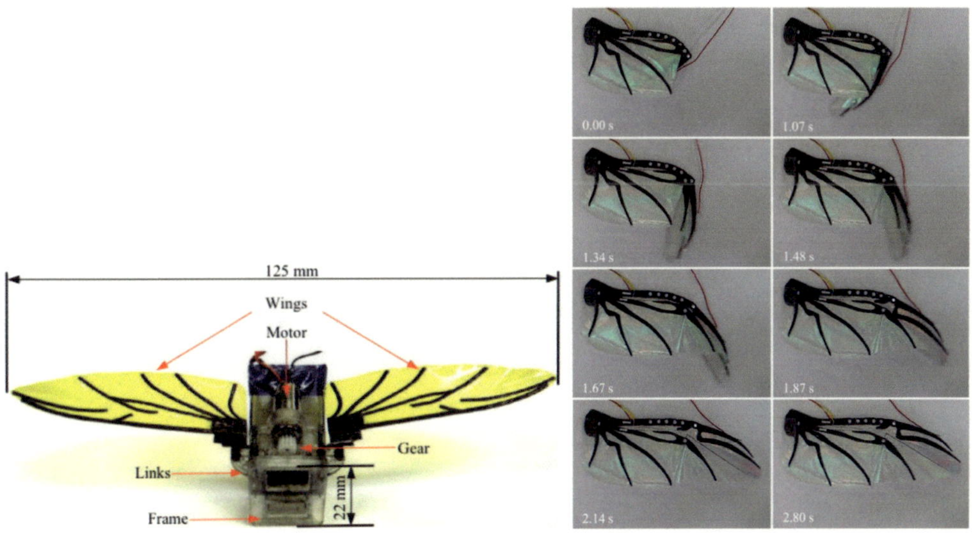

Figure 5. Prototype of the robotic beetle and detail of the unfolding of the artificial wing [51].

3.2.2. BATMAV-a biologically inspired micro-air vehicle for flapping flight: artificial-muscle based actuation

The BATMAV is a biologically inspired bat-like Micro-Aerial Vehicle (MAV) with flexible and foldable wings capable of flapping flight [2]. The robot features bat-inspired wings with a large number of flexible joints that allow mimicking the kinematics of a real bat flyer. Figure 6 details the overall structure of the robot, and the main connections of the SMA-like muscles.

BATMAV is the first robot that uses the SMA wires to play a dual role: first, as muscle-like actuators that provide the flapping and morphing wingbeat motions of the robot, and second, as super-elastic flexible hinges that join the wing's bone structure. Most of the experiments in [?] were carried out with a two-degree of freedom wing capable of flapping at 3Hz. Despite the fact that their robot is able to achieve accurate bio-inspired trajectories, the results presented lack experimental evidence of aerodynamics measurements that might demonstrate the viability of their proposed design.

3.3. Ground

Ground bio-inspired robots have been divided in two categories: the ones that uses actuated appendices (i.e, legged robots) and those that actuate the whole body, i.e. crawling robots such as snakes and worms.

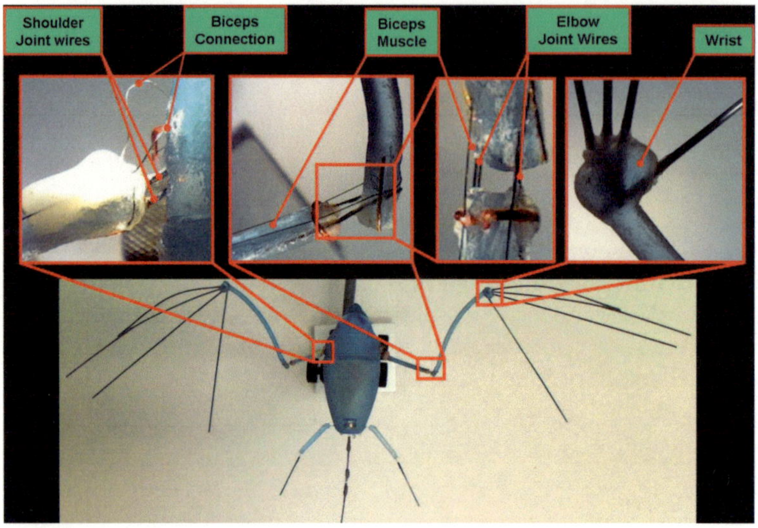

Figure 6. BATMAV. Dual Role of Shape Memory Alloy wires: as actuation muscles, and super elastic joints. (Picture from `https://sites.google.com/site/gheorghebunget/research/batmav`, with permission).

3.3.1. *Sensor fusion in a SMA-based hexapod bio-mimetic robot*

In [44] SMABOT is presented, a hexapod biomimetic robot with two SMA actuators that allow for the motion of the two degree-of-freedom robot. Each SMA actuator produces $300gram - force$ of pull force. Figure 7 shows the SMABOT IV. SMABOT IV incorporates two-dimensional inertial navigation system for position control. The average speed when moving with tripod gait is $30cm/min$. Its maximum power consumption is about $25W$ (the mass is $290g$).

Figure 7. SMABOT IV, a SMA based hexapod robot with the IMU module, compass sensor and step touch sensors [44].

3.3.2. Omegabot: Crawling robot inspired by Ascotis Selenaria

In [38] a robot inspired by the inchworm Ascotis Selenaria is presented. The robot, called Omegabot, is named after the omega (Ω) shape of the crawling motion of the inchworm. Figure 8 shows the Omegabot platform. Previous work about this robot can be also found in [37].

Experimental results report the first step for establishing an inchworm-like robot that can crawl on various terrains where conventional robots cannot move. The Omegabot uses a SMA coil actuator that requires a current of $200mA$ for activation. The frequency of motion is about $1Hz$, limited by the response time of the SMA wires. The inchworm robot is manually controlled by an IR remote operation, and it achieves a maximum linear velocity of $5mm/s$. The robot travels a distance of $5mm$ per stroke.

Figure 8. Omegabot, a biomimetic inchworm robot, grasps the branch of a wood, raises its head, and turns right. Bottom right: Proleg of Omegabot [38].

3.3.3. An earthworm-like micro robot using shape memory alloy actuator

In [34] a bio-mimetic micro earthworm-like robot with wireless control is proposed. The actuation mechanism consists on a SMA spring that contract and extend the earthworm muscle respectively. The proposed mechanism is simple but effective when traveling in narrow and rough environments, such as human digestive organs, bended long pipeline and so on. Also, this micro robot incorporates both control and power supply onboard. The theoretical speed of the micro robot is approximately $3.4mm/cycle$, where the total time per cycle is $8s$ (the contraction time of the SMA is $2s$, whereas the recovery time is $6s$). The fabricated micro robot can move with the velocity of $10mm/min$ during 8 minutes. The stroke per cycle is $2.0mm$.

3.3.4. Other peristaltic motion concepts

A concept similar to the one depicted in Figure 9 has been proposed in [48]. A SMA-spring has been used for changing the axial length of worm's modules, and consequently changing

their width. The robot has a length and diameter of $1cm$. SMA springs of 50 and $100\mu m$ wires have been tested. The resulting prototype, made of 4 modules, was able to achieve a speed up to $0.22mm/sec$, with a power consumption of $600mA$. The previous prototype was composed of a number of identical segments attached in series, each of which alternately contract axially and expand radially. To conclude this section two prototypes are worth mentioning that use SMAs structures for peristaltic motion.

The proposed structure in these cases is a tubular mesh made of SMA wires that convert radial contraction into longitudinal lengthening. Such a structure has been proposed independently in [61] (see also [36]) and in the SoftWorm project [4].

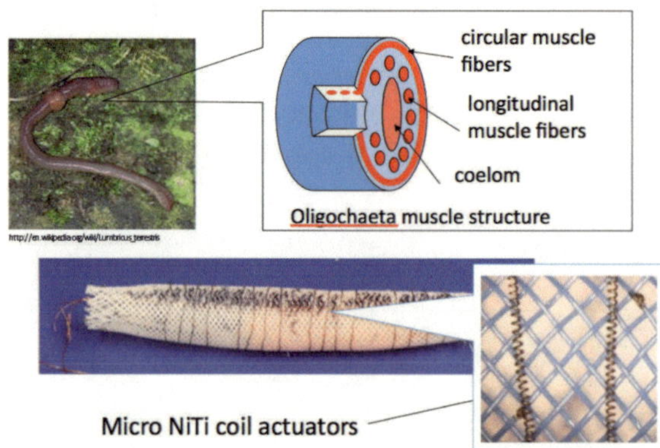

Figure 9. Arrangement of Antagonistic circular muscles in Oligochaeta [48] (Image of the real Oligochaeta from http://en.wikipedia.org/wiki/Lumbricus_terrestris).

3.4. Other SMA-based actuation systems

In this section we present two works that do not address the development a full robot, but rather studying and developing appendices to be added to future full robotic systems.

3.4.1. Research on Development of a Flexible Pectoral Fin Using Shape Memory Alloys

In [83], experimental research on pectoral fin structure is presented. The design of the pectoral fin actuator is based on SMAs wires composed by a couple of plates with the opposite functions. Figure 10 shows the biomimetic pectoral fin.

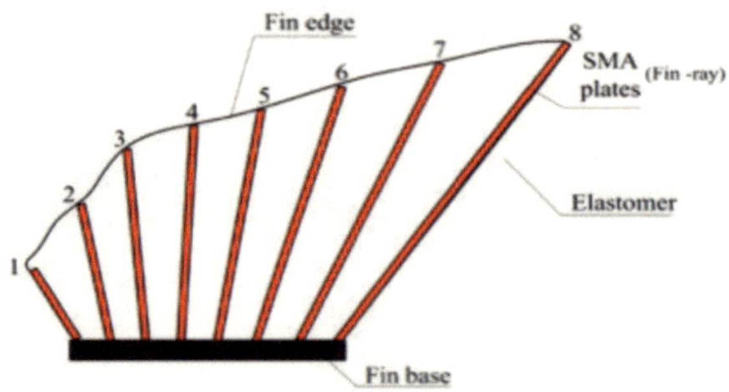

Figure 10. Biomimetic pectoral fin driven by eight couples of SMA plates [83].

This pectoral fin was the first and today continues being the only designed with only SMA wires. In fact, most research on fish-like robots is focused on studying propulsion (how to generate thrust), while maneuvers is largely unexplored.

3.4.2. Development of a dexterous tentacle-like manipulator using SMA-actuated hydrostats

Novel design principles and technologies for a new generation of high dexterity soft-bodied robots inspired by the morphology and behavior of the octopus are being developed in the framework of the OCTOPUS-IP project[1].

The imitation of the internal muscular structure of octopuses' tentacles is being studies and imitated. Longitudinal cables and transverse SMA imitate the arrangement of muscle fibers, controlling contractions as soft actuators within the robot arm [42]. Moreover, this manipulator is surrounded by a sensitive skin, with contact sensors embedded into silicone rubber, equipped with passive suckers that allow the grasping of objects. SMA actuators are used to change the section of the tentacle in several locations, inducing its bending.

Figure 11. The SMA-based tentacle (See http://www.octopusproject.eu/).

3.4.3. Development of a Shape-Memory-Alloy actuated biomimetic hydrofoil

The development and testing of a biomimetic active hydrofoil using Shape Memory Alloy (SMA) actuators is presented in [16]. This work describes the development and testing of a six-segment demonstration foil and the control schemes used.

4. Review on recent advances: iTuna and BaTboT

In this section, we report our most recent results on two SMA-actuated bio-inspired robots. The first, called iTuna, ia an underwater robot that according to our classification falls into the "full-actuated-boy" category. The second is an aerial robot, which implements the concept of morphing wings by means of SMA-based muscles.

[1] http://www.octopusproject.eu/

4.1. iTuna: a bending structure swimming robotic fish

The iTuna [59] is a swimming fish-like robot that apart from the external appearance, imitates some key features of the internal morphology of fishes.

This mechatronic concept takes inspiration from the arrangement of the red or slow-twitch muscles (see inset in Figure 12). In live fishes, such muscles are used for bending a flexible but nearly incompressible axis. Such axis is either composed of a (visco) elastic beam (notochord) or a series of vertebrae connected through intervertebral discs. The main structure of the iTuna robot fish is inspired by the former solution, and is composed by a continuous flexible backbone. The backbone is composed of polycarbonate of $1mm$ thickness actuated by SMA muscles acting as red muscles.

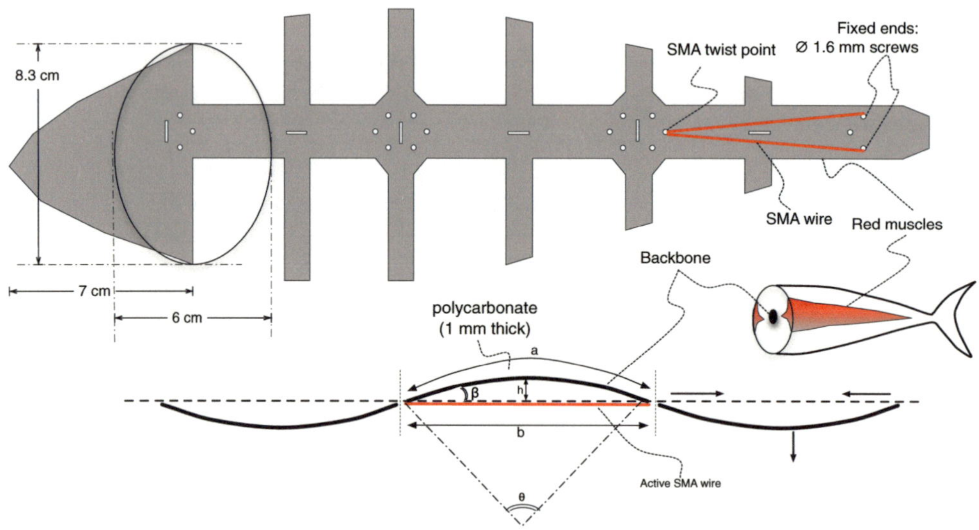

Figure 12. Main structure of the iTuna robot fish. a=8.5 cm. Under nominal operatrion, $b \cong 96\%$ a = 8.16 cm, h=1.02 cm, b=28° [59].

Six SMA-based actuators whose length is 1/3 of the body length are positioned in pairs, parallel to the body in such a way to produce an antagonistic movement on three body segments of $8.5cm$ length. This antagonistic configuration of SMA wires has some advantages in terms of increasing the range of controllable actuation, since both directions of motion (contraction and elongation) can be actively controlled. Figure 12 shows the location of the SMA wires within the skeleton structure of the prototype.

A V-shaped configuration of the wires, where each artificial muscle is composed of a single V-shaped SMA wire, twisted around the tension screw, allows to double the pull force without a significant increase of power consumption. NiTi SMA wires with a diameter size of $150\mu m$ have been adopted. These have a pull force of $230grams - force$ at consumption of $250mA$ at room temperature, and a nominal contraction time of 1 second.

Under nominal operation such SMAs can bend the body segments up to 28 degrees (angle β of Figure 12), even if SMA wires only contract approximately 4% of their length. By increasing the input electrical current and including a suitable control that handles an overloaded SMA operation, contraction time of $0.5s$ was achieved, and strain could be increased up to 6%, corresponding to a bending of approximately 36° (Fig. 13).

4.1.1. SMA control in the iTuna

After identification, a low-level PID controller has been designed to address two main limitations of SMAs: slack in the fibers, and limited actuation speed. Slack issues appear when SMA wires develop a two-way memory effect during operation [11]. Limitation in actuation speed occurs due to the large switching time between cooling and heating phases. To address such problems, a pre-heating mechanism has been developed that works in conjunction to the antagonistic arrangement. The pre-heating avoids the temperature on both wires drops below the 10% of the maximum applied electrical current, preventing the inactive alloy from complete cooling. On the other hand, the antagonistic arrangement provides an external stress to the cooling wire (provided both by the elastic backbone and by the active antagonistic wire). Working with an already-warm wire allows for a faster stretch and slack issues are avoided. Note that the PID controller is based on the experimental observation that the hysteresis on the electrical resistance curve was smaller than the hysteresis on the temperature curve. Resistance measurements are used as a feedback signal for closed-loop control (see [59] and [60] form more details).

The control developed allows overloading the SMA with up to $350mA$ peak current (note that power signals are sinusoidal, hence overloading only lasts a brief period of time). Overloading has allowed for achieving a $1Hz$ oscillation time (i.e. 0.5 seconds contraction and cooling times) and a bending angle of 36 degrees of each body segment.

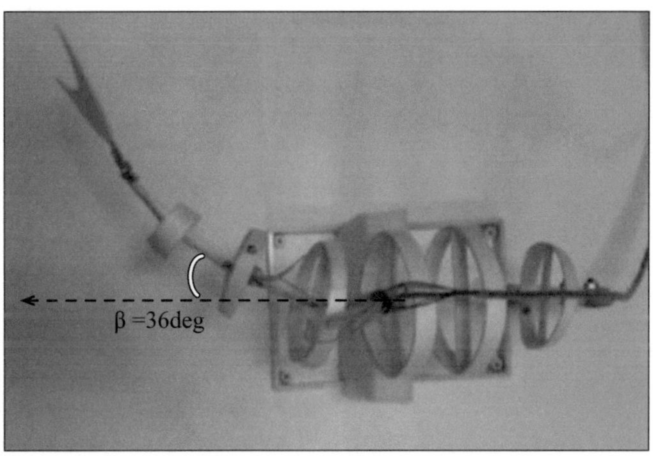

Figure 13. Bending under SMA overloading [59].

4.1.2. Control architecture

A key feature of SMAs is the possibility to develop closed loop control systems without the need of external sensor hardware. The feed back signal is provided by the detection of inner electrical resistance, that allows an indirect measurement of the temperature.

The main components are described in the following. A micro controller implements the PID algorithm. The PID controller receives the input reference position (set point) and the feedback of SMA's voltage and current that allows calculating the heating current to drive the SMA actuator. The digital output of the PID controller is converted to a reference current in

two steps. First, it is converted into an analog signal using a 2-wire serial 8-Bit DAC (Digital to Analog Converter) with Rail-to-Rail outputs. Then, a Voltage Controlled Current Source (VCCS) transforms the DC voltage in a constant current that feeds the SMA. This stage has a power consumption of less than $10mA$.

The DAC output ranges from 0 to 5 Volts with a resolution of $0.02V$. The measured voltage (VSMA) and current (ISMA) on the SMA fiber are fed-back to the micro controller in order to close the control loop. The hardware used (16F690-PIC) had a 12-bits A/D converter with a resolution of $0.537mV$ (considering the maximum voltage measured at the SMA $VSMA = 0.55V$.

On the other hand, taking into consideration the maximum current through the wire ($500mA$), SMA resistance variations about 1.074mΩ can be measured. Therefore, since the maximum variation in the SMA length is $0.34cm$, and the maximum variation of the resistance is 1.6Ω, the theoretical position error of the system based on the SMA length is 0.067%. i.e.,$0.12mm$.

4.2. BaTboT: a biologically-inspired bat-like aerial robot

BaTboT is a bio-inspired bat robot that uses Shape Memory Alloys (SMAs) as artificial muscles for powering the morphing motion of the wings. The morphing motion is related to the capacity of the robot to modulate its wings by contracting and extending the membrane in sync with the flapping motion. It is precisely this characteristic what makes biological bats more agile to maneuver than any other flying creature within the same Reynolds number range ($10^3 - 10^4$) [66], [22]. In addition, biological studies in [32], [58] have revealed that real bats are able to maneuver because of the inertial changes produced by the wings' modulations. Attempting to mimic this functionality using an artificial counterpart -BaTboT- mainly presents a twofold challenge: i) biomechanical design of the wings, and ii) proper control/actuation to module BaTboT's wings.

Prior work in [5] presented experimental results regarding both challenges. The investigation carried out in [5] does not only describes the biomechanical design of BaTboT's wings, but also focuses on evaluating the implications of using SMAs as artificial muscles to power the change of wing's morphology. Figure 14 shows the design-flow process to evaluate key issues of SMA performance and their implications to the application at hand.

The use of SMAs as artificial muscles has been concretely evaluated in terms of two issues:

- Functionality: SMA Power-to-Force.
- SMA Performance: actuation speed and fatigue.

4.2.1. Functionality

The SMA actuators shown in Fig. 14 (step-1) are the commercial Migamotor NanoMuscle model RS-70-CE [49]. Each NanoMuscle consists of several short strips of SMA NiTi wire with a thickness of $150\mu m$ attached to opposite ends of six metal strips stacked in parallel. Each SMA segment pulls the next strip about $0.67mm$ relative to the previous strip, and the relative movements sum to make a stroke of $4mm$. As depicted, two Migamotors muscles have been arranged into an antagonistic configuration working as artificial biceps and triceps that provide the rotation motion of the wing elbow's joint θ_3. The range of motion of the joint

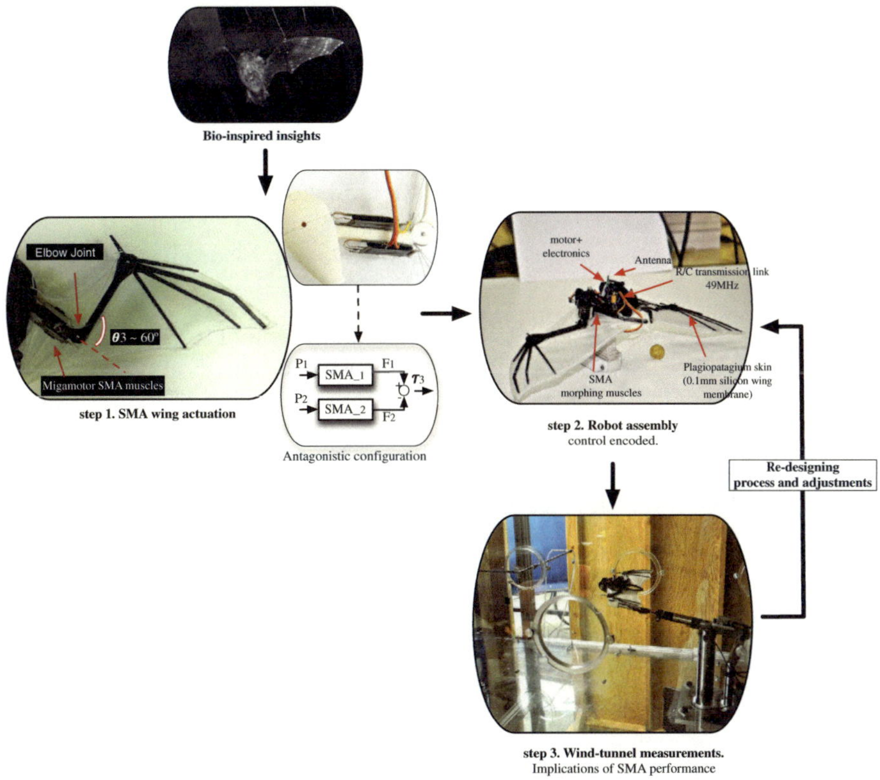

Figure 14. Flow-process for SMA evaluation in the BaTboT prototype [5].

is about 60°. The joint motion is achieved when each SMA actuator contracts upon heating, thus generating a pull-force (F_{sma}). Because both actuators are connected to the joint in an antagonistic fashion, the pull-force F_{sma} generates a joint torque, denoted as τ_3. Therefore, each actuator requires an input heating power P, to produce and output torque τ_3. The input heating power (P) is proportional to the input electrical current (I): $P = I^2R$, being R the nominal electrical resistance of the NiTi wires, $R = 8.5\Omega$.

In [5] simulations and experiments have been carried out aimed at quantifying the Power-to-Force tradeoff of the SMA muscles working under two operation modes: nominal and overloaded. To the application at hand, nominal-mode implies an input heating current (I) between $175mA$ and $350mA$, whereas overloaded-mode, between ~$400mA$ and $600mA$. Overloading allows for increasing the heating-time of the SMA, therefore increasing the contraction speed and the overall actuation frequency. However, overloading must be monitored to avoid overheating issues that may cause physical damage of the shape memory effect. In [5], SMA limitations were found under simulation and validated under experimentation.

Fig.15 shows the Power-to-Force tradeoffs: i) simulation using a SMA phenomenological model, ii) SMA response using an identified model, and iii) experimental measurements of output torque. Herein, our goal is not focused on describing the models, but on comparing

the correlations between models (i)-(ii) and experimental data (iii). Further details about the models can be found in [5].

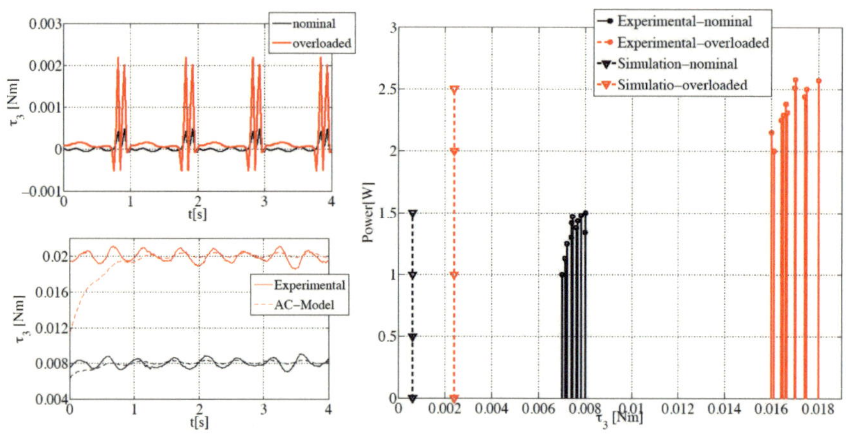

Figure 15. SMA Power-to-Force response. The input power is of the form: $a + b\sin(2\pi ft)$. a) Simulation of elbow's torque response under nominal and overloaded SMA operation, b) AC small-signal response of SMAs at f=~2Hz, c) maximum peak-values of output torque as a function of the input power

In order to simulate how the SMA muscles response upon electrical heating, a phenomenological model was used. This model is composed by thermo-mechanical equations that describe how the temperature, strain, and stress of the SMA change during its hysteresis loop, going from austenite to martensite and vice versa. Likewise, the SMA phenomenological actuation model is coupled with the dynamics equations that govern the motion of the robot's wing (refer to [5] for further details).

Figure 15a shows simulation results of joint torques achieved by applying nominal and overloaded values of input power of the form: $a + b\sin(2\pi ft)$, where a is the mean input power P, the term b is the small-magnitude of the sinusoidal component, and f is the commanded frequency for SMA contraction. Simulation data determines that by applying a nominal input power of $P = 1.04W$, the SMAs are able to generate a torque around the elbow joint of $\tau_3 = 0.0007Nm$, whereas by applying an overloaded input power of $P = 2.57W$, the output torque can be increased up to $\tau_3 = 0.0022Nm$. This implies that by increasing the input power about 2.5 times, the output torque can be increased about three times.

In order to verify the accuracy of simulation results, Fig.15c shows experimental measurements of output torques. In this trial, several profiles of input power were applied, and the respective forces were measured. Note that under nominal-mode, the output torque varies between 0.007 - 0.008Nm, whereas under overloading-mode, from 0.016 - 0.018Nm. Likewise, at the left side of the plot, the maximum peaks of nominal and overloaded torques corresponding to the simulation in Fig. 15 a, are also shown. One can note there is almost an order of magnitude of difference between simulation and experimental data. The error is produced because the phenomenological model does not take into account the joint friction, and even most important, the anisotropic loading of the silicone-based membrane. However, despite the error in magnitude, note that by doubling the input heating power, the measured

	Parameter	Theoretical[1]	Simulation[2]	Experimental[3]	AC-model[4]
Nominal	Input power P [W]	0.26	1.04	0.87-1.5	1.36
	Input current I [mA]	175	350	320-420	400
	SMA Pull-force F [N]	0.012	0.007	0.07 - 0.08	0.08
	Output torque tau_3 [Nm]	0.0012	0.0007	0.007 - 0.008	0.008
Overloaded	Input power P [W]	–	2.57	2 - 2.57	3.06
	Input current I [mA]	–	550	485 - 550	600
	SMA Pull-force F [N]	–	0.022	0.16 - 0.18	0.2
	Output torque tau_3 [Nm]	–	0.0022	0.016 - 0.018	0.02

Table 2. SMA-muscle Power-to-Force performance. [1] Nominal values can be found in Migamotors website [49]. The manufacturer does not provide overloading values. [2] Simulation data corresponds to Fig. 15 a. Please refer to Fig.9 in [5] for details about the SMA phenomenological model. [3] Experimental data corresponds to Fig. 15 c. [3] Identified AC Power-to-force model response. Data corresponding to Fig. 15 b.

output torque increases about 2.5 times. This scaling factor is quite similar compared to the simulation prediction.

The use of a phenomenological SMA model can provide a useful insight into the implications of overheating, by increasing the input power beyond the limits of SMA stress, and thus defining the upper threshold for overloading when using the real system. Unfortunately, the phenomenological model has not proven to be accurate for modeling the response of the SMAs when applying small-signal power inputs. Therefore, the next step is to identify the Power-to-Force AC behavior of the SMA actuators.

As observed by [11], [71] the AC response of NiTi SMA wires behaves as a first order low-pass filter. In [40], [19], [20] similar first-order behavior in their SMA actuator response had also been observed. Therefore in [5], we have followed a similar frequency response analysis procedure for finding a transfer function model that matches the measured gain and phase data of the frequency range of interest. Fig. 16 shows the results.

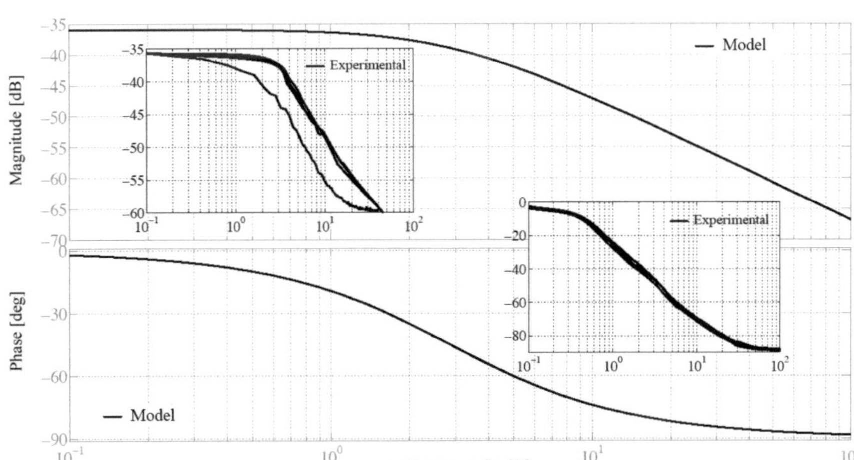

Figure 16. (Experiment VS model) Bode magnitude and phase plots for NiTi 150μm SMA Migamotor actuator. Input heating power of the form: $a + b\sin(2\pi ft)$.

The suitable transfer function that fits the experimental data from Fig. 16 is: $\tau_3(s) = 0.016(0.35s+1)^{-1}P(s)$. The response of the AC model is shown in Fig. 15 b. The AC model is compared against the experimental response of the SMA actuator to the AC small-signal. Note that by applying an input power of ~1.36W (I = 400mA), the output torque τ_3 stabilizes around ~0.008Nm. This response corresponds to the real nominal mode of the SMA actuator. Increasing the input power up to ~3.06W (overloaded mode, $I = 600mA$), the registered output torque stabilizes around ~0.02Nm. In both experiments the anisotropic loading of the 0.1mm silicone-based membrane has been taken into account.

Note how well the AC model correlates against the real behavior of the SMA muscles. A detailed comparison of the obtained data is summarized in Table 2. In conclusion, to properly evaluate the functionality of using SMA muscles based on the criteria of the application at hand:

- First of all, identify the required force the SMA muscle must provide to drive the motion of the robot. This quantification requires a dynamic model of the robot.
- Use a phenomenological model for defining the limits of SMA actuation. The best models in the literature can be found in [69],[1],[7],[8],[10].
- Classify the response of the SMA in terms of nominal and overloaded input power. Verify that under overloaded operation, SMAs are safe from overheating, and yet achieving the desired output torque - to - input power tradeoff.
- Using the real system, apply an AC input power of the form $a + b\sin(2\pi ft)$, being a the mean power (nominal or overloaded),b the small-signal amplitude of the sinusoidal component, and f the desired actuation frequency. Subsequently, measure the small changes of pulling-force generated by the SMA when contracting. Note that forces can be mapped to torques (if using a rotational joint).
- Follow the frequency response analysis procedure described in [11], [71]. Note the AC behavior of a NiTi SMA wire resemble a first order low-pass filter. Find the transfer function that matches the data. The Power-to-Force model of the SMA will be of the form: $F(s) = k(\tau s+1)^{-1}P(s)$. Use the identified model for control tuning. Note that a simple PID control technique is suitable for driving SMA actuation. Depending on the available sensors, force control or position control could be applied. The following section provides a brief insight about the advantages and implications of using both control schemes: force or position feedback. For detailed insights about SMA force control, refer to [71], whereas for SMA position control, refer to [5], and [59].

4.2.2. *SMA Performance: actuation speed and fatigue*

As concluded in [29], [57], [11], [17], [76], PID technique is sufficient for allowing accurate and faster SMA actuation. The key issue for achieving an outstanding SMA performance does not necessarily depend on the control technique, but in complementary mechanisms that monitor the lower and upper limits of input power.

In [11], and [71], these mechanisms are introduced: i) anti-slack, and ii) anti-overload. The former deals with the two-way shape memory effect [39], improving accuracy and speed,

whereas the latter limits the amount of input heating power to prevent physical damage when SMAs are overloaded. As a result, the force controller presented in [71] was capable of tracking fast and accurate force references when compared with other works reported in [20], [84].

The improvement in accuracy and speed are due to two factors: i) by avoiding wire entanglement, and ii) by ensuring the passive wire of the antagonistic configuration does not cool completely. Wire entanglement can be produced when the SMA wires extended upon cooling. The passive SMA wire can develop a few millimeters of slack as it cools, which consequently affects the accuracy of the control. This slack-phenomenon is only presented in the antagonistic configuration, due to the two-way shape memory effect produced by each actuator.

To avoid the aforementioned issues, the anti-slack mechanism defines a minimum threshold of input heating power that ensures the inactive wire does not cool completely. The improvement in actuation speed is due to the fact that the already-warm SMA wire can begin to pull as soon as the heating current is raised, whereas a cold wire would first need to be raised to its operating temperature. It has been observed from experimental results in [71] and [5] that a suitable minimum threshold of input heating power is about 10% of the power applied.

On the other hand, the anti-overload mechanism is in charge of ensuring that the maximum input power does not increase above an upper limit. This approach avoids overheating the SMAs in case the controller delivers a large amount of power to the wires. As mentioned before, this upper limit can be found using a phenomenological model, or by performing real measurements of SMA temperature and stress on the wires.

Both anti-slack and anti-overload mechanisms are key for improving on SMA performance under a force control architecture. The advantages of using a force-control scheme are twofold: i) high-bandwidth response, and ii) SMA fatigue avoidance. High-bandwidth response requires the use of force sensors capable of providing the force feedback. It has also been demonstrated in [71], that by using high-bandwidth force feedback, limit cycles of SMA operation are eliminated.

Nevertheless, for some systems, the use of force sensors could be a hardware limitation. In [5], it has been demonstrated that both anti-slack and anti-overload mechanisms can be implemented in a position control scheme. The position feedback can be achieved by measuring the electrical resistance of the SMA wires, which is a linear function of the strain. The key disadvantage of using a position scheme that forces the SMA to behave in overloaded operation mode relies on fatigue. As experimentally observed in [5], overloaded operation mode could be maintained only for about five minutes of SMA continuous operation before decreasing performance to nominal mode. For the application at hand, overloaded mode implied an actuation frequency of $2.5Hz$, while nominal mode, an actuation frequency of $1.3Hz$. In this case fatigue issues caused a decrease in actuation speed performance about 56%. Further investigations shall be devoted to quantifying the lifetime of SMAs when subjected to higher stresses and larger heating currents within a position control scheme.

5. Conclusions

SMA technology allows the development of a wide variety of robotics designs, exploiting their advantages in terms of weight, volume and sensing capabilities. Their particular features open a wide range of possibilities hardly achievable with classical technology: hydrofoils, morphing shapes, and hydrostats are just some of the new concepts that have been made possible to develop thanks to SMAs. The prototypes described in this chapter have been classified first according to the operation environment (water, air and ground) and second according to their application in the robotics system.

Researchers have successfully overcome SMA disadvantages in terms of power consumption, actuation speed and low strain, by:

- Finding suitable niches of application
- Designing dedicated mechatronics
- Developing ad-hoc control strategies

Biomimetics has emerged as a very promising field together with bio-medical applications. Even if SMAs cannot substitute classical servomotor and hydraulic technology in general, niches can be found where they can effectively compete with, and even outperform standard actuation technology. Clearly, merely replacing servomotors with SMA-based actuators in classical linear-actuation mechanic setups would make little sense. But dedicated mechatronic design, such as embedding SMA fibers in silicone obtaining morphing materials, bending of continuous structures and hydrostats make the best out of this technology, regardless the relatively small strain achievable from the fibers. Finally, cleverly designed control strategies, that exploit the knowledge of the physics of the material and of its behavior over time, coupled with dedicated mechanic setups can dramatically reduce response time.

Our experience with SMA-based actuators has revealed a relatively little considered drawback, which relates to fatigue. In fact, extensive testing with our prototypes has demonstrated that even if SMA performance can be improved by overloading, the effects of overloading disappear quite rapidly. We believe that further investigations shall be devoted to study the behavior of the alloys when exposed to high stresses and large heating currents over a large period of time.

In conclusion, we can say that Shape Memory Alloys have demonstrated great capabilities, as it has been demonstrated by this review of the state of the art. Most importantly, they have the potential to be used in future robotic systems: the more researchers will apply and study them, the more their limitations are smoothed and their employment becomes more effective. After all, they are being extensively investigated only since a short time in comparison with electric motors and hydraulic actuators, that can boast more than a century of history.

Author details

William Coral, Claudio Rossi, Julian Colorado, Daniel Lemus and Antonio Barrientos
Centre for Automation and Robotics (CAR) of the Technical University of Madrid, Spain
CSIC (National Council for Scientific research)

Acknowledgements

This work was supported by the Robotics and Cybernetics Group at the *Centre for Automation and Robotics (CAR) UPM-CSIC*, and funded under the project *ROBOCITY 2030* sponsored by the Community of Madrid (S-0505/DPI/000235) and The project *ROTOS (MultiRobot Systems for Protection of large outdoor infrastructures)* sponsored by the Ministry of Science And Innovation of Spain (DPI2010-17998)

6. References

[1] Brinson, L. C. [1993]. One-Dimensional Constitutive Behavior of Shape Memory Alloys: Thermomechanical Derivation with Non-Constant Material Functions and Redefined Martensite Internal Variable, *Journal of Intelligent Material Systems and Structures* 4(2): 229–242.
URL: *http://jim.sagepub.com/cgi/doi/10.1177/1045389X9300400213*

[2] Bunget, G. & Seelecke, S. [2010]. BATMAV: a 2-DOF bio-inspired flapping flight platform, *Proceedings of SPIE* 7643: 76433B.

[3] Caldwell, D. G. & Taylor., P. [1988]. Artificial muscles as robotic actuators, *IFAC Robot Control*, Karlsruhe, pp. 401–406.

[4] *Case Western Reserve University. Continuous Wave Peristaltic Locomotion, In: Center For Biologically Inspired Robotics Research, 2012* [2012].
URL: *http://biorobots.cwru.edu/projects/softworm/*

[5] Colorado, J., Barrientos, A., Rossi, C. & Breuer, K. [2012]. Biomechanics of smart wings in a bat robot: morphing-wings using SMA actuators, *Bioinspiration & Biomimetics* 7(3).
URL: *http://iopscience.iop.org/1748-3190/7/3/036006*

[6] Dynalloy [2012, Available from: http://www.dynalloy.com/TechDataWire.php]. Dynalloy, Inc. FLEXINOL®Actuator Wire In: Technical and Design Data.
URL: *http://www.dynalloy.com/TechDataWire.php*

[7] Elahinia, M. H. [2004]. *Effect of System Dynamics on Shape Memory Alloy Behavior and Control*, PhD thesis, Virginia Polytechnic Institute and State University.

[8] Elahinia, M. H. & Ahmadian, M. [2005]. An enhanced SMA phenomenological model: I. The shortcomings of the existing models, *Smart Materials and Structures* 14(6): 1297–1308.
URL: *http://stacks.iop.org/0964-1726/14/i=6/a=022?key=crossref.356645df188b7494fe51e5de94b4f281*

[9] Eren, Y., Mavroidis, C. & Nikitczuk, J. [2002]. B-Spline Based Adaptive Control of Shape Memory Alloy Actuated Robotic Systems, *Dynamic Systems and Control*, Vol. 2002, ASME, pp. 471–478.
URL: *http://dx.doi.org/10.1115/IMECE2002-33425*

[10] Esfahani, E. T. & Elahinia, M. H. [2007]. Stable Walking Pattern for an SMA-Actuated Biped, 12(5): 534–541.

[11] Featherstone, R. [2008]. An Architecture for Fast and Accurate Control of Shape Memory Alloy Actuators, *The International Journal of Robotics Research* 27(5): 595–611.
URL: *http://ijr.sagepub.com/cgi/doi/10.1177/0278364908090951*

[12] Featherstone, R. & Teh, Y. H. [2004]. Improving the Speed of Shape Memory Alloy Actuators by Faster Electrical Heating, *9th International Symposium on Experimental Robotics* pp. 1–10.

[13] Fujita, H. [1989]. Studies of micro actuators in Japan, *International Conference on Robotics and Automation*, IEEE Comput. Soc. Press, pp. 1559–1564.
URL: *http://ieeexplore.ieee.org/lpdocs/epic03/wrapper.htm?arnumber=100200*

[14] Fukuda, T. & Asaka, K. [2003]. A new type of fish-like underwater microrobot, *IEEE/ASME Transactions on Mechatronics* 8(1): 136–141.
URL: *http://ieeexplore.ieee.org/lpdocs/epic03/wrapper.htm?arnumber=1187383*

[15] G. K. Batchelor, U. o. C. [2000]. *An Introduction to Fluid Dynamics*, cambridge edn, Cambridge University Press, Cambridge.
URL: *http://www.cambridge.org/gb/knowledge/isbn/item1161783/?site_locale=en_GB*

[16] Rediniotis, O. K., Wilson, L. N., Lagoudas, D. C. & Dimitris, C. [2002]. Development of a Shape-Memory-Alloy Actuated Biomimetic Hydrofoil, *Intelligent Materials Systems and Structures* 13(1): 35–49.

[17] Gorbet, R. B. & Russell, R. A. [1995]. A novel differential shape memory alloy actuator for position control, *Robotica* 13(4): 423–430.
URL: *http://www.journals.cambridge.org/abstract_S0263574700018853*

[18] Granito, M. [2011]. *S.M.A. Shape Memory Alloy Cooling System by Peltier Cells: A Cooling System for Shape Memory Alloy Based on the Use of Peltier Cells*, LAP Lambert Acad. Publ.
URL: *http://books.google.es/books?id=IFvSygAACAAJ*

[19] Grant, D. [1999]. *Accurate and rapid control of shape memory alloy actuators*, PhD thesis.

[20] Grant, D. & Hayward, V. [1997]. Variable structure control of shape memory alloy actuators, *Control Systems Magazine IEEE* 17(3): 80–88.
URL: *http://ieeexplore.ieee.org/xpls/abs_all.jsp?arnumber=588180*

[21] Hashimoto, M., Tabata, T. & Yuki, T. [1999]. Development of electrically heated SMA active forceps for laparoscopic surgery, *IEEE International Conference on Robotics and Automation Cat No99CH36288C* 3(May): 2372–2377.
URL: *http://ieeexplore.ieee.org/lpdocs/epic03/wrapper.htm?arnumber=770460*

[22] Hedenström, A., Johansson, L. C. & Spedding, G. R. [2009]. Bird or bat: comparing airframe design and flight performance., *Bioinspiration biomimetics* 4(1): 015001.
URL: *http://www.ncbi.nlm.nih.gov/pubmed/19258691*

[23] Ho, M., McMillan, A. B., Simard, J. M., Gullapalli, R. & Desai, J. P. [2011]. Toward a Meso-Scale SMA-Actuated MRI-Compatible Neurosurgical Robot: Robotics, IEEE Transactions on, *Robotics IEEE Transactions on* PP(99): 1–10.

[24] Homsy, G., Aref, H., Breuer, K.S., Hochgreb, S., Powell, K., Munson, B. & Robertson, C. [2008]. *Multimedia Fluid Mechanics DVD-ROM*, second edi edn, Cambridge University Press.

[25] Honma, D., Miwa, Y. & Iguchi. [1985]. Micro Robots and Micro Mechanisms Using Shape Memory Alloy to Robotic Actuators, *Robotic Systems* 2(1): 3–25.

[26] Hunter, I. W. & Lafontaine, S. [1992]. A comparison of muscle with artificial actuators, *SolidState Sensor and Actuator Workshop 5th Technical Digest IEEE*, pp. 178–185.
URL: *http://ieeexplore.ieee.org/xpls/abs_all.jsp?arnumber=228297*

[27] Hunter, I. W., Lafontaine, S., Hollerbach, J. M. & Hunter, P. J. [1991]. Fast reversible NiTi fibers for use in microrobotics, *Micro Electro Mechanical Systems*, Ieee, pp. 166–170. URL: *http://ieeexplore.ieee.org/lpdocs/epic03/wrapper.htm?arnumber=114789*

[28] Ihálcz, I. M. [2001]. Fundamental characteristics and design method for nickel-titanium shape memory alloy, *Periodica Polytechnica Mechanical Engineering* 45(1): 75–86. URL: *http://www.pp.bme.hu/me/2001_1/pdf/me2001_1_10.pdf*

[29] Ikuta, K. [1990]. Micro/miniature shape memory alloy actuator, *International Conference on Robotics and Automation*, Vol. 3, IEEE Comput. Soc. Press, pp. 2156–2161.

[30] Ikuta, K., Tsukamoto, M. & Hirose, S. [1988]. Shape memory alloy servo actuator system with electric resistance feedback and application for active endoscope, *International Conference on Robotics and Automation*, Vol. 1, IEEE, pp. 427–430. URL: *http://ieeexplore.ieee.org/xpls/abs_all.jsp?arnumber=12085*

[31] Ikuta, K., Tsukamoto, M. & Hirose, S. [1991]. Mathematical model and experimental verification of shape memory alloy for designing micro actuator, *IEEE Micro Electro Mechanical Systems*, pp. 103–108.

[32] Iriarte-Díaz, J., Riskin, D. K., Willis, D. J., Breuer, K. S. & Swartz, S. M. [2011]. Whole-body kinematics of a fruit bat reveal the influence of wing inertia on body accelerations., *Journal of Experimental Biology* 214(Pt 9): 1546–1553. URL: *http://www.ncbi.nlm.nih.gov/pubmed/21490262*

[33] Kim, B., Kim, D.-h., Jung, J. & Park, J.-o. [2005]. A biomimetic undulatory tadpole robot using ionic polymer-metal composite actuators, *Smart Materials and Structures* 14(6): 1579–1585. URL: *http://stacks.iop.org/0964-1726/14/i=6/a=051?key=crossref.f252f55e78ad58af50f72847f0cd41a3*

[34] Kim, B., Lee, M. G., Lee, Y. P., Kim, Y. & Lee, G. [2006]. An earthworm-like micro robot using shape memory alloy actuator, *Sensors and Actuators A: Physical* 125(2): 429–437. URL: *http://linkinghub.elsevier.com/retrieve/pii/S0924424705003067*

[35] Kim, B., Park, S. & Park, J.-O. [2009]. Microrobots for a capsule endoscope, *2009 IEEE/ASME International Conference on Advanced Intelligent Mechatronics* pp. 729–734. URL: *http://ieeexplore.ieee.org/lpdocs/epic03/wrapper.htm?arnumber=5229926*

[36] Kim, S., Hawkes, E., Choy, K. & Joldaz, M. [2009]. Micro artificial muscle fiber using NiTi spring for soft robotics, *Robots and Systems* pp. 2228–2234. URL: *http://ieeexplore.ieee.org/xpls/abs_all.jsp?arnumber=5354178*

[37] Koh, J. [2009]. Omegabot: Biomimetic inchworm robot using sma coil actuator and smart composite microstructures (scm), *Robotics and Biomimetics (ROBIO), 2009* pp. 1154–1159. URL: *http://ieeexplore.ieee.org/lpdocs/epic03/wrapper.htm?arnumber=5420752 http:// ieeexplore. ieee.org/xpls/abs_all.jsp?arnumber=5420752*

[38] Koh, J.-s. & Cho, K.-j. [2010]. Omegabot: Crawling robot inspired by Ascotis Selenaria, *International Conference on Robotics and Automation*, IEEE, pp. 109–114. URL: *http://ieeexplore.ieee.org/lpdocs/epic03/wrapper.htm?arnumber=5509425*

[39] Kohl, M. [2004]. *Shape Memory Microactuators*, Springer. URL: *http://www.amazon.com/Shape-Memory-Microactuators-Manfred-Kohl/dp/3540206353*

[40] Kuribayashi, K. [1986]. A New Actuator of a Joint Mechanism Using TiNi Alloy Wire, *The International Journal of Robotics Research* 4(4): 47–58.

URL: *http://ijr.sagepub.com /cgi/doi/10.1177/027836498600400404*
http://ijr.sagepub.com/cgi/content/abstract/4/4/47

[41] Kuribayashi, K. [1991]. Improvement of the Response of an SMA Actuator Using a Temperature Sensor, *The International Journal of Robotics Research* 10(1): 13–20.
URL: *http://ijr.sagepub.com/cgi/doi/10.1177/027836499101000102*

[42] Cecilia Laschi, Matteo Cianchetti, Barbara Mazzolai, Laura Margheri, Maurizio Follador & Paolo Dario [2012]. Soft Robot Arm Inspired by the Octopus, *Advanced Robotics* 26: 709-727.

[43] Li, W. D., Guo, W., Li, M. T. & Zhu, Y. H. [2006]. Design and Test of a Capsule Type Endoscope Robot with Novel Locomation Principle, *2006 9th International Conference on Control Automation Robotics and Vision* 20(02): 1–6.
URL: *http://ieeexplore.ieee.org/lpdocs/epic03/wrapper.htm?arnumber=4150137*

[44] Liu, S.-H. & Yen, J.-Y. [2008]. Sensor fusion in a SMA-based hexapod bio-mimetic robot, *Workshop on Advanced robotics and Its Social Impacts*, IEEE, pp. 1–6.
URL: *http://ieeexplore.ieee.org/lpdocs/epic03/wrapper.htm?arnumber=4653611*

[45] Liu, X., Luo, H.-Y., Liu, S.-P. & Wang, D.-F. [2011]. Pilot study of SMA-based expansion device for transanal endoscopic microsurgery, *2011 International Conference on Machine Learning and Cybernetics*, IEEE, pp. 1420–1424.
URL: *http://ieeexplore.ieee.org/lpdocs/epic03/wrapper.htm?arnumber=6016875*

[46] M. Hashimoto, Takeda, M., Sagawa, H., Chiba, I. & Sato, K. [1985]. Application of shape memory alloy to robotic actuators, *Journal of Robotic Systems* 2(1): 3–25.

[47] McNaney, J. [2003]. An experimental study of the superelastic effect in a shape-memory Nitinol alloy under biaxial loading, *Mechanics of Materials* 35(10): 969–986.
URL: *http://linkinghub.elsevier.com/retrieve/pii/S0167663602003101*

[48] Menciassi, A., Gorini, S., Pernorio, G. & Dario, P. [2004]. A SMA Actuated Artificial Earthworm, *IEEE International Conference on Robotics and Automation 2004 Proceedings ICRA 04 2004* 4: 3282–3287 Vol.4.
URL: *http://ieeexplore.ieee.org/lpdocs/epic03/wrapper.htm?arnumber=1308760*

[49] Migamotors [n.d.]. Migamotors Company, 2012, Avaliable from: http://www.migamotors.com/. URL: *http://www.migamotors.com/*

[50] M.W.M. van der Wijst [1998]. *Shape control of structures and materials with shape memory alloys*, PhD thesis, Universiteit Eindhoven.
URL: *http://en.scientificcommons.org/17599133*

[51] Nguyen, Q., Park, H., Byun, D. & Goo, N. [2010]. Recent progress in developing a beetle-mimicking flapping-wing system, *World Automation Congress (WAC)*, pp. 1–6.

[52] Pons, J. L. [2005]. *Emerging Actuator Technologies: A Micromechatronic Approach*, Wiley.

[53] Pornsin-sirirak, T., Tai, Y., Nassef, H. & Ho, C. [2001]. Titanium-alloy MEMS wing technology for a micro aerial vehicle application, *Sensors and Actuators A: Physical* 89(1-2): 95–103.
URL: *http://linkinghub.elsevier.com/retrieve/pii/S0924424700005276*

[54] Punning, A., Anton, M., Kruusmaa, M. & Aabloo, A. [2004]. A Biologically Inspired Ray-like Underwater Robot with Electroactive Polymer Pectoral Fins, *IEEE Confrence "Mechatronics and Robotics*.

[55] Rabinovich, M., Varona, P., Selverston, A. & Abarbanel, H. [2006]. Dynamical principles in neuroscience, *Reviews of Modern Physics* 78(4): 1213–1265.
URL: *http://link.aps.org/doi/10.1103/RevModPhys.78.1213*

[56] Raynaerts, D. & Brussel, H. V. [1991]. Development of a SMA high performance robotic actuator, *Fifth International Conference on Advanced Robotics Robots in Unstructured Environments*, Ieee, pp. 61–66.
URL: *http://ieeexplore.ieee.org/lpdocs/epic03/wrapper.htm?arnumber=240475*

[57] Reynaerts, D. & Brussel, H. V. [1998]. Design aspects of shape memory actuators, *Mechatronics* 8(6): 635–656.
URL: *http://linkinghub.elsevier.com/retrieve/pii/S0957415898000233*

[58] Riskin, D. K., Iriarte-Díaz, J., Middleton, K. M., Breuer, K. S. & Swartz, S. M. [2010]. The effect of body size on the wing movements of pteropodid bats, with insights into thrust and lift production., *Journal of Experimental Biology* 213(Pt 23): 4110–4122.
URL: *http://www.ncbi.nlm.nih.gov/pubmed/21075953*

[59] Rossi, C., Colorado, J., Coral, W. & Barrientos, A. [2011]. Bending continuous structures with SMAs : a novel robotic fish design, *Bioinspiration biomimetics* 045005(4): 045005.
URL: *http://iopscience.iop.org/1748-3190/6/4/045005*

[60] Rossi, C., Coral, W. & Barrientos., A. [2010]. SMA Control for Bio-mimetic Fish Locomotion, *International Conference on Informatics in Control, Automation and Robotics (ICINCO)*, Madeira.

[61] Seok, S., Onal, C. D., Wood, R., Rus, D. & Kim, S. [2010]. Peristaltic locomotion with antagonistic actuators in soft robotics, *2010 IEEE International Conference on Robotics and Automation* 0060: 1228–1233.
URL: *http://ieeexplore.ieee.org/lpdocs/epic03/wrapper.htm?arnumber=5509542*

[62] Shi, Z., Liu, D., Ma, C. & Zhao, D. [2011]. Accurate controlled Shape Memory Alloy Actuator for Minimally Invasive Surgery, *Self* pp. 817–822.

[63] Shin, B., Kim, H.-Y. & Cho, K.-J. [2008]. Towards a biologically inspired small-scale water jumping robot, *2008 2nd IEEE RAS & EMBS International Conference on Biomedical Robotics and Biomechatronics*, IEEE, pp. 127–131.
URL: *http://ieeexplore.ieee.org/lpdocs/epic03/wrapper.htm?arnumber=4762896*

[64] Shin, D. D., Mohanchandra, K. P. & Carman, G. P. [2004]. High frequency actuation of thin film NiTi, *Sensors And Actuators* 111(2-3): 166–171.
URL: *http://linkinghub.elsevier.com/retrieve/pii/S0924424703004771*

[65] Suleman, A. & Crawford, C. [2008]. Design and testing of a biomimetic tuna using shape memory alloy induced propulsion, *Computers and Structures* (86): 491?499.

[66] Swartz, S. M., Bishop, K. L. & Ismael-Aguirre, M. F. [2005]. Dynamic complexity of wing form in bats: implications for flight performance, *Functional and evolutionary ecology of bats* pp. 110–130.

[67] Tadesse, Y., Thayer, N. & Priya, S. [2009]. Tailoring the Response Time of Shape Memory Alloy Wires through Active Cooling and Pre-stress, *Journal of Intelligent Material Systems and Structures* 21(1): 19–40.
URL: *http://jim.sagepub.com/cgi/doi/10.1177/1045389X09352814*

[68] Tadesse, Y., Villanueva, A., Haines, C., Novitski, D., Baughman, R. & Priya, S. [2012]. Hydrogen-fuel-powered bell segments of biomimetic jellyfish, *Smart Materials*

and Structures 21(4): 045013. URL: *http://stacks.iop.org/0964-1726/21/i=4/a=045013?key=crossref.fb72cd34852f3dfd4f1efbce ff7c630b*

[69] Tanaka, K. [1986]. A thermomechanical sketch of shape memory effect: one-dimensional tensile behavior, *Res Mechanica* 2(3): 59–72.
URL: *http://www.sciencedirect.com/science/article/pii/0749641986900161*

[70] Teh, Y. H. [2003]. *A Control System For Achieving Rapid Controlled Motions From Shape Memory Alloy (SMA) Actuator Wires*, PhD thesis, Final Year Project, Australian National University.

[71] Teh, Y. H. [2008]. *Fast , Accurate Force and Position Control of Shape Memory Alloy Actuators*, PhD thesis, The Australian National University.

[72] Teh, Y. H. & Featherstone, R. [2004a]. A new control system for fast motion control of SMA actuator wires, *The 1st International Symposium on* .
URL: *http://axiom.anu.edu.au/ roy/SMA/smart04paper.pdf*

[73] Teh, Y. H. & Featherstone, R. [2004b]. Experiments on the Audio Frequency Response of Shape Memory Alloy Actuators, *Information Sciences* .

[74] Teh, Y. H. & Featherstone, R. [2004c]. Experiments on the Performance of a 2-DOF Pantograph Robot Actuated by Shape Memory Alloy Wires, *Information Sciences* pp. 1–8.

[75] Teh, Y. H. & Featherstone, R. [2007]. Accurate Force Control and Motion Disturbance Rejection for Shape Memory Alloy Actuators, *Information Sciences* (April): 10–14.
URL: *http://ieeexplore.ieee.org/lpdocs/epic03/wrapper.htm?arnumber=4209783*

[76] Troisfontaine, N., Bidaud, P. & Dario, P. [1997]. Control experiments on two SMA based micro-actuators, *The Fifth International Symposium on Experimental Robotics V Barcelona Spain* pp. 490–500.

[77] Villanueva, A., Smith, C. & Priya, S. [2011]. A biomimetic robotic jellyfish (Robojelly) actuated by shape memory alloy composite actuators., *Bioinspiration & biomimetics* 6(3): 036004.
URL: *http://www.ncbi.nlm.nih.gov/pubmed/21852714*
http://stacks.iop.org/1748-3190/6/i=3/a =036004

[78] Wang, Z., Hang, G., Li, J., Wang, Y. & Xiao, K. [2008]. A micro-robot fish with embedded SMA wire actuated flexible biomimetic fin, *Sensors and Actuators A: Physical* 144(2): 354–360.
URL: *http://linkinghub.elsevier.com/retrieve/pii/S0924424708001283*

[79] Wang, Z., Wang, Y., Li, J. & Hang, G. [2009]. A micro biomimetic manta ray robot fish actuated by SMA, *2009 IEEE International Conference on Robotics and Biomimetics (ROBIO)*, IEEE, pp. 1809–1813.
URL: *http://ieeexplore.ieee.org/lpdocs/epic03/wrapper.htm?arnumber=5420423*

[80] Waram, T. [1993]. *Actuator Design Using Shape Memory Alloys*, T.C. Waram.
URL: *http://books.google.es/books?id=fjZBAgAACAAJ*

[81] Westphal, A., Rulkov, N. F., Ayers, J., Brady, D. & Hunt, M. [2011]. Controlling a lamprey-based robot with an electronic nervous system, *Smart Structures and Systems* 8(1): 39–52.

[82] Yang, Y., Ye, X. & Guo, S. [2007]. A New Type of Jellyfish-Like Microrobot, *2007 IEEE International Conference on Integration Technology*, number 2006, IEEE, pp. 673–678. URL: *http: //ieeexplore.ieee.org/lpdocs/epic03/wrapper.htm?arnumber=4290404*
http://ieeexplore.ieee.org/xpls/abs_all.jsp?arnumber=4290404

[83] Zhang, Y., He, J. & Yang, J. [2006]. Initial research on development of a flexible pectoral fin using Shape Memory Alloy, *Automation, Proceedings of* pp. 255–260.
URL: *http://ieeexplore.ieee.org/xpls/abs_all.jsp?arnumber=4026090*

[84] Zhong, Z. W. & Yeong, C. K. [2006]. Development of a gripper using SMA wire, *Sensors and Actuators A: Physical* 126(2): 375–381.
URL: *http://linkinghub.elsevier.com/retrieve/pii/S0924424705005935*

Chapter 4

Shape Memory Actuators for Medical Rehabilitation and Neuroscience

Simone Pittaccio and Stefano Viscuso

Additional information is available at the end of the chapter

http://dx.doi.org/10.5772/50201

1. Introduction

Actuators based on shape memory alloys (SMA) proved to be particularly advantageous with respect to other actuation technologies when they are embedded in applications requiring strict compliance to a set of compatibility (e.g. mechanical, biological, weight, ...) and environmental constraints. Most noteworthy are the uses in miniaturised components, lightweight systems, sensing-actuating systems (e.g. actuators interacting with changing environmental variables), low-noise or low-impact appliances (e.g. actuators with reduced interaction with the environment) and self-sensing controllable devices. With particular regard to these preferred applications, SMA can also play a role in solving specific actuation problems in the fields of Medical Rehabilitation and Neuroscience. The main characteristics expected of SMA actuators for these fields are light weight and portability; self-adjustment to evolving needs; applicability of actuation in shielded environments (with bioimaging and diagnostic devices: magnetic resonance imaging (MRI, fMRI), magnetoencephalography (MEG), and the like).

Taking the move from these key ideas, this Chapter will discuss some innovative uses and some implemented examples of biomedical devices based on the Shape Memory Effect (SME), explaining their advantages and limitations. In particular, it will address the following main topics: analysis of the relevant literature and background; the derivation of technical constraints and required material properties given the expected biomechanical and or clinical cases; the mathematical design of sample actuators and their implementations; ideas for their clinical application and control. It is important to notice that these same concepts can be easily transferred to other important industrial engineering fields.

1.1. Biomedical background

Neuromuscular Rehabilitation is the science and practice of supporting the recovery of lost or impaired motor function, particularly due to damage of the Central Nervous System

(CNS). The established approach to treatment is physiotherapeutic, pharmacological and surgical, the choice of a specific line of action depending on the exact diagnosis, the general state of the patient, the severity of the damage, the time distance from the acute event, the evolution of sequelae, the presence of co-morbidities, etc.

In the general case, a patient undergoing neuromuscular rehabilitation was struck by a primary event (e.g. a stroke, a traumatic brain injury, an infection, hypoxia, a tumour,...) producing severe neuronal loss. According to the affected area of the CNS different scenarios may result, including a state of paresis (essentially, tetra- or hemi-plegia), and possibly other types of impairment, such as cognitive, sensory and so on. The recovery process from these initial insults often does not proceed independently of a series of interconnected secondary phenomena, which, if left uncontrolled, can undermine all hopes of an optimal final outcome. These phenomena include muscular contracture, spasticity, CNS remodelling and learned non-use.

With no ambition to be exhaustive, it can be pointed out here that the main mechanisms by which the human body, and in particular the neuromuscular system, react to the primary event are mediated by two functional disruptions of the physiological status, namely, *immobility* and *disuse*. Immobility is the absence of movement, especially of a limb or body joint; disuse is the lack of muscular activation and consequent mechanical loading of the skeleto-muscular structure. It is evident that both conditions are tightly connected to and often caused by the state of paresis. *Contracture* (shortening of the muscles) can occur due to prolonged (days or weeks) permanence of the body segment in a given position. This can be the case e.g. for a flexed elbow when patients sitting on wheelchairs keep their hands in their laps; or for an ankle joint that is maintained in an overly extended posture by gravity during continuous rest in bed. Muscular contracture produces, as a major consequence, a forceful and rigid ill-positioning of the adjacent joint, reduction of the available articular range of motion, and malformation. It can lead, in time, through reduced mobility, to structured calcifications of joints and retraction. Alongside contractures, the development of *spasticity* is one of the most common and severe secondary events in the sub-acute phase of paresis. Spasticity is an abnormal and speed-dependent stiffening of the muscles due to hyperactive stretch reflex. This exaggerated response to muscle lengthening exacerbates the impossibility of movement and subjects the patient to discomfort and pain as soon as the limb is mobilised. In turn, contrasting immobility becomes an even harder challenge. The picture resulting from the instauration of contracture and spasticity is given the name of *spastic paresis*, which is a painful, disfiguring and severely disabling condition, and constitutes a serious obstacle to patients' recovery [1-2]. In fact, the increased rigidity of muscles and joints further hinders any voluntary or involuntary motion or use of the limb, thus creating a vicious circle towards greater disruption of neuromuscular functionality.

Besides producing peripheral modifications, the prolonged immobility and disuse of the limb appears to affect negatively the plastic reorganisation and remapping of brain structures, by which live neurons take over the role of lost ones. The manner of such remapping is bound to have a paramount influence on ultimate functional healing, and this

is one more motivation for trying and avoiding those conditions in the first place. Their removal could shun the worsening of muscle stiffness and hypertone, while a sustained proprioceptive (sensorial) stimulation through movement may support appropriate cortical reorganisation. However, a complete picture of brain remodelling after a CNS insult and the effect of rehabilitation in modulating such a reorganisation is still not available. Studies of Neuroscience investigating in particular rehabilitation processes are certainly of great interest for physicians, neurologists, researchers and physiotherapists, as they can shed some light on those phenomena.

Recent trends in neuromuscular rehabilitation rely on pharmacological treatments against mild and severe spasticity, such as denervating neurotoxins (BOTOX) or myorelaxation agents (Baclofen) [3]; in most severe cases of contracture the approach can resort to surgical interventions such as tendon lengthening or transfer. The most widely used technique for all severities and types of spastic paresis is however physical therapy, which consists in static repositioning by means of splints and casting, controlled and gradual mobilisation of the limbs, muscular stretching and active exercising. It has been widely acknowledged that active work-out is a particularly important part of the rehabilitation program [4-5]. However, if autonomous control by the patient is absent or insufficient, passive mobilisation is utilised to try and maintain the viscoelastic properties of muscles and periarticular tissues. Furthermore, it is reasonable to think that minimising immobility and disuse, even by means of passive motion, can be advantageous for the prevention of the other adverse sequelae of CNS damage or to contrast their worsening.

Robotic tools are becoming very popular, as they make it possible to extend the duration of therapeutic sessions without the need for an individual involvement of human therapists, and reduce costs. In fact, though time-extensive physiotherapy is known to influence positively the rehabilitation outcome, organisational issues may limit the practical application of this principle, especially for passive mobilisation, in which case the physiotherapist has to move each affected joint one by one. In most situations, robotic tools provide repetitive and repeatable passive mobilisation of the whole limb [6], or joint by joint [7]. Moreover, several rehabilitation robots can assist active work-out, by completing patient's efforts to move [8] with the aim of improving muscle recruitment, by perturbing limb trajectories to enhance active control, or by contrasting movement [9] to strengthen muscles. Some drawbacks can be identified concerning the current use of robots in rehabilitation. First, as many of these machines are large, massive and costly, in a typical clinical environment the number of robots is limited and their use at home is ruled out. Second, most part of them require minimal skills by the patients, e.g. trunk control, equilibrium or cognitive abilities: that excludes very early or severe subjects, although they could benefit from extended sessions of passive mobilisation as well.

Robotic devices have been used for the mobilisation of body parts also in the field of Neuroscience [10-15]. Machines to manage the motion of the effectors (muscles) can in fact be used to support the measurement and visualisation of the interconnections between the activities of peripheral segments and the brain structures involved in their control. In this

type of applications tight compatibility with Bioimaging and Biosignal technologies is mandatory [13].

1.2. Overview of shape memory alloys

Robotic devices for Rehabilitation, like all robotic devices, are based on one or more actuators to transfer motion and displace a load. The load, in the case of Rehabilitation, will often consist of body limbs and joints, comprehensive of their mechanical properties, inertia and gravity. As was reported in Section 1.1, there is still a want of appropriate robots, meeting the needs of frailer patients and organisational requirements of the clinical structure. Much investigation in this respect is devoted to the manufacture of light-weight, portable and/or wearable devices. To the ears of the materials scientist or technologist these properties and the capability for actuation are typical features of a special class of metallic alloys, namely Shape Memory Alloys (SMA).

SMA are a compositionally heterogeneous class of materials, comprising a number of intermetallic compounds made up of transition and post-transition metals of different groups. It is beyond the scope of this Chapter to relate about all the different available compositions and their properties, which can be found in reference texts and scientific literature [16-19]. The most relevant compound of this class, employed in most technological applications to date is nearly-equiatomic NiTi, which is also supplied commercially in many semi-finished forms. The general name of NiTi (or TiNi) or Nitinol refers to several grades of the alloy found in the quasi-stoichiometric range from 49%at Ni to 51%at Ni. There are also some important ternary compositions based on NiTi, such as NiTiCu, NiTiHf, NiTiPt, NiTiNb, NiTiCr, with optimised characteristics for dedicated uses.

The Shape Memory Effect (SME) is an athermal reversible martensitic transformation producing macroscopic strain recovery upon heating above a certain characteristic temperature, generally referred to as A_f in the specialised literature. In the case of binary NiTi, the stable low-temperature phase (below the M_f temperature) is a B19′ martensite, while the high-temperature structure is a B2 parent phase. M_f and A_f are separated by a temperature difference of around 20°C in the solution-treated state. The effect of cold working is to strengthen and embrittle the material, and suppress the transformation. A controlled work-hardening however produces beneficial effects on the mechanical properties. In this state the material forward and backward transformations occur across spread-out temperature ranges and the hysteresis is also increased. Ageing treatments are necessary to optimise mechanical and functional properties and to adjust the characteristic temperatures. Typically, NiTi is aged at 350-650°C and water quenched, and this process also shape-sets the material in the shape "to be remembered". Treatment temperature and duration have to be honed to the application and size of the specimen. The mechanical properties, as described by a stress-strain tensile loading curve, may vary but, in the most representative cases, show an initial linear-elastic range, followed by a long flat plateau and a final increase. Most of the macroscopic shape change happens along the plateau region (up to 6% long) and corresponds to the microstructural phenomenon of martensite detwinning.

There are two phenomenological varieties of the unloading curve, according to the temperature of the test. If the test is carried out below M_f, unloading occurs along a line and strain decreases only minimally, so no macroscopic shape recovery is often observable. Full recovery is obtained by successive heating above the A_f temperature. This is called the SME proper. However, if the test is carried out above A_f the unloading curve is different and shows a long recovery plateau, at lower stresses than the loading one, and a final linear decrease towards zero-strains. The curve, in this case, is hysteretic and the shape recovery is attained without any need to heat the material. When the test is at room temperature T_r ,and A_f lies below T_r, SME is given the special name of Pseudoelastic Effect (PE). In all cases where the test temperature falls between M_f and A_f the behaviour is intermediate. Cold-work, ageing and precipitation of second phases can have very significant effects on all aspects of the mechanical curves, and in particular, on the height and separation of the plateaux, and cycling stability. So has the precise test temperature: first of all, in the case of SME loading occurs through the deformation of preformed martensite, resulting in lower stress values of the plateau and lower linear elastic modulus than in the case of PE, where austenite is initially loaded until stress starts inducing the formation of detwinned martensite; furthermore, through the Clausius-Clapeyron effect, stress is proportional to temperature and thus plateau stress levels are increased by an increase in test temperature, in particular in the case of PE. Finally, the compositions in the range 49%at-50.5%at Ni (Ti-rich) tend to show SME proper, while the Ni-rich ones (50.6%at-51%at Ni) have a pseudoelastic behaviour at room temperature. As a general trend, the higher the Ni content, the lower are the transformation temperatures.

1.3. State of the art of SMA in rehabilitation and neuroscience

Many efforts were made in the last 20 years for developing SMA-rehabilitation tools. Figure 1 shows the evolution of the number of papers, patents and conference contributions over this period. It is evident that in the last five years several groups focused on studying the matter, issuing some two thirds of the overall production. All papers dealing with implanted devices were not included in the search, for a number of reasons: because the typical fields of use of implantable devices are hardly connected to Neuromuscular Rehabilitation; because the design of parts to be utilised inside the human body is based on assumptions and limitations rather different from those distinctive of external actuators; because the domain of implantable devices in NiTi is well covered by extensive and comprehensive reviews and is now an established application. A choice was made to include only papers addressing directly the issue of Neuromuscular Rehabilitation or, indirectly, suggesting manners of applying static or dynamic external forces to reposition or move body parts. Particular attention was paid to applications including SMA *actuators,* i.e. devices exploiting the SME to transform heat into mechanical work.

1.3.1. Repositioning

Repositioning is the set of procedures aiming at contrasting ill-postures and malformations caused by contractures and spasticity. Static orthoses are often used to impart these

treatments and may be as rigid as castings or partially compliant. Many authors imagined the use of SMA in orthotics as passive components that apply static corrective forces [20-30]. In these applications SMA are mostly employed for their pseudoelastic properties, even though some designs are also based on SME. A series of papers [20-25] showed applications of pseudoelastic NiTi for spastic limb repositioning (elbow, ankle). The authors suggest that pseudoelasticity can be an interesting solution to the disuse and immobility problems during the orthotic repositioning therapies, in that it safeguards residual motor capabilities (voluntary or reflex) and decreases contact pressure by yielding under muscular jerks. They also showed possible advantages of this type of devices for correcting equinus gait [25]. In [26], pseudoelastic NiTi was utilised to remodel deformed auricles. Though not strictly correlated to the main subject of Neuromuscular Rehabilitation, this paper presents an approach in which the anatomic and biomechanical constraints are integrated in the design procedure. In [27], PE is employed to try and correct flat foot malformation and provide increased stability.

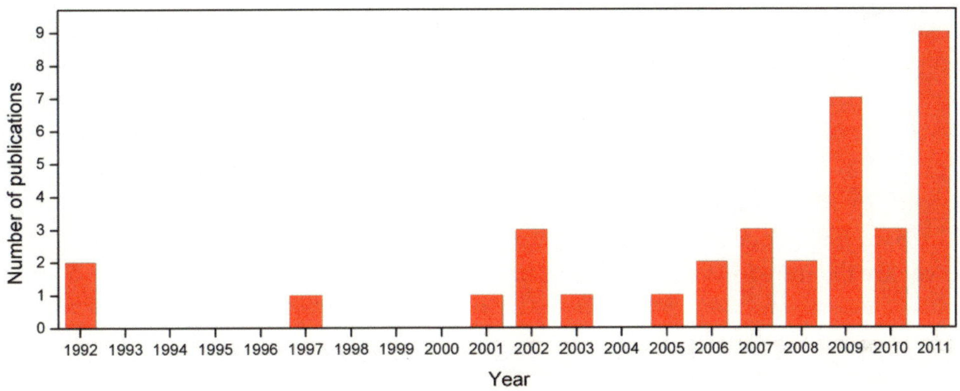

Figure 1. Histogram of the number of publications about SMA-based rehabilitation devices in the last 20 years.

On the other hand, [28] used the Shape Memory Effect to stretch gradually spastic wrist and fingers of paretic patients. The authors of this particular report also carried out measurements of joint resistance to movement, and designed the NiTi elements on the basis of these results. Thanks to the biomechanical design, the joints could be positioned at the neutral angle when the SMA was activated. The main characteristic of SMA exploited for this application was its ability to change shape, which eased putting on the device on malformed limbs. In an international patent [29], SMA are cited as a possible means to produce adjustable degrees of knee joint distraction and assuage contact overload between femur and tibia in unicompartmental osteoarthritis. Another international patent [30] discloses the use of SMA elements as possible actuators in a dynamically adjustable shoe to adapt to congenital or acquired deformities of the foot.

1.3.2. Functional exercise

Passive physiotherapy consists in repetitions of movements, muscular stretching and proprioceptive stimulation imparted by the hands of a therapist or by robots. Some authors

[31-41] investigated the possibility of employing SMA to make portable or wearable devices to act as functional exercising robots.

Some papers focused on finger movement, in particular the review [31] reported an example of rehabilitation glove fully actuated by SMA wires showing the two-way Shape Memory Effect. Another attempt to move fingers is reported in [32,33]. This paper describes the characterisation of bending wire actuators on a dummy finger and a temperature-controlled switch, and clearly evidences the issue of the force-speed trade-off. In [34] another device for passive mobilisation of the flaccid fingers is presented. This paper dealt with biomechanical constraints and design issues connected to the use of SMA springs as actuators, also proposing several modelling equations. Interestingly, both [33,34] employed the composition NiTiCu, which is known to have a narrower thermal hysteresis and lower detwinning stress than binary NiTi.

Other papers [35-41] concentrated on the lower limb. In [35], a concept for a paediatric boot is described. In the intention of the authors, a control system would activate two NiTi wires producing slow movement of the ankle and provide the possibility for home passive mobilisation. Two different implementations of a wearable ankle mobiliser were published by another group [36-41]. In the first paper [36] a description of the design procedure is presented, including mechanical and power dimensioning of SMA linear actuators based upon biomechanical considerations. Pre-clinical trials are also reported in the text. The following publications [40-41] dealt with an amagnetic rotary actuator and its use for the construction of a biosignal-controlled ankle exerciser that could be used in two different modes: fully passive and haptic-assistive, according to the evolving state of the patient during recovery. Trials on healthy subjects are described in [41].

A different use of SMA is proposed in [42], where a NiTi-wire was utilised to produce fingertip tactile stimulation with the purpose of providing a means for sensory feedback in haptic rehabilitation. Several implementations are presented and also qualitative tests on healthy volunteers are reported.

1.3.3. Muscle toning

Some robotic tools provide resisting forces that contrast active exercising by the patients. This could help increase muscle strength and control abilities. A few publications propose SMA-based devices for manifestly therapeutic uses and for generic muscle strength training. In the international patent [43], a set of wearable pseudoelastic garments are proposed to favour muscle toning during daily life activities. Though this application was not described as a therapeutic tool, it shows similarities with techniques that are used in Neuromuscular Rehabilitiation. Two papers by another group [32,33] suggest the use of SMA actuators for applying resisting forces to active movement. Unfortunately, the authors did not discuss either the level of force required or the consequent dimensioning of the actuator.

1.3.4. *Assistive robotics*

Other proposed applications of SMA in Rehabilitation comprise a number of studies that present active tools to support [44-47] or take over [48-51] impaired motor functions.

An M.Sc. thesis and a following paper by the same authors [44,45] report about the design and control of a downscaled bio-inspired mechanical model of an ankle joint wearing a SMA-actuated orthosis to support the walking capabilities of plegic subjects. The thesis discusses the biomechanical constraints and both publications clearly explain the dimensioning of the device. The authors also discuss the relationship between the size and number of SMA elements and the actuation properties. Another group issued a paper [46] and a patent [47] describing the use of SMA actuators to manufacture soft wearable orthoses for the knee joint. The garments were intended to provide sufficient torque to support the impaired gait of neurological patients, but the authors eventually found that the cycling speed was somewhat slower than appropriate.

An international patent [48] describes an orthotic device for restoring the grasping function in paretic subjects. The SMA actuation is intended to substitute entirely the lost function by allowing index and middle finger tips to reach jointly the tip of the thumb. A more complex multi-joint device with the same general purpose was developed in a Ph.D. thesis [49], where motors (possibly SMA-based ones) are however not mounted on the patient's arm, but lie at a distance, being connected to the limb by Bowden cables. This manuscript also includes biomechanical measurements and a control system. Another patent [50] imagines orthotic devices activated by SMA elements aiming at accurately simulating the mobile capabilities of human limbs. An unusual application is provided by [51], where SMA wires are employed as actuators to regain lost facial mobility and the ability to smile. The activation of the wires, aligned along the main affected muscle groups, is triggered by bioelectric signals acquired on the contralateral healthy side of the face.

1.3.5. *Neuroscience*

The scope of limb-moving actuators in the field of Neuroscience is mainly to provide proprioceptive, kinaesthetic or tactile stimulation in order to investigate neural responses to those stimuli. Increasing interest has being expressed during the last few years [10-15] to the effect of finding specialised tools that can be compatible with the instrumentation used to conduct Neuroscience experiments, e.g. magnetic resonance imaging (MRI, fMRI), magnetoencephalography (MEG), electroencephalography (EEG), near-infrared spettroscopy (NIRS).

SMA actuators have found little application in this field as yet: only two examples of SMA-based devices for use in Neuroscience were identified in the literature [37-41]. These publications describe a linear and a rotary actuator designed to comply with electromagnetic noise constraints, and be compatible in particular with MEG and fMRI. Most other devices were implemented using alternative technologies such as pneumatic and piezoelectric. The main reported advantage of SMA actuators is again portability and wearability.

1.3.6. Final remarks

Most authors recognised as a major advantage of using SMA in orthotics the light weight and compactness of the actuators. These factors, in fact, suggest that these materials can be used to build portable and/or wearable devices, and oriented the concepts of many designs towards compact robots, light static or dynamic splints, or even soft wearable garments, often for use in the home environment. Going from concept to actual implementation of the devices, the most challenging issue in this field appears to be the dimensioning of the actuators to address at the same time long-strokes and fairly-high-force requirements. In particular, human joints, bones and muscles move or deform across broad ranges of angular motion and several centimetres in length, and possess non-negligible masses. Understanding this is a first fundamental step of an engineering approach to the problem of actuation for Rehabilitation. Some papers did not find an optimal solution to the design question, because there is a serious trade-off between size and power variables to be overcome. In particular, high forces require large cross-sections of SMA: big diameters or many units working in parallel, or both. Any one of these choices, however, brings along some drawbacks: large diameters generally imply larger currents (for Joule's heating) and longer heating and especially cooling times; thin actuators are faster, but building large bundles or arrays of those require cross-insulation of the parallel fibres, larger volumes or surfaces to store them on board and the right strategy of electric connection (in series to minimise the current but requiring high voltages, or *vice versa* for the parallel configuration). It must be kept in mind, as an extra constraint, that patient's safety issues would tend to favour solutions with both low currents and voltages. Furthermore, when precise timing of the movements, matching some physiological rhythm (such as gait), or speed control are parts of the desired functionality of the device, actuation frequency is also an important factor, imposing tighter limitations on the acceptable surface-volume ratio and thermal exchange properties (viz. heating and cooling times) of the SMA actuators. The most successful approaches appeared to be characterised by a precise preliminary analysis of the biomechanics of the joint to be moved, and an optimal and simultaneous modelling solution of the overall functioning and sizing of the SMA actuator: this method made it possible to balance advantages and drawbacks against all standing trade-offs.

In terms of control systems, some designs implemented fine closed-loop architectures to set, maintain or change speed and position of the effector. These systems attracted the attention mostly of designers of dynamic tools, such as those to support the gait or the motion of the fingers. Simpler approaches were used to obtain automatic switching on and off of the actuators, based on some measurable variable (position, temperature, time,…). A few papers employed biosignals (particularly muscular bioelectric activity measured from the skin surface) to trigger actuation. We find this idea particularly interesting, dealing with Neuromuscular Rehabilitation, because it may bring the residual abilities of the patients into the design picture.

Actuators for Neuroscience share a number of characteristics with those described for the field of Rehabilitation. Besides the constraints depending on biomechanics and SMA

properties, special requirements must be fulfilled in order for these devices to be able to be used in electromagnetically shielded environments. In particular, if the Joule's Effect is chosen as a means to heat SMA elements and produce phase transformation, care will have to be taken in controlling any unwanted stray magnetic fields generated by the current flow.

2. Uses of SMA actuators in medical rehabilitation

This section will provide guidelines for designing SMA actuators aiming at moving human joints. General statements and suggestions will be accompanied by the discussion of a case study taken from the authors' experience, i.e. an ankle mobiliser conceived as a lightweight and portable tool for providing neurologic patients (stroke, traumatic brain injury, paresis) with motor support and sensorial stimulation not only during the very early phase of rehabilitation when they are still unable to exercise, but also during the subsequent phase of motor relearning, thanks to a specially-designed biofeedback-based control system.

2.1. Biomechanical, bioengineering and clinical specifications

The initial step of a biomechanical design is identifying the anatomical compartment, the pathology of interest and aim of the medical device. This is however just the basic set of information to tackle the problem and provides a very rough description of the actual boundary conditions affecting the following steps. Further human-related design constraints that ought to be considered in developing rehabilitation devices span from the mechanical properties of the joint, to the typical characteristics of target patients, and as far as considering the psychological impact.

2.1.1. Tissues and joints

Any biomedical device, either implantable or external, should take into account the peculiar characteristics of living tissues. Focusing on the musculoskeletal system, a great variety of tissues can be described: cortical bone, cancellous bone, muscular fibres, tendon, ligament, cartilage, synovial fluid, fibrous tissues… All these tissues show non-linear, viscoelastic and hysteretic properties. Some of them, such as the muscular tissues, also display active mechanical behaviour. Moreover, tissue characteristics may vary significantly with age, sex, ethnicity, fitness, pathology or malformation. The properties of biological tissues work together to produce the observed features of joints and limbs, which therefore also display non-linear, viscoelastic and patient-dependent behaviour.

Human joints may differ significantly from one another in terms of dimensions, degrees of freedom, neutral position, range of motion, associated muscular power and weight of the adjacent limb segments, neighbouring structures, etc. A "general purpose" actuator is hard to imagine, but dedicated devices can be tailored to target specific joints and pathologies. Furthermore, once the application has been selected, a statistical analysis of the target population could be carried out and possibly *ad-hoc* assessment of joint characteristics, biometric parameters, classic associated malformations or impairments and comorbidities, so that an appropriate average set of biomechanical parameters is acquired for a "typical user".

Modelling of the joint behaviour is often an essential step in the design process. The desired application can direct modelling choices as to whether a detailed or simplified description of the joint should be included. For example, it may be a purpose of the new device to activate only one degree of freedom (d.o.f.) of a di- or tri-axial joint; in this case, the extra d.o.f's can be excluded from the joint model, but it must be realised that the neglected mobility can produce unwanted stresses or sliding, or pain when the actuator is in use, and it is up to the designer to decide if not controlling those events is acceptable, or else they ought to be controlled by some other means (e.g by fixing and preventing movements along the extra d.o.f.'s). Joint characteristics cannot be considered as unchangeable data, but have to be connected to the current design case: for instance, in judging what a suitable range of motion (ROM) for the actuator should be, the designer must consider the joint type, but also the application (what is the required movement for the rehabilitation exercise?), the pathology (is it likely to change the ROM? Does the actuator have to work to restore the physiological ROM?), the relationships with posture (e.g. certain positions of adjacent joints can affect the available ROM of the target one), the influence of other d.o.f.'s, etc. In relation to this, it can also be decided if joint mechanical properties can be linearised within the ROM of interest.

2.1.2. *Expected loads*

Considering the force or torque generated by the actuator, it is important to keep in mind that joints have passive and active torque-angle characteristics that can be activated by the patient voluntarily or unconsciously.

A passive response is elicited from the muscles and periarticular structures when the joint is moved from its neutral angle, i.e. a resting position where forces producing joint flexion are balanced by those producing joint extension (including or not including gravity, according to limb position in space). As mentioned above, this response is generally nonlinear, viscoelastic and hysteretic, even in physiologic conditions. It can be evaluated by means of a motor-dynamometer-encoder system, in such a manner that the joint can be mobilised passively at controlled speed across a set range of motion, and resisting torque and position can be acquired simultaneously throughout the movement. By changing movement speed the viscoelastic characteristics can be evidenced, while conducting tests both in the flexion and extension direction (or inversion and eversion; or adduction and abduction; etc.) can reveal the hysteresis. The obtained torque-angle curves can be used as load curves in the design of the actuator. The absolute values of passive joint torques strongly depend not only on pathology (viz. contracture, retraction, fibrosis, etc.), but also on age, sex and muscular trophism, and of course change from joint to joint in the body. For this reason we can only give here some examples to clarify the order of magnitude of these torques. Maximum physiological passive torques can be as low as 50-60Ncm for finger joints [52], 100-300Ncm for the elbow [53] and up to 500-1000Ncm for the ankle joint [53-54], also depending on the angle of the knee. In healthy conditions, joint passive stiffness is not prevalently affected by movement speed. The same may not be true in some pathological cases. Muscle contracture and joint retraction can cause the reported values of joint passive stiffness to increase several times and can even make joints completely rigid to any practical purpose.

Apart from a passive response, joints can resist imposed movement also through the activation of the associated muscle groups. Muscular power can be delivered voluntarily (active motion) or by reflex activation. Involuntary active motion can also be produced by neurological conditions such as cloni, dystonia, chorea, etc. During limb and joint passive mobilisation muscle stretching can occur: a dedicated physiologic reflex produces instantaneous muscular contraction when the muscle fibres are elongated over a certain limit. This so called *stretch reflex* is much exasperated in spastic syndromes and may produce an involuntary, uncontrolled increase of resisting loads. Spasticity strongly depends on the speed of stretching; preconditioning the limb by a few cycles of slow manual stretching can help reduce momentarily the intensity of stretch reflex. Very severe spasticity, however, may correspond to the absolute impossibility of mobilising a joint.

Furthermore it must be remembered that the limbs adjacent to a joint possess a mass, and are therefore subjected to gravity and inertial forces connected to accelerations produced even in different parts of the body. One important case is gait. Where and when any or all of these components of the resisting or facilitating loads for the actuation must be taken into account has to be decided for each new application separately, in particular considering the typical position held by the target patient while the actuator is functioning, whether the patient will be prone to uncontrollable movements (jerks, dystonia, convulsions, etc.), or, on the contrary, whether some physiological movements will be typically impeded or suppressed (paresis, coma, castings, etc).

2.1.3. Interfaces

The forces and strokes generated by the actuator must be suitably transmitted to the patient's limb, so connective or fastening elements have to be carefully considered in relation to local anatomical constraints, such as the size of the limbs or body parts and the available surrounding space. In particular, when designing a wearable device, additional dimensional and weight constraints must be taken into account. Generally, the device should be attached to the limb segments that articulate at the target joint, in order to promote their reciprocal movement. Designing such connections should take into account the dimensions of limb segments (that strongly depend on age, sex and pathology) and the available space (determined primarily by joint position and orientation relative to the rest of the body, sex and malformation, etc.). For example, a device to move the shoulder should be designed taking into account the sex of the patient, especially if a frontal strap is required; in the case of a knee device, on the other hand, the lateral aspect of the joint could provide more free space for attachment than the medial one, so that interference with the other leg can be avoided. Another important factor linked to the transmission of forces and movement is the device-skin interface. This must be carefully designed so that no excessive pressure, friction or relative motion occurs during activation of the actuator. Some of these problem can arise due to misalignment of the actuator rotation axis with respect to the corresponding anatomic joint axis. In case of wearable applications, also the weight and stability of the device on the limb should be taken into account.

2.1.4. Other issues

Safety is a mandatory constraint when designing medical devices in general. For the particular case of rehabilitation actuators based on SMA, most risks could arise from electric, mechanical or thermal hazards. IEC 60601-1 is a widely accepted standard for medical electric equipment, and many countries tend to require compliance with IEC 60601-1 before commercialization of an electrical medical device is permitted [55]. Even if the medical device is intended for use only in the scientific laboratory, reference to this set of prescriptions may suggest precautions and guide designers and engineers in the development of safe devices. In connection with this latter standard, international standard ISO 14971 specifies a process for identifying the hazards associated with the use of medical devices and carrying out a suitable risk analysis [56]. Consequently, good practice requires to try and minimise all the identified risks. Moreover, when the medical device is approaching clinical validation, another norm to be considered is ISO 14155, which addresses the issues of designing, conducting, recording and reporting of clinical investigations carried out on human subjects [57]. All of these guidelines are valid also for SMA-based devices and may help consider further implications of new designs for the human body.

Dealing with patients, there are still other questions to be considered. Even if the device is functional and efficient and the actuator fulfils all the technical requirements, they will not be employed in the clinical setting if they are not acceptable to the patient. At the basis of *acceptability* there is a set of physical and psychological factors. Of course, the medical device should not produce nuisance or pain during short or prolonged use. Those problems, in the particular case of rehabilitation appliances and actuators, may depend on badly-designed interfaces with the skin or on movements imparted too fast. In fact, as already stated, contracted and spastic muscles are very responsive to stretch, especially when it is applied rapidly. During mobilisation, the increased muscle tone due to spastic reflexes may result in much discomfort for the patient. The use of SMA actuators in general has the advantage of providing low movement rates; however, severe spasticity may be hard to treat even with slow active devices. Other acceptability issues are often connected to psychological aspects. In particular, wearable devices should not embarrass the patient with noisy, bulky or cumbersome structures. For children, shape and colour can also be important features. SMA actuators are likely to be silent and miniaturised, and thus suitable in this respect, but all other components may not (i.e. power supply). Other impediments for patients employing rehabilitation devices actively or for assistive therapies may arise from their possible co-existing cognitive impairment: this consideration supports the use of very clear instructions and user interfaces, especially using non-verbal pictorial communication, lights, vibrations or sounds.

2.2. Actuator design

2.2.1. Strokes and loads

Rehabilitation devices in most cases are intended to generate a mutual rotation of limb segments around an anatomical joint. By employing rotary actuators, it is possible to

provide directly angular strokes and torques. On the other hand, linear actuators can be coupled with suitable lever arms and achieve the same results indirectly. The first step in designing an SMA-based rehabilitation actuator is to define, by careful analysis of the biomechanical constraints and clinical requirements, the appropriate neutral position, range of motion and resisting loads. The neutral position and range of motion will help determine a correct linear or angular stroke for the actuator; the resisting loads will set the linear force (and lever arm) or torque requisites. Of course, the two quantities will influence each other, for example different choices of lever arm may affect the linear stroke.

Once the required stroke and load have been determined, the geometrical parameters of SMA element can be worked out. Different types of actuators need different formulas for determining the parameters: for example, wire actuators have only two parameters, length and cross-section that directly depend on stroke and load respectively, while springs (helical, torsion or flat) parameters comprise coil dimensions, wire diameter and number of turns, which all depend on both stroke and load. Moreover material properties influence actuator parameters, especially in springs where anisotropy may play an important role [34]. Attention should be paid in particular to the critical stresses needed to detwin martensite both in the heated (above A_f) and cooled (below M_f) states. This consideration can guide the choices about material composition and thermomechanical treatment, because those factor have great impact on the height of the plateaux and the mechanical hysteresis. In most actuation applications, in fact, it must be guaranteed that the load to be displaced is low enough that stress in the material is lower than the critical stress for the heated state; on the other hand, sufficient loading should be provided to detwin martensite and ensure actuator deformation and position reset during cooling for most cyclic applications. Whatever the SMA geometry, it can be stated in general terms that cross-section depends mainly on the maximum desired work output, through the mean level of stress acting on it. The question of choosing the level of stress is quite complex, and also involves the expected fatigue life of the actuator and the strains recovered upon activation. For example, it is reported that commercial NiTi-wire actuators can remain stable for more than 10^5 cycles at 200MPa and 4% of strain in traction [58], but other optimal compromises can be found, provided that stress does not exceeds 400MPa and strains 5%. The use of parallel actuators can help reduce required work from each actuator, i.e. given a stress level, cross-section can be reduced. Dimensioning cross-section should also take into account cooling rates, as will be discussed in the next section.

2.2.2. *Power*

Heating and cooling are key points concerning SMA actuation. The general equation that drives these processes is

$$P_{in}(t) = c_p m \frac{dT}{dt} + m\Delta H \left|\frac{d\xi(T)}{dt}\right| + \frac{dW(\theta(t),t)}{dt} + P_{diss}(t). \tag{1}$$

Considering the heating (actuation) phase, $P_{in}(t)$ is the heating power, $c_p m\frac{dT}{dt}$ is the power needed to increase SMA element temperature (c_p is the constant-pressure specific heat and m

is the mass), $m\Delta H\left|\frac{d\xi(T)}{dt}\right|$ is the power involved in phase transformation (ξ is the fraction of austenite and increases in time as transformation goes on, ΔH is the transformation enthalpy), *W(θ(t),t)* is the mechanical output (useful work, inertia and frictional losses) which varies along the movement trajectory *θ(t)*, and $P_{diss}(t)$ is the power dissipated by thermal convection, thermal conduction and (less important in this case) thermal irradiation. Considering the cooling phase, and imagining that a bias force acts to restore the starting actuator configuration (by deforming the SMA element), $P_{in}(t)$ is again a heating power that could be used to help control the cooling rate, $c_p m\frac{dT}{dt}$ is the power to be dissipated in order to decrease SMA element temperature, $m\Delta H\left|\frac{d\xi(T)}{dt}\right|$ is the power involved in phase transformation (enthalpy in this case is negative), *W(θ(t),t)* accounts for the mechanical energy stored in detwinned martensite plus inertial and loss terms, and $P_{diss}(t)$ is the power dissipated. The contribution of $P_{diss}(t)$ is present both during the heating and cooling phases of an actuation cycle, but it may differ significantly if any active cooling system is included. Equation 1 gives a simplified description of the process, but permits to focus on some key points. First, SMA mass is an important factor when designing power consumption. Second, as all the diverse forms of dispersed energy depend on the amount of exposed surface, surface/mass ratio is the driving contribution to determine cooling time. Therefore, thin actuators (maximising the surface/mass ratio) should be preferred to bulky ones if cyclic, especially high-frequency actuation is desired. Having in mind which use is intended for the device provides a guidance for the choice of heating source and consequently gives some hints for the actuator design. The best strategy for providing controllable, cyclic actuation is heating by Joule's effect, but, especially for non-cyclic and one-shot activation, also direct thermal transfer may be suitable. Focusing on Joule's effect, the energy provided depends on the injected current and the electric resistance of the SMA element according to Equation 2:

$$P_{in}(t) = P_{Joule}(t) = I(t)^2 \cdot R_{SMA}(\xi) \tag{2}$$

where the electric resistance $R_{SMA}(\xi)$ can be expressed as

$$R_{SMA}(\xi) = \rho(\xi) \cdot \frac{L(t)}{S} \tag{3}$$

with $\rho(\xi)$ representing the SMA resistivity, S the cross-sectional area of the SMA element, L(t) the length over which the current flows, that obviously changes as actuation goes on. Given a length of SMA wire (which depends directly on the stroke required from the actuator), the thinner the wire, the higher is the electrical resistance. Moreover, mass-dependent contributions to Equation 1 are also minimised, and heating efficiency increased.

SMA actuators are subjected also to the Clausius-Clapeyron relation, that can be expressed as:

$$\theta^* = \theta + \frac{\sigma}{C} \tag{4}$$

where $\boldsymbol{\theta}$ is one of the transformation temperatures, which is effectively increased to $\boldsymbol{\theta}^*$ when a stress σ is applied to the material. This is a very important issue when purchasing shape memory materials or dimensioning SMA actuators powering: the transformation

temperatures as provided by DSC analysis will be shifted when the material is loaded. Typical values of *C* for NiTi range from 5 to 9 MPa/°C, i.e. an actuator designed to bear 200MPa would have transformation temperatures increased by as much as 40°C. However, actuation temperatures cannot be increased indefinitely: in fact, above a certain temperature (M_d), part of the deformation cannot be recovered and material undergoes plastic slip.

2.2.3. *Choice of the SMA element*

SMA actuators can be provided in a number of shapes, which may differ significantly from one another in terms of available stroke and force. The first distinction is based on the deformation mode imposed onto the material. Wires are subjected mainly to traction, which produces a uniform loading of the material cross-section: this fact makes it possible to produce high force actuators, but, on the other hand, the available stroke is limited by the fact that linear deformation can be as high as 8% only, 4% in most cyclic operating modes. Other simple SMA actuators are linear springs (tension or compression springs), whose principal mode of deformation is torsion. Springs of this kind are helical coils of SMA wire arranged around a central axis: this allows for large elongations along the central axis depending on spring ratio and the number of turns. On the contrary, force is limited by the fact that torsional loading is not uniform in the material cross-section. Some authors proposed also bending actuators, made of SMA ribbons [27] or SMA wires [33]. For the same purpose, also torsion springs and flat springs can be employed, which share the same principal mode of deformation, i.e. flexion. The advantage of this actuation is that angular stroke is directly available, but the non-uniform cross-section loading limits the available torque.

The problem of moving human joints generally involves large angular strokes coupled with fairly high levels of force or torque. The simple configurations just described most of the time provide solutions only for very long wires or thick cross-section springs, ribbons or bars, which are often deemed impractical for a number of reasons. Long wires have to be housed in a suitable manner, to limit length into a compact three-dimensional structure. A way to do so is coiling the SMA wire along a series of pulleys, whose diameter should be sufficiently large with respect to the wire diameter, to avoid strain concentrations. On the other hand, large cross-section actuators are impractical because they need high electric power to reach the transformation temperature. Moreover, cooling down bulky wires or ribbons may require too long a time for most cyclic applications. A possible solution for increasing force output in springs and ribbons is arranging many actuators in parallel or in bundles. Provided space around the joint is sufficient, attention should be posed to electrical insulation and electrical connections, in order to provide suitable solution to tension-current requirements for actuation, keeping in mind also the safety issues.

2.3. Actuator control

The problem of controlling actuation is very important for many applications in Rehabilitation, as patient's needs and responses may vary during the evolution of therapy or even during the same session of exercise. Control strategies apply mainly to assistive robots,

but may also be employed in passive mobilisation devices. As the focus of this Chapter is not on control systems, only a brief overview of possible applications to SMA rehabilitation devices will be discussed.

2.3.1. *Control of passive mobilisation cycles*

Two major strategies exist to control cyclic heating and cooling of SMA actuators: in *open-loop control* parameters are predetermined by a set routine, whereas *closed-loop control* employs a feedback signal of any nature to adjust heating and cooling parameters (timing, current or voltage, active cooling systems, etc). The main drawback of open-loop control is that any perturbation to the trajectory would not be compensated, as the current profile is predetermined. This could happen, for example, if patient exerts any unexpected active contraction or in the case of unpredicted changes in environmental conditions. However, there are applications in which open-loop control is still feasible, for example repetitive passive motion of flaccid limbs. Closed-loop controls need monitoring of a feedback variable, which could be the SMA temperature, the joint angle, actuation speed, SMA force or any other measurable physical quantity. Apart from rare examples of feedback loops based on monitoring of SMA resistance [59], closed-loop controls rely on dedicated sensors that are better included from the initial steps of the device design.

Another distinction can be made on different types of control of movement trajectory. In many applications an ON/OFF actuation is suitable: the focus is only on the initial and final positions, while the detailed trajectory is controlled only mechanically and the speed only in average terms. In this case, open-loop control may be practical, and heating can be achieved with a very simple current profile, e.g. a step or a ramp. Experimental tests are a viable solution to adjust the heating and cooling parameters. In particular, attention should be paid to the experimental conditions, including the orientation of the actuator in the gravitational field, as thermal convection is strongly dependent on that.

On the other hand, when a precise trajectory of movement over time is required for the application, alternatively open-loop or closed-loop control strategies can be applied. In the open-loop approach, Equation 1 should be solved by imposing the desired mechanical output $W(\theta(t),t)$ and by calculating the time-varying current profile to be injected. However, the most reliable way to provide trajectory control is closed-loop control, in which the current input is continuously adjusted to match the prescribed position, thus counterbalancing perturbations.

2.3.2. *Assistive control*

As rehabilitation implies that patient's capabilities evolve during therapy, the characteristics of therapeutic tools could be devised to change accordingly. A way to implement this is to trigger SMA actuators only when the patient makes some effort to move the paretic limb: thus, a measure of the "effort" should be chosen to implement such a biofeedback-based closed-loop control. In some cases, mechanical variables could be employed: for example, the SMA actuator completes movement only if the patient generates at least a certain level of

voluntary muscular force or flexes the joint over a minimal extent. On the other hand, possible biosignals for closed-loop control can be the surface electromyographic signals from the muscles that control the movement to be rehabilitated. The level of the biofeedback variable above which actuators are switched on may depend on the patient, pathology and progression of therapy, and therefore should be adjusted to match patient's capabilities with rehabilitation goals.

Besides the biofeedback signal, another key point for assistive rehabilitation is how to encourage the patient to participate actively in the rehabilitative session. Visual or verbal or acoustic feedback are the most common way to prompt, instruct and reward the patients during the exercise, but when selecting the communication format attention should be paid to their possible co-existing cognitive impairment.

2.3.3. Gait triggering

A very peculiar example of precise trajectory control is gait, during which any wearable active device should be able to activate according to the walking rhythm of the patient. A closed-loop control is desirable, and can be implemented with mechanical sensors (which detect angles at different joints or contact forces when the foot contacts the ground) and/or electromyographic monitoring of the leg muscles. Gait is a quite demanding application for SMA in terms of frequency of activation, as even slow walk takes place at over 1Hz. This influences dramatically actuator design, i.e. diameter and number of SMA elements. For gait applications, movement trajectory must be controlled fully within the step and from step to step: in other words, both heating and cooling phases should have adjustable parameters so that actuation can adapt to changing stride or external perturbations. The possibility of modulating input current profiles has already been discussed and may be sufficient for the control of heating; cooling, on the other hand, shall not be left to natural convection in this case, but an active system for accelerating and tuning heat transfer will often have to be adopted.

2.4. Case study: Passive/assistive SMA device for ankle dorsiflexion

2.4.1. Design aims and clinical constraints

Ankle is one of the body joints that often suffer from the sequelae of stroke, i.e. immobility, disuse, contracture and spasticity. After stroke, a rehabilitation program should be carried out as soon as the patient's conditions are sufficiently good, but in the acute phase the patient is unable to exercise actively. In these circumstances, moving the limb passively can help maintaining joint flexibility and normal muscle tone, while also contrasting contracture, spasticity and sensory deprivation. A rehabilitation device for promoting repetitive dorsiflexion of the ankle in the first few weeks after the acute event could help improve clinical outcomes. The fact of limiting mobilisation to just one degree of freedom (sagittal plane of the joint) is appropriate because dorsi-plantarflexion is the principal motion involved in gait. It is regarded as a matter of importance that the device be able to manage patient's changing conditions during recovery. As needs evolve during patient's improvements, this therapeutic

device should be able to guide and sustain gradual recovery by providing commensurate aid. This includes exploiting even initial attempts at voluntary motion and turn those into effective workout. To this end appropriate control will have to be included.

The device is intended for patients in the first weeks after a stroke, i.e. we should expect adult subjects having a flaccid ankle (no active control of the muscles, mild or no hypertone) and possibly increased ankle stiffness. Figure 2 shows the passive viscoelastic ankle characteristic of a healthy subject at rest (no spasticity, no contracture, no voluntary movement) and a typical chronic flaccid patient with mild contracture. The ankle joint can move in the sagittal plane in the approximate range -30°/+20° (negative being towards plantarflexion) even though the range of -15° to +10° is the functional range of utmost interest in gait. These ranges become reduced with evolving contracture and spasticity, and in particular the positive degrees are progressively lost. In the range of interest (-15°/+10°) the characteristics of the presented healthy and paretic joints differ mostly for dorsiflexed angles (greater than around +2°, in this case), where increased stiffness in the paretic ankle can be observed. The typical acute flaccid patient will have intermediate characteristics between the healthy one and the chronic.

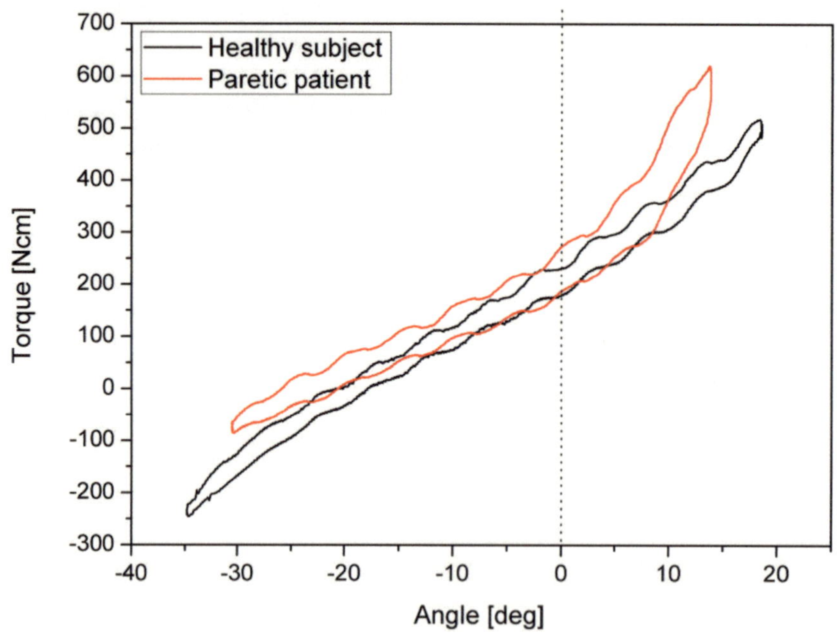

Figure 2. Measured passive characteristic of the ankle in a healthy subject vs. a paretic patient with mild contracture and ROM limitation. The main difference appears for positive dorsiflexion angles

Given this information, the biomechanical constraints on the design can be summarised as follows:

1. the resisting forces which have to be considered in actuator dimensioning are the viscoelastic passive characteristic of the joint and the foot weight. No active contribution from the muscles are expected.

2. Choosing a reduced range for the device, biased towards dorsiflexion, has the advantage that the calf muscles are kept under a relative stretch (physiological neutral angle is around -25°); furthermore it is appropriate to avoid forcing the joint excessively towards the extreme dorsiflexion position (conservative approach to limit insurgence of spastic reflexes). A target ROM for mobilisation was therefore was chosen as -5°/+10°. Of course, the actual working range will depend on the passive characteristic of the specific ankle, but is not expected to be much different from the target one.
3. For a target population represented by this type of subjects it is reasonable to assume that passive resisting torque does not exceed 400Ncm in the range of interest.

Considering the general conditions of the patients in the first weeks after a stroke, it would be wise to design a device that can be employed in the bed. However, the knee joint should be sustained in a flexed position, so that the bi-articular *gastrocnemius* muscle is not pre-stretched and full ankle range of motion is available. As a way of compromise, an angle of 10° for the leg rest can be assumed, with the foot positioned lower than the knee. It should be taken into account that, in such a configuration, the contribution of foot weight to the resisting force is limited. Figure 3 shows the total resisting torque, comprising different foot weights, and the same viscoelastic resistance from a typical patient. The expected maximal resisting torque can be approximately calculated in 400Ncm. As the curve hysteresis is not large for this joint (cf. Figure 2), the same values used for dorsiflexion can be also utilised when considering the movement towards plantarflexion and the loading level associated to martensite detwinning.

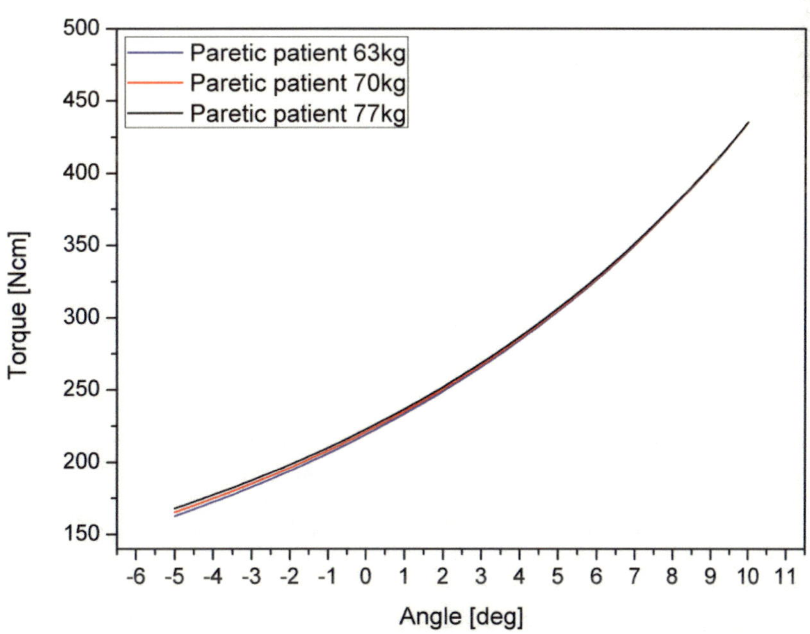

Figure 3. Influence of foot weight on the total resisting torque at the ankle. Differences in foot weight (proportional to the body weight) account for little change due to the lying position of the limb.

As the device is not intended for walking, there are no direct constraints on cycle duration, provided that a suitable number of cycles can be delivered to the patient's ankle in a therapeutic session (at least 100 cycles/hour), and movement speed is not too elevated, lest spastic responses are aroused. The actuator should provide repeatable movement and maintain the same characteristics for a number of sessions, as it may be impractical to adjust or change the SMA actuator too often. For this reason, maximal strain should be limited to 3% and stress to 300MPa.

2.4.2. *General concept of the device*

Figure 4 shows a simple embodiment of the concept. Two thermoplastic shells lined with soft foam are modelled on a prototype human lower limb of average size. These shells are hinged together and strapped by Velcro® bands respectively on the frontal aspect of the shin and the foot, in such a manner that the ankle and the hinge are perfectly aligned along the same axis (this minimises unwanted friction and sliding between the orthosis and the skin). The choice of planar hinges was also made to control (fixate) ad/abduction and in/eversion, although this solution makes the device unsuitable for patients that have already developed severe malformations out of the sagittal plane (unusual in the acute phase). With this structural configuration, two linear actuators are fixed on the front of the leg shell and transfer actuation pull to the foot shell through inextensible threads.

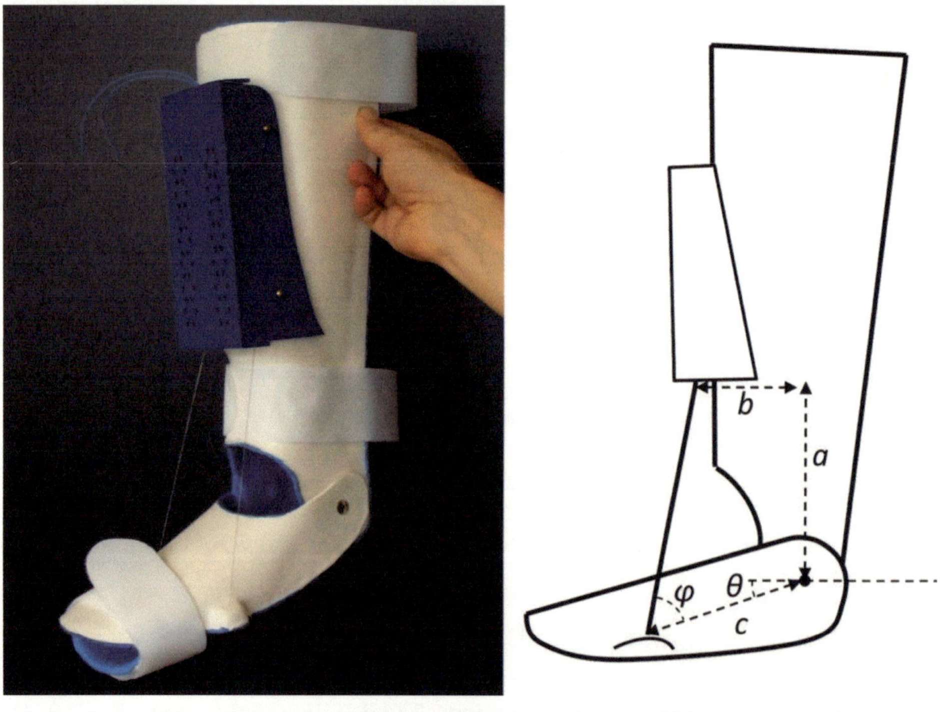

Figure 4. Implementation of an exerciser for the ankle joint with linear SMA actuators. The device is designed to provide cyclic joint flexion in the rehabilitation of neurological patients in the acute phase. On the left, schematic representation of the main design dimensions to calculate the lever arm.

The linear stroke and force output required of the SMA actuators depend, through the lever arm, on the fixation points for the actuators (on the shin), and the distal ends of the inextensible threads (foot shell). Actuators produce dorsiflexion, while plantarflexion and actuators position reset is left to gravity and viscoelastic resistances.

Power supply is to be provided by a dc-generator, which is not on board the wearable device, as the intended use is not for walking. This choice helps limiting weight and keep the device stable on the patient's leg during use.

An open-loop control strategy is adopted for the use as passive exercise device, whereas a simple closed-loop control for assistive rehabilitation employs the electromyographic activity from *tibialis anterior* as control signal, picked up by surface electrodes. As for the power supply, it was decided that all the components needed for control are not mounted on board the wearable device, even though integration would be possible by developing *ad-hoc* electronics.

2.4.3. Actuator design and characterisation

With the selected set of geometrical parameters (Figure 4: a=16cm, b=8.5cm, c=15cm), a linear stroke of 2.5cm is required to produce an angular movement across the range -5°/+10°. Considering a linear deformation of 3%, the required amount of wire for linear actuation would be 83cm. However, in our case it would be impractical to have a free-standing wire of such a length as a linear actuator. For this reason, the NiTi wire should be confined into a more compact actuator. In dimensioning the length of wire now we must take into account also the local deformations that inevitably will be imposed on the SMA wire when coiling it. The actuator was designed as an insulated cartridge wherein the necessary reference length of NiTi wire is led back-and-forth between two arrays of ten pulleys. An end of the wire is connected to an inextensible thread that transmits the force to the foot, while the other one is fixed to the housing. A pseudoelastic spring is connected to the moving end of the NiTi wire in order to keep it just taut, and its pull is negligible during actuation.

Considering the force requirement, each actuator should bear half of the resisting torque T_r showed in Figure 3, and the amount of force varies at varying ankle angles θ (defined between the foot direction and the horizontal, Figure 4):

$$F(\theta) = \frac{T_r(\theta)}{c \cdot \sin\varphi} \tag{5}$$

where φ is the angle between the inextensible thread and the foot and depends on θ according to:

$$\varphi(\theta) = \frac{\pi}{2} - \theta - \arccos\left[\frac{c \cdot \sin\theta + a}{\sqrt{(c \cdot cos\theta - b)^2 + (c \cdot sin\theta + a)^2}}\right] \tag{6}$$

With a maximum torque of 200Ncm, Equations 5 and 6 give a maximum load of 13N on each actuator. By selecting a 250µm-diameter commercial NiTi wire, stabilised for actuation, the maximum stress on the wire cross-section is 266MPa. In order to complete

plantarflexion, full martensite detwinning should be achieved. It is certainly possible to find stabilised wires for which martensite detwinning occurs with less than 100MPa, corresponding to around 5N pulling on a 250μm-diameter wire. With the expected resisting torque in Figure 3 and the geometrical parameters, force requirement for martensite detwinning is always satisfied in the range -5°/+10° and thus cyclic actuation will take place in that range. Considering a diameter of the pulleys of 12mm, the maximum amount of localised strain is 2%. In order to respect the 3% limit for the maximal deformation, only 1% of strain is available for actuation and can be employed for calculating the length of wire to be coiled within each actuator, that is to say, 250cm.

Having limited strains to 3% and stress to 300MPa, a cycling life of 30k-100k cycles is expected [58].

2.4.4. *Power dimensioning*

In order to evaluate the parameters for Joule's effect heating, various experimental tests were carried out on samples of NiTi wire. First, the transformation temperatures were investigated by means of differential scanning calorimetry (on DSC 220 SSC/5200 - Seiko Instruments, Tokyo, Japan), showing A_f = 351K and M_f = 274K. However, calorimetry gives information about the material with no loads applied, whereas Equation 4 suggests that transformation temperature depends on the load applied. For this reasons, tensile tests on the material at different temperatures were conducted using an MTS 2/M thermo-mechanical test machine (MTS Systems, Eden Prairie, MN, USA) equipped with a 2kN load cell. The material was deformed up to an engineering strain of 5%, at 365K, 380K, 390K and 400K. The pseudoelastic plateau values varied as a function of temperature with a ratio of 8.402 MPa/K, which is exactly the constant C in Equation 4. With the estimated stress level of 266MPa, full transformation and strain recovery can be achieved at 383K.

Activation tests on the wire were conducted injecting different currents (0.65A, 0.7A, 0.75A, 0.8A) in the wire at a constant strain of 3% for a set period of 13s, in order to evaluate what current value is most appropriate to provide the working load of 13N, minimizing current expenditure and heating time. As expected, the higher the current the faster is actuation. Patient safety considerations should be taken into account, as well. A compromise solution can be accepted with 0.7A, which allows for reaching 13N within 6s. With the selected current, voltage is limited to 35V, considering an average resistance of 49.5 Ω (45-54Ω during transformation from fully austenitic to fully martensitic). Notice that the rehabilitation device mounts two actuators, which need a 0.7A current each. The electrical configuration of the two actuators could be a series or a parallel circuit: in the first case, the dc-generator should be able to provide 0.7A at 70V, in the latter case 1.4A at 35V. According to IEC 60601-1 specifications, dc tension preferably does not exceed 60V, i.e. in our case the two linear actuators are better connected in parallel.

Heating parameters can be extracted by testing a free-standing NiTi wire, and by applying 35V to the ends of the cartridge actuator suspending a 13N weight. It was demonstrated that nominal linear stroke is reached in around 6s (mean dorsiflexion speed 2.5°/s). On the other hand, cooling times depend strictly on the geometrical arrangement of the wire in space. A

compact actuator may have a considerably slower cooling and position reset time, as tests on the actuator confirms. Basically, if the vertical free-standing wire cools down in approximately 10s, cooling and actuator position reset takes place in 30s. The full cycle thus lasts 36s, which makes it possible to deliver 100 cycles/hour to the patient's ankle, as required by the design specifications.

2.4.5. Control strategy

An open-loop control is adopted for using the device as a passive exerciser. A computer routine was written in LabView (National Instruments, Austin, TX, USA) to control an electronic relay (NI9481 - National Instruments, Austin, TX, USA) that closes the electric circuit between the dc-generator and the actuators.

For the assistive therapy, a closed-loop architecture is preferred, with surface electromyographic signal (sEMG) from *tibialis anterior* as control variable representing the patient's attempt to move his ankle. Three Ag/AgCl electrodes (positive, negative and reference) were used to pick up the signal. Analogical waveforms were acquired, pre-amplified and band-pass filtered (18-478Hz) before being digitalised (sampling at 1000Hz) using an NI9205 (National Instruments, Austin TX, USA) connected to an ordinary laptop computer running a LabView routine. Pre-amplifying and filtering stages were assembled using conventional 8-pin PDIP components, and included also a feed-back loop towards the body (akin to the driven right leg stage used in electrocardiographers). Further signal manipulation is built into the control software, including rectification and low-pass filtering (3Hz), so that the obtained waveform is sufficiently smooth that it can be employed as a measure of instantaneous muscular activation.

Through the user interface of the control routine two patient-specific sEMG threshold values can be selected. The lower one is set to the minimal required level of exercise (which can lie even lower than the muscular motor threshold, in some cases); the upper one is set to an appropriate activation representing the ultimate (or a higher) therapeutic goal, i.e. an effective motion.

At the start of the assistive exercise session, a visual cue is presented to the patient to dorsiflex the ankle. Then the system is set on hold waiting for the sEMG from *tibialis anterior* to cross one of the threshold values. If the lower threshold is reached, then the system waits a few milliseconds for the upper threshold also to be reached. If this latter event does not occur before the time-out, then power is supplied to the actuators and the motor task is completed for the patient. If, on the contrary, the upper threshold is reached, then a visual feedback is provided to the patient that the higher goal was hit, while the device does not intervene to support the movement.

2.4.6. Experimental tests on the device

The assembled device was tested with a static load of 40N fixed on the foot shell at a distance of 10cm from the hinge axis, while measuring the angular displacement by means

of electrogoniometer SIM-HES-EG 042 (Signo Motus, Messina, Italy). The device was tested for 4000 cycles, with the actuators powered with a 6s step of 1.4A (0.7A per cartridge) and 30s allowed for cooling. The results of these tests (Figure 5) demonstrated that stroke is quite stable across cycling. Plantarflexion occurs more rapidly after cycling because of the instauration of two-way shape memory. Following this result, by employing cycled wire or wire stabilised for two-way shape memory, cycle duration could be reduced from 36s to 30s without changing the stroke, if desired (reaching 120 cicles/hour).

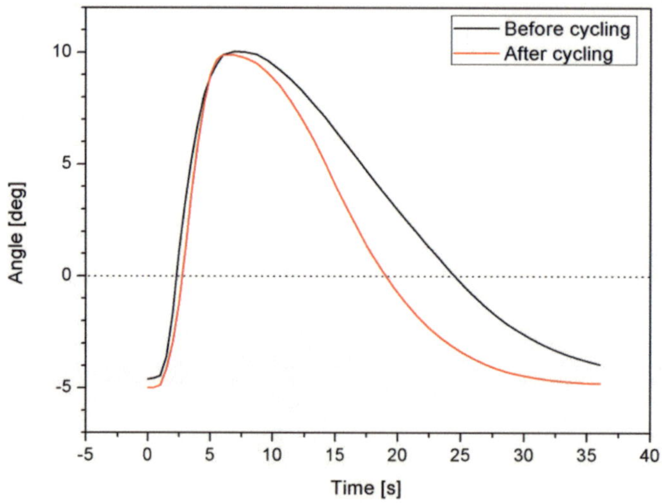

Figure 5. Angular stroke of the SMA-based device for ankle dorsiflexion: first activation and after 4000 cycles against a resisting torque of 400Ncm.

The closed loop control strategy was tested on three healthy volunteers (28.17±6.08 years old). Ag/AgCl electrodes were placed on the belly of the *tibialis anterior* muscle of the dominant side, the corresponding distal muscle-tendon junction, and the internal malleolus (driven electrode). Subjects were asked to perform a maximal isometric contraction at the ankle neutral position, then to sustain the minimum voluntary activation they could manage. Subsequently, values were set for the lower (110% of minimum individual contraction) and upper (60% of individual maximum isometric contraction) thresholds. Then, the control routine was launched and subjects were asked to follow on-screen instructions (graphic and written) trying to respond with just a supra-minimal contraction when cued to dorsiflex the ankle.

The measured joint angle and sEMG time courses for a representative subject are shown in Figure 6. When the lower threshold was crossed, the system triggered the powering of the orthosis, which completed the movement of dorsiflexion (assisted active session). It can be appreciated how passive mobilisation can be triggered by a very subtle muscular contraction, which may correspond only to a very slight movement (~1°). The movement produced by the orthosis as a consequence of a minimal contraction brings along some degree of reflex sEMG activity: this may also be thought of as an interesting result to the effect of rehabilitative exercise.

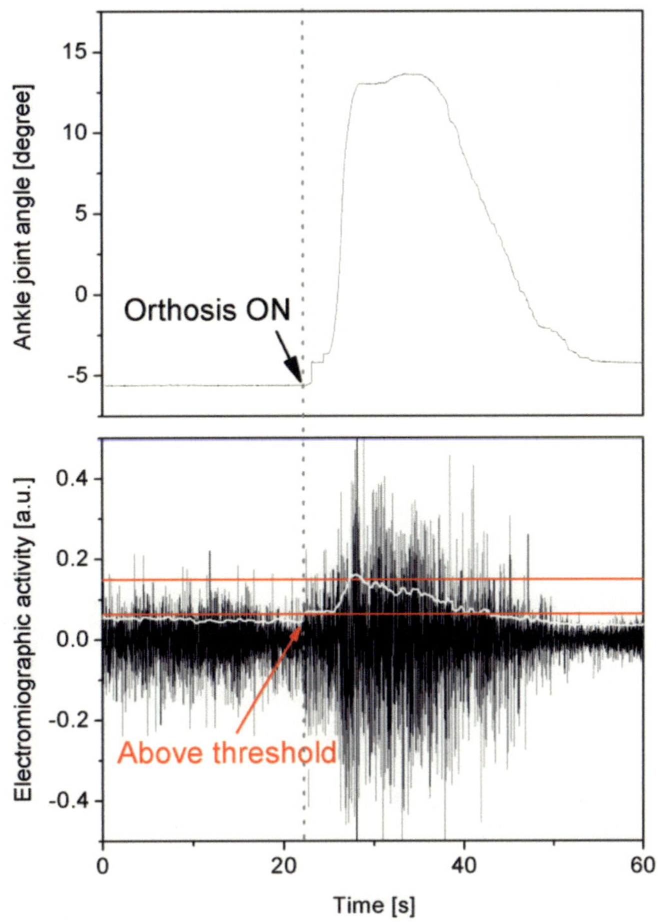

Figure 6. Top: angle timecourse measured during an EMG-triggered activation of the device for a healthy volunteer. Bottom: EMG recording during the trial. When the subject's muscular activity crosses the lower threshold, actuators are triggered to complete the movement of dorsiflexion.

3. SMA actuators for neuroscience

The development of SMA devices for use in Neuroscience is a very challenging task as diagnostic equipment utilised in this field of research is generally very sensitive to electromagnetic noise. Whereas some techniques (e.g. electroencephalography or near-infrared spectroscopy) are less affected by environmental conditions, magnetic resonance imaging (MRI, fMRI) and magnetoencephalography (MEG) acquisitions can be carried out only in highly shielded Faraday cages, to abate electromagnetic noise. Therefore, materials and devices must comply with a number of requirements, in order to be allowed into the acquisition room. In the next paragraph, these two diagnostic techniques will be discussed separately, as their peculiar characteristics demand different compatibility constraints.

3.1. Electromagnetic constraints

3.1.1. Functional magnetic resonance imaging (fMRI)

fMRI is a magnetic resonance procedure utilised in Neuroscience for measuring brain activity by detecting associated haemodynamic responses. MRI equipment uses a strong static magnetic field to polarise the nuclear spins of atoms in living tissues: the higher the magnetic field, the higher the polarisation of the spins, thus, the higher signal-to-noise ratio. Commercial scanners can generate polarisation fields up to 3T, but efforts are being made to increase the available magnetic field far beyond [13]. In addition to this static magnetic field, the MR scanner uses rapidly-varying magnetic field gradients for spatial encoding during the imaging sequence. The use of strong polarisation magnetic fields entails safety precautions, as high forces can be exerted especially on objects made of ferromagnetic materials, which may become dangerous projectiles for the patient, medical personnel, and the instruments. Besides safety issues, ferromagnetic materials generally affect image quality, as they alter the homogeneity of the static magnetic field. Another potential source of artefacts is the presence of objects made of conductive materials inside the acquisition room. In fact, the magnetic field gradients can induce electrical fields and currents flowing through the conductive materials (eddy-currents): these currents, in turn, induce magnetic fields which inevitably interact with the MR fields and affect the quality of the images. In much the same way as eddy-currents, it can be inferred that any current flowing inside conductive materials could generate image artefacts.

Given these strict constraints, a possible solution for moving the limbs of a patient during fMRI investigations by an actuator is to keep the actuator and control system outside the scanning room and transfer the mechanical energy to the subject via pneumatic, hydraulic, or mechanical (pulleys, ropes, etc.) means. These systems are usually complex and suffer from long transmission lines accompanied by dissipative and delay effects. It would be desirable to place actuators directly inside the acquisition room, but this limits the possibilities of using many categories of actuators. Of course, electromagnetic actuators are, in general, not compatible with the MRI environment, but also pneumatic or hydraulic motors could hardly get into the acquisition room, because of their bulky metallic components. Generally speaking, in order to be utilised in an MRI acquisition room, actuators should not include ferromagnetic and conductive materials. However, it is reported that small parts made of MR-incompatible materials do not compromise safety or generate image artefacts as long as they are sufficiently small and appropriately positioned relative to the imaged area [13]. This clears the way to the use of SMA actuators in this field, provided that some specific design rules are respected and that they are applied to body segments sufficiently distant from the head, which is the area of principal interest in Neuroscience.

3.1.2. Magnetoencephalography (MEG)

MEG is a technique for investigating neuronal activity in the living human brain by recording magnetic fields produced by the electrical currents flowing in cortical neurons. These weak magnetic fields (ranging 10^{-14}T-10^{-12}T), can be detected by employing arrays of

SQUID (Superconducting Quantum Interference Device) gradiometers, which convert the magnetic flux threading a pick-up coil into voltage. Since the magnetic fields to be measured are extremely small as compared to the Earth's magnetic field (10^{-5}-10^{-4}T), MEG measures are carried out in a shielded room that minimises interference. Of course, any electromagnetic noise should be avoided inside the acquisition room: there is no safety issue for the patient connected with this constraint, but if not respected the measure would be impossible because SQUID channels would saturate rapidly and would remain unusable as long as the pick-up coils are in the superconductive state. Unlike the case of MRI, theoretically any material could enter an MEG acquisition room. In practice, objects made of conductive materials should remain still inside the shielded room, as any movement would generate an artefact. These constraints are very demanding for most actuation technologies, but design solutions are possible for SMA actuators.

3.2. Compatible design guidelines

Some design indications could be useful for devices intended for either fMRI or MEG studies. In particular, the SMA element should have a shape or should be arranged in a way that limits to a great extent the magnetic fields induced by eddy-currents and by the current used for Joule's heating. For example, the coil of a spring is not particularly indicated, as the intensity of the magnetic field is proportional to the number of turns. Moreover, the SMA actuator should be supplied with a dc-current rather than other types of time-varying currents (pulse-width modulation, sinusoidal current…). In fact, this could help reduce the magnetic field induced by currents flowing into the SMA elements. Power generators and control systems should be positioned outside the acquisition room and wires passed through. As these cables could act as antennas that radiate electromagnetic noise, some countermeasures should be adopted: shielded or twisted cables can limit pick-up of stray frequencies, while using proper filters when connecting the wires inside and outside the acquisition room can help rejecting the time-varying components of the power signal. Coming to the control strategy, an open-loop may be preferred (for its simplicity), as both fMRI and MEG data analysis generally need a precise windowing of signals that matches different phases of movement, in order to extract features of interest with statistical significance. So, closed-loop architectures could be used e.g. to control precise abidance to set movement speed evolution, but might not be strictly necessary if the experimental protocol only requires repeatable ranges of motion in a given time. In fact, within the very protected environments of the shielded rooms, often no major disturbances are expected to intervene and perturb the movement.

Additional recommendations apply only to devices for use in the MRI acquisition room: in particular, ferromagnetic materials should be avoided for safety reasons, even if small parts could be tolerated in some cases. On the other hand, as we said before, during MEG acquisition conductive materials should not move inside the shielded room. Translating this information into design specifications, a first recommendation would be that the moving parts should not mount conductive materials, including the SMA actuators. However, the principle of SMA actuation is that metal moves! The geometry of the SMA element should

be conceived so that macroscopic changes of space occupancy are avoided, thus limiting the amount of movement. For example, in SMA helical springs the linear length can vary considerably during actuation, spanning in many cases more than 300% the unloaded length. On the other hand, traction wires, having a limited elongation, change their space occupancy less dramatically. Interestingly, it should be noticed that the slow actuation rates typical of SMA could be an advantage for limiting artefacts influencing MEG measurements. In fact, it is reasonable that artefacts have mostly the same time frequency components as the movement that generated them, i.e. typically less than 1Hz for SMA actuation. Cortical oscillations range 1-100Hz, but when investigating sensorimotor representation (as expected in the case of Neuroscience for Neuromuscular Rehabilitation), the range of interest reduces to 8-100Hz, giving the possibility to devise suitable signal filtering stages.

3.3. Case study: amagnetic rotary actuator for ankle dorsiflexion during MEG and fMRI acquisitions

3.3.1. Concept

A possible device for ankle dorsiflexion could be devised as a leg part and a foot part connected by planar hinges parallel to the ankle joint, in a similar way to the device presented in section 2.4.2. Two SMA rotary actuators mounted externally with respect to the hinges promote the dorsiflexion of the foot part with respect to the leg part, while plantarflexion and position reset is left to viscoelastic resistance and foot weight (Figure 7). It should be noticed that the requirements for the ranges of movement are different in this application, as the purpose is to produce a clearly detectable movement without muscular stretching. This can be done maintaining joint angle negative or slightly positive (i.e. extended). The objective of the study was to test healthy subject, so it can be estimated that maximum resisting torque will be in the range 200-250Ncm (see Figure 2).

A rotary SMA actuator can be generally described as a structure in which the SMA wire connects parts that can rotate relative to one another about a central axis: when the elongated SMA wire is heated above A_f and recovers its deformation, the linear stroke ΔL is converted into a rotation $\Delta\theta$ of the moving parts. However, it is likely that a conspicuous length of wire is needed to produce a suitable amount of rotary stroke $\Delta\theta$ for ankle dorsiflexion. As we already said, in such cases coiling the wire could improve conveniently overall compactness of the actuator. Unfortunately, this design decision contrasts with the aim of fabricating an amagnetic actuator, as the electric current flowing in coils of wire produces, according to Ampère's Law, an overall magnetic field along the central axis of the solenoid, which is proportional to the number of turns. A very simple solution can be found exploiting the same physical laws of electromagnetic fields. In fact, the magnetic field vector is directed according to the right hand rule with respect to the direction in which the current flows that induced it. Thus, theoretically, given two identical and concentric solenoids traversed by the same electric current in opposite directions, the magnetic fields generated by the coils will mutually annihilate. Moreover, no ferromagnetic materials should be employed in the implementation of all other components of such actuator, as well as the assembled device.

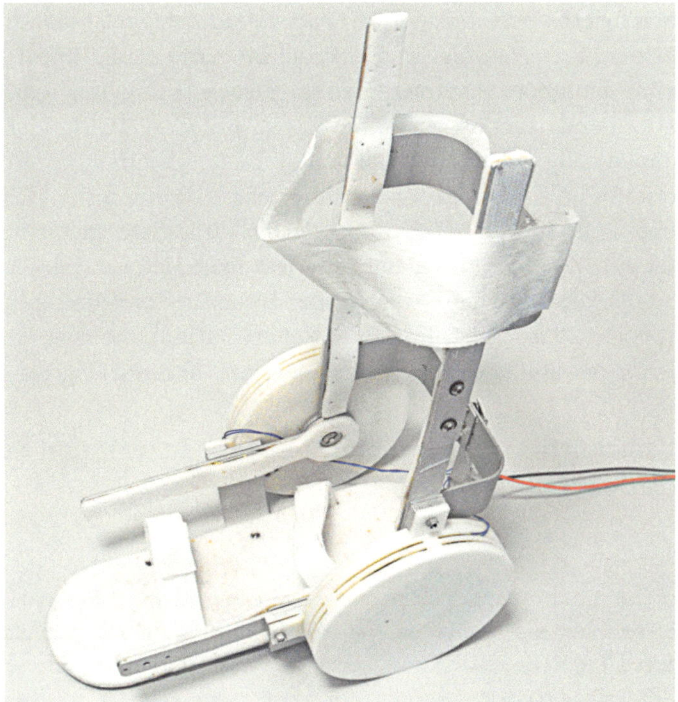

Figure 7. Implementation of an amagnetic device for ankle dorsiflexion with rotary actuators

3.3.2. Actuator implementation and mechanical characterisation

An implementation of the invented amagnetic actuator is shown in Figure 8. Two discs of approximately 10cm in diameter are connected through a central shaft allowing reciprocal rotation, leaving a clearance of 2cm between the two discs. One of the two discs mounts, along its circumference, 6 pins bearing 3 pulleys each, to create a spiralling sequence, along which the SMA wire can be coiled suitably. These pulleys have a double triangular groove, which makes the double coiling as compact as possible. The two ends of the SMA wire are fixed to one disc, while the mid-point of the wire connects to the other disc. The two halves of the wire are would along the pulleys thus creating two concentric coils very close to one another. By providing sufficient current flowing between the two ends of the SMA wire, shape recovery occurs generating a linear stroke ΔL that is converted into a reciprocal rotation $\Delta\theta$ of the two discs. Notice that the torque generated is given by twice the cross-section of the wire, while the induced magnetic field is self-compensated to a large extent, as the two coils are traversed by the same electric current in opposite directions [40].

As already described for the linear actuator in paragraph 2.4.3, wire length depends on the expected angular stroke and the working strain level, while torque output is connected to the wire diameter and mean stress on the cross-section. Moreover, the geometrical parameters of the assembled actuator affect both wire length and torque output. The rotary actuator was designed to provide an angular stroke up to 40° against resisting torques in the

range 120Ncm-250Ncm. By limiting the linear strain to 3.8%, the required length of NiTi wire can be calculated in 219cm. The selected NiTi wire is 250μm in diameter, corresponding to stresses of 130-270MPa in the above-mentioned range of torques. Taking into account the localized strain on the wire resting on the pulleys, the total maximal strain reaches 4.42%, which in combination with the chosen stress level should guarantee suitable fatigue life (several thousands of cycles). Commercial stabilised NiTi wire was utilised, displaying large deformability (4.5%) at room temperature for stress levels as low as 150MPa [58]. The power dimensioning procedure was similar to the one described in paragraph 2.4.3 and led to utilizing 0.7A at 30V for each actuator. Control is achieved with an open-loop strategy, by which acquisition windows can be synchronised to the ankle movement. Power supply and control appliances are left outside the acquisition room, and shielded cables are let into the shielded room through suitable access vents.

Figure 8. Amagnetic SMA rotary actuator, consisting of two coils of wire traversed by the same current in opposite directions. This arrangement produces self-annihilation of induced magnetic fields.

In the present implementation, ferromagnetic materials were avoided. However, it was not possible to exclude completely metallic components, such as the bearing balls, the shafts and the pins. After careful consideration, it was decided that the rotating discs and the frame of the device could be made of an aluminium alloy, in order to minimise structural sections. This choice did not affect the acquisition of biosignals, as will be discussed in the next paragraphs.

Technical tests were carried out on the assembled device, to assess its characteristics. Increasing loads were attached to the foot part of the orthosis 13cm from the axis of rotation,

while the device was held aloft by a static support. The weights used in this test produced resisting torques in the range 28-250Ncm. A direct current injection at 30V for 7s was applied to the actuators, connected in parallel; then 30s were allowed for position reset through natural cooling and the action of the weights. The resulting angular upward and downward strokes were measured by means of electrogoniometer SIM-HES-EG 042 (Signo Motus, Messina, Italy). Figures 9 shows the results. At lower values of resisting torques (i.e. in the range 28Ncm-120Ncm), angular stroke is not stabilised and steadily increases from 24° to 36°. For torques above 120Ncm, angular stroke is quite stable at 36°. Curves steadily shift towards negative angles (zero being the horizontal position, and negative in the direction of plantarflexion) with increasing torques. This increase in stroke is caused by the incomplete detwinning of martensite at lower torques: it is worth noticing that anyway, even at the lower values of measured angular stroke, suitable angular strokes are obtained.

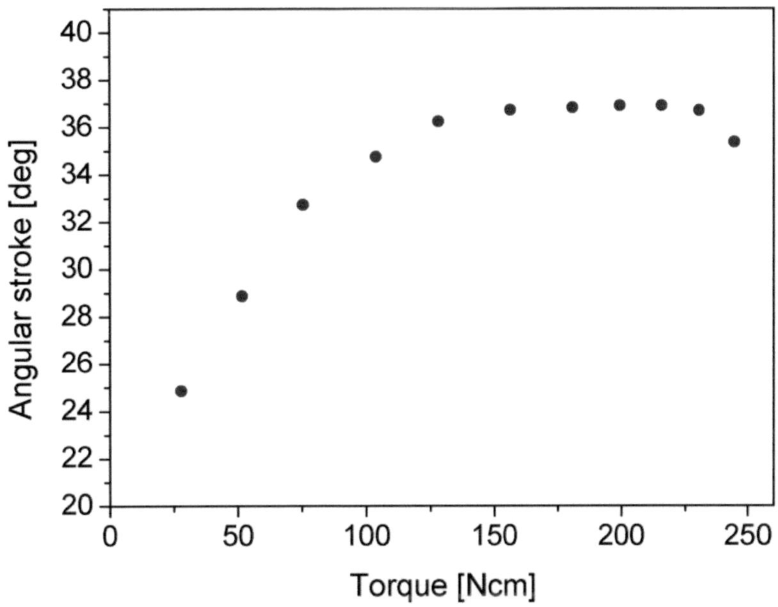

Figure 9. Angular stroke of the SMA-based exerciser for different resisting static torques. The dotted line is just a guide for the eye.

3.3.3. *Tests in a MEG shielded room*

In order to assess compatibility with the MEG technology, neural activity from a healthy subject's brain during repeated passive mobilisation of the ankle by the device was carried out with an MEG system composed of 153 SQUID channels covering the whole surface of the scalp. The electric power generator and the programming computer were kept out of the shielded room, and cables were passed through suitable vents in the shims. The NiTi wire was activated by a current pulse (ramp to 0.7A in 1s, then flat for 9s).

These tests revealed no significant noise on SQUID channels. Figure 10 (left) shows the signal recorded by one representative channel during MEG testing of the healthy volunteer.

At t=0s, the actuators on board the orthosis were switched on. There is no artefact at that moment, indicating that the level of noise in the acquisition room has not changed. Furthermore, the right graph in the same Figure demonstrates that the frequency components in the actuator-OFF and actuator-ON states are the same. The variation in intensity of the signal power spectral density (PSD) can be totally accounted for by a change in the cortical reactivity of the subject under testing: this variation is an important part of the measured quantities of interest.

Incidentally, the use of conductive materials in the construction of the device did not affect the acquisition. This is probably due to the slow movement provided by the actuators, which generates low-frequency artefacts that can be suitably eliminated during routine filtering and windowing of the MEG data.

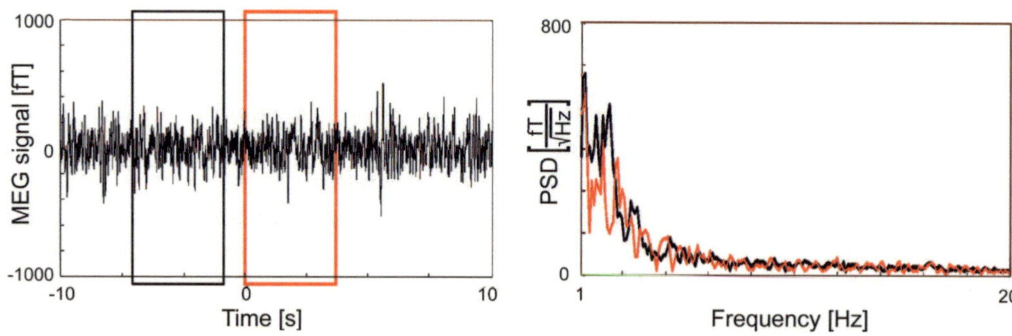

Figure 10. Left: MEG signal recorded by a representative SQUID channel from a healthy subject during passive mobilisation of the ankle by the amagnetic exerciser. Right: spectral components of the recorded MEG signal, before and after switching on the actuators at t=0s (black and red line, respectively).

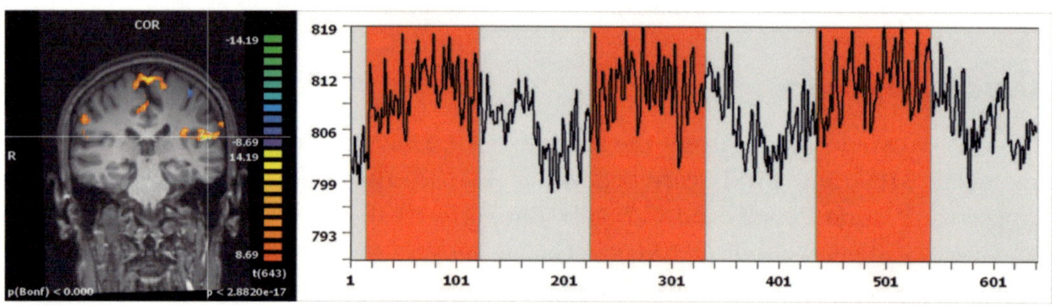

Figure 11. Left: fMRI image of metabolic activity in a healthy brain during passive mobilisation of the ankle. Right: time-course of the BOLD signal (arbitrary units) in the area of the brain highlighted in the left figure.

3.3.4. Tests in a MRI acquisition room

Magnetic resonance images were collected in a Philips 3T Achieva X-Series Magnetic Resonance scanner. fMRI signals were treated to extract the Blood Oxygen Level Dependent (BOLD) image, which depicts changes in neural metabolism related to neural activity. fMRI images (Figure 11) were not affected by any artefact. The BOLD signal was clearly collected

from all parts of the brain and in some areas seemed to be temporally dependent on the movements of the ankle. The use of conductive materials in the implementation of the device did not affect the acquisition. The distance between the gantry (or head coil) and the ankle probably helped limit any influence on image encoding.

4. Conclusions

This Chapter showed some innovative applications of Shape Memory Alloys requiring deep understanding of the interaction of the material characteristics with the complex constraints imposed by the human body. In particular, it was explained how the design plan should be laid considering the many aspects connected to the state of the target patient, and the technical requirements should be chosen to meet very well identified needs. The field of Medical Rehabilitation is an interesting domain for exploiting the functional properties of NiTi-based alloys in making new lightweight and portable actuators. The Neuroscience applications introduced in this Chapter, on the other hand, albeit representing a niche sector *per se*, both make the most of the typical design techniques employed for rehabilitation devices, and provide a development ground for interesting industrial actuators with amagnetic characteristics. It is hoped that the SMA-based design strategies presented here will be of inspiration to engineers interested in utilising shape memory actuation for biomedical, robotic, aerospace or automotive applications with tight and mandatory external constraints where compactness, light weight or wearability are desired features of the device.

Author details

Simone Pittaccio* and Stefano Viscuso
National Research Council of Italy, Institute for Energetics and Interphases, Lecco, Italy

Acknowledgement

The authors would like to thank the staff of ITAB, University of Chieti, where the tests employing MEG and fMRI were conducted: in particular Gian Luca Romani, Filippo Zappasodi, Vittorio Pizzella and Gabriella Tamburro for the MEG measurements, Cosimo Del Gratta, Antonio Ferretti and Mauro Gianni Perrucci for the fMRI trials. The financial support from Regione Lombardia (Mind in Italy Project) is most gratefully acknowledged.

5. References

[1] Gracies J-M (2005) Pathophysiology of spastic paresis I: paresis and soft tissue changes, Muscle Nerve 31: 535–551.

[2] Gracies J-M (2005) Pathophysiology of spastic paresis II: emergence of muscle overactivity, Muscle Nerve 31: 552–571.

* Corresponding Author

[3] Simon O, Yelnik AP (2010) Managing spasticity with drugs. Eur J Phys Rehabil Med. 46: 401-410.

[4] European Stroke Organisation (ESO) Executive Committee, ESO Writing Committee (2008) Guidelines for management of ischaemic stroke and transient ischaemic attack, Cerebrovasc. Dis. 25: 457-507.

[5] Duncan PW, Zorowitz R, Bates B, Choi JY, Glasberg JJ, Graham GD, Katz RC, Lamberty K, Reker D (2005) Management of Adult Stroke Rehabilitation Care: a clinical practice guideline, Stroke. 36: e100-e143.

[6] Hesse S (2008) Treadmill training with partial body weight support after stroke: a review. NeuroRehabilitation. 23: 55-65.

[7] Krebs HI, Rossi S, Kim S, Artemiadis PK, Williams D, Castelli E, Cappa P. (2011) Pediatric anklebot. IEEE Int. Conf. Rehabil. Robot 1-5.

[8] Staubli P, Nef T, Klamroth-Marganska V, Riener R (2009) Effects of intensive arm training with the rehabilitation robot ARMin II in chronic stroke patients: four single-cases. J Neuroeng Rehabil. 17: 6:46.

[9] Fasoli SE, Krebs HI, Stein J, Frontera WR, Hogan N (2003) Effects of robotic therapy on motor impairment and recovery in chronic stroke. Arch. Phys. Med. Rehabil. 84: 477-482.

[10] Alary F, Simões C, Jousmäki V, Forss N, Hari R (2002) Cortical activation associated with passive movements of the human index finger: an MEG study. Neuroimage. 15: 691-696.

[11] Boonstra TW, Clairbois HE, Daffertshofer A, Verbunt J, van Dijk BW, Beek PJ (2005) MEG-compatible force sensor. J Neurosci Methods. 144: 193-196.

[12] Suminski AJ, Zimbelman JL, Scheidt RA (2007) Design and validation of a MR-compatible pneumatic manipulandum. J Neurosci Methods. 163: 255-266.

[13] Tsekos NV, Khanicheh A, Christoforou E, Mavroidis C (2007) Magnetic resonance-compatible robotic and mechatronics systems for image-guided interventions and rehabilitation: a review study. Annu Rev Biomed Eng. 9: 351-387.

[14] Onishi H, Oyama M, Soma T, Kubo M, Kirimoto H, Murakami H, Kameyama S (2010) Neuromagnetic activation of primary and secondary somatosensory cortex following tactile-on and tactile-off stimulation. Clin Neurophysiol. 121: 588-593.

[15] Hollnagel C, Brügger M, Vallery H, Wolf P, Dietz V, Kollias S, Riener R (2011) Brain activity during stepping: a novel MRI-compatible device. J Neurosci Methods. 201: 124-130.

[16] Funakubo H (1987) Shape memory alloys. New York: Gordon and Breach Science Publishers. 275p.

[17] Otsuka K, Wayman CM (1998) Shape Memory Materials, Cambridge: Cambridge University Press. 287 p.

[18] Otsuka K, Ren X (2005). Physical metallurgy of Ti–Ni-based shape memory alloys, Progr. in Mater. Science. 50: 511-678.

[19] Lagoudas DC (2008) Shape Memory Alloys: Modeling and Engineering Applications. Berlin: Springer. 435 p.

[20] Viscuso S, Pittaccio S, Caimmi M, Gasperini G, Pirovano S, Besseghini S, Molteni F (2009) Pseudoelastic alloy devices for spastic elbow relaxation. In: Lim CT, Goh JC, editors. ICBME 2008 – 13th International Conference on Biomedical Engineering, IFBME Proceedings 23. Berlin: Springer. pp. 1584–1587.

[21] Pittaccio S, Viscuso S (2009) The use of muscle "creep" as opposed to relaxation in stretching braces: a pseudoelastic device. In: Dössel O, Schlegel WC, editors. World Congress on Medical Physics and Biomedical Engineering 2009, IFMBE Proceedings 25/9. Berlin: Springer. pp. 178-181.

[22] Pittaccio S, Viscuso S (2009) Customizable neuro-mechanical model of a hemiplegic elbow interacting with a pseudoelastic dynamic orthosis. In: Dössel O, Schlegel WC, editors. World Congress on Medical Physics and Biomedical Engineering 2009, IFMBE Proceedings 25/9. Berlin: Springer. pp. 182-185.

[23] Viscuso S, Pittaccio S, Caimmi M, Gasperini G, Pirovano S, Villa E, Besseghini S, Molteni F (2009) Pseudoelastic Nitinol-Based Device for Relaxation of Spastic Elbow in Stroke Patients. J Mater Eng Perform 18: 805–813.

[24] Patent Number: WO/2011/137999. Joint For Articulations With Pseudo-Elastic Elements.

[25] Pittaccio S, Viscuso S, Beretta E, Turconi AC, Strazzer S (2010) Pilot studies suggesting new applications of NiTi in dynamic orthoses for the ankle joint. Prosthet Orthot Int 34: 305-318.

[26] Hanafusa A, Isomura T, Sekiguchi Y, Takahashi H, Dohi T (2002) Orthosis Design System for Malformed Ears Based on Spline Approximation. In: Dohi T, Kikinis R, editors. Medical Image Computing and Computer-Assisted Intervention — MICCAI 2002. Berlin: Springer. pp. 227-234.

[27] Shimizu Y, Kobayashi (2011) The Development of Foot Orthotics using Shape Memory Alloy for Preventing Falls. In: Werner B, editor. 2011 IEEE 11th International Conference on Bioinformatics and Bioengineering. Washington: Conference Publishing Services. pp. 117-120.

[28] Takami M, Fukui K, Saitou S, Sugiyama I, Terayama K (1992) Application of a shape memory alloy to hand splinting. Prosthet Orthot Int. 16: 57-63.

[29] Patent Number: US 2008/0294079 A1. Orthotic Apparatus And Method Of Operation.

[30] Patent Number: US 2011/0131838 A1. Dynamically Adjustable Orthotic Device.

[31] Machado LG, Savi MA (2003) Medical applications of shape memory alloys. Braz J Med Biol Res. 36: 683-691.

[32] Lai YJ, Yeh LJ, Chiu MC (2010) An Assessment of the Body Joint Bending Actuator using a Shape Memory Alloy. J Appl Sci, 10: 1973-1977.

[33] Lai YJ, Yeh LJ, Chiu MC (2012) An experimental investigation on shape memory alloy dynamic splint for a finger joint application. Sens Actuator A-Phys 173: 210–218.

[34] Torri M, Viscuso S, Pittaccio S, Nespoli A, Besseghini S (2006) Biomechanical design of a shape memory alloy spring for the activation of a flaccid hand rehabilitation device. In: Venugopalan R, Wu MH, editors. Medical Device Materials III: Proceedings of the Materials & Processes for Medical Devices Conference. Materials Park: ASM International. pp.237-242.

[35] Van Kuren MB, Gillette S, Mejia P, Stoever T, Walker A,(2005) Design considerations for a wearable pediatric rehabilitative boot. In: ICORR 2005, 9th International Conference on Rehabilitation Robotics. pp. 400- 403. Available: http://ieeexplore.ieee.org/stamp/stamp.jsp?tp=&arnumber=1501128. Accessed 2012 Mar 28.
[36] Pittaccio S, Viscuso S, Rossini M, Magoni L, Pirovano S, Villa E, Besseghini S, Molteni F (2009) SHADE: A shape-memory-activated device promoting ankle dorsiflexion. J Mater Eng Perform 18: 824–830.
[37] Viscuso S, Pittaccio S, Tecchio F, Zappasodi F, Rossini M, Magoni L, Pirovano S, Besseghini S, Molteni F (2009) Assessment of the peripheral performance and cortical effects of SHADE, an active device promoting ankle dorsiflexion. In: Lim CT, Goh JC, editors. ICBME 2008 – 13th International Conference on Biomedical Engineering, IFBME Proceedings 23. Berlin: Springer. pp. 961-965.
[38] Viscuso S, Pittaccio S, Zappasodi F, Tecchio F (2009) A magnetically quasi-transparent tool for ankle passive mobilization in investigations on cortical involvement using MEG. In: Dössel O, Schlegel WC, editors. World Congress on Medical Physics and Biomedical Engineering 2009, IFMBE Proceedings 25/7. Berlin: Springer. pp. 410-413.
[39] Pittaccio S, Zappasodi F, Viscuso S, Mastrolilli F, Ercolani M, Passarelli F, Molteni F, Besseghini S, Rossini PM, Tecchio F (2011) Primary sensory and motor cortex activities during voluntary and passive ankle mobilization by the SHADE orthosis. Hum Brain Mapp 32: 60-70.
[40] Patent Number: WO/2011/141183. Rotational Actuator.
[41] Pittaccio S, Viscuso S (2011) An EMG-Controlled SMA Device for the Rehabilitation of the Ankle Joint in Post-Acute Stroke. J Mater Eng Perform 20: 666-670.
[42] Merrett GV, Metcalf CD, Zheng D, Cunningham S, Barrow S, Demain SH (2011) Design and Qualitative Evaluation of Tactile Devices for Stroke Rehabilitation. Presented at IET Assisted Living 2011, London, UK. Available: http://eprints.soton.ac.uk/271802/. Accessed 2012 Mar 28.
[43] Patent Number: US 2001/0029224 A1. Exercise Kit And Method Of Using Same.
[44] Esfahani ET (2007) Developing an Active Ankle Foot Orthosis Based On Shape Memory Alloys. M.Sc. thesis. Available: http://etd.ohiolink.edu/send-pdf.cgi/Tarkesh%20 Esfahani%20Ehsan.pdf?toledo1197561946. Accessed 2012 Mar 28.
[45] Esfahani ET, Elahinia MH (2007) Stable Walking Pattern for an SMA-Actuated Biped IEEE/ASME Trans Mechatron 12: 534-541.
[46] Stirling L, Yu CH, Miller J, Hawkes E, Wood R, Goldfield E, Nagpal R (2011) Applicability of Shape Memory Alloy Wire for an Active, Soft Orthotic. J Mater Eng Perform 20: 658-662.
[47] Patent Number: WO 2011/008934 A2. Actively Controlled Orthotic Device.
[48] Patent Number: US 5103807. Shape Memory Alloy Orthotic Device.
[49] Abolfathi PP (2007) Development of an Instrumented and Powered Exoskeleton for the Rehabilitation of the Hand. Ph.D. thesis. Available: http://ses.library.usyd.edu.au/bitstream/2123/3690/1/01pp-abolfathi-2008-thesis.pdf. Accessed 2012 Mar 28.

[50] Patent Number: US 6379393 B1. Prosthetic, orthotic, and other rehabilitative robotic assistive devices actuated by smart materials.

[51] Jayatilake D, Gruebler A, Suzuki K (2010) Robot Assisted Smile Recovery. In: Kordic V, editor. Cutting Edge Robotics 2010. Rijeka: InTech. pp. 333-350.

[52] Esteki A, Mansour JM (1996) An experimentally based nonlinear viscoelastic model of joint passive moment. J Biomech. 29: 443-450.

[53] Given JD, Dewald JPA, Rymer WZ (1995) Joint dependent passive stiffness in paretic and contralateral limbs of spastic patients with hemiparetic stroke. J Neurol Neurosurg Psychiatry 59: 271-279.

[54] Lin CCK, Ju MS, Chen SM, Pan BW (2008) A Specialized Robot for Ankle Rehabilitation and Evaluation. J Med Biol Eng 28: 79-86.

[55] IEC 60601-1. Medical electrical equipment - Part 1: General requirements for basic safety and essential performance.

[56] ISO 14971. Medical devices - Application of risk management to medical devices.

[57] ISO 14155. Clinical investigation of medical devices for human subjects - Good clinical practice.

[58] Fumagalli L, Butera F, Coda A (2009) SmartFlex® NiTi Wires for Shape Memory Actuators. J Mater Eng Perform 18: 691-695.

[59] Liu SH, Huang TS, Yen JY (2010) Tracking Control of Shape-Memory-Alloy Actuators Based on Self-Sensing Feedback and Inverse Hysteresis Compensation. Sensors 10: 112-127.

Chapter 5

1D Phenomenological Modeling of Shape Memory and Pseudoelasticity in NiTi Alloys

Carmine Maletta and Franco Furgiuele

Additional information is available at the end of the chapter

http://dx.doi.org/10.5772/51283

1. Introduction

A one-dimensional phenomenological approach to simulate both the mechanical and functional properties in shape memory alloys (SMAs) is described in the following sections. In fact, shape-memory alloys exhibit unique mechanical and functional features, due to reversible transformations in crystal structure. In particular, on the macroscopic scale, SMAs are able to remember a geometrical shape and can return to that shape by activating the phase transition mechanisms. Many kinds of SMAs have been exploited in the last decades, such as the copper-zinc-aluminum (ZnCuAl), copper-aluminum-nickel (CuAlNi), nickel-manganese-gallium (NiMnGa), nickel-titanium (NiTi), and other ones made by alloying zinc, copper, gold, iron, etc. Among these alloys the near equiatomic NiTi binary system shows the most exploitable characteristics due to the high stress and strain recovery capabilities associated with their functional properties, namely pseudoelastic effect (PE) and shape memory effect (SME). These properties are due to a reversible solid state phase transformation between a parent phase (austenite) and a product phase (martensite), the so called thermoelastic martensitic transformation (TMT), that can be activated either by temperature (Thermally Induced Martensite, TIM), or by applied stress (Stress Induced Martensite, SIM) [1]. Due to these features NiTi alloys are currently used in an increasing number of applications in many fields of engineering [2], for the realization of smart sensors and actuators, joining devices, hydraulic and pneumatic valves, release/separation systems, consumer applications and commercial gadgets. However, thanks to their good mechanical properties and biocompatibility the most important applications of NiTi alloys are in the field of medicine, where pseudoelasticity is mainly exploited for the realization of several components, such as cardiovascular stent, embolic protection filters, orthopedic components, orthodontic wires, micro surgical and endoscopic devices. As a direct consequence of their interesting features NiTi alloys have attracted the interest of scientific and engineering community in the last years. However, despite the increasing interest and the efforts of many researchers to better understand these unusual mechanisms, the use of NiTi alloys is currently

limited to high-value applications (i.e. medical devices, MEMS, etc.), due to the high cost of the raw material as well as to the complex component manufacturing; in fact, an accurate control of the processing parameters must be carried out as the functional and mechanical properties of NiTi alloys are significantly affected by the thermo-mechanical loading history experienced during manufacturing [3–8]. On the other hand, the design of complex shaped NiTi-based components needs an accurate knowledge of the mechanical and functional response of the material, as well as how this evolves during subsequent thermo-mechanical processes. Within this context the use of numerical modeling techniques, to simulate both mechanical and functional behavior of SMAs, is of major concern and, consequently, many studies have been focused on this topic in the last few years [9, 10], with the aim to model the non-linear hysteretic behavior that describes the phase transformation, and the related functional properties. Some of these models are based on microscopic and mesoscopic approaches [10], where the thermo-mechanical behavior is modeled starting from molecular level and lattice level, respectively; other models are based on macroscopic approaches, where only phenomenological features of the SMAs are used [11–24]. In this field, some authors proposed one-dimensional models based on an assumed polynomial-free energy potential [11, 12] while other models are based on an assumed phase transformation kinetic and consider simple mathematical functions to describe the phase transformation behavior of the material [13–15]. These models are probably the most popular in the literature due to their phenomenological approaches, which allow easy developments without considering the underlying physics of the transformation kinetic. Furthermore, other models are based on the elastoplasticity theory [16–22] which are capable of describing the functional behavior of the material using plasticity concepts. Finally, some researchers used the Galerkin method to describe thermo-mechanical behaviors of shape memory alloys [23, 24]. More recently, a 1-D phenomenological approach to simulate both the shape memory effect [27–29] and pseudoelastic effect [30] in NiTi-based shape memory alloys has been developed and it is described in the following sections. In particular, the temperature-strain and stress-strain hysteretic behavior of SMAs, associated with the thermally induced and stress-induced phase transition mechanisms, are modeled from a phenomenological point of view, *i.e.* without considering the underlying physics of the problem, by using Prandtl-Ishlinksii hysteresis operators [25, 26]. The main features of this approach is a simple implementation together with a good accuracy and efficiency in predicting the stress-strain hysteretic behavior of 1D components. Unfortunately, the one dimensional nature of the proposed model, represents one of the major drawback with respect to some of the pre-existing phenomenological models, which are based on more thermodynamically consistent frameworks and, consequently, are able to capture several behaviors of NiTi alloys, such as detailed stress-strain distribution in 2D and 3D components. However, the high computational efficiency of the proposed model allows its use for real time simulation and control o 1D SMA components. The parameters of the phenomenological model are identified by simple and efficient numerical procedures, starting from a set of experimentally measured hysteresis loops. The identification procedures have been developed in the commercial software package *Matlab*^{TM}, while the computed parameters are used in *Simulink*^{TM} models, which are able to simulate the whole path dependent hysteretic behavior of the SMAs, *i.e.* for generic complete and incomplete stress-induced and/or thermally induced phase transition mechanisms. The models are also able to capture the hysteresis modifications due to complex loading conditions, *i.e.* they are able to predict the change of the transformation stresses and temperatures according to the

Clausius-Clapeyron relation [1]. The unique thermo-mechanical features of SMAs are firstly illustrated in the following section 2, while the numerical approach is described in section 3 together with some case studies, involving both shape memory and pseudoelasticity, and the results show good accuracy and small computational time.

2. Thermo-mechanical properties of NiTi alloys

2.1. Thermoelastic martensitic transformation

Nickel-Titanium (NiTi) based shape memory alloys exhibit unique thermomechanical properties due to a reversible solid state phase transformation between a high temperature parent phase (B2 - austenite) and a low temperature product phase (B19′ - martensite), the so called Thermoelastic Martensitic Transformation (TMT). In particular, TMT can be activated by a temperature variation (TIM, Thermally Induced Martensite) or by the application of an external mechanical load (SIM, Stress Induced Martensite), and it allows the crystal lattice structure to accommodate to the minimum energy state for a given temperature and/or stress. Figure 1 schematically shows the crystal structures of the two phases.

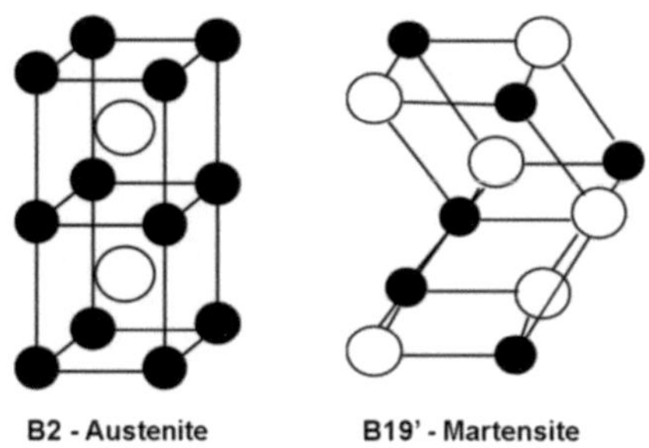

Figure 1. Schematic depiction of the lattice structures of austenite and martensite.

The austenitic phase is characterized by a Body Centered Cubic structure (BCC), with a nickel atom at the center of the crystallographic cube and titanium atoms at the cube's corners, while the lattice structure of the martensitic phase consists of a rhombus alignment with an atom at each of the rhombus corners. On the macroscopic scale the two crystal structures exhibit different engineering properties, such as Young's modulus, electrical resistance, damping behavior, etc. As a consequence, the transition between the two phases gives the possibility to obtain variable and/or tunable properties, *i.e.* NiTi alloys are able to change and or/adapt their response as a function of external stimuli. In addition, phase transition mechanisms are also associated with high strain recovery capabilities resulting from both thermally-induced and/or stress-induced transformations as described in the following section.

2.1.1. *Thermally-induced martensitic transformation*

When cooling the austenitic structure a thermally-induced martensitic transformation (B2→B19′) occurs in the temperature range between martensite start temperature (M_s), and martensite finish temperature (M_f). When the martensitic structure is heated the reverse transformation between martensite and austenite (B19′→B2) occurs in the range between

austenite start temperature (A_s) and austenite finish temperature (A_f). These characteristic temperatures, the so called phase transition temperatures (PPTs), can be regarded as material parameters, which depend on the alloy composition and on the thermomechanical processing conditions [1], and can be easily detected by Differential Scanning Calorimetry (DSC) investigation, as schematically shown in Figure 2.

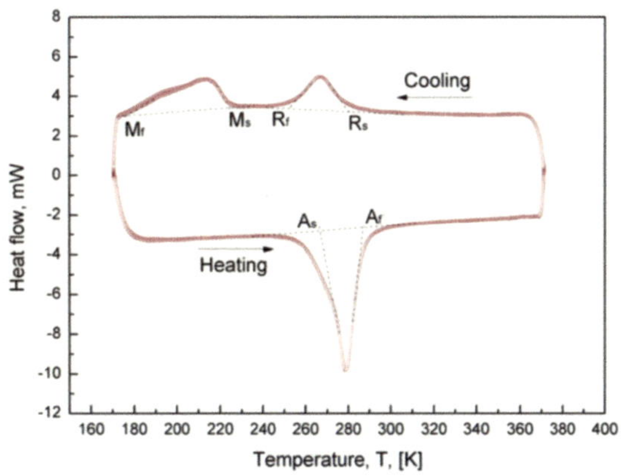

Figure 2. Differential scanning calorimetry thermogram of a NiTi alloy

In addition, transformation from B2 cubic austenite into monoclinic B19′ martensite could occur either directly or via an intermediate rhombohedral phase (R-phase), as shown in Figure 2. In particular, the R-phase transformation (B2→R) can be observed during cooling from A_f to M_s prior to martensitic transformation, resulting in a sequential transformation B2→R→B19′. However, it is worth noting, that the B2→R transformation is observed only under specific processing conditions of the alloy [33]. In addition, marked differences are normally observed between direct and reverse transformation temperatures as a direct consequence of the thermal hysteretic behavior of the alloy, as illustrated in Figure 3. In particular, this figure shows the thermal hysteresis describing the evolution of the volume fraction of martensite (ξ_M) together with the characteristic transformation temperatures.

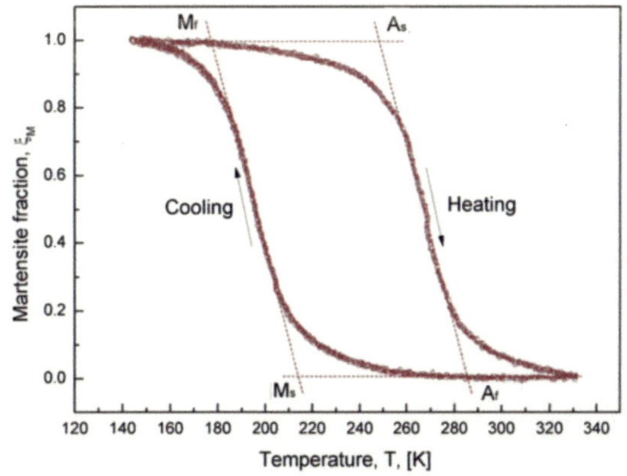

Figure 3. Thermal hysteresis of a NiTi alloy describing the evolution of volume fraction of martensite

2.1.2. *Stress-induced martensitic transformation*

When a mechanical load is applied to the austenitic structure the stress-induced B2→B19′ transformation occurs, corresponding to a plateau in the stress-strain curve of the alloy. If the mechanical load is removed the reverse B19′→B2 transformation occurs which is related to another stress plateau and allows an almost complete strain recovery. Figure 4 illustrates an example of stress-strain curve of a NiTi alloy exhibiting stress-induced phase transformation mechanisms, together with the characteristic transformation stresses of the alloy, *i.e.* the stresses for direct B2→B19′ transformation (σ_s^{AM}, σ_f^{AM}) and the stresses for reverse B19′→B2 transformation (σ_s^{MA}, σ_f^{MA}).

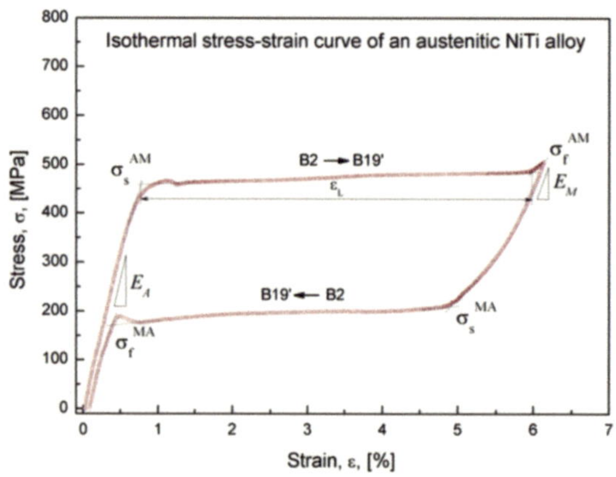

Figure 4. Stress-strain curve of an austenitic NiTi alloy with characteristic transformation stresses.

Figure 4 also illustrates the recovery strain ε_L due to the stress-induced transformation mechanisms, together with the Young's moduli of the two phases (E_A and E_M). Another stress-induced microstructural change occurs when a mechanical load is applied to the martensitic structure, *i.e.* for $T < M_f$, the so called detwinning. This mechanism can be regarded as a variant reorientation process and, on the macroscopic scale, it causes large plastic-like deformations which corresponds to a plateau in the stress-strain curve of the alloy. This mechanism is responsible for the shape memory effect as described in the following section. In addition, it is worth noting that NiTi SMAs exhibit a marked temperature dependent stress-strain response, as schematically depicted in Figure 5. In particular, the temperature dependence of transformation stresses is given by the Clausius-Clapeyron relation of equations 1:

$$\frac{d\sigma^{AM}}{dT} = C_M; \frac{d\sigma^{MA}}{dT} = C_A \quad (1)$$

where C_M (direct martensitic transformation) and C_A (reverse austenitic transformation) are in the range between 5 and 10 $MPaK^{-1}$.

2.2. Shape memory effect

Shape Memory Effect (SME) is the ability of a SMA to remember a predetermined shape and to recover this shape even after being subjected to large mechanical deformations (up to 10%). In NiTi alloys this property is observed under martensitic conditions, *i.e.* when $T < M_f$,

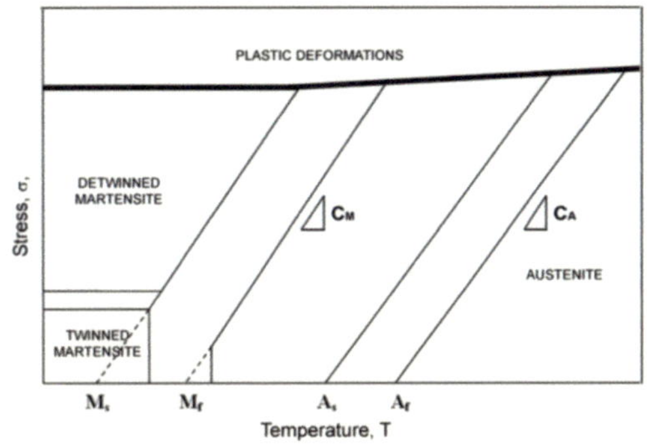

Figure 5. Relation between transformation stresses and temperature according to the Clausius-Clapeyron relation.

and it can be attributed to the combination of two microstructural changes: *i*) detwinning of martensitic variants and *ii*) thermally induced phase transformation. Figure 6 shows a schematic depiction of the SME together with the associated phase transition mechanisms (Figure 6.a) and the corresponding stress-strain-temperature response (Figure 6.b).

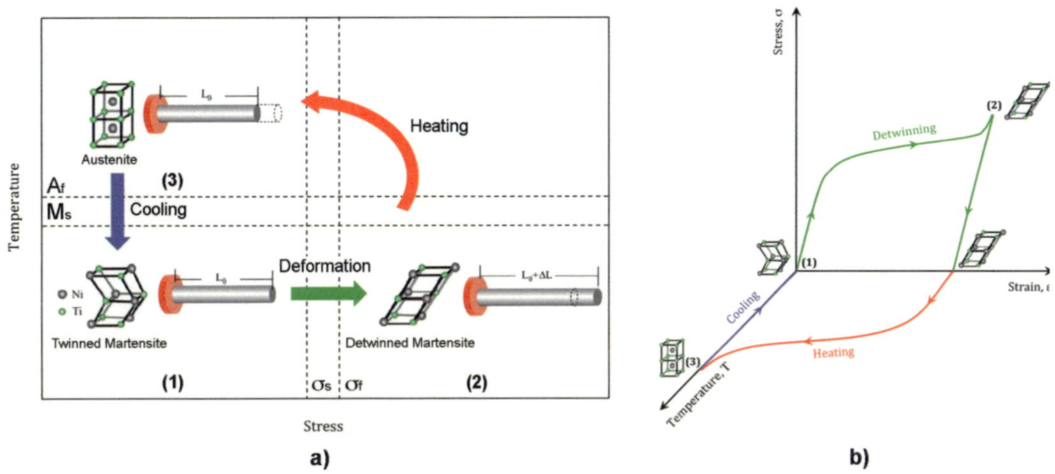

Figure 6. Schematic depiction of the shape memory effect: a) phase transition mechanisms and b) stress-strain-temperature response

In particular, Figs. 6 show that if a mechanical load is applied to the twinned martensitic structure (1), *i.e.* for $T < M_f$, detwinning occurs at a given critical stress value which corresponds to large plastic-like deformations (up to 10%) through a plateau in the stress-strain curve (2). In fact, these deformations persist after complete unloading as only elastic recovery of the detwinned structure is observed. However, if the material is heated up to the austenite finish temperature ($T > A_f$) a complete thermally induced phase transformation occurs from the detwinned martensitic structure to the austenitic one (3) and, on the macroscopic scale, this transformation allows a complete shape recovery. Finally, if the material is cooled down to the martensite finish temperature ($T < M_f$) it is able to remember

its original twinned martensitic structure (1). This unusual functional property is also known as one-way shape memory effect (OWSME) as it defines the ability of material to remember just one shape, the cold one ($T < M_f$), and to recover this shape after being mechanically deformed. However, under specific thermo-mechanical treatments NiTi alloys could exhibit another shape memory mechanism, the so called two-way shape memory effect (TWSME), *i.e.* they are able to remember a cold shape, linked to the martensitic structure, and a hot shape, linked to the austenite. As a consequence, during repeated heating and cooling, the material changes its shape in a reversible way, through a hysteresis loop, as schematically illustrated in Figure 7.

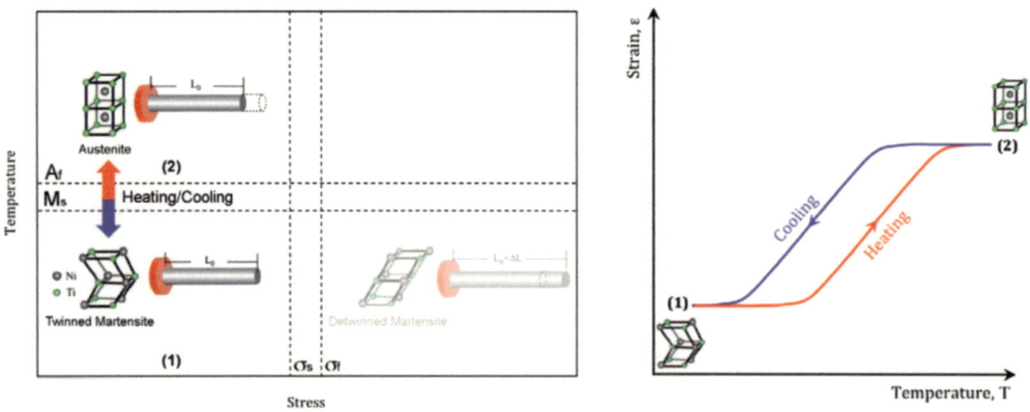

Figure 7. Schematic depiction of the two-way shape memory effect: a) phase transition mechanisms and b) strain-temperature hysteretic response

In particular, TWSME can be induced by proper thermomechanical procedures, the so-called training, which usually involve repeated deformations and transformations between austenite and martensite. This thermomechanical process produces a dislocation structure and, consequently, creates an anisotropic stress field that benefits the formation of preferentially oriented martensite variants [32], resulting in a macroscopic shape change between the phase transition temperatures.

Figure 8 shows an example of the thermomechanical cycle, which is composed of four subsequent steps: 1) strain controlled uniaxial loading up to a training deformation ε_{tr}, 2) complete unloading at the same rate and recording of the residual strain ε_r, 3) heating up to the austenite finish temperature A_f, in stress free conditions, to activate SME and measuring the recovery deformation ε_{re} and permanent strain ε_p, and 4) cooling down to the martensite finish temperature M_f, in stress free conditions, and recording the induced two-way shape memory strain ε_{tw}. Experimental measurements have been carried out in [28] where several training cycles have been executed with a training deformation $\varepsilon_{tr} = 5.5\%$. Each training cycle starts from the end of the cooling stage of the previous one, so that the total deformation at the $i-th$ cycle, $\varepsilon_{tot(i)}$, can be defined as follows:

$$\varepsilon_{tot(i)} = \varepsilon_{tr(i)} + \varepsilon_{p(i-1)} + \varepsilon_{tw(i-1)} \quad (2)$$

Figure 9 reports the measured ε_{tw}, ε_p, ε_{pe}, and ε_{tot} *vs* the number of training cycles. The figure clearly shows that the two-way shape memory strain ε_{tw} increases with increasing the number

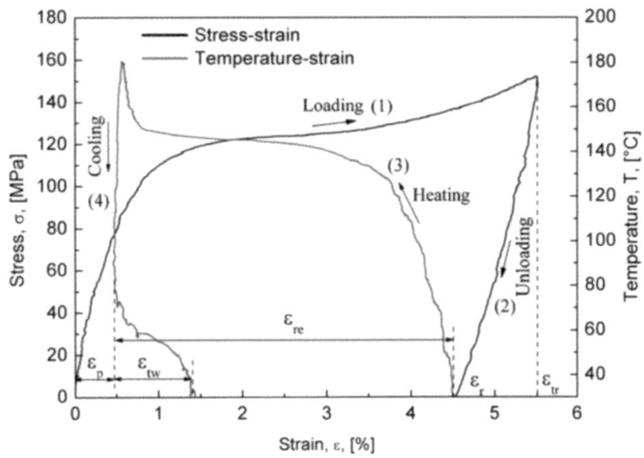

Figure 8. Example of training cycle: 1) loading, 2) unloading, 3) heating up to A_f, and 4) cooling down to M_f [28].

of training cycles, and a similar behavior is observed for the permanent deformation ε_p, the strain recovery ε_{re}, and the total deformation ε_{tot}. In particular, ε_{tw} increases from 1.0% at the first training cycle to 2.8% after six cycles; ε_p and ε_{re}increase from 1.5% to 4.2% and from 4% to 6%, respectively, and, consequently, ε_{tot} increases from 5.5% to 11.8%. In Figure 10, the

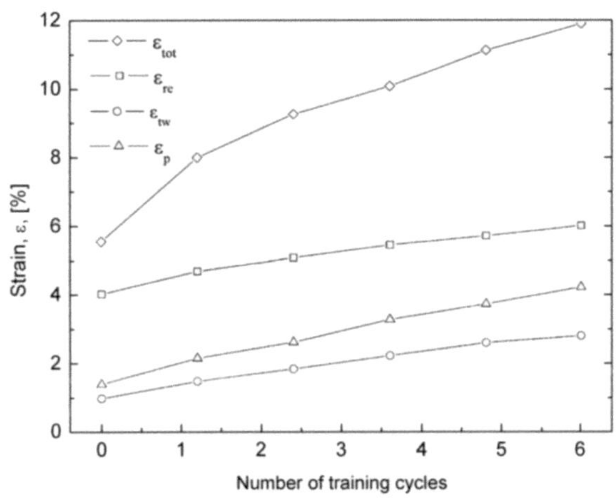

Figure 9. Measured deformations versus number of training cycles: ε_{tw}, ε_p, ε_{pe}, and ε_{tot} [28].

measured thermal hysteresis behavior strain *vs* temperature, describing the TWSME of the trained material, is shown. In particular, Figure 10.a illustrates the stress-free hysteresis loop, together with the PTTs, while in Figure 10.b the stress-free thermal hysteresis loop is compared with those obtained under a tensile stress of 50 MPa and 100 MPa. The comparison clearly shows and a systematic increase in ε_{tw}, as well as in all PTTs, when increasing the applied stress σ. In particular, the increase of ε_{tw} is attributed to *i*) the variation of Young's modulus in the thermal hysteresis behavior between martensite and austenite and *ii*) the increased volume fraction of favorably oriented martensite variants with increasing external stress.

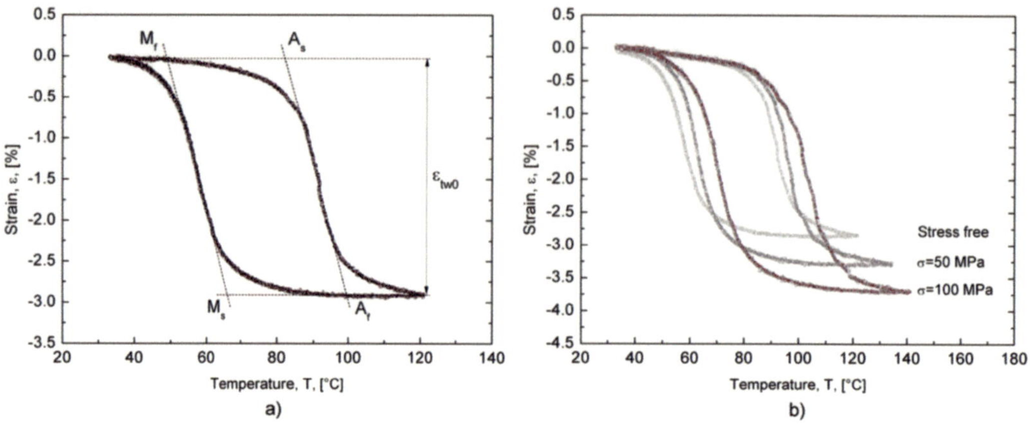

Figure 10. Thermal hysteresis behavior of the trained material strain *vs* temperature: a) stress-free hysteresis loop with an highlight of the PTTs and b) effects of the applied tensile stress [28].

2.3. Pseudoelastic effect

The pseudoelastic (PE) effect in NiTi alloys consists in the high strain recovery capability (up to 10%) observed during isothermal loading/unloading cycles carried out at temperature $T > A_f$, *i.e.* when the alloy is in austenitic conditions. This functional property can be directly attributed to the reversible stress-induced martensitic transformations as discussed in section 2.1.2. In particular, Figure 11 illustrates that if a mechanical load is applied to austenitic structure (1) the B2→B19′ transformation occurs and, on the macroscopic scale, large mechanical deformation are achieved through a stress-strain transformation plateau (2). However, if the mechanical load is removed the reverse B19′→B2 transformation occurs and, consequently, the material is able to recover its original shape through an unloading plateau in the stress strain curve. However, the reverse transformation occurs at lower stress values resulting in a marked stress-strain hysteretic behavior.

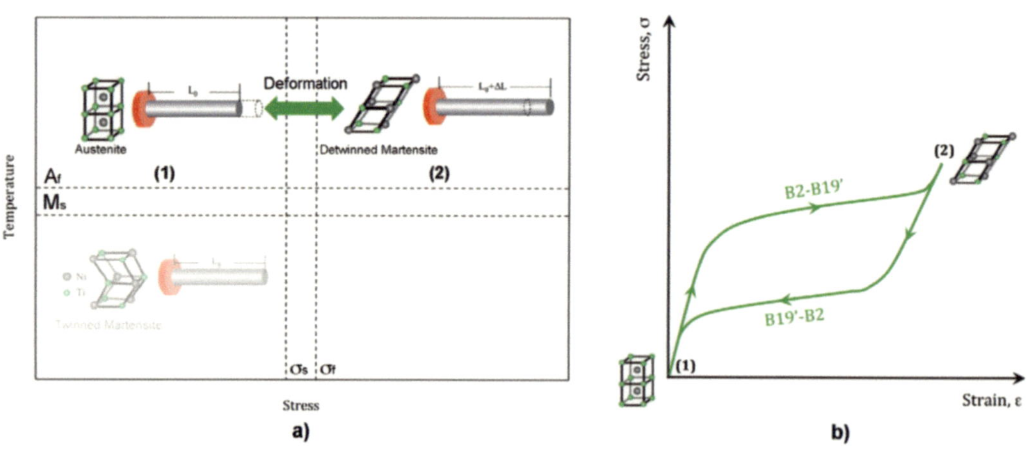

Figure 11. Schematic depiction of the pseudo elastic effect: a) stress-induced phase transition mechanisms and b) stress-strain hysteretic response

It is worth noting that Figure 11.b illustrates the nominal stress-strain behavior of a pseudoelastic SMA, while NiTi alloys exhibit a marked evolution of the stress-strain hysteretic behavior in the first mechanical cycles, up to a stable response, due to the formation of stabilized martensite [32], which causes a reduction of the pseudoelastic recovery of the SMA. In particular, Figure 12 illustrates the evolution of the stress-strain hysteretic behavior of the material in the first 25 mechanical cycles for a fixed value of total strain $\varepsilon_{tot} = 3.5\%$. These data were obtained from experimental testing of a commercial pseudoelastic NiTi alloy [30]; the figure clearly shows a marked reduction of the pseudoelastic recovery, from 3.5% to about 3%, but the stress-strain loops becomes stable after 20 cycles. Furthermore, as reported in section

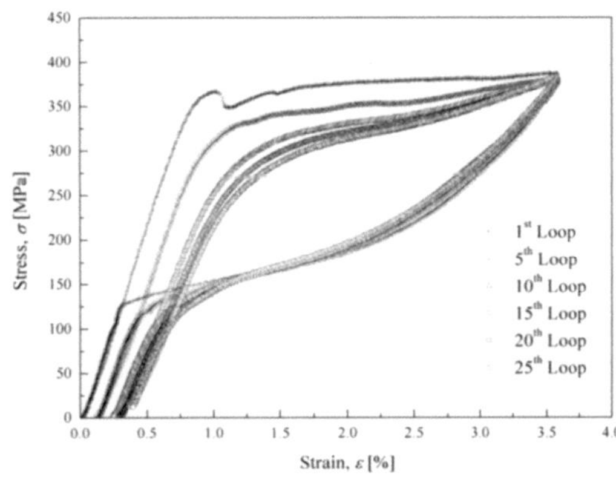

Figure 12. Evolution of the stress-strain response of a commercial pseudoelastic NiTi alloy during the first 25 mechanical cycles for a maximum applied deformation $\varepsilon_{tot} = 3.5\%$ [30]

2.1.2, the critical stresses for direct and inverse phase transformation are significantly affected by the temperature, according to the Clausius-Clapeyron constant (Equation 1), as illustrated in the experimentally measured curves in Figure 13 [30]. These curves have been obtained from isothermal loading unloading cycles, carried out at increasing values of the testing temperature for $T > A_f$ ($303K - 328K$), by using a SMA with a stable pseudoelastic response. The analysis of the data in Figure 13 allows to obtain the value of the Clausius-Clapeyron constant ($C_M = C_A = 8.7MPaK^{-1}$).

3. Hysteresis modeling: basics

A one-dimensional numerical approach to simulate the stress-strain and strain-temperature hysteresis behavior of SMAs is described in this section. In particular, in a purely phenomenological way, the hysteresis loop is modeled by a Prandtl-Ishlinskii hysteresis operator H [26]; the basic idea of this approach consists in modeling the non-linear hysteretic behavior by a weighted superposition of many elementary hysteresis operators, such as the backlash operators Hr, as schematically illustrated in Figure 14.

$$H = \{w\}^T \{H_r\} \tag{3}$$

where $\{H_r\}$ is the vector of backlash operators and $\{w\}$ is the corresponding vector of weights. As shown in Figure 14.a, each backlash operator H_{ri} is characterized by its dead band width dw_i, while the corresponding weight w_i represents the slope of the oblique lines of the operator. As illustrated in Figure 14.b, which represent a generic $x(t) - y(t)$ hysteretic

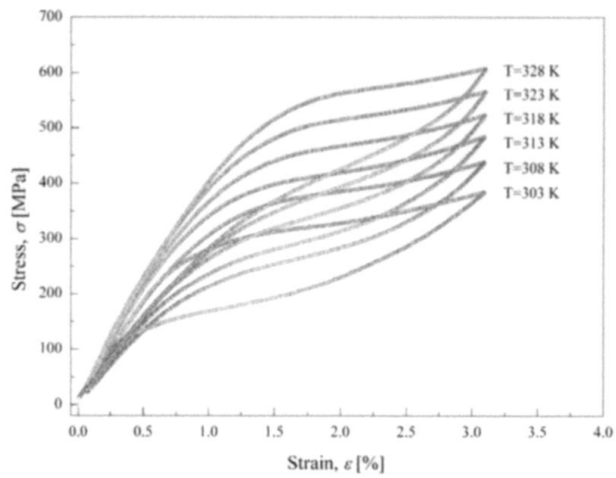

Figure 13. Isothermal stress-strain hysteresis loops of a commercial pseudoelastic NiTi alloy as a function of the testing temperature [30].

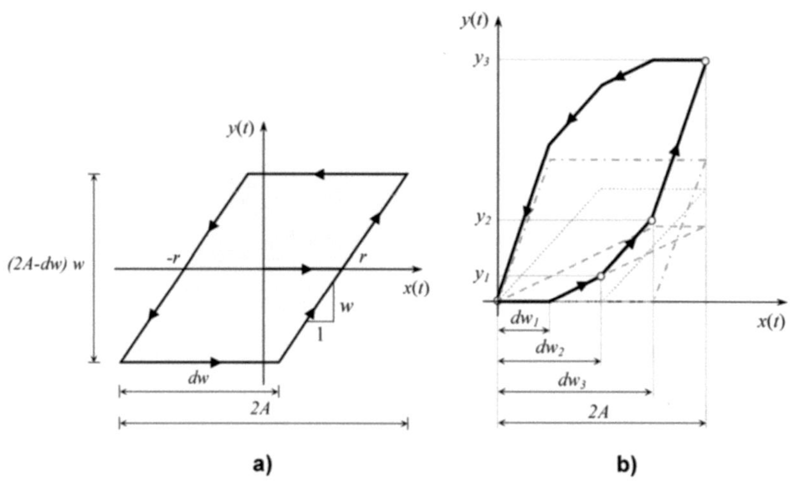

Figure 14. Schematic depiction of the hysteresis operators: a) elementary backlash operator H_r and b) Prandtl-Ishlinskii hysteresis operator H given by weighted superposition of elementary hysteresis operators.

behavior, the proposed approach consists of modeling the hysteretic loop by a linear piecewise discretization. The accuracy of the model can be improved by increasing the total number of linear pieces, which represent the total number of the backlash operators. The problem of modeling the hysteretic behavior, starting from the experimental measurements, is now reduced to the determination of the deadband width vector $\{dw\}$ of the backlash operators and the associated gain vector $\{w\}$. In particular, the parameters of the model can be easily identified by the outer loop of the hysteretic region by using the following simple relation:

$$y_k = \sum_{i=1}^{k} (dw_{k+1} - dw_i) w_i \tag{4}$$

where y_k is the output value of the lower branch of the loop in the generic point of discontinuity k, as shown in Figure 14.b. The vector $\{dw\}$ is a user defined discretization of

the total amplitude of the input signal. Equation 4 can be rewritten in matrix form as follows:

$$\{y\} = [A]\{w\} \tag{5}$$

where the matrix $[A]$ is constructed, for a given $\{dw\}$ vector, by using equation 4; the unknown vector $\{w\}$ can be found by solving a system of N linear equations, where N is the total number of backlash operators, as follows:

$$\{w\} = [A]^{-1}\{y\} \tag{6}$$

The main drawback of the Prandtl-Ishlinskii approach consists in the fact that only loops with an odd symmetry to the relative center can be modeled; in fact the symmetry of the elementary hysteresis operator, with respect to the center of the loop, persists under linear superposition. However, this drawback can be overcome by using a modified Prandtl-Ishlinksii operator, as described in details in [26], in which a weighted superposition of saturation operators is combined with the hysteresis operator. The parameters of this sub model, such as saturation limits $\{S\}$ and associated gains $\{w_s\}$, can be identified by using a procedure similar to that described above.

3.1. Modeling of pseudoelastic effect

The numerical model described in this section is able to simulate the pseudoelastic effect of a shape memory alloy [30], *i.e.* the the stress-strain ($\sigma - \varepsilon$) hysteretic behavior, based on the Prandtl-Ishlinksii operator and on the assumptions reported in the following.

3.1.1. Basic assumptions

Figure 15 shows the stabilized stress-strain hysteretic behavior of a commercial NiTi alloy, *i.e.* the response of the material after the first training cycles (see Figure 12), for different values of the applied deformation. The figure also illustrates the Young's moduli of austenite and detwinned martensitic structures, E_A and E_M, together with the generic young's modulus of the alloy, $E(\xi_M)$, corresponding to an incomplete stress induced martensitic transformation, *i.e.* as a function of the martensite fraction ξ_M ($0 < \xi_M < 1$). In particular, E_A represents the slope of the early stage of the stress-strain loading curve, E_M is measured from the unloading curve of a complete martensitic transformation ($\xi_M = 1$), while $E(\xi_M)$ is obtained from the unloading path of an incomplete phase transformation.

The total strain ε can be decomposed in elastic and a transformation strain components, ε_{el} and ε_{tr}, respectively:

$$\varepsilon = \varepsilon_{el} + \varepsilon_{tr} \tag{7}$$

where the elastic strain can be expressed as a function of the applied stress, σ, and of the Young's modulus, $E(\xi_M)$, of the material:

$$\varepsilon_{el} = \frac{\sigma}{E(\xi_M)} \tag{8}$$

As schematically shown in Figure 15 the Young's modulus changes during stress-induced phase transformation between austenite and martensite, *i.e.* it decreases from E_A to E_M, and

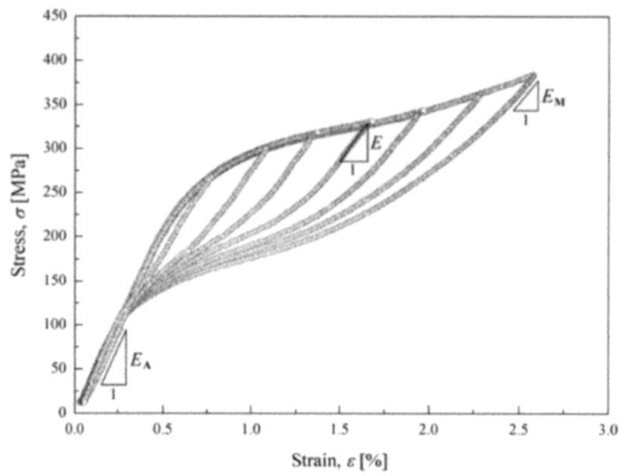

Figure 15. Stress-strain hysteretic behavior of a commercial NiTi alloy together with the Young's moduli of austenite and detwinned martensite [30].

it is assumed to be dependent on the volume fraction of martensite ξ_M according to the Reuss formula [20]:

$$\frac{1}{E(\xi_M)} = \frac{\xi_M}{E_M} + \frac{1-\xi_M}{E_A} \tag{9}$$

The evolution of martensite is assumed to be a linear function of the stress in the stress-strain transformation curves, *i.e.* in the range $(\sigma_s^{AM}, \sigma_f^{AM})$ in the loading stage and $(\sigma_s^{MA}, \sigma_f^{MA})$ in the unloading stage. In particular, the evolution rule for a complete transformation, can be expressed as follows:

$$\xi_M = \begin{cases} \frac{\sigma-\sigma_s^{AM}}{\sigma_f^{AM}-\sigma_s^{AM}} & \text{Loading path} \\ 1-\frac{\sigma-\sigma_s^{MA}}{\sigma_f^{MA}-\sigma_s^{MA}} & \text{Unloading path} \end{cases} \tag{10}$$

The assumptions given by equations (9) and (10) have been validated by experimental measurements of the Young's modulus, as reported in Figure16. In particular, the figure shows the measured values of the Young's modulus as a function of the applied stress, on the direct stress-strain transformation plateau, together with the predictions of the Reuss formula, and a satisfactory agreement is observed.

However, it's worth noting that the evolution of martensite is characterized by a hysteretic behavior, *i.e.* it is stress path dependent, and equations (9) and (10) do not correctly predict the effects of incomplete transformations. These effects are simulated mathematically, by using a backlash operator, as schematically illustrated in Figure 17; in particular, Figure 17.a shows the stress *vs* time path, while Figure 17.b illustrates the evolution of ξ_M and $1/E(\xi_M)$ *vs* the applied stress. Furthermore, the continuous lines in Figure 17.b are relative to a complete martensitic transformation, *i.e.* ξ_M increases from 0 to 1, while the dashed lines show the effects of an incomplete transformation. Due to the modification in Young's modulus during the phase transformation between austenite and martensite, as shown in Figure 17.b, both elastic and transformation strain components are represented by a hysteretic behavior and it can be calculated using equations (7-10); as an example in Figure 18 a typical $\sigma - \varepsilon$ hysteresis loop for a complete phase transformation is compared with the corresponding computed elastic strain ($\sigma - \varepsilon_{el}$) and transformation strain ($\sigma - \varepsilon_{tr}$) hysteresis loops.

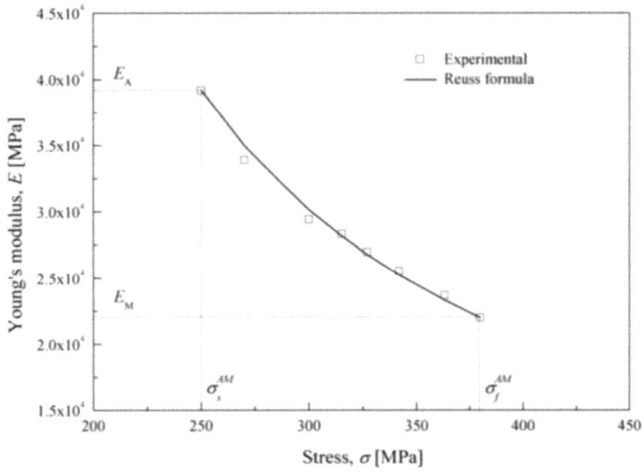

Figure 16. Evolution of the Young's modulus as a function of the applied stress: experimental measurements *vs* simulations [30].

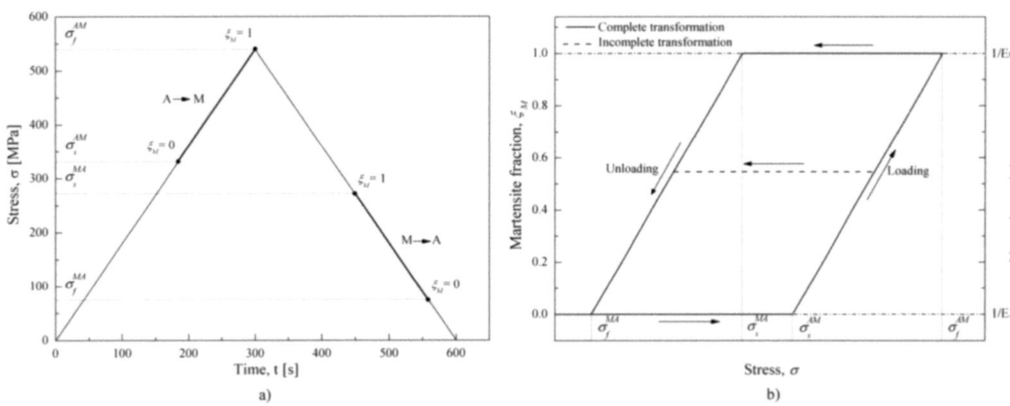

Figure 17. Evolution rule of the martensite fraction ξ_M in the tension cycle: a) example of stress-time path and b) simple hysteresis model to predict the Young's modulus $E(\xi_M)$ [30].

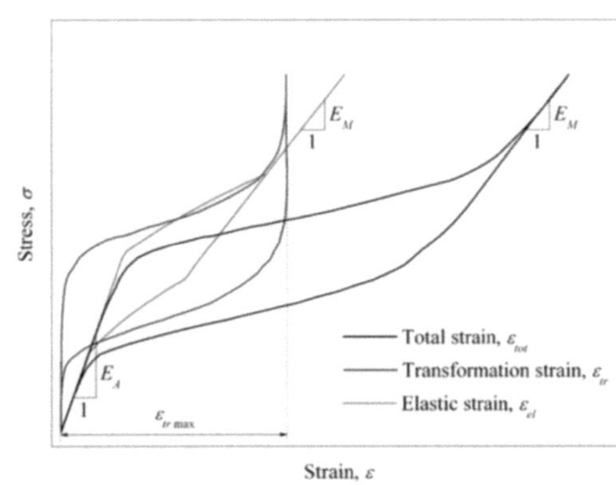

Figure 18. Stress strain ($\sigma - \varepsilon$) hysteresis loops: elastic strain ε_{el} and transformation strain ε_{tr} [30].

3.1.2. Numerical flowchart

Based on the Prandtl-Ishlinskii hysteresis modeling approach together with the model assumptions described in the previous section, a one-dimensional numerical model can be easily developed by using commercial software packages. In particular, Figure 19 illustrates an implementation of the model in the *MatlabSimulink*^{TM} platform, in which the two strain components, ε_{el} and ε_{tr}, are treated separately. As clearly illustrated in Figure 19, both stress

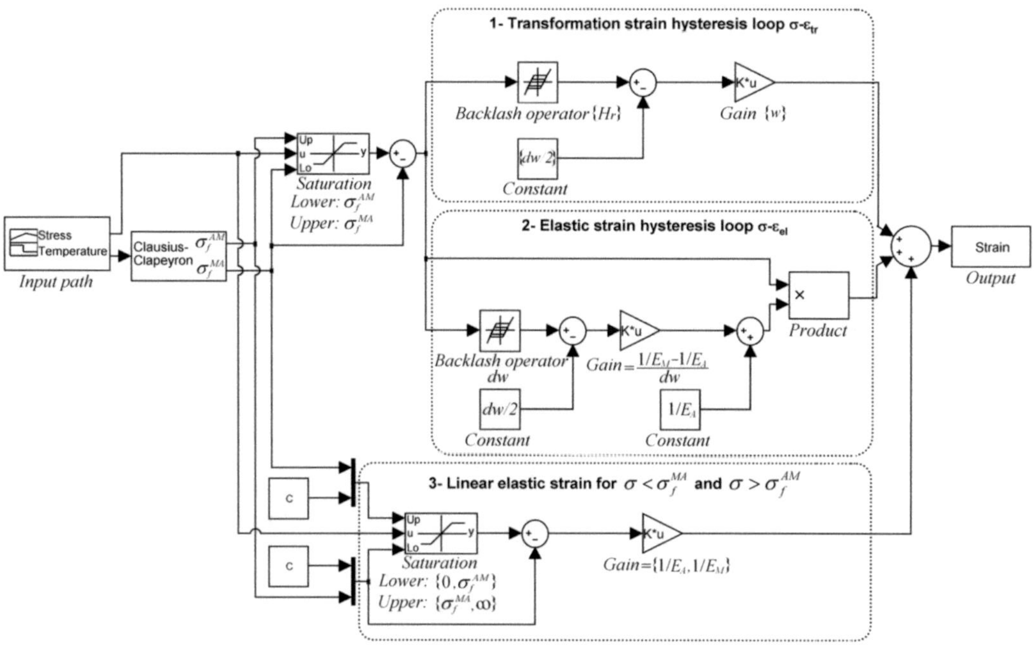

Figure 19. Flowchart of a *MatlabSimulink*^{TM} model to simulate the pseudoelastic effect in SMAs [30].

and temperature are taken as input variables, and the critical stresses are calculated, based on the current value of the temperature, by using the Clausius-Clapeyron relation (equation 1). In the model the saturation operators are used to identify the stress range where the material exhibits the hysteretic behavior ($\sigma_f^{MA} < \sigma < \sigma_f^{AM}$), and the dead zones of transformation where only elastic deformation of austenite ($\sigma < \sigma_f^{MA}$) or martensite ($\sigma > \sigma_f^{AM}$) occurs. Three different sub-models are highlighted in Figure 19 which simulate the $\sigma - \varepsilon_{tr}$ and $\sigma - \varepsilon_{el}$ hysteresis loops, and the linear elastic response of the material in austenitic and martensitic conditions. In particular, a Prandtl-Ishlinskii hysteresis operator, was used to model the $\sigma - \varepsilon_{tr}$ hysteretic behavior, a single backlash operator was adopted to model the $\sigma - \varepsilon_{el}$ loop (see Figure 17.b), and a gain operator was used to describe the linear elastic response for $\sigma < \sigma_f^{MA}$ and for $\sigma > \sigma_f^{AM}$. The parameters of the Prandtl-Ishlinskii operator describing the $\sigma - \varepsilon_{tr}$ hysteretic behavior, *i.e.* the deadband width vector $\{dw\}$ and the associated gain vector $\{w\}$ are determined from an experimentally measured stress-strain hysteresis loop by using the procedure described in section 3. In particular, as shown in Figure 20, the generic input and output variables (x and y) can be regarded as the stress and strain values (σ and ε), respectively. As a consequence, the vector $\{dw\}$ represents a user defined discretization of

the stress amplitude while the vector $\{w\}$ can be determined by using equation (6), where the vector $\{y\}$ is the vector of strain values $\{\varepsilon\}$. The computational procedure to obtain the aforementioned model parameters from an experimentally measured hysteretic loop can be easily implemented, which generates the vector $\{dw\}$, by a partition of the input stress amplitude, and calculates the unknown vector of weight $\{w\}$ by solving a system of N linear equations, where N is the number of backlash operators.

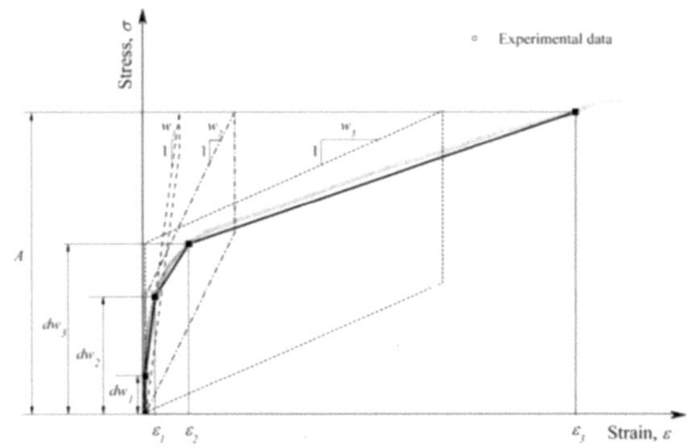

Figure 20. Loading branch of the stress-strain hysteresis loop and linear piecewise discretization obtained by a weighted superposition of three backlash operators [30].

3.1.3. Numerical results vs experiments

The accuracy of the numerical method is illustrated by comparisons with experimentally measured hysteresis loops, by using he thermo-mechanical parameters reported in [30] (E_A=39 GPa, E_M=20 GPa, σ_s^{AM}=260 MPa, σ_f^{AM}=385 MPa, σ_s^{MA}=250 MPa, σ_f^{MA}=125 MPa, $C_A = CM$=10.3 MPa/K). Figure 21 shows the stress-strain hysteretic behavior of the SMA for a stress path which involves several incomplete stress-induced martensitic transformations ($A \rightarrow M$), by repeated isothermal tension cycles ($T = 303K$) carried out between $\sigma_{min} = 0$ and decreasing values of $\sigma_{max} < \sigma_f^{AM}$. In particular, the comparison between numerical simulations and experimental measurements, illustrated in Figure 21.a, clearly shows a satisfactory accuracy of the model in predicting the non-linear stress-strain hysteretic behavior of the material; the figure also shows that the model is able to capture the modification of Young's modulus in the stress-strain transformation curve, as it correctly predicts the change in the slopes of the unloading curves. Furthermore, Figure 21.b shows the evolutions of the transformation strain, ε_{tr}, and elastic strain, ε_{el}, for the same input stress path of Figure 21.a. As illustrated in section 3.1.1, both ε_{tr} and ε_{el} are characterized by hysteretic behaviors, which are due to the mismatch between the critical stresses in the stress induced transformations and the modification of Young's modulus in the stress-strain transformation curve, respectively.

Figures 22 show comparisons between numerical predictions and experimental measurements for two input isothermal stress paths ($T = 303K$) which involve incomplete $M \rightarrow A$ transformations (Figure 22.a) and both incomplete $A \rightarrow M$ and $M \rightarrow A$ transformations (Figure 22.b). In particular, Figure 22.a shows the hysteretic response of the material for a stress path which is composed of some subsequent tension cycles between increasing values of $\sigma_{min} < \sigma_f^{MA}$ and $\sigma_{max} = const$, while Figure 22.b is relative to a stress

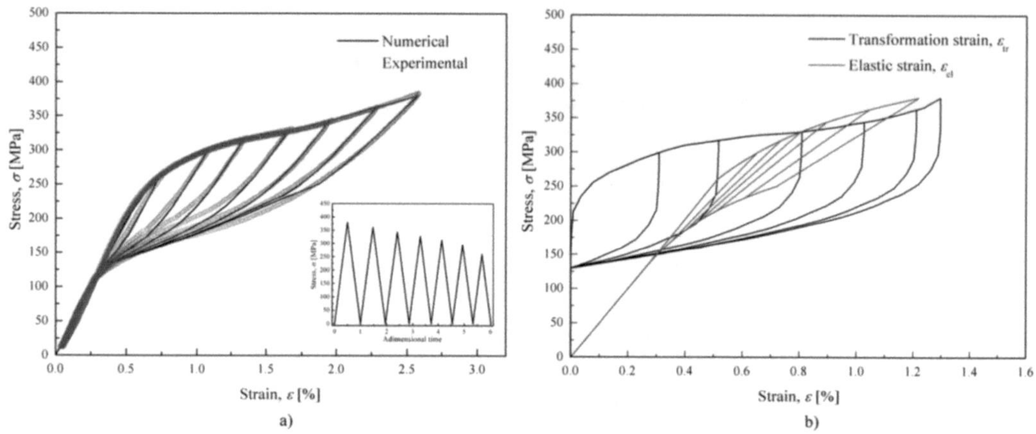

Figure 21. Numerical simulation for isothermal stress cycles with incomplete $A \to M$ transformations: a) comparison with experimentally measured loops and b) evolution of transformation strain, ε_{tr}, and elastic strain, ε_{el} [30].

path which involves different subsequent tension cycles carried out between increasing values of $\sigma_{min} > \sigma_f^{MA}$ and decreasing values of $\sigma_{max} < \sigma_f^{AM}$; both figures shows good agreements between experiments and numerical simulations.

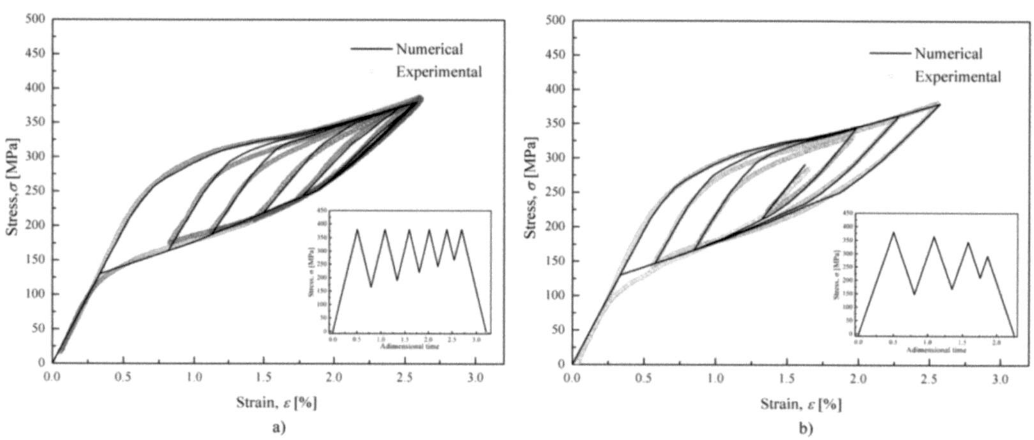

Figure 22. Comparison between numerical simulations and experimentally measured hysteresis loops for isothermal stress cycles with: a) incomplete $M \to A$ transformations and b) incomplete $A \to M$ and $M \to A$ transformations [30].

3.2. Modeling of two-way shape memory effect

The numerical model described in this section is able to simulate the two-way shape memory effect of a trained NiTi based shape memory alloy, *i.e.* the the strain-temperature ($\varepsilon - T$) hysteretic behavior [27–29]. Furthermore, the model is able to capture the effects of applied stress on the thermal recovery of the alloy.

3.2.1. Basic assumptions

In a pure phenomenological way, the variation of the two-way shape memory strain, ε_{tw}, with increasing the applied stress (see Figure 10.b) can be attributed to two different mechanisms: *i*) the variation of Young's modulus in the thermal hysteresis behavior between martensite and austenite, *ii*) the increased volume fraction of favorably oriented martensite variants with increasing external stress.

$$\Delta\varepsilon_{tw} = \varepsilon_{tw} - \varepsilon_{tw0} = \left(\frac{1}{E_M} - \frac{1}{E_A}\right)\sigma + c\sigma \tag{11}$$

where subscript 0 indicate the stress-free condition, E_M and E_A represent the Young's moduli of martensite and austenite, respectively, while c can be assumed as a material constant. In particular, the first term at the right end side of equation 11, indicated as $\Delta\varepsilon_{mech}$, describe the first effect while the last terms, namely $\Delta\varepsilon_{mem}$, take into account the second effect. Young's moduli E_M and E_A can be measured by isothermal tensile tests carried out at temperatures $T < M_f$ and $T > A_f$, respectively, while the parameter c can be obtained from experimental measurements of the two-way shape memory strain carried out at different value of the applied stress. In the following the values of the parameters reported in [28] have been used ($E_M = 36 * 10^3 MPa$, $E_A = 67 * 10^3 MPa$, $c = 8.5 * 10^{-5} MPa^{-1}$). Two simple mathematical functions, obtained by a numerical fitting of the experimental data, are used to describe the phase transformation kinetics. In particular, as reported in the model by Tanaka and Nagaki [13], the heating and cooling branches of the hysteresis loop can be represented by two exponential curves:

$$\varepsilon = \begin{cases} \varepsilon_{tw} e^{a_A(A_f - T + \delta T_M)} & \mathrm{M} \rightarrow \mathrm{A} \\ \varepsilon_{tw}[1 - e^{a_M(M_f - T + \delta T_A)}) & \mathrm{A} \rightarrow \mathrm{M} \end{cases} \tag{12}$$

where ε_{tw}, M_s, and A_s are functions of the applied stress, as reported in Equations 11 and 1; a_M, a_A, δT_M, and δT_A, which define the shape of the heating and cooling branches of the loop, can be identified by a numerical fitting of the experimental data. Figure 23.a shows a comparison between experimental measurement and exponential curves; in the figure, points P_1 and P_2 represent the range where the numerical fitting is executed to identify the parameters of the heating branch of the loop, while points P_3 and P_4 are relative to the cooling branch. In Figure 23.b, a linear fitting between points P_1 and P_2 of the experimental data in the $T - log\varepsilon$ plane is shown, where the slope of the line defines the parameter a_A and the intersection with the $log\varepsilon$ axis allows us to obtain the parameter δT_M. If the loop is characterized by an odd symmetry with respect to its center, as is quite well observed in the investigated material, the same values can be assumed for the constants a_A and a_M and δT_A, and δT_M. The two exponential curves describe the two branches of the hysteresis loop in a parametric way for a generic value of the applied stress by using Equations 1 and 11. Starting from the curves $\varepsilon - T$, the numerical method based on the Prandtl-Ishlinksii operator can be developed, which is able to predict the output response for a generic temperature path, as decribed in the following section.

3.2.2. Numerical flowchart

The numerical method described above can be easily implemented in a *MatlabSimulink*^{TM} model, as shown in Figure 24, by a modified Prandtl-Hishlinkii hysteresis operator. The

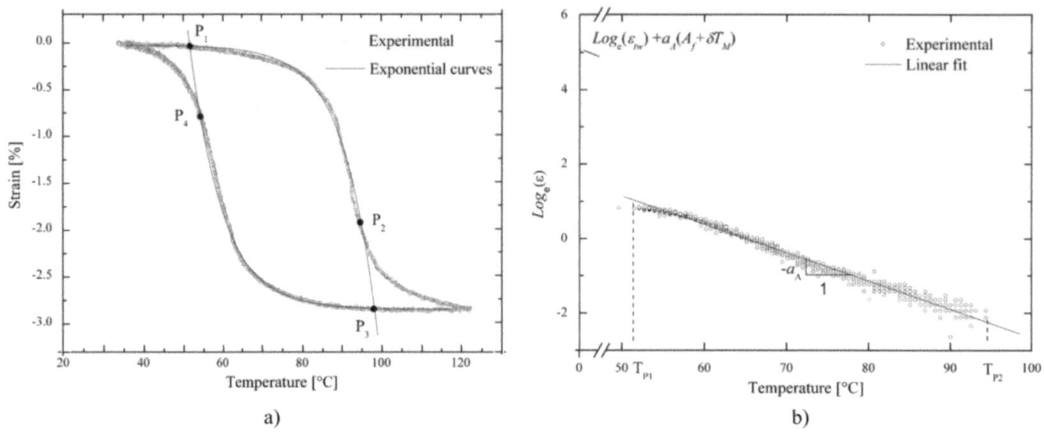

Figure 23. Numerical fitting of the experimental data: a) comparison between exponential curves and experimental measurements; b) numerical fitting in the $T - log\varepsilon$ plane to identify the parameters of the exponential curve in the heating branch of the hysteresis loop [28]

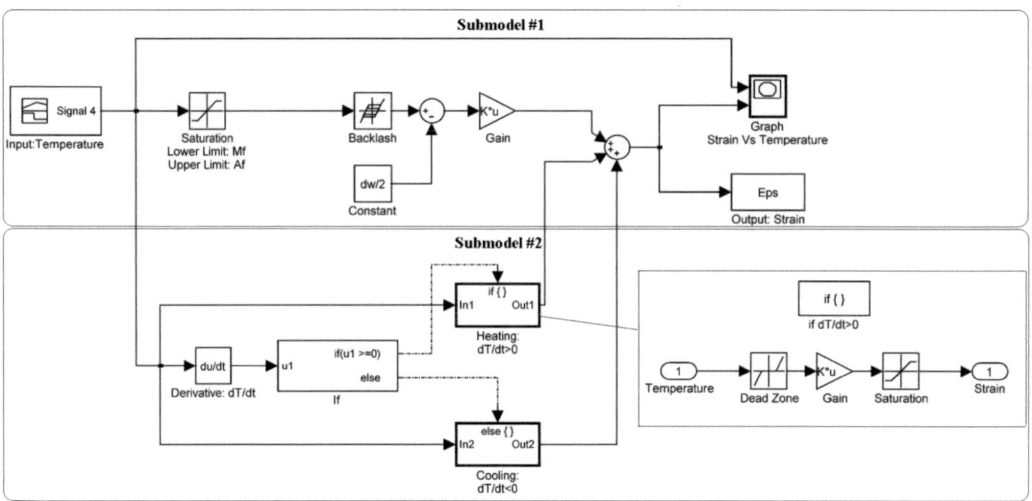

Figure 24. Flowchart of a $MatlabSimulink^{TM}$ model to simulate the two way shape memory effect in SMAs.

Prandtl-Hishlinkii operator is implemented in the submodel #1 of Figure 24, by a weighted superposition of several backlash operators, and the corresponding parameters, *i.e.* the deadband width vector $\{dw\}$ and the associated gain vector $\{w\}$, are determined from the exponential curves of equation 12, which in turn are obtained from a fitting of experimental data $(T - \varepsilon)$ as illustrated in Figure 23. In particular, the generic input and output variables (x and y) can be regarded as the temperature and strain values (T and ε), respectively. As a consequence, the vector $\{dw\}$ represents a user defined discretization of the temperature amplitude while the vector $\{w\}$ can be determined by using equation (6), where the vector $\{y\}$ is the vector of strain values $\{\varepsilon\}$ obtained from equation 12 ($\varepsilon_i = \varepsilon(T_i)$). The saturation operator in the submodel #1 is used to simulate the dead zones of transformation, *i.e.* the

material behavior when the temperature is above A_f during heating and below M_f in the cooling branch of the hysteresis loop. In particular, this operator imposes upper and lower bounds on the temperature, which are A_f and M_f, respectively, so that when the temperature is outside these bounds the signal is clipped to the upper or lower bounds. In Figure 25.a, the response of the submodel #1, in terms of strain-temperature loop, is illustrated for a thermal cycle between the temperatures $T_0 < M_f$ and $T_1 > A_f$; the figure clearly shows that the hysteretic behavior of the material is properly described in the range of temperatures between M_f and A_f, and the dead zone of transformation, when the temperature exceeds A_f or falls below M_f, are also simulated.

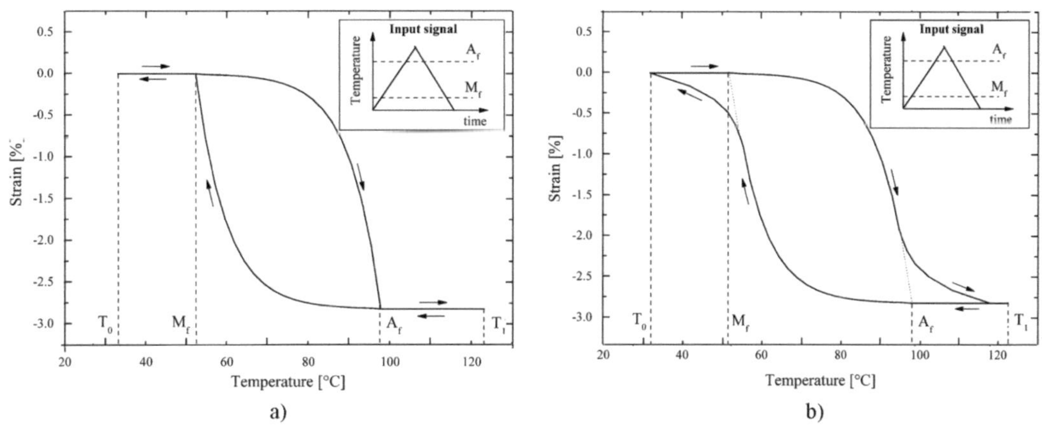

Figure 25. Numerically simulated loop for a thermal cycle between the temperatures $T_0 < M_f$ and $T_1 > A_f$ obtained by: a) Prandtl-Ishlinskii model and b) modified Prandtl-Ishlinskii model [28].

Unfortunately, when comparing the experimental results with the numerically simulated loops, high errors are observed in the extremity of the hysteretic region, *i.e.* when the temperature is below M_f during cooling and above A_f during heating. To overcome this limitation, a modified *Simulink* model can be implemented by including the submodel #2 of Figure 24. This latter uses two subsystems, for the heating and cooling branches of the loop, which modifies the output response of the system when the temperature is near M_f and A_f. In particular, each subsystem implements a weighted superposition of several dead band operators, which is executed by a series of a dead band block and a gain block, while the saturation block assures that the correction is carried out only in a limited range of temperatures near M_f and A_f. Figure 25.b shows a simulated hysteresis loop, obtained by the modified model, between the temperatures $T_0 < M_f$ and $T_1 > A_f$; the figure clearly shows that the model allows a better simulation of the extremity of the hysteretic region with respect to Figure 25.a.

3.2.3. Numerical results vs experiments

In this section, the accuracy and efficiency of the 1D numerical model are illustrated by comparing some experimentally measured hysteresis loops with the corresponding numerical predictions. The simulations have been carried out by using a model with 20 backlash operators and 5 dead zone operators to modify the loops in the extremity of the hysteretic region. Figure 26.a shows a comparison between the experimentally measured hysteresis loop

for a stress-free martensitic transformation, between the temperatures $T_0 < M_f$ and $T_1 > A_f$, and the numerically simulated one; the figure clearly shows a good accuracy of the numerical model with very small errors. In Figure 26.b, a comparison between numerical predictions and experimental results when the material is subjected to a tensile stress $\sigma = 100 MPa$ is shown. Also, in this case a satisfactory agreement is observed, but the errors increase with respect to the stress free condition. However, it is important to point out that the model parameters were identified by using the measured hysteresis loop under stress-free conditions, and by applying Equations 1 and 11 to modify both the PTTs and ε_{tw}.

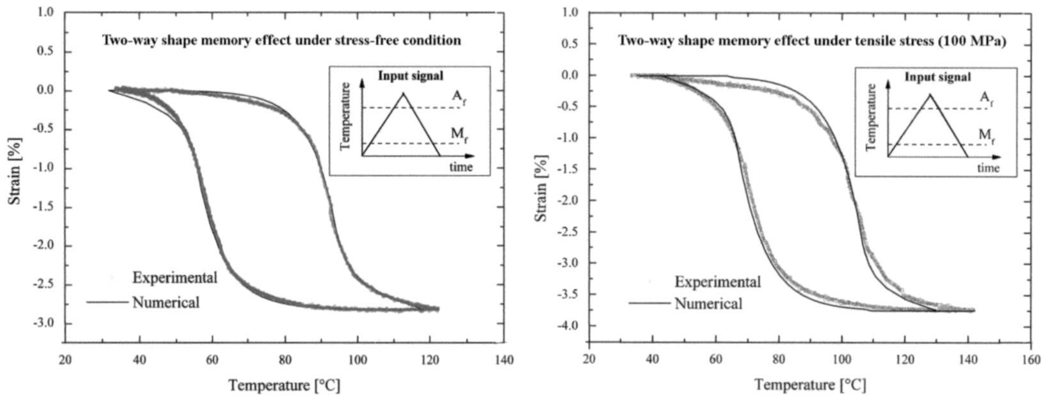

a) b)

Figure 26. Comparison between experimental measurements and numerical predictions for a thermal cycle between the temperatures $T_0 < M_f$ and $T_1 > A_f$ under a) stress-free condition and b) tensile strees of 100 MPa [28].

The accuracy of the model was also analyzed when the material is subjected to partial thermal cycles, *i.e.* to incomplete martensitic transformations. Figures 27.a and 27.b show the hysteretic behavior of the material for two different temperature-time paths in stress-free conditions; in particular, Figure 27.a shows incomplete $A \to M$ transformations, while Figure

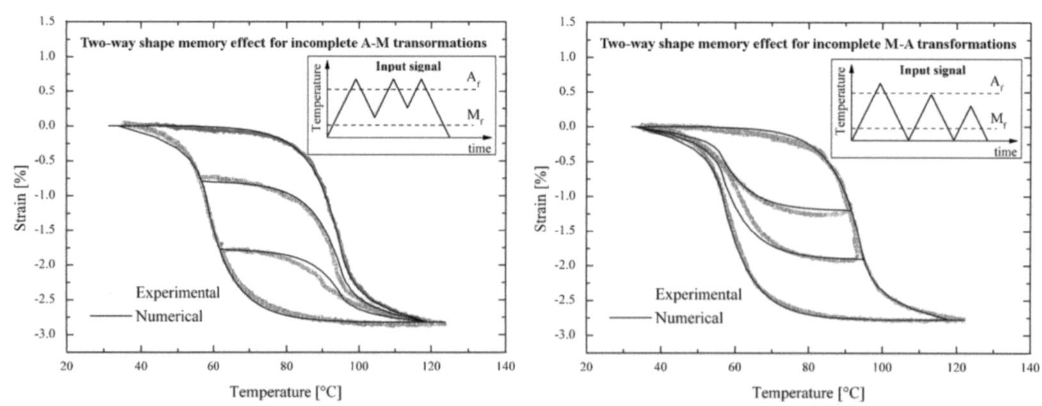

Figure 27. Comparison between experimental measurements and numerical predictions for two different temperature-time paths: a) incomplete $A \to M$ transformations and b) incomplete $M \to A$ transformations [28].

27.b illustrates incomplete $M \rightarrow A$ transformations. As shown in the figures, the comparison between experimental measurements and numerical predictions show a good accordance in both cases. It is worth noting that the same model parameters of the first example (Figure 26.a) were used in these numerical simulations.

Author details

Maletta Carmine
Assistant professor of Machine Design at University of Calabria, Department of Mechanical Engineering, Rende (CS), Italy

Furgiuele Franco
Full professor of Machine Design at University of Calabria, Department of Mechanical Engineering, Rende (CS), Italy

4. References

[1] Otsuka K, Ren X, (2005) Physical Metallurgy of TiÐNi-Based Shape Memory Alloys. Prog. Mater. Sci. 50: 511-678.
[2] Otsuka K, Wayman CM (1998) Shape Memory Materials. Cambridge University 559 Press, Cambridge.
[3] Wu Ming H (2002) Fabrication of Nitilon Materials and Components. Mater. Sci. Forum. 394-395:285-292.
[4] Schlossmacher P, Haas T, Shussler A (1997) Laser-Welding of a Ni-Rich Tini Shape Memory Alloy: Mechanical Behavior. J. Phys. IV. 7(5):251-256.
[5] Tuissi A, Besseghini S, Ranucci T, Squatrito F, Pozzi M (1999) Effect of Nd-YAG Laser Welding on the Functional Properties of the Ni-49.6 at. %Ti. Mater. Sci. Eng. A. 273-275:813-817.
[6] Theisen W, Schuermann A (2004) Electro Discharge Machining of Nickel-Titanium Shape Memory Alloys. Mater. Sci. Eng. A. 378:200-204.
[7] Falvo A, Maletta C, Furgiuele F M (2005) Laser Welding of a Niti Alloy: Mechanical and Shape Memory Behavior, Mater. Sci. Eng. A. 412:235-240.
[8] Falvo A, Maletta C, Furgiuele F M (2006) Functional Behavior of a Niti Welded Joint: Two-Way Shape Memory Effect. Mater. Sci. Eng. A. 481-482:647-650.
[9] Gall K A, Seitoglu H, Chumlyakov Y (2000) Niti Experiments Versus Modeling: Where Do We Stand?, Proc. SPIE. 3992:536Ð547.
[10] Paiva A, Savi M A (2006) An Overview of Constitutive Models for Shape Memory Alloys. Math. Probl. Eng. art. no. 56876.
[11] Falk F (1980) Model Free-Energy, Mechanics and Thermodynamics of Shape Memory Alloys. Acta Metall. 28(12):1773-1780.
[12] Falk F (1983) One-Dimensional Model of Shape Memory Alloys. Arch. Mech. 35(1):63-84.
[13] Tanaka K, Nagaki S (1982) Thermomechanical Description of Materials With Internal Variables in the Process of Phase Transitions. Ing. Arch. 51:287-299.
[14] Liang C, Rogers C A (1990) One-Dimensional Thermomechanical Constitutive Relations for Shape Memory Materials. J. Intell. Mater. Syst. Struct. 1:207-234.

[15] Brinson L C (1993) One Dimensional Constitutive Behavior of Shape Memory Alloys: Themomechanical Derivation With Non-Constant Material Functions and Redefined Martensite Internal Variable. J. Intell. Mater. Syst. Struct. 4:229-242.
[16] Bertran A (1982) Thermomechanical Constitutive Equations for the Description of Shape Memory Effects in Alloys. Nucl. Eng. Des. 74(2):173-182.
[17] Souza A C, Mamiya E, Zouain N (1998) Three-Dimensional Model for Solids Undergoing Stressinduced Phase Transformations. Eur. J. Mech. A/Solids. 17(5):789-806.
[18] Boydand J G, Lagoudas D C (1996) A Thermodynamical Constitutive Model for Shape Memory Materials. Part I. The Monolithic Shape Memory Alloy International Journal of Plasticity. 12(6):805-842.
[19] Auricchio F, Lubliner J (1997) A Uniaxial Model for Shape Memory Alloys. Int. J. Solids Struct. 34(27):3601-3618.
[20] Auricchio F, Sacco E (1997) A One-Dimensional Model for Superelastic Shape Memory Alloys With Different Elastic Properties Between Austenite and Martensite. Int. J. Non-Linear Mech. 32(6):1101-1114.
[21] Auricchio F, Taylor R L, Lubliner J (1997) Shape-Memory Alloys: Macro and Numerical Simulations of the Superelastic Behavior. Comput. Methods Appl. Mech. Eng. 146(3-4): 281-312.
[22] Marfia S, Sacco E, Reddy J N (2003) Superelastic and Shape Memory Effects in Laminated Shape-Memory-Alloy Beams. AIAA J. 41:100-109.
[23] Liew K M, Ren J Reddy J N (2005) Numerical simulation of thermomechanical behaviors of shape memory alloys via a non-linearmesh-free Galerkin formulation. Int. J. Numer. Meth. Engng 63:1014-1040.
[24] Amalraj J, Bhattacharyya A, Faulkner M G (2000) Finite element modeling of phase transformation in shape memory alloy wires with variable material properties. Smart Mater. Struct. 9:622-31.
[25] Krejci P, Kuhnen K, (2001) Inverse Control of Systems With Hysteresis and Creep, IEE Proc.: Control Theory Appl. 148(3):185-192.
[26] Kuhnen K, Janocha H (2001) Inverse Feedforward Controller for Complex Hysteretic Nonlinearities in Smart Material Systems. Control Intell. Syst. 29(3):74-83.
[27] Falvo A, Maletta C, Furgiuele F M (2007) Two-Way Shape Memory Effect of a Ti Rich Niti Alloy: Experimental Measurements and Numerical Simulations. Smart Mater. Struct. 16:771-778.
[28] Falvo A, Maletta C, Furgiuele F M (2008) A Phenomenological Approach for Real-Time Simulation of the Two-Way Shape Memory Effect in NiTi Alloys. ASME Journal of Engineering Materials and Technology. 130:771-778.
[29] Falvo A, Maletta C, Furgiuele F (2008) Hysteresis modeling of two-way shape memory effect in NiTi alloys. Meccanica. 43:165-172.
[30] Falvo A, Maletta C, Furgiuele F (2009) A phenomenological model for superelasticity in NiTi alloys. Smart Mater. Struct. 18:025005.
[31] Raniecki B, Lexcellent C H and Tanaka K 1992 Thermodynamic Models of Pseudoelastic Behavior of Shape Memory Alloys .Arch. Mech. 44 261-284.
[32] Hamilton R F, Sehitoglu H, Chumlyakov Y, Maier H J (2004) Stress Dependence of the Hysteresis in Single Crystal Niti Alloys. Acta Mater. 52:3383-3402.

[33] Sittner P, Landa M, Lukas P, Novak V (2006) R-phase transformation phenomena in thermomechanically loaded NiTi polycrystals. Mechanics of Materials 38:475-492

Chapter 6

Modeling, Compensation and Control of Smart Devices with Hysteresis

Daniele Davino, Alessandro Giustiniani and Ciro Visone

Additional information is available at the end of the chapter

http://dx.doi.org/10.5772/51388

1. Introduction

The chapter aims to provide a review on the basic issues concerning the techniques to handle innovative devices employing materials with hysteresis and the related control strategies.

Nowadays, the great availability of materials designed to gather different physical characteristics (electric, magnetic, mechanic, etc.) working together to provide improved capabilities, has determined a huge rise of devices able to integrate different "coupled" functionalities into the same device. In this respect, the set of these "smart" or "multi-functional" materials and the devices employing them is really huge in number and quality. With no claim of exhaustivity, we can recall, for sake of example, thermo-electric, [11], or magneto-caloric [33] effects, where a coupling between Entropy/temperature and electric or magnetic fields is observed, and which promise new devices for heat recovery or refrigeration tasks, through the exploitation of new and more effective materials.

Other materials, conversely, are able to couple mechanical quantities (i.e. stress and strain) to electric or magnetic fields so amplifying the application range for actuation or sensing aims. The most known and widespread are the piezo-electric alloys ($Pb[Zr_xTi_{1-x}]O_3$, $PbTiO_3$), which show increased piezoelectric properties so allowing to make smart devices, otherwise unrealizable, [34], [32], [2]. A similar behavior is shown by the electro-active polymers, [3] which, exploiting the characteristics of polymers, allow to foresee really innovative applications. Materials with magneto-elastic coupling show a complementary behavior with respect to the former and enable to further increase the already huge set of potential smart applications. It is worth to recall the magnetostrictive materials, such as Terfenol-D or Galfenol, [21], exploited for actuation or sensing purposes, or the Ni-Mn-Ga alloys, which presents different magneto-elastic behavior and, even if less assessed than classical magnetostrictives, promise very interesting application, [37].

Omitting electro- or magneto-rheological materials and many others, we would finally mention Ni-Ti alloys showing a thermo-mechanical coupling by the exploitation of the Austenite-Martensite phase transition that enables to realize further and interesting smart devices, [31], [4]. In summary, the interest and attention of researchers and inventors is so

focused on these materials and their applications, that everyone can experience, navigating on the web, an impressive "blooming" of ideas and proposals. The key element is the consciousness that smart materials achieve new functionalities going well beyond the "sum" of the individual properties and allowing to develop new and really innovative devices.

Among them, those materials coupling mechanical to other physical quantities (thermal, electric or magnetic) received a special attention since their suitability for sensing or actuation goals. The latter involved many researchers in a multi-disciplinary frame in designing and producing really smart actuators, able to provide high forces, or high precision *micropositioning*, or high speed responses, in dependence of the selected application.

In particular, as we can find in the scientific literature, the employ of Piezo-electric materials, [23], or magnetostrictives, [21], which are able to cover complementary sets of applications, play a primary role and many devices suited for micropositioning, active vibration control, smart actuation, ultrasonic generators, are available.

Similar conclusions can be drawn for Shape Memory Alloys (SMA) which show huge deformations driven by temperature variations and therefore are generally quite slow [45]. The Ni-Mn-Ga materials, also referred to as Ferromagnetic Shape Memory Alloys (FSMA), [40] overcome this limitation still preserving high deformations as SMAs.

However, all of them, share *rate independent* memory properties which strongly affect the global behavior of the device and its performances. Rate independence means that the observed memory behavior doesn't arise from "dynamics" and therefore is still kept also for "quasi-static" input variations, as happens in the well-known behavior of ferromagnetic materials. The common way to refer to it is *hysteresis*.

In the sequel, a strategy to "handle" devices employing smart materials with hysteresis in general working conditions will be outlined and several applications employing magnetostrictives will be discussed to check the validity of the proposed techniques.

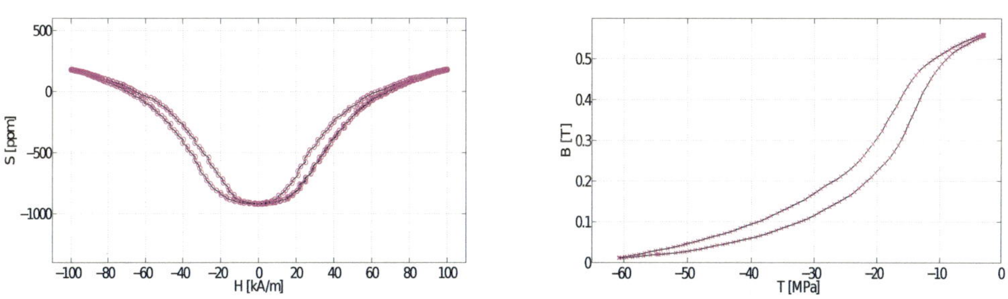

(a) Magnetostriction at constant stress (b) Inverse Magnetostriction at constant field

Figure 1. Elastic and magnetic response of a Terfenol-D sample to a magnetic (a) or mechanical (b) input

2. Hysteresis operators and their basic properties

The general approach to materials showing nonlinear coupling and hysteresis between the involved variables, can be written as

$$y_1 = \mathcal{F}_1(x_1, x_2), \tag{1}$$

$$y_2 = \mathcal{F}_2(x_1, x_2), \tag{2}$$

being (x_1, x_2) the couple of input variables, and (y_1, y_2) the couple of output variables. Of course, the choice of input and output variables is, in principle, arbitrary and is performed according to the specific application needs. Moreover, $\mathcal{F}_1$ and $\mathcal{F}_2$ represents arbitrary and general operators with memory, which "link" the input to the output functions. This is a quite generic picture, by which we would just acquaint the reader on the complex phenomena lying below the behavior of such materials. In particular, we would evidence that, generally speaking, equations above define a *multi-variate operator* with memory which could be handled by means of *vector operators* with hysteresis, [28], [25]. Unfortunately, they require huge efforts in identification procedures and high computational weight, so resulting unsuitable for *sensing* or *actuation* tasks. As a consequence, other approaches have been considered with the aim to gather good generality and modeling capabilities to a relative simple handling and computational effort, [28], [1], [42]. There, the basic concern was to fully exploit the "machinery" of the classical hysteresis operators, well suited to link together *one input* to *one output*. They are usually referred to as SISO (Single Input-Single Output) systems. As a starting point, any functional material can be modeled as a SISO system whenever only one of the output variables is of concern and, moreover, one of the inputs is kept constant. For the sake of example, Fig.1-(a) and -(b) sketches the typical elastic response driven by magnetic field or the flux density induced by applied stress, respectively. In each of these pictures, a specific characteristic with memory is described and can easily be modeled through a classical SISO hysteresis operator. In a more general framework, as shown in Fig.2-(a) and -(b), a SISO modeling is still applicable whenever the interest is only on one of the output variables (i.e. the *strain* in picture -(a)), and when *stress* can be assumed as stationary (same picture). Of course, the same reasoning can be applied when the *flux density* as output variable is of concern (picture -(b)). A quite unifying approach to define a wide class of hysteresis operator

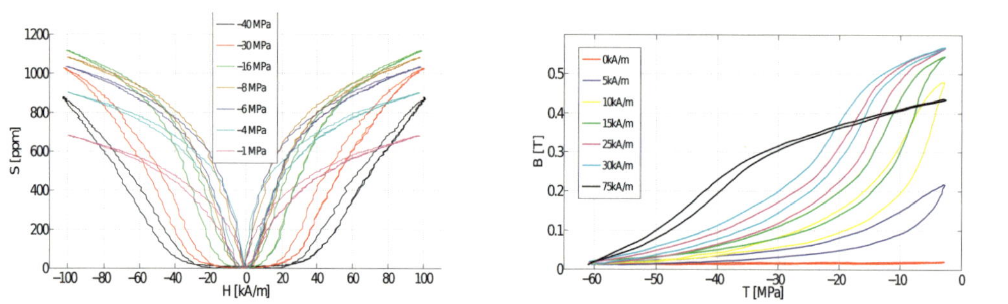

(a) Magnetostriction as function of stress

(b) Inverse Magnetostriction as a function of magnetic field

Figure 2. Elastic and magnetic response of a Terfenol-D sample to a magnetic (a) or mechanical (b) input at various magnetic fields and stresses

is based on the *Preisach memory* updating rules, [28], [25] and we refer to them in the sequel. For the sake of simplicity and assuming some basic knowledge on the Preisach operator, we can introduce the Preisach hysteresis operator as the superposition of a continuous set of ideal relay operators, $\hat{\gamma}_{rs}$ defined by the parameters $r \in \mathbb{R}^+$ and $s \in \mathbb{R}$ as specified in the Fig. 3. Here it should be noticed that the *relay* can be represented by means of the r and s parameters, or by its switching fields a and b ($a \geq b$), equivalently. In the latter case the formalism is the one exploited by I.D. Mayergoyz in [28], while the former yields to the formalism used in the books [7, 26] and one can be derived from the other by a simple 45° axis rotation, [44]. In the

sequel we decided to use the second approach which allows the reader to refer to those books so exploiting several results there addressed. On the contrary, this choice has the drawback to discuss some tool of the Preisach formalism (i.e. identification, or inversion algorithms) in a less natural way. For this reason we would further evidence the aim of this chapter, focused in providing a sufficiently complete discussion concerning the modeling and compensation of systems with hysteresis and referring for details to the cited references.

Each *relay*, referred to $R_{r,s}[x]$, has $r \geq 0$ and $s \in \mathbb{R}$. We assume now that they are distributed into the $\mathbb{R}^+ \times \mathbb{R}$ half plane with a prescribed distribution function $\mu(r,s)$ and the output of this set is the linear weighted superposition of the response of each *relay*. In symbols, [28], [25], [26]:

$$y(t) = \iint_{\mathbb{R}^+} \mu(r,s) R_{rs}[x](t) dr ds \tag{3}$$

Of course the past input history of each relay of the distribution results in a non trivial way to compute the integral. However, this problem can be overcome by giving a geometric interpretation to it. First of all, let us associate to each point of the half plane (r,s) the relay having the same parameters. Moreover, recall that each relay of the distribution switches from the state -1 to $+1$ if the input field exceeds the upper switching point, that is $x(t) \geq s + r$, while the opposite transition from positive to negative output is performed if the field is below the lower switching point or, in symbols: $x(t) \leq s - r$. This yields to the following conclusion: the relay operators switched to the $+1$ state are those below the straight line of equation $s + r = x$; conversely, those switched from the state $+1$ to -1 lie above the line $s - r = x$. If all the relays start from the state -1 (usually referred as the state of *negative saturation*) and applying the input field $x(t)$, the line separating the $+1$ from the -1 elements is sketched in Fig. 3-(b) and represents the state of the system. When, from such initial state,

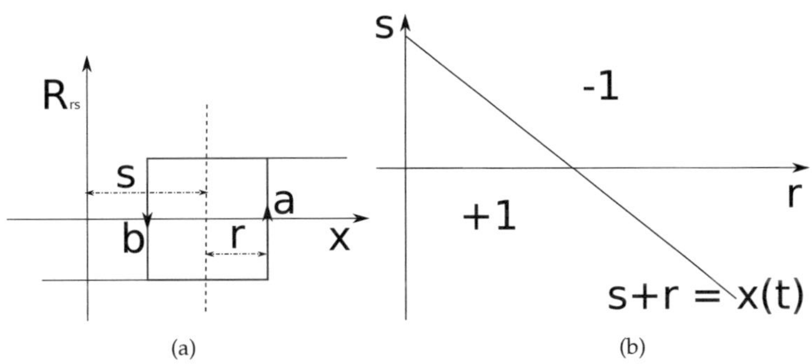

Figure 3. (a) The ideal relay with parameters r and s; (b) geometrical interpretation of the relay switching

the input is decreased, the relays above the line $s - r = x(t)$ switch down to -1. The result is sketched Fig. 4-(a), while in Fig. 4-(b) it is represented the state corresponding to an arbitrary *non-monotone* input variation. The *demagnetized* or *virgin state* is represented in the $\mathbb{R}^+ \times \mathbb{R}$ half plane as the horizontal axis ($s = 0$). We will refer to it as the h_0 state.

After this brief resume, it should be quite easy to realize that the above mentioned rules define a specific relation between the set of continuous input functions, $x(t)$ and the piece-wise linear function $h(r,t)$, as follows:

$$\Theta : x \in C^0[0,T] \rightarrow h(r,t) \in \Psi_0, \tag{4}$$

with, h the staircase of the classical Preisach model (cfr.[28]), and Ψ_0 the set of admissible Preisach states. The properties of this operator, referred to as Preisach Hysteresis Generator (PHG), can be easily sketched. To this aim, let us consider the functions $x(t)$ and $z(t)$, shown in Fig. 5-(a), sharing the same sequence of input extrema and linked by the equation $x(t) = z \circ \rho(t)$. The function ρ is a time rescaling and $\circ$ the composition operator. In other words the two functions differ only by the rate of variation. It is easy to verify that the staircases corresponding to x and z coincide (Fig. 5-(b)). The memory operator, therefore, is not affected by the input rate, but only by the sequence of the input extrema.

Let us now consider the memory operator with initial state equal to the h_{-1} *virgin state*, i.e. $h_0 = h_{-1}$ undergoing the input $x(t)$, shown in Fig. 5-(a). The vertex of the staircase (x_3, x_4) are wiped out as soon as the input exceeds x_3 as they were never existed, as shown in Fig. 5-(b). The final state generated by the sequence of input extrema $(x_1, x_2, x_3, x_4, x_5)$ coincides with the one generated by the *reduced* input sequence (x_1, x_2, x_5), [41]. The latter is the basic rule for memory updating which is universally referred to as *Preisach deletion rule*. In conclusion, the Preisach-memory operator is *rate independent* and fulfills the *wiping-out* property. What we have just introduced is the machinery enabling the classical Preisach operator, defined in eqn. (3), to take into account *rate-independent* memory phenomena. To

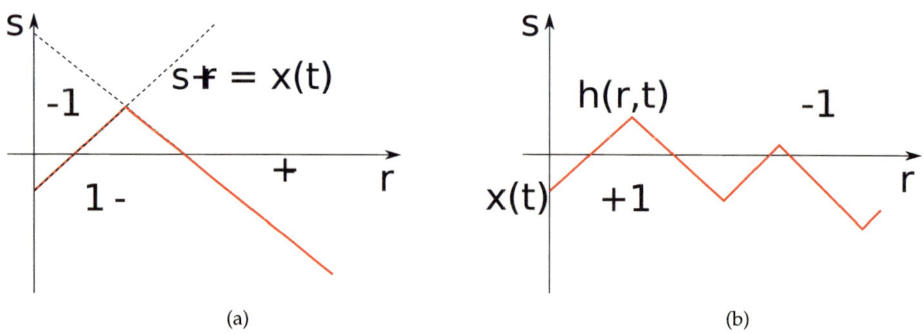

Figure 4. (a) staircase in the case of non-monotone input variation (b) Staircase corresponding to a non-monotone input $x(t)$

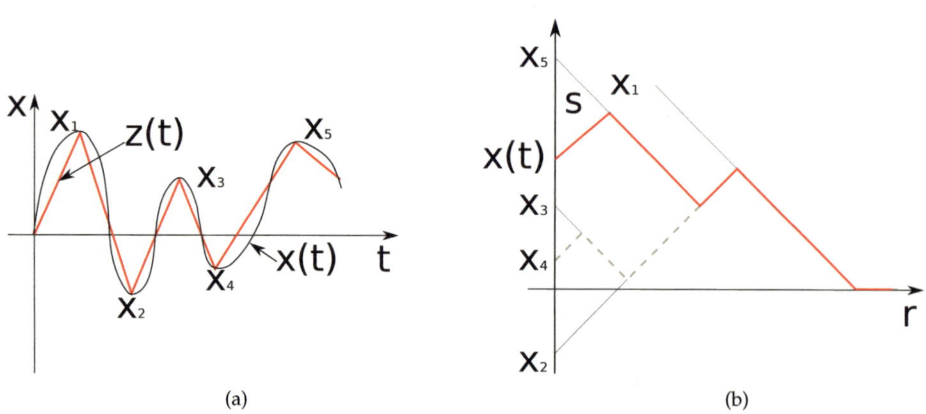

Figure 5. (a) staircase in the case of non-monotone input variation (b) Staircase corresponding to a non-monotone input $x(t)$

fully exploit such operator, it is possible to define a larger class of operators with hysteresis, based on the Preisach memory updating rules, as proposed in [7], where any operator with Preisach memory can be represented as follows:

$$\mathcal{W} = \mathcal{F} \circ \Theta, \tag{5}$$

where Θ is the PHG, while

$$\mathcal{F} : h(r,x) \in \Psi_0 \to y(t) \in \mathbb{R}. \tag{6}$$

represents a memoryless functional by which a wide class of hysteresis operators, based on the Preisach memory, can be defined. It should be noted that this formalism, adopted in [7, 26], is slightly different with respect to [28], representing the Preisach plane by means of a rotated coordinate system, [44]. For the sake of example, the following functional

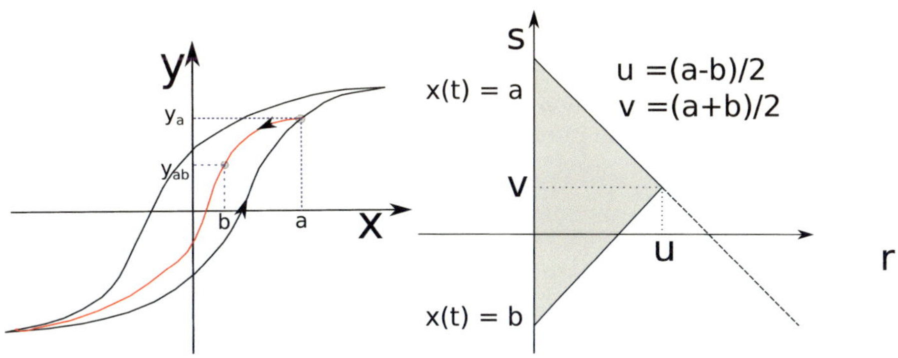

(a) Definition of first order reversal curve (b) Triangular domain over which the Everett integral is defined

Figure 6. Identification procedure of a Preisach model, based on the exploitation of first order reversal curves

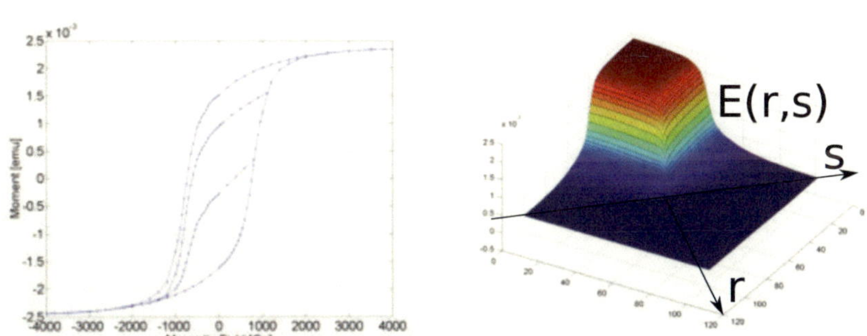

(a) Characteristic of a floppy disk (b) Everett function of a floppy disk

Figure 7. Example of the identification of the Everett function from experimental data coming from a floppy disk

$$\mathcal{F} : h \in \Psi_0 \to \int_0^{+\infty} dr \left(\int_{-\infty}^{h} \mu(r,s)ds - \int_h^{+\infty} \mu(r,s)ds \right), \tag{7}$$

defines the Classical Preisach operator, and $\hat{\mu}(r,s)$ is a density function to be determined. The functional

$$F : h \in \Psi_0 \rightarrow \int_0^{+\infty} \mu(r,s)h(r,t)dr, \tag{8}$$

conversely, defines the well known Prandtl-Ishilinskii operator [7]. Such models have been widely exploited in last years by several authors to compensate the memory of real devices, so allowing a model-based control approach. A non exhaustive review can be found in [44]. It should be quite evident that these models are phenomenological in nature and so require well-defined identification procedures to link their parameters (i.e. the mapping functional, F) to the real material, through a suitable set of measured data. It should be said that such a procedure is not addressed for any arbitrary functional, while, in the case of the Preisach operator, a simple and well-behaved identification procedure is available and, for convenience it is reported with some detail hereafter.

In particular, it could be observed that in order to compute the output of a classical Preisach model, it is not necessary to know the Preisach Distribution Function, $\mu(r,s)$ [28], but it is sufficient the knowledge of the integrals of $\mu(r,s)$ over any triangle of the $\mathbb{R}^+ \times \mathbb{R}$ half plane and vertex (u,v), according to the procedure described below. Preliminary, let us define the following function known as *Everett function*, as follows:

$$\mathcal{E}(u,v) = \int_0^u dr \left(\int_{v-u+r}^{v} \mu(r,s)ds + \int_v^{v+u-r} \mu(r,s)ds \right). \tag{9}$$

The link between the actual measured data and the Everett function can be outlined as follows. Let us start from negative saturation and increase the input until the value $x(t) = a$ and the corresponding output is y_a are attained. Then decrease the input until the value $x(t) = b$ is reached; the corresponding output is referred to as y_{ab}. The hysteresis branch (Fig. 6-(a)) traced from the reversal point y_a until, again, the negative saturation is known as *first order reversal curve*, or *for*-curve. It is easy to realize (cfr. also [28]) that the following relation linking the output measured data and the Everett integral holds:

$$\mathcal{E}(u,v) = \frac{y_a - y_{ab}}{2} \tag{10}$$

with

$$u = \frac{a-b}{2} \tag{11}$$

$$v = \frac{a+b}{2}. \tag{12}$$

Let us subdivide the input interval into N equal intervals of width $\Delta x = x_{max} - x_{min}$ and assume $x_1 = x_{max} = N\Delta x$, $x_k = (N-k+1)\Delta x$. The corresponding reversal curve (Fig. 6-(b)) so has $N-k+1$ measured points, $y_{a_k b_n}$. Hence the total number of measured points to be used for the identification procedure, are $N \times (N-1)/2$, which correspond to the number of nodes in the grid in the $\mathbb{R}^+ \times \mathbb{R}$ half plane. Now, if we associate the value of the Everett integral to each point of the grid, a discrete version of the Everett function is easily available. The details on its construction and use can be found in [28]. An example of the reversal curves for a real material, and the corresponding Everett function is shown in Fig. 7.

3. Compensation algorithms

The definition of suitable control algorithms to be embedded in actuators employing smart materials is sensibly affected by the memory phenomena involved. This because, roughly speaking, the controller has no memory of past history which, conversely, affects the plant behavior, and will be not able to provide suitable correction to track a desired trajectory. The system therefore will show poor performances. A simple way to overcome this problems, proposing quite a simple control strategy, is based on the following well known issues. As a preliminary step, let us assume that the plant is adequately modeled by the series connection of a hysteresis operator and a linear dynamic component. In Fig. 8-(a) it is shown a control scheme employing a generic nonlinear controller which gather both dynamics and *rate-independent* properties, even if specifying its details is non-trivial. Let us assume that the *rate independent* behavior of the plant can be modeled by the aid of a hysteresis operator $W[x]$, as defined in section 2. If the inverse of the hysteresis operator partly modeling the plant were available, their series connection in the control scheme bears to a linear system, allowing the employment of all standard linear control strategies, as shown in Fig. 8-(b). In summary, in order to simplify the controller design, it is suitable to compensate nonlinearity and *rate-independent* memory of the actuator. This approach is generally referred to as *model-based* control strategy. The issue is now to address the definition of the compensator of a hysteresis

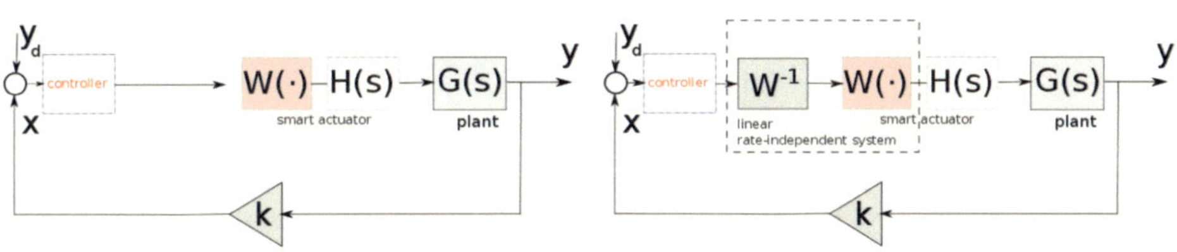

(a) The general controller should take into account memory effects

(b) Model-based control scheme: a suitable modeling of the plant is performed which enable to provide a compensator

Figure 8. General control scheme (a) - Model-based control scheme (b)

operator and to define the conditions under which such an operator admits an inverse or *compensator*,[7, 27]. In particular:

Definition 1. *A hysteresis operator $\mathcal{W}^{-1}$ with initial state $\phi_{-1} \in \Psi_0$, is called a* compensator *(or inverse) of the operator $\mathcal{W}$, with initial state ψ_{-1} if, for any state $\phi \in \Psi_0$, there is a state $\psi \in \Psi_0$, such that $\mathcal{W}_\psi \circ \mathcal{W}_\phi^{-1} x(t) = \mathcal{W}_\phi^{-1} \circ \mathcal{W}_\psi x(t) = x(t)$, for every input function x(t).*

Of course, the above definition lead us asking the conditions making this issue sound. This requires to specify the hysteresis operator with more detail. To this aim, we refer to the class of operators based on the Preisach deletion rule, stated in eqn. (5). We refer to this operators, to as $\mathcal{P}_0$-operators. In particular, we can define a piece-wise strictly monotone operator of this class, [7], according to the following

Definition 2. *The operator $\mathcal{W} = \mathcal{F} \circ \Theta$ with initial state ψ_{-1} mapping the set of continous functions, $C[0,T]$, into itself, is piece–wise strictly increasing if*

$$[\mathcal{F}(\Theta(x,\psi) - \mathcal{F}(\psi)] \, [x - \psi(0)] > 0, \quad x \neq \psi(0), \tag{13}$$

where $\psi(0)$ is the past input, while the symbol $\Theta(x, \psi)$ is the staircase attained by starting from ψ and applying the value of input x at time t.

Such a property guarantees the invertibility of $\mathcal{P}_0$-operators and is connected to the specific memoryless mapping functional $\mathcal{F}$. It turns to the classical definition of monotonicity for the (single valued) hysteresis branch when the state (i.e. the staircase in the Preisach plane) is specified. In the particular case of the classical Preisach operator, with mapping functional specified in eqn. (7), the following theorem holds [6]:

Theorem 1. *The Preisach operator, $\mathcal{H} \in \mathcal{P}_0$, in eqn. 3 with distribution function μ, admits an inverse if and only if $\mu \geq 0$, and $E(u,v) \geq 0$, for every $(u,v) \in \mathbb{R}^+ \times \mathbb{R}$.*

Before discussing an efficient algorithm to implement the inverse of Preisach operator, it is important asking why the inverse operator cannot be employed by just exchanging the role of input and output.[1] In principle, this is not accurate, due to a specific property of the Preisach mapping functional, known as *congruency*. Let us refer to Fig. 9-(a), where it is sketched the state updating corresponding to the same *up and forth* input variation (from x_- to x_+) for two different states h_1 and h_2. It is quite evident that two minor loops are generated, each corresponding to an initial state, and they are *congruent* by vertical translation, Fig. 9-(b). The inverse operator describe a characteristic which is placed symmetrically to the direct characteristic, with respect to the $y = x$ line. As a consequence, any minor loop described by the model is transformed into one which is symmetrically placed with respect the same line.This implies that two congruent minor loops generated by the model, are transformed by the inverse operator into two minor loops which are congruent by a horizontal translation. Fig. 10-(a) illustrates such a situation. Conversely the operator which just exchanges the

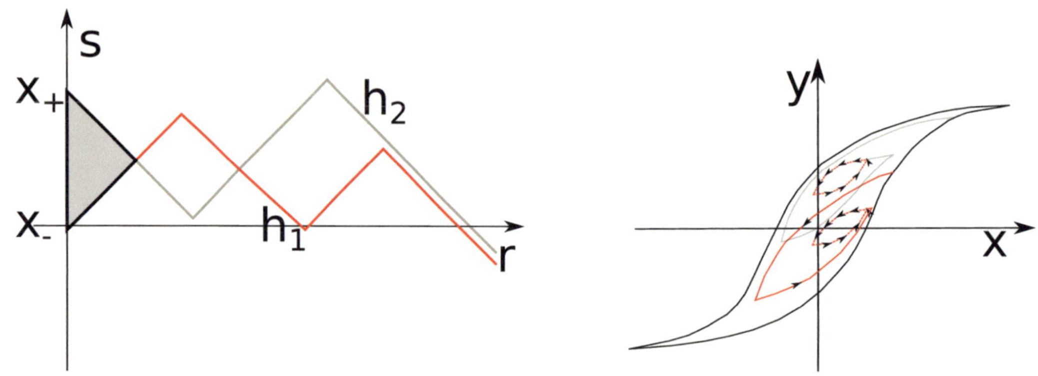

(a) States corresponding to the same up and forth input variation (b) Congruent minor loops

Figure 9. Illustration of the Congruency property of the Classical Preisach Problem

role of the variables that we call *pseudo-compensator* is still a Preisach operator and therefore generates vertically congruent minor loops, as shown in Fig. 10(b). It is so manifest that this operator cannot be strictly considered as the inverse of a Preisach operator, due to the limitations imposed by the congruency property. However this observation doesn't limit the possibility to adopt pseudo-compensator as a hysteresis model by itself which is able to

[1] In reality this is a procedure to define and approximation of the inverse, referred to as *pseudo-compensator* as detailed in [30]

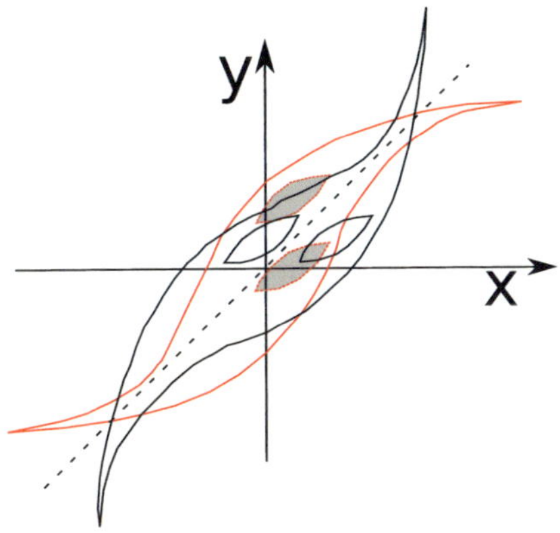

(a) States corresponding to the same up and forth input variation

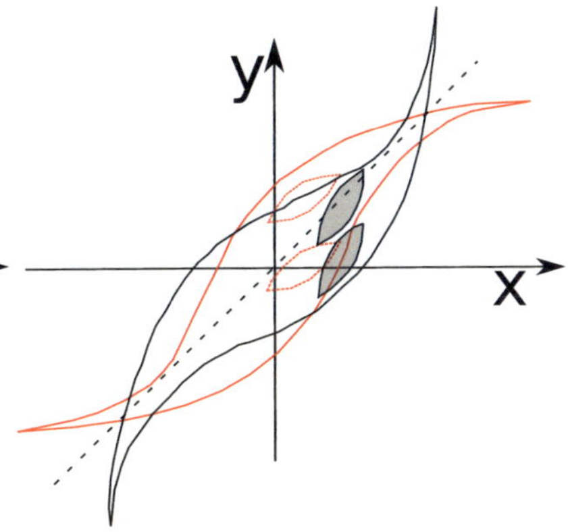

(b) Congruent minor loops

Figure 10. Illustration of the Congruency property of the Classical Preisach Problem

guarantee good performances, as widely described in [20, 29]. In summary, the *inverse* of a classical Preisach operator *is not* a Preisach operator; the pseudo-compensator, conversely, *is* a Preisach operator and cannot rigorously represent the Preisach compensator.

It should be stressed that even if the inverse of the Preisach operator exists, it is not possible, generally speaking, to represent the compensator in a closed form, except than in very specific cases, [44], [26], [27], [43]. So it is mandatory the definition of well-defined and computationally efficient inversion algorithms in general working conditions. To this aim we will briefly sketch the issue, by referring to [12, 44] for details. When the conditions of invertibility of the Preisach operator are fulfilled the problem of finding its inverse has sense and the inversion procedure is based on the algorithm of state updating which assumed the output $y(t)$ as known. In other words, the algorithm described hereafter, allows to provide the input field of the operator at time instant t, corresponding to a given and known output $y(t)$, once the initial state is assigned. This procedure, since no numerical iterations are needed, allows to compute the inverse with the same computational efficiency and accuracy of the classical algorithm used to compute the Preisach operator. For this reason we refer to it as a "fast" inversion algorithm, [13] [14]. If the Preisach plane is divided into a $\Delta x \times \Delta x$ mesh, with $\Delta x = \frac{x_{max}-x_{min}}{N}$ then the Everett function is uniformly sampled into a $(N+1) \times N/2$ mesh, as sketched in the Fig. 11. Therefore the solution of the above equation can be performed by a simple inspection of a look-up table, according to the rules described below. Moreover, let us assume that the Everett integral is limited to the triangle ABC. This is a quite reasonable hypothesis, since the function must vanish at infinity. In the same plane, it can be evidenced the initial state of the operator, $h_0(r,t)$. Let us further assume the output $y(t) = y_0 + \Delta y$, is known. In the ordinary fashion of handling the Preisach operator, the output variation Δy coincides with the integral of the distribution function over the area S, that is:

$$\iint_S \mu(r,s)dr\,ds = \mathcal{E}(r,s) - \iint_{S^*} \mu(r,s)dr\,ds. \tag{14}$$

By exploiting the wiping out property, all the vertex of the staircase such that the constraint

$$\Delta y \geq \mathcal{E}_k(r,s) - \iint_{S^*} \mu(r,s)dr\,ds. \tag{15}$$

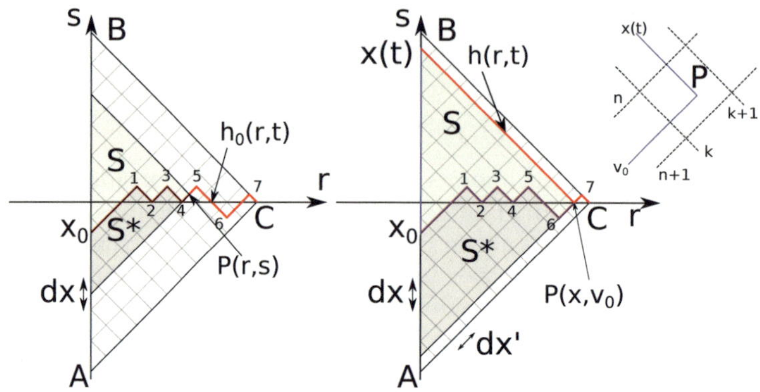

(a) State updating algorithm (b) Input computation by interpolation over the grid points

Figure 11. The algorithm of inversion of the Preisach operator

is fulfilled, are cancelled. The suffix k of he Everett function evidence the vertex of the triangle. In the case reported in Fig.11-(a) the initial state h_0 represented by the sequence of dominant extrema $\{0,1,2,3,4,5,6,7\}$ is updated and the vertex $\{0,1,2,3,4\}$ cancelled. So the state updating algorithm consists in applying the above condition and cancel all the vertex until it is verified. Once the last vertex has been cancelled, the following condition hold:

$$\Delta y = \mathcal{E}(x, v_0) - \iint_{S^*} \mu(r,s) dr\, ds. \tag{16}$$

being x the unknown value of the input field to be determined. In Fig. 11-(b) it is shown how $x(t)$ wipes out the vertex of the state function, and the operator attains the new state $h(r,t) = \{x, P, 7, C\}$ evidenced in red in the same figure where it is also shown a zoom of the new state around the point $P(x, v_0)$ lying within a square determined by the points $(n+1, k)$, $(n+1, k+1)$, (n, k) and $(n, k+1)$. As a first step, the Everett integrals at the points $(k+1, v_0)$ and (k, v_0) are found by linear interpolation:

$$\mathcal{E}(m, v_0) = \mathcal{E}(m, n) - (\mathcal{E}(m, n) - \mathcal{E}(m, n+1)) \frac{v_0 - n\Delta x'}{\Delta x'} \tag{17}$$

being $m = k+1$ and $m = k$. Finally, once these values have been determined, and exploiting eqn.(16) the final output value can be written as:

$$x' = \frac{\Delta y + \iint\limits_{S} \mu(r,s) dr\, ds - \mathcal{E}(k, v_0)}{\mathcal{E}(k+1, v_0) - \mathcal{E}(k, v_0)} \Delta x' + k\Delta x', \tag{18}$$

where $x' = x/\sqrt(2)$ and $\Delta x' = \Delta x/\sqrt(2)$. For this reason the equation does not change when the variables without index are concerned. Moreover, it should be stressed that the integral in the previous equation is a known quantity. Such an approach requires only three linear interpolations at most and does not need any lengthy iteration. The procedure only requires the knowledge of the matrix storing the samples of the Everett integrals and its precision is linked to the mesh adopted for the Preisach plane or, in other words, to the number of measured samples used for the identification. Moreover, the algorithm, as already mentioned,

does not increase the computational weight of the algorithm which is equivalent to that of the direct operator.

4. Multi-variate systems with hysteresis

Smart materials, in general application conditions, cannot be considered as SISO systems, since each output quantity is a function of at least two input functions. For sake of example, in a magnetostrictive material the strain depends both on the mechanical load (stress) and magnetic field. A similar behavior can be experienced for the magnetic *flux density*. The behavior is illustrated in Fig. 12-(a) where the strain is affected by both the stress and field variation. Moreover, not only the H-field affects the internal state of the system, which is also altered by the applied stress, as shown by the hysteresis branches sketched in Fig. 12-(b). The experimental evidence witnessed by Figs. 2 and 12, allows to conclude

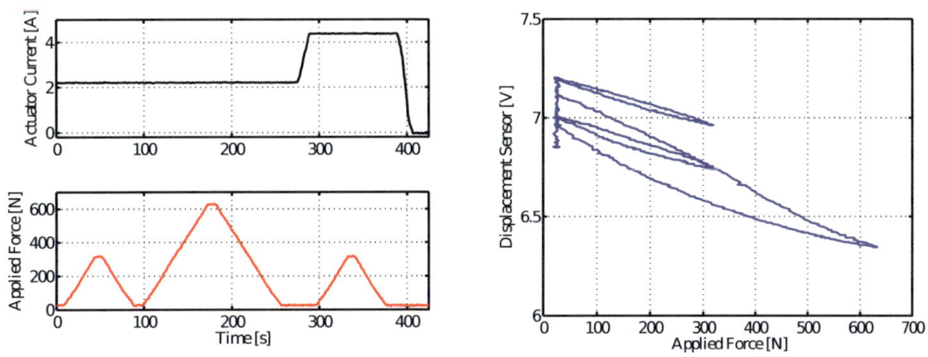

(a) (b)

Figure 12. Applied current and force vs. time to a magnetostrictive sample (a); displacement/force hysteresis branches (b).

that the modeling of these systems requires a generalization of all the classical approaches to hysteresis phenomena. This need started two decades ago [1], [46] and yielded to models able to couple together two different input variables (i.e. ε and σ). This effort become of fundamental importance whenever the coexistence of non stationary stress and fields cannot be overlooked.

The simplest way to define more general models is to keep the usual SISO structure of systems with hysteresis, as the ones discussed in the previous section. To this aim, let us consider the mapping:

$$\zeta : (u, v) \in C^0[0,T] \times C^0[0,T] \to x \in C^0[0,T], \tag{19}$$

where ζ is the operator relating the couple of continous functions (u, v), affecting the system behavior, to the input, x, of the memory operator, Θ. For the sake of example, in the case of a magnetostrictive material, ζ "links" the stress σ and magnetic field H to the input of the memory operator. Therefore

$$\mathcal{W} = \mathcal{F} \circ \Theta \circ \zeta, \tag{20}$$

represents a hysteresis operator with two inputs which generalizes the definition in [5] to a wider class of hysteresis operators. Before proceeding in the description of multi-variate systems with hysteresis in control applications, it is mandatory to provide some information

concerning its identification. To this aim, let us now consider the following operator, specified for a magneto-elastic material

$$y = \digamma \circ \Theta \circ \zeta(x_1, x_2) + q(x_2), \tag{21}$$

where $y = \varepsilon$, $x_1 = H$, and $x_2 = \sigma$, being ε, H, and σ the strain, magnetic field and stress experienced by the material, respectively. Moreover, $q(\sigma)$ is a pure elastic response to be identified with almost trivial mechanical measurements at zero magnetic field. Even it is not strictly necessary, the latter assumption allows to take into account separately the effects of vertical translation of the magnetostrictive response due to pure mechanical reasons. Let us preliminary assume that the set of $j = 1, .., M$ anhysteretic[2] curves, e_j shown in Fig. 13-(b) is available. Moreover, assume that the interval of input variation ($x_2 = \sigma$) is $[\sigma_{min}, \sigma_{max}]$ and that the magnetostrictive response experiences a monotonic decrease for increasing compressive stresses, that is $\sigma_{min} = \sigma^*$. With reference to the figure, this corresponds to a 1.05 MPa applied mechanical stress. Finally, each curve is discretized into N samples such that $H_{i+1} - H_i = \Delta H$, with $i = 1, ..N$. Assuming further that $\zeta(H, \sigma^*) = H$ and $\lambda(\sigma^*) = 0$, if

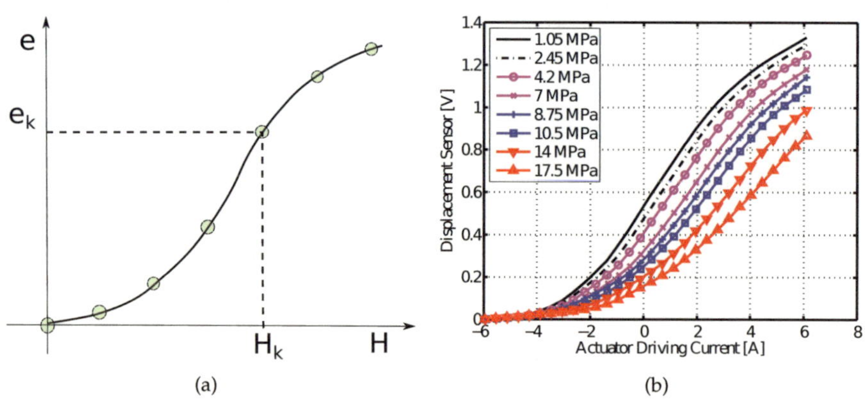

Figure 13. Example of *Anhysteretic* curve (a) Experimental Anisteretic curves for a magnetostrictive material (b).

$\sigma = \sigma^*$, then the operator takes the following classical form

$$e(t) \equiv \varepsilon(t) - q(\sigma) = \digamma \circ \Theta[H], \tag{22}$$

which could be easily identified, according to the procedure sketched in sect. 2 and detailed in [28]. Once the Preisach distribution function is known, the procedure to identify the $\zeta(\cdot, \cdot)$ function can start. Recalling now that in suitable conditions $\digamma \circ \Theta$ admits an inverse and that an efficient inversion algorithm is available, as shown in section 3, the eq. (22) can easily be re-arranged as:

$$\Gamma^{-1}[\varepsilon - q(\sigma)] = f(H, \sigma). \tag{23}$$

By this procedure, a set of $M \times N$ measured samples is now available, which enable us finding the samples $\zeta(H_i, \sigma_k)$ of the unknown function.

[2] It is defined as *anhysteretic*, the curve in the input-output plane, as in Fig. 13-(a), each point of which corresponds to the output of the $\mathcal{P}_0$ operator in the state represented by the horizontal curve $s = H_k$, that can be constructed by applying to the system the input $x(t)$ slowly converging to H_k

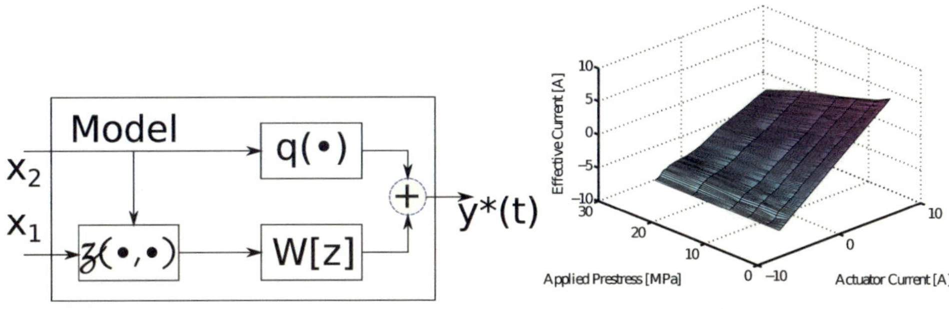

(a) (b)

Figure 14. Block scheme of the 2-variables operator (a); Example of reconstruction of the ζ function, named as "effective current" in the vertical axis of the plot (b).

In order to show the effectiveness of the described procedure, it is important now to show the performances of a multi-variable model as the one described in eqn. (23). In particular, in Fig. 14-(b) it is shown the reconstructed ζ function, while in Fig. 14-(a), the scheme of the model is shown. It undergoes the input variable x_1 and x_2, according to eqn. (23), which in the specific case are the magnetic field and the stress, respectively. The output $y^*(t)$ is compared to the actual measured strain function and the results are sketched in Fig. 15. In particular, in Fig. 15-(a) it is shown the time variation of the input variables, i.e. $x_1(t) = H(t)$ and $x_2(t) = \sigma(t)$, in the first two frames, while in the third, it is described the output of the ζ-function, to which the hysteresis operator $\mathcal{W}$ is subject. In Fig. 15-(b) it is shown the system response, compared to the time behavior of the measured strain. Here the performances of the model described so far are compared to those of a model proposed in [10], where the dependence of memory on the stress is not taken into account.

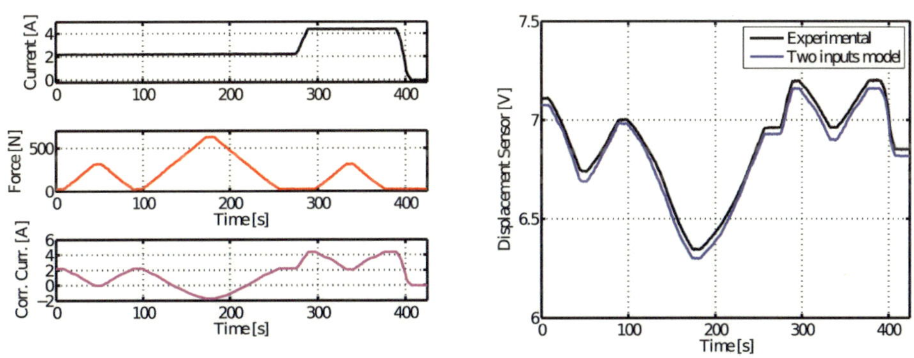

(a) (b)

Figure 15. Comparison of the model's behavior to experiments for a prescribed history. Stress (load), magnetic field (curr) and ζ (corr-curr) vs. time (a); output displacement compared to the experimental data (b).

In conclusion of this section, it is worth to be further stressed that the procedure described so far is of general breath and can be exploited in modeling and compensation of any material/system with multi-variable hysteresis as most of multifunctional materials behave.

5. Model-based control strategies for smart actuators

After a complete review on the modeling and compensation strategies of actual systems with hysteresis, in general working conditions, has been settled, it is possible to discuss different approaches to their control through the definition of specific application frameworks. As a first step, let us start from the classical *model based* control approach, already sketched in Fig. 8, sect. 3

There, the employment of the compensation algorithm allows a easy design of the controller, which could be, as a first attempt, a classical PI system, with all its limits, but with the great advantage of its implementation neatness. This viewpoint has been widely used in smart actuation tasks where Piezo, SMAs or magnetostrictives were employed, [23, 24, 27, 35, 39]. In Fig. 16-(a) the whole control system is shown. In this figure, the block $D(s) = H(s)G(s)$ merge together the linear dynamic part of the actuator and the linear dynamic plant. So the series connection of $\mathcal{W}$ and $D(s)$ represents the *smart actuator* and the plant (SmA). This is a quite general scheme allowing to discuss the control system of the actuator both when loaded by the plant or *unloaded*. The part (b) of the same Fig. 16 shows the same system actuator-plant when it is affected by two inputs (x_1, x_2). The detailed discussion of this scheme will be provided later, at the end of this section.

In the example shown in Fig. 17 a triangular waveform with period $T = 0.02$ s is tracked by the control system sketched in Fig. 16-(a), with and without the compensation algorithm described in the previous sections, [12]. The effects of compensation evidence the increase of the tracking performances of the actuator. In [9] it is also discussed the improvements of the control performances also in term of stability.

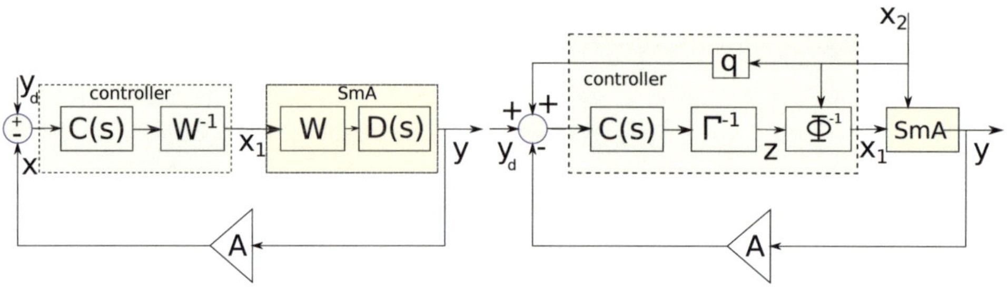

(a) Model-based control scheme (b) Control scheme with two variables feeding the Sma

Figure 16. Model-based control scheme (a) and its generalization to a multi-variable, model-based controller (b). In both cases, the smart actuator and the plant (SmA) are represented by the connection of $\mathcal{W}$ and $D(s)$

The examples considered until now refers to *slow* actuation applications, such as micro positioning tasks (cfr. also [8]). When the assumption of low input variation cannot be assumed as the general working condition, a more accurate discussion should be provided. To this aim, let us again consider the Fig 2, showing how the mechanical response of the material is strongly affected by the applied constant stress (i.e. the *pre-stress*). Normally the magnetostrictive material is subject to a prescribed (and optimal) stress, with the aim to maximize such response. When the actuator is working to actuate a prescribed mechanical load, the total stress experienced by the material has very low fluctuations with respect to the applied stress, whenever slow variations of the applied field are concerned. For high frequencies application tasks this assumption no longer holds and the material experiences

a simultaneous variation of stress and field. In that case the employment of a compensation algorithm with two variable is necessary. To this aim the *multi-variable* hysteresis modeling described in the section above could be an interesting starting point to define suitable multi-variable compensation and control strategies. A first attempt to tackle such a quite demanding problem was settled in [10], where a simple generalization of the Preisach model, taking into account the presence of a second variable (i.e. the stress) in the distribution function was stated and discussed. Such approach had the drawback that only a correction of the operator's output was concerned while no effects on the state were considered. It should be said, however, that even so the system showed promising characteristics, which yielded to further improvements and to the definition of the model described above (eqn. (21)). In the following we will sketch how the latter model can be employed to define suitable and well behaved multi-variable control strategies, [15–17]. However, a preliminary discussion to settle the existence conditions of the compensator for such a kind of multi-variable hysteresis operator is required. Let us now refer to the model specified in eqn. (21) which is re-arranged

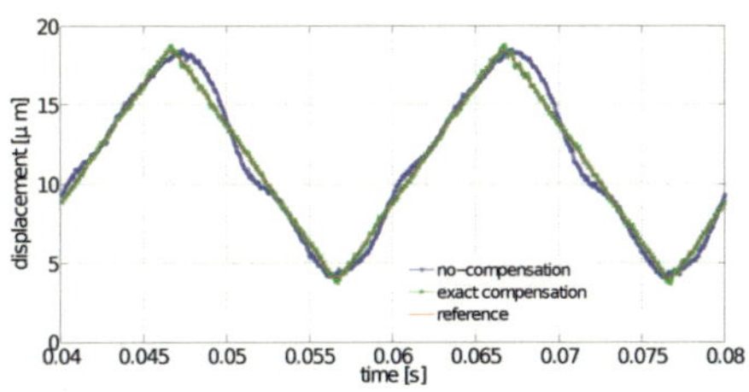

(a) Model-based control scheme

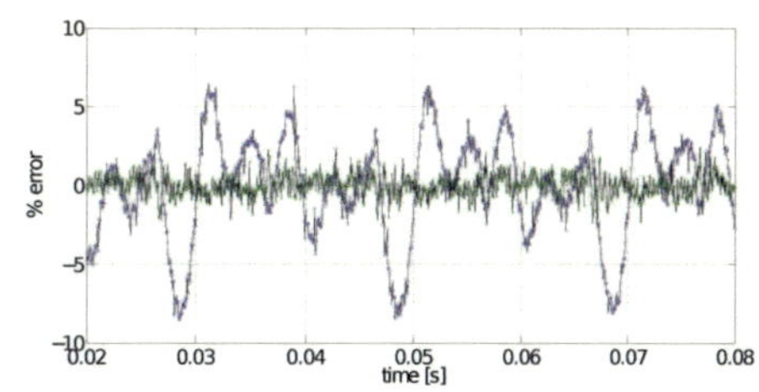

(b) Green: compensated control; Blue: uncompensated control

Figure 17. Model-based control scheme (a) and its performance in tracking a reference signal (b)

as follows:

$$z \equiv \mathcal{W}^{-1}[y-q] = \zeta(x_1, x_2), \tag{24}$$

where

$$\zeta : (x_1, x_2) \in \Omega \longrightarrow \mathbb{R}$$

is a continuous function and Ω the set where it is defined. Moreover, consider now the set where the above eqn. (24) admits at least one solution:

$$S = \left\{ (\bar{x}_2, \bar{z}) \in \mathbb{R}^2 : \exists x_1 \,\text{with}\, \zeta(x_1, \bar{x}_2) \equiv \bar{z} \right\} \tag{25}$$

and the function:

$$\Phi : x_1 \in X \subseteq \mathbb{R} \longrightarrow (x_2, z) \in S. \tag{26}$$

It should be noticed that the latter defines a *single valued* function, being $\zeta(x_1, x_2)$ a *single valued* function. Moreover, recall that $e = y - q$ is the model's output without the pure mechanical response represented by $q(x_2)$.

This function could be either injective or not, depending on the characteristics of the real material, as evidenced hereafter. To this aim, let us consider the material, showing the relation between the output $y - q$ and the input variables x_1 and x_2 as shown in Fig. 18-(a). There with z it has been indicated the input variable corresponding to the output $y - q$ when $x_2 = x_2^*$. If now $x_2 = x_2'$ two distinct values of x_1 are obtained. This implies that the function Φ is *not injectve* (Fig 18-(b)). Conversely, if we refer to Fig. 19-(a), the value of x_1 corresponding

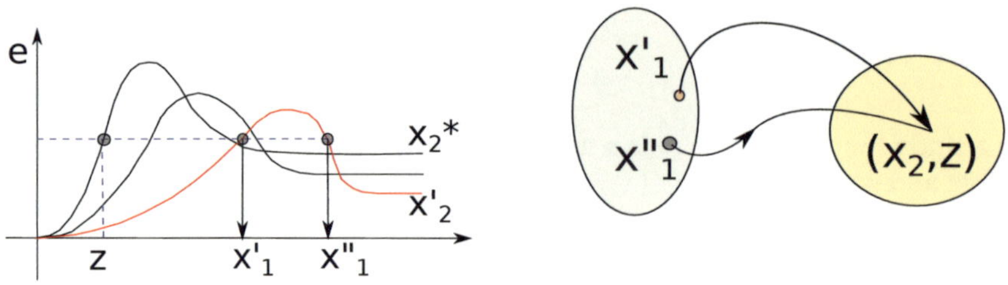

(a) Non monotone response characteristic of a general smart material.

(b) Example of non-injective Φ function

Figure 18. Example of the Φ function reconstructed from input-output available data. It can be stressed that the non-monotonicity of the response functions, as sketched in (a), yields to $\frac{\partial \zeta}{\partial x_1}$ that can assume any sign for assigned x_2

to a given z (which, as before, refers to the assigned value of output $y - q$ on the curve corresponding to $x_2 = x_2^*$), is now unique, when it exists. In fact in the same figure, the possibility that no x_1 values corresponding to a given couple (x_2, z) is also considered. This example evidences that not all values (x_2, z) belongs to the domain of definition of the function Φ. In this case the function is *injective* in S. Moreover, it would appear quite evident that the properties of the Φ function are strictly related to the monotonicity of the functions $(y - q) \leftrightarrow z$ or, in other words to the derivatives $\frac{\partial \zeta(x_1, \bar{x_2})}{\partial x_1}$, as detailed hereafter.

The function

$$z = \zeta(x_1, x_2)$$

can be rearranged by the aid of the function: $g : \mathbb{R}^2 \to \mathbb{R}^2$, having components:

$$\begin{cases} z = \zeta(x_1, x_2); \\ x = x_2, \end{cases} \tag{27}$$

and *Jacobian*:

$$J_g(x_1, x_2) = \begin{vmatrix} \frac{\partial \zeta}{\partial x_1} & \frac{\partial \zeta}{\partial x_2} \\ 0 & 1 \end{vmatrix} = \frac{\partial \zeta}{\partial x_1}. \tag{28}$$

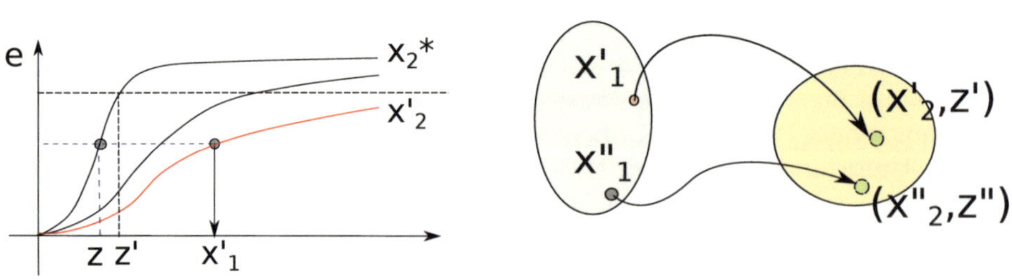

(a) Monotone response characteristic of a general smart material.

(b) Example of injective Φ function

Figure 19. Example of the Φ function reconstructed from input-output available data. It can be stressed that the strict monotonicity of the response functions, as sketched in (a), yields to $\frac{\partial \zeta}{\partial x_1} > 0$ for assigned x_2.

If now we define the set:

$$A = \{(P \equiv (x_1, x_2) \in \Omega : J(P) \neq 0\}, \tag{29}$$

it is possible to conclude that whenever A is a non-empty set, for every $P \in A$ the g function g is locally invertible and so the equation $\zeta(x_1, x_2) = z$ has only one x_1 solution. The sets

$$S^* = \left\{(\bar{x}_2, \bar{z}) \in \mathbb{R}^2 : (x_1, x_2) \in A\right\} \tag{30}$$

is a subset of S, i.e. $S^* \subseteq S$. As a consequence, Φ is injective in S^* and finally it admits the inverse function

$$\Phi^{-1} : S^* \longrightarrow X^*,$$

being X^* a subset of X. This line of reasoning shows the conditions which guarantee the existence of the Φ^{-1} function and hence the possibility to define the compensator of the 2-variables operator with hysteresis. This is a crucial point since only with this condition a generalization of the classical *model-based* control strategy can be settled for multi-variable systems with rate independent memory.

The availability of a procedure enabling the inversion of a hysteresis operator depending on two input variables opens the opportunity to extend *model-based* control strategies to systems having in realty two inputs (for example, desired strain and mechanical stress). This need arises in specific dynamic working conditions of magnetostrictives, (i.e. active vibration suppression of structures), where the device could experience forces of comparable magnitude of the applied stress, but could be extended to similar conditions involving different smart materials.

The simplest control strategy exploiting the results concerning the inversion of multi-variable hysteresis operators, is sketched in Fig. 16-(b). There we can observe the compensation algorithm described so far. In particular the operator Γ^{-1} which, in open loop, takes the variable $e = y - q$ and provides the corresponding input function for the reference case (i.e. $x_2 = x_2^*$), that is $x_1 = z$. The inverse of the Φ function, takes z and x_2 and provides the control variable x_1. If a given desired signal is provided, in presence of a time varying $x_2 = \sigma(t)$, the global behavior of the actuators drastically changes and, without a suitable *stress-dependent* compensation algorithm, the controller performances suddenly worsen. For the sake of example, in Fig. 20 it is shown the relative error in tracking a $\sin(t)/t$ function provided a time variation of the applied force over the full simulation time interval of about 400N. It is evident the improved behavior of the 2-variables control (2VC) system with respect to the case when stress corrections are not provided, [18]. The present approach allows to compensate the effects of nonlinearity and hysteresis in more general working conditions, guaranteeing a close-to-linear behavior of the actuator and making so easier the controller design. Unfortunately, the effects of the compensation algorithm together with the PI controller are beneficial for the whole control system whenever the desired signal is slowly varying, but its performances decrease for faster actuation purposes. Further, multi-variable compensation and control is mostly required exactly when faster signal to track are involved. Thus, different control schemes able to guarantee sufficient performances in a wider frequency range are necessary, [19, 22]. As a first attempt to address the issue, let us recall that closed loop control represents a necessary solution, in presence of disturbances model uncertainties, guaranteeing performances otherwise unachievable by open loop solutions. On the contrary, feedforward controls are easier to implement, do not affect the stability of the overall system

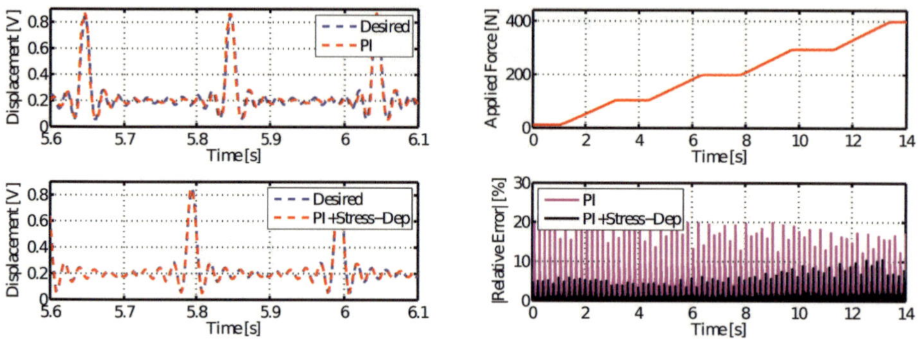

(a) Desired and measured output vs. time. Up: PI controller; bottom; PI+2VC (b) Applied stress and relative tracking error vs. time

Figure 20. Example of real time control task, by considering the effects of compensation with one or two variables

and show faster behavior with respect to feedback solutions since its action depends only on the feedforward controller and not on process dynamics. Modern control systems often employ both actions with the aim to obtain the advantages of both configuration, avoiding, as far as possible, their drawbacks so yielding to a controller with higher performances. Such a configuration is called *two-degrees-of-freedom*, 2-DoF, control system since in this way it is possible to design a controller that weights independently the reference signal and the measurements coming from the process [36, 38]. Examples of 2-DoF control systems for

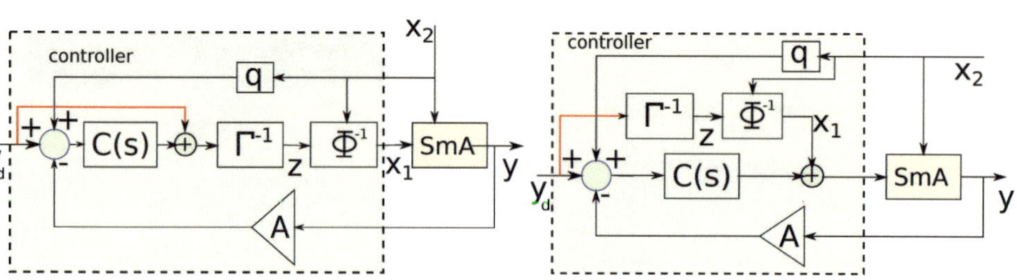

(a) (b)

Figure 21. Block schemes of 2-DoF control systems. Hysteresis compensation on the feedback loop (a); Hysteresis compensation on the feedforward action (b)

actuators with memory, proposed in [19], are shown in Fig. 21. In both of the diagrams it is possible to locate a feedback and a feedforward action. In part -(a) of the figure, the feedforward action trivially consists in adding the desired output to the output of the linear controller $C(s)$. The compensation, is located on the feedback action. In that case both the compensator and actuator with hysteresis experiences exactly the same input history and so they share the same internal state. Conversely, the scheme reported on the part -(b) of the same figure shows the compensation located on the feedforward action, while the feedback loop is without compensation. In such a scheme it is quite evident that the real actuator and its compensator doesn't experiences the same input history. For this reason we could expect a different behavior with respect to the previous system.

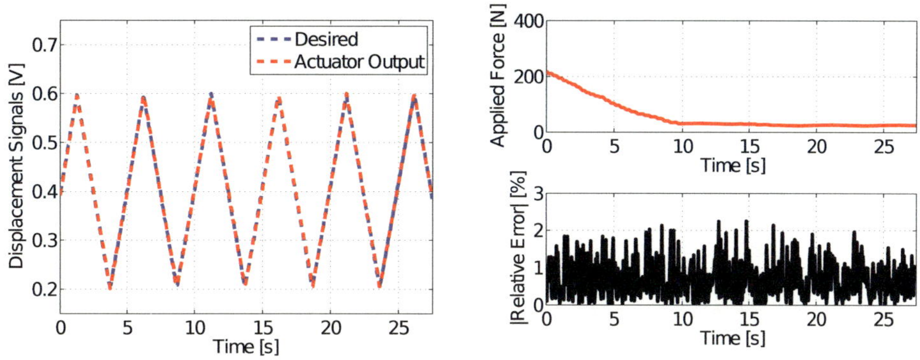

(a) (b)

Figure 22. Tracking performances of a PI control system with 2VC for low frequency application. In this case the period of the desired output is $T = 5$s

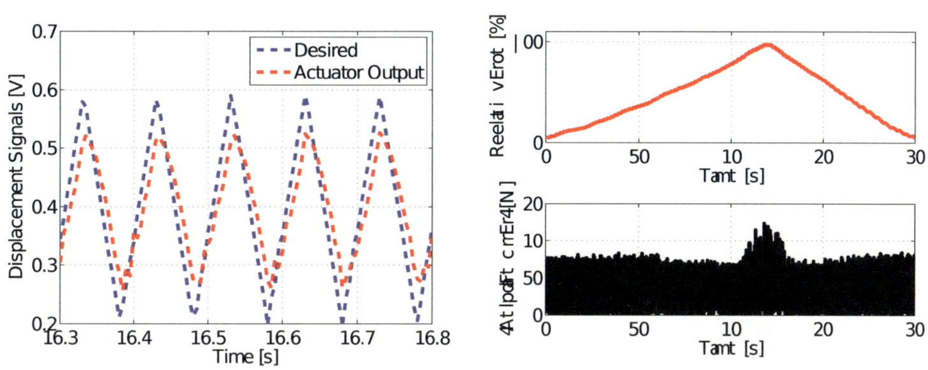

(a) (b)

Figure 23. Tracking performances of a PI control system with 2VC for higher frequency application. In this case the period of the desired output is $T = 0.1$s

In order to evidence the drawbacks of a classical feedback loop employing a simple PI controller, and the improvements provided by the 2-DoF control system, let us discuss the experiments performed by a *real-time* environment where the compensation algorithm and the PI controller are implemented in a PC with Matlab, and a xPC Target toolbox with a 16 bit acquisition/generation card are also employed. Further details are discussed in [19]. In particular, in Fig. 22 it is shown the performances of the feedback system, sketched in Fig. 16, in tracking a triangular waveform, with period of $5\,s$. Here the relative error lies below 3%. When the frequency is increased to 10Hz, the problems in tracking the same waveform dramatically arise and tracking error goes well beyond 15%, as evidenced in Fig 23. The performances results of the scheme reported in Fig. 21-(b) are shown in Fig. 24, with an increase of tracking performances. In Fig. 25, conversely, the performances of the 2-DoF control system presented in Fig. 21-(a) are shown. It can be observed a further increase of the tracking performances of the system with quite low tracking error, since the system provides the same state evolution of both the actuator and compensator.

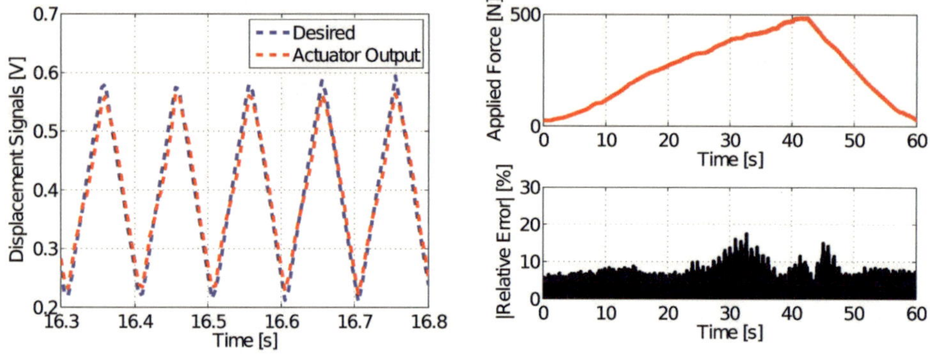

(a) (b)

Figure 24. Tracking performances of a 2-DoF control system with 2VC for higher frequency application. In this case the period of the desired output is $T = 0.1$s and the controller is specified in Fig. 21-(b)

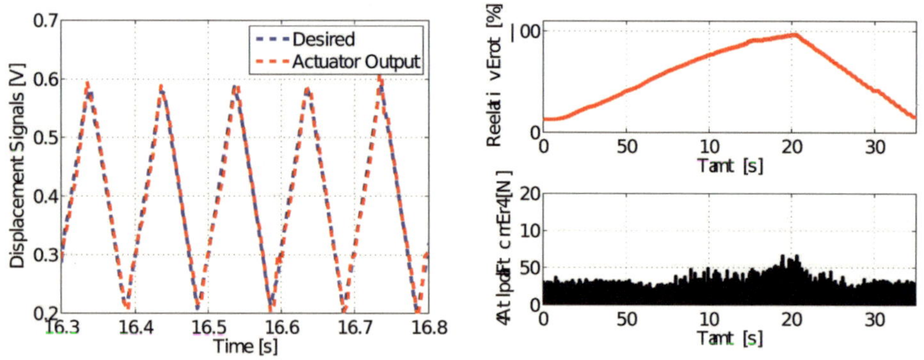

(a) (b)

Figure 25. Tracking performances of a 2-DoF control system with 2VC for higher frequency application. In this case the period of the desired output is $T = 0.1$s and the controller is specified in Fig. 21-(a)

In conclusion the availability of suitable and well assessed tools for the modeling and compensation of actuators with hysteresis enables to define control algorithms that fully exploits this high modeling capabilities. The result is the availability of a model-based control system which gather good performances and quite low computational effort, guaranteeing its suitability for real time and fast actuation tasks.

6. Conclusion

The chapter aimed to provide a view on the basic problems involved when modeling and control of smart materials, normally affected by memory phenomena, are concerned. So, after a brief review of some of the contributions provided in this field, the attention was focused specifically on materials showing a magnetostrictive behavior, such as Terfenol-D and Galfenol. However, all the ideas and tools introduced and discussed could be widely exploited for other smart materials with non-negligible memory effects.

In Sect. 2 a brief description of the phenomena involved in magnetostrictive materials is drawn and the ideas of their modeling are also provided. In this respect, the basic operators describing systems with hysteresis is presented, with a specific attention to the classical Preisach model. Although the most known and applied approach is the one in the 'rotated' system of coordinates, according to Mayergoyz's book, [28], the adopted formalism allover the chapter is that preferred in the books [7] and [26], showing some interesting characteristic from the modeling viewpoint. In any case, the relationship between them is explicitly pointed out.

In the next section 3 the attention is focused on the main modeling and compensation issues which are normally adopted for control purposes. Here, a specific attention is devoted to the congruency property of Preisach operator, which forces a specific attention in the inversion of the operator. Finally, a basic point is discussed with some detail, that is, the definition of compensation algorithms which doesn't require further computational effort with respect to the 'direct' operator, i.e. the Preisach operator. This specific point becomes crucial when a *model-based* real time control approach is of concern.

In order to put the issues discussed in the chapter in the current framework of smart materials and devices, in section 5 the need to manage two independent variables in controlling the device is emphasized, by proposing a well-behaved procedure to handle the stress and magnetic field simultaneously. The effectiveness of this approach for a more precise description and control of smart devices is also discussed by comparison to measured data. Such discussion paves the way to the last section of the chapter, where some application in *real time* controlling smart materials are presented. In particular, 'standard' and 2-DoF control strategies are presented, all fulfilling the constraint to keep a low computational complexity of the whole control system, still ensuring good tracking and stability performances.

Author details

Daniele Davino, Alessandro Giustiniani, Ciro Visone
University of Sannio - Engineering Department, 82100 Benevento (BN), Italy

7. References

[1] Adly, A.A., Mayergoyz, I.D. & Bergqvist, A. (1991). Preisach modeling of magnetostrictive hysteresis *AIP Journal of Applied Physics* 69, 5777.
[2] Anton, S.R. & Sodano, H.A. (2007). A review of power harvesting using piezoelectric materials (2003-2006), *IOP Smart Materials and Structures*, vol. 16, 3, pp. R1-R21.
[3] Bar-Cohen, Y. (2002). Electro-active polymers: Current capabilities and challenges, *Proceedings of SPIE*, San Diego, CA.
[4] Bellouard, Y. (2008). Shape memory alloys for microsystems: A review from a material research perspective *ELSEVIER Materials Science and Engineering A* 481-482, pp. 582-589.
[5] Bergqvist, A. & Engdahl, G. (1991). A stress-dependent magnetic Preisach hysteresis model, *IEEE Transactions on Magnetics*, Vol. 27, No. 6, pt 2, pp. 4796-4798.
[6] Brokate, M. (1989). Some mathematical properties of the Preisach model of hysteresis, *IEEE Transactions on Magnetics*, 25, pp. 2922-24.
[7] Brokate, M. & Sprekels, J. (1996) *Hysteresis and Phase Transitions*,Springer.
[8] Cavallo, A., Natale, C. Pirozzi, S. & Visone, C. (2003). Effects of Hysteresis Compensation in Feedback Control Systems, *IEEE Transactions on Magnetics* Vol. 39, No. 3, pp.1389-1392.

[9] Cavallo, A., Natale, C., Pirozzi, S., Visone, C. & Formisano, A. (2004). Feedback Control Systems for Micropositioning Tasks With Hysteresis Compensation, *IEEE Transactions on Magnetics*, Vol. 40, No. 2, pp. 876-879.

[10] Cavallo, A., Davino, D., De Maria, G., Natale, C., Pirozzi, S. & Visone, C. (2008). Hysteresis compensation of smart actuators under variable stress conditions, *ELSEVIER Physica B*, 403, 2-3, pp. 261-265.

[11] Chung, D.Y., Hogan, T., Brazis, P., Rocci-Lane, M., Kannewurf, C., Bastea, M., Uher, C. & Kanatzidis, M. G. (2000). $CsBi_4Te_6$: A High-Performance Thermoelectric Material for Low-Temperature Applications, *Science* 287, pp. 1024-27

[12] Davino, D., Natale, C., Pirozzi, S. & Visone, C. (2005). A fast compensation algorithm for real-time control of magnetostrictive actuators, *ELSEVIER Journal of Magnetism and Magnetic Materials*, 290-291, pp. 1351-54.

[13] Davino, D., Giustiniani, A. & Visone, C. (2008). Fast Inverse Preisach Models in Algorithms for Static and Quasistatic Magnetic-Field Computations, *IEEE Transactions on Magnetics*, 44 (6) , pp. 862-865.

[14] Davino, D., Giustiniani, A. & Visone, C. (2008). Properties of Hysteresis Models Relevant in Electromagnetic Fields Numerical Solvers, *ELSEVIER Physica B: Condensed Matter*, 403 , pp. 414-417.

[15] Davino, D., Giustiniani, A. & Visone, C. (2008). A Magnetostrictive Model with Stress Dependence for Real-Time Applications, *IEEE Transactions on Magnetics*, 44 (11-2) , pp. 3193-3196.

[16] Davino, D., Giustiniani, A. & Visone, C. (2009). Experimental Properties of an Efficient Stress-Dependent Magnetostriction Model, *AIP Journal of Applied Physics*, 105 (07D512) , pp. 1-3.

[17] Davino, D., Giustiniani, A. & Visone, C. (2010). Design and Test of a Stress Dependent Compensator for Magnetostrictive Actuators, *IEEE Transactions on Magnetics*, 46 (2) , pp. 646-649.

[18] Davino, D., Giustiniani, A. & Visone, C. (2011). Compensation and control of two-inputs systems with hysteresis, *IOP Journal of Physics: Conference Series*, 268, doi:10.1088/1742-6596/268/1/012005.

[19] Davino, D., Giustiniani, A., Iannelli, L. & Visone, C. (2012). Comparison of real-time control strategies with hysteresis compensation for magnetostrictive actuators, *IOS International Journal of Applied Electromagnetics and Mechanics*, in press.

[20] Dlala, E. & Arkkio, A. (2006). Inverted and forward Preisach models for numerical analysis of electromagnetic field problems, *IEEE Transactions on Magnetics*, Vol. 42, No. 8, pp. 1963-1973.

[21] Engdhal, G. (editor) (2000) *Handbook of Giant Magnetostrictive Materials*, Academic Press.

[22] Esbrook, A., Guibord, M., Tan, X. & Khalil, H.K. (2010). Control of systems with hysteresis via servocompensation and its application to nanopositioning, *Proceedings of the 2010 American Control Conference, ACC 2010*, Article number 5531422, Pages 6531-6536.

[23] Ge, P. & Jouaneh, M. (1996). Tracking Control of a Piezoceramic Actuator, *IEEE Transactions On Control Systems Technology*, 4, pp. 209.

[24] Hasegawa, T. & Majima, S. (1998). A control system to compensate the hysteresis by Preisach model on SMA actuator, *Proceedings of IEEE International Symposium on Micromechatronics and Human Science*, 171.

[25] Krasnoselskii, M. A. & Pokrovskii, A. V. (1989). Systems with Hysteresis, Springer-Verlag.

[26] Krejci, P. (1996). Hysteresis, convexity and dissipation in hyperbolic equations, *Gakuto Int. Series Math. Sci. & Appl.*, 8, Gakkotosho, Tokyo.

[27] Krejci, P. & Kuhnen, K. (2001). Inverse control of systems with hysteresis and creeps, *Proceedings of Control Theory and Applications*, Vol. 148, No. 3, pp. 185-192.

[28] Mayergoyz, I. D. (1991) *Mathematical Models of Hysteresis*, Springer.

[29] Miano, G., Serpico, C. & Visone, C. (1997). A new model of magnetic hysteresis, based on Stop hysterons: an application to magnetic field discussion, *IEEE Transactions on Magnetics*, Vol. 32, No. 3, pp. 1132-1135.

[30] Natale, C., Velardi, F. & Visone, C. (2001). Identification and compensation of Preisach hysteresis models for magnetostrictive actuators, *ELSEVIER Physica B*, 306, pp. 161-165.

[31] Otsuka, K. & Wayman, C.M. (1998) *Shape Memory Materials*, Cambridge University Press.

[32] Panda, P. K. (2009). Review: environmental friendly lead-free piezoelectric materials, *SPRINGER Journals of Matererials Science*, 44, pp. 5049-5062, DOI 10.1007/s10853-009-3643-0.

[33] Pecharsky, V.K. & Gschneidner, K.A. (2006). Advanced magnetocaloric materials: What does the future hold? *International Journal of Refrigeration* 29 pp. 1239-1249.

[34] Polla, D.L. & Francis, L.F. (1998). Processing And Characterization Of Piezoelectric Materials And Integration Into Microelectromechanical Systems, *Annual Review Matererials Science*, 28, pp. 563-597.

[35] Schafer, J. & Janocha, H. (1995). Compensation Of Hysteresis In Solid-State Actuators, *ELSEVIER Sensors And Actuators A* 49, pp. 97-102.

[36] Skogestad, S. & Postlethwaite, I. (2005). Multivariable Feedback Control: Analysis and Design, *Wiley-Interscience*.

[37] Söderberg, O., Ge, Y., Sozinov, A., Hannula, S.P. & Lindroos, V.K. (2005). Recent breakthrough development of the magnetic shape memory effect in Ni-Mn-Ga alloys *IOP Smart Materials and Structures* 14, 5.

[38] Sugie, T. & Yoshikawa, T. (1986). General solution of robust tracking problem in two-degree-of-freedom control systems *IEEE Transactions on Automatic Control*, Vol. 31 (6), pp. 552-554.

[39] Tan, X., Venkataraman, R. & Krishnaprasad, P.S. (2001). Control of hysteresis: Theory and experimental results, SPIE Modeling, Signal Processing, and Control in Smart Structures (Rao V. S., Ed.), 4326, pp. 101-112.

[40] Tellinen, J., et al., (2002). Basic properties of magnetic shape memory actuators, *Proceedings of 8th ACTUATOR Conference*, Bremen, Germany.

[41] Visintin, A. (1991) *Differential Models of Hysteresis*, Springer.

[42] Visone, C. & Serpico, C. (2001). Hysteresis operators for the modeling of magnetostrictive materials, *ELSEVIER Physica B: Condensed Matter*, 306 (1-4) , pp. 78-83.

[43] Visone, C. & Sjostrom, M. (2004). Exact invertible hysteresis models based on play operators, 2004, *ELSEVIER Physica B: Condensed Matter*, 343, pp. 148Ð52.

[44] Visone, C. (2008). Hysteresis Modelling and Compensation for Smart Sensors and Actuators, *IOP Journal of Physics Conference Series*.

[45] Webb, G.V., Lagoudas, D.C. & Kurdila, J. (1998). Hysteresis modeling of SMA actuators for control applications, *SAGE Journal of Intelligent Material Systems and Structures*, Vol. 9, No. 6, pp. 432-48.

[46] Webb, G., Lagoudas, D. & Kurdila, A. (1999). Adaptive hysteresis compensation for SMA actuators with stress-induced variations in hysteresis, *SAGE Journal of Intelligent Material Systems and Structures*, Vol. 10, No. 11, pp. 845-854.

Section 2

Smart Polymers

Chapter 7

A Technology for Soft and Wearable Generators

Thomas G. McKay, Benjamin M. O'Brien and Iain A. Anderson

Additional information is available at the end of the chapter

http://dx.doi.org/10.5772/50540

1. Introduction

Never again needing to recharge the batteries of portable electronic devices is an exciting prospect. The uptake of portable electronics is increasing and recharging or replacing batteries is not only an inconvenience, but also contributes to an environmental hazard. Furthermore, with technologies such as GPS, pressure sensors, and cameras becoming smaller and cheaper, smart clothing such as the Adidas miCoach range are now a reality [1]. For seamless integration of smart devices into clothing, the inconvenience of battery replacement or recharging needs to be eliminated.

Conveniently there is an alternative source of energy in the exact location where portable and wearable devices operate: biomechanical energy from human movements. Interestingly, the device often credited as being the first ever power generator was a biomechanical energy harvester. 17th century engineer Otto von Guericke produced a sulfur globe which was charged through the triboelectric effect[1] when it was rubbed with a dry hand [2]. However, four centuries later the ability to harvest appreciable amounts of energy from biomechanical motions remains a research challenge. In this chapter we will discuss recent progress towards harvesting biomechanical energy using a soft and wearable electro-active polymer technology.

To provide sufficient power for portable devices a biomechanical energy harvester needs to supply in the order of a few Watts. For instance, an insulin pump consumes approximately 5 Watts [3] and the Nexus One smartphone (HTC Corporation, Taiwan) consumes 25 milli-Watts in standby, 330 milli-Watts in idle, and 750 milli-Watts during a phone call [4]. So to eliminate the need to recharge batteries, a similar or greater quantity of power needs to be sourced.

[1] The triboelectric effect refers to the transfer of electrons between two materials when they come into contact. Some materials are more susceptible to donating electrons when the contact is separated and some are more likely to hold on to electrons so the materials remain charged when separated.

Starner and Paradiso provide a good review of the energy available for scavenging from human movement and identify walking as a rich source of energy [5]. They highlight that 13Watts of power is available from the heel-strike of a 68Kg person if the sole of their shoe is compressed by 1cm when walking at 2 steps per second. Furthermore, a typical running shoe midsole dissipates a relatively large amount of energy as heat. Shorten's analysis of the energetics of a shoe midsole worn by a 76Kg runner suggests that a running shoe will dissipate between 2 and 10 Joules per step [6]. A well designed energy harvesting shoe could instead turn this energy into electricity, without altering the comfort or energy expenditure of the person wearing the shoe.

The most prevalent energy harvesting technology, electromagnetic generators, have been used to harvest energy from human gait, but additional mechanisms are required to condition the mechanical energy [5, 7]. Donelan et al. developed a knee-brace generator for harvesting energy from human gait. Their generator contained auxiliary components including a gear train, bearings, and a separate input shaft to convert the mechanical energy to suit their electromagnetic generator. Their system cost on average 59W of metabolic power to carry without harvesting energy, whereas an additional 5W of metabolic power was required to produce 4.8W of electrical power [7]. Although their harvesting mechanism was extremely efficient, they could have achieved larger efficiency gains if their system's mass was reduced and the device did not alter gait patterns. This begs the question: why are auxiliary mechanical components required for an electromagnetic biomechanical energy harvester?

The system by Donelan et al. was relatively heavy because it required additional componentry for it to work efficiently: electromagnetic generators produce more energy during a single rotation or stroke as velocity increases and are poorly suited for the low induced velocities associated with walking unless augmented mechanically. Thus the viability of harvesting energy from human motions could be improved by utilizing a different energy harvesting technology.

We propose the key characteristics of an ideal technology for harvesting biomechanical energy below:

1. Efficient operation at low biomechanical "walking" speeds.
2. Ability to couple directly to large walking motions.
3. A high energy density.
4. Good mechanical impedance matching to muscle.
5. Cheap to produce and maintain.

The first two criteria eliminate the need for additional mechanisms to condition the mechanical energy when it is transferred to the generator. The first three criteria therefore provide low mass/bulk. By having good impedance matching to muscle, the generator will be comfortable to wear, reducing its effects on the person wearing it. For the mass market, it is essential that the generator is low cost because consumers are unlikely to pay a premium for generators that produce power of approximately 1 Watt.

The research efforts of the authors of this chapter have focused on an energy harvesting technology called dielectric elastomer generators (DEG) which have been identified as a

highly promising technology for biomechanical energy harvesting [8-11]. Furthermore, DEG fit the five listed criteria particularly well, so this chapter will focus on our recent developments towards portable and wearable DEG. For the interested reader, Anderson et al. provide a good, broad review of dielectric elastomer transducers [12].

2. The dielectric elastomer generator

DEG are a class of variable capacitance generators that are fabricated from a rubbery dielectric material sandwiched between stretchable electrodes. They have excellent impedance matching to natural muscle [13], can be fabricated from a wide range of low cost materials (commonly acrylic or silicone membranes sandwiched between carbon-based electrodes), have demonstrated extremely high energy densities, can undergo strains in excess of 100% [14], and have the ability to work over a wide frequency range without sacrificing efficiency [15].

Mechanical energy can be converted to electrical energy by cyclically deforming a DEG and placing charges on its flexible electrodes in the deformed state. DEG are typically produced from incompressible polymers so that an area stretch results in a decreased thickness. Relaxation of the charged, deformed DEG forces the opposite charges apart and packs like charges closer together and this transfers the mechanical energy to the electrical charges. The energy flows during an ideal DEG cycle are highlighted by the red arrows in Figure 1. The system receives 1 unit each of mechanical and electrical energy and converts the unit of mechanical energy into electrical, so that it outputs two units of electrical energy.

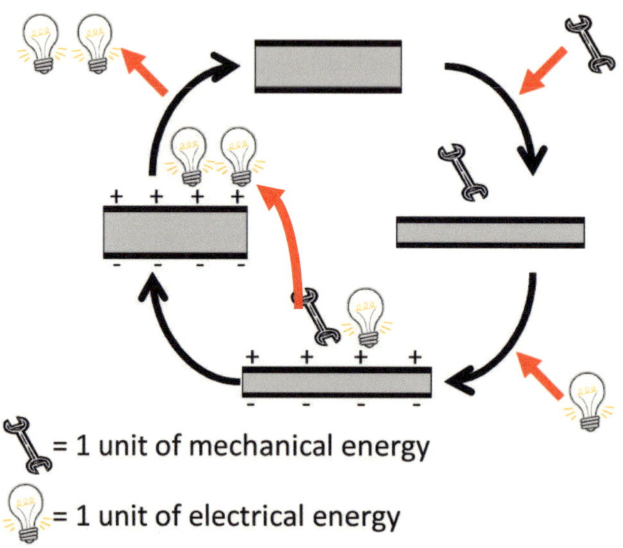

Figure 1. Schematic of the DEG states during a generation cycle. The grey area represents the dielectric and the black area represents the electrodes. From top moving clockwise, mechanical energy is input to the system deforming the DEG; an electrical energy input then charges the stretched DEG; the mechanical energy is then transferred to the charges by separating opposite and compressing like charges together. The electrical energy is then extracted and the cycle repeats (From [16]).

The major advantage that DEG hold for biomechanical energy harvesting is their ability to directly harvest low frequency motions without any gear mechanisms. To illustrate why this is true we will briefly describe their fundamental energy harvesting mechanism. As illustrated in Figure 1, DEG convert mechanical energy to electrical when the deformation of a stretched charged DEG is relaxed. During this relaxation period the thickness of the dielectric increases and the electrode area decreases, both resulting in a reduction of the capacitance. If charge is trapped on the generator during this relaxation phase, the voltage on the DEG will increase and there will be an increase in energy given by equation 1 where C_d and V_d are the capacitance and voltage of the DEG in its deformed state, and C_r and V_r are the relaxed DEG's capacitance and voltage, respectively.

$$E = 0.5C_r V_r^2 - 0.5C_d V_d^2 \tag{1}$$

Since we are considering the case where the charge on the DEG is fixed during the relaxation period, and that the charge on a capacitor can be calculated from its voltage and capacitance (Q=CV), we can relate the voltage of the deformed DEG to the voltage of the generator in its relaxed state using equation 2. We substitute equation 2 into equation 1 to get equation 3.

$$V_R = \frac{C_d}{C_r} V_d \tag{2}$$

$$E = 0.5\left(\frac{C_d^2}{C_r} - C_d\right) V_d^2 \tag{3}$$

These equations emphasise that along with driving voltage, C_r and C_d are the key parameters that influence the energy output of a DEG. The capacitance is dependent on the geometry of the DEG and the driving voltage is controlled by the generator's associated electronics. Neither of these parameters are dependent on the velocity at which the generator is deformed, so the fundamental mechanism of DEG is not dependent on driving velocity.

Although we have highlighted that DEG have a highly suitable mechanism and characteristics for harvesting biomechanical energy, Figure 1 highlights that DEG need an electrical circuit that will control the flow of charge onto and off of the generator at appropriate stages of the energy harvesting cycle. Such circuitry can add weight and is typically composed of stiff and bulky parts. Furthermore the requirement for the electronics to replace charge delivered to the load has traditionally reduced the portability of DEG because they have been either tethered to the grid or used batteries that need periodic replacement [10]. This chapter will focus now on recent developments that eliminate the need for a secondary power source and reduce external DEG circuitry mass and stiffness.

3. Portable dielectric elastomer generator electronics

A passive DEG circuit for controlling the charge state of the DEG appropriately as it is mechanically cycled called the Self-Priming Circuit (SPC) [17, 18] has been developed to

eliminate the need for a secondary power source and thus improve DEG portability. The circuit works as a charge pump that provides energy in a higher charge form than the energy supplied to it. The self-priming circuit is configured so that it harvests energy from a DEG and then supplies that energy in a higher charge form to a load or to the DEG when it requires priming. This effectively boosts the charge of the generated energy before it is used, thus a secondary power source is not necessary because the generated energy is used to replace circuit charge losses and charge delivered to the load. A schematic of the simplest form of an SPC is given in Figure 2, showing that an SPC is simply a capacitor bank that has diodes to convert the topology of the circuit between a low capacitance when it is charged (Figure 2c, capacitors in series) and a high capacitance when it discharges (Figure 2d, capacitors in parallel). This toggling of state provides the SPC with a higher output capacitance which converts the energy to a higher charge form. This can be explained using equation 4, which provides two expressions for the energy (E) stored on a capacitor. If the energy is conserved when the capacitance (C) of the SPC increases, then the charge (Q) must increase too, and since the SPC is an adiabatic process the increased charge is accompanied by a decreased voltage (V). Thus the SPC outputs energy in a higher charge, lower voltage form than the input.

$$E = \frac{Q^2}{2C} = \frac{CV^2}{2} \tag{4}$$

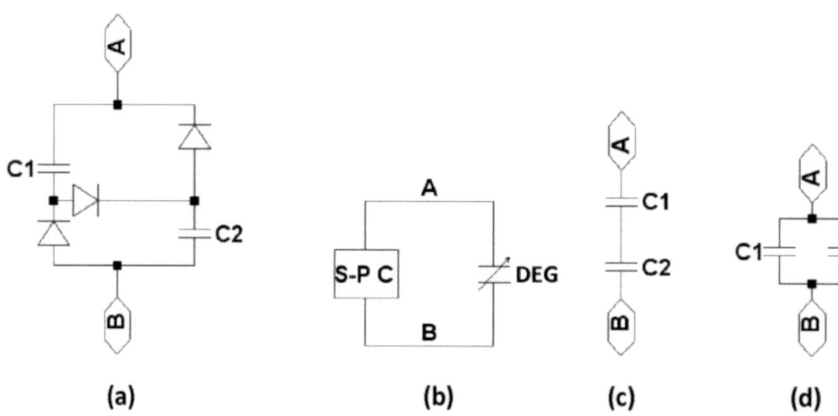

Figure 2. Schematic of a self priming circuit (a) which connects to a DEG in parallel (b), the diodes control current flow so that the SPC capacitor bank takes on the form given in (c) when energy is transferred to it from the DEG, when energy is transferred off the SPC the diodes configure it to take on the higher charge form given in (d).

The SPC has additional benefits for portable DEG. First, the SPC is passive and requires no active switching or control, so it does not require any power to drive it. Second, the SPC accumulates charge from cycle to cycle, this means that the system voltage climbs (see Figure 3). The ability to boost its own voltage is highly desirable because as demonstrated in equation 3, generated energy climbs with priming voltage. This high voltage has

traditionally been supplied by a high voltage power supply or converter, so the SPC eliminates the need for these typically high cost components.

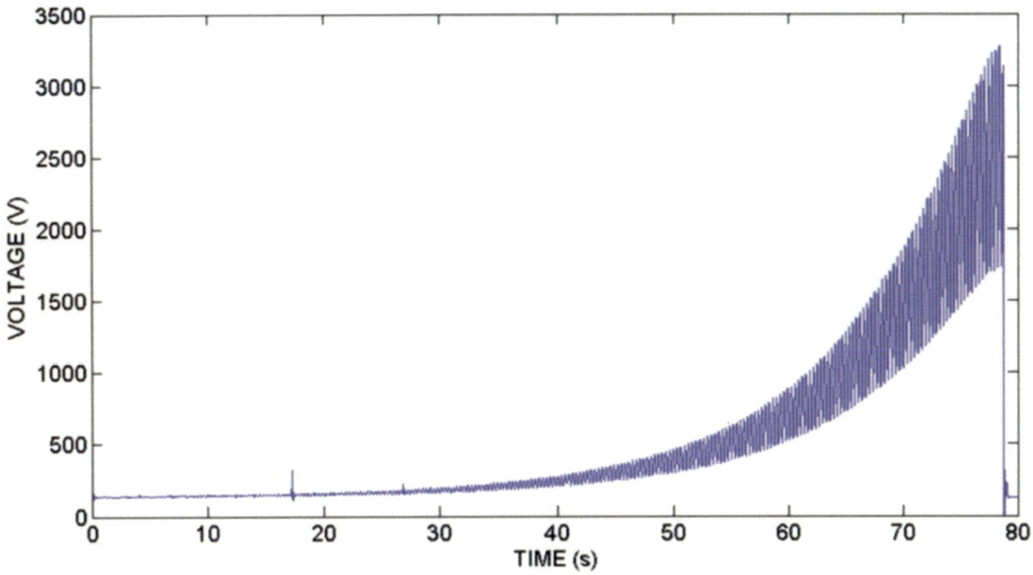

Figure 3. The output voltage waveform of a self-priming generator mechanically deformed at 3 Hz. The voltage climbs from cycle to cycle because the generated energy accumulates in the form of extra charge stored on the generator and priming circuit (From [19]).

Although the SPC is low cost, low complexity, and autonomous it still adds considerable mass and stiffness to the DEG system. The SPC consists of diodes and capacitors. The function of the capacitors is to store priming charges and the diodes control the transfer of charge to and from the DEG, so that an appropriate generation cycle is achieved. We will now discuss how these functions have been integrated onto the DEG membrane to produce a generator that can be fabricated entirely from soft elastomers.

4. Dielectric elastomer generators with integrated soft electronics

The DEG membrane is essentially a soft capacitor and one can take advantage of this, using the DEG to provide the energy storage function of the SPC. This means the generator can be fabricated with its external circuitry consisting solely of a few diodes [16, 20]. One configuration that has been used to integrate the SPC storage function into the DEG membrane is given in Figure 4 where DEG1 is electrically configured as an SPC [16]. Because the generator's elastomer membranes are integrated into the self-priming circuit, this system has been referred to as the integrated self-priming circuit [21]. The generator was configured into two DEG membranes which were deformed so that as one membrane was stretched the other was relaxed. When the voltage of DEG1A and DEG1B connected electrically in parallel exceeded the voltage of DEG2, energy was transferred from the SPC to DEG 2 through the path shown in Figure 4c; when the voltage of DEG2 exceeded that of

DEG1A and DEG1B connected in series, energy was transferred onto the SPC from DEG2 through the path shown in Figure 4d.

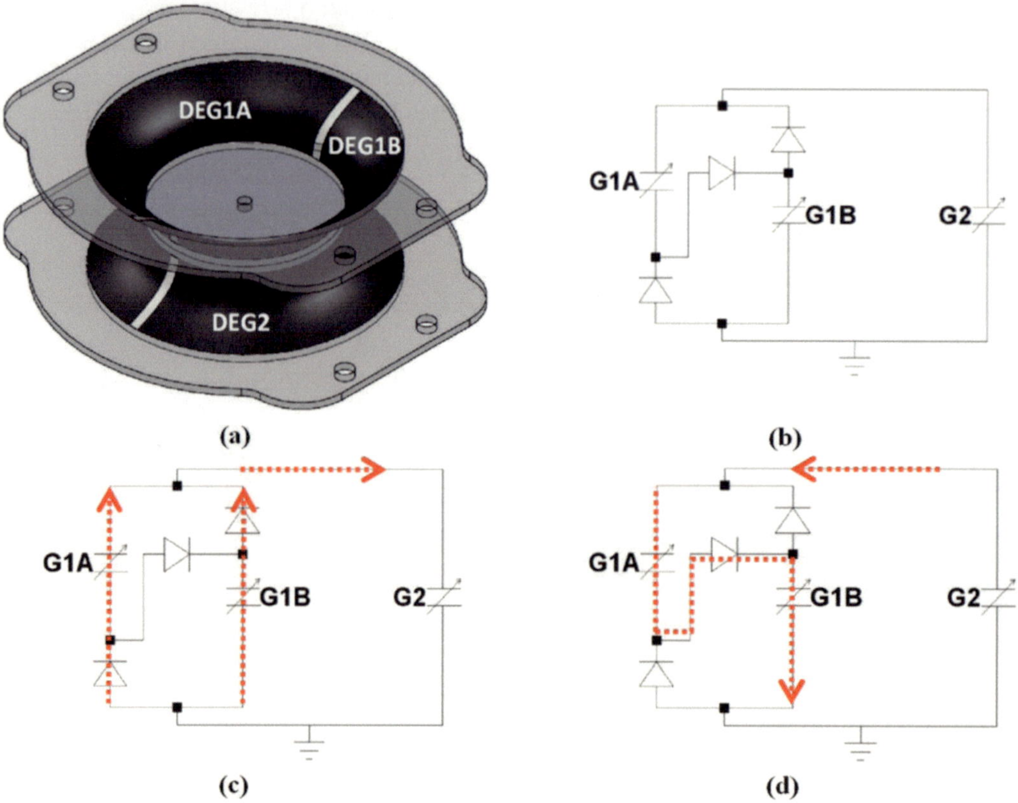

Figure 4. A segmented generator which consists of two membranes that are antagonistically deformed when the inner hub displaces up and down (a). The top membrane DEG1 can be electrically interconnected with diodes to form a self-priming circuit (b). The paths along which current flows off (c) and onto (d) the SPC are also illustrated.

For a small scale portable and potentially wearable generator the diodes of the SPC can represent a significant mass. For instance, the prototype of the generator shown in Figure 4 consisted of DEG membranes with a combined mass of 0.35 grams, whereas the diodes weighed 0.63 grams. If we could remove the diodes, 64% of the total mass (ignoring the mass of the frame) could be eliminated.

The function of the diodes is to control the charge transfer between the DEG and SPC. The diodes simply allow current to flow along one path when one diaphragm is stretched, and along an alternative path when that diaphragm is relaxed. This means that the diodes can be replaced by switches that are toggled at the appropriate time. Since this timing is dependent on the material stretch state, the diodes can be replaced by stretch-sensitive switches coupled to the DEG.

Stretch sensitive electronics called Dielectric Elastomer Switches (DES) have been used to integrate the functionality of the diodes into the DEG membrane. DES consist of piezoresistive electrodes fabricated directly onto a highly stretchable dielectric elastomer membrane. They exhibit very large changes in resistance with stretch. O'Brien et al. first presented the concept [22] and characterized M-shaped DES as illustrated in Figure 5 [23]. These DES had a resistance of several MΩ in their rest state, which increased to several GΩ when they were stretched to approximately 1.4 times their original length (see Figure 5b) [23].

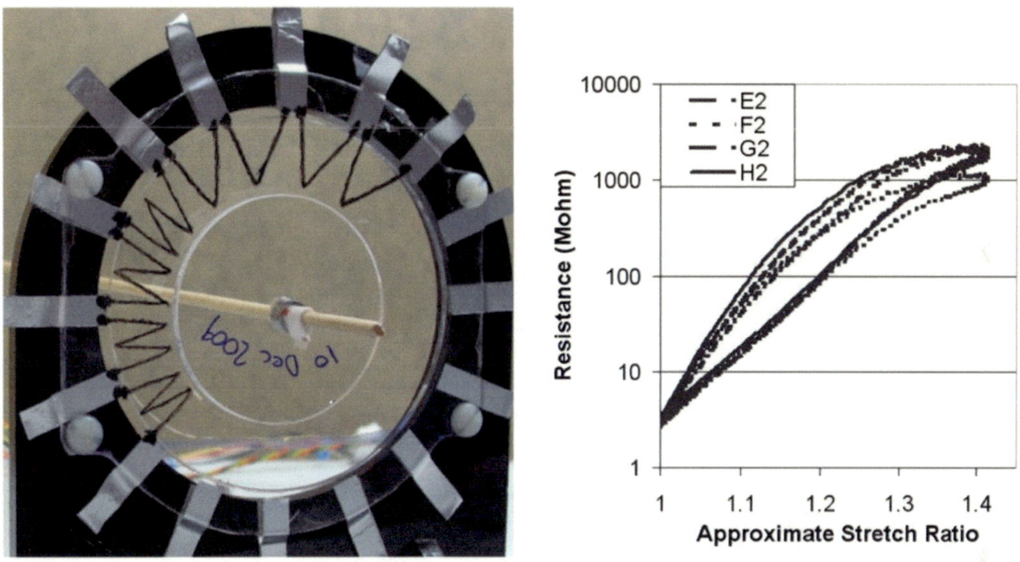

(a) (b)

Figure 5. Carbon powder-based DES applied to a dielectric elastomer diaphragm (black "M" and "V" shaped tracks) for characterisation by O'Brien et al. (a), and a plot of measured DES resistance versus approximate radial stretch ratio showing that the resistance climbed several orders of magnitude when the DES were stretched (b). Images taken from O'Brien [23].

The generator illustrated in Figure 4 was redesigned by placing DES onto the membrane to replace the hard diodes as illustrated in Figures 6 and 7. The SPC capacitors were fabricated onto diaphragm 1. When diaphragm 1 was relaxed the switches Q1 and Q2 were also relaxed and therefore conducted, configuring the SPC capacitors into a parallel topology (high charge form). When diaphragm 1 was stretched so too were Q1 and Q2, so they no longer conducted, but diaphragm 2 was simultaneously relaxed, causing Q3 to conduct, connecting the SPC capacitors in series. This generator can be fabricated entirely from soft elastomers, so we refer to it as the soft generator.

The output voltage of a prototype soft generator is given in Figure 8. When compared to the output of the self-priming circuit given in Figure 3, the soft generator's voltage climbs very rapidly. But the most profound advantages are the ability to produce DEG that maintain the

advantages of high flexibility, softness, low volume, low cost, low component count, and low mass at the system level.

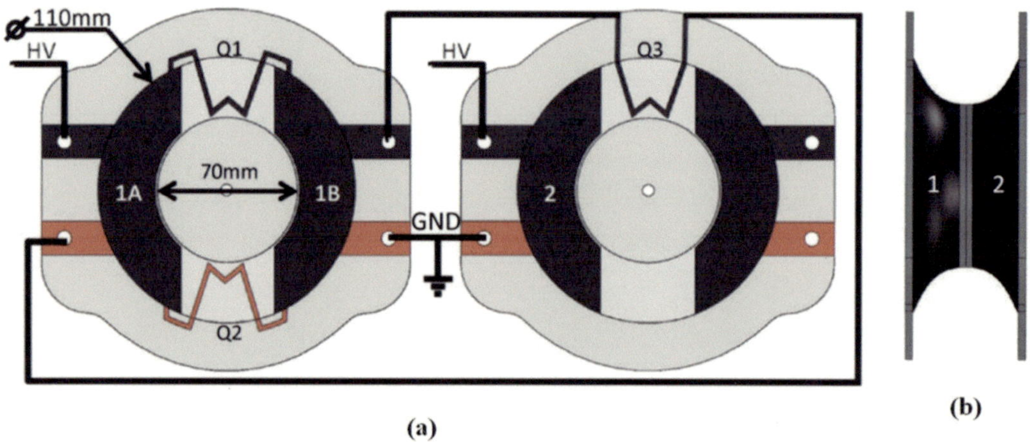

Figure 6. A schematic of the soft generator. The switches Q1-Q3 control the charge flow within the self-priming circuit in a similar manner to the diodes in the integrated self-priming circuit. The two diaphragms in (a) are connected together to form the antagonistic pair shown in (b). (From [24]).

Figure 7. A photograph of a soft generator prototype. The large black regions are the generator electrodes and the thin "M-shaped" electrodes are the piezo-resistive switches.

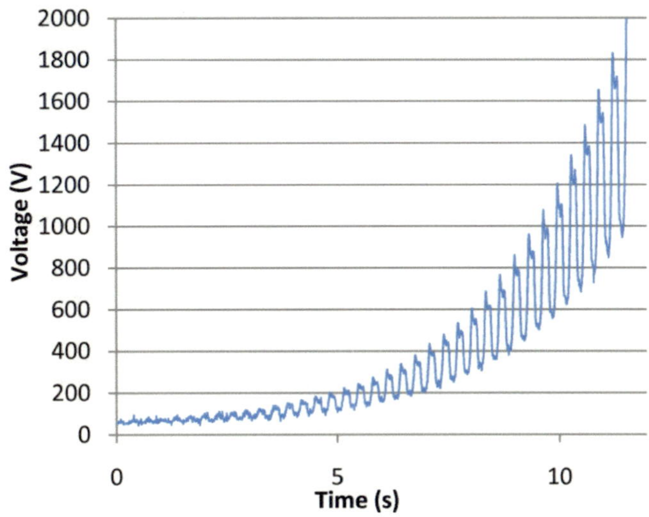

Figure 8. The output voltage waveform of a soft generator initially primed to 10 Volts then mechanically deformed at 3 Hz showing that the generator rapidly boosts its operating voltage through accumulation of generated energy (From [24]).

The total mass of biomechanical energy harvesters has a great influence on the metabolic cost of their use and can be easily compared in a quantitative manner. In Figure 9 we compare the energy density of each DEG system described in this chapter with the mass of their associated electronics included in the calculations (see equation 5).

$$EnergyDensity = \frac{EnergyGenerated}{Mass_{dielectric} + Mass_{ExternalCircuit}} \quad (5)$$

The energy and energy densities produced by similarly sized DEG membranes (~0.3 grams) mechanically cycled at a rate of 3 Hz and operating at 2 kV are compared in Figure 9. The soft generator energy density was superior to both the integrated and external SPC generators because their respective external circuit masses are approximately 0 grams, 0.6 grams, and 13.4 grams. The soft generator's energy density of 9.5 mJ/g is highly competitive with the predicted practical maxima of electromagnetics (4 mJ/g) and piezoelectrics (17.7 mJ/g) at the ~1cm3 scale [25], demonstrating the utility of DEG for small-scale energy harvesters.

The recent developments discussed in this chapter have provided progress towards wearable, soft power generators becoming a reality, but there are still issues that need to be addressed:

1. The switching technology used in the soft generator, DES, are in their infancy. Material and process developments are required to create more reliable DES with resistances that can be tuned to their application.

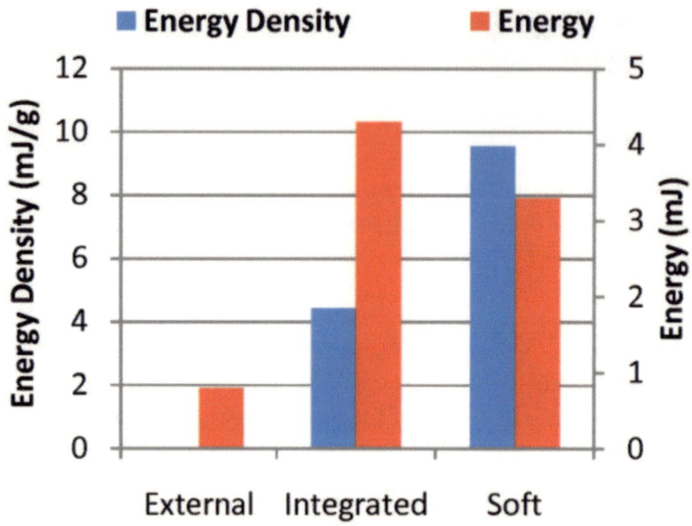

Figure 9. The energy and energy density produced by the external SPC, integrated SPC, and soft generators.

2. The prototypes discussed in this chapter are 1 to 2 orders of magnitude smaller than those required to produce sufficient power to drive the more power hungry portable electronics such as smart phones, which begs the question: How do we expect them to scale? Figure 9 illustrates that the integrated SPC generator produced more energy than the soft generator, so as the generator is scaled up and the diode mass becomes insignificant the integrated generator may outperform the soft generator in terms of energy density. When scaled up to produce an order of a Watt, are the soft generator's advantages sufficient to provide a worthwhile alternative to more complex DEG control strategies that will provide superior energy production [26-29].
3. The soft and integrated generators have an antagonistic pair of membranes. What is the best configuration for harvesting energy from a given biomechanical source?
4. DEG provide the best energy production when operated at high voltages (typically kV range). Compact energy-efficient circuitry is required to reduce this voltage down to consumer electronics levels (~5 Volts) or perhaps the high voltage energy can be used to drive another high voltage device such as a dielectric elastomer actuator.

With these challenges comes an exciting future: the emergence of smart, wearable soft devices. The DEG membranes are multifunctional; they can be operated as actuators or generators and simultaneously sense strain [12]. Furthermore DES have been used to fabricate rubber-based NAND-gates and memory elements, the two primitives required to build a computer [30-32]. Perhaps the future will not only include soft generators to power portable electronic devices, but it may also include a new breed of smart, multifunctional self-powered portable soft devices. These possibilities open the door for smart devices to be integrated directly into clothing. So soft wearable generators will not only revolutionize the use of today's portable devices, but they will be an integral part of a distributed body-area

network of sensors and smart devices that will improve future prosthetic devices, sports monitors, and video gaming interfaces.

Author details

Thomas G. McKay and Benjamin M. O'Brien
The Biomimetics Laboratory of the Auckland Bioengineering Institute, The University of Auckland, Level 6, Auckland, New Zealand

Iain A. Anderson
The Biomimetics Laboratory of the Auckland Bioengineering Institute, The University of Auckland, Level 6, Auckland, New Zealand
Department of Engineering Science, School of Engineering, The University of Auckland, Level 3, Auckland, New Zealand

Acknowledgement

This work was supported by a University of Auckland Bioengineering Institute FRDF postdoctoral fellowship and a Rutherford Foundation postdoctoral fellowship from the Royal Society of New Zealand. The authors acknowledge the support of the Auckland Bioengineering Institute, their colleagues in the Biomimetics lab, Kristal Murray, Michelle O'Brien, and Suzie Pilkington.

5. References

[1] *Adidas miCoach.* 2012 [cited 2012]; Available from: www.micoach.com/.

[2] Schiffer, M.B. and C.L. Bell (2003) *Draw the Lightening Down: Benjamin Franklin and Electrical Technology in the Age of Enlightenment.* Berkeley: University of California Press, Ltd.

[3] Jia, D. and J. Liu (2009) *Human power-based energy harvesting strategies for mobile electronic devices.* Frontiers of Energy and Power Engineering in China. 3(1): p. 27-46.

[4] Carroll, A. and G. Heiser (2010) *An Analysis of Power Consumption in a Smartphone. USENIX ATC '10.*

[5] Starner, T. and J.A. Paradiso (2004) *Human Generated Power for Mobile Electronics* in *Low Power Electronics Design* C. Piguet Editor CRC Press. p. 45-1 - 45-35.

[6] Shorten, M.R. (1993) *The energetics of running and running shoes.* Journal of Biomechanics. 26, Supplement 1(0): p. 41-51.

[7] Donelan, J.M., Q. Li, V. Naing, J.A. Hoffer, D.J. Weber, and A.D. Kuo (2008) *Biomechanical Energy Harvesting: Generating Electricity During Walking with Minimal User Effort.* Science. 319(5864): p. 807-810.

[8] Jean-Mistral, C. and S. Basrour (2010) *Scavenging energy from human motion with tubular dielectric polymerElectroactive Polymer Actuators and Devices (EAPAD) 2010* 7642: p. 764209-12.

[9] Jean-Mistral, C., S. Basrour, and J.J. Chaillout (2008) *Dielectric polymer: scavenging energy from human motion.* Proc. SPIE. 6927: p. 692716.

[10] Pelrine, R., R.D. Kornbluh, J. Eckerle, P. Jeuck, S. Oh, Q. Pei, and S. Stanford (2001) *Dielectric elastomers: generator mode fundamentals and applications.* Proc. SPIE. 4329: p. 148.

[11] Prahlad, H., R. Kornbluh, R. Pelrine, S. Stanford, J. Eckerle, and S. Oh 2005 *Polymer Power: Dielectric Elastomers and Their Applications in Distributed Actuation and Power Generation* in *Proc. ISSS*: Bangalore, India. p. 100.

[12] Anderson, I.A., T.A. Gisby, T.G. McKay, B.M. O'Brien, and E.P. Calius (2012) *Multi-functional dielectric elastomer artificial muscles for soft and smart machines.* Journal of Applied Physics. 112(4): p. 041101-20.

[13] Bar-Cohen, Y. (2004) *Electroactive Polymer Actuators as Artificial Muscles: Reality, Potential, and Challenges* SPIE: New York.

[14] Pelrine, R., R. Kornbluh, P. Qibing, and J. Joseph (2000) *High-speed electrically actuated elastomers with strain greater than 100%.* Science. 287(5454): p. 836-9.

[15] Pelrine, R. and H. Prahlad (2008) *Generator Mode: Devices and Applications* in *Dielectric Elastomers as Electromechanical Transducers* F. Carpi, et al. Editors Elsevier: Oxford, UK. p. 146-155.

[16] McKay, T.G., B.M. O'Brien, E.P. Calius, and I.A. Anderson (2011) *Realizing the Potential of Dielectric Elastomer Generators. Proc. SPIE* 7976: p. 7976-10.

[17] McKay, T., O'Brien, B., Calius, E., Anderson, I. (2010) *Self-priming dielectric elastomer generators.* Smart Materials and Structures. 19(5).

[18] Pelrine, R.E., R.D. Kornbluh, J.S. Eckerle, and H. Prahlad 2008 *Circuits for electroactive polymer generators* SRI International.

[19] McKay, T.G., B. O'Brien, E. Calius, and I. Anderson (2010) *Self-priming dielectric elastomer generators.* Smart Mater. Struct. 19(055025): p. 055025.

[20] McKay, T., O'Brien, B., Calius, E., Anderson, I. (2010) *An integrated, self-priming dielectric elastomer generator.* Applied Physics Letters. 97(6).

[21] McKay, T., B. O'Brien, E. Calius, and I. Anderson (2010) *An integrated, self-priming dielectric elastomer generator.* Applied Physics Letters. 97(062911): p. 062911.

[22] O'Brien, B.M., E.P. Calius, T. Inamura, S.Q. Xie, and I.A. Anderson (2010) *Dielectric elastomer switches for smart artificial muscles.* Appl. Phys. A.: p. 1-5.

[23] O'Brien, B. Simulation, Fabrication, and Control of Biomimetic Actuator Arrays PhD Auckland Bioengineering Institute 2010 The University of Auckland.

[24] McKay, T.G., B.M. O'Brien, E.P. Calius, and I.A. Anderson (2011) *Soft generators using dielectric elastomers.* Applied Physics Letters. 98(14): p. 142903-3.

[25] Roundy, S., P.K. Wright, and J.M. Rabaey (2004) *Energy Scavenging For Wireless Sensor Networks*Boston, MA: Kluwer Academic Publishers. 212.

[26] Due, J., S. Munk-Nielsen, and R.O. Nielsen (2010) *Energy harvesting with Di-Electro Active Polymers.* IET Conf. Pub. 2010(CP563): p. WE244-WE244.

[27] Graf, C., J. Maas, and D. Schapeler (2010) *Energy harvesting cycles based on electro active polymers. Proc. SPIE* 7642: p. 764217-12.

[28] Kaltseis, R., C. Keplinger, R. Baumgartner, M. Kaltenbrunner, T. Li, P. Machler, R. Schwdiauer, Z. Suo, and S. Bauer (2011) *Method for measuring energy generation and efficiency of dielectric elastomer generators.* Appl. Phys. Lett. 99(16): p. 162904.

[29] Koh, S.J.A., X. Zhao, and Z. Suo (2009) *Maximal energy that can be converted by a dielectric elastomer generator.* Appl. Phys. Lett. 94(26): p. 262902.

[30] O'Brien, B.M. and I. Anderson (2011) *An Artificial Muscle Ring Oscillator.* Mechatronics, IEEE/ASME Transactions on. PP(99): p. 1-4.

[31] O'Brien, B.M., E.P. Calius, T. Inamura, S.Q. Xie, and I.A. Anderson (2010) *Dielectric Elastomer Switches for Smart Artificial Muscles.* Applied Physics A: Materials Science & Processing. 100: p. 385-389.

[32] O'Brien, B.M., T.G. McKay, S.Q. Xie, E.P. Calius, and I.A. Anderson (2011) *Dielectric Elastomer memory*. Proc. SPIE. 7976: p 797621-7.

Chapter 8

Large Scale Capacitive Skin for Robots

P. Maiolino, A. Ascia, M. Maggiali, L. Natale, G. Cannata and G. Metta

Additional information is available at the end of the chapter

http://dx.doi.org/10.5772/50489

1. Introduction

In order to allow robots to share our space and chores, tactile sensing is crucial. Indeed it allows safe interaction of robots with people and objects, because it provides the most direct feedback to control contact forces both in voluntary and involuntary interactions. Furthermore, it allows improving performance in tasks that require controlled physical interactions in uncontrolled environments where the location and the characteristics of contact cannot be predicted or modeled in advance and more complex forms of interactions are required. Therefore, a tactile sensor system capable of measuring contact forces over large areas is needed. Tactile sensing in robotics has been widely investigated in the past 30 years and many examples of engineering solutions to tactile sensing have been presented in the literature [1]. Research in this field has focused largely on transduction principles and transduction technologies [2]; however, various technical issues have limited the transition from a single tactile element (or a small matrix prototype) to a large scale integrated solution: it is easy to understand that a sensitive robot skin cannot be achieved by simply aggregating a large number of single sensors. In fact, the concept of robot skin entails a number of system level problems that simply do not appear when focusing on small tactile sensors or small area arrays:

- Modularity and Conformability: the system should be modular and it should be simple to tailor it to the shape of the surface of the robot.
- Infrastructure and Networking: embedded electronics and distributed computation are necessary to facilitate the integration in the robot, otherwise a large number of wires would impede mobility and dexterity of the robot.
- Coverage: The sensors must cover the largest possible area of the robot without increasing the number of wires or the size of the ancillary electronics.
- Compliance: The sensors must be covered by a soft layer that allows safe interaction with humans and the environment and that forms a protective layer for the sensors themselves.

- Dynamic range and sensitivity: The system should be able to detect a wide range of contact pressures.
- Total Weight and installation space: In order to cover the whole robot body surface the weight of the sensors must be small as well as the required space for installation.
- Costs: The system must be as cheap as possible and off-the-shelf components should possibly be used to decrease the manufacturing cost.
- Manufacturing and deployment: the ease and speed of production have to be taken into account. The deployment procedures must be reproducible on robots with different characteristics.

Several tactile systems have been integrated into robots and described in the literature. Some of them are modular and include hierarchical data processing. Yet, the modules are usually big and cannot be installed on small robot parts. In many cases the spatial resolution is low; in others the sensory modules connectivity is cumbersome. This is clearly in contrast with the requirements above. Small sensor modules are typically necessary in order to allow the communication between them; size constraints are then very critical as not enough space typically can be found on the sensor modules to the local integration of microcontrollers or other networking electronics. One of the first examples of full-body skin for robots was proposed by Inaba et al. [3] that describes a tactile sensor system composed of a layered structure of electrically conductive fabric implementing a matrix of pressure sensitive switches. However, the complexity of the manufacturing process and the limited sensing capabilities represent the most significant limits of this approach. Force detectable surface covers have been introduced by Iwata et al. [4]. Information originating from both resistive and force sensors to correlate pressure information and exerted force is exploited. Although the presented system can be used to cover large surfaces, the actual design is strictly dependent on the actual robot surface. However, a principled discussion about design issues is not carried out. Ohmura et al. [5] proposed a conformable and truly scalable robot skin system formed by self-contained modules supported by a flexible printed circuit boards (PCBs) that can be interconnected. Each module contains 32 tactile elements consisting of a photo-reflector covered by urethane foam. In order to adjust the distance between each tactile sensor element, a band-like bendable substrate that can be easily folded (or even cut) is adopted. The system integrates a microcontroller based architecture for data acquisition and networking, however, the dimension of the system are too big for small robot parts and the spatial resolution featured is quite low [6]. The tactile system that has been developed for the robot RI-MAN also uses flexible PCBs with a tree-like shape to conform to curved surfaces [7]. The tactile elements are commercially available piezoresistive semiconductor pressure sensors, and the measurements have less hysteresis than the one shown by Ohmura et al. [5]; to reduce the number of wires, the sensor modules include multiplexers, but this approach requires to connect each module individually to a controller board. The robot ARMAR-III [8] uses skin patches based on piezoresistive sensor matrices with embedded multiplexers. The patches have a flat or a cylindrical shape and are specifically designed for the different parts of the robot; smaller patches are used for the fingers. In Kotaro [9], tactile sensing is achieved by using flexible bandages formed by two flexible PCBs with an intermediate layer of pressure sensitive conductive rubber. Each bandage has 64 taxels, but no integrated data acquisition electronics is mentioned. Piezoelectric transducers are used for the humanoid

robots Robovie-IIS [10] and CB2 [11]. The transducers were placed individually on the robots and the sensitive skin has a low spatial resolution.

In Yoshikai et al. [12] the construction method for soft stretchable enclosing type tactile sensing exterior for humanoid robots is presented. A cardigan knit sensor which encloses an upper body of a small humanoid has been developed using the proposed method. The overall sensor is constituted by layers of conductive fabrics that are knitted for improved stretchability. The presented cardigan knit has, however, only 75 electrodes.

Shimojo et al. [13] have developed a mesh of tactile sensors that satisfy the requirements of a high speed and reduced number of wires and that is able to cover surfaces of arbitrary curvatures. The tactile sensors mesh is arranged as a net, where only nearby taxels are connected through wires. Since the shape of a patch must be specifically designed, modularity issues are not properly addressed, but scalability is possible thanks to the smart wiring infrastructure: only 4 wires actually exit from the patch.

Mittendorfer et al. [14] have presented a tactile sensor system made by small hexagonal PCB modules equipped with multiple discrete off-the-shelf sensors for temperature, acceleration and proximity. Each module contains a local controller that pre-processes the sensory signals and actively routes data through a network of modules towards the closest PC connection. The sensory system is embedded into a rapid prototyped elastomer skin material and redundantly connected to neighboring modules by just four ports. The functionality of some modules on a KUKA robot arm has been demonstrated, however a complete integration has not been shown.

The ROBOSKIN tactile system (Patent No. I0128764) is our proposal for a skin that is able to cover large area of a robot body. It incorporates a distributed pressure sensor based on capacitive technology. The transducer consists of a soft dielectric sandwiched by electrodes. When pressure is applied to the sensor, the distance between the electrodes changes, and the capacitance changes accordingly (capacitance is a function of distance). The ROBOSKIN system is made up of a number of tactile elements (taxels) geometrically organized in interconnected modules of triangular shape. Above the flexible PCB, there is a layer of silicone foam (Soma Foama 15 from Smooth-On) that covers the 12 taxels and acts as a deformable dielectric. On top of the silicone foam is placed a deformable conductive layer made of electrically conductive lycra-like fabric. This layer is connected to ground and enables the sensor to respond to contacts with objects of any material.

As we mentioned earlier, manufacturing and deployment processes constitute an important aspect of the robot skin design since the skin has in fact to be integrated on robots with different characteristics. Flexibility in this sense is a good property for a robot skin system. In addition, it is desirable that the skin production process is reproducible so to guarantee resilience against impact and forces and the general wear and tear of the material. In particular, the ROBOSKIN tactile system has been integrated on three different types of robots: iCub [15], KASPAR [16] and NAO [17]. The three robots have different sizes and shapes, and the tactile feedback has been used for different purposes. Nevertheless, the methods that have been used to implement the skin were nearly the same, which demonstrate the portability of the sensor system technology. This chapter is organized as follows: firstly, the ROBOSKIN tactile system is presented in details in section 2; section 3 presents the methods and procedures for ROBOSKIN manufacturing and integration; in particular, section 3.1 is

related to the iCub platform; section 3.1.1 presents a new implementation of iCub fingertip with experimental data related to their behavior characterization; section 3.2 is related to KASPAR and section 3.3 presents the ROBOSKIN integration on robot NAO; finally section 4 is dedicated to conclusion.

2. ROBOSKIN technology

As we told before, ROBOSKIN tactile system, incorporates a distributed pressure sensor based on capacitive technology. The transducer consists of a soft dielectric sandwiched by electrodes. When pressure is applied to the sensor, the distance between the electrodes above and below the dielectric changes, and the capacitance changes accordingly (capacitance is a function of distance).

The basis of the sensor is a flexible printed circuit board (PCB). It includes the electronics to obtain 12 measurements of capacitance and send them over a serial bus. In particular, each PCB includes 12 round pads, one for each taxel, and a capacitance to digital converter (CDC) (AD7147 from Analog Devices).

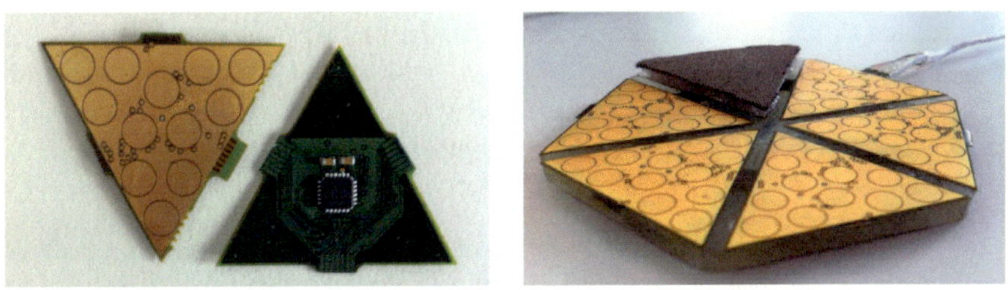

(a) (b)

Figure 1. The triangular module: (a) in the front view it is possible to see the 12 round taxels, in the back view the CDC chip from Analog Devices is shown; (b) An hexagonal patch glued on a cover with the foam elastomer and electrically conductive lycra layers.

The chip can measure the capacitance of all taxels with 16 bit resolution, but we use only 8 bit measurements, as any higher resolution is covered by noise. As a result, one measurement unit corresponds to 2.88 fF. The CDC has an I^2C serial interface and each chip can be assigned with a 2 bit address; therefore up to 4 chips can communicate over the same serial bus. The shape of the PCB is in most cases a triangle (only for the fingertips of the robot iCub we used a unique solution, which we will discuss in section 3.1.1), see figure 1(a). The triangular shape was chosen in analogy to polygonal modeling in 3D computer graphics, which uses triangles to describe the shapes of objects. The triangles can conform up to a certain degree to generic smooth curved surfaces (see figure 2(a) and figure 2(b)).

The triangular PCBs also include the electronics to communicate between themselves: three communications ports placed along the sides of the triangle (one for input and two for output) relay the signals from one triangle to the adjacent one. Up to 16 triangles can be connected in this way (4 serial buses with 4 different addresses each) and only one of them needs to be connected to a microcontroller board (MTB, see figure 3). This is a critical advantage since it reduces the amount of wires and electrical connections that are required. The MTB can also

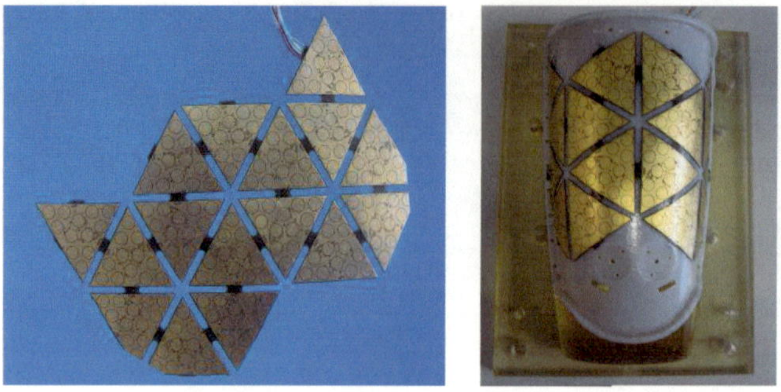

(a) (b)

Figure 2. (a) A ROBOSKIN patch; (b) ROBOSKIN patch conformed to a curved body parts.

be used to program each CDC to either measure its 12 taxels independently at 50 Hz or an average of them at about 500 Hz.

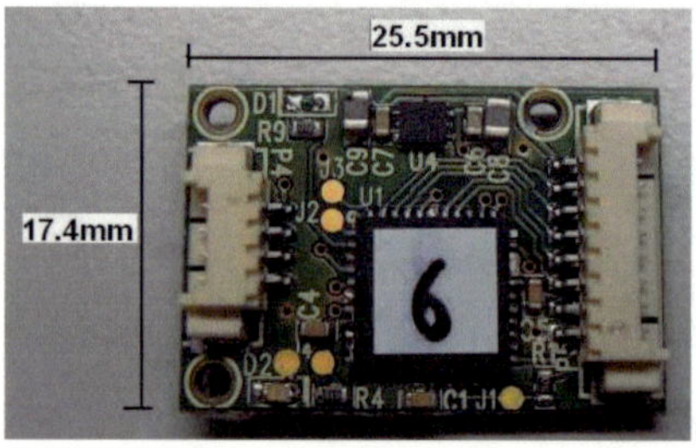

Figure 3. The microcontroller board (MTB). We designed a small microcontroller board, which collects the measurements from up to 16 CDC chips and sends the measurements over a CAN bus.

Above the flexible PCB there is a layer of silicone foam (Soma Foama 15 from Smooth-On). It is 2 mm thick for the hands of the robot iCub, and 3 mm in all other cases. It covers the 12 pads and acts as a deformable dielectric for the capacitive pressure sensor. The foam also makes the skin compliant. On top of the silicone foam there is a second conductive layer made by electrically conductive Lycra. This layer is connected to ground and enables the sensor to respond to objects irrespective of their electrical properties. It serves as the common electrode above the silicone foam for all the taxels, see figure 1(b). When pressure is applied to the sensor, this layer gets closer to the round pads on the PCB, and the sensor measures the distance. This layer also reduces the electronic noise coming from the environment, in particular the stray capacity, which can be a problem for capacitive pressure sensor systems [2]. For the KASPAR and NAO, we didn't use the conductive layer on top of the silicone foam, as in these cases the robot is intended to interact only with humans, in which case the human constitutes the ground plane (like in many consumer products, which are responsive to humans, but are not responsive to insulators, for example).

3. Integration process

In order to integrate the ROBOSKIN sensor on a robot, this general procedure must be followed:

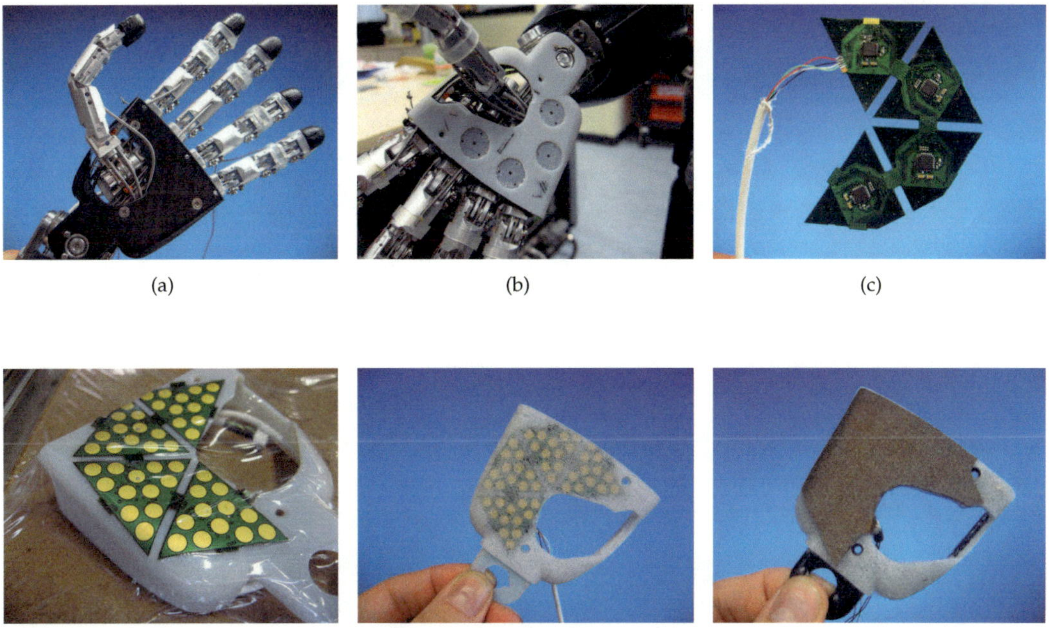

(a) (b) (c)

(d) (e) (f)

Figure 4. The production steps for the palm of iCub. For a description of each step, please refer to the text.

- Identification of the part to be covered, see Fig 4(a). If no CAD model is available, obtain the shape with a 3D laser scanner (as for example for the hands of KASPAR).
- Manufacturing of the part (or of a cover) with a 3D printer (Eden 3D printer from *Objet*). The resulting parts look for example like in Fig. 4(b). Round holes provide space for the CDC chip and the other electronic components which are soldered on the PCB.
- Identification and wiring of the mesh of flexible PCBs that is needed to cover the part, see Fig. 4(c).
- Bonding of the PCBs on the part with bi-component glue and the help of a vacuum system, see Fig. 4(d).
- Covering the PCBs with silicone foam, see Fig. 4(e). To this aim we employ a specific purpose-built mold for each part.
- Covering of the silicone foam with a conductive lycra as ground plane, as shown in Fig. 4(f).

3.1. Implementation on iCub robot

The integration of the ROBOSKIN tactile sensor on iCub, has involved forearms, upper arms, torso and hands (palms and fingertips). The current implementation allows obtaining a number of distinguished taxels equal to 2400. In figure 5 the final result is shown.

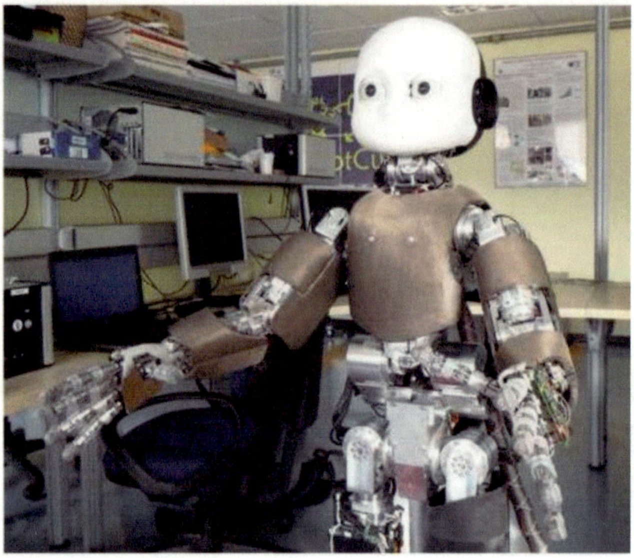

Figure 5. The iCub covered with artificial skin

3.1.1. Implementation on the iCub hands

As it is possible to see in figure 4(a), the palm of iCub is made from carbon fiber; since this is a structural part, we decided not to modify it, but instead, we added another cover above the carbon fiber part as a basis for the sensor: it has a thickness of 1.2 *mm* and provides space for the CDC chip and the other electronic components which are soldered on the PCB. The implementation steps are reported in Fig. 4. While for all the iCub parts the standard ROBOSKIN solution for the PCB has been used, for the iCub fingertips it has been necessary to design a specific solution (see figures 7(a), 7(b), 7(c)), since they have small size and round shape (each fingertip is 14.5 mm long and 13 mm wide and high and has a round shape that resembles a human fingertip).

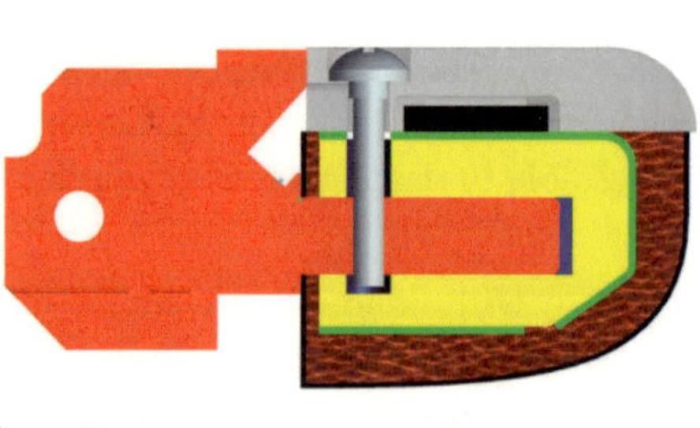

Figure 6. Crosssection of the fingertip.

The structure of the fingertip is illustrated schematically in Fig. 6: the inner support of the fingertip is shown in yellow, and the flexible PCB, that is wrapped around, is depicted in green. To mechanically attach the fingertip to the hand, the last phalanx of each digit (shown in red) has a stick that fits inside a hole in the inner support. A screw is used to secure the fingertip and, in addition, the screw fixes a fingernail on top of the fingertip that covers the

PCB. The dielectric, made of silicone rubber foam, is depicted in brown; around the foam there is the conductive lycra layer shown in black as well as the AD7147 chip.

(a) (b) (c) (d) (e) (f)

Figure 7. The production steps for the fingertip iCub

The fingertip production protocol involves the following steps:

- The flexible PCB is wrapped around an inner support that was printed with a 3D printer. (see figures. 7(a), 7(b), 7(c)). As we are using an I^2C serial bus, only 4 wires have to be connected to the PCB (Vcc, ground, serial data line and serial clock). They travel along the side of the fingers to small boards at the back of the hand. These boards relay the data from all five fingertips (and the four triangular modules in the palm) to one microcontroller board, which is located in the forearm of iCub.
- A first layer, made with a foam elastomer (Smooth-on Soma Foama 15), is deposited over the PCB; mechanical deformation of this soft dielectric material leads to capacitance variations; therefore, it is possible to detect pressures applied on the fingertip, see figure 7(d).
- The second layer, that is made with conductive lycra, is glued over the silicone foam substrate, allowing the development of a single ground plane above all electrodes placed on the PCB thus enabling the detection of each type of object (conductive and non conductive) within noise reduction. This layer is connected to the digital ground of the CDC by one flat pad on the PCB.
- Finally the third protective layer, made again with Smooth-on Soma Foama 15, is deposited; it is used such as a protection, thus intrinsically increases the lifetime of the fingertip sensor.

Figure 8. Fingertip sensor placed on the support used for measurements together with schematic view of the line where measurements have been taken

With respect to the first implementation of the fingertips [18], where it was used a self-made mixture of silicone (CAF4 from *Rhodia-Silicones*) and carbon-black particles (Vulcan XC72 from *Cabot*) as a ground plane, and, as protective layer, silicone glue (Sil-Poxy from *Smooth-On*) sprayed above it, the new implementation made with conductive lycra, has been chosen because it increases the durability of the fingertip to usury due to friction forces appearing during grasp tasks. In order to characterized the behavior of the new fingertip several experiments have been performed.

The characterization setup consists of a cartesian robot (TT-C3-2020 from IAI) which moves one non-conductive probe against the fingertip. The non-conductive probe is fixed at the top of an off-center load cell (AS1 form Laumas) which measures independently the force applied to the fingertip by the probe during the test. A microcontroller records the CDC output and the load cell circuit output. Therefore the Data are stored in a computer by a dedicated graphic user interface made in Matlab.

The characterization protocol was the following:

- One non-conductive probe of 4 mm diameter has been used. By the use of the Cartesian robot, the non-conductive probe applied different pressures to 13 positions, on the fingertip, separated by 0.5 mm along a 6 mm long straight line of two capacitors, indicated with C1 and C2(see figure 8).
- In each position, different pressures have been applied to the fingertip by vertical displacement of the non-conductive probe using the cartesian robot.
- Each pressure has been applied for 2 seconds, with intervals of 2 seconds. In order to investigate sensor repeatability, each pressure has been repeated fifty times. It must be observed that, along the straight line the probe was perpendicular to the fingertip.

Only the steady-state responses at external pressures of the sensor have been taken into account during post elaboration. Mean values of the fifty steady-state responses of C1 and C2, at different positions of the probe, are reported from figure 9(a) to figure 9(e). It is possible to observe a non linear trend of the capacitor variation as a function of the applied pressure. Therefore, the least squares method has been used to fit a polynomial model on each steady state response at each position. Figure 10(a) shows an example of the choice of the order of the polynomial function for CDC-C1 in the case of the position 0 mm, while, in figure 10(b), it is shown the norm of residual with respect to the polynomial order, on the basis of which,

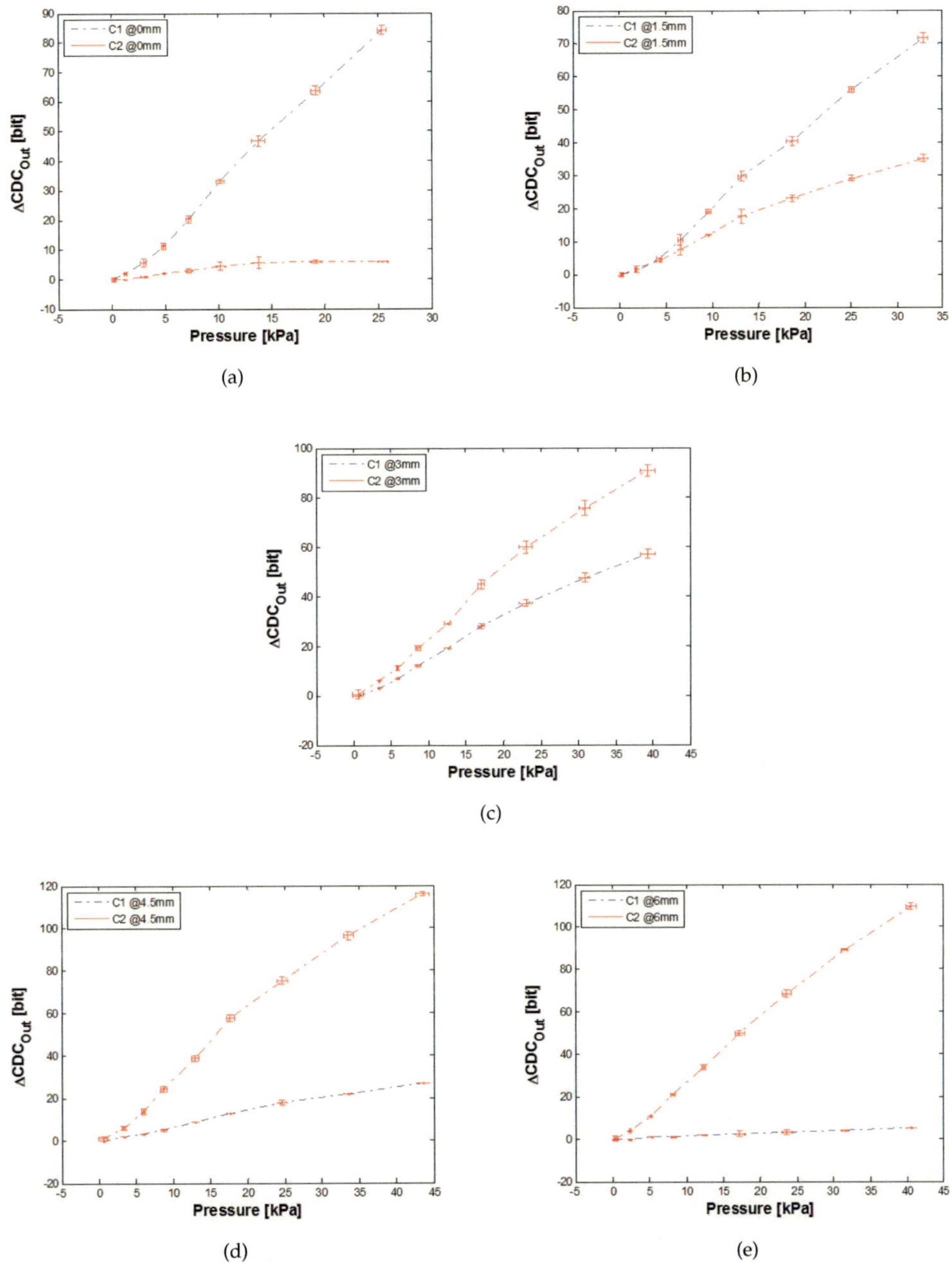

Figure 9. steady-state response at 0 mm (a), 1.5 mm (b), 3 mm (c), 4.5 (d) mm and 6 mm (e) position point of the straight line of measurement

a 4th polynomial model order has been chosen. Polynomial functions thus obtained, were used to calculate the variation of capacitors C1 and C2 due to a fixed pressure as a function of position along the measurement line. Results are presented from figure 11(a) and figure 11(e), which report the variation of capacitors C1 and C2 as a function of the probe position for five different pressure values from 5kPa to 25 kPa.

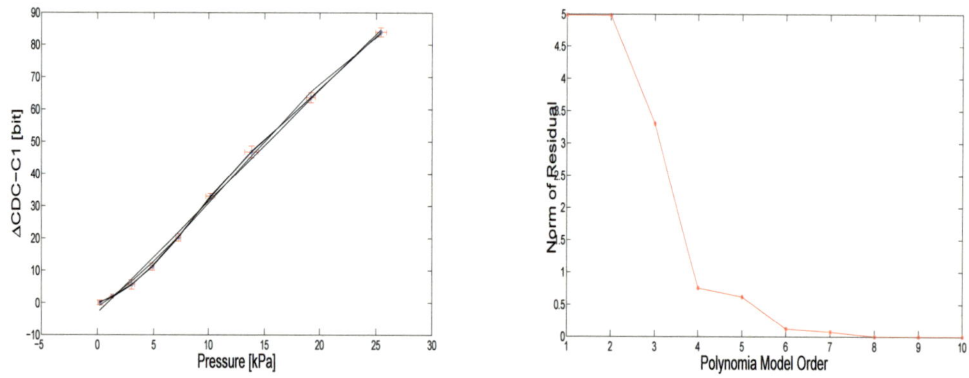

(a) (b)

Figure 10. a) An example of the choice of the order of the polynomial function for CDC-C1 in the case of the position 0 mm; b) Norm of the residual with respect to polynomial order

3.1.2. Implementation on the iCub arms

For the iCub arms, the skin was integrate directly into the covers, so new covers were designed. The production steps are the same as for the palm and the other iCub parts and are illustrated in fig. 12, for the forearm, and fig. 13, for the upper arm. On the two iCub arms we have 1464 contact points: 6 patches (61 triangular module) and 7 MTBs for each arm.

3.1.3. Implementation on the Torso of iCub

For covering the iCub Torso, five patches has been used, obtaining 528 contact points. In figure 15 it is possible to see the front of the torso covered with the glued patches and the silicone foam substrate (figure15(a)), the torso with the conductive lycra substrate on top (figure 15(b)) and finally the back of the torso cover with the 4 MTBs that are needed for the five patches (figure 15(c)).

3.2. Implementation of artificial skin on KASPAR

Despite the other two robot, that have an iconic appearance, KASPAR has a more human-like appearance, in fact in order to have a "natural" shape a child-sized mannequin was used as a basis and the legs, torso and the hands were kept. In order to invite children to touch the hands (which is more like touching a doll), the maintenance of human-like appearance has been an important parameter that has been kept under consideration in the covering process. For this purpose the hands have been covered with coloured foam silicone rubber (it has been

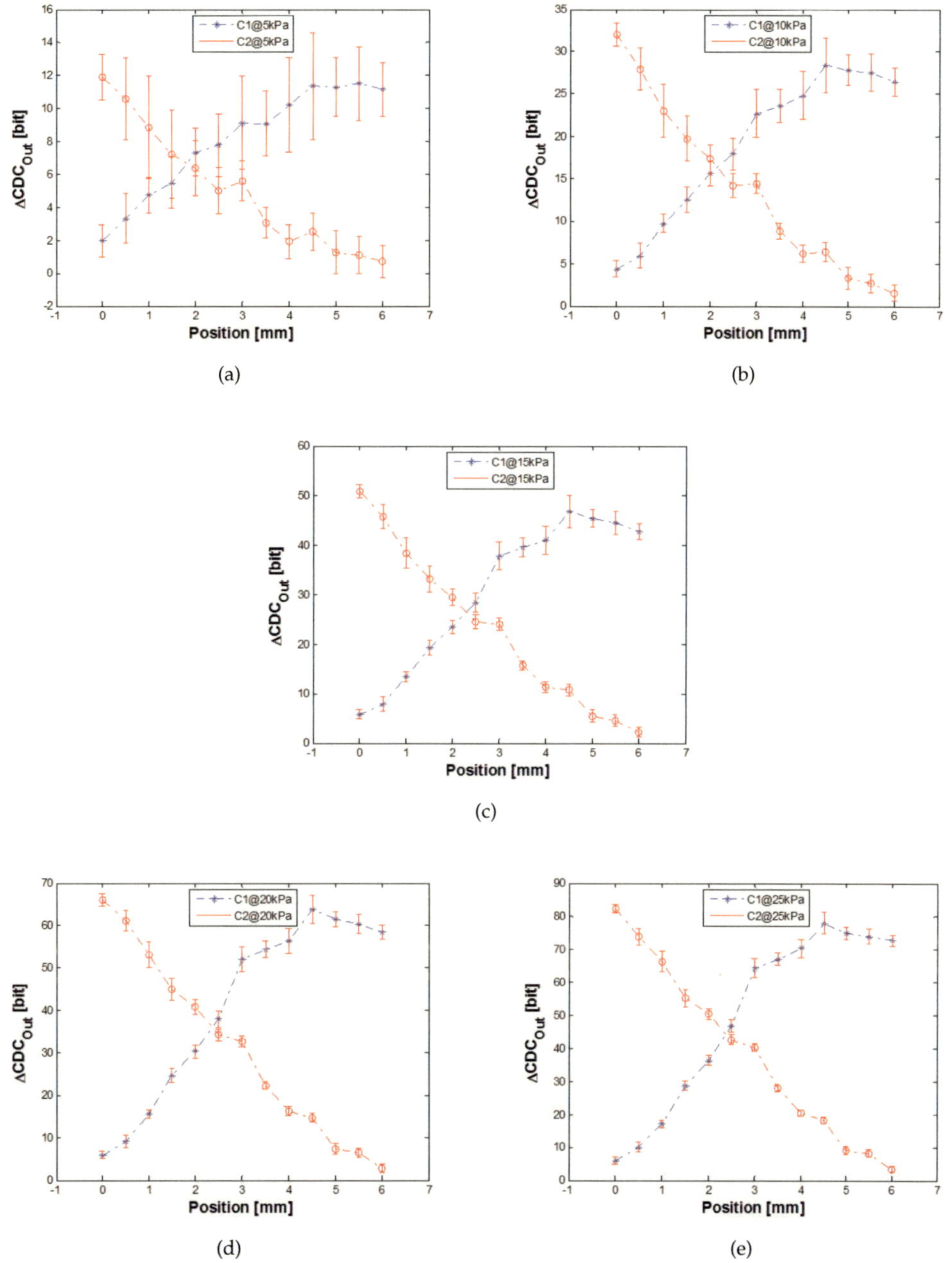

Figure 11. Calculated variations of C1 and C2 due to a fixed pressure applied by the probe along different positions of the straight line of measurement

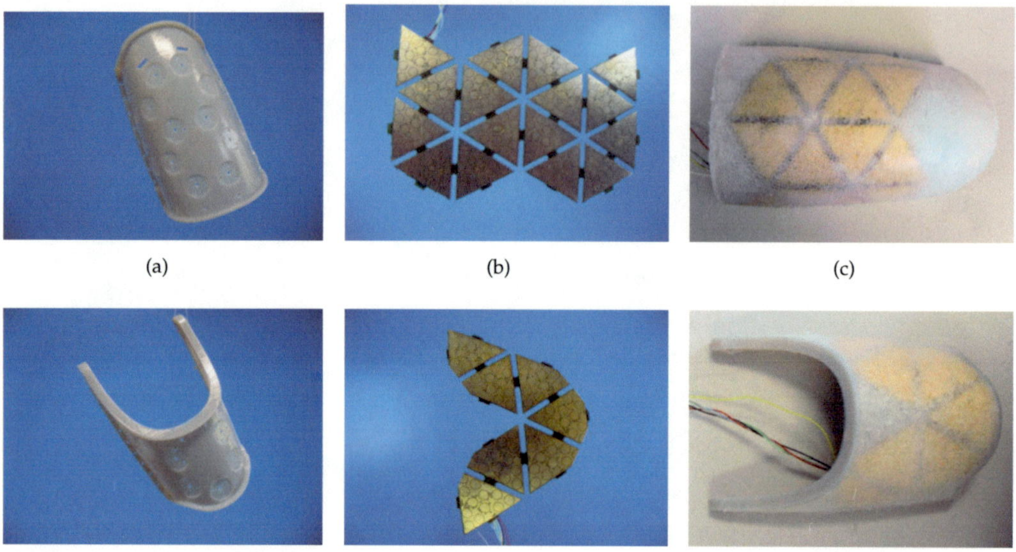

(a) (b) (c)

(d) (e) (f)

Figure 12. The process steps for iCub forearm: two covers are needed which are mounted together on the robot forearm.

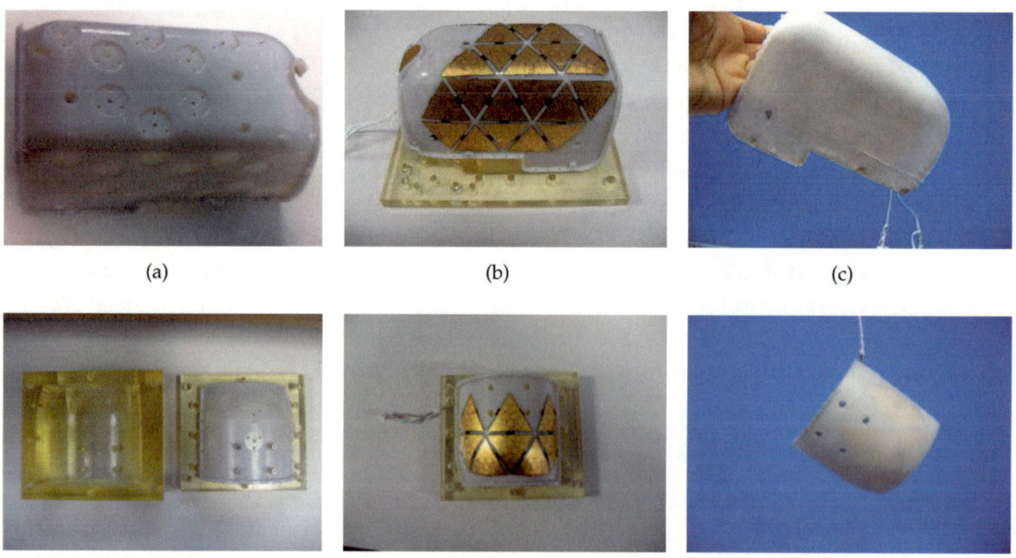

(a) (b) (c)

(d) (e) (f)

Figure 13. The process steps for iCub upper arm: two covers are needed which are mounted together on the robot upper arm.

Figure 14. The covers of iCub arms covered with conductive lycra

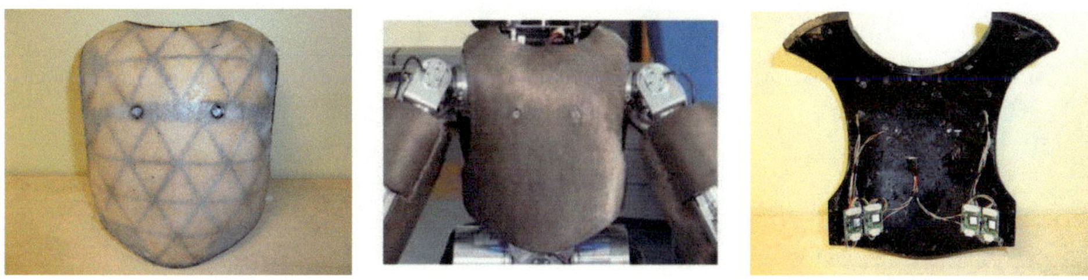

(a) (b) (c)

Figure 15. The process steps for iCub Torso

obtained adding color pigments to the rubber mixture). Furthermore it has been decided to not cover the silicone rubber substrate with ground plane (the KASPAR parts come in contact only with human being that constitute the ground plane during tactile contact). The hands have been scanned with a laser scanner and completely rebuild by a 3D printer in order to have the housing for the electronics (see figures 16(a) and 16(b)); The procedure used was the same explained before for the iCub parts. In figure 16(d) the final result for KASPAR hands is shown. For the other parts (cheeks, torso, upperarms and feet) we didn't take in consideration the human-like appearance because this parts, in the final setup of KASPAR, are covered by clothes. For the upper arms a cover with the housing for the electronics was build by a 3D printer and the patch was covered by a neoprene substrate. For the cheeks, torso and the feet it has not been possible to rebuild the part with the 3D printer (more expensive), so Artificial Skin patches have been applied on the puppet surfaces by glue. In total on KASPAR there are 12 MTB boards and 68 triangles that corresponds to 1128 contact point.

3.3. Implementation on NAO

For NAO the goal was to sensorize the forearms and the upper arms. Therefore, we designed 6 covers (2 for each forearm and 1 for each upper arm) to mount over the robot structural covers as for the iCub palm.

Fig. 18 shows the steps to cover NAO's forearm with skin. In total NAO has 18 triangles that correspond to 216 contact points for each lower arm and 9 triangles that correspond to 108 contact points for each upper arm.

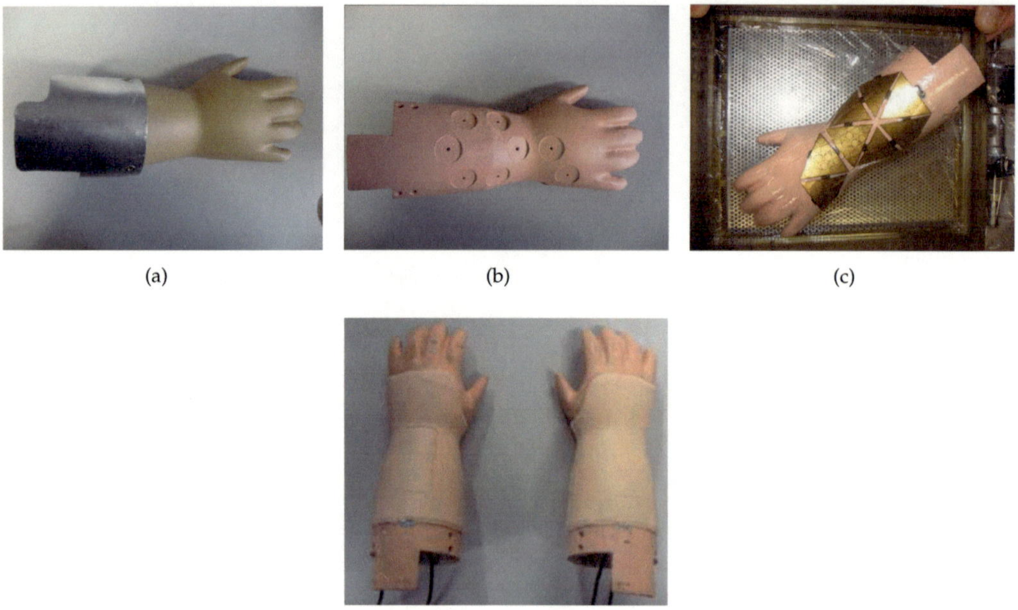

(a) (b) (c)

(d)

Figure 16. The process steps for KASPAR hands

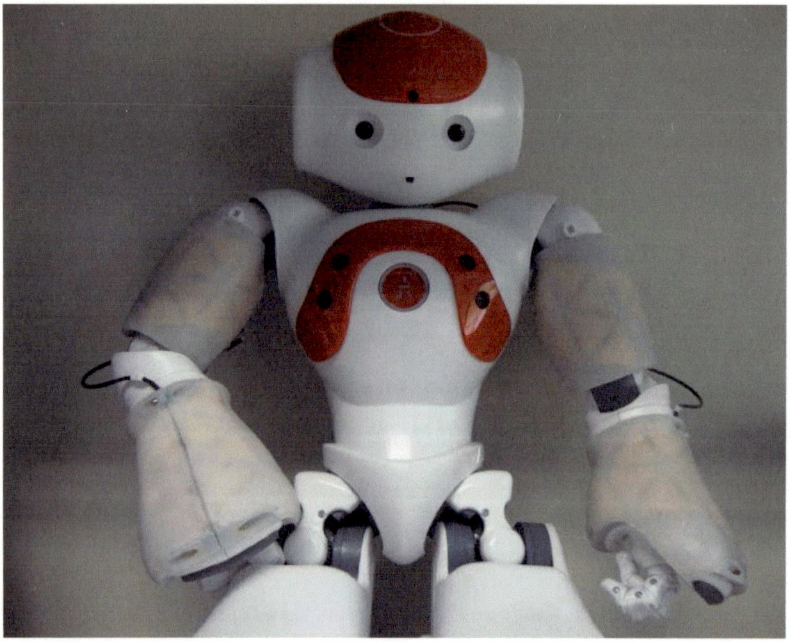

Figure 17. NAO with his forearms covered with the artificial skin.

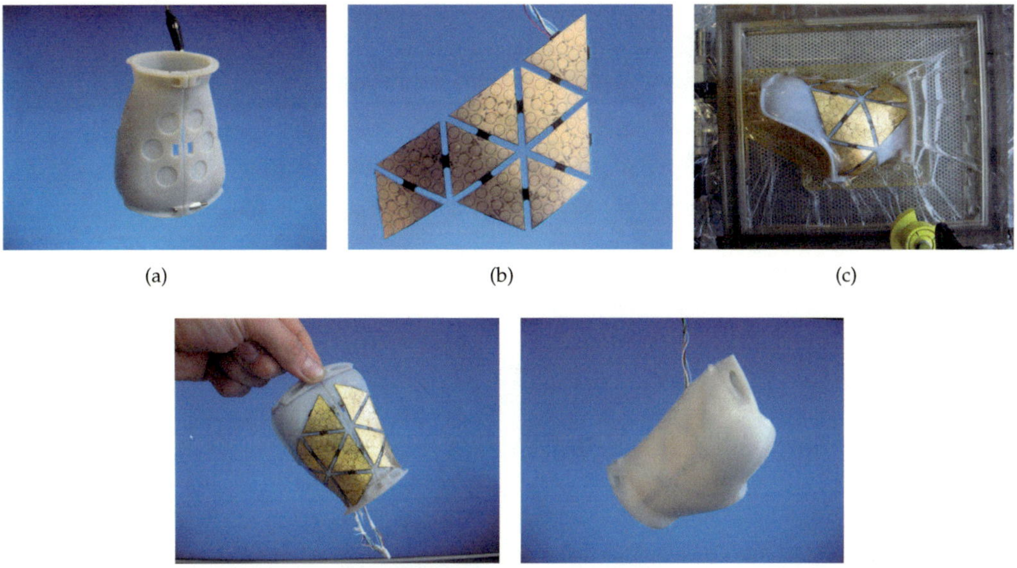

(a) (b) (c)

(d) (e)

Figure 18. **The steps for implementing the skin on NAO's hands.** (a) The two parts of the cover for NAO's forearm. (b) The triangles for one half of the forearm. (c) The PCBs are glued to the cover. (d) The cover with the PCBs. (e) The silicone foam has been molded onto the forearm. (e) The final result.

4. Conclusion

In the chapter the ROBOSKIN tactile system has been presented; in particular the procedures and methods for its integration in three different robotic platforms have been shown, proving its portability and its capability to cover large area of the robots body. Indeed the three robots have very different characteristics, but the integration steps were the same. A new implementation for the iCub fingertips has been shown and behavior characterization data have been presented. The new implementation allows to have a better robustness increasing the durability of the fingertip during grasp operations.

Acknowledgements

The research leading to these results has received funding from the European Community Seventh Framework Programme(FP7/2007-2013) under Grant Agreement no. 231500 (Project ROBOSKIN).

Author details

P.Maiolino and G.Cannata
Department of Communication, Computer and System Sciences (DIST), University of Genova, Genova, Italy

A.Ascia, M.Maggiali, L.Natale and G.Metta
Fondazione Istituto Italiano di Tecnologia (Italian Institute of Technology), Genova, Italy

5. References

[1] Webster J (1988) Tactile Sensors for Robotics and Medicine. Wiley, NewYork.

[2] Lee M H, Nicholls H R (1999) Tactile sensing for mechatronics: a state of the art survey. Mechatronics. 9:1Ű31.

[3] Inaba M, Hoshino Y, Nagasaka K, Ninomiya T, Kagami S, Inoue H (1996) A full-body tactile sensor using electrically conductive fabric and strings. Proc. of 9th Int. Conf. on Intelligent Robots and Systems. 450-457.

[4] Iwata H, Hoshino H, Morita T, Sugano S (2001)Force Detectable Surface Covers for Humanoid Robots. Proceedings of the 2001 IEEE/ASMA International Conference on Advanced Intelligent Mechantronics (AIM'01).

[5] Ohmura Y, Kuniyoshi Y, Nagakubo A. (2006) Conformable and scalable tactile sensor skin for curved surfaces. Proc. IEEE Int. Conf. on Robotics and Automation (ICRA).

[6] Ohmura Y, Kuniyoshi Y (2007) Humanoid robot which can lift a 30kg box by whole body contact and tactile feedback. Proc. IEEE/RSJ Int. Conf. on Intelligent Robots and Systems (IROS).

[7] Mukai T, Onishi M, Hirano S, Luo Z (2008) Development of the tactile sensor system of a human-interactive robot ŞRI-MANŤ. IEEE Transactions on Robotics. 24:505Ű512.

[8] Asfour T, Regenstein K, Schroder J, Dillmann R (2006) ARMAR-III: A humanoid platform for perception-action integration. Proc. Int. Workshop on Human-Centered Robotic Systems.

[9] Mizuuchi I, Yoshikai T, Nishino T, Inaba M (2006) Development of musculoskeletal humanoid Kotaro. Proc. IEEE Int. Conf. on Robotics and Automation (ICRA).

[10] Tajika T, Miyashita T, Ishiguro H (2006) Automatic categorization of haptic interactions - what are the typical haptic interactions between a human and a robot?. Proc. IEEE-RAS Int. Conf. on IEEE Conference on Humanoid Robots (Humanoids).

[11] Minato T, Yoshikawa Y, Ishiguro H, Asada M (2007) CB2: A child robot with biomimetic body for cognitive developmental robotics. Proc. IEEE-RAS 7th Int. Conf. on Humanoid Robots (Humanoids).

[12] Yoshikai T, Fukushima H, Hayashi M, Inaba M (2009)Development of Soft Stretchable Knit Sensor for Humanoids Whole-Body Tactile Sensitivity. Proceedings of the 2009 IEEE Conference on Humanoid Robotics (HUMANOIDS 2009).

[13] Shimojo M, Araki T, Ming A, Ishikawa A(2010) A High Speed Mesh of Tactile Sensors Fitting Arbitrary Surfaces. IEEE Sensors Journal. 10:822-831.

[14] Mittendorfer P, Cheng G (2011) Humanoid Multimodal Tactile-Sensing Modules. Robotics, IEEE Transactions on. 27:401-410.

[15] Metta G, Sandini G, Vernon D, Natale L, Nori F (2008) The iCub humanoid robot: an open platform for research in embodied cognition. In Proceedings of the 8th Workshop on Performance Metrics for Intelligent Systems (PerMIS '08).

[16] Dautenhahn K, Nehaniv C L, Walters M L, Robins B, Kose-Bagci H, Assif Mirza N, Blow M (2009) KASPAR Ű a minimally expressive humanoid robot for humanŰrobot interaction research. Applied Bionics and Biomechanics. 6:3-4.

[17] Gouaillier D, Hugel V, Blazevic P, Kilner C, Monceaux J, Lafourcade P, Marnier B, Serre J, Maisonnier B (2009)Mechatronic design of NAO humanoid. Robotics and Automation.IEEE International Conference on (ICRA). 769-774.

[18] Schmitz A, Maggiali M, Natale L, Bonino B, Metta G (2010) A tactile sensor for the fingertip of the humanoid robot iCub. Intelligent Robots and Systems (IROS), 2010 IEEE/RSJ International Conference on. 2212-2217.

Chapter 9

Ionic Polymer Metal Composite Transducer and Self-Sensing Ability

Doan Ngoc Chi Nam and Ahn Kyoung Kwan

Additional information is available at the end of the chapter

http://dx.doi.org/10.5772/51310

1. Introduction

During the recent decade, ionic polymer metal composite (IPMC), as a smart material, is gaining great importance in the use for both sensors and actuators. An ion polymer metal composite (IPMC) is an Electro-Active Polymer (EAP) that bends in response to a small applied electrical field as a result of mobility of cations in the polymer network [1] and vice versa. A typical IPMC sheet is constructed with a thin (200 μm) ionic polymer membrane like Nafion or Flemion and two metalic electrode layers (10 μm thick) outside such as platinum or gold. When a low voltage electrical field (1 – 5V) is applied, the transport of hydrated cations within the solvated IPMC and the associated electrostatic interactions lead to bending motions of the IPMC sheet. Thus, an IPMC can work as a small size actuator. Figure 1 illustrates the mechanism of the IPMC as actuator. On the other hand, when an IPMC is mechanically bent, it will generate a low voltage (order of milivolts) between the two electrodes. The generated voltage is due to the non-uniform concentration of ions in the IPMC membrane. Figure 2 illustrates the behavior of the IPMC as sensor.

Since firstly introduced, the IPMC has been applied in many research fields such as biomedical system, biomimetic robot, and MEMS devices. In these applications, the IPMC may give advantages thank to the low driven voltage, flexible operation, and simple actuating structure. Some applications using IPMC material have been a snake-like robot with IPMC actuator [7], a micro pump [8], a scale biped walking robot [9], an underwater micro robot [10], etc. Because of the increasing importance of IPMC in the smart material field, it is important to clearly understand the behavior inside the IPMC actuator.

Recently, it has been found that the IPMC possesses an attractive characteristic which is so called the "self – sensing" behavior. The self – sensing ability originates from the variation of surface resistance of the two electrode layers during operating period [11 – 12]. If one can capture these variations of surface resistance, the bending curvature can be figured out. The self – sensing ability has been one of the most interesting features of the IPMC transducer.

By employing self – sensing feature, the sensing problem will be easier for the field of biomedical devices, biomimetic and micro robots.

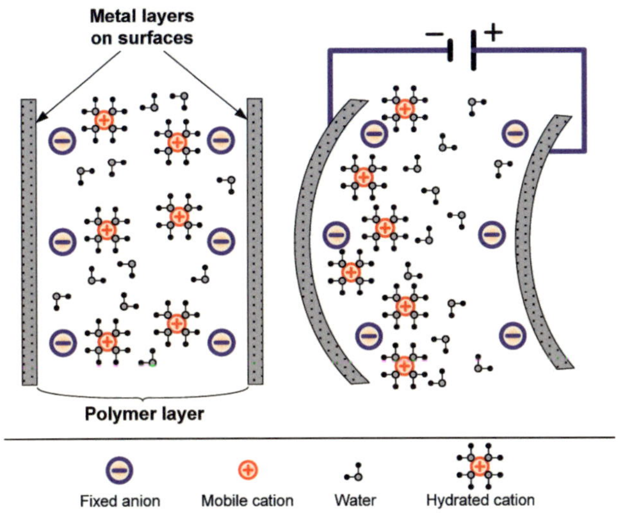

Figure 1. Operating principle of IPMC as actuator

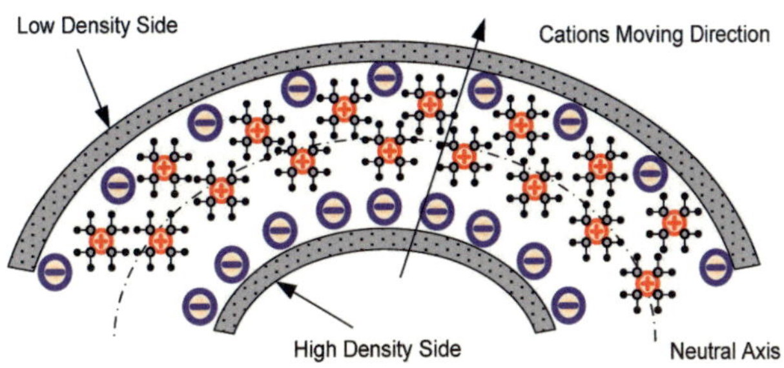

Figure 2. Fundamental of IPMC as a motion sensor

In order to investigate the behavior of IPMC and its self- sensing ability, this chapter will cover physical analysis on an IPMC actuator and its self – sensing model. Firstly, the mechanism inside the IPMC actuator under an applied input voltage signal is carried out. The analysis starts from the governing partial differential equation (PDE) for charge distribution in an IPMC, which was introduced firstly by Nasser et al. in [13] and then used in [14] and [15] for investigating the actuation and sensing response. The effect of distributed surface resistance is also considered to obtain accurate boundary conditions for the PDE. Solving the PDE in asymmetric boundary conditions, the charge density inside the IPMC is obtained. Consequently, the induced stress of the IPMC is obtained. Using this induced stress model, a visual simulation technique is proposed for investigating the actuation capabilities of IPMC actuator via ANSYS environment [18].

Next, the self – sensing ability of the IPMC actuator is analyzed. The self sensing technique is inspired by the variation of surface resistance along the IPMC. An equivalent circuit of

IPMC actuator is present and analyzed. Consequently, a simple technique to implement the self – sensing behavior of IPMC is carried out. In this technique, the variation of surface resistance along the IPMC is figured out using half clamped configuration. Four voltage feedback signals are then employed to estimate the bending behavior. Experimental results show that this technique has a strong ability to apply the self – sensing technique into real application, i.e. micro robot, biomimetic robot.

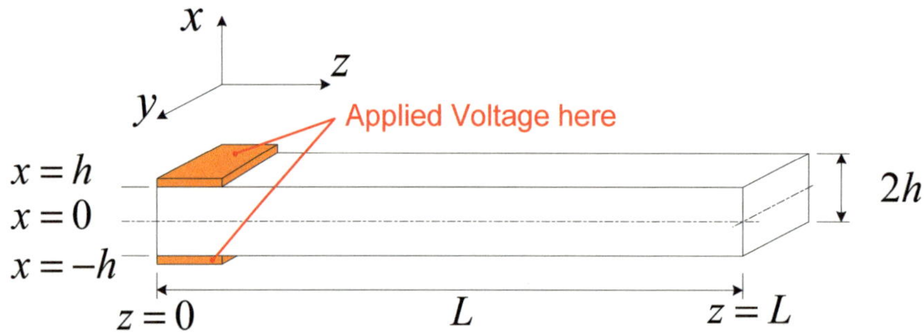

Figure 3. Geometric description of IPMC transducer

2. Electromechanical model of IPMC actuator

The electromechanical model developed for the Nafion based IPMC relies on field and continuity equations. Let D, E, ϕ , and ρ denote the electric displacement, the electric field, the electric potential, and the charge density, respectively, the field equations and continuity equation using in this analysis are as follows:

$$E = \frac{D}{\kappa_e} = -\nabla\phi \tag{1}$$

$$\nabla \cdot D = \rho = F\left(C^+ - C^-\right) \tag{2}$$

$$\nabla \cdot J = -\frac{\partial C^+}{\partial t} \tag{3}$$

where, κ_e is the effective dielectric constant of the polymer, F is Faraday's constant, J is the ion flux vector, and C^+ and C^- are the cation and anion concentrations, respectively. The ion flux is considered to be composed of three primary terms: the diffusion, migration and convection terms:

$$J = -d\left(\nabla C^+ + \frac{C^+F}{RT}\nabla\phi + \frac{C^+\Delta V}{RT}\nabla p\right) + C^+v \tag{4}$$

where, d is the ionic diffusivity, R is the gas constant, T is the absolute temperature, p is the fluid pressure, v is the free solvent velocity field, and ΔV is the volumetric change.

Considering the geometric description of IPMC as in figure 3, since the thickness of an IPMC is much smaller than its length or width, the first assumption is claimed as: inside the polymer, D, E, and J are all restricted to the thickness direction (x-direction). This assumption enables to drop the boldface notation for these variables. Rewrite equation (2) as:

$$C^{+} = \frac{1}{F}\rho + C^{-} \tag{5}$$

Because the anions are constrained to the polymer backbone, the concentration of anions inside the IPMC is considered to be spatially and temporally constant. Therefore, the cation gradient can be rewritten specifically in terms of the electric field E:

$$\nabla C^{+} = \frac{\kappa_e}{F}\nabla^2 E \tag{6}$$

Employing a modified version of Darcy's Law [29], the free solvent velocity field can be calculated as:

$$v = k'\left[C^{-}FE - \nabla p\right] \tag{7}$$

where k' represents the hydraulic permeability. The second assumption is that the convection term can be neglected: $v \to 0$. Then, equation (7) becomes:

$$\nabla p = C^{-}FE \tag{8}$$

By employing equations (5), (6) and (8), the original flux equation (4) can be rewritten as:

$$J = -d\left(\frac{\kappa_e}{F}\nabla^2 E - \frac{\left(1 - C^{-}\Delta V\right)}{RT}E\left(\kappa_e \nabla E - FC^{-}\right)\right) \tag{9}$$

By taking the gradient of (9) and employing (3), the nonlinear PDE in term of electrical field is obtained as:

$$-\frac{\kappa_e}{F}\frac{\partial(\nabla E)}{\partial t} = -d\left(\frac{\kappa_e}{F}\nabla^3 E - \frac{\kappa_e\left(1 - C^{-}\Delta V\right)}{RT}\left(\nabla^2 E.E + (\nabla E)^2\right) - \frac{FC^{-}\left(1 - C^{-}\Delta V\right)}{RT}\nabla E\right) \tag{10}$$

Consider the properties of the IPMC actuator as shown in table 1. It can be seen that the coefficient ratio of the linear term and the nonlinear term:

$$ratio = \frac{FC^{-}}{\kappa_e} \tag{11}$$

is around the order of 5×10^{10}. Hence, it can be concluded that:

$$\frac{\kappa_e\left(1-C^{-}\Delta V\right)}{RT}\left(\nabla^2 E.E+\left(\nabla E\right)^2\right) << \frac{FC^{-}\left(1-C^{-}\Delta V\right)}{RT}\nabla E \tag{12}$$

Neglecting small nonlinear term, the PDE for the electric field can be obtained as:

$$\frac{\partial}{\partial x}\left[\frac{\partial(\kappa E)}{\partial t}-d\frac{\partial^2(\kappa E)}{\partial x^2}-\frac{F^2 dC^{-}}{\kappa RT}\left(1-C^{-}\Delta V\right)(\kappa E)\right]=0 \tag{13}$$

or in the form of charge density ρ :

$$\frac{\partial \rho}{\partial t}-d\frac{\partial^2(\rho)}{\partial x^2}-\frac{F^2 dC^{-}}{\kappa RT}\left(1-C^{-}\Delta V\right)\rho=0 \tag{14}$$

where, C^{-} is the anion concentration and κ is the effective dielectric constant of the polymer.

$F=96487\ C/mol$	$R=8.3143\ J/mol\cdot K$	$T=300K$	$w=0.02mm$
$h=100\,\mu m$	$d=1.030\times10^{-10}m^2/s$	$C^{-}=1200mol/m^3$	$\kappa=1.88\mu F/m$

Table 1. Parameters of IPMC charging model

The solutions of (13) and (14) are analyzed as follows:

It is noted that the IPMC transducer consists of two noble metallic layers outside as electrodes and one ion-exchange polymer membrane with fixed anions at both sides and movable cations. Because of the fixed anion layers, an applied step voltage will lead to an asymmetry in the ion redistribution [16 – 17].

Considering the steady state where $J=0$, the PDE (13) becomes:

$$\frac{\partial^2 E}{\partial x^2}-KE=0 \tag{15}$$

where:

$$K \triangleq \frac{F^2 C^{-}}{\kappa RT}\left(1-C^{-}\Delta V\right) \tag{16}$$

Since fixed anions locate at both sides of the polymer membrane (see figure 1), the charge density in the anode boundary layer keeps a constant value, given by:

$$\rho_{(1)}=-C^{-}F \tag{17}$$

Whereas, in the remaining part, the charge density is given as:

$$\rho_{(2)}=\left(C^{+}-C^{-}\right)F \tag{18}$$

Thus, the charge density ρ, electric field E, and electric potential field ϕ in the anode boundary layer are then given by:

$$\begin{cases} \rho_{(1)}(x) = -C^{-}F \\ E_{(1)}(x) = \dfrac{1}{\kappa}\left(-C^{-}Fx + E_0\right) \qquad \text{for}\left(-h \le x \le -h + w\right) \\ \phi_{(1)}(x) = \dfrac{1}{\kappa}\left(\dfrac{C^{-}Fx^2}{2} - E_0 x\right) + A_0 \end{cases} \tag{19}$$

where, w is the thickness of the anode boundary layer, E_0, and A_0 are an integration constants. In the remaining region, solution of the ODE (5) can be obtained as:

$$\begin{cases} E_{(2)}(x) = C_1 \exp(\sqrt{K}x) + C_2 \exp(-\sqrt{K}x) \\ \phi_{(2)}(x) = \dfrac{1}{\sqrt{K}}\left(-C_1 \exp(\sqrt{K}x) + C_2 \exp(-\sqrt{K}x)\right) + C_3 \quad \text{for}\left(-h + w \le x \le h\right) \\ \rho_{(2)}(x) = \kappa\sqrt{K}\left(C_1 \exp(\sqrt{K}x) - C_2 \exp(-\sqrt{K}x)\right) \end{cases} \tag{20}$$

where, C_1, C_2, and C_3 are integration constants. Consequently, the function of charge density in steady state is obtained for an IPMC actuator under a step voltage input signal:

$$\rho^*(x) = \begin{cases} -C^{-}F & \text{for}\left(-h \le x \le -h + w\right) \\ \kappa\sqrt{K}\left(C_1 \exp(\sqrt{K}x) - C_2 \exp(-\sqrt{K}x)\right) & \text{for}\left(-h + w \le x \le h\right) \end{cases} \tag{21}$$

3. Effect of distributed surface resistance in an IPMC actuator

In order to find correct value for the charge density ρ, electric field E, and electric potential field ϕ, a sufficient set of boundary conditions must be considered. In this section, the effect of distributed surface resistance is investigated to serve as a boundary condition for solving the distribution of charge density, electric field, and electric potential along the z axis. Since the chemical reduction of salt is typically used to build the electrode layers, the surfaces of an IPMC typically consist of aggregated nano particles of noble metal. Thus, the affect of surface resistance is non-negligible in the behavior of IPMC during sensing and actuating process.

Considering the effect of distributed surface resistance, the equivalent circuit of the IPMC is shown as in figure 4. Let the electrode resistance per unit length be r_1 in z direction and r_2 in x direction. As shown in figure 4, $i_p(z,t)$ is the distributed current per unit length going through the polymer due to the ion movement, $i_k(z,t)$ represents the leaking current per unit length, and $i_s(z,t)$ is the surface current on the electrodes. R_{dc} denotes the leakage resistance per unit length. The surface current $i_s(0,t)$ at $z=0$ is the total actuation current. The following equations hold for the impedance model:

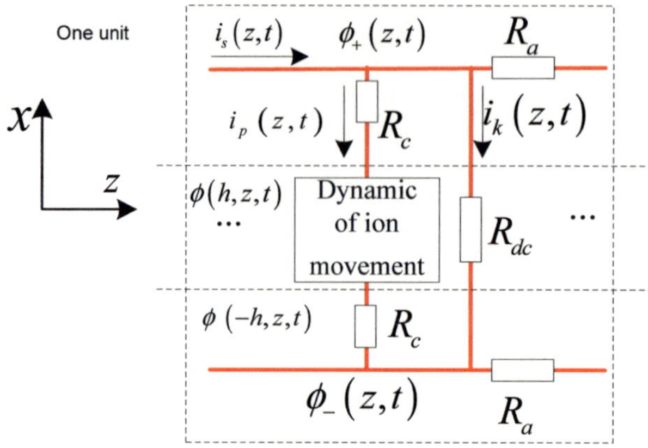

Figure 4. Equivalent circuit of IPMC in a unit length

$$\frac{\partial \phi_{\pm}(z,t)}{\partial z} = \mp R_a i_s(z,t) \tag{22}$$

$$\frac{\partial i_s(z,t)}{\partial z} = -\left(i_p(z,t) + i_s(z,t)\right) \tag{23}$$

Under the effect of distributed surface resistance, the first two boundary conditions for the PDE are:

$$\frac{\partial \phi(h,z,t)}{\partial z} = \int_0^z R_a i_s(\tau,t)d\tau - R_c i_p(z,t) \tag{24}$$

$$\frac{\partial \phi(-h,z,t)}{\partial z} = \phi_0 - \int_0^z R_a i_s(\tau,t)d\tau - R_c i_p(z,t) \tag{25}$$

Other boundary conditions are described as:

$$C^{+}(-h+\mathrm{w}) = 0 \tag{26}$$

$$E_{(1)}(-h+\mathrm{w}) = E_{(2)}(-h+\mathrm{w}) \tag{27}$$

$$\phi_{(1)}(-h+\mathrm{w}) = \phi_{(2)}(-h+\mathrm{w}) \tag{28}$$

These constraints are continuity, and the overall charge-balance conditions. By employing the above constraints, one can solve for E_0, A_0, C_1, C_2, and C_3 to obtain the distribution of charge density, electric field, and voltage potential along the z axis.

Approximately, the dynamic of charging model can be express as:

$$\rho(x,t)=\left(1-\exp\left(\frac{-t}{\tau}\right)\right)\rho^{*}(x),\quad \tau=\frac{\kappa RT}{dC^{-}F} \tag{29}$$

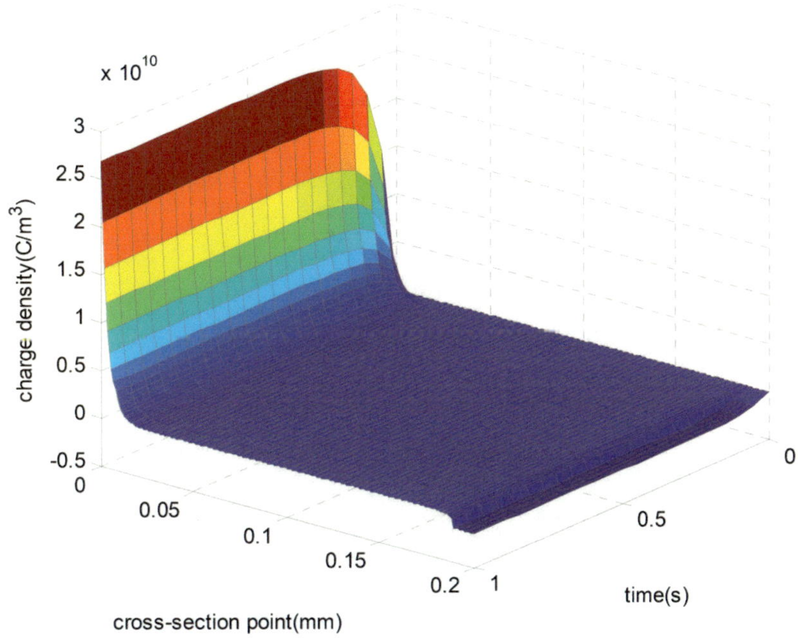

Figure 5. Simulation result for the charge density profile

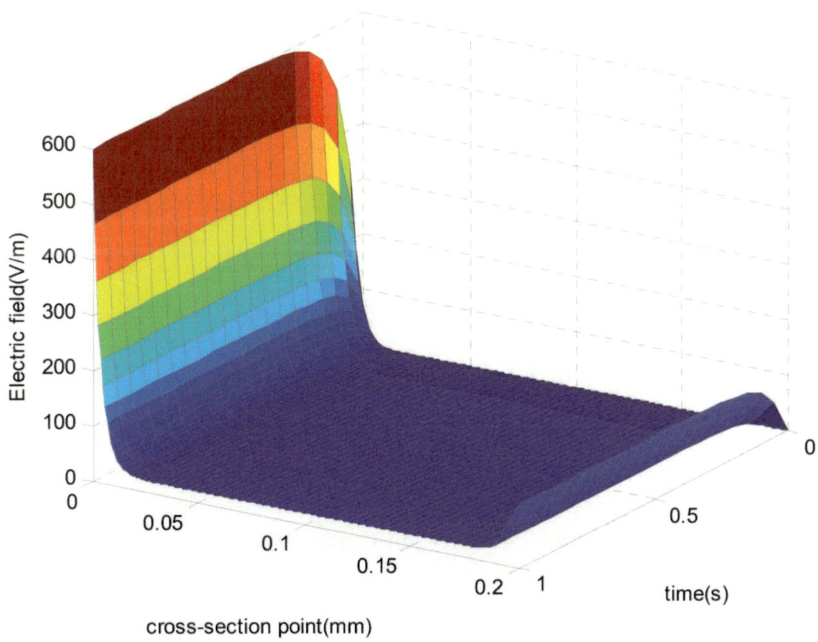

Figure 6. Simulation result for the distributed electric field

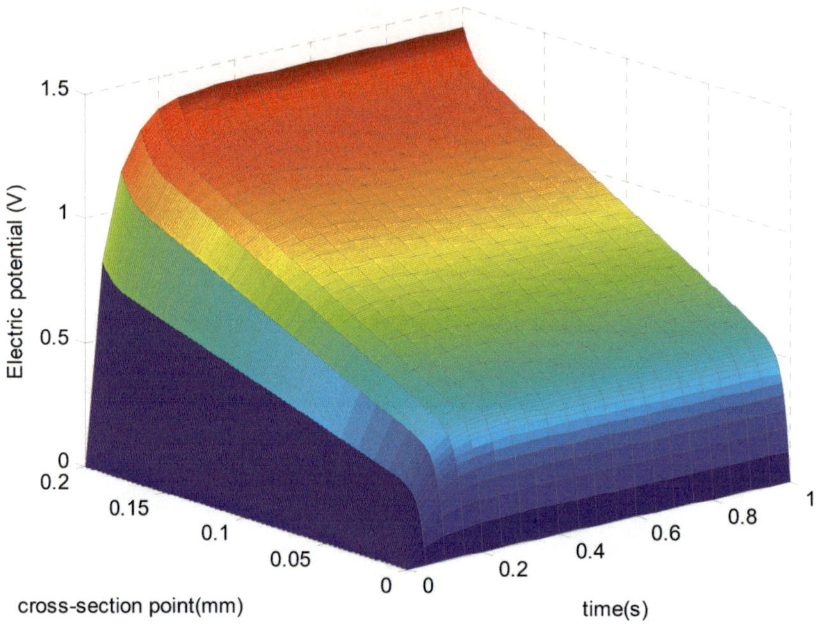

Figure 7. Simulation result for the distributed voltage potential

Here, τ is the natural time scale which characterizes relative speed of cation redistribution. Applying the above analysis, the charging behavior, the distributed electric field, and the distributed electric potential can be trivially obtained. Figure 5, figure 6, and figure 7 are the step response of the charge density, electric field, and electric potential when a 1.5V step input voltage signal is applied to the IPMC actuator.

4. Visual analysis of IPMC actuator

Since the charge distribution is obtained, the induced stress inside the IPMC actuator is available. The key assumption for analysis the electromechanical behavior inside the IPMC transducer is that the induced stress inside the IPMC σ is proportional to the electric charge density ρ :

$$\sigma(x,y,z,t)=\alpha_0\rho(x,y,z,t) \tag{30}$$

where, α_0 is the coupling constant to be determined by experiment. The internal stress σ produces the deformation of the IPMC actuator. Generally, the ideal linear beam theory is employed to analyze the bending behavior of a beam configured IPMC actuator. The bending moment is obtained by integration of induced stress along the x axis. However, this is a complicated work since the induced stress σ is not in simple form. Moreover, if the configuration of IPMC is complex (i.e. round shape, triangular shape), the above technique is unavailable.

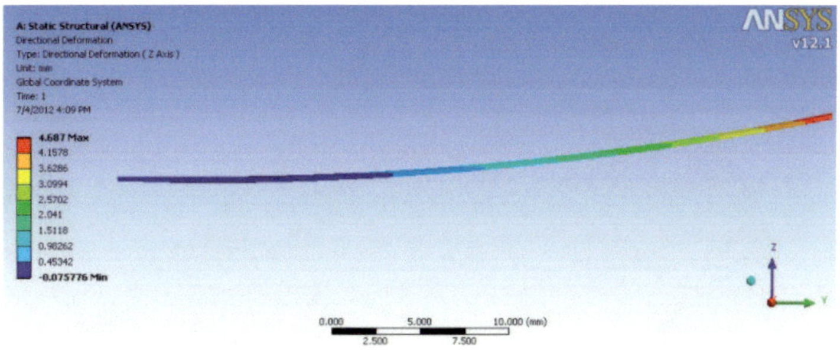

(a) Displacement of cantilever IPMC actuator (iso view)

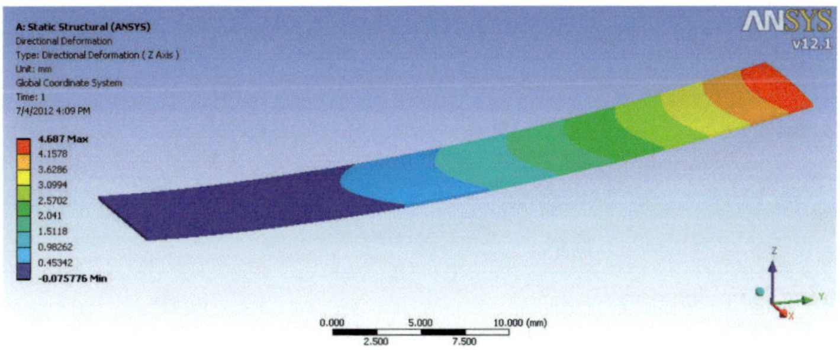

(b) Displacement of cantilever IPMC actuator (front view)

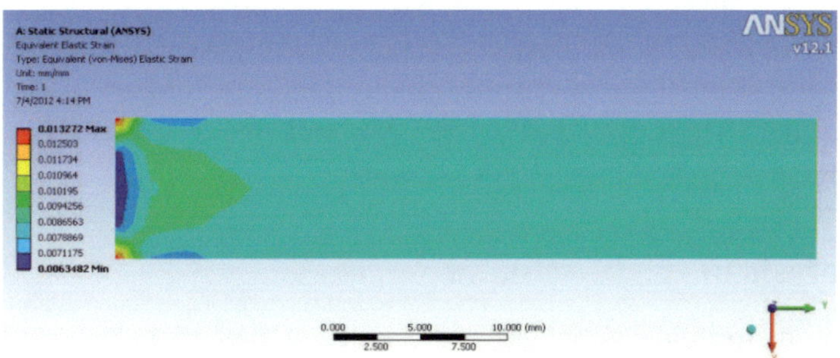

(c) Strain of cantilever IPMC actuator (top view)

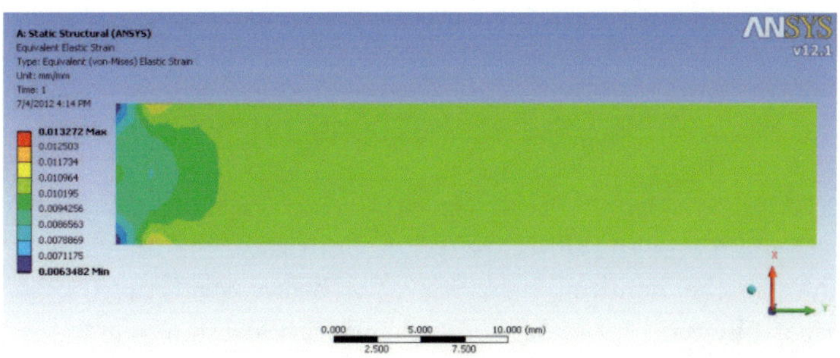

(d) Strain of cantilever IPMC actuator (bottom view)

Figure 8. Simulation of a cantilever IPMC actuator

For ease of analysis, this chapter proposes a novel visual simulation technique for the IPMC actuator using a commercial mechanical analysis tool, ANSYS 12.1 (2009) [18]. In this technique, a finite element model (FEM) is created to model an IPMC actuator. By using the command "INISTATE", the induced stress, obtained from equation (30), is added to the FEM as the deformation mechanism. Next, the constraints at the clamped edges of the IPMC actuator are considered as boundary conditions (fixed support to the clamped edges). Displacement constraints are employed as boundary conditions for neutral plane inside the IPMC actuator.

An example of visual simulation is shown as figure 8 for the cantilever IPMC actuator with a constant 3V input signal. The material properties of the IPMC actuator, employed for the FEM, are shown as in Table 2. A FEM with total 400 elements was created. As shown in figure 8a and figure 8b, the deformation of the IPMC can reach 4.69 mm at maximum under 3V input signal. Figure 8c and figure 8d show the internal strain ratio of the IPMC actuator under operation (maximum strain ratio is 0.013). Figure 9 shows the response of the IPMC beam under a step input voltage signal, and figure 10 shows the displacement versus input voltage graph at the steady state. Modeling results show that the visual simulation technique possesses a strong ability to be applied in analyzing operation of IPMC. It is obvious that the proposed visual simulation technique helps to analyze the behaviors of IPMC actuators more conveniently. In high input voltage signal range, the experimental displacements of IPMC are smaller than simulations. The reasons may be caused by the incompetence of the key assumption (see equation 30), the neglecting of nonlinear term in the PDE, or the variation of surface resistance along the z axis when the IPMC deforms.

	Elastic modulus (Pa)	Poisson's ratio	Size of IPMC actuator (mm)	Diaphragm radius (mm)	Thickness (mm)
Specification	5.71×10^{8}	0.487	$30\times6\times0.2$	5	0.2

Table 2. Material properties of IPMC diaphragm

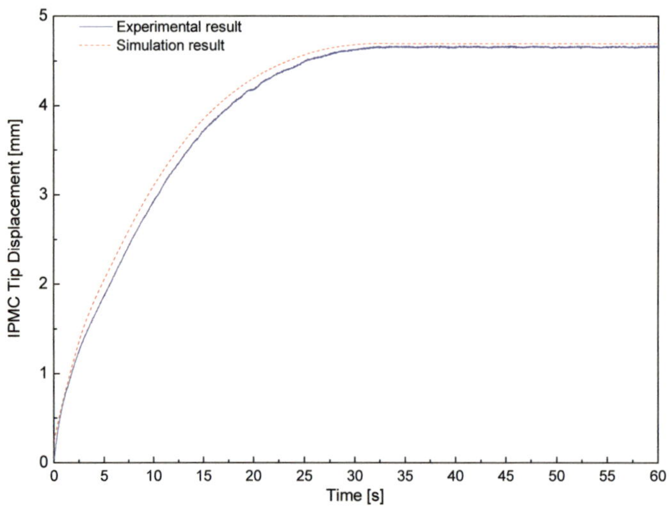

Figure 9. IPMC actuation with respect to a 3V step input signal

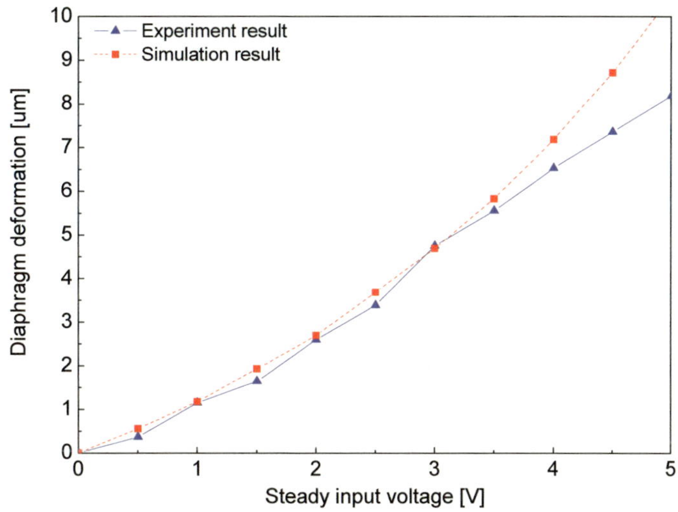

Figure 10. Displacement vs. input voltage

5. Variation of surface resistance and self sensing ability

5.1. Physical model of self sensing ability

During the bending operation, surface resistances of the two electrode layers may change due to the non – uniform material distribution between the two electrode layers as reported in [11 – 12]. During the deforming process, the surface resistance on the stretched side of IPMC increases corresponding to the reduction of conductive material density. Meanwhile, the surface resistance on the compressed side of IPMC decreases because the density of conductive material on the surface layer increases. In order to investigate the bending curvature - surface resistance correlation, a simple model of an IPMC sheet is derived out as in figure 11. Here, half of the IPMC sheet is fixed by a clamp while the other half sheet is free

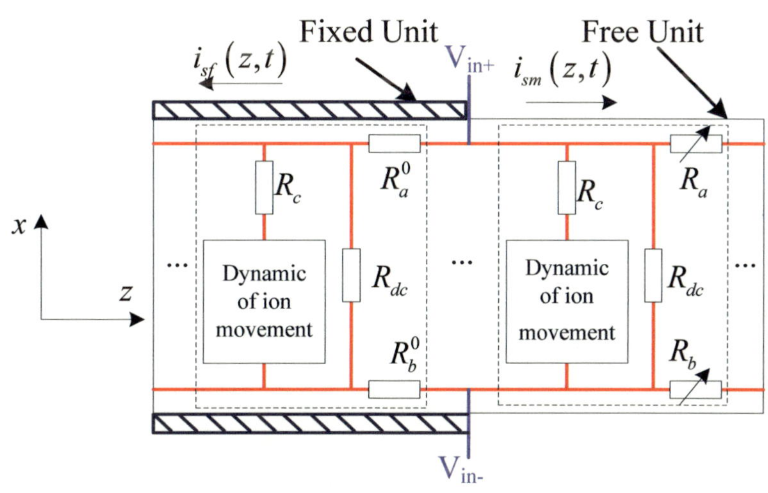

Figure 11. Equivalent circuit of an IPMC actuator

for moving. The actuator is then represented by the equivalent circuit in which affecting factors are defined as follows:

- R_a^0, R_b^0 : the fixed surface resistances at both sides of the clamped part.
- R_a, R_b : the variable surface resistances at both sides of the movable part.
- $i_{sf}(z,t)$: the surface current of the movable part.
- $i_{sm}(z,t)$: the surface current of the movable part.

The analysis of self – sensing behavior deals with the following assumptions:

- The properties of the material are uniform. Consequently, surface resistances on both sides of the clamped part are equal:

$$R_a^0 = R_b^0 \tag{31}$$

- The variation of surface resistances is proportional to the unit bending displacement (the bending behavior corresponding to a unit length along the z axis). Given the unit bending displacement $\psi(z,t)$, the surface resistances on the movable part are calculated as follows:

$$\begin{cases} R_a(z,t) = \beta_a \psi(z,t) + R_a^0 \\ R_b(z,t) = \beta_b \psi(z,t) + R_b^0 \end{cases} \tag{32}$$

where, and β_a, β_b are coupling constants to be determined by experiment.

- The neutral axis of the resistance variation is the center line of the IPMC along the z axis. This assumption allows one to treat β_a, β_b at same value:

$$\beta_a = -\beta_b = \beta \tag{33}$$

With the above assumption, the self – sensing ability of IPMC is discussed as follows. Define:

$$\begin{cases} \Delta R_a(z,t) = \beta_a \psi(z,t) \\ \Delta R_b(z,t) = \beta_b \psi(z,t) \end{cases} \tag{34}$$

Employing equation (33), the variant of surface resistance in equation (34) is rewritten as:

$$\Delta R_a(z,t) = -\Delta R_b(z,t) = \beta \psi(z,t) \tag{35}$$

A simplified circuit for the self – sensing actuator is carried out (figure 12)

Based on the equation (35), the following statements hold:

$$Z_F = Z_M = Z_{eq} \tag{36}$$

$$i_{sf}(z,t) = i_{sm}(z,t) = i_s(z,t) \tag{37}$$

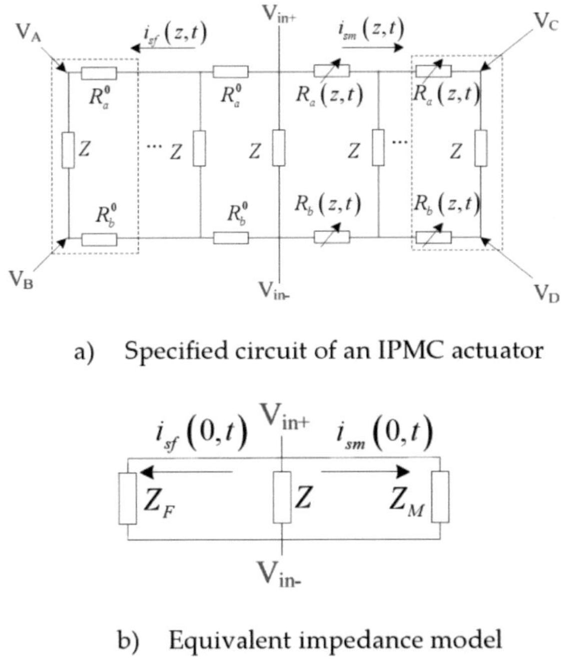

Figure 12. Simplified equivalent circuit of an IPMC actuator

where, Z_F, Z_M are equivalent impedance of fix part and movable part, respectively. The voltage signal at 4 points *A, B, C,* and *D* are in the forms:

$$V_A(t)=V_{in+}(t)-\int_0^{L/2} R_a^0 i_s(\tau,t)d\tau \tag{38}$$

$$V_B(t)=V_{in-}(t)+\int_0^{L/2} R_b^0 i_{sf}(\tau,t)d\tau \tag{39}$$

$$V_C(t)=V_{in+}(t)-\int_0^{L/2} R_a(\tau,t)i_{sm}(\tau,t)d\tau \tag{40}$$

$$V_D(t)=V_{in-}(t)+\int_0^{L/2} R_b(\tau,t)i_{sm}(\tau,t)d\tau \tag{41}$$

The following analyses are carried out:

$$\begin{aligned} V_{AC}(t)=V_A(t)-V_C(t)&=\int_0^{L/2} R_a(\tau,t)i_s(\tau,t)d\tau-\int_0^{L/2} R_a^0 i_s(\tau,t)d\tau \\ &=\int_0^{L/2}\left(R_a^0+\Delta R_a(\tau,t)\right)i_s(\tau,t)d\tau-\int_0^{L/2} R_a^0 i_s(\tau,t)d\tau \\ &=\int_0^{L/2}\Delta R_a(\tau,t)i_s(\tau,t)d\tau \\ &=\beta\int_0^{L/2}\psi(\tau,t)i_s(\tau,t)d\tau \end{aligned} \tag{42}$$

Denote $V_{in+A}(t)=V_{in+}(t)-V_A(t)$, the derivatives of equation (38) and equation (42) give:

$$i_s(z,t) = \frac{1}{R_a^0}\frac{\partial V_{in+A}(t)}{\partial z} \tag{43}$$

$$\psi(z,t) = \frac{1}{\beta i_s(z,t)}\frac{\partial V_{AC}(t)}{\partial z} \tag{44}$$

$$\psi(z,t) = \frac{R_a^0}{\beta}\frac{\frac{\partial V_{AC}(t)}{\partial z}}{\frac{\partial V_{in+A}(t)}{\partial z}} \tag{45}$$

Note that R_a^0 is a constant. Equations (43), (44) and (45) imply that if one can observe the change of voltage signal along the length of IPMC actuator, the unit bending displacement can be obtained accurately. Consequently, the tip displacement of the IPMC can be obtained as follows:

$$d(t) = \int_0^{\frac{L}{2}} \psi(z,t)dz \tag{46}$$

In practice, the bending curvature can be estimated by using the feedback voltage signals at several points along the z axis. The following section presents the bending estimation of IPMC actuator using four points feedback configuration.

5.2. Experimental results for self sensing of an IPMC actuator

The self – sensing test rig using an IPMC actuator is shown in figure 13. Here, the actuator is a sheet of IPMC clamped by two plastic plates to which electrodes (V_{in+}, V_{in-}, A, B, C, and D) are attached and also contacted with two sides of the IPMC. The IPMC sheet in size of 30 x 6 x 0.2 mm manufactured by Environmental Robots Inc. can operate in both wet and dry environments. A processing system is built on a personal computer (Intel ® CoreTM2 Duo 1.8 GHz) within Simulink environment combined with Real-time Windows Target Toolbox of MATLAB. Two multi-function data acquisition Advantech cards, A/D 1711 and D/A 1720, are installed on PCI slots of the PC to perform the peripheral buses. In addition, a CCD laser displacement sensor, LK-081, from Keyence Corp. is installed on the rig base to measure the IPMC tip displacement.

Employing the technique described in section 5.1, the voltage signals at four points A, B, C, D are collected. In ideal case when the assumption in equations (31), (32), and (33) hold, the estimation can be done with only two feedback voltage signals at A and C. However, for more accurate result, the two more voltage signals at B and D are obtained. Since this implementation uses only four feedback signals, the equation (45) is rewritten as:

$$\psi(z,t) = \frac{R_a^0}{2\beta}\left(\frac{V_{AC}(t)}{V_{in+A}(t)} + \frac{V_{BD}(t)}{V_{in-}(t) - V_B(t)}\right) \tag{47}$$

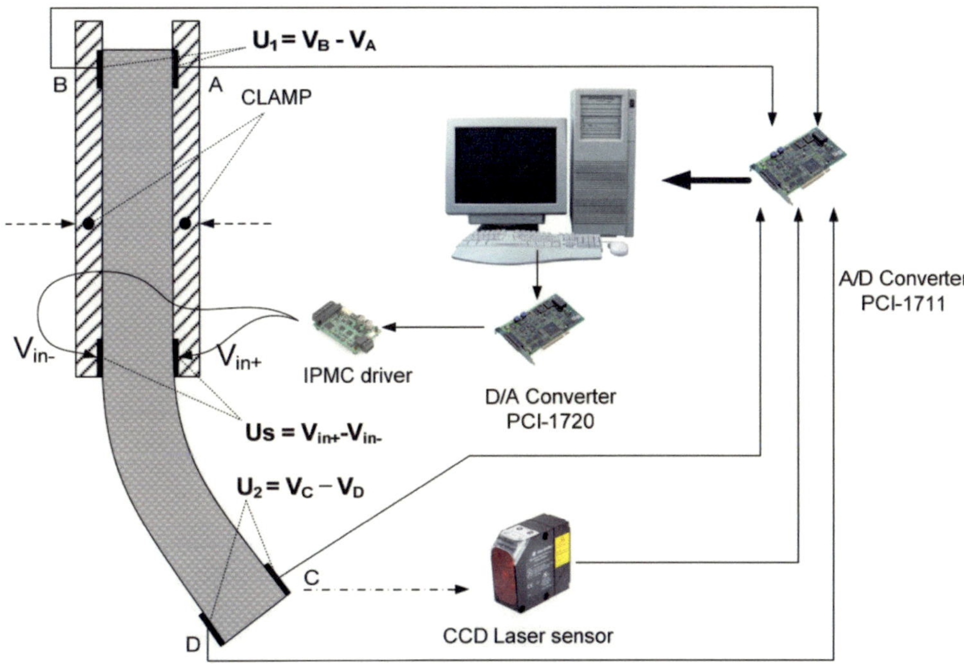

a) Configuration of IPMC in self – sensing application

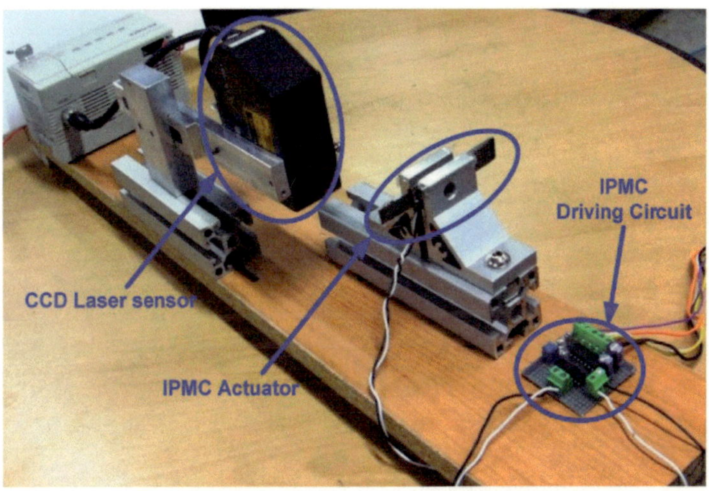

b) Experimental self – sensing IPMC actuator

Figure 13. Configuration of IPMC in self – sensing application

By applying equations (46) and (47), the self sensing has been carried out corresponding to four cases: 2V, 4V, 4.5V and 5V driving square input voltages signal, 0.05 Hz in frequency. Consequently, the experimental results are then shown in figure 14. Experimental results prove that the developed self – sensing technique has the ability to measure the bending behavior of the IPMC actuator. In order to increase the accuracy of the self sensing

performance, one can increase the number of captured voltage signals. However, the accuracy requirements depend on the practical application.

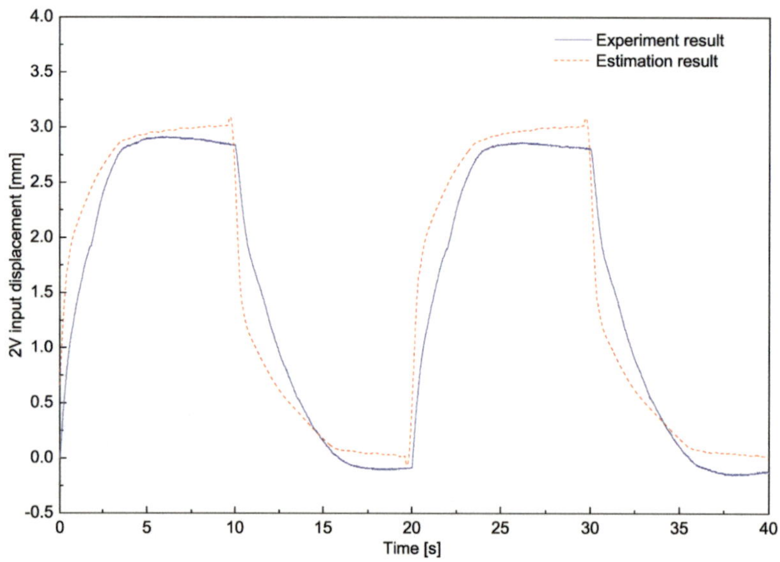

a) Estimation result – 2V square input signal

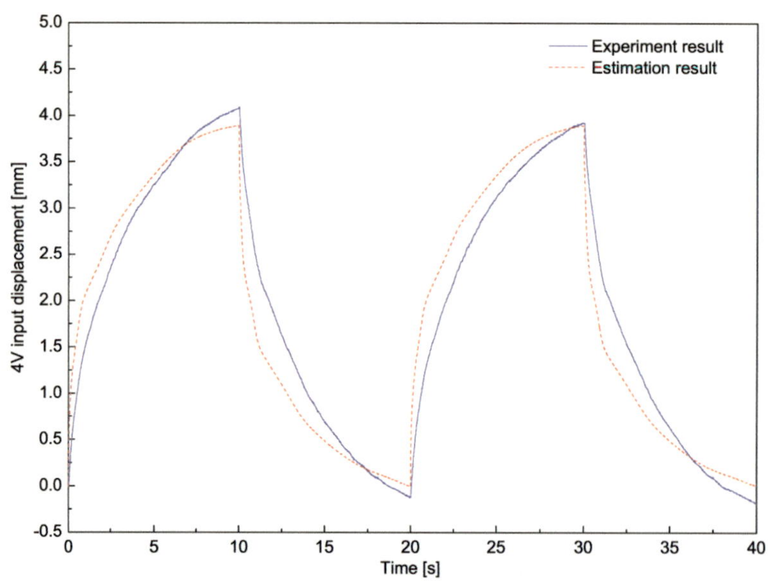

b) Estimation result – 4V square input signal

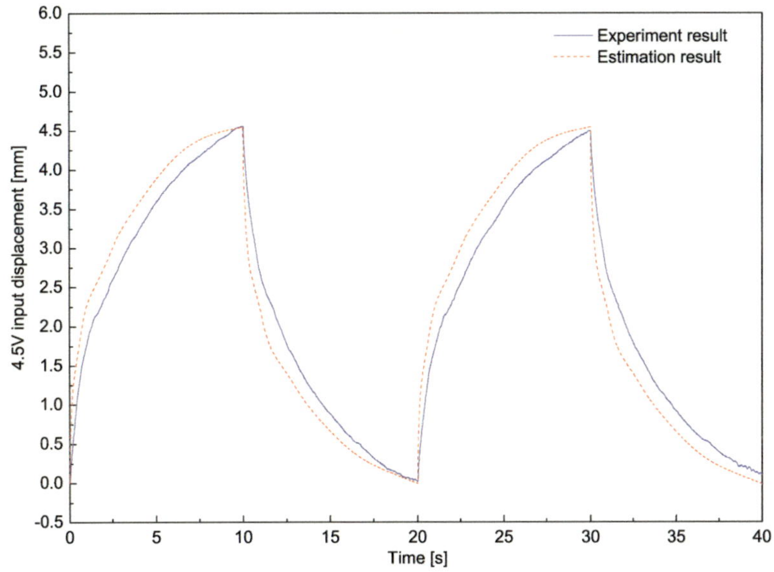

c) Estimation result – 4.5V square input signal

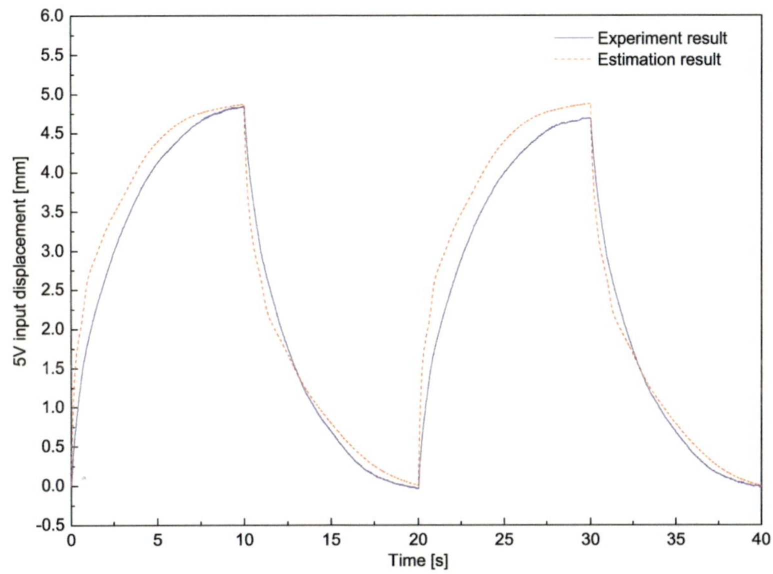

d) Estimation result – 5V square input signal

Figure 14. Self - sensing behavior of an IPMC actuator

6. Conclusion

This chapter gives a physical view in modeling the IPMC actuator and its self – sensing ability. The modeling technique provides a powerful tool to investigate the characteristic of

IPMC actuator via incorporation of physical induced stress model and visual ANSYS environment. The self – sensing model gives a powerful methodology to measure the displacement of IPMC actuator. This is a great advantage for biomedical, biomimetic, micro or mobile robotic applications where external sensing systems are usually unavailable.

Author details

Ahn Kyoung Kwan*
School of Mechanical and Automotive Engineering, University of Ulsan, Korea

Doan Ngoc Chi Nam
Graduate School of Mechanical and Automotive Engineering, University of Ulsan, Korea

Acknowledgement

This work was supported by the Korea Science and Engineering Foundation (KOSEF) grant funded by the Korea government (MEST) (No. 2009-0080924)

7. References

[1] Ba-Cohen Y. (2001) Electro-Active polymer (EAP) actuators as artificial muscles - Reality, potential, and challenges, Spie press .

[2] Abadi S., Dehghani A., Nelson E.A. (2008) A soft sensing method for monitoring ambulatory activities of patients with venous ulceration, Proc. of the 4th IET Int. Conf. on Adv. in Medical, Sig. and Inf. Processing, MEDSIP. pp.1 – 4.

[3] Paola B., Fortuna L., Giannone P., Graziani S., Strazzeri S. (2008) IPMCs as Vibration Sensors, Proc. of IEEE Int. Instrumentation and Measurement Tech. Conf. pp.2065 – 2069.

[4] Bonomo C., Fortuna L., Giannone P., Graziani S. (2004) A sensor-actuator integrated system based on IPMCs [ionic polymer metal composites], Sensors, Proc. of IEEE conf. pp. 489 – 492.

[5] Fang B.K., Lin C.-C.K., Ju M.-S. (2010) Development of Sensing/Actuating Ionic Polymer Metal-Composite (IPMC) for Active Guide-Wire System, Sensors & Actuators: A. Physical, v 158, pp. 1-9.

[6] Chen Z., Kwon K.Y., Xiaobo T. (2008) Integrated IPMC/PVDF sensory actuator and its validation in feedback control, Sen. and Act. A: Physical 144(2), pp. 231-241.

[7] Yamakita M., Kamamichi N., Kozuki T., Asaka K., Zhi-Wei L. (2005) A snake-like swimming robot using IPMC actuator and verification of doping effect, Proc. of IEEE/RSJ Int. Conf. pp.2035 – 2040

[8] Tung N.T., Khanh N.V., Youngtai Y., Goo N.S. (2006) A Novel Polymeric Micropump based on a Multilayered Ionic Polymer-Metal Composite, Proc. of IECON - 32nd Annual Conference, pp.4888 – 4893.

* Corresponding Author

[9] Yamakita M., Kamamichi N., Kozuki T., Asaka K., Zhi-Wei L. (2005) Control of Biped Walking Robot with IPMC Linear Actuator, Proc. of IEEE/ASME International Conference, Sep. pp.48 – 53.

[10] Guo S., Shi L., Asaka K. (2008) IPMC actuator-sensor based a biomimetic underwater microrobot with 8 Legs, Proc. of ICAL. IEEE International Conference, Sep. 2008, pp.2495 – 2500.

[11] Punning A., Kruusmaa M. and Aabloo A. (2009) Surface resistance experiments with IPMC sensors and actuators Sen. and Act. A: Physical 133(1). Pp. 200-209.

[12] Punning A., Kruusmaa M. and Aabloo A. (2007) A self-sensing ion conducting polymer metal composite (IPMC) actuator Sen. and Act. A: Physical 136(2). Pp. 656-664.

[13] Nemat-Nasser S. and Li J. (2000) Electromechanical response of ionic polymer-metal composites, Journal of Applied Physics, 87 (7), pp. 3321–3331.

[14] Farinholt K. M. (2005) Modeling and characterization of ionic polymer transducers for sensing and actuation, Ph.D. dissertation, Virginia Polytechnic Institute and State University.

[15] Chen Z. and Tan X. (2008) A control-oriented and physics-based model for ionic polymer –metal composite actuators, IEEE/ASME Trans. Mechatronics. 13, pp. 519–529.

[16] Chen Z., Hedgepeth D. R. and Tan X. (2009) A nonlinear, control-oriented model for ionic polymer–metal composite actuators, Smart Mater. Struct. 18055008 (9pp).

[17] Nemat-Nasser S. (2002) Micromechanics of Actuation of Ionic Polymer-metal Composites, Journal of Applied Physics, 92 (5) (21pp).

[18] http://www.ansys.com/Support/Documentation

Ionic Polymer-Metal Composite Artificial Muscles in Bio-Inspired Engineering Research: Underwater Propulsion

Zheng Chen, T. Um and Hilary Bart-Smith

Additional information is available at the end of the chapter

http://dx.doi.org/10.5772/51292

1. Introduction

Electroactive polymers (EAPs), known as artificial muscles, can generate large deflections under an electrical stimulus (Bar-Cohen 2000, 2004). Due to the similarities with biological muscles, in terms of achievable stress and strain, EAPs have great potential to be used as actuators in bio-inspired robots, bio-medical devices, and micro/nano manipulation systems. Ionic polymer-metal composites (IPMCs) are an important category of ionic EAPs (Shahinpoor & Kim, 2001, 2005; Kim & Shahinpoor, 2003, 2003; Shahinpoor et al, 2003; Kim et al, 2007; Nemat-Nasser & Li, 2000; Nemat-Nasser, 2002), which can work well under a low actuation voltage (1 to 2 Volts) in a sodium salt-water environment.

As shown in Fig. 1, an IPMC consists of one ion exchange membrane, such as Nafion (DuPont), sandwiched in between two thin metal electrodes (Shahinpoor & Kim, 2001). When the IPMC is hydrated, the positive ions in the Nafion polymer, such as sodium and calcium ions, bring water molecules and migrate to the cathode side under an electric field, whereas the negative ions are permanently fixed to the carbon chain (Nemat-Nasser & Li, 2000). The ion transportation and water migration introduce swelling in the cathode side and shrinking in the anode side, which eventually causes bending motion of the IPMC (Shahinpoor & Kim, 2001).

IPMCs can also be used for electrolysis. Electrolysis of water occurs when a relatively high voltage (> 1.6 V) is applied between the cathode and anode (Shaaban, 1994; Shimizu et al, 2005). The water molecules are decomposed into hydrogen and oxygen gases. IPMC enhanced electrolysis of water is due to several reasons; (i) ion transport brings high dense of water close to cathode side (Shahinpoor & Kim, 2001); (ii) the strong electrical fields

established on the boundary (Nemat-Nasser & Li, 2000; Nemat-Nasser, 2002) breaks down of the water molecules; and (iii) the porous platinum electrodes (Shahinpoor & Kim, 2001) on the surface acts as a stable electrolysis catalyst (Millet et al, 1989).

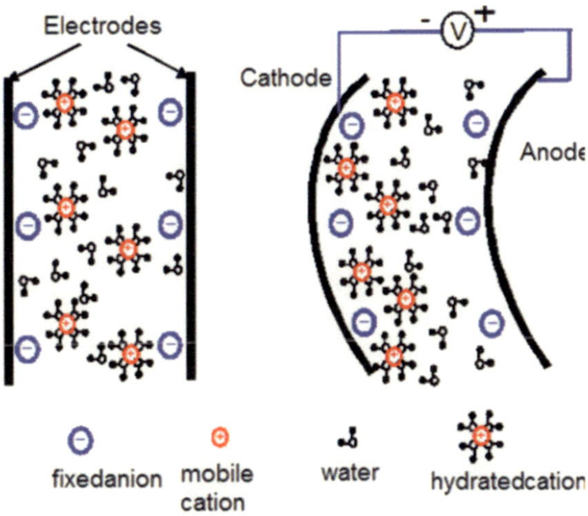

Figure 1. Actuation mechanism of IPMC (Chen & Tan, 2008).

In this chapter, the advantages and challenges of IPMCs as applied to bio-inspired engineering are discussed. A bio-inspired robotic manta ray integrated with IPMCs as artificial muscles and a novel buoyancy control device using IPMCs as an electrolysis generator are presented.

1.1. Introduction to bio-inspired robotic manta ray

Aquatic animals (e.g., fishes, cetaceans, etc.) are ultimate examples of superior swimmers as a result of millions of years of evolution, endowed with a variety of morphological and structural features for moving through water with speed, agility, and efficiency (Chen et al, 2010). The manta ray (*Manta birostris,* shown in Fig. 2) demonstrates excellent swimming capabilities; generating highly efficient thrust via flapping of dorsally flattened pectoral fins (Rosenberger, 2001). Many efforts have been directed towards building a bio-inspired pectoral fin structure to mimic the swimming behavior of the ray. Examples include rigid plates or tensegrity structures actuated by servo-motors (Gao et al, 2007; Moored et al, 2008; Moored & Bart-Smith, 2009; Zhou & Low, 2012) and flexible membrane actuated by shape memory alloy (SMA) (Wang et al, 2009). However, these methods are not suitable for small-scale robots (on the order of 5-10 cm) (Shahinpoor, 1992; Mojarrad & Shahinpoor, 1996; Tan et al, 2006; Guo et al, 2003; Punning et al, 2004) because of either the limitations in scaling or high power consumption. To construct a free swimming and small-scale robotic manta ray, there is a need for a bio-inspired actuating material that is lightweight, compliant, resilient, and capable of generating 3 dimensional (3D) deformations with portable power consumption.

Figure 2. Bio-inspiration: The Manta Ray (courtesy of www.elasmodivec.com).

In the past, IPMC actuators have been used as a caudal fin in bio-inspired robotic fishes (Shahinpoor, 1992; Mojarrad & Shahinpoor, 1996; Tan et al, 2006; Guo et al, 2003), where the propulsive fin mimics the bending motion observed in many biological fishes. In the propulsion mechanism of rays, undulatory and oscillatory flapping motions of the pectoral fin play an important role in generating highly efficient propulsion and maneuvering (Rosenberger, 2001). To fabricate an actuating membrane capable of generating complex deformations, lithography-based (Chen & Tan, 2010) and surface machining-based approaches (Kim et al, 2011) have been explored to pattern the electrodes of the IPMC. To create active and passive areas in a Nafion membrane, the electrodes on the membrane were separated. By independently controlling the bending of each active area, 3-dimensional deformations of the membrane have been achieved. However, the stiffness of the Nafion in the passive area limited its capability of generating large twisting motions. Punning et al, developed pectoral fins for ray-like underwater robot by assembling separated IPMC beams with a latex foil (Punning et al, 2004). However, this robot did not achieve free-swimming due to high power consumption and low propulsion efficiency. The challenge in this study is to fabricate a compliant actuating membrane cable of complex 3D kinematic motions capable of generating energy efficient locomotion. This will then be integrated in a small-scale robotic ray capable of free swimming.

1.2. Introduction to bio-inspired depth control device

A simple and efficient method for buoyancy control is critical in the design of an autonomous underwater device. The most common method used today to achieve depth control is to utilize a piston-cylinder assembly connected to a servomotor (Lin et al, 2010). A cylinder, usually containing air, is compressed and expanded to adjust the volume of the system. Another similar approach is to use a water-pump powered bladder to compress the air in a blaster tank (Zhou & Low, 2012). These two methods usually yield reliable results with relatively fast response times. However, in addition to the significant noise generated by the servomotors, there are limitations in scaling the servomotor. Consequently, this

solution is not feasible for implementation in small devices. In order to build more efficient buoyancy control devices, researchers have turned to biology for inspiration for the next generation of autonomous underwater vehicles.

Biology has many novel and effective depth control mechanisms suitable for a variety of environments. For example, sperm whales (Fig. 3(a)) achieve buoyancy control by using their spermaceti oil. An adult sperm whale contains about 4 tons of spermaceti oils in their spermaceti organ, which represents approximately 8% of its total mass (Shibuya et al, 2006). The spermaceti oil has a low melting point and its density depends largely on the temperature of the oil. By manipulating the arterial blood flow through the spermaceti organ, the sperm whales can regulate the temperature of the oil and are thus able to control their buoyancy. There have been recent demonstrations of buoyancy control concepts manipulating temperature to change the density of oil (Shibuya et al, 2006) or wax (McFarland et al, 2003). However, the response times are slow (on the order of 10 minutes), and it is inefficient for small devices because a constant power must be supplied to maintain the temperature of the oil while cruising at a certain depth. Ray-finned fishes, such as one depicted in Fig. 3(b), change the buoyancy of their body using a swim bladder (Bond, 1996). Expansion of the bladder results in increased volume, thus making the body more positively buoyant and vice versa. Inspiration for the artificial bladder presented in this chapter comes from these ray-finned fishes. The challenges arise from how to generate and release the gas to control the volume of the bladder underwater.

(a) (b)

Figure 3. (a) Sperm whale (*Physeter macrocephalus*); (b) Golden fish (*Carassius auratus*).

The rest of this Chapter is organized as the follows. Section 2 is focused on the development of a bio-inspired robotic manta ray powered by IPMC artificial pectoral fin. Fabrication and characterization of the artificial pectoral fin and design of the robotic manta ray are presented. Section 3 is focused on bio-inspired depth control device enabled by IPMC enhanced water electrolysis, where the buoyancy control mechanism, device design, and open loop testing are demonstrated. Conclusions and future work on both studies are discussed in Section 4.

2. Bio-inspired robotic manta ray powered by IPMC artificial pectoral fin

In this section, we present a bio-inspired, free-swimming robotic manta ray propelled by artificial pectoral fins. In our previous work, we developed an assembly-based fabrication process to bond four IPMC actuators with Polydimethylsiloxane (PDMS) elastomer to create an actuating membrane capable of 3D kinematic motions (Chen et al, 2011). However, the control strategy for four IPMCs is too complicated to be implemented on-board. In this study, a pectoral fin, with the planform shape of a manta ray pectoral fin and consisting of one IPMC artificial muscle in the leading edge and a passive PDMS membrane in the trailing edge, has been designed. When the IPMC is actuated, the passive PDMS membrane follows the bending of IPMC with a phase delay, which causes an undulatory motion of the fin. The pectoral fin has been characterized by key factors as they relate to the function of the robot: tip deflection; twist angle; and power consumption. Since only one IPMC is needed for actuation in the pectoral fin, the control strategy is greatly simplified. To test the performance with a free-swimming robot, a light, compact on-board control unit with a lithium ion polymer battery has been developed. Experimental results have shown that the robot is capable of free swimming.

2.1. Fabrication of artificial pectoral fin

The proposed artificial pectoral fin must be able to generate oscillatory motion with a twist angle as observed in swimming of the manta ray, under hydrodynamic loads. An artificial pectoral fin was fabricated by combining one IPMC actuator with a PDMS elastomer in a mold to create a predefined planform shape. The design of pectoral fin is shown in Fig. 4(a). The outline shape of the fin mimics that of the manta ray. The fin is divided into two areas: IPMC beam on the leading edge and PDMS passive membrane on the trailing edge. Note that the size and shape of the IPMC is chosen to generate enough bending moment with limited power consumption. Optimal design of the fin will be focused in the future work.

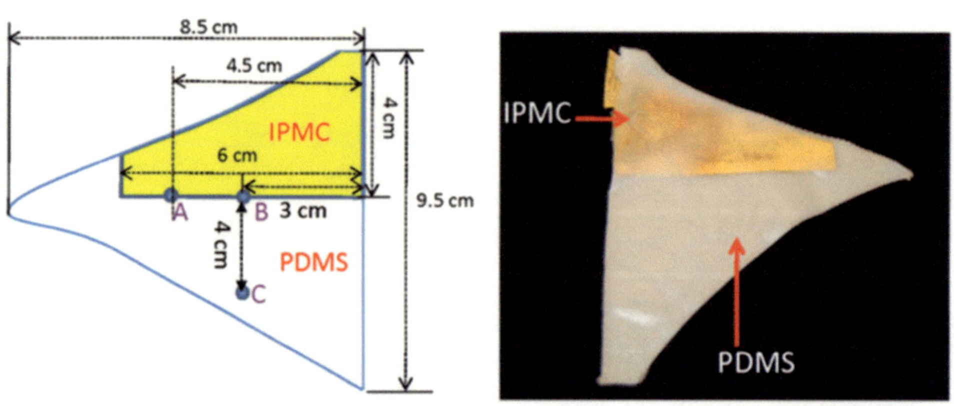

(a) Pectoral fin design (b) Fabricated pectoral fin

Figure 4. IPMC powered artificial pectoral fin.

The first step in creating the artificial pectoral fin is to fabricate the IPMC actuator. Many groups have developed different IPMC fabrication processes for various purposes (Kim & Shahinpoor, 2003; Chung et al, 2006; Lee et al, 2006; Kim & Shahinpoor, 2002). In our fabrication method, we followed most of the procedure outlined by K. Kim and M. Shahinpoor (Kim & Shahinpoor, 2003) but added additional platinum/gold plating process to reduce the surface resistance of the electrodes (Lee et al, 2006). The following supplies are used to fabricate the IPMC beams: (1) Nafion ion exchange membrane Nafion 1110 (258 μm thick, DuPont); (2) tetraammineplatinum chloride 98% (Aldrich); (3) sodium borohydride ($NaBH_4$, Aldrich); (4) dilute ammonium hydroxide solution (NH_4OH 29% solution); 5) dilute hydrochloric acid (HCl aq, 1.0 N solution); and 6) de-ionized water. Fig. 5 shows the IPMC fabrication process. The major fabrication steps are as follows:

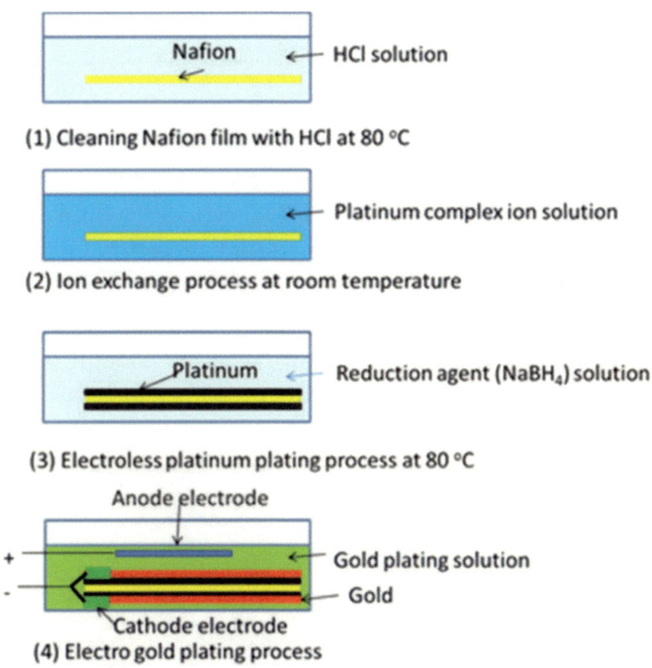

Figure 5. IPMC fabrication process.

1. Treatment of Nafion with HCl (cleaning). This step is to remove metal particles and other impurities from the film. The Nafion film (10 cm by 5 cm) is boiled in 1.0 N hydrochloride acid (HCl) at 80 °C for 30 min. The film is then rinsed with DI water to remove acid residue.

2. Ion exchange process. The purpose of the ion exchange process is to enable the Nafion film to absorb enough platinum complex ions $Pt(NH4)_2^{+2}$,which will be reduced to platinum particles to form electrodes in step 3. The membrane is then immersed in 50 ml tetraammineplatinum Chloride hydrate solution (3 mg/ml) mixed with 1 ml ammonium hydroxide (29%) to make a weak base environment. The ion exchange process takes about 3 hours to allow the Nafion film to fully absorb enough platinum complex ions.
3. Electroless platinum plating. In this step, the platinum complex ions are reduced to platinum particles to create the metal electrodes on the outer surfaces of the Nafion film. The ion-exchanged Nafion film is placed in a DI water bath. Then the bath is gradually heated to 40 °C. A sodium boron hydride solution ($NaBH_4$) (20 mg/ml) is prepared, which is used as a reducing agent. 5 ml $NaBH_4$ solution is added to the bath and the temperature of the bath is increased to 65 °C. 5 ml $NaBH_4$ solution is added to the bath every 30 min. After 3 h plating, approximately 6 μm of platinum is deposited on the surface with good polymer-metal adhesion.
4. Electro gold plating: To further improve the conductance of the electrode, about 1 μm thick gold was deposited on the surface via electroplating process. The sample was then submerged in a sodium solution (1 N) for one day to exchange H^+ with Na^+ to enhance the actuation performance of the IPMC.

After fabrication of the IPMC actuator, the next step is to bond the IPMC with a PDMS elastomer membrane. The PDMS bonding process (shown in Fig. 6) is described in the following 5 steps. (1) A Delrin polymer (McMaster) mold was made using a Computer Numerical Control (CNC) rapid milling machine (MDX-650, Roland). The mold has two concaved areas to house the PDMS passive membrane and IPMC actuator. The thicknesses of PDMS membrane and IPMC are 600 μm and 280 μm, respectively. (2) The IPMC was cut into the shape shown in Fig. 3(a). (3) About 3% glass bubbles (Glass bubble K37, 3M Inc) were added into PDMS gel (Ecoflex 0030, Smooth-on Inc.) to gain a neutrally buoyant pectoral fin. (4) The IPMC and the PDMS gel were then clamped with the mold and the PDMS was cured at room temperature for 3 hours. (5) The IPMC/PDMS artificial pectoral fin (shown in Fig. 4(b)) was removed from the mold. Note that there is about 150 μm thick PDMS covered on the top of IPMC to guarantee good bonding between passive and active areas. The additional layer of PDMS stiffens the IPMC active area. However, since the PDMS (Ecoflex) is an extremely compliant material (Young's modulus, 0.06 MPa), compared to the Nafion (Young's modulus 114 MPa), the stiffening effect can be neglected.

2.2. Characterization of pectoral fin

The pectoral fin was characterized in terms of tip deflection, twist angle, and power consumption. These characteristics are useful in providing comparison data in the design of the bio-inspired robot. To characterize the actuating response of the pectoral fin, three testing points (A, B, C) are defined on the membrane (shown in Fig. 4(a)).

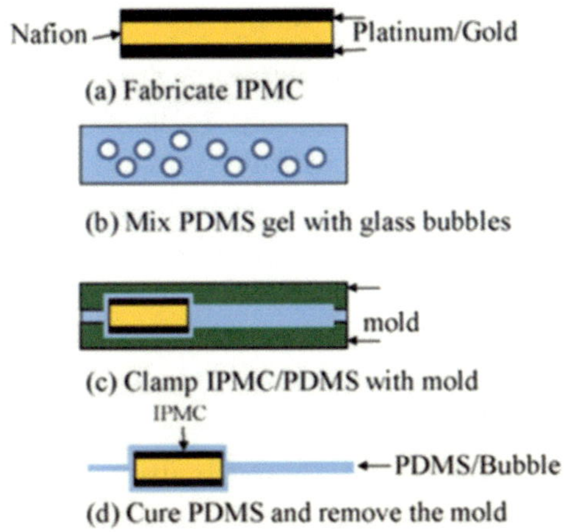

Figure 6. Fabrication process of pectoral fin.

2.2.1. *Tip displacement*

To measure the tip deflection, the fin was actuated in a transparent tank containing water. A laser sensor (OADM 20I6441/S14F, Baumer Electric) was fixed outside of the tank to measure the bending displacement at point A. Note that due to large deflection of the fin, the laser sensor was unable to capture the bending displacement of the tip. So we moved the measurement point from the tip to the point A. The tip deflection was normalized by dividing the bending displacement by the length at the point A. Fig. 7 shows the tip deflection when a square wave actuation voltage (0.09 Hz, 4 V) was applied to the IPMC. It shows a peak-to-peak deflection of 100% in the span-wise direction. One can achieve a larger deflection by applying a higher actuation voltage. But there is a limit to the size of voltage applied across the IPMC—anything greater than 6 V risks dielectric breakdown through the IPMC (Lee et al, 2006).

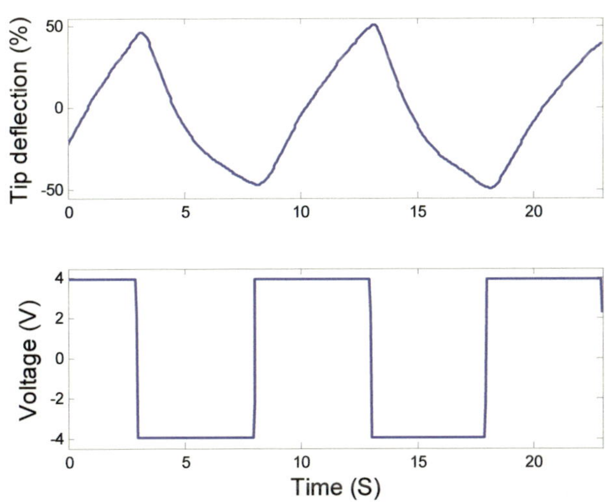

Figure 7. Tip deflection of the pectoral fin.

2.2.2. *Bode plot*

To capture the Bode plot of the pectoral fin, a series of sinusoidal actuation signals with amplitude of 4 V and frequencies ranging from 0.05 Hz to 0.9 Hz were applied to the IPMC. The tip deflection at the point B and the actuation voltage were measured. The magnitude and phase shift of the tip deflection over the actuation voltage were calculated. The Bode plot (Fig. 8) demonstrates that the actuation dynamics of the pectoral fin behaves as a low-pass filter with a 0.4 Hz cut-off frequency. This is to be expected as the ions in the IPMC cannot move very rapidly (Nemat-Nasser, 2002) and hydrodynamic force dumps high frequency vibration (Chen & Tan, 2010).

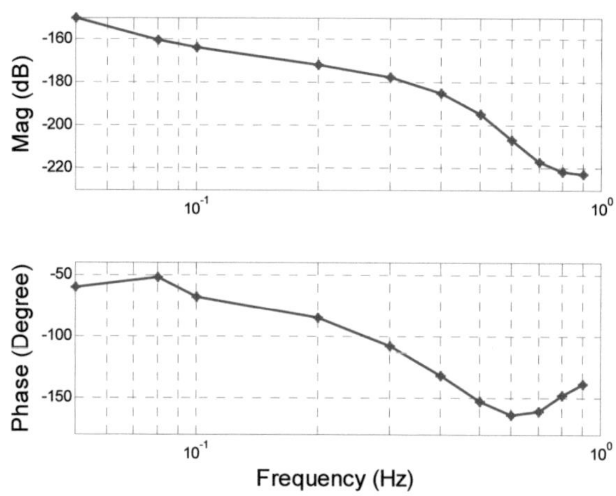

Figure 8. Bode plot of pectoral fin.

2.2.3. *Characterization of twisting motion*

In order to characterize the 3-dimensional (3D) kinematics of the fin, two laser sensors (OADM 20I6441/S14F, Baumer Electric) were used to measure the bending displacements at the points B and C, d_B and d_C respectively. The twist angle was calculated by

$$\alpha = \tan^{-1}\left(\frac{d_B - d_C}{BC}\right) \tag{1}$$

where BC=20 mm.

A series of square wave voltage signals were generated via LabVIEW (National Instruments), amplified using power amplifiers, and then applied to the IPMC actuator. All signals have the same amplitude of 4 V but varying frequencies from 0.06 Hz to 1 Hz. As the IPMC actuator is being used to generate thrust in underwater vehicles, the 3D kinematics was quantified in water. The fin was placed in a water tank and the laser sensors were fixed outside of the tank. Fig. 9 shows the twisting and flapping motion on the fin at f=0.09 Hz. The upper figure shows the bending displacements d_B and d_C, which indicates a phase delay

and an amplified magnitude between the bending motions at the points B and C. The lower figure shows the calculated twist angle. The peak-to-peak twist angle is 40°. Fig. 10 shows the plot of twist angle versus operating frequency. The twist angle decreases as the frequency increases. The cut-off frequency of twist angle is around 0.1 Hz. Since the twist angle plays an important role in generating a thrust, the optimal operating frequency of the robot will be set around 0.1 Hz.

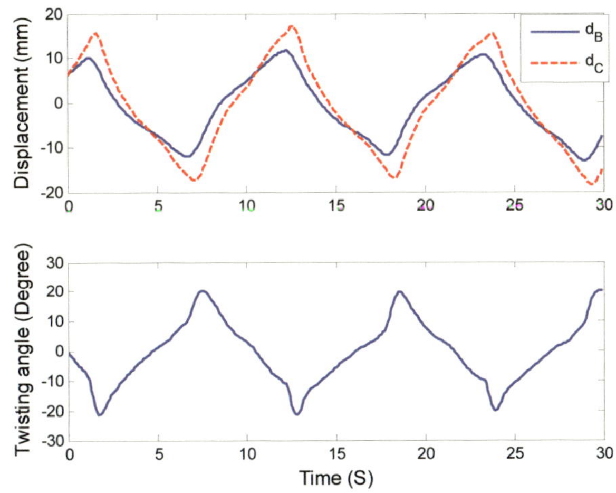

Figure 9. Twisted flapping motion on the fin.

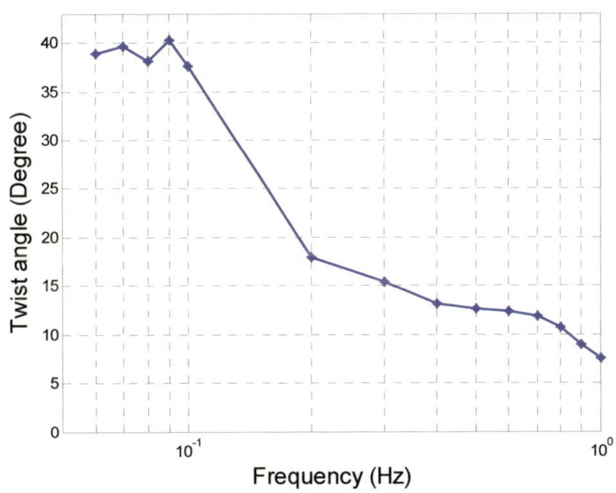

Figure 10. wist angle versus operating frequency.

2.2.4. *Power consumption*

For un-tethered bio-inspired robot applications, key questions regarding power consumption and optimal power management must be addressed. In this section, we study the power consumption of the pectoral fin under a square wave actuation voltage, which is

easy to generate on board. The power consumed by the IPMC was calculated using the following equation:

$$P = \frac{1}{T}\int_0^T i(t)v(t)dt \tag{2}$$

where i(t), v(t), and T are measured actuation current, voltage, and period, respectively. A 4 V, 0.1 Hz square wave voltage was applied to the IPMC of the pectoral fin, which was fixed under water. Fig. 11 shows the actuation voltage and current. The power consumption at 0.1 Hz is 1.09 W.

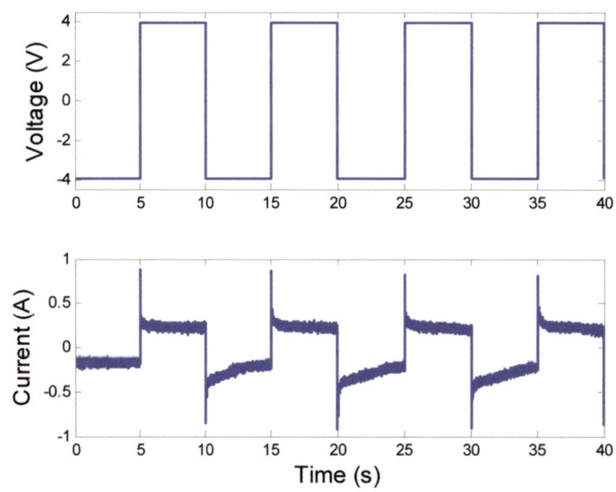

Figure 11. Actuation current and voltage under a 0.1 Hz square wave signal.

Under an electrical voltage signal, the IPMC behaves as a pseudo capacitor. When the applied voltage changes its polarization, the current reaches its peak, and then drops down to a steady-state current. The steady-state current is due to the DC resistance of the polymer and the water electrolysis on the electrodes. To reduce the power consumption of the pectoral fin, it would be better to eliminate the water electrolysis by coating a waterproof protecting layer on the IPMC, which will be our future focus. To study the relationship between power consumption and operating frequency, a series of square wave actuation signals with amplitude of 4 V and frequencies ranging from 0.1 Hz to 1 Hz were applied to the fin. Fig. 12 shows the power consumption versus operating frequency. It demonstrates that they are positively related and the power consumption is below 1.5 W.

2.3. Design of robotic manta ray

In our previous work, a free swimming and IPMC enabled robotic manta ray was developed for the first time (Chen et al, 2011). We used four IPMC beams bonded with thin PDMS membrane to create artificial pectoral fin. By independently controlling the bending of each IPMC beam, the fin can generate undulatory motions. However, in the free-swimming test of the robot, only a flapping motion was generated due to the complexity of generating the

four control signals on-board. In this paper, we present an IPMC-enabled robotic manta ray capable of free swimming. Since there is only one IPMC beam in the leading edge, a single signal is generated on-board, greatly simplifying the control strategy.

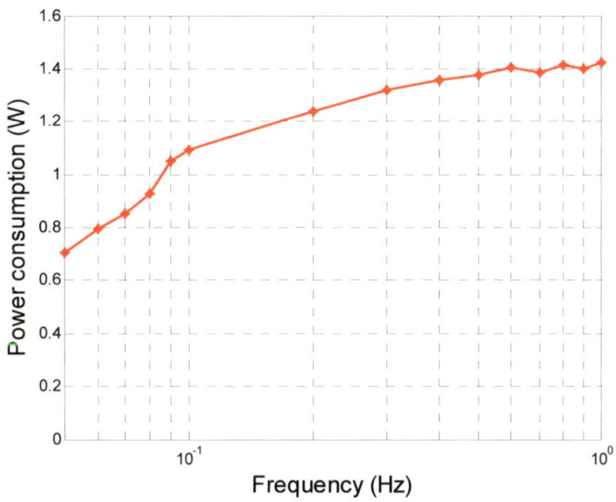

Figure 12. Power consumption versus operating frequency.

2.3.1. Control circuit

The on-board circuit, which was developed in our previous work (Chen et al, 2011), provided a square wave voltage signal to the IPMC actuator in the pectoral fin. Fig. 13(a) shows the schematic of the circuit. A 555-timer was used to generate a frequency adjustable square wave. The amplitude of the voltage signal, V_P, was controlled by an adjustable voltage regulator. An H-bridge driver was used to draw up to 2 A output peak current. A rechargeable 7.3 V Lithium Ion Polymer battery (400 mAh, AA Portable Power Corp) was selected as the power source for the robot. Fig. 13(b) shows the picture of PCB board and battery.

Figure 13. Control circuit and battery (Chen et al, 2011).

2.3.2. *Free swimming robot design*

Fig. 14 shows the overall shape of the robot. Two acrylic frames with gold electrodes were made to clamp the artificial wings to the body support. Gold electrodes were used to minimize corrosion. Polymer foam was put into the middle of the frame to make the robot slightly positively buoyant. A simple hinged plastic box was used to house the circuit board and battery. Once closed, only two wires were exposed outside the box. PDMS was used to seal the box waterproof. The plastic box was then glued on the top of the robot. With the robot was put under the water, the cover of the box was above the water level thus water could not leak into the box. The fully assembled robot was 11 cm long, 21 cm wide, and 2.5 cm thick with a mass of 55 grams. The free-swimming robot with the control unit is shown in Fig. 15. The total cost of the robot is about $200.

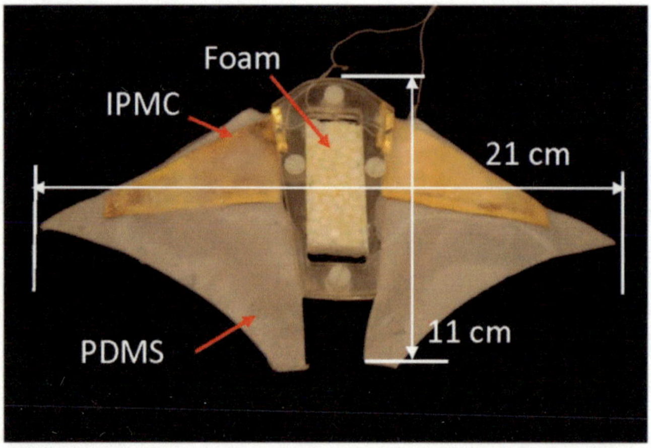

Figure 14. Robotic manta ray body.

Figure 15. Free swimming robotic manta ray body.

2.3.3. *Free-swimming test*

The robot was tested in a water tank (1.5 m wide, 4.7 m long, and 0.9 m deep). The robot was positively buoyant. The operating frequency of the square wave actuation voltage was tuned to 0.167 Hz and the amplitude was set to 4 V. Since each pectoral fin consumed less than 1.2 W (shown in Fig. 11), the overall power consumption of the robot was under 2.5 W. A digital video camera (VIXIA HG21, Canon) was used to capture the motion of the swimming robot. Fig. 16 presents six snap shots of the swimming robot from top view, where the trajectories of left and right fins are plotted in blue and red, respectively.

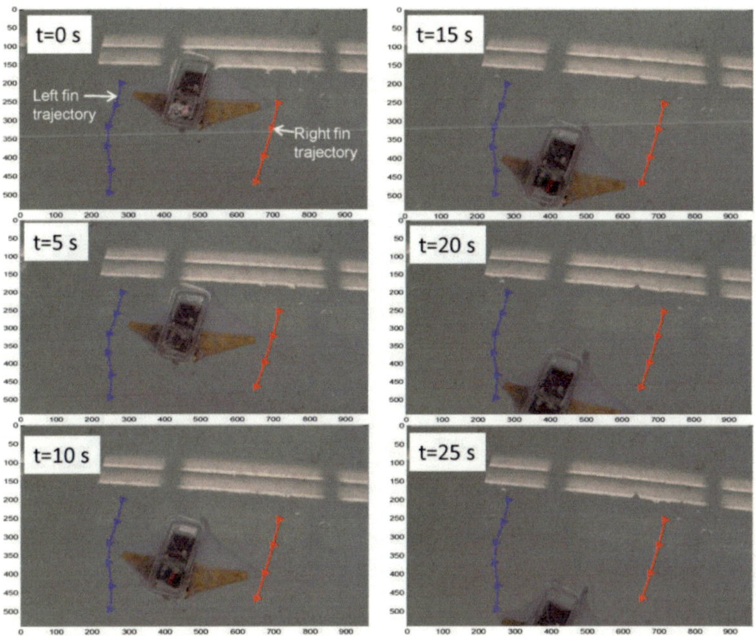

Figure 16. Snap shots of free-swimming robotic manta ray.

Each snap shot was taken every 5 seconds. A swim speed of 0.74 cm/s was calculated from the movie using the Edge Detection program in the LabVIEW. Since the body length was 11 cm, the speed in body length per second (BL/s) was 0.067.

2.4. Comparison of ray-like swimming robots

In our previous work, we developed an artificial pectoral fin that consists of four IPMCs bonded by PDMS membrane. The robot swam at 0.42 cm/s (0.053 BL/s) consuming 1 W in power. Many other research groups have been developed ray-like robots, using IPMCs, servomotors, and shape memory alloys as actuators. The comparison of ray-like swimming robots is shown in Table 1. It indicates that the IPMC powered ray-like robots are lighter and consume less power than the robots actuated by a servomotor or SMA. But they swim at lower speed for the reasons that IPMC is unable to flap at high frequency and generated force is also very low.

Reported Article	Speed (BL/s)	Power (W)	Weight (g)	Actuator
This chapter	0.067	2.5	55	IPMC
Chen et al. 2010	0.053	1	55	IPMC
Punning, et al. 2004	0.038	2.2	60	IPMC
Zhou et al, 2012	0.8	10.8	7300	Servo Motor
Wang et al, 2009	0.43	Not given	354	SMA

Table 1. Comparison of Ray-like Robots.

The advantages of using IPMCs as artificial muscle in ray-like swimming robots are: 1) low actuation voltage; 2) works well in wet conditions; 3) no gears and motors, 4) simple mechanical structure; 5) low noise; 6) able to be shaped in bio-inspired engineering design. However, IPMC can only generate low force with slow response time, which limits the swimming speed of the robot. To accommodate the disadvantages and utilize the advantages, an optimal design of the pectoral fin must be conducted, where the dimensions and location of the IPMC actuator are optimized. The challenge arises from modelling the IPMC actuating membrane to produce optimal 3D kinematic motions, which is a future focus. Another focus will be on modelling of the fluid dynamics introduced by the 3D kinematic motions of the fin, to understand how the thrust force is generated.

3. A novel buoyancy control device enabled by IPMC

In this study, a novel depth control device has been designed and built. The proof-of-concept device utilizes the principles of electrolysis of water, enhanced by the inclusion of an ionic polymer-metal composite (IPMC) membrane as a medium. The device design incorporates an artificial bladder where the volume of gas generated by electrolysis is controlled by a solenoid valve, thus changing the device's buoyancy. A set of gold electrodes, separated by an IPMC film, is used as a lightweight and compact electrolysis generator. IPMC acts as stable, low power, highly efficient and environmentally friendly gas generator. Experimental results using open-loop control show that the device is capable of controlling its buoyancy efficiently with no noise and low power consumption. Applications for this technology include integration into bio-inspired, unmanned underwater vehicles.

3.1. Mechanism of buoyancy control

3.1.1. *Gas generation by water electrolysis*

When an electric current is applied between positively and negatively charged electrodes in ionized water, a chemical reaction occurs, in which pure oxygen and hydrogen gases are produced. Hydrogen gas appears at the cathode electrode (3) and oxygen gas appears at the anode electrode (4). Equation (5) shows the net reaction (Shaaban, 1994).

$$2H_2O_2 + 2e^- \rightarrow H_2 + 2OH^- \tag{3}$$

$$4OH^{-} \rightarrow O_2 + 2H_2O + 4e^{-} \tag{4}$$

$$2H_2O(l) \xrightarrow{current} O_2(g) + 2H_2(g) \tag{5}$$

Two thin gold plates, each approximately 2 cm by 2 cm, are used as electrodes (Fig. 17). Typically, the two plates are placed 1-2 cm apart to allow the current to travel through the water. Most electrolysis experiments are performed in ionic solutions, which are usually prepared by adding salt, acid, or base. However, adding electrolytes is not feasible in this case because the device must be able to generate gas in regular tap water.

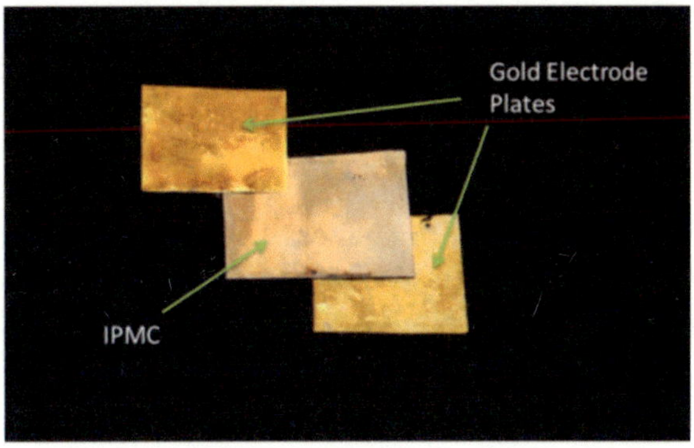

Figure 17. Gold electrodes and a sample of an ionic polymer-metal composite.

Electrolysis in tap water is much slower because of the limited number of ions present. In order to enhance the electrolysis process, an IPMC (150 μm thick) is placed directly in between the electrodes (Fig. 17). In this study, the IPMC is used only as a medium to contain electrolytes and provide channels for ion movement in the electrolysis process. By allowing the current to flow through the electrodes and IPMC, oxygen and hydrogen gases are generated. These gases are collected in a gas chamber, displacing the water in the artificial bladder. Because the gas mixture has a much lower density than that of water, the device becomes more positively buoyant. Care must be taken because the mixture contains a proportion of constituents such that it can burn instantly when ignited.

3.1.2. Gas releasing mechanism by solenoid valve

The mechanism to control the release of gas, and thereby depth, uses a two-way solenoid valve (Hargraves Tech. Corp, Part # 75M06U2.A005S). This valve is closed in its non-actuated state and is suitable for the size and power constraints of this proof-of-concept design. This particular valve can be opened by applying 6 VDC voltage and has a power consumption and mass of 0.5 W and 5 g, respectively. When the valve is actuated, the gas formed during electrolysis escapes from the device and water supersedes. As a result, the density of the device increases, causing it to become more negatively buoyant. When the valve is closed, water cannot enter through a bottom opening because the pressure inside the device is equal to that of outside. Thus the device is able to maintain the same depth by

having a constant buoyant force. In order to travel up or downwards, gas is either produced or released in a controlled manner.

3.2. Device design

The device consists of three parts: bottom chamber, middle seal, and top chamber. The overall schematic of the device is shown in Fig. 18. The parts are drawn using Autodesk Inventor 2010 and printed using a Fused Deposition Modeling (FDM) machine (uPrint Plus by Dimension). The device is approximately 15 cm tall, 6.5 cm in diameter, and has a mass of 114 g.

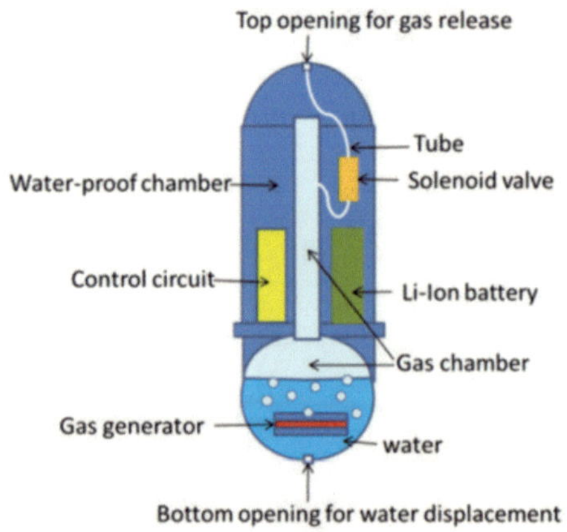

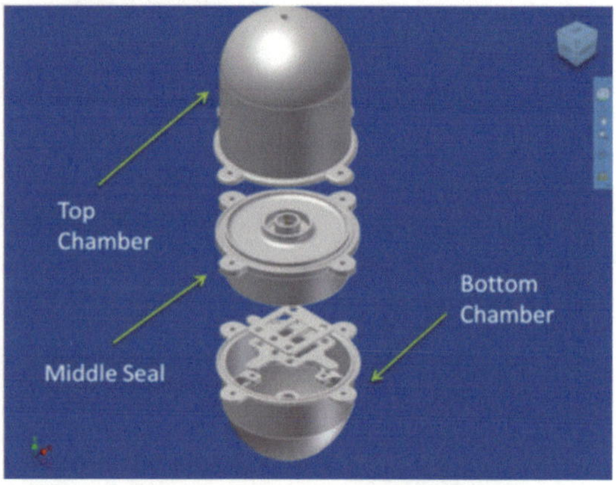

Figure 18. Schematic of the device (left); Computer drawing of the depth control device (right).

3.2.1. *Bottom chamber*

The bottom chamber encases the gas generator and holds water (Fig. 19). The water level varies depending on the depth location of the device. When the device is below the surface, the gas generator is submerged under water to ensure that the electrolysis can take place to allow the device to float up. The bottom chamber also has room to secure up to 8 metal washers as deadweights such that the device is initially about 80-90% submerged. On the bottom of the chamber, there is a small opening to allow water to enter or escape the device. This opening ensures that the inside and outside pressures are kept equal.

3.2.2. *Middle seal*

The purpose of the middle seal (Fig. 20) is to provide waterproofing of the electronics in the top chamber (Fig. 21). The bottom inner surface is concaved to direct the gas products to the gas chamber.

3.2.3. *Top chamber*

The top chamber is partitioned into a waterproof chamber and a gas chamber. The waterproof chamber contains all the electronics, Li-ion battery and Solenoid valve (Fig. 21).

The gas chamber contains a mixture of air and gases produced at the electrodes in the bottom chamber. A tube connects the gas chamber to the device surroundings via solenoid valve.

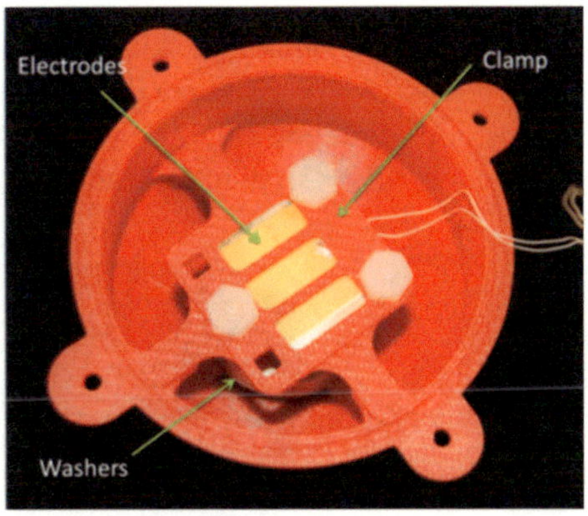

Figure 19. Bottom Chamber with electrodes assembly (Top View).

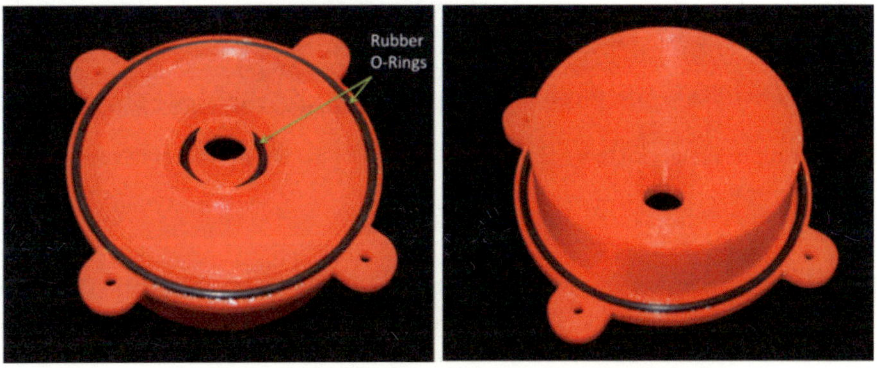

Figure 20. Middle Seal with rubber O-rings (Left: Top view; Right: Bottom view).

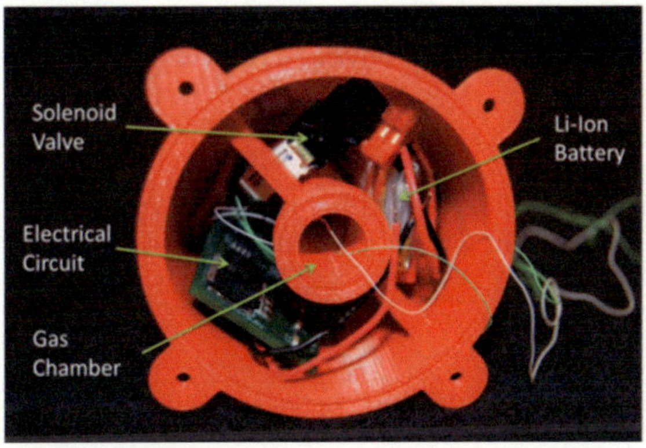

Figure 21. Top Chamber with electronic components and gas chamber.

3.2.4. On-board electrical circuit design

The on-board circuit (Fig. 22) provides actuation voltage signals to the electrodes and solenoid valve. A rechargeable 7.4 V, 400 mAh AA Portal Power Corp Lithium Ion Polymer battery is used as a power source and PIC12F508 microcontroller is used to generate two square wave control signals, S_1 and S_2. A square wave is chosen for simplicity. Since the microcontroller draws only 25 mA, two H-bridge drivers are used to provide up to 2 A peak current output to the electrodes and solenoid valve, which draw up to 500 mA and 80 mA, respectively. Amplitude of the voltage that controls the valve, V_{p1}, is 7.4 V. A voltage regulator sets the amplitude of the voltage applied to the electrodes and microcontroller, V_{p2}, to 5 V. Mass of the circuit and the battery are 11.5 g and 19.1 g, respectively.

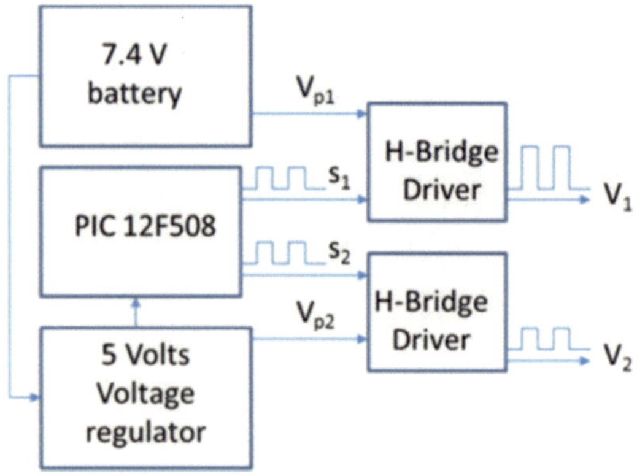

Figure 22. Circuit schematics.

3.3. Experimental results

3.3.1. Gas generation rate

An experiment is set up to measure the gas generation rate at different voltage levels. Voltages ranging from 2.0 V to 6.0 V at 0.5 V intervals are applied to the electrodes using an Agilent DC Power Supply (Model #E3646A). The hydrogen and oxygen gases generated are collected using water displacement technique with 50 mL graduated cylinder (Fig. 23(left)). A constant voltage is applied for 60 s then the average current and displaced volumes are recorded. The gas generation rate can be found by dividing the displaced volume by 60 s. Gas generation rate vs. power consumption is plotted (Fig. 23(right)) and a least squares regression line is fitted to find a correlation. The results indicate a fairly strong linear relationship between gases generated and power consumption. The proportionality constant is approximately 0.032 mL/J.

The on-board circuit is set up such that the output voltage is 5V. However the actual voltage measured across the electrodes is 4 V due to limited capacity of the battery. At this voltage, the average current and power consumption based on 5 trials are 0.3 A and 1.2 W, respectively. The average gas generation rate was 0.048 mL/s. The linear model predicts 0.040 mL/s at this power so the model and the result are in a fair agreement.

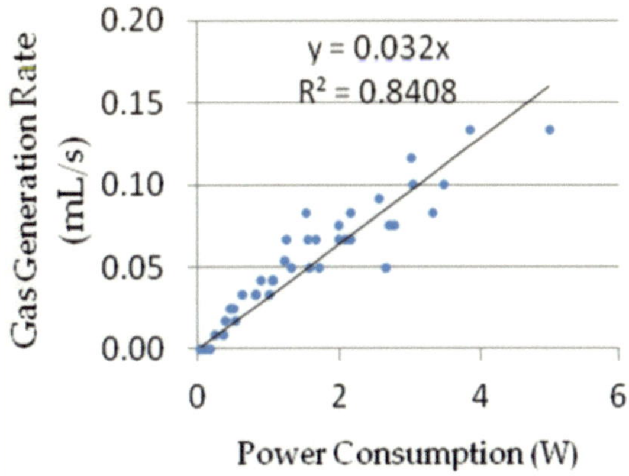

Figure 23. Gas generation rate experiment set-up; using water displacement technique to collect the gas generated (left); Gas generation rate vs. power consumption (right).

3.3.2. *Diving test*

The depth control device is tested in a water tank (1.5 m wide, 4.7 m long, and 0.9 m deep). The tank is filled with tap water at a room temperature. A critical mass is when the device is about 95% submerged under water, at which a slight decrease in buoyancy causes the device to sink. The critical mass of the device is experimentally found to be 283 g. Because the mass of the device is 114 g, 169 g of metal washers are added to the bottom chamber as deadweights. This mass is also equal to the payload for this particular device. The microcontroller in the circuit is programmed such that there is an initial 3-minute delay to allow time for assembly and fastening of the bolts. After the initial delay, the solenoid valve turns on for 12 s to allow gas to escape such that the device will sink to the bottom of the tank. Then the solenoid valve turns off and 5 V is applied to the electrode plates for up to 15 minutes. Gases generated by the electrolysis fill the gas chamber, which displaces the water inside the bottom chamber. Thus, the device becomes more positively

buoyant. The time it takes to rises back up to the top is measured. Fig. 24 shows the timing of the control signals.

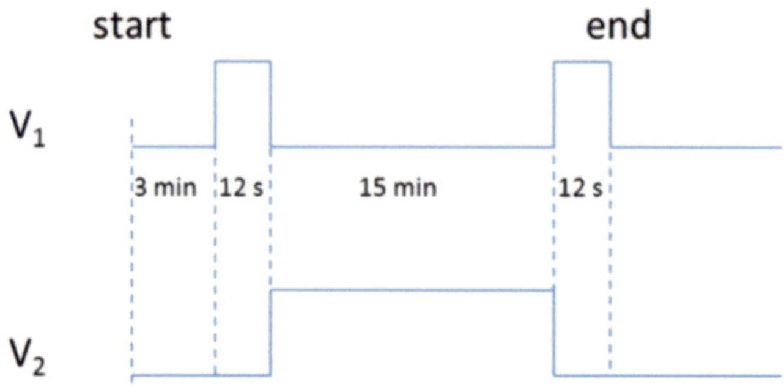

Figure 24. Timing of the control signals.

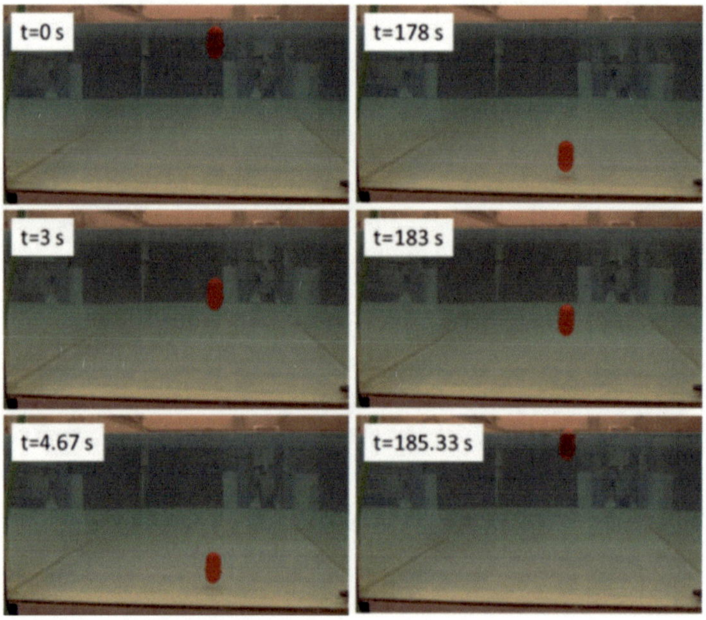

Figure 25. Snapshots of the device during open-loop testing.

It took approximately 4.7 s for the device to sink to the bottom of the tank and 180 s to rise back up to the top. The power consumption for sinking and rising are 0.5 W and 1.2 W, respectively. Fig. 25 shows snapshots of a successful demonstration of the open-loop proof-of-concept depth control device. In order to reduce the rising response time, multiple sets of electrodes can be implemented and actuated simultaneously at a higher voltage. To decrease the current rising time by a factor of 1/3 (60 s), a gas generation rate must triple to be at least 0.14 mL/s. This is achievable with two sets of electrodes with approximately 6 V actuation voltages.

3.3.3. *Discussion*

The advantages of using IPMC as electrolysis generator in depth control device are: 1) quiet gas generation; 2) low activation voltage; 3) compact design. The challenge arises from the low gas generation rate. In the close-loop depth control (Chen et al, 2012b), a fast gas generation rate is required to compensate the depth drop caused by disturbances and the volume decrease caused by water pressure, rapidly. A multi-IPMC electrodes design might be a good solution to increase the gas generation. However, it also increases the power consumption. In our future work, an additional volume-fixed gas storage chamber will be used to collect the generated gas with high pressure. Another solenoid valve will be used to release the high pressure gas into the gas chamber, which controls the volume of the device rapidly. One may be aware of hydrogen and oxygen explosive gas in the gas storage chamber. A two-gas-chamber design can solve this issue by separating oxygen and hydrogen gases. If a fuel cell is embedded, these gases can be recycled as fuel to recharge the battery, which makes the depth control more power efficient.

4. Chapter conclusion

In this Chapter, we presented two studies on IPMC artificial muscles in bio-inspired engineering research. In the first study, we developed a bio-inspired robotic manta ray powered by IPMC pectoral fin. We developed a novel synthesis process to fabricate IPMC pectoral fin that is capable of 3D kinematic motion and characterized the pectoral fin in terms of tip deflection (up to 100%), bode plot (0.4 Hz cut-off frequency), twist angle (up to 40o), and power consumption (below 1.5 W). A small size free-swimming robotic manta ray has been developed and experimental results have demonstrated its free-swimming capability with speed at 0.067 body length per second and 2.5 W power consumption. In the second study, we developed a novel buoyancy control device as an artificial swimming bladder. IPMC acted as an efficient, environmentally friendly water electrolysis generator to gain volume of the device while a solenoid valve was used as gas releasing controller. A compact and low power device has been assembled with an on-board open loop controller. Experimental results have shown that the device was able to control its 0.9 m depth within 3 minutes.

In both studies, advantages and challenges of IPMC in bio-inspired engineering research have been addressed. In this first study, the advantages of using IPMC as artificial muscle are: 1) low actuation voltage; 2) working well in wet condition; 3) low noise; 4) simple mechanical structure; 5) able to be shaped into bio-inspired engineering design. The disadvantages are: 1) low generated force; 2) slow response time. The challenge in this study comes from optimal design of the pectoral fin. In future research, we will focus on modeling an IPMC powered pectoral fin and modeling of the fluid dynamics introduced by the 3D kinematic motions of the fin. In the second study, the advantages of using IPMC as electrolysis generator are: 1) low noise; 2) compact design; 3) low

activation voltage and low power consumption. The disadvantage arises from 1) slow gas generation rate that limits its capability of feedback depth control; 2) unstable nonlinear dynamics. The challenge in this study is feedback depth control that requires high gas generation rates and complex nonlinear control algorithms. In future work, we will focus on nonlinear feedback control and improved device design that can gain gas volume rapidly.

Author details

Zheng Chen, T. Um and Hilary Bart-Smith*
University of Virginia, USA

Acknowledgement

This research was supported in part by the Office of Naval Research (ONR) under the Multidisciplinary University Research Initiative (MURI) Grant N00014-08-1-0642 and the David and Lucille Packard Foundation.

5. References

Bar-Cohen, Y. (2004). Electric flex. *IEEE Spectrum,* Vol.41, No.6, pp. 29-33, ISSN 0018-9235

Bar-Cohen, Y. (2000). Electroactive Polymers as Artificial Muscles - Capabilities, Potentials and Challenges, In: *Handbook on Biomimetics,* Y. Osada, (Ed.), 936-950, NTS Inc.

Bond, C. (1996). Swim Bladder. In: *Biology of Fishes* (2nd edition). pp. 283-290, Saunders College Publishing.

Chen, Z. & Tan, X. (2008). A Control-oriented and Physics-based Model for Ionic Polymer-Metal Composite Actuators. *IEEE/ASME Transactions on Mechatronics*, Vol. 13, No. 5, pp. 519–529, ISSN 1083-4435.

Chen, Z. ; Shatara, S. & Tan, X. (2010). Modeling of Biomimetic Robotic Fish Propelled by an Ionic Polymer-Metal Composite Caudal Fin. *IEEE/ASME Transactions on Mechatronics,* Vol. 15, No. 3, pp. 448-459, ISSN 1083-4435.

Chen, Z & Tan, X. (2010). Monolithic Fabrication of Ionic Polymer-Metal Composite Actuators Capable of Complex Deformation, *Sensors and Actuators A: Physical,* Vol. 157, No. 2, pp. 246–257, ISSN 0924-4247.

Chen, Z.; Um, T. I. & Bart-Smith, H. (2011). A Novel Fabrication of Ionic Polymer-Metal Composite Membrane Actuator Capable of 3-dimensional Kinematic Motions. *Sensors and Actuators A: Physical,* Vol. 168, No. 1, pp. 131–139, ISSN 0924-4247.

Chen, Z.; Um, T. I. & Bart-Smith, H. (2012a). Bio-inspired Robotic Manta Ray Powered by Ionic Polymer-Metal Composite Artificial Muscles. *International Journal of Smart and Nano Materials,* to appear, ISSN 1947-5411.

---

* Corresponding Author

Chen, Z; Um, T. I. & Bart-Smith, H. (2012b). Modelling and Control of Artificial Bladder Enabled by Ionic Polymer-Metal Composite. *Proceedings of the 2012 American Control Conference*, to appear, June 27-29 2012, Montreal, Canada, 2012, pp. 3340-3345.

Chung, C.; Fung, P.; Hong, Y.; Ju, M.; Lin, C. & Wu, T. (2006). A Novel Fabrication of Ionic Polymer-Metal Composites (IPMC) Actuator with Silver Nano-powders. *Sensors and Actuators B: Chemical*, Vol. 117, No. 2, pp. 367–375, ISSN 0925-4005.

Gao, J.; Bi, S.; Xu, Y. & Liu, C. (2007). Development and Design of A Robotic Manta Ray Featuring Flexible Pectoral Fins. *Proceedings of the 2007 IEEE International Conference on Robotic and Biomimetics*, pp. 519-523, ISBN 978-1-4244-1761-2, Sanya, China, December 15-18, 2007.

Guo, S.; Fukuda, T. & Asaka, K. (2003). A New Type of Fish-like Underwater Microrobot. *IEEE/ASME Transactions on Mechatronics*, Vol. 8, No. 1, pp. 136–141, ISSN 1083-4435.

Kim, K. J. & Shahinpoor, M. (2002). A Novel Method of Manufacturing Three-dimensional Ionic Polymer-Metal Composites (IPMCs) Biomimetic Sensors, Actuators, and Artificial Muscles. *Polymer*, Vol. 43, No. 3, pp. 797–802, ISSN 0032-3861.

Kim, K. J. & Shahinpoor, M. (2003). Ionic polymer-metal composites: II. Manufacturing techniques. *Smart Materials and Structures*, Vol. 12, No. 1, pp. 65–79, ISSN 0964-1726.

Kim, K. J.; Pugal, D. & Leang, K. K. (2011). A Twistable Ionic Polymer-Metal Composite Artificial Muscle for Marine Applications. *Marine Technology Society Journal*, Vol. 45, No. 4, pp. 83-98, ISSN 0025-3324.

Kim, S. J.; Lee, I. T. & Kim. Y. H. (2007). Performance Enhancement of IPMC Actuator by Plasma Surface Treatment. *Smart Materials and Structures*, Vol. 16, No. 1, pp. N6–N11, ISSN 0964-1726.

Lee, S. J.; Han, M. J.; Kim, S. J.; Jho, J. Y.; Lee, H. Y. & Kim, Y. H. (2006). A New Fabrication Method for IPMC Actuators and Application to Artificial Fingers. *Smart Materials and Structures*, Vol. 15, No. 5, pp. 1217–1224, ISSN 0964-1726.

Lin, L.; Xu, H.; Xie, H. & Shen, L. (2010). Bionic Bladder Based Depth Control for Bionic Underwater Robots. *Proceedings of 2010 8th IEEE International Conference on Control and Automation*, pp. 932-936, ISBN 978-1-4244-5195-1, June 9-11, Xiamen, China, 2010.

McFarland, D.; Gihespy, I. & Honary, E. (2003). Divebot: A Diving Robot with A Whale-like Buoyancy Mechanism. *Robotica*, Vol. 21, No. 4, pp. 385-398, ISSN 0263-5747

Millet, P.; Pineri, M. & Durand, R. (1989). New Solid Polymer Electrolyte Composites for Water Electrolysis. *Journal of Applied Electrochemistry*, Vol. 19, No. 2, pp. 162-166, ISSN 0021-891X.

Mojarrad, M. & Shahinpoor, M. (1996). Noiseless Propulsion for Swimming Robotic Structures Using Polyelectrolyte Ion-exchange Membranes. *Proceedings of SPIE Conference on Smart Structures and Materials*, Vol. 2176, pp. 183, ISSN 0277-786X, February 26, San Diego, California, 1996.

Moored, K.; Smith, W.; Hester, J.; Chang, W. & Bart-Smith, H. (2008). Investigating the Thrust Production of A Myliobatoid-inspired Oscillating Wing. *Advances in Science and Technology*, Vol. 58, pp. 25–30, ISSN 1662-0356.

Moored, K. & Bart-Smith, H. (2009). Investigation of Clustered Actuation in Tensegrity Structures. *International Journal of Solids and Structures*, Vol. 46, No. 17, pp. 3272–3281, ISSN 0020-7683.

Nemat-Nasser, S. & Li, J. (2000). Electromechanical Response of Ionic Polymer–Metal Composites. *Journal of Applied Physics*, Vol. 87, No. 7, pp. 3321–3331, ISSN 0021-8979.

Nemat-Nasser, S. (2002). Micromechanics of Actuation of Ionic Polymer–Metal Composites. *Journal of Applied Physics*, Vol. 92, No. 5, pp. 2899–2915, ISSN 0021-8979.

Punning, A.; Anton, M.; Kruusmaa, M. & Aabloo, A. (2004). A Biologically Inspired Ray-like Underwater Robot with Electroactive Polymer Pectoral Fins. *Proceedings of the IEEE International Conference on Mechatronics and Robotics 2004*, pp. 241-245, ISBN 3-938153-50-X, September 13-15, Aachen, Germany, 2004.

Rosenberger, L. J. (2001). Pectoral Fin Locomotion in Batoid Fishes: Undulation versus Oscillation. *The Journal of Experimental Biology*, Vol. 204, pp. 379–394, ISSN 0022-0949.

Shahinpoor, M. (1992). Conceptual Design, Kinematics and Dynamics of Swimming Robotic Structures Using Ionic Polymeric Muscles. *Smart Materials and Structures*, Vol. 1, No. 1, pp. 91-94, ISSN 0964-1726.

Shahinpoor, M. & Kim, K. J. (2001). Ionic polymer-metal composites: I. Fundamentals. *Smart Materials and Structures*, Vol. 10, No. 4, pp. 819–833, ISSN 0964-1726.

Shahinpoor, M.; Kim, K. J. & Leo, D. (2003). Ionic Polymer-Metal Composites as Multifunctional Materials. *Polymer Composites*, Vol. 24, No. 1, pp. 24-33, ISSN 0272-8397

Shahinpoor, M & Kim K. J. (2005). Ionic polymer–metal composites: IV. Industrial and medical applications. *Smart Materials and Structures*, Vol. 14, No. 1, pp. 197–214 ISSN 0964-1726.

Shaaban, A. H. (1994). Pulsed DC and Anode Depolarization in Water Electrolysis for Hydrogen Generation. In: *StormingMedia*, 01.08.1994, Avalable from http://www.stormingmedia.com.

Shimizu, N.; Hotta, S.; Seekiya, T. & Oda, O. (2005). A Novel Method of Hydrogen Generation by Water Electrolysis Using an Altra-short-pulse Power Supply. *Journal of Applied Electrochemistry*, Vol. 36, No. 4, pp. 419-423, ISSN 0021-891X.

Tan, X.; Kim, D.; Usher, N.; Laboy, D.; Jackson, J.; Kapetanovic, A.; Rapai, J.; Sabadus, B, & Zhou, X. (2006). An Autonomous Robotic Fish for Mobile Sensing. *Proceedings of the 2006 IEEE/RSJ International Conference on Intelligent Robots and Systems*, pp. 5424-5429, ISBN 1-4244-0259-X, Oct. 9-15, Beijing, China, 2006.

Um, T. I.; Chen, Z. & Bart-Smith, H. (2011). A Novel Electroactive Polymer Depth Control Device for Bio-inspired Underwater Vehicles. *Proceedings of the IEEE International Conference on Robotics and Automation*, pp. 172-177, ISBN 978-1-61284-380-3, May 13-19, Shanghai, China, 2011.

Wang, Z.; Wang, Y.; Li, J. & Hang, G. (2009). A Micro Biomimetic Manta Ray Robot Fish Actuated by SMA. *Proceedings of the 2009 IEEE International Conference on Robotics and Biomimetics*, pp. 1809-1813, ISBN 978-1-4244-4774-9, Guilin, China, December 19-25, 2009.

Zhou, C. & Low, K. H. (2102). Design and Locomotion Control of A Biomimetic Underwater Vehicle with Fin Propulsion. *IEEE/ASME Transactions on Mechatronics*, Vol. 17, No. 1, pp. 25-35, ISSN 1083-4435.

Chapter 11

Sensing and Rheological Capabilities of MR Elastomers

Weihua Li, Tongfei Tian and Haiping Du

Additional information is available at the end of the chapter

http://dx.doi.org/10.5772/50501

1. Introduction

Magnetorheological elastomers (MREs) are smart materials where polarized particles are suspended in a non-magnetic solid or gel-like matrix. Two kinds of MREs, namely anisotropic and isotropic, are fabricated either under a magnetic field or without a field [1,2]. In anisotropic MREs, polarised particles are arranged in chains within a polymer media such as silicon rubber or natural rubber. The shear modulus of MRE can be controlled by the external magnetic field, which has led to many applications, such as tuned vibration absorbers, dampers and sensors [3].

Additives are used to adjust the mechanical properties or electrical performance of MREs. Silicone oil is an additive which is used to increase the gaps between the matrix molecules and to decrease the gaps between the conglutination of molecules. Apart from increasing the plasticity and fluidity of the matrix, the additives can average the distribution of internal stress in the materials, which makes them ideal for fabricating MRE materials. Graphite powder is a kind of additive which can affect the magnetorheology and electrical conductivity of MREs [4,5]. By introducing graphite microparticles into the elastic matrix, MREs exhibit a lower electrical conductive and a different magnetorheological response.

When the material in the matrix is magnetic, the polarization of the particles is less effective and the magnetorheological response is therefore smaller. The addition of magnetically active additives (other than MR particles) also decreases the magnetorheology [6, 7]. The overall properties of the elastomer composite are also influenced by the additives, as the filler material causes the volume to increase, so the previous effect also increases. Lokander et al. [7, 8] have shown that the absolute effect of MR (the difference between the zero-field modulus and modulus measured under an external magnetic field) is independent of the matrix material. However, the zero-field modulus can be much higher for hard matrix material

(for instance, where materials with high volume fraction of iron specifically possess a high zero-field modulus), which means that the relative magnetorheological effect is quite low.

MREs feature viscoelastic properties and magnetorheology [9]. The magnetorheology of MREs is described as a reversible change in modulus in an applied magnetic field. Aligned MREs have mostly been characterized at relatively low frequencies (1 to 20 Hz) to measure the changes in the dynamic shear modulus induced by the external magnetic field [9, 10]. Ginder et al. found a substantial magnetorheology over the entire frequency range studied. The increase in the shear modulus varied initially with the strength of the magnetic field but saturated at higher field strength. When the magnetic field was increased from 0 to 0.56 Tesla the consequent increase in shear modulus was nearly 2 MPa and the frequency of the resonance was shifted upward by over 20% [10]. Zhou et al. [9] stated that the changes of dynamic shear storage modulus can be over 50%, while Gong et al. [11] said that it can be over 100%. Lokander et al. studied the dynamic shear modulus for isotropic MR elastomers with different filler particles and matrix materials. They measured the magnetorheology as a function of the amplitude of strain and found that the magnetorheology decreases rapidly with increasing strain within the measured range, and is not dependent on the frequency of testing. The fact that the absolute magnetorheology is independent of the matrix material means that softer matrix materials will show a greater relative magnetorheology [7, 8].

The effect of additives on the sensing capabilities were studied by a few groups. Kchit and Bossis [7] found that the initial resistivity of metal powder at zero pressure is about 108 Ωcm for pure nickel powder and 106 Ωcm for silver coated nickel particles. The change in resistance with pressure was found to be an order of magnitude larger for a MRE composite than for the same volume fraction of fillers dispersed randomly in the polymer. Wang et al. [8] proposed a phenomenological model to understand the impedance response of MREs under mechanical loads and magnetic fields. Their results showed that MRE samples exhibit significant changes in measured values of impedance and resistance in response to compressive deformation, as well as applied magnetic field. Bica [9] found that MRE with graphite micro particles (~14%) is electroconductive. These MREs possess an electric resistance whose value diminishes with both the increase of the intensity of the magnetic field and with the compression force. Such a variation of resistance with magnetic field intensity is due to the compression of MRE with graphite microparticles. In the approximation of the perfect elastic body, the sums of the main deformations and the compressibility module of MRE with graphite microparticles, depend on the magnetic field intensity. Li et al. [10] introduced graphite into conventional MREs and found that a MRE sample with 55% carbonyl iron, 20% silicon rubber and 25% graphite powder exhibits the best performance. The test result showed that at a normal force of 5 N, the resistance decreases from 4.62 kΩ without a magnetic field to 1.78 kΩ at a magnetic field of 600 mT. The decreasing rate is more than 60%. This result also demonstrated the possibility of using MREs to develop a sensor for measuring magnetic fields. This result indicates that the detection is very sensitive to the normal force. When the normal force is 15N, the field-induced resistance only has less than 28% change from 0.65 kΩ at 0 mT to 0.47 kΩ at 600 mT.

Depending on an elasto-plastic asperity microcontact model for contact between two nominally flat surfaces, Kchit and Bossis developed a model to analyse the contact of two

rough surfaces. They used two kinds of magnetic particles: nickel and nickel coated with silver which are dispersed in a silicone polymer as the polarised particles [12]. To understand the complex conductivity of particle embedded composites, quantitative or semi-quantitative models can be used [16]. The Maxwell–Wagner and the Bruggeman symmetric and asymmetric media equations were introduced by McLachlan [17] to model the electrical behaviour of conductor-insulator composites. The microstructures for which these effective media equations apply are considered in simulating the measured impedance and modular spectra of these composites. Woo et al. [18] developed a universal equivalent circuit model in modelling the impedance response of composites with insulating or conductive particles or fibres. Based on the microstructure of MREs, Wang et al. proposed an equivalent circuit model to interpret the impedance measurement results [13].

This chapter is organized as follows. Section 2 describes the fabrication of graphite based MRE samples with various weight fractions. The morphology of the MRE samples and rheological properties of graphite based MREs were presented in Section 3 and 4, respectively. Section 5 presents the theoretical development based on a representative volume for investigating MRE electric properties. The main findings are summarized in Section 6.

2. Fabrication of graphite based MREs

The materials used for the Graphite based Magnetorheological Elastomers (Gr-MREs) are: silicone rubber (Selleys Pty. LTD), silicone oil, type 378364 (Sigma-Aldrich Pty. LTD), carbonyl iron particles, type C3518 (Sigma-Aldrich Pty. LTD) and graphite powder, type 282863 (Sigma-Aldrich Pty. LTD). The particle sizes of graphite powder are less than 20 μm, the iron particles' diameter is between 3 μm and 5 μm.

The MREs with various graphite weight fractions (Gr %) were fabricated to compare the effect of graphite on to MREs. Table 1 shows the compositions of all Gr MRE samples. all samples contain 10g carbonyl iron particles, 3g silicone rubber and 3g silicone oil. The only difference is the graphite weight fraction, which is from 0% to 23.81%. For each composition, the mixture is fabricated under a constant magnetic field of 165 mT to form anistropic MRE samples.

3. Microstructure observation

To see microstructure of Gr-MREs, LV-SEM (JSM 6490LV SEM) is used. Figures 1- 3 show the surface imaging for MRE microstructures. Specifically, Fig. 1 shows the microstructure of isotropic and anisotropic Gr-MREs without graphite. The carbonyl iron particles are arranged in chains in the anisotropic sample, where they are dispersed randomly in the isotropic sample. According to Fig. 2 (a) and Fig. 3 (a), in the anisotropic Gr-MREs, the carbonyl iron particles are arranged in chains, whereas the graphite powders dispersed randomly in the matrix. The reason of this phenomenon is that the magnetic field only affects the carbonyl iron particles, but not the graphite. So upon magnetisation, the carbonyl iron particle chains tend to orient along the same direction as the magnetic field in the matrix.

By comparing Fig. 1 (a), Fig. 2 (a) and Fig. 3 (ba), we can see that the carbonyl iron chains in the sample without graphite have the best alignment performance. Further, the carbonyl iron chains in Fig. 2 (a) are aligned better than those in Fig. 3 (a). The reason is that when the mixture of carbonyl iron, silicone rubber, silicone oil and graphite is curing under the magnetic field, the graphite powders in Gr-MREs affect the carbonyl iron particles' movement. The more graphite in the mixture, the more effects are applied on to the carbonyl iron chains, which influence the magnetorheology of MREs.

	Graphite based MREs							
Sample No.	1	2	3	4	5	6	7	8
Carbonyl iron	10g	10g	10g	10g	10g	10g	10g	10g
Silicone oil	3g	3g	3g	3g	3g	3g	3g	3g
Silicone rubber	3g	3g	3g	3g	3g	3g	3g	3g
Graphite	0g	1g	2g	3g	3.5g	4g	4.5g	5g
Graphite weight fraction (Gr %)	0%	5.88%	11.11%	15.79%	17.95%	20%	21.95%	23.81%

Table 1. Components of Gr MRE samples

4. MRE rheology

A rotational Rheometer (MCR 301, Anton Paar Companies, Germany) and a Magneto Rheolgical Device (MRD 180, Anton Paar Companies, Germany) were used to measure the MREs' mechanical properties. The Magneto Rheolgical Device is equipped with an electromagnetic kit which can generate a magnetic field perpendicular to the direction of the shear flow. Specifically, a 20mm diameter parallel-plate measuring system with 1 mm gap was used. The samples were sandwiched between a rotary disk and a base placed in parallel. In this study both a steady-state rotary shear and an oscillatory shear were applied for the experiments.

The schematic of the experimental setup is shown in Fig. 4. The stress and stain signals were output from two ports which were detected through a Data Acquisition (DAQ) board (Type: LABVIEW PCI-6221, National Instruments Corporation. U.S.A) and transferred to a computer by which the data was processed.

In this experiment the magnetic flux density of the sample of MRE (B_{MRE}) in the measuring gap depends not only on the current (I) applied to the samples, but also on the magnetic properties of MRE materials. The relationship between B_{MRE} versus I is found to be: $B_{MRE} = 220\ I$, where the units of B_{MRE} and I are in mT and Ampere (A), respectively.

In the following test, the test current varies from 0A to 2A with the increment 0.5A, for which the intensity of magnetic field is 0 mT to 440 mT with the increment 110 mT.

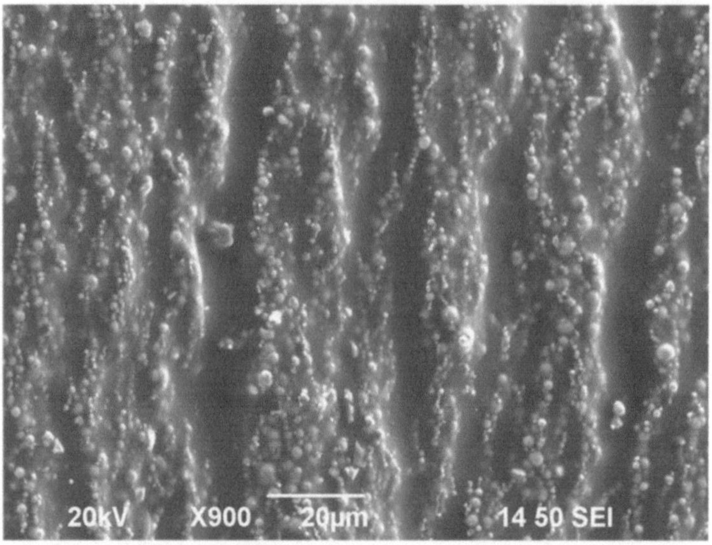

(a)

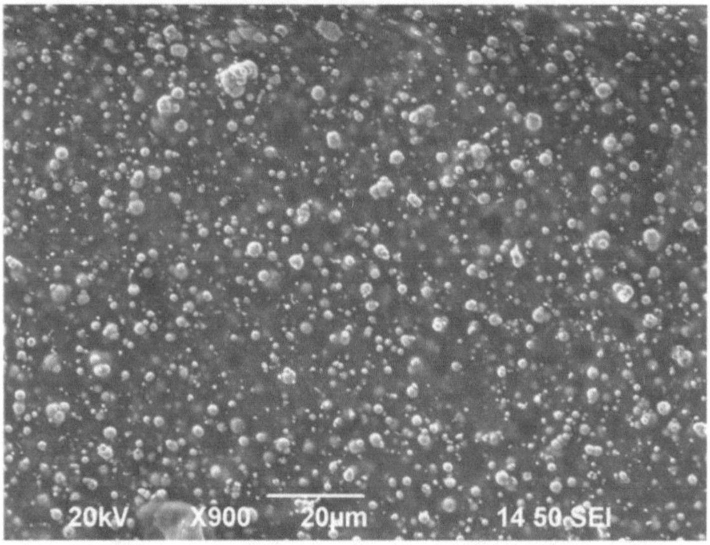

(b)

Figure 1. Microstructure of Gr-MREs (Gr 0%) (a) anisotropic (b) isotropic

4.1. Steady state results

Under rotary shear the shear stress and shear strain of MREs under fields varying from 0~440mT were measured. The MR effect was evaluated by measuring the shear strain-stress curve of the sample with and without a magnetic field applied.

Figures 5 and 6 show the strain-stress curve of different samples at 5 different magnetic field intensities ranging from 0 to 440mT.

The slope of the strain-stress curve is the shear modulus of the material. As can be seen in the figures, all the samples' shear modulus show an increasing trend with magnetic field before they reach magnetic saturation at high field strength, which proves that all the MRE samples exhibit obvious MR effects.

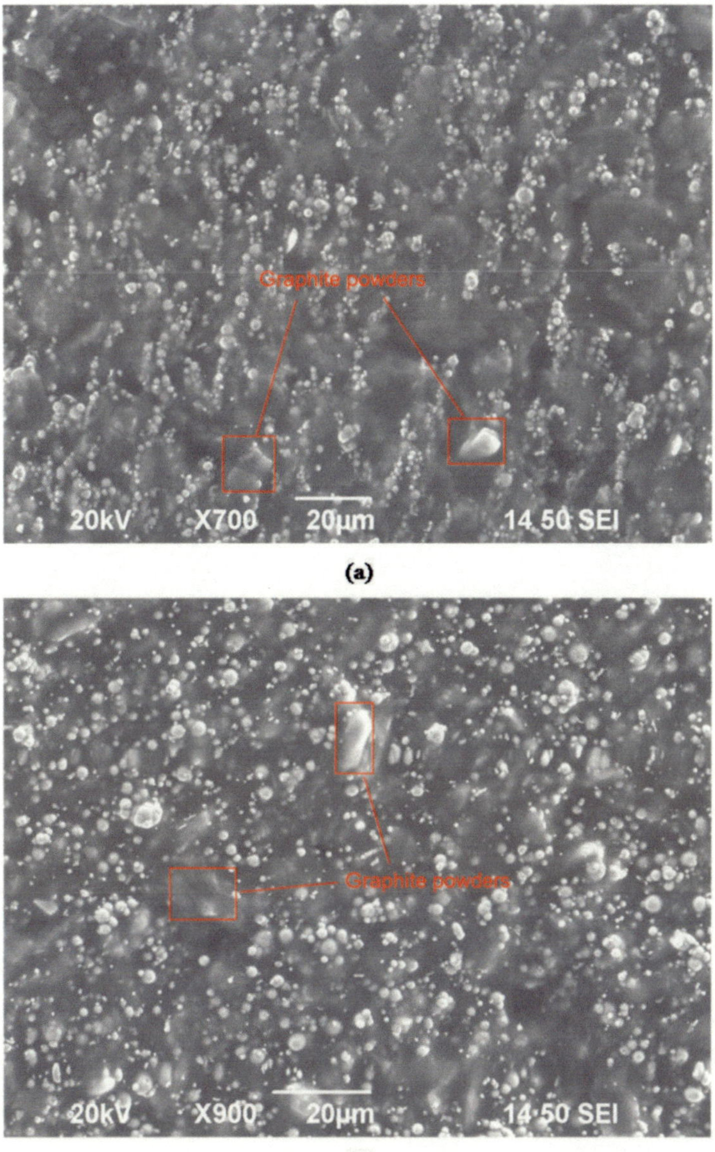

Figure 2. Microstructure of Gr-MREs (Gr 11.11%) (a) anisotropic (b) isotropic

From Figures 5 and 6, the shear stress shows a linear relationship with the shear strain when the strain is within a range. This means the MRE acts with linear viscoelastic properties when the strain is below a certain limit. For conventional MREs, the limitation is around

50% shear strain. When the graphite weight fraction increases from 0 to 15.79%, the range of linearity decreases from 50% to around 10%. For the samples with higher graphite weight fraction alike 23.81%, the linearity ranges are only 6% and 4% for isotropic and anisotropic samples, respectively. When the strain is above the limitation, the shear modulus reaches a saturation (maximum value) and decrease steadily. This could be due to sliding effect. The higher magnetic field intensity leads to higher steady shear stress. Fig. 7 shows the linear ranges of different samples at various magnetic field intensities.

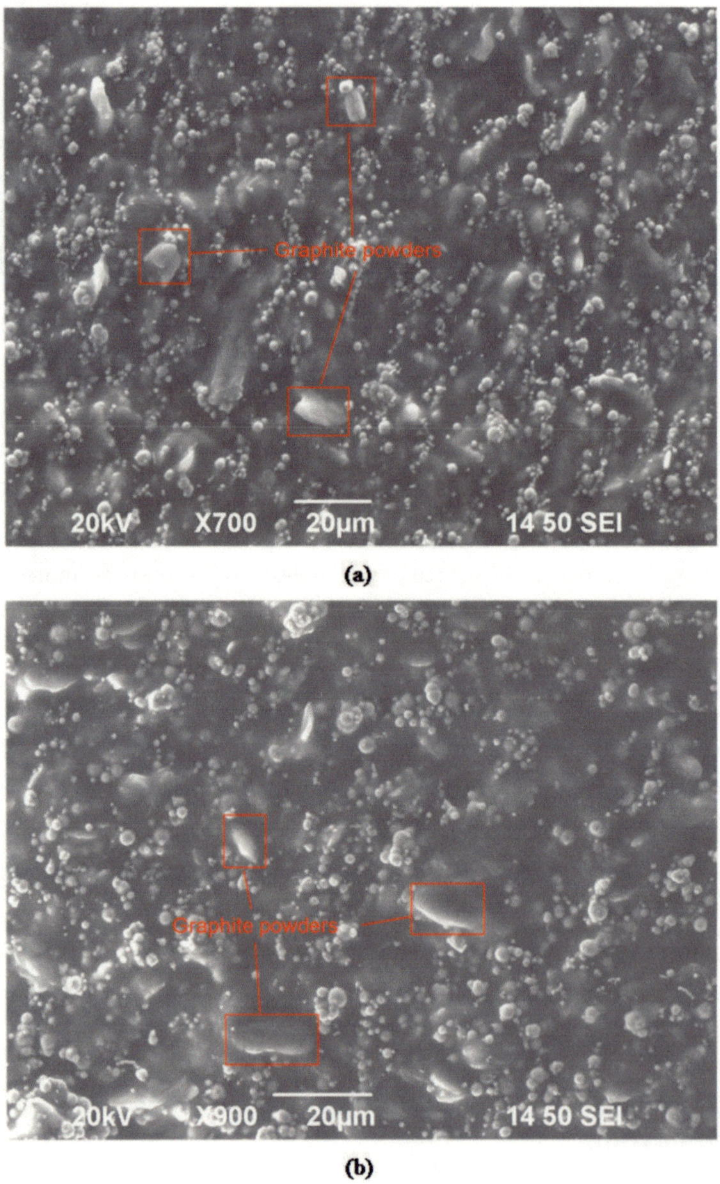

Figure 3. Microstructure of Gr-MREs (Gr 20%) (a) anisotropic (b) isotropic

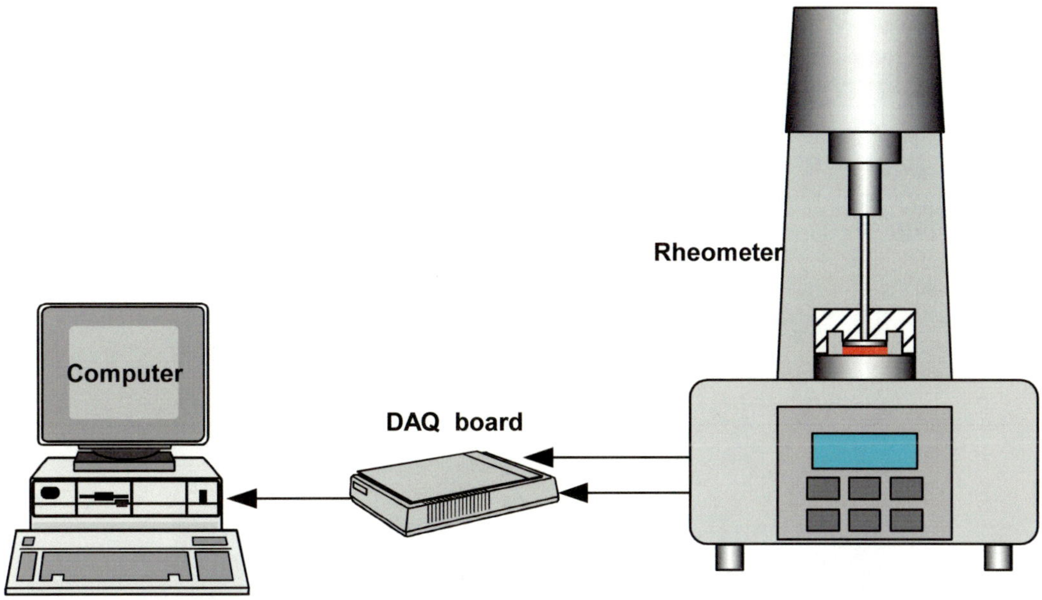

Figure 4. A diagram of the experimental setup

For the isotropic and anisotropic samples with same compositions, the isotropic samples always have the bigger linearity ranges and steady shear stress than chose of anisotropic samples.

For each curve, the slope equals the ratio between peak shear stress and relevant shear strain. By analyzing the slope of these curves, it is easy to see that the more graphite in the material, the smaller increment in slopes occurs when the magnetic field increase from 0 to 440mT. This is due to the contributions of graphite powders to the stiffness of the samples. The graphite increases the initial stiffness of Gr-MREs, thus the stiffness change induced by the MR effect can not be comparable to that conventional MREs. Fig. 8 shows the peak stresses of different samples for a magnetic field intensity equalling 220mT, 330mT and 440mT, respectively.

The relative MR effect ($\Delta G_{max}/G_0$) of several samples is shown in Fig. 9. G_0 denotes the MRE samples' zero-field modulus, ΔG_{max} denotes the saturated field-induced modulus, and $\Delta G_{max}/G_0$ denotes the relative MR effect.

It can be seen from Fig. 9 that G_0 is enhanced with the increase in graphite powders content. This result indicates that graphite powders can modify particle properties and, consequently, influenced the MR effect. From the abovementioned results, the exhibited MR effects correspond well with the microstructures of Gr-MREs.

4.2. Dynamic tests result

The dynamic mechanical behavior of MREs was studied by using the strain amplitude sweep tests. Five sets of data were collected for different amplitudes of oscillation, according

to various magnetic fields applied to the samples of MREs. Similar to the steady state tests, five different magnetic field intensities, 0, 110, 220, 330 and 440mT, were used in this experiment. A constant frequency of 5Hz was selected for the strain amplitude sweep tests.

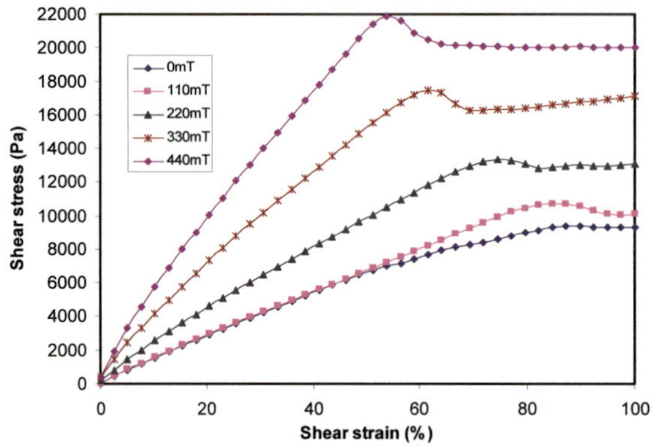

(a)

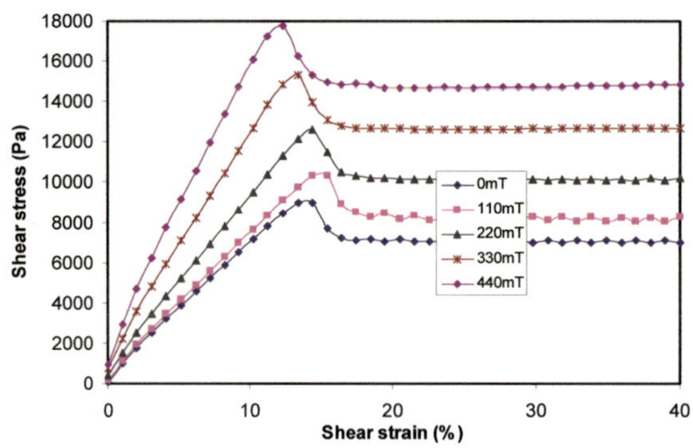

(b)

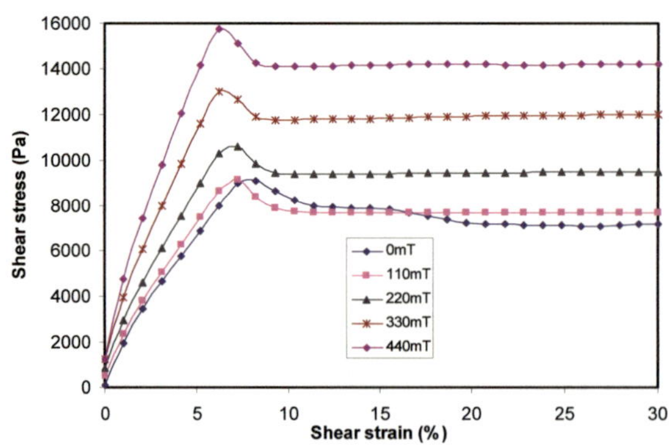

(c)

Figure 5. Strain-stress curve versus magnetic field (isotropic MRE) (a) Gr 0% (b) 15.79% (c) Gr 23.81%)

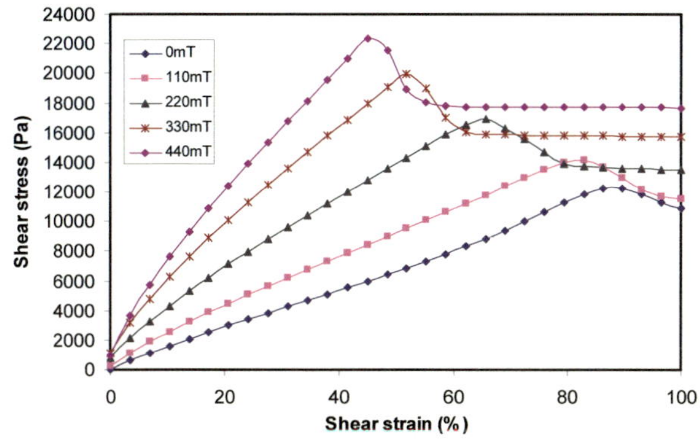

(a)

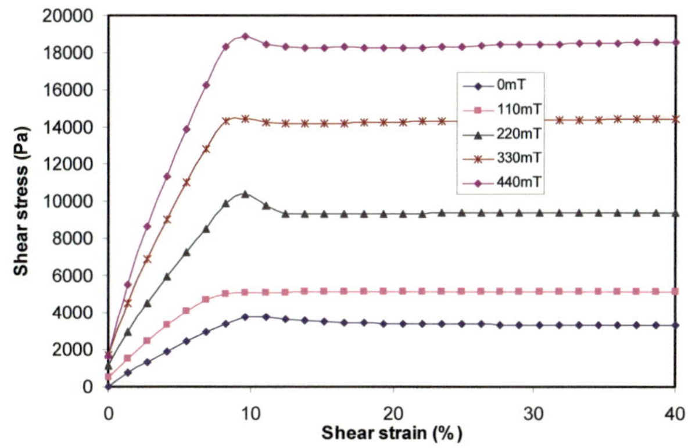

(b)

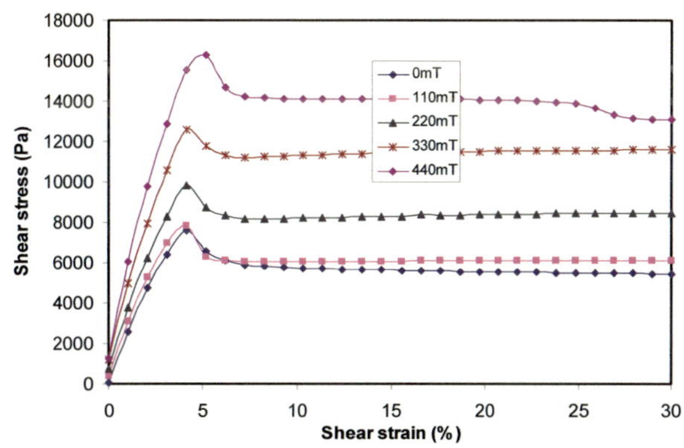

(c)

Figure 6. Strain-stress curve versus magnetic field (anisotropic MRE) (a) Gr 0% (b) 15.79% (c) Gr 23.81%)

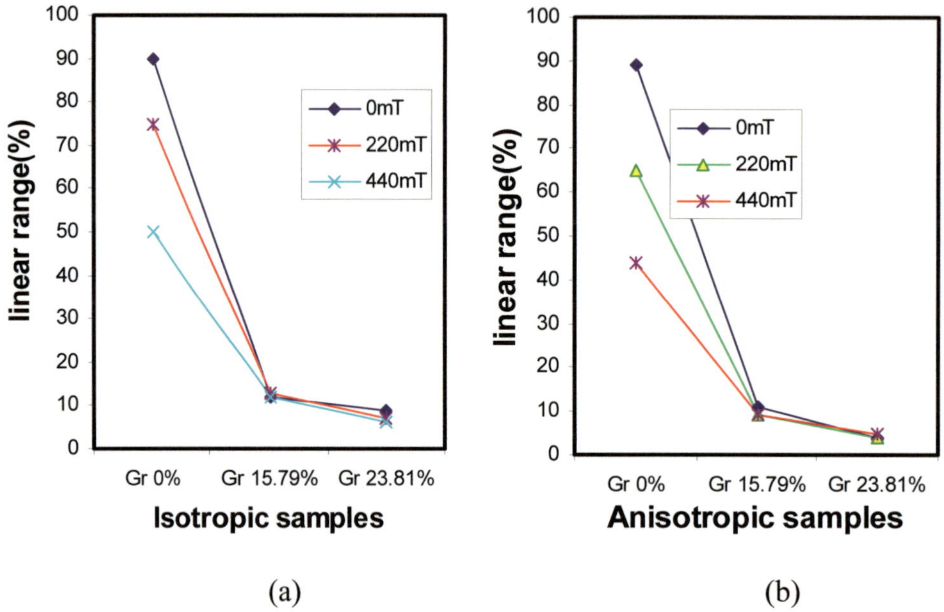

Figure 7. Linear ranges versus different samples (a) isotropic samples (b) anisotropic samples

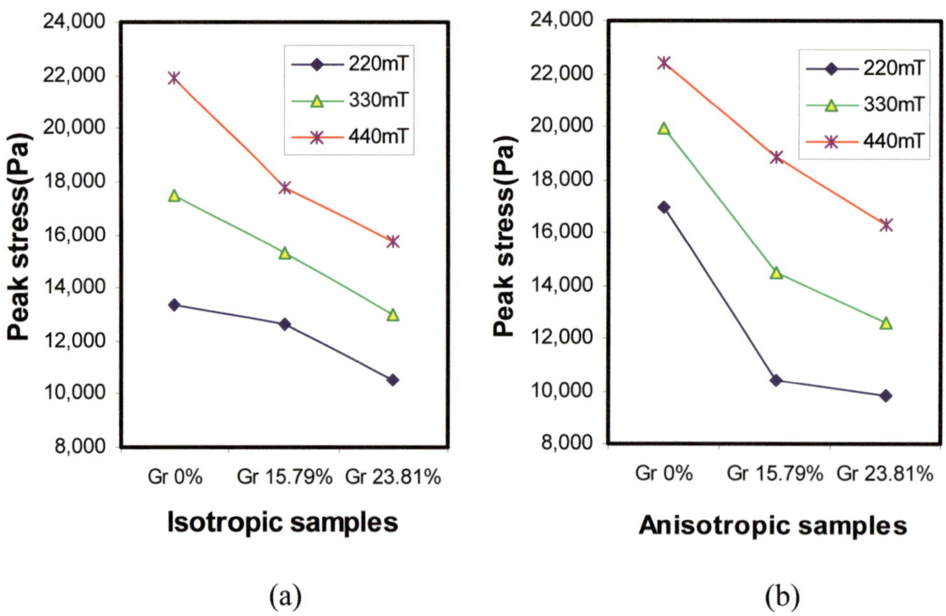

Figure 8. Peaks stresses versus different samples (a) isotropic samples (b) anisotropic samples

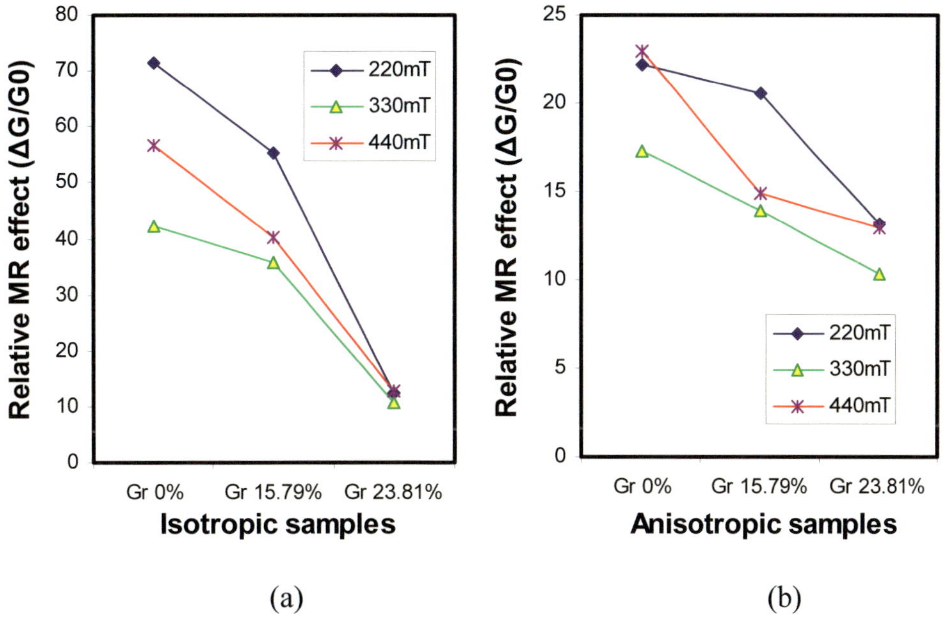

Figure 9. Relative MR effects versus different samples (a) isotropic samples (b) anisotropic samples

In the strain sweep test, the storage and loss moduli were measured by varying strain from 0.01% to 100% at different magnetic fields. Figs. 10-13 show the variation of storage and loss moduli with the strain amplitude sweep.

As shown in Figures from 10 to 13, the overall trend of storage modulus is decreasing with strain amplitude. The storage modulus goes down smoothly until 10% shear strain and begins to drop significantly beyond 10% shear strain. Except for isotropic MREs without graphite, the Loss modulus has almost the same trend as the storage modulus. This means that at the higher shear strains, the storage and loss moduli are much smaller than those at lower shear strains.

Figures 14 and 15 show the storage modulus of different samples at 0mT, 220mT and 440mT magnetic field. The data are collected at 10% and 87.5% shear strain, respectively.

As can be seen from Fig. 14, the storage modulus of all samples shows an increasing trend with graphite weight fraction at 10% shear strain which is in the linear range for most of the samples. In Fig. 15, it turns to a diminishing trend with graphite weight fraction at 87.5% shear strain which is out of the linear range. This means that in the linear range of shear strain, the samples with higher graphite weight faction have the bigger storage modulus.

Figures 16 and 17 show the storage modulus vs. magnetic field at 0.1% and 10% shear strain, respectively. The two shear strains are the beginning and the end of the linear range. From these figures we can see that the storage modulus shows an increasing trend with the intensity of magnetic field. The ratio of storage modulus at 440mT to that at 0mT is indicative of the MR effect. The MR effect of isotropic MREs with 0% graphite is around 4.5,

when the graphite weight fraction increases to 15.79% and 23.81%, the MR effect decreases down to around 2.8 and 2.8, respectively. For anisotropic samples, the MR effects of 0%, 20% and 23.81% Gr-MREs are 4.8, 3.2 and 2.2, respectively. This proves again that with the increase of graphite weight fraction, the MR effect decreases.

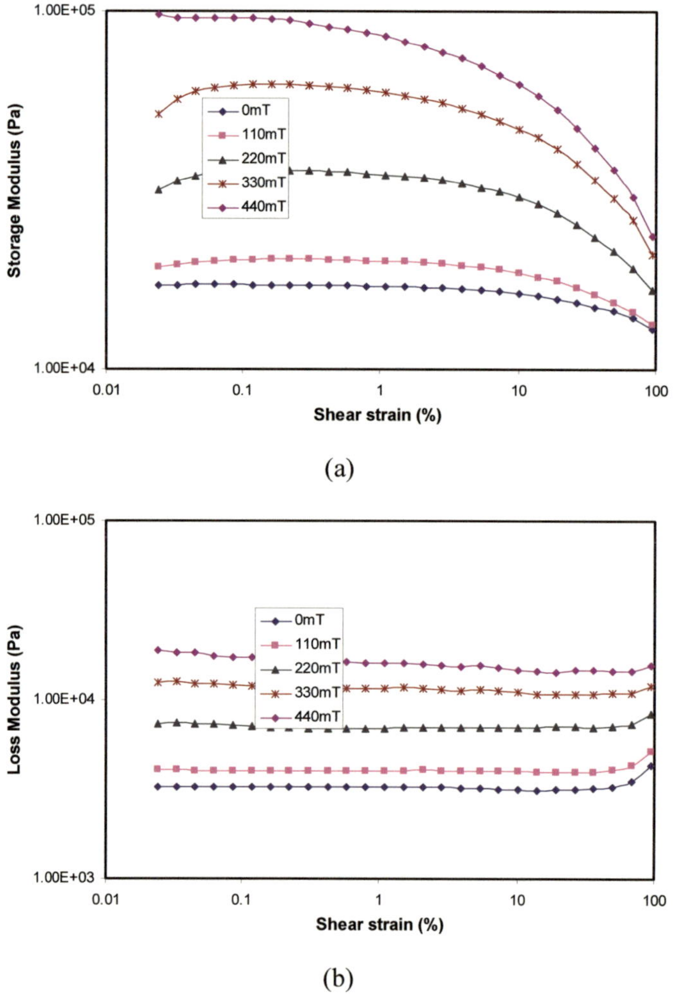

Figure 10. Storage and Loss Modulus versus strain amplitude sweep (isotropic MRE Gr 0%) (a) Storage modulus vs. shear strain (b) Loss modulus vs. shear strain

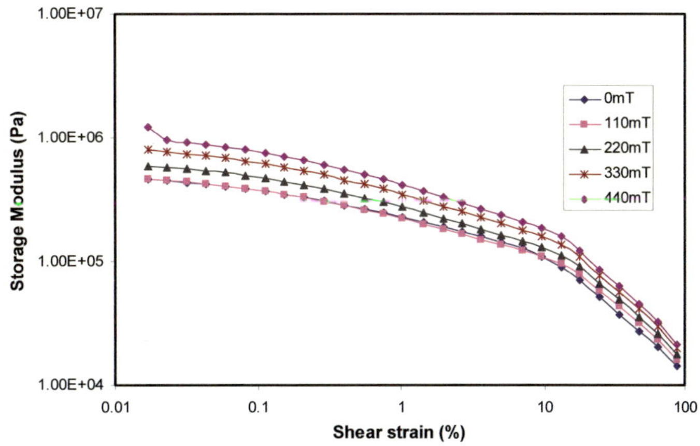

(a)

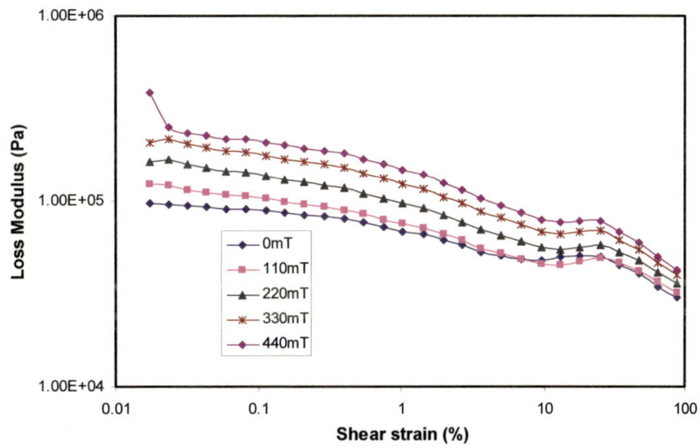

(b)

Figure 11. Storage and Loss Modulus versus strain amplitude sweep (isotropic MRE Gr 20%) (a) Storage modulus vs. shear strain (b) Loss modulus vs. shear strain

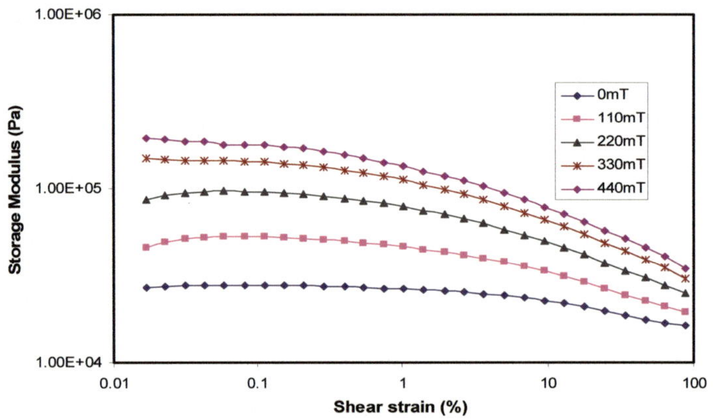

(a)

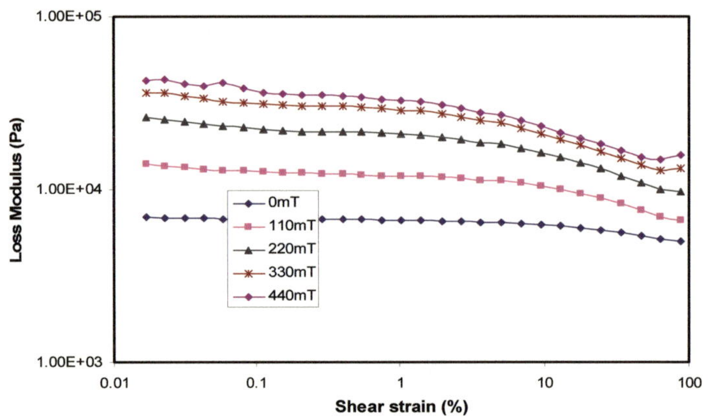

(b)

Figure 12. Storage and Loss Modulus versus strain amplitude sweep (anisotropic MRE Gr 0%) (a) Storage modulus vs. shear strain (b) Loss modulus vs. shear strain

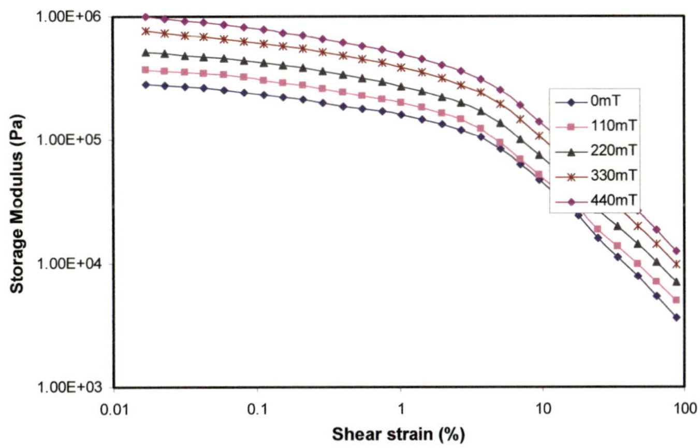

(a)

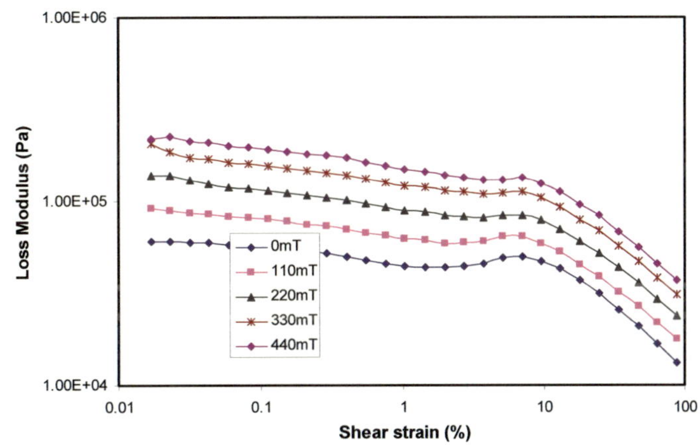

(b)

Figure 13. Storage and Loss Modulus versus strain amplitude sweep (anisotropic MRE Gr 20%) (a) Storage modulus vs. shear strain (b) Loss modulus vs. shear strain

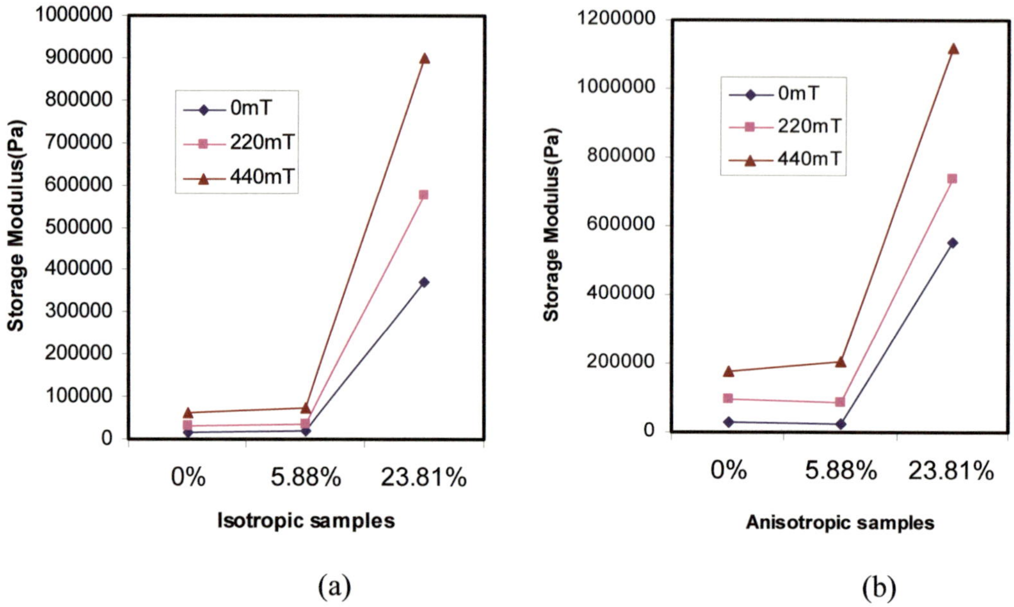

Figure 14. Storage Modulus of different samples at 10% shear strain (a) isotropic samples (b) anisotropic samples

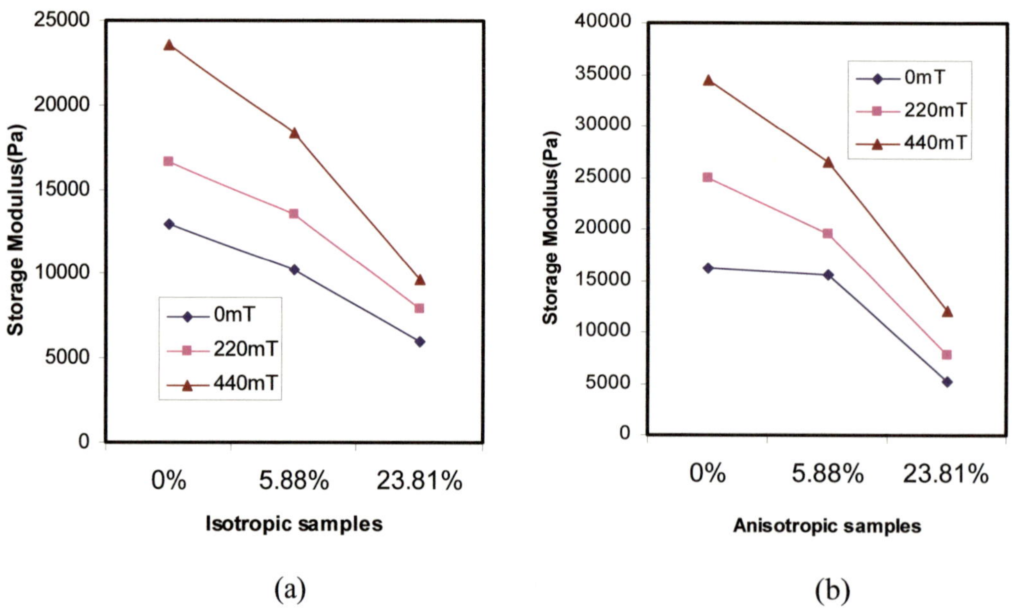

Figure 15. Storage Modulus of different samples at 87.5% shear strain (a) isotropic samples (b) anisotropic samples

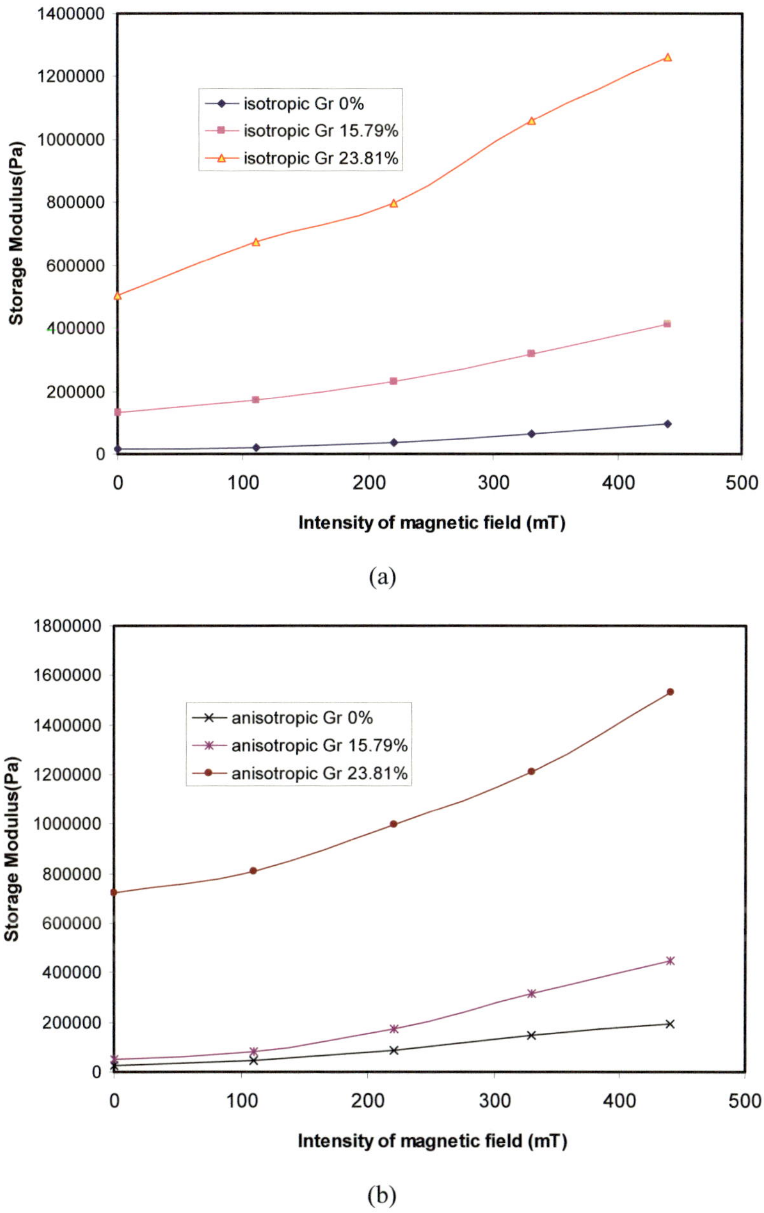

Figure 16. Storage Modulus versus magnetic field at 0.1% shear strain (a) isotropic samples (b) anisotropic samples

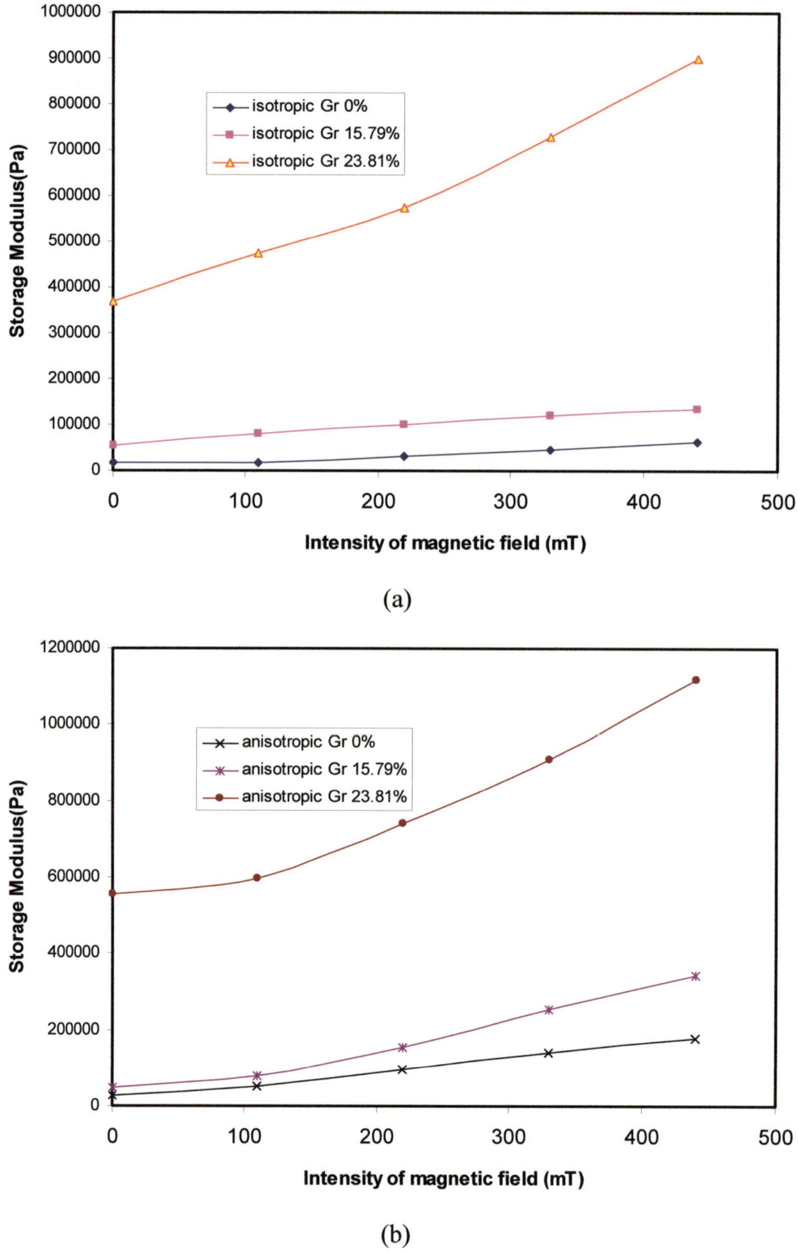

Figure 17. Storage Modulus versus magnetic field at 10% shear strain (a) isotropic samples (b) anisotropic samples

5. MRE sensing capabilities

5.1. Measurement of the relationship between the current and magnetic field

Figure 18 shows a schematic of the experimental device used. In this setup a long plastic plate is used to hold the weight which is applied load to the Gr-MRE samples. A Gaussmeter (HT201, Hengtong magnetoelectricity CO., LTD) is employed to test the intensity of the magnetic field. A multimeter (Finest 183, Fine Instruments Corporation) measures the resistance in the Gr-MRE samples.

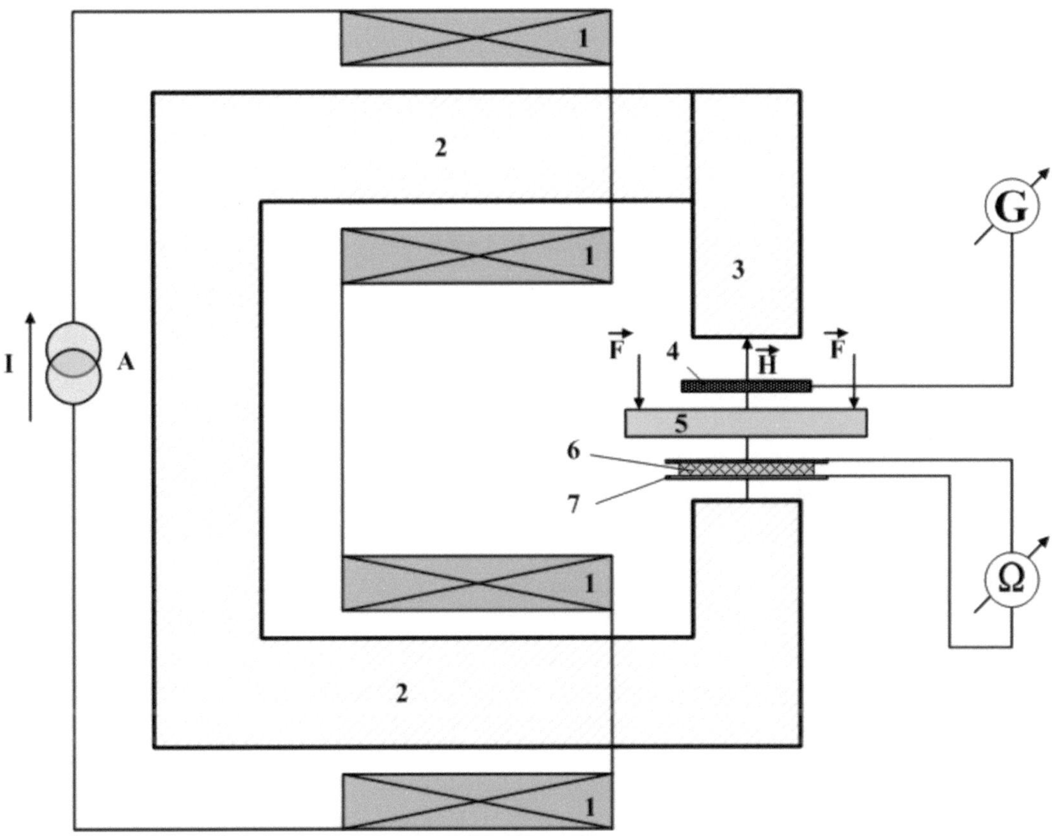

Figure 18. Sketch of the experimental device 1- coils; 2- electrical magnetic; 3- moving magnetic pole; 4- Gaussmeter probe; 5- plastic plate; 6- Gr MRE sample; 7- measuring copper plates; G- Gaussmeter; Ω- Multimeter; F- external force applied on Gr MRE samples

During the testing of the resistance of all the anisotropic MRE samples only the samples whose graphite weight fraction is higher than 15% are detectable by Finest 183 multimeter. Thus, among the tested anisotropic MRE samples only three of them are considered in Table 1, namely those whose graphite weight fractions are 20%, 21.95% and 23.81%. The results of these tests are shown in Figures 19, 20, and 21, respectively for the following data of anisotropic MRE with graphite weight fraction 20%, 21.95% and 23.81%.

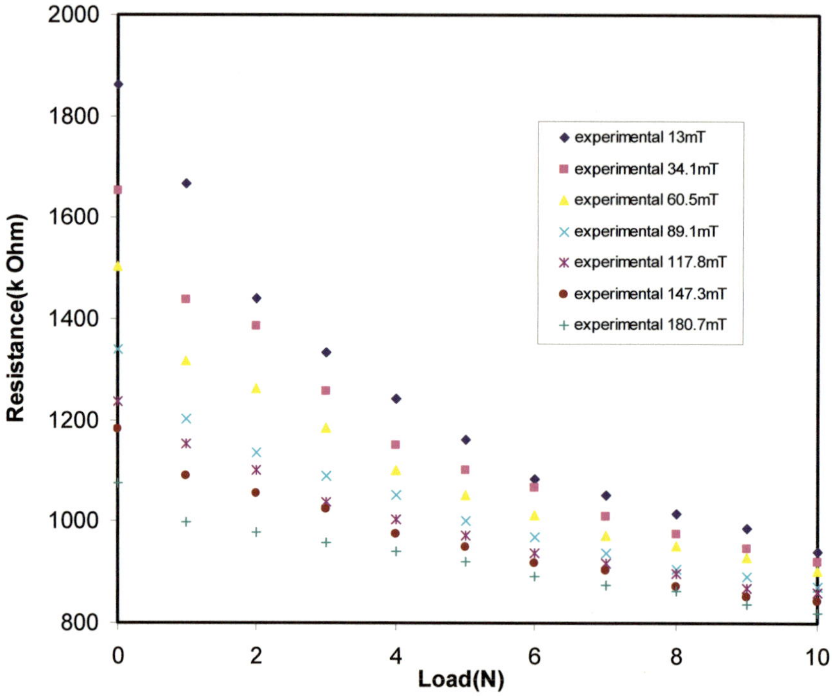

Figure 19. Resistance versus load (anisotropic MRE Gr 20%)

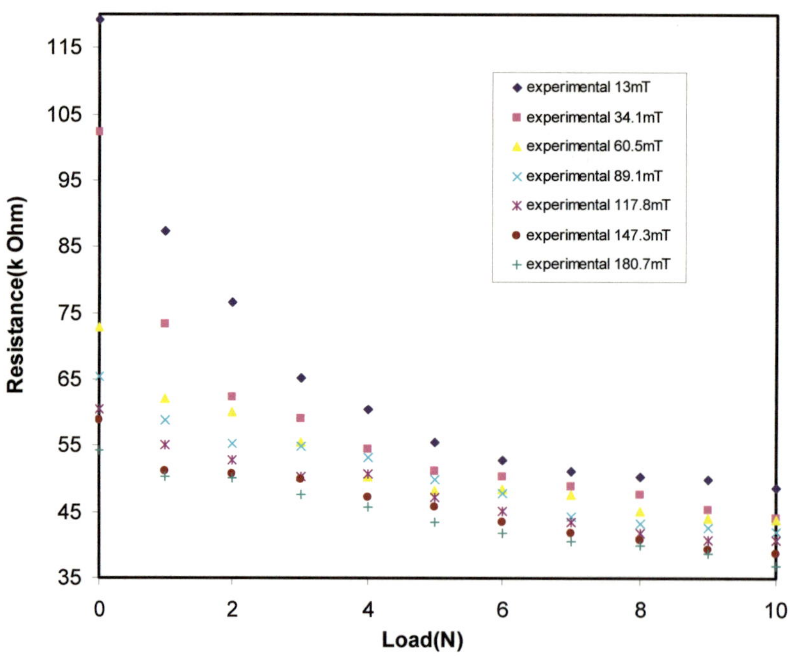

Figure 20. Resistance versus load (anisotropic MRE Gr 21.95%)

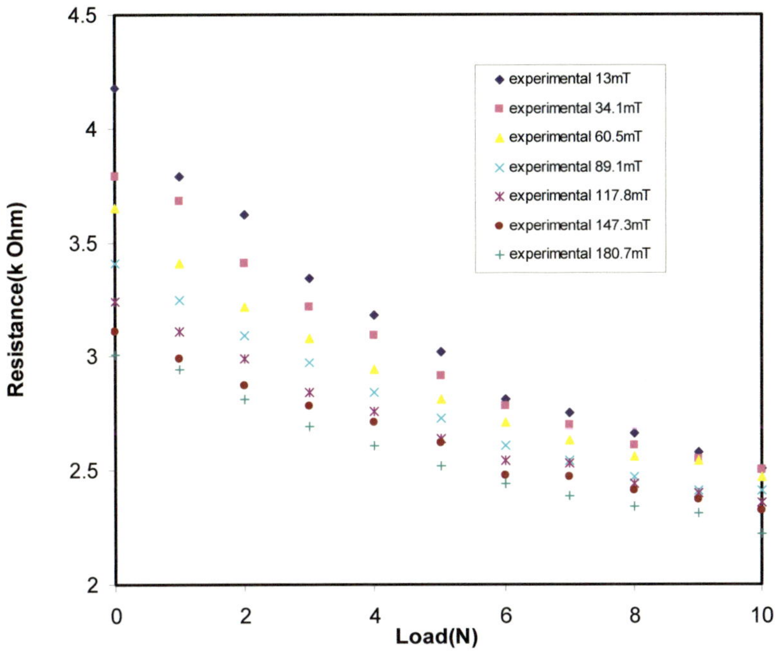

Figure 21. Resistance versus load (anisotropic MRE Gr 23.81%)

As shown in the plots similar trends are observed for the three samples. Specifically, in a fixed magnetic field, when the external load increases from 0 to 10 N, the resistance reduces in all three samples. With small loads, the resistance changes significantly but it decreases slowly when the load is more than 5 N. According to the absolute values, the sample with higher graphite weight fraction shows the higher electrical conductivity and the smaller decline in resistance. For instance, the resistance of the sample with graphite weight fraction 20% drops from 1862 kΩ at 0 N to 942 kΩ at 10 N, whereas a decline from 4.18 kΩ at 0 N to 2.51 kΩ at 10 N for the sample with graphite weight fraction 23.81%.

Besides, the resistance of each sample at a fixed external load decreases with increase in magnetic field intensity. Considering the sample with 21.95% graphite weight fraction as an example, at 5 N external force, the resistance 55.4 kΩ without the magnetic field decreases to 43.5 kΩ at a 440 mT magnetic field. This trend is shown in detail in Fig. 22.

5.2. A representative volume unit based mathematical model for investigating the magnetic field dependent sensing capabilities

5.2.1. Theoretical approach

From the Dipole Model a Representative Volume Unit (RVU) is derived. A RVU consists of two neighbouring hemispheres and the surrounding polymer matrix, which can be regarded as the minimum volume element in the conventional MRE. Fig. 23 shows the position of RVU in carbonyl iron particle chains and a longitudinal section of the unit is shown in Fig. 24.

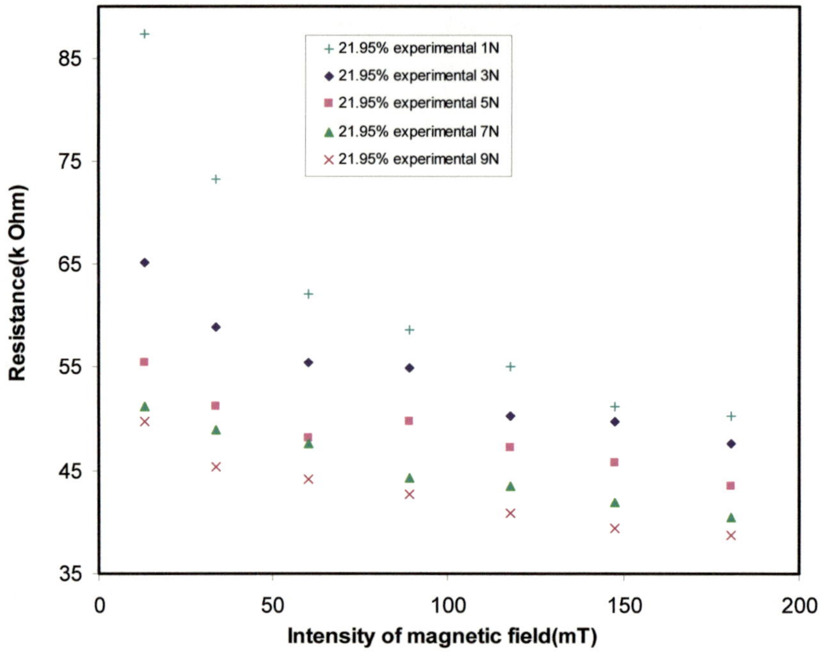

Figure 22. Resistance versus magnetic field (anisotropic MRE Gr 21.95%)

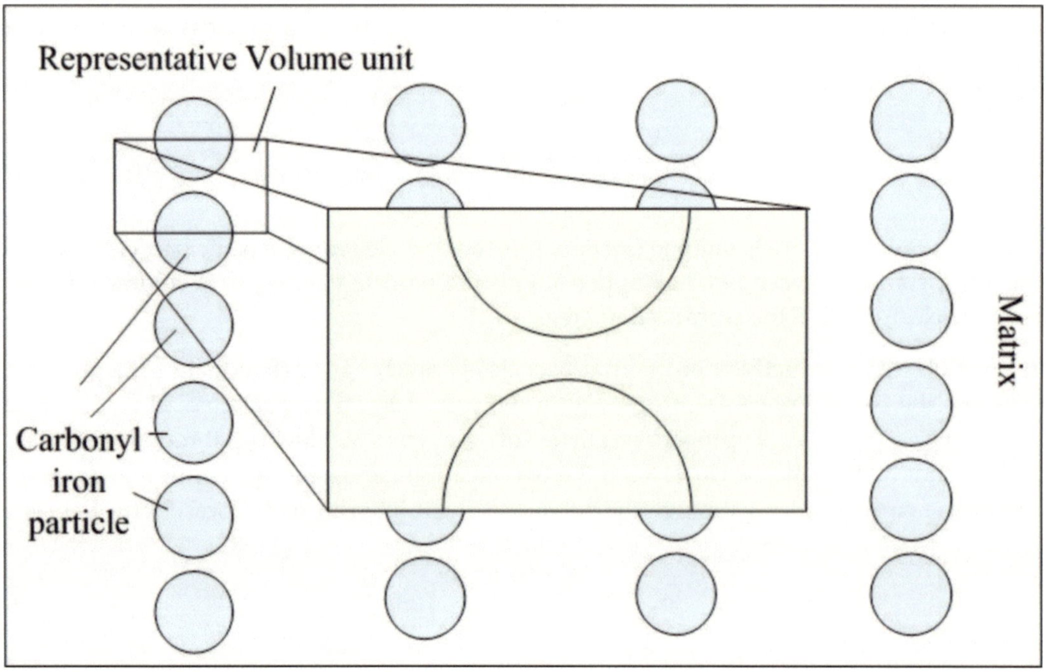

Figure 23. RVU in carbonyl iron chains

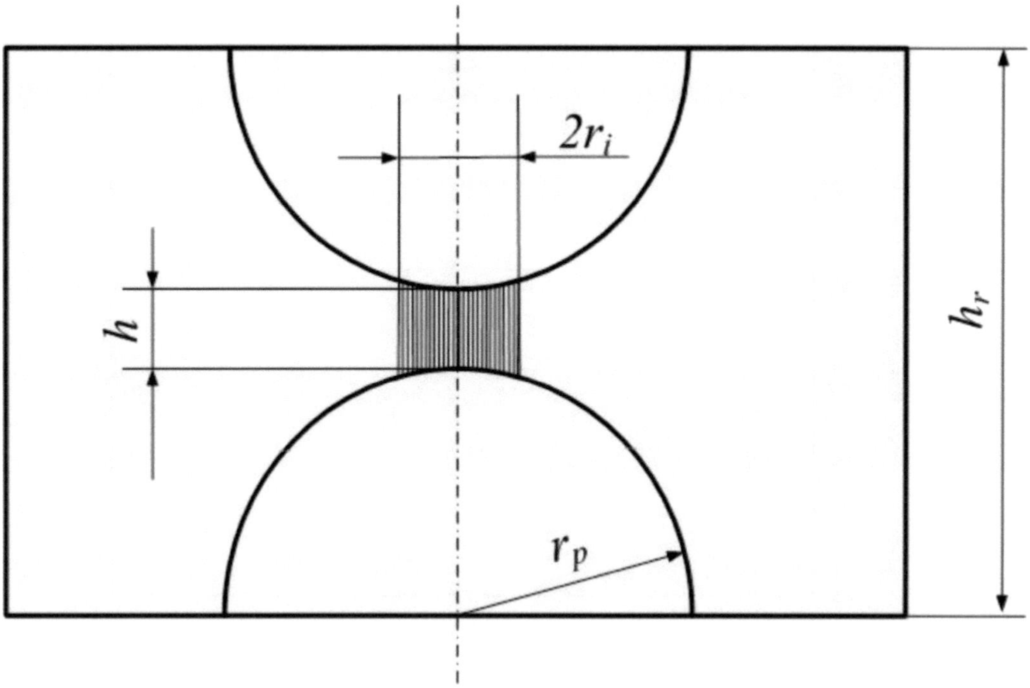

Figure 24. The longitudinal section of RVU

The RVU is a model for ideal anisotropic MREs which is supposed to have a chain structure.

Given the iron particle volume fraction ϕ in the RVU, the area of cross section can be expressed as

$$S_r = \frac{V_r}{h_r} = \frac{V_s}{\phi h_r} = \frac{4\pi {r_p}^3}{3\phi\left(2r_p + h\right)} \approx \frac{2\pi {r_p}^2}{3\phi} \tag{1}$$

where ϕ is the iron particle volume fraction, r_p is the particle radius, h is the particle distance, h_r is the distance between two half spheres, V_s is the particle volume, V_r is volume between two particles, and S_r is the cross section area.

In the RVU, the conductivity of the iron particles is much higher than that of the polymer, therefore the electric potential drops within the particles can be neglected. Due to system geometry and electrical properties, most of the current flowing through the RVU concentrates on the small area between the two adjacent hemispheres. Specifically, given the intensity of current flowing through the polymer is E, the intensity of the local electric field is

$$E_{loc} = \frac{h_r E}{h} = \frac{(2r_p + h)E}{h} \approx 2r_p \frac{E}{h} \tag{2}$$

Owing to the magnetic attraction during MRE preparation, the insulating polymer film between two neighbouring iron particles (here described by h) is very thin, and it is across

this film that the electrical field induced tunnel current can occur. The Fowler-Nordheim equation [19-21] can be used to express the tunnel current. In addition, the iron particles dispersed in the polymer matrix contribute to the conductivity of the polymer and then the total current density is the sum of the tunnel density j_t and the conduction density of the polymer j_c

According to the Fowler-Nordheim equation, the relation between the tunnel density j_t and electric field intensity E is

$$j_t = \alpha E_{loc}^2 \exp\left(-\frac{\beta}{E_{loc}}\right) \tag{3}$$

in which α and β are constants determining the tunnel current. So the total current density j is

$$j = j_t + j_c = \alpha E_{loc}^2 \exp\left(-\frac{\beta}{E_{loc}}\right) + \sigma_f E_{loc} \tag{4}$$

in which σ_f is the conductivity of the polymer film.

The total current density j is for the current flowing through the small area between the tips of two adjacent iron particles, however, the overall density of RVU j_r should be derived from its entire cross section. So from the total density of RVU j_r and the electric field intensity E, the global conductivity of typical MRE σ_r can be represented as

$$\sigma_r = \frac{j_r}{E} = \frac{\pi \cdot r_i^2 \cdot j}{S_r \cdot E} = 3\phi \cdot r_i^2 \left[\frac{2\alpha}{h^2} E \exp\left(-\frac{h\beta}{2r_p E}\right) + \frac{\sigma_f}{r_p h}\right] \tag{5}$$

in which r_i is the radius of the circular section across which the tunnelling current flow, as shown in Fig. 24.

MRE also exhibits piezoresistivity. When a MRE sample is compressed, its conductivity increases. This phenomenon is explained by two factors, one of which is the increments of the conductive area induced by the deformation of MREs and the other is the reduction of the thickness of the polymer membrane between the two adjacent iron particles. Because of the large ratio of r_i/h, it is difficult to compress the membrane further and thus the increment of the conductive area is the significant reason for the conductivity increasing.

From the Hertz Theory [22-24], when the initial pressure applied on MRE is σ_0 through a constant loading, corresponding to which there is an initial contact area radius r_{i0}

$$r_{i0} = \left[\frac{3\pi\sigma_0\left(1-v^2\right)}{2E_p}\right]^{1/3} \cdot r_p \tag{6}$$

Where ν is the Poisson's ratio, E_p is the Young's Modulus.

So the radius r_i increases along with the increment of pressure

$$r_i = r_{i0} + r_p\left(\left(\sigma_0 + \sigma\right)^{1/3} - \sigma_0^{\ 1/3}\right)\cdot\left(\frac{3\pi\left(1-\nu^2\right)}{2E_p}\right)^{1/3} \tag{7}$$

The magnetic field also contributes to the electrical resistance of MREs. When the external magnetic field is applied to MREs, the carbonyl iron particles are attracted by the poles of magnetic field, with closer magnetic pole providing more powerful magnetic force than the other pole. So the attraction from the farther magnetic pole can be neglected.

For the two iron particles in each RVU, the magnetic attraction from the pole applied to the farther particle compresses the thin film between the two adjacent iron particles. Similar to the piezoresistivity, the increment of the conductive area is the main cause for the conductivity increasing.

Thus, the radius r_i can be updated as

$$r_i = r_{i0} + r_p\left(\left(\sigma_0 + \sigma_1 + \sigma_2\right)^{1/3} - \sigma_0^{\ 1/3}\right)\cdot\left(\frac{3\pi\left(1-\nu^2\right)}{2E_p}\right)^{1/3} \tag{8}$$

in which, σ_1 is the stress due to pressure, σ_2 is the stress from the magnetic attraction.

So the dependence of the conductivity of MREs on electric field intensity and the stress due to pressure is

$$\sigma_m = 3\phi\cdot\left[\frac{2\alpha}{h^2}E\exp\left(-\frac{h\beta}{2r_pE}\right)+\frac{\sigma_f}{r_ph}\right]\cdot\left[r_{i0} + r_p\left(\left(\sigma_0 + \sigma_1 + \sigma_2\right)^{1/3} - \sigma_0^{\ 1/3}\right)\cdot\left(\frac{3\pi\left(1-\nu^2\right)}{2E_p}\right)\right]^2 \tag{9}$$

When the initial condition σ_0 and r_{i0} are set, Apart from E and σ, the other parameters in this equation are all constants. So the conductivity of MREs σ_m is dependent on the intensity of the electric field E and the stress due to pressure σ.

In response to the effects of graphite and carbonyl iron on the resistance model, two parameters λ_g and λ_i are introduced to show the effects of graphite volume fraction ϕ_g and carbonyl iron volume fraction φ_i to the conductivity of new MREs. Two parameters λ_g and λ_i show the contribution of the carbonyl iron particles and graphite powder to the resistance respectively. The higher are values for λ_g and λ_i, the less resistance the Gr-MREs have.

In the RVU, the iron particle volume fraction is set as ϕ = 0.4, which means that the volume of two hemispheres is 40% of the whole volume of RVU.

The shape of Gr MRE samples is fixed. The thickness l is l= 1mm=0.001m, the diameter D is D=0.021m, so the cross section area is $A=\pi\cdot (D/2)^2$= 0.000346 m².

In the tunneling equation $\left[\frac{2\alpha}{h^2}E\exp\left(-\frac{h\beta}{2r_pE}\right)+\frac{\sigma_f}{r_ph}\right]$, α and β are pre-exponential and exponential terms of the standard Fowler-Nordheim Tunneling which are both constants. In this case, value α is set as 2, value β is set as 1. h is the height of the RVU as two times of iron particle's radius. So h=0.000004 m. The iron particle's radius r_p is 0.000002 m. σ_f is the conductivity of the polymer film namely silicone rubber. Because of the high resistance of silicone rubber, the value of conductivity of silicone rubber σ_f is set as $1*10^{-10}$. The electric field E is from the function file of the multimeter used in the test. The value of E is 9V. So the whole equation can be calculated as $\frac{2\alpha}{h^2}E\exp\left(-\frac{h\beta}{2r_pE}\right)+\frac{\sigma_f}{r_ph}\approx 2.01*10^{12}$. At the normal condition, the MRE Young's modulus E_p is set as 4000 Pa. The Poisson's ratio of particles v is set as 0.1. Substitute these parameters into Equation (9), the final resistance of Gr-MREs is

$$R_g=\frac{4.78*10^{-13}\times\lambda_g\lambda_i}{10^{-8}+0.000002\left(\left(2000+\sigma_1+\sigma_2\right)^{1/3}-2000^{1/3}\right)\times 0.001166} \tag{10}$$

where σ_1 is the compressive stress from the external force and σ_2 is the compressive stress from the external magnetic field [25]. In this study, the weight of plastic plate is 10 g and the weight increment is 110 g. Thus, the force increment is 1 N. The stress, σ_1, is then calculated by dividing the force with the area A (0.000346 m²), and shown in Table 2.

External load(N)	Real load(N)	Compressive stress σ_1 (Pa)
0	0.1	288.7165
1	1.1	3175.881
2	2.1	6063.045
3	3.1	8950.21
4	4.1	11837.37
5	5.1	14724.54
6	6.1	17611.7
7	7.1	20498.87
8	8.1	23386.03
9	9.1	26273.2
10	10.1	29160.36

Table 2. Relation of weight, load and compressive stress

The anisotropic MRE with graphite weight fraction 21.95% (contains 10g iron particles, 3g silicone rubber, 3g silicone oil and 4.5g graphite powder) can be used as an example. Table 3 shows the volume fractions of all the ingredients for this sample.

	Mass(g)	Density(g/cm³)	volume(cm³)	Volume fraction
Iron particle	10	7.86	1.272265	16.50%
Silicone rubber	3	3.18	0.943396	12.24%
Silicone oil	3	0.96	3.125	40.54%
Graphite powder	4.5	1.9	2.368421	30.72%

Table 3. Volume fractions of all the ingredients (anisotropic MRE Gr 21.95%)

The factors showing the efforts of graphite and iron particles to the resistance in Gr-MREs are λ_g and λ_i respectively and λ_g and λ_i depend on the graphite volume fraction ϕ_g and iron particles volume fraction ϕ_i. For each sample there are different values for λ_g and λ_i. Fig. 25 shows the product of λ_g and λ_i versus Gr-MREs (Gr 20%, 21.95% and 23.81%). The data is from the ratio of experimental result and theoretical prediction.

If we set $\lambda_g = \exp(a \cdot \phi_g)$ and $\lambda_i = \exp(b \cdot \phi_i)$ and substitute the data in Table 4 and Fig. 8 to λ_g and λ_i, the parameters λ_g and λ_i can be obtained and so the parameter a and b were derived as -65 and 250, respectively. So the relationship of λ_g and λ_i to ϕ_g and ϕ_i are $\lambda_g = e^{-65*\varphi_g}$, $\lambda_i = e^{250\varphi_i}$

	Gr 20%	Gr 21.95%	Gr 23.81%
Iron particle (ϕi)	17.09%	16.5%	15.96%
Graphite powder(ϕg)	28.27%	30.7%	33.01%

Table 4. Volume fractions of iron and graphite (anisotropic MRE Gr 20%, 21.95% & 23.81%)

After having determined all parameters, Equation 10 is ready to calculate the final resistance R_g of the anisotropic MRE with graphite weight fraction 20%.

5.2.2. *Comparison and analysis*

To show the comparison between the experimental result and the theoretical prediction, anisotropic MRE with graphite weight fraction 21.95% is considered. Comparison between experimental and theoretical result is provided in Fig. 26.

Figure 26 shows that the experimental and theoretical result do not match each other perfectly. However, the trends of both experimental result and theoretical result are the same.

Similarly some other comparison can be made from the data. At the fixed external force (which is expressed as load) when the magnetic field intensity increases the resistance of the

sample decreases. The data under three external loads such as 1 N, 5 N and 10 N were chosen to compare the experimental and theoretical result in Fig. 27.

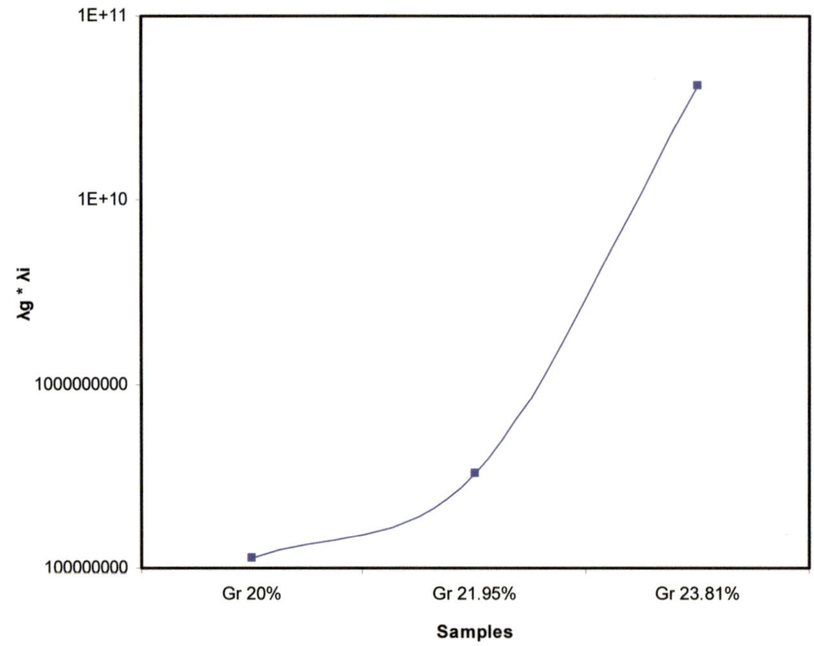

Figure 25. $\lambda_g * \lambda_i$ versus graphite weight fraction

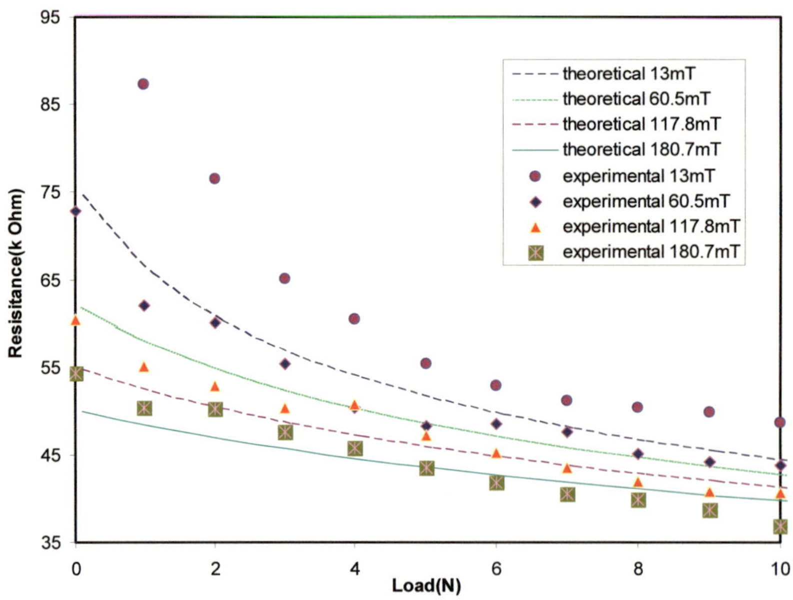

Figure 26. Comparison between experimental result and theoretical result (anisotropic MRE Gr 21.95%)

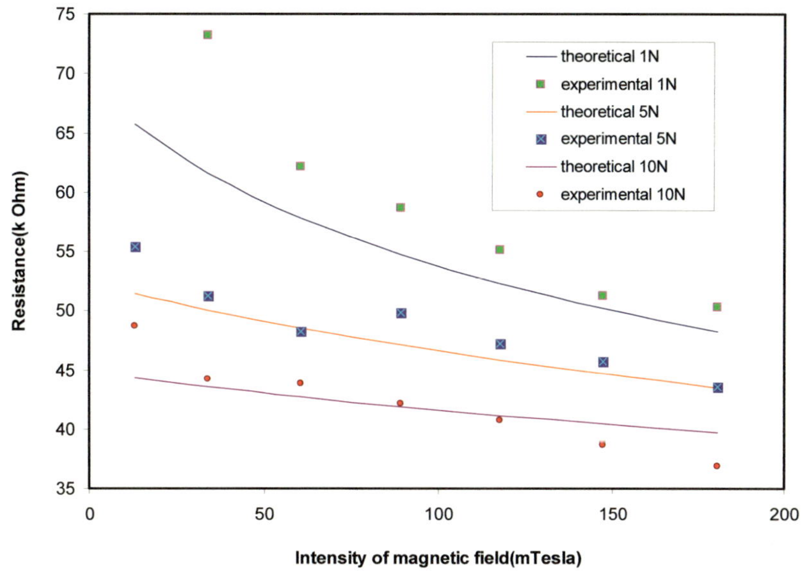

Figure 27. Resistance changing at a fixed load (anisotropic MRE Gr 21.95%)

The Fig. 27 shows the change in the resistance for given external forces. Along with the raising of magnetic field intensity, the sample's resistance decreases. The higher external load applied leads to lower resistance of Gr-MREs.

The next two figures Fig. 28 and 29 show the resistance variation between anisotropic MREs with graphite weight fraction 20%, 21.95%, and 23.81%. The magnetic field is fixed in Fig. 28 and the external load is fixed in Fig. 29.

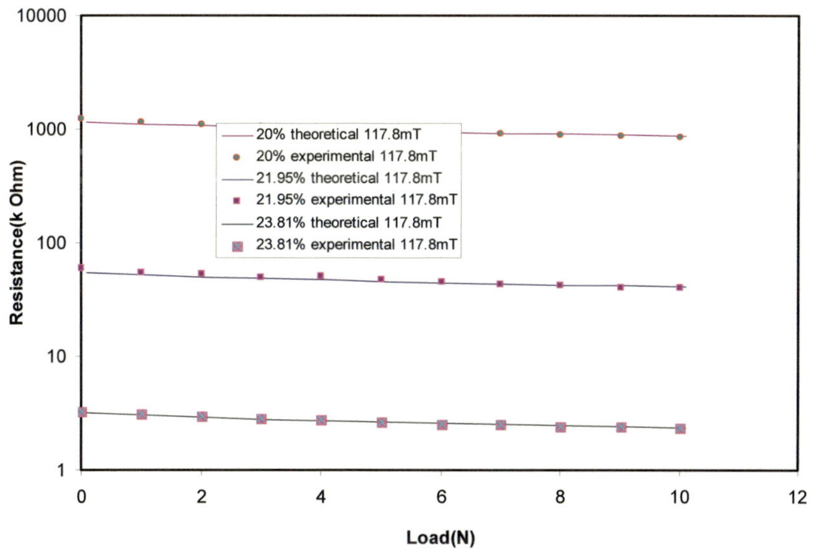

Figure 28. Resistance between different sample at 117.8mT magnetic field (anisotropic MRE with graphite weight fraction 20%, 21.95% & 23.81%

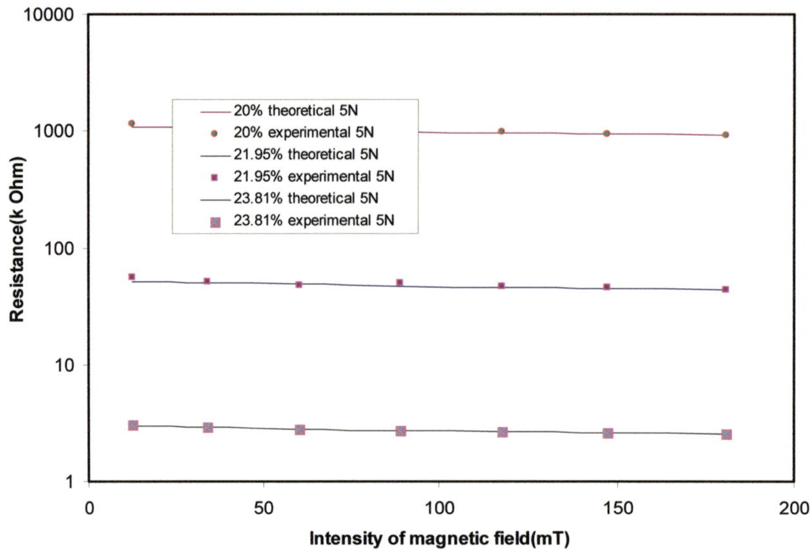

Figure 29. Resistance between different samples at 5.1N external load (anisotropic MRE with graphite weight fraction 20%, 21.95% & 23.81%)

Because of the order of values for the three samples it is easy to see from the Fig. 28 and 29 that theoretical and experimental results can match each other very well. This again provides a proof of the theory and when either external load or magnetic field intensity increases the resistance decrease.

Therefore when the magnetic field intensity and resistance are measured, Equation 10 can be used to calculate the value of the external load which means the Gr-MREs have the potential to work as the key component in a force sensor.

6. Conclusion

Both isotropic and anisotropic samples of graphite-based magnetorheological elastomers (Gr-MREs) with various graphite weight fractions ranging from 0% to 23.8% were fabricated in this study. The microstructural observation of these samples shows that the presence of graphite powder affects the forming of carbonyl iron chains. The sample with less graphite shows better-aligned carbonyl iron chains which influence the magnetorheology of MREs. In addition by connecting two iron chains in parallel and connecting the disconnected iron chains, the graphite contributes to the conductivity of MREs.

Steady state and dynamic tests such as strain amplitude sweep and angular frequency sweep were used to test the magnetorheology of Gr-MREs. With the help of graphite in MREs, the Storage and Loss Moduli are both changed. The steady state tests showed that the graphite can diminish the viscoelastic linear range of MREs. The dynamic tests proved that the samples with higher graphite weight fraction show higher initial Storage and Loss Moduli and lower MR effects. Additionally, the resistance of each MRE sample exhibits a decreasing trend with the graphite weight fraction.

Based on a Dipole model, a representative volume unit was presented to show the resistance of ideal anisotropic MREs. The current flowing through the ideal chain structure were derived by taking into account both the tunnel current and conductivity current. In the mathematical model, two parameters, λ_i and λ_g, were introduced to reflect the effect of the iron particles and graphite, which are both exponential functions of particle volume fractions. The exponential parameters were identified and then used to reconstruct modelling predictions. The comparison between experimental results with modelling predictions indicate that the proposed mathematical model can well investigate the sensing capabilities of the graphite based MRE elastomers.

Author details

Weihua Li and Tongfei Tian
School of Mechanical, Materials & Mechatronic Engineering, University of Wollongong, Wollongong, NSW, Australia

Haiping Du
School of Electrical, Computer and Telecommunications Engineering, Wollongong, NSW, Australia

Acknowledgement

This work is partially supported by the University of Wollongong through a UIC funding scheme.

7. References

[1] G. Y. Zhou and Z. J. Jiang, "Deformation in magnetorheological elastomer and elastomer-ferromagnet composite driven by a magnetic field", *Smart Mater. Struct.*, 13: 309-316 (2004)

[2] X. L. Gong, X. Z. Zhang and P. Q. Zhang, "Fabrication and Characterization of Isotropic Magnetorheological Elastomers", *Polymer Testing*, 24(5): 669–676 (2005)

[3] H. X. Deng, X. L. Gong and L. H. Wang, "Development of an adaptive tuned vibration absorber with magnetorheological elastomer", *Smart Materials & Structures*, 15 (5): N111-N116 (2006)

[4] Zou, H., Zhang, L.Q., Tian, M., Wu, S.Z. and Zhao, S.H., 2009. Study on the Structure and Properties of Conductive Silicone Rubber Filled with Nickel-Coated Graphite. Journal of Applied Polymer Science, 115(5): p. 2710-2717.

[5] Li, W.H., Kostidis, K., Zhang, X.Z., Zhou, Y. and Ieee. 2009. 'Development of a Force Sensor Working with MR Elastomers'. in 2009 Ieee/Asme International Conference on Advanced Intelligent Mechatronics, Vols 1-3. New York: Ieee.

[6] Ginder, J.M., Clark, S.M., Schlotter, W.F. and Nichols, M.E., 2002. Magnetostrictive phenomena in magnetorheological elastomers. International Journal of Modern Physics B, 16(17-18): p. 2412-2418.

[7] Lokander, M. and Stenberg, B., 2003. Performance of isotropic magnetorheological rubber materials. Polymer Testing, 22(3): p. 245-251.

[8] Lokander, M. and Stenberg, B., 2003. Improving the magnetorheological effect in isotropic magnetorheological rubber materials. Polymer Testing, 22(6): p. 677-680.

[9] Zhou, G.Y., 2003. Shear properties of a magnetorheological elastomer. Smart Materials and Structures, 12(1): p. 139.

[10] Ginder, J.M., Schlotter, W.F. and Nichols, M.E. 2001. 'Magnetorheological elastomers in tunable vibration absorbers'. in Smart Structures and Materials 2001: Damping and Isolation. Newport Beach, CA, USA: SPIE.

[11] Gong, X.L., Zhang, X.Z. and Zhang, P.Q., 2005. Fabrication and characterization of isotropic magnetorheological elastomers. Polymer Testing, 24(5): p. 669-676.

[12] Kchit, N. and Bossis, G., 2009. Electrical resistivity mechanism in magnetorheological elastomer. Journal of Physics D-Applied Physics, 42(10): p. 5505-5505.

[13] Wang, X.J., Gordaninejad, F., Calgar, M., Liu, Y.M., Sutrisno, J. and Fuchs, A., 2009. Sensing Behavior of Magnetorheological Elastomers. Journal of Mechanical Design, 131(9): p. 6.

[14] Bica, I., 2009. Influence of the transverse magnetic field intensity upon the electric resistance of the magnetorheological elastomer containing graphite microparticles. Materials Letters, 63(26): p. 2230-2232.

[15] Li, W.H., Kostidis, K., Zhang, X.Z., Zhou, Y. and Ieee, Development of a Force Sensor Working with MR Elastomers, in 2009 Ieee/Asme International Conference on Advanced Intelligent Mechatronics, Vols 1-3. 2009, Ieee: New York. p. 233-238.

[16] Zhao, Y., Maietta, D.M. and Chang, L., 2000. An Asperity Microcontact Model Incorporating the Transition From Elastic Deformation to Fully Plastic Flow. Journal of Tribology, 122(1): p. 86-93.

[17] McLachlan, D.S., 2000. Analytical functions for the dc and ac conductivity of conductor-insulator composites. Journal of Electroceramics, 5(2): p. 93-110.

[18] Woo, L.Y., Wansom, S., Hixson, A.D., Campo, M.A. and Mason, T.O., 2003. A universal equivalent circuit model for the impedance response of composites. Journal of Materials Science, 38(10): p. 2265-2270.

[19] Weinberg, Z.A., 1982. On tunneling in metal-oxide-silicon structures. Journal of Applied Physics, 53(7): p. 5052-5056.

[20] Serdouk, S., Hayn, R. and Autran, J.L., 2007. Theory of spin-dependent tunneling current in ferromagnetic metal-oxide-silicon structures. Journal of Applied Physics, 102(11): p. 113707-1-113707-5.

[21] Dahlke, W.E. and Sze, S.M., 1967. Tunneling in metal-oxide-silicon structures. Solid State Electronics, 10(8): p. 865-873.

[22] Zhupanska, O.I. and Ulitko, A.F., 2005. Contact with friction of a rigid cylinder with an elastic half-space. Journal of the Mechanics and Physics of Solids, 53(5): p. 975-999.

[23] Etsion, I., Levinson, O., Halperin, G. and Varenberg, M., 2005. Experimental investigation of the elastic-plastic contact area and static friction of a sphere on flat. Journal of Tribology, 127(1): p. 47-50.

[24] Maouche, N., Maitournam, M.H. and Dang Van, K., 1997. On a new method of evaluation of the inelastic state due to moving contacts. Wear, 203-204: p. 139-147.
[25] Chang, W.R., Etsion, I. and Bogy, D.B., 1988. Static Friction Coefficient Model for Metallic Rough Surfaces. Journal of Tribology, 110(1): p. 57-63.

Chapter 12

Simultaneous Smart Actuating-Sensing Devices Based on Conducting Polymers

José G. Martínez, Joaquín Arias-Pardilla and Toribio F. Otero

Additional information is available at the end of the chapter

http://dx.doi.org/10.5772/51733

1. Introduction

Towards the end of 1970s new artificial organic materials, conducting polymers (CP), were discovered (Chiang et al., 1977; Inzelt, 2011; Shirakawa et al., 1977). Since then, most of the scientists working on CP became interested by the fact that their conductivity can shift, in a reversible way, over several orders of magnitude by oxidation/reduction (also called doping/dedoping) processes. The availability of these new organic semiconductors has opened up possibilities to rebuilt electronics and microelectronics producing flexible devices (Guo et al., 2010; Klauk, 2006; Perepichka & Perepichka, 2009; So, 2010).

Besides conductivity other properties such as stored charge, stored chemicals, volume, porosity or colour also change during doping/dedoping, under electrochemical control, in parallel with the material composition (counterion content) along the redox reactions (Otero, 2008). Any intermediate oxidation state determines a chemical equilibrium characterized by an equilibrium potential. Any physical (temperature, pressure, applied current) or chemical (electrolyte concentration) magnitude affecting the chemical equilibrium modifies the electrical potential of the material that therefore can be used as a sensor of that magnitude (Otero, 2009).

These properties, whose value changes with the material composition, are allowing the development of different devices (Onoda et al., 1999a; Otero et al., 1992a; Otero et al., 1992c; Pei & Inganäs, 1992a). Volume variations driven by oxidation/reduction reactions are being used to generate new electrical motors having different configurations of the polymeric actuators (Alici et al., 2007; Smela et al., 1993; Wu et al., 2005).

We will present devices constructed with conducting polymers based on an electrochemical reaction and working, simultaneously, as an actuator and as several sensors of the surrounding conditions. During the movement they store and release charge working as a battery: they are multifunctional devices.

2. Electrochemical reactions in conducting polymers

Once generated (chemical or electrochemically), conducting polymers can be considered as both, classical unreactive materials having either a constant composition and a constant value of the magnitude of its physical and chemical properties; or new reactive materials varying their composition (polymeric chains, ions and solvent in different ratios) and properties along several orders of magnitude in a reversible way under electrochemical reaction (Otero, 1999), or promoted by redox agents present in solution (Kuttel et al., 2009).

2.1. p-doping

When the polymer chains are oxidized, consecutive electrons are removed from each chain generating an excess of positive charges (holes) along the chains. This excess of positive charges (lack of electrons) promotes the repulsion between the polymeric chains and the generation of free volume between them. This free space is occupied by anions arriving from the solution to compensate the emerging positive charges (keeping the electroneutrality) and solvent molecules to keep osmotic pressure balance (Huang et al., 1986; Otero, 1999; Tsai et al., 1988).

When the polymer is generated in the presence of small anions, they can be exchanged by other small anions present in solution by electrochemical reactions so a prevailing exchange of anion occurs during reaction:

$$\left(Pol^0\right)_s + n\left(A^-\right)_{sol} + m\left(Solv\right) \rightleftarrows \left[\left(Pol^{n+}\right)_s \left(A^-\right)_n \left(Solv\right)_m\right]_{gel} + n\left(e^-\right)_{metal} \tag{1}$$

where the different subscripts mean: *s*, solid; *sol*, in solution; *gel* indicates that the material is a gel formed by oxidized polymeric chains (Pol^{n+}) generated after the extraction of *n* electrons (e^-) through the metal (indicated by subscript *metal*) from neutral polymer chains (Pol^0), *n* anions (A^-) coming from the solution to keep the gel electroneutrality and m molecules of solvent (*Solv*) required to keep osmotic pressure balance.

When the polymer is generated in the presence of a macroanion, due to its volume and the interaction with polymer chains, this macroanion cannot be exchanged by the electrochemical reaction keeping trapped inside the polymer film. So, in order to keep the electroneutrality, smaller cations are exchanged with the solution during the reaction:

$$\left[\left(Pol^0\right)\left(MA^-\right)_n \left(C^+\right)_n \left(Solv\right)_m\right]_{gel} \rightleftarrows \left[\left(Pol^{n+}\right)\left(MA^-\right)_n\right]_{gel} + n\left(C^+\right)_{sol} + m\left(Solv\right) + n\left(e^-\right)_{metal} \tag{2}$$

where MA^- is the macroanion trapped inside the polymer film and C^+ are cations exchanged in order to keep the electroneutrality.

Usually, the real redox process is not as easy as expressed by reactions (1) and (2): anions and cations are exchanged simultaneously (Hillman et al., 1989; Inzelt, 2008). Usually one of the previous interchanges prevails supporting the greater percentage of charge balance (Kim et al., 2010; Lyutov et al., 2011; Orata & Buttry, 1987; Torresi & Maranhao, 1999).

2.2. n-doping

Some CP such as PEDOT (Ahonen et al., 2000), polythiophene (Arbizzani et al., 1995) or polyfluorenes (Ranger & Leclerc, 1998) have an electronic affinity high enough to allow transitions from the neutral state to a reduced state, storing negative charges (by electron injection) on the chains at high cathodic potentials. In this case, very stable solvents and salts are required, as electrolytes, to perform this redox reaction:

$$\left(Pol^{0}\right)_{s} + n\left(C^{+}\right)_{sol} + m\left(Solv\right) + n\left(e^{-}\right)_{metal} \rightleftarrows \left[\left(Pol^{n-}\right)_{s}\left(C^{+}\right)_{n}\left(Solv\right)_{m}\right]_{gel} \quad (3)$$

where Pol^{n-} represents the reduced polymer chains after insertion of n electrons. Here, in an analogous way compared to reaction (1), an excess of negatives charges promotes repulsion between polymeric chains generating the free volume that will be occupied by cations (exchanged with solution to keep electroneutrality) and solvent (exchanged with solution to keep pressure and osmotic balance).

2.3. Double doping

Some conducting polymers can be doped both, by p-doping and by n-doping. Thus, from their neutral state they can be reduced (suffering n-doping) or oxidized (suffering p-doping). In those polymers the energy difference between both processes is the electrochemical bandgap (Arias-Pardilla et al., 2010; Otero et al., 2011).

3. Reversible change of the electrochemical properties

Electrochemical reactions 1, 2 and 3 are reversible reactions moving through *n* consecutive oxidation states (Otero et al., 2012) having different content of the counterion. The value of the magnitude of those properties being a function of the material composition (electrochemical properties) can also be shifted in a reversible way under faradic control. As previously indicated, conducting polymers can be oxidized and/or reduced from their neutral state, with the entry/exit of ions and solvent. The most studied electrochemical properties of conducting polymers are: volume, colour, stored charge, stored chemicals, porosity or permselectivity, sensing responses or wetability, among others. The progressive and reversible variation of the value of these electrochemical properties allows the development of different devices and products. The change in volume will be reviewed in detail below: this is the most important property of conducting polymers for the development of actuators or artificial muscles.

3.1. Volume variations

A chain of conducting polymer in solution can be considered as an electrochemical molecular motor (Balzani et al., 2005; Davis, 1999; Otero, 2011): movements are produced by reversible conformational changes in chains originated by oxidation/reduction reactions.

The reversible conformational movement from a coil like structure to a rod like structure is produced by extraction (oxidation) or injection (reduction) of *n* electrons through *n* consecutive steps of one electron per step, together with movement of balancing counterions. This results in length variation of a free polymer chain in solution but, in polymer films three dimensional changes of volume are observed. The entanglement of the polymer chains in the film gives reversible swelling or shrinking changes of volume under reversible electrochemical stimulation (Fig. 1).

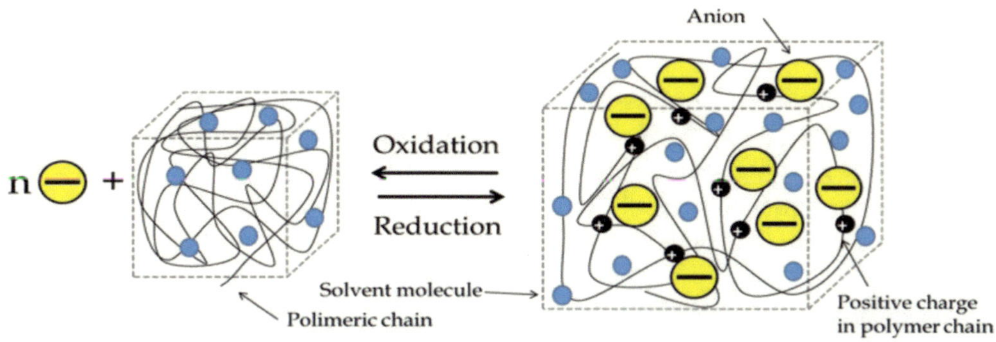

Figure 1. Schematic representation of the reversible volume change associated with the electrochemical reaction in conducting polymer chains during oxidation/reducction during p-doping exchanging anions.

Some mechanical test machines have been developed following length or thickness variations produced by submitting the film to different potential (Bay et al., 2003; Kiefer et al., 2007; Mazzoldi et al., 1998; Spinks et al., 2002) or current (Otero et al., 2006; Otero et al., 2007c) programs. In situ Atomic Force Microscopy (AFM) technique follows film thickness variation during reverse oxidation/reduction processes (Bieńkowski et al., 2011; Cho et al., 2011; Smela & Gadegaard, 2001). In this way, it has been possible to measure a volume difference between reduced and oxidized state up to 35% (Smela & Gadegaard, 1999). The volume change depends on multiple factors such as type of polymer, synthesis conditions (potential or current applied), electrolyte and solvents used.

4. Simultaneous sensing and actuating properties in conducting polymers

The driving reversible electrochemical reaction supports the development of simultaneous sensing and actuating properties by conducting polymer materials and by any electrochemical device based on those materials. The electrochemical working device (based on the electrochemical reaction) senses changes of any physical or chemical variable acting on the polymer reaction rate while working. This is the Le Chatelier principle applied outside equilibrium conditions (Smith, 2004). Therefore, for increasing values of the electrolyte concentrations or of the temperature working under flow of a constant current (constant reaction rate), lower values of the device potential are observed (the reaction is easier) during the transition between the same initial and final oxidation states of the materials. When a

greater mechanical work is required (moving faster, applying a higher current, or a higher mechanical stress is needed to move the actuator) the reaction gives increasing potential when the device moves between the same initial and final oxidation states.

Those sensing abilities are intrinsic properties of the reaction. They are characteristics of the material reaction and of any device based on this electrochemical reaction. So, the dual and simultaneous sensing-actuation property is expected to be quantified from electrochemical equations.

The evolution of the conducting polymer film potential with time E(t) during the movement from the same initial oxidation/reduction state to the same final oxidation/reduction state driven by flow of a constant anodic current is given by a stair function (Otero et al., 2012).

$$E(t)=\sum E_n(t)p_n(t)=E_1(t)p_1(t)+E_2(t)p_2(t)+...+E_n(t)p_n(t) \tag{4}$$

where:

$$p_n(t)=u(t-t_n)-u(t-t_{n+1})=\begin{cases}1, t\in[t_n,t_{n+1}]\\0, t\notin[t_n,t_{n+1}]\end{cases} \tag{5}$$

Being t_n the time while the nth electron is removed from every polymeric chain and $E_n(t)$:

$$E_n(t)=E_0+(n-1)\Delta E+\frac{RT}{(1-\alpha)F}\left\{\ln\left(\frac{i_a}{FV}\right)-d\ln\left[A^-\right]-e\ln\left(\left[Pol^*\right]_{initial}-\frac{i_a t}{FV}\right)-\ln k_{a0}\right\} \tag{6}$$

where E_0 is the standard potential, i_a is the applied current; n is the number of consecutive electrons extracted from a chain; ΔE is the increment observed in the potential when a new electron is extracted from a polymeric chain, R is the universal gas constant (R = 8.314 J K^{-1} mol^{-1}); α is the symmetry factor; F is the Faraday constant (F=96485 C mol^{-1}); V, the volume of the film; $[A^-]$ the concentration of anions (counterion) in solution; t, the time of current flow; T is the experimental temperature; d and e are the reactions orders related with the concentration of anions in solution or to that of the active centres $[Pol^*]$ in the film (sites of the polymer where a positive charge will be stored after oxidation) and k_{a0} is the rate constant or rate coefficient for $E=E_0$.

Therefore, Eqs. 4 and 6 are the sensing equations: the evolution of the device potential during actuation is a function of either, driving (current) and working (temperature, electrolyte concentration and film volume) variables.

Being electrical machines, by integration of Eq. 6 the evolution of the electrical energy consumed by the electrochemical device (U_a) during the actuation time is attained:

$$U_a(t)=i_a\int E(t)dt=i_at\{E_0+(n-1)\Delta E\}+\frac{RTi_at}{(1-\alpha)F}\left\{\ln\left(\frac{i_a}{FV}\right)-d\ln\left[A^-\right]-\ln k_{a0}\right\}+$$
$$\frac{RTVe}{(1-\alpha)}\left\{\ln\left(\left[Pol^*\right]_{initial}-\frac{i_a t}{FV}\right)-1\right\}\left\{\left[Pol^*\right]-\frac{i_a t}{FV}\right\} \tag{7}$$

The consumed energy (U_a) after any constant time (t) of current flow is also a sensing function of the same variables. Fig. 2 shows the good agreement between experimental and theoretical results for the consumption of three different charges (from the same initial oxidation/reduction state, three different final oxidation/reduction states are obtained) at different experimental temperatures by flow of a constant anodic current for three different times of current flow.

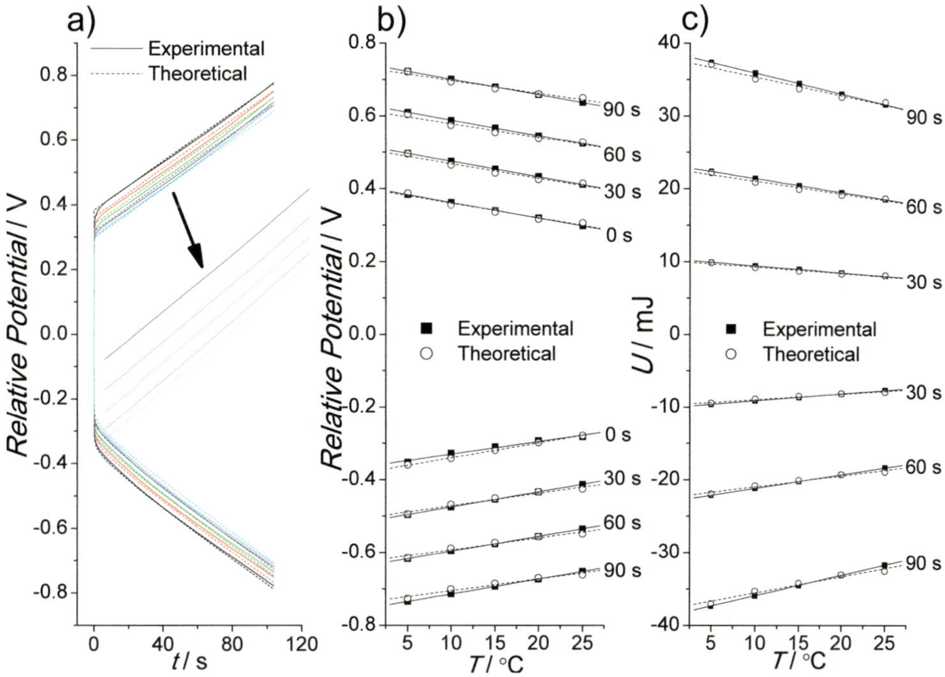

Figure 2. a) Anodic and cathodic experimental (full lines) and theoretical (dotted lines) chronopotentiograms obtained by flow of ±0.75 mA through a 1.6 mg pPy film (10.77 mm x 5.09 mm x 19 μm) at different temperatures (black line: 5ºC; red line: 10ºC; green line: 15ºC; blue line: 20ºC and cyan line: 25ºC) in 1 M $LiClO_4$ aqueous solution. b) Achieved potential after different times of anodic (positive) or cathodic (negative) current flow. c) Consumed electrical energy after the same times of current flow. Reproduced from (Otero et al., 2012), with permission of American Chemical Society).

5. Artificial muscles

In general, artificial muscles are devices attempting to reproduce composition, characteristics and capabilities of natural muscles. Different materials have been used as piezoelectrics (PZT), shape memory alloys (SMA) (Ouyang et al., 2008), carbon based materials as nanotubes (Baughman et al., 1999) or graphene (Huang et al., 2012) and polymeric gels. Also the pneumatic approach has been studied with contractile or extensional devices operated by pressurized air (Daerden & Lefeber, 2002).

Natural muscles transform chemical energy into mechanical energy and heat. Their actuation involves: a) aqueous media, b) an electric pulse arriving from the brain to the

muscle through the nervous system, c) liberation of calcium ions inside the sarcomere, d) chemical reactions (ATP hydrolysis), e) conformational changes along natural polymeric chains (actin and myosin) with change of the sarcomere volume and f) water exchange. Among of the above mentioned materials for artificial muscles, conducting polymers are the most similar to natural devices including: electric pulses, polymeric chains, chemical reactions, aqueous solutions (ions and water), volume variations, and strain and stress changes. Two families of electrical polymeric actuators can be differentiated (Otero et al., 2007a; Otero, 2000; Otero, 2007):

- **Electromechanical actuators**: the polymer responds to electric fields (E) and the polymer chains do not participate in chemical reactions (the structure of its intramolecular chemical bonds doesn't change) during actuation. In this type of actuator the dimension variations are proportional to E^2 for electrostrictive actuators and to E for electrostatic and electrokinetic (movement of ions and/or solvent molecules) actuators.
- **Electrochemomechanical actuators**: the polymer chains respond to electric currents and participate in electrochemical or chemical reactions, changing the structure of its chemical bonds, varying their composition and originating volume changes. In this case dimensional variations are under control of the electrochemical reaction becoming proportional to consumed charge. Conducting polymers can be used for the production of this kind of actuators. Whether carbon nanotubes and graphenes based actuators are electromechanical or electrochemomechanical is still under discussion (Gimenez et al., 2012; Mukai et al., 2011). Any electrochemically reactive material that can be laminated as stable films can be used as part of electrochemomechanical actuators.

From now on we will focus on electrochemomechanical actuators. These actuators have been built using different configurations as discussed below, always containing one or several films of conducting polymers, where reaction 1, 2 or 3 takes place. Volume variations generated by reactions 1, 2 or 3 are almost isotropic, while natural muscles are anisotropic devices. So to produce anisotropy, only volume changes following length variation of the films are used, with the consequent efficiency reduction.

5.1. Bending artificial muscles

Historically the first way to transduce reversible length variations in films of conducting polymers into macroscopic movements was through a bilayer, or bimorph structure, i.e. CP/passive layer (Otero et al., 1992b; Otero et al., 1992a; Otero et al., 1992c; Pei & Inganäs, 1992a) (Fig. 3). The variation of the mechanical stress gradient generated across the bilayer interface by swelling/shrinking processes induced by the electrochemical reactions in the conducting polymer film develops a macroscopic movement of the bilayer free end by the progressive bending of the device. The direction (clockwise or anticlockwise) of the movement depends on the prevalent ionic exchange (anions or cations) of the conducting polymer film. Conducting polymers with a prevalent exchange of anions swell by oxidation, pushing the bilayer free end meanwhile conducting polymers with prevalent cation

exchange shrink during oxidation, trailing the device. Different materials have been used as passive layer, for example a tape (Otero et al., 1992b; Otero et al., 1992a; Otero et al., 1992c; Pei & Inganäs, 1992a), a sputtered metal (Jager et al., 2000a; Jager et al., 2000b; Smela et al., 1993), a piece of paper (Deshpande et al., 2005b), a non conducting plastic (Higgins et al., 2003), a solid state electrolyte film (Alici et al., 2011; Baughman, 1996) or a thin film of any flexible material which is metal coated (i.e. by sputtering) (Deshpande et al., 2005a).

In a similar way it is possible to obtain bending movement from asymmetrical monolayers of the same conducting polymer, having an internal asymmetry capable of producing asymmetric swelling or shrinking across the film under the same electrochemical process (Okamoto et al., 2000; Onoda et al., 1999a; Onoda & Tada, 2004; Onoda et al., 1999b; Shakuda et al., 1993; Takashima et al., 2003; Takashima et al., 1997; Wang et al., 2002). Here half of the film has a prevalent anionic exchange, while the second half experiences a prevalent cationic exchange. These asymmetrical monolayers are obtained in two separate stages of electrogeneration using different salts and the same monomer. Other ways are being studied to produce asymmetric monolayers by physical means, for example, by growing the conducting polymer on adsorbed and porous materials (Li et al., 2004), or by electrochemical means generating a film of conducting polymer with a counterion concentration gradient (Okuzaki & Hattori, 2003; Sansiñena et al., 2003), conductivity (Nakano & Okamoto, 2001; Onoda et al., 2005) or morphology gradients (Han & Shi, 2006; Okamoto et al., 2001) by crosslinked networks.

All these bilayer devices require a counter-electrode in order to close the electrical circuit allowing the current flow. In this electrode (usually a metal) different electrochemical reactions as solvent oxidation, must occur during current flow consuming a major fraction of the electrical energy, resulting in pH variations and generating new chemicals, which can deteriorate progressively the bilayer device.

Trying to avoid the counter electrode and its associated problems, a three layer structure was proposed (Otero et al., 1992c). Initially, it was produced by using a central passive film (two sides tape) each side coated with a conducting polymer film (Fig. 3). The triple layer was immersed in an electrolyte allowing the current flow. One of the conducting polymer films acts as the anode while the second film acts as the cathode (Garcia-Cordova et al., 2011; John et al., 2008; Yao et al., 2008). But using this configuration it is also possible to obtain movement outside a liquid electrolyte media using an ionic conducting membrane to separate the two films of conducting polymers. This membrane can be obtained by solvent evaporation and UV irradiation (Blonsky & Meridian, 1997; Heuer et al., 2002; Sansinena et al., 1997; Song et al., 2002), or by formation of interpenetrated networks (Cho et al., 2007; Plesse et al., 2005; Vidal et al., 2009; Vidal et al., 2003). In this case, the two conducting polymer films are generated by chemical polymerization on the external part of the membrane. Using this approach, multilayer devices were constructed and characterized (Ikushima et al., 2009; Zainudeen et al., 2008). The three-layer configuration provides greater efficiencies of the consumed energy: the same current is used two times to produce opposite electrochemical reactions and volume variations in the conducting

polymer films; the anode swells and pushes the device and the cathode shrinks and trails the device.

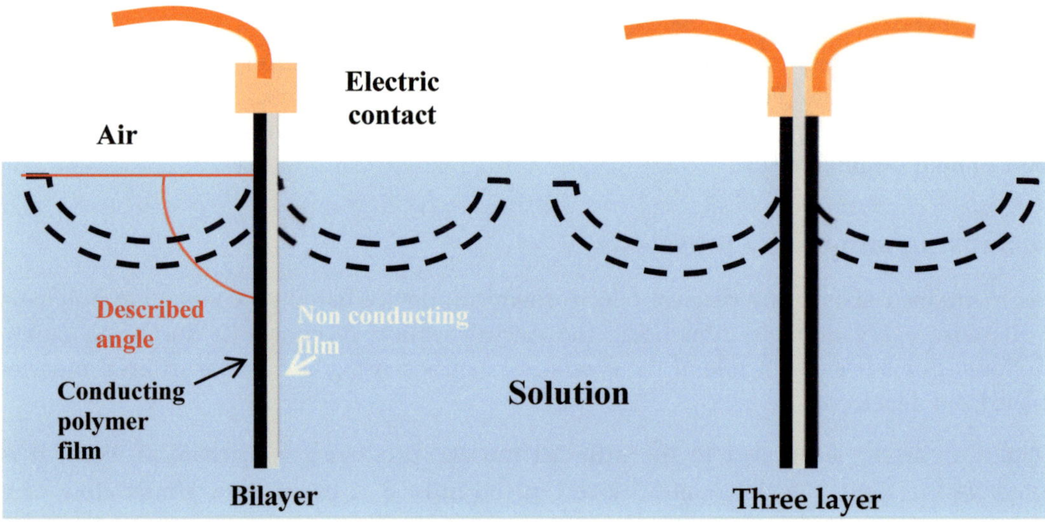

Figure 3. Bilayer and three layer devices in solution, formed by conducting polymer films and non-conductive films. Described angle is also shown.

5.2. Longitudinal or linear artificial muscles

Freestanding conducting polymer films are the simplest longitudinal actuators (DellaSanta et al., 1997). Its actuation principle is based on longitudinal expansion and contraction of the polymer during ionic exchange, although expansion and contraction occur in all three dimensions, as previously indicated. To improve the performance of these actuators overcoming problems of fragility, multilayered actuators were proposed (Hara et al., 2006; Kaneto et al., 2008; Kaneto et al., 2009), in which several thin conducting polymer films and an electrolyte (ionic liquid-soaked paper) are used to develop a compact and scalable longitudinal actuator with a high work output (Ikushima et al., 2009). Also folded films with Origami shapes provide good linear movements (Okuzaki, 2008).

Fibres of conducting polymer also can be considered as longitudinal actuators. Fibres can be obtained by extrusion (Mazzoldi et al., 1998) or by chemical polymerization over a fibre-shaped substrate (Ismail et al., 2011; Lu et al., 2002), or hollow fibre solid polymer electrolyte (Plesse et al., 2010) making it possible to obtain two concentric CP films separated by the electrolytic medium, allowing its movement in air (Dobbelin et al., 2010; Plesse et al., 2009; Vidal et al., 2010; Vidal et al., 2009). Microrods (Cho et al., 2011) or nanorods (Park et al., 2009; Vlad et al., 2012) of conducting polymers and bundles of films or fibres were investigated to produce vertical displacements of weights (Lu et al., 2002). Also, conducting polymer tubes were generated using springs and helical metallic wires (Ding et al., 2003; Hara et al., 2003; Hara et al., 2005; Spinks et al., 2003b) or zigzag metal wires (Hara et al.,

2004; Morita et al., 2010) as substrates, looking for uniform potential and current distribution. When individual fibres, bundles or tubes are used, a counter electrode is necessary, for the same reasons given above in the case of the bilayers.

Finally, it is possible to obtain linear displacement by combination of different bending structures as bilayers (Fuchiwaki et al., 2009; Naka et al., 2010; Otero & Broschart, 2006) or trilayers (Mutlu & Alici, 2010; Otero et al., 2007b) achieving longitudinal displacements over 60% of their original length.

6. Sensing and tactile muscles

As mentioned above, any reactive (electrochemical) device based on conducting polymers will sense every variable influencing the electrochemical reaction rate during actuation. Following this basic principle of the chemical kinetics sensing and tactile artificial muscles have been developed.

While a current is applied to the artificial muscle, producing a mechanical work, it is possible to follow the potential achieved in the muscle at every time. Under flow of a constant current (constant reaction rate) the achieved potential is lower when a variable which favours the electrochemical reaction increases. This is the case for temperature (Garcia-Cordova et al., 2011; Ismail et al., 2011; Otero & Cortes, 2003b; Valero Conzuelo et al., 2010; Valero et al., 2010) or electrolyte concentration (Arias-Pardilla et al., 2011; Garcia-Cordova et al., 2011; Otero & Cortes, 2003b; Otero et al., 2007b; Valero Conzuelo et al., 2010). On the other hand, the potential shifts to higher values when a variable makes the reaction harder: the muscle moves larger masses (Garcia-Cordova et al., 2011; Otero et al., 2007b; Valero Conzuelo et al., 2010) or moves the mass faster by applying now rising currents (Garcia-Cordova et al., 2011; Ismail et al., 2011; Valero et al., 2011).

Being the potential evolution a sensor of the working variables, the electrical energy (U) consumed by the device during actuation and obtained by integration is also sensor:

$$U(t) = \int E(t) I dt \tag{8}$$

Where t is the elapsed time, E(t) is the potential evolution during the actuation time and I is the constant applied current.

When a free muscle moves driven by a constant current finds an obstacle, the potential steps to higher values, trying to produce more energy and to shift the obstacle. The potential increment detects the object and its mechanical resistance to be shifted. Related with this property, artificial muscles with tactile sensitivity have been developed too (Otero & Cortes, 2003a).

7. Multi-devices: Actuators-sensors-battery-electrochromic

As mentioned above, different properties can be tuned simultaneously and in a reversible way during electrochemical reactions in conducting polymers. Electrochemical devices

created with conducting polymers may change several of those properties, such as volume, stored charge or color, at the same time. For example, using the same configuration, changing the quantity of conducting polymer used in the device, it's possible to obtain different optical or mechanical devices (Vidal et al., 2010).

Also, as the configuration of a three layer is the same of a battery with two conducting polymer films as electrodes, electrical charge is trapped in the actuator during electrochemical reaction. A way to recover that charge may be developed in order to recover that energy, acting now as a battery and moving the actuator in the opposite direction until a uniform oxidation state is attained in both polymer films (vertical position in Fig. 3).

8. Theoretical models

Several models have been proposed to characterize the electrochemomechanical behaviour of artificial muscles. At the moment, there exist different approaches from different fields as mathematics, physics or chemical-physics.

8.1. Faradic control of the movement

As mentioned above, artificial muscles are electrochemomechanical machines controlled by the electrochemical reaction occurring while actuating. As any electrochemical reaction, actuation in conducting polymers is controlled by the charge consumed during actuation. Volume changes are not an exception: according to reactions 1, 2 or 3, volume variations will be given by the number of extracted (or inserted) charges from (or to) the polymeric chains, promoting the swelling/shrinking and the ionic exchange during reaction.

For bending bilayer or trilayer artificial muscles, it has been experimentally proven that the described angle (α in rad) follows a linear relationship with the consumed charge (Q in C):

$$\alpha(t) = kQ(t) \tag{9}$$

where k is a constant (rad C^{-1}) depending on every actuator system: the device (conducting polymer and isolating tape) and the electrolyte where is moving.

By definition of the angular rate of the movement (ω):

$$\omega(t) = \frac{d\alpha(t)}{dt} = k\frac{dQ(t)}{dt} = kI(t) \tag{10}$$

This expression confirms the faradic control of the movement: the angular rate of the movement is a linear function of the applied current. Any increment of the current produces (immediately, without any relaxing time) a faster movement of the actuator, by stopping the current flow the movement stops (the driving reaction and the film volume variation stops). Eq. 9 also indicates that the actuator is a positioning device: the same charge produces the

same displacement and the charge consumed during description of a movement of one degree ($\alpha/Q = k$) is constant (independent of the applied current).

The above expressions can be normalized by mass unit of active conducting polymer reacting during actuation. This allows predicting the behaviour of every artificial muscle moving in a known electrolyte made of the same material whatever the geometry of the device is (shape, thickness, surface area, etc.). That means that the same change of the specific composition (according with reactions 1, 2 or 3) per unit time produce the same angular rate in devices having different geometry. This means that experiments from one muscle are only required in order to obtain this faradic characteristic of the CP.

The faradic control of the movement has been checked with different artificial muscles made of different polymers, exchanging both anions (Otero & Cortes, 2004) or cations (Valero et al., 2011).

8.2. Bending beam method

The bending beam method (Gere & Goodno, 2009; Timoshenko, 1925) is based on the analogy existing between a bending artificial muscle and a solid-state bending beam: the study of the forces generated at the interface between the non-conductive layer, keeping its volume constant during actuation, and the conducting polymer film, varying its volume locally.

This mechanical model assumes several characteristics related with the study of traditional mechanical bending beam: (I) the thickness of the beam is small compared to the minimum radius of curvature, (II) a linear relationship exists between stress and strain of the material and (III) the Young's modulus, Y, and the actuation expansion coefficient of the conducting polymer, α, keep constant: they do not depend on spatial location inside each layer.

The actuator curvature radius (R_∞ is the radius at the equilibrium and R_0 is the initial radius) is related to either, the Young's modulus (Y) and the thicknesses (h) of the conducting and non-conductive films (indicated by subscripts 1 and 2 respectively), and to the volume changes locally produced at the interface between both films $\alpha(t)$ (Pei & Inganäs, 1993b; Pei & Inganäs, 1993a; Pei & Inganäs, 1992b):

$$\frac{1}{R_\infty}-\frac{1}{R_0}=\frac{6\alpha_{(t)}}{\dfrac{\left(Y_1h_1^2-Y_2h_2^2\right)^2}{Y_1Y_2h_1h_2\left(h_1+h_2\right)}+4\left(h_1+h_2\right)} \tag{11}$$

Christophersen et al. (Christophersen et al., 2006) expanded the model by including strain and modulus variations along the direction of film thickness. Actuator's position, rate of the movement and force generated by the actuator (Alici & Huynh, 2006; Alici et al., 2006b) were simulated and applied to the design of biomimetic device (propulsion fins) (Alici et al., 2007). Du et al. (Du et al., 2010) have developed a general model for a multilayer system (N layers) to link the actuation strain of the actuator to the bending curvature.

8.3. Finite element method

The finite elements methodology is a well know mathematical treatment for engineering designs that can be applied to solve the movement of the artificial muscles too. Alici et al. (Alici et al., 2006a; Metz et al., 2006) developed a model based on a lumped-parameter mathematical model for trilayer actuators employing the analogy between thermal strain and the real strain (due to the insertion/extraction of ions inside the polymeric film) in polypyrrole actuators actuating in air. An optimization of the geometry was required, in order to obtain the greater output properties from a determined input voltage. Shapiro et al. (Shapiro & Smela, 2007) developed a two dimensional model (along a full area) to obtain curvature and angular moment from bilayer and trilayer actuators. Thus, they combined the results from the previous method (bending beam method) with finite element method to attain a solution. Another example of the employment of this method was carried out by Gutta et al. who applied it to the study the movement of a cylindrical ionic-polymer metal composite actuator (Gutta et al., 2011).

8.4. Equivalent transmission line model

Electrochemical systems, as many other systems, can be assimilated to electrical circuits and electrochemomechanical actuators have been treated by the equivalent transmission line method. This resource is a practical tool due to the great number of facilities available to the study of electrical circuits through different steps or modules. Such treatments are employed by engineers and physicists, or electrochemists, in order to explain the claimed capacitive behaviour of CP (Albery & Mount, 1993; Bisquert et al., 2000; Paasch, 2000). Ren et al. (Ren & Pickup, 1995) proposed equivalent electrical circuits to model the electron transport and electron transfer in composite pPy-PSS films based on Albery's works. Fang et al. (Fang et al., 2008; Yang et al., 2008) have developed a scalable method including dynamic actuation performance under a given voltage input, joining three different modules for different aspects of the actuator: electrochemical dynamics, stress-generation by charge and mechanical dynamics. Shoa et al. (Shoa et al., 2011) developed a dynamic electromechanical method for electrochemically driven conducting polymer actuators based on a 2-D impedance model using an RC transmission line equivalent circuit to predict the charge transfer during actuation. Besides, a mechanical model (based on the bending beam model) is considered after the equivalent circuit that simulates ion "diffusion" through the thickness and electronic resistance along the length (Shoa et al., 2010). If the angular movement is not linear along the full geometry of the actuator, the bending beam method has to be modified, for example for cantilever type conducting polymer actuators (Alici, 2009).

From all these kind of models, it is possible to employ only one or several of them at the same time in order to obtain the best required model (Woosoon et al., 2007).

9. Actuators applications

The investigation of these devices is mainly performed in academic laboratories. Nevertheless a rising number of applications and products are emerging with pioneering

companies that are being incorporated by large multinationals. So Creganna Tactx Medical and Bayer MaterialScience recently acquired a pair of companies working in the field, indicating the potential of these technologies. Also EAMEX from Japan is developing actuators for biomedical and robotic applications. NASA and ESA space agencies consider polymeric actuators as preferential technologies, and the European Scientific Network for Artificial Muscles (ESNAM) has started funded by the European Union. Many different applications can be found in literature. The following is a summary of a few of them, both macroscopic and microscopic.

- **Smart electronic textiles**: Although different strategies have been used to overcome problems like short lifetimes and high response times (Carpi & De Rossi, 2005; De Rossi et al., 2009; Wallace et al., 2007), much work has to be done in order to endow textiles with efficient and reliable actuating functions.
- **Braille displays**: Conducting polymers is one of the materials explored to develop low-cost and efficient displays for Braille text. Actual displays are close to achieve the required specs, but they have to overcome short cycling life (Ding et al., 2003; Spinks et al., 2003a).
- **Micropump**: It is possible to find in literature different types of micropumps made with conducting polymer actuators able to transport fluids at a microflow rate with high precission with low energy consumption (Kiefer et al., 2008; Kim et al., 2008; Morita et al.,2010 ; Naka et al., 2010; Ramírez-García & Diamond, 2007; Wu et al., 2005).
- **Propulsion or locomotion devices**: Systems based on trilayer configuration were proposed in aquatic devices (Alici et al., 2007; James et al., 2007; McGovern & et al., 2009; McGovern et al., 2010) for creating enough thrust for propulsion. Also bending actuators like cilia can be used in mini-robotic devices (Alici & Gunderson, 2009).
- **Microcatheters**: By depositing two conducting polymer films around a passive catheter and using a fast conductor solid state electrolyte, it is possible to control the movement of the catheter tip (DellaSanta et al., 1996; Shoa et al., 2008). This kind of microcatheters can be used in Optical Coherence Tomography to enable high-resolution 3D imaging (Lee et al., 2009).
- **Microactuators**: The electrochemical synthesis of conducting films and their electrochemical actuation (Wilson et al., 2007) are suitable for the construction different microdevices (He et al., 2007; Jager et al., 2001; Pede et al., 1998; Roemer et al., 2002) and microtools constituted by bilayers (Jager et al., 2000a; Jager et al., 2001; Jager et al., 2000b; Jager et al., 1999; Smela, 1999; Smela & Gadegaard, 1999; Smela et al., 1995; Smela et al., 1993) or trilayers (Gursel et al., 2009; Kiefer et al., 2008) using microelectronic technologies. These microactuators can be integrated in CMOS chips containing all the elements required to transport, contain and nurture biological cells (Jager et al., 2002; Smela et al., 2007).

10. Published papers and patents evolution

An indication of the initiation and fast growth of the field is given by bibliographic and patents evolution. As can be observed, the literature related to conducting polymers

actuators starts from 1990 (Fig. 4.a) reaching a maximum in 2006 with 45 papers, an excellent value for such a specific topic. After that, around 30 papers have been published every year. Fig. 4.b shows exponential evolution of citations with a value of almost 2000 in 2010. A similar evolution is observed in Fig. 4.c for patents, reaching a maximum in 2009 with 180 patents.

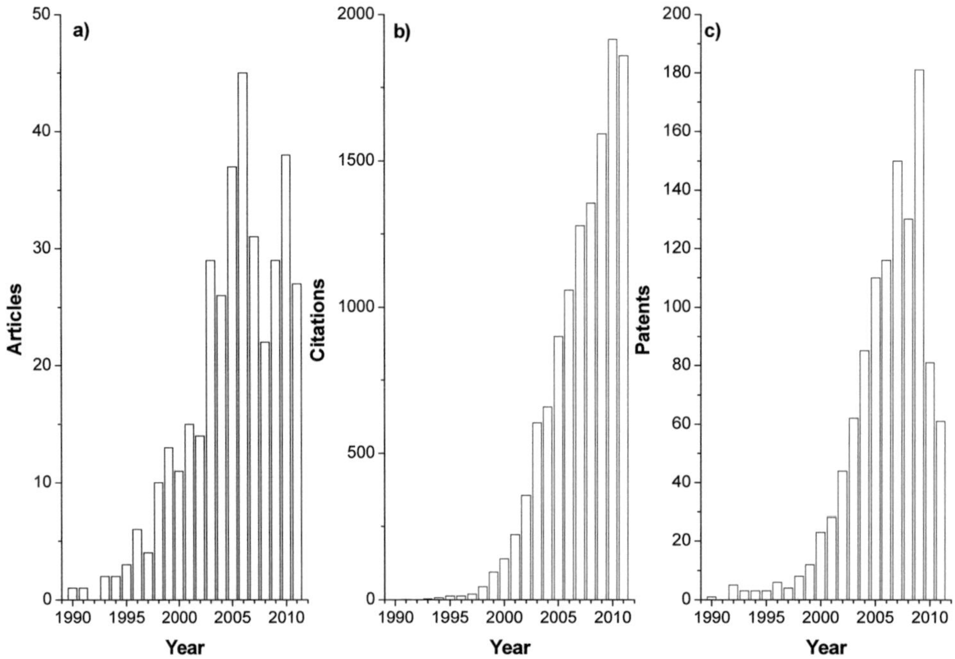

Figure 4. Evolution of a) published papers, b) citations and c) patents for "conducting polymer & actuator", from the ISI web of Knowledge® and Scopus®.

11. Future and challenges

Although a hard work has been performed, it does not exist yet an uncontroversial model for the description of the new reactive polymers and devices based on these reactive polymer gels whose composition and properties mimic those from artificial organs in mammals. New models should include concepts from very different fields of knowledge as electrochemistry, mechanics, polymer science and thermodynamics. In those systems, electrochemical reactions produce structural (conformational) changes in the polymeric chains and macroscopic volume changes related to the composition change of the material. Those aspects are not considered in the classical chemical kinetic models. Conducting polymer films should be used as models for the study and quantification of chemical kinetics under structural control with the aim to develop a new structural chemical model, able to quantify conformational changes and structural information in conducting polymers and in biological reactions originating life in living cells.

This model will allow the synthesis of new conducting polymers providing a more precise control of structural changes and intermolecular forces (polymer-ion interactions) with the

reaction. Those materials should be used in a new generation of polymeric actuators able to overcome current limitations, opening possibilities for new applications.

Other factors, such as configuration, manufacture of the actuators, design of the electric contacts or electrolytic media (solid or liquid) are also very important for the improvement of the electrochemomechanical actuators. This is a multidisciplinary field, a lot of work performed by specialists from different disciplines is required in order to attain a good control and modelling of both, devices and soft robots.

Author details

José G. Martínez, Joaquín Arias-Pardilla and Toribio F. Otero
Universidad Politécnica de Cartagena, Spain

Acknowledgement

Authors acknowledge financial support from Spanish Government (MCINN) Projects MAT2008-06702, MAT2011-24973, Seneca Foundation Project 08684/PI/08 and to the European network for Artificial Muscles ESNAM. Jose G. Martinez acknowledges Spanish Education Ministry for a FPU grant (AP2010-3460).

12. References

Ahonen, H. J., Lukkari, J. & Kankare, J. (2000). n- and p-doped poly(3,4-ethylenedioxythiophene): Two electronically conducting states of the polymer. *Macromolecules*, Vol. 33, No. 18, pp. 6787-6793, ISSN 0024-9297.

Albery, W. J. & Mount, A. R. (1993). Application of A Transmission-Line Model to Impedance Studies on A Poly(Vinylferrocene)-Modified Electrode. *J.Chem.Soc.Faraday T.*, Vol. 89, No. 2, pp. 327-331, ISSN 0956-5000.

Alici, G. & Huynh, N. N. (2006). Predicting force output of trilayer polymer actuators. *Sens. Actuat. A-Phys*, Vol. 132, No. 2, pp. 616-625, ISSN 0924-4247.

Alici, G., Metz, P. & Spinks, G. M. (2006a). A methodology towards geometry optimization of high performance polypyrrole (PPy) actuators. *Smart Mater.Struct.*, Vol. 15, No. 2, pp. 243-252, ISSN 0964-1726.

Alici, G., Mui, B. & Cook, C. (2006b). Bending modeling and its experimental verification for conducting polymer actuators dedicated to manipulation applications. *Sens. Actuat. A-Phys*, Vol. 126, No. 2, pp. 396-404, ISSN 0924-4247.

Alici, G., Spinks, G., Huynh, N. N., Sarmadi, L. & Minato, R. (2007). Establishment of a biomimetic device based on tri-layer polymer actuators-propulsion fins. *Bioinspir. Biomim.*, Vol. 2, No. 2, pp. S18-S30, ISSN 1748-3182.

Alici, G. (2009). An effective modelling approach to estimate nonlinear bending behaviour of cantilever type conducting polymer actuators. *Sens. Actuat. B-Chem*, Vol. 141, No. 1, pp. 284-292, ISSN 0925-4005.

Alici, G. & Gunderson, D. (2009). A bio-inspired robotic locomotion system based on conducting polymer actuators. In: *2009 IEEE/ASME International Conference on Advanced Intelligent Mechatronics*. pp. 998-1004, IEEE, ISBN 978-1-4244-2852-6, New York.

Alici, G., Punning, A. & Shea, H. R. (2011). Enhancement of actuation ability of ionic-type conducting polymer actuators using metal ion implantation. *Actuat. B-Chem*, Vol. 157, No. 1, pp. 72-84, ISSN 0925-4005.

Arbizzani, C., Catellani, M., Mastragostino, M. & Mingazzini, C. (1995). N-doped and p-doped polydithieno[3,4-b-3',4'-d] thiophene - a narrow-band gap polymer for redox supercapacitors. *Electrochim. Acta*, Vol. 40, No. 12, pp. 1871-1876, ISSN 0013-4686.

Arias-Pardilla, J., Walker, W., Wudl, F. & Otero, T. F. (2010). Reduction and Oxidation Doping Kinetics of an Electropolymerized Donor-Acceptor Low-Bandgap Conjugated Copolymer. *J. Phys. Chem. B*, Vol. 114, No. 40, pp. 12777-12784, ISSN 1520-6106.

Arias-Pardilla, J., Plesse, C., Khaldi, A., Vidal, F., Chevrot, C. & Otero, T. F. (2011). Self-supported semi-interpenetrating polymer networks as reactive ambient sensors. *J. Electroanal. Chem.*, Vol. 652, No. 1-2, pp. 37-43, ISSN 1572-6657.

Balzani, V., Credi, A., Ferrer, B., Silvi, S. & Venturi, M. (2005). *Artificial molecular motors and machines: Design principles and prototype systems*, Springer-Verlag, ISBN 0340-1022, Berlin.

Baughman, R. H. (1996). Conducting polymer artificial muscles. *Synth. Met.*, Vol. 78, No. 3, pp. 339-353, ISSN 0379-6779.

Baughman, R. H., Cui, C. X., Zakhidov, A. A., Iqbal, Z., Barisci, J. N., Spinks, G. M., Wallace, G. G., Mazzoldi, A., De Rossi, D., Rinzler, A. G., Jaschinski, O., Roth, S. & Kertesz, M. (1999). Carbon nanotube actuators. *Science*, Vol. 284, No. 5418, pp. 1340-1344, ISSN 0036-8075.

Bay, L., West, K., Sommer-Larsen, P., Skaarup, S. & Benslimane, M. (2003). A conducting polymer artificial muscle with 12% linear strain. *Adv. Mater.*, Vol. 15, No. 4, pp. 310-313, ISSN 0935-9648.

Bieńkowski, K., Strawski, M. & Szklarczyk, M. (2011). The determination of the thickness of electrodeposited polymeric films by AFM and electrochemical techniques. *J. Electroanal. Chem.*, Vol. 662, No. 1, pp. 196-203, ISSN 1572-6657.

Bisquert, J., Belmonte, G. G., Santiago, F. F., Ferriols, N., Yamashita, M. & Pereira, E. C. (2000). Application of a distributed impedance model in the analysis of conducting polymer films. *Electrochem.Commun.*, Vol. 2, No. 8, pp. 601-605, ISSN 1388-2481.

Blonsky, P. M. & Meridian, I. (1997). Structurally stable gelled electrolytes, US Patent 5648011.

Carpi, F. & De Rossi, D. (2005). Electroactive polymer-based devices for e-textiles in biomedicine. *IEEE T. Inf. Technol. B*, Vol. 9, No. 3, pp. 295-318, ISSN 1089-7771.

Chiang, C. K., Fincher, C. R., Park, Y. W., Heeger, A. J., Shirakawa, H., Louis, E. J., Gau, S. C. & Macdiarmid, A. G. (1977). Electrical-Conductivity in Doped Polyacetylene. *Phys. Rev. Lett.*, Vol. 39, No. 17, pp. 1098-1101, ISSN 0031-9007.

Cho, M., Seo, H., Nam, J., Choi, H., Koo, J. & Lee, Y. (2007). High ionic conductivity and mechanical strength of solid polymer electrolytes based on NBr/ionic liquid and its

application to an electrochemical actuator. *Sens. Actuat. B-Chem*, Vol. 128, No. 1, pp. 70-74, ISSN 0925-4005.

Cho, M. S., Choi, J. J., Kim, T. S. & Lee, Y. (2011). In situ three-dimensional analysis of the linear actuation of polypyrrole micro-rod actuators using optical microscopy. *Sens. Actuat. B-Chem*, Vol. 156, No. 1, pp. 218-221, ISSN 0925-4005.

Christophersen, M., Shapiro, B. & Smela, E. (2006). Characterization and modeling of PPy bilayer microactuators - Part 1. Curvature. *Sens. Actuat. B-Chem*, Vol. 115, No. 2, pp. 596-609, ISSN 0925-4005.

Daerden, F. & Lefeber, D. (2002). Pneumatic Artificial Muscles: actuators for robotics and automation. *Eur. J. Mech. Environ. Eng.*, Vol. 47, No. 1, pp. 10-21, ISSN 1371-6980.

Davis, A. P. (1999). Nanotechnology - Synthetic molecular motors. *Nature*, Vol. 401, No. 6749, pp. 120-121, ISSN 0028-0836.

De Rossi, D., Carpi, F. & Galantini, F. (2009). Functional Materials for Wearable Sensing, Actuating and Energy Harvesting. In: *Biomedical Applications of Smart Materials, Nanotechnology and Micro/Nano Engineering*. pp. 247-256, Trans Tech Publications Ltd, ISBN 978-3-908158-23-3, Stafa-Zurich.

DellaSanta, A., Mazzoldi, A. & de Rossi, D. (1996). Steerable Microcatheters Actuated by Embedded Conducting Polymer Structures. *J. Intel. Mat. Syst. Str.*, Vol. 7, No. 3, pp. 292-300, ISSN 1045-389X.

DellaSanta, A., DeRossi, D. & Mazzoldi, A. (1997). Performance and work capacity of a polypyrrole conducting polymer linear actuator. *Synth. Met.*, Vol. 90, No. 2, pp. 93-100, ISSN 0379-6779.

Deshpande, S. D., Kim, J. & Yun, S. R. (2005a). New electro-active paper actuator using conducting polypyrrole: actuation behaviour in LiClO4 acetonitrile solution. *Synth. Met.*, Vol. 149, No. 1, pp. 53-58, ISSN 0379-6779.

Deshpande, S. D., Kim, J. & Yun, S. R. (2005b). Studies on conducting polymer electroactive paper actuators: Effect of humidity and electrode thickness. *Smart Mater. Struct.*, Vol. 14, No. 4, pp. 876-880, ISSN 0964-1726.

Ding, J., Liu, L., Spinks, G. M., Zhou, D., Wallace, G. G. & Gillespie, J. (2003). High performance conducting polymer actuators utilising a tubular geometry and helical wire interconnects. *Synth. Met.*, Vol. 138, No. 3, pp. 391-398, ISSN 0379-6779.

Dobbelin, M., Marcilla, R., Pozo-Gonzalo, C. & Mecerreyes, D. (2010). Innovative materials and applications based on poly(3,4-ethylenedioxythiophene) and ionic liquids. *J. Mater. Chem.*, Vol. 20, No. 36, pp. 7613-7622, ISSN 0959-9428.

Du, P., Lin, X. & Zhang, X. (2010). A multilayer bending model for conducting polymer actuators. *Sens. Actuat. A-Phys*, Vol. 163, No. 1, pp. 240-246, ISSN 0924-4247.

Fang, Y., Tan, X. O., Shen, Y. T., Xi, N. & Alici, G. (2008). A scalable model for trilayer conjugated polymer actuators and its experimental validation. *Mater. Sci. Eng., C*, Vol. 28, No. 3, pp. 421-428, ISSN 0928-4931.

Fuchiwaki, M., Tanaka, K. & Kaneto, K. (2009). Planate conducting polymer actuator based on polypyrrole and its application. *Sens. Actuat. A-Phys*, Vol. 150, No. 2, pp. 272-276, ISSN 0924-4247.

Garcia-Cordova, F., Valero, L., Ismail, Y. A. & Otero, T. F. (2011). Biomimetic polypyrrole based all three-in-one triple layer sensing actuators exchanging cations. *J. Mater. Chem.*, Vol. 21, No. 43, pp. 17265-17272, ISSN 0959-9428.

Gere, J. M. & Goodno, B. J. (2009). *Mechanics of Materials 4th Ed.*, Cengage Learning, ISBN 0534553974, Toronto.

Gimenez, P., Mukai, K., Asaka, K., Hata, K., Oike, H. & Otero, T. F. (2012). Capacitive and faradic charge components in high-speed carbon nanotube actuator. *Electrochim. Acta*, Vol. 60, No. 177-183, ISSN 0013-4686.

Guo, Y., Yu, G. & Liu, Y. (2010). Functional Organic Field-Effect Transistors. *Adv. Mater.*, Vol. 22, No. 40, pp. 4427-4447, ISSN 1521-4095.

Gursel, A., Valerie, D., Philippe, R. & Geoff, S. (2009). Conducting polymer microactuators operating in air. *J. Micromech. Microeng.*, Vol. 19, No. 2, pp. 025017, ISSN 0960-1317.

Gutta, S., Realmuto, J., Woosoon, Y. & Kim, K. J. (2011). *Dynamic model of a cylindrical ionic polymer-metal composite actuator.* Proceedings of 8th International Conference on Ubiquitous Robots and Ambient Intelligence (URAI), Incheon.

Han, G. & Shi, G. (2006). Electrochemical actuator based on single-layer polypyrrole film. *Sens. Actuat. B-Chem*, Vol. 113, No. 1, pp. 259-264, ISSN 0925-4005.

Hara, S., Zama, T., Sewa, S., Takashima, W. & Kaneto, K. (2003). Polypyrrole-metal coil composites as fibrous artificial muscles. *Chem. Lett.*, Vol. 32, No. 9, pp. 800-801, ISSN 0366-7022.

Hara, S., Zama, T., Ametani, A., Takashima, W. & Kaneto, K. (2004). Enhancement in electrochemical strain of a polypyrrole-metal composite film actuator. *J. Mater. Chem.*, Vol. 14, No. 18, pp. 2724-2725, ISSN 0959-9428.

Hara, S., Zama, T., Tanaka, N., Takashima, W. & Kaneto, K. (2005). Artificial fibular muscles with 20% strain based on polypyrrole-metal coil composites. *Chem.Lett.*, Vol. 34, No. 6, pp. 784-785, ISSN 0366-7022.

Hara, S., Zama, T., Takashima, W. & Kaneto, K. (2006). Tris (trifluoromethylsulfonyl)methide-doped polypyrrole as a conducting polymer actuator with large electrochemical strain. *Synth.Met.*, Vol. 156, No. 2-4, pp. 351-355, ISSN 0379-6779.

He, X. M., Li, C., Chen, F. G. & Shi, G. Q. (2007). Polypyrrole microtubule actuators for seizing and transferring microparticles. *Adv. Funct. Mater.*, Vol. 17, No. 15, pp. 2911-2917, ISSN 1616-301X.

Heuer, H. W., Wehrmann, R. & Kirchmeyer, S. (2002). Electrochromic window based on conducting poly (3,4-ethylenedioxythiophene)poly(styrene sulfonate). *Adv. Funct. Mater.*, Vol. 12, No. 2, pp. 89-94, ISSN 1616-301X.

Higgins, S. J., Lovell, K. V., Rajapakse, R. M. G. & Walsby, N. M. (2003). Grafting and electrochemical characterisation of poly-(3,4-ethylenedioxythiophene) films, on Nafion and on radiation-grafted polystyrenesulfonate-polyvinylidene fluoride composite surfaces. *J. Mater. Chem.*, Vol. 13, No. 10, pp. 2485-2489, ISSN 0959-9428.

Hillman, A. R., Loveday, D. C., Swann, M. J., Eales, R. M., Hamnett, A., Higgins, S. J., Bruckenstein, S. & Wilde, C. P. (1989). Charge Transport in Electroactive Polymer-Films. *Faraday Discuss. Chem. Soc.*, Vol. 88, No. 151-163, ISSN 0301-7249.

Huang, W. S., Humphrey, B. D. & MacDiarmid, A. G. (1986). Polyaniline, A Novel Conducting Polymer - Morphology and Chemistry of Its Oxidation and Reduction in Aqueous-Electrolytes. *J. Chem. Soc., Faraday Trans. I*, Vol. 82, No. 2385-2400, ISSN 0300-9599.

Huang, Y., Liang, J. & Chen, Y. (2012). The application of graphene based materials for actuators. *J. Mater. Chem.*, Vol. 22, No. 9, pp. 3671-3679, ISSN 0959-9428.

Ikushima, K., John, S., Yokoyama, K. & Nagamitsu, S. (2009). A practical multilayered conducting polymer actuator with scalable work output. *Smart. Mater. Struct.*, Vol. 18, No. 9, pp. 9, ISSN 0964-1726.

Inzelt, G. (2008). Redox Transformations and Transport Processes. In: *Conducting Polymers*, Scholz, F. pp. 169-224, Springer-Verlag Berlin, ISBN 9783540759294, Heidelberg (Germany).

Inzelt, G. (2011). Rise and rise of conducting polymers. *J. Solid State Electrochem.*, Vol. 15, No. 7-8, pp. 1711-1718, ISSN 1432-8488.

Ismail, Y. A., Martínez, J. G., Harrasi, A. S. A., Kim, S. J. & Otero, T. F. (2011). Sensing characteristics of a conducting polymer/hydrogel hybrid microfiber artificial muscle. *Actuat. B-Chem*, Vol. No. 0, pp. ISSN 0925-4005.

Jager, E. W. H., Smela, E., Inganäs, O. & Lundstrom, I. (1999). Polypyrrole microactuators. *Synth. Met.*, Vol. 102, No. 1-3, pp. 1309-1310, ISSN 0379-6779.

Jager, E. W. H., Inganäs, O. & Lundstrom, I. (2000a). Microrobots for micrometer-size objects in aqueous media: Potential tools for single-cell manipulation. *Science*, Vol. 288, No. 5475, pp. 2335-2338, ISSN 0036-8075.

Jager, E. W. H., Smela, E. & Inganäs, O. (2000b). Microfabricating conjugated polymer actuators. *Science*, Vol. 290, No. 5496, pp. 1540-1545, ISSN 0036-8075.

Jager, E. W. H., Inganäs, O. & Lundstrom, I. (2001). Perpendicular actuation with individually controlled polymer microactuators. *Adv. Mater.*, Vol. 13, No. 1, pp. 76-79, ISSN 0935-9648.

Jager, E. W. H., Immerstrand, C., Peterson, K. H., Magnusson, K.-E., Lundström, I. & Inganäs, O. (2002). The Cell Clinic: Closable Microvials for Single Cell Studies. *Biomed. Microdev*, Vol. 4, No. 3, pp. 177-187, ISSN 1387-2176.

James, T., Patrick, A., Timothy, F., Angela, C., Mike Del, Z. & Ian, H. (2007). The application of conducting polymers to a biorobotic fin propulsor. *Bioinspir. Biomim.*, Vol. 2, No. 2, pp. S6, ISSN 1748-3190.

John, S., Alici, G. & Cook, C. (2008). *Frequency response of polypyrrole trilayer actuator displacement*. Proceedings of Electroactive Polymer Actuators and Devices (Eapad) 2008, San Diego (USA).

Kaneto, K., Suematsu, H. & Yamato, K. (2008). Training effect and fatigue in polypyrrole-based artificial muscles. *Bioinspir. Biomim.*, Vol. 3, No. 3, pp. 035005, ISSN 1748-3182.

Kaneto, K., Suematsu, H. & Yamato, K. (2009). Conducting Polymer Soft Actuators based on Polypyrrole-Training Effect and Fatigue. In: *Artificial Muscle Actuators Using Electroactive Polymers*, Vincenzini, P., Bar-Cohen, Y. & Carpi, F. pp. 122-130, Trans Tech Publications Inc., ISBN 1662-0356, Stafa-Zuerich, Switzerland.

Kiefer, R., Chu, S. Y., Kilmartin, P. A., Bowmaker, G. A., Cooney, R. P. & Travas-Sejdic, J. (2007). Mixed-ion linear actuation behaviour of polypyrrole. *Electrochim. Acta*, Vol. 52, No. 7, pp. 2386-2391, ISSN 0013-4686.

Kiefer, R., Mandviwalla, X., Archer, R., Tjahyono, S. S., Wang, H., MacDonald, B., Bowmaker, G. A., Kilmartin, P. A. & Travas-Sejdic, J. (2008). *The application of polypyrrole trilayer actuators in microfluidics and robotics*. Proceedings of Electroactive Polymer Actuators and Devices (Eapad).

Kim, J. H., Lau, K. T., Shepherd, R., Wu, Y., Wallace, G. & Diamond, D. (2008). Performance characteristics of a polypyrrole modified polydimethylsiloxane (PDMS) membrane based microfluidic pump. *Sensors and Actuators A: Physical*, Vol. 148, No. 1, pp. 239-244, ISSN 0924-4247.

Kim, L. T. T., Sel, O., Biemme-Chouvy, C., Gabrielli, C., Laberty-Robert, C., Perrot, H. & Sanchez, C. (2010). Proton transport properties in hybrid membranes investigated by ac-electrogravimetry. *Electrochem.Commun.*, Vol. 12, No. 8, pp. 1136-1139, ISSN 1388-2481.

Klauk, H., Ed. (2006). *Organic Electronics: Materials, Manufacturing, and Applications* Wiley-VCH Verlag, ISBN 978-3-527-31264-1, Weinheim.

Kuttel, C., Stemmer, A. & Wei, X. (2009). Strain response of polypyrrole actuators induced by redox agents in solution. *Sens. Actuat. B-Chem*, Vol. 141, No. 2, pp. 478-484, ISSN 0925-4005.

Lee, K. K. C., Munce, N. R., Shoa, T., Charron, L. G., Wright, G. A., Madden, J. D. & Yang, V. X. D. (2009). Fabrication and characterization of laser-micromachined polypyrrole-based artificial muscle actuated catheters. *Sens. Actuat. A-Phys*, Vol. 153, No. 2, pp. 230-236, ISSN 0924-4247.

Li, W. G., Johnson, C. L. & Wang, H. L. (2004). Preparation and characterization of monolithic polyaniline-graphite composite actuators. *Polymer*, Vol. 45, No. 14, pp. 4769-4775, ISSN 0032-3861.

Lu, W., Fadeev, A. G., Qi, B. H., Smela, E., Mattes, B. R., Ding, J., Spinks, G. M., Mazurkiewicz, J., Zhou, D. Z., Wallace, G. G., MacFarlane, D. R., Forsyth, S. A. & Forsyth, M. (2002). Use of ionic liquids for pi-conjugated polymer electrochemical devices. *Science*, Vol. 297, No. 5583, pp. 983-987, ISSN 0036-8075.

Lyutov, V., Tsakova, V. & Bund, A. (2011). Microgravimetric study on the formation and redox behavior of poly(2-acrylamido-2-methyl-1-propanesulfonate)-doped thin polyaniline layers. *Electrochim. Acta*, Vol. 56, No. 13, pp. 4803-4811, ISSN 0013-4686.

Mazzoldi, A., Degl'Innocenti, C., Michelucci, M. & De Rossi, D. (1998). Actuative properties of polyaniline fibers under electrochemical stimulation. *Mater. Sci. Eng., C*, Vol. 6, No. 1, pp. 65-72, ISSN 0928-4931.

McGovern, S. & et al. (2009). Finding NEMO (novel electromaterial muscle oscillator): a polypyrrole powered robotic fish with real-time wireless speed and directional control. *Smart Mater. Struct.*, Vol. 18, No. 9, pp. 095009, ISSN 0964-1726.

McGovern, S. T., Abbot, M., Emery, R., Alici, G., Truong, V. T., Spinks, G. M. & Wallace, G. G. (2010). Evaluation of thrust force generated for a robotic fish propelled with polypyrrole actuators. *Polym. Int.*, Vol. 59, No. 3, pp. 357-364, ISSN 0959-8103.

Metz, P., Alici, G. & Spinks, G. M. (2006). A finite element model for bending behaviour of conducting polymer electromechanical actuators. *Sens. Actuat. A-Phys*, Vol. 130, No. 1-11, ISSN 0924-4247.

Morita, T., Chida, Y., Hoshino, D., Fujiya, T. & Nishioka, Y. Fabrication and characterization of a polypyrrole soft actuator having corrugated structures. *Mol. Cryst. Liq. Cryst.*, Vol. 519, No. 121-127, ISSN 1542-1406.

Morita, T., Chida, Y., Hoshino, D., Fujiya, T. & Nishioka, Y. (2010). Fabrication and Characterization of a Polypyrrole Soft Actuator Having Corrugated Structures. *Mol. Cryst. Liq. Cryst.*, Vol. 519, No. 121-127, ISSN 1542-1406.

Mukai, K., Asaka, K., Hata, K., Otero, T. F. & Oike, H. (2011). High-Speed Carbon Nanotube Actuators Based on an Oxidation/Reduction Reaction. *Chem.-Eur. J.*, Vol. 17, No. 39, pp. 10965-10971, ISSN 1521-3765.

Mutlu, R. & Alici, G. (2010). A Multistable Linear Actuation Mechanism Based on Artificial Muscles. *J. Mech. Des.*, Vol. 132, No. 11, pp. 111001, ISSN 1050-0472.

Naka, Y., Fuchiwaki, M. & Tanaka, K. (2010). A micropump driven by a polypyrrole-based conducting polymer soft actuator. *Polym. Int.*, Vol. 59, No. 3, pp. 352-356, ISSN 1097-0126.

Nakano, T. & Okamoto, Y. (2001). Synthetic Helical Polymers: Conformation and Function. *Chem. Rev.*, Vol. 101, No. 12, pp. 4013-4038, ISSN

Okamoto, T., Tada, K. & Onoda, M. (2000). Bending machine using anisotropic polypyrrole films. *Jpn. J. Appl. Phys. 1*, Vol. 39, No. 5A, pp. 2854-2858, ISSN 0021-4922.

Okamoto, T., Kato, Y., Tada, K. & Onoda, M. (2001). Actuator based on doping/undoping-induced volume change in anisotropic polypyrrole film. *Thin Solid Films*, Vol. 393, No. 1-2, pp. 383-387, ISSN 0040-6090.

Okuzaki, H. & Hattori, T. (2003). Electrically induced anisotropic contraction of polypyrrole films. *Synth. Met.*, Vol. 135-136, No. 45-46, ISSN 0379-6779.

Okuzaki, H. (2008). A biomorphic origami actuator fabricated by folding a conducting paper. *J. Phys. Conf. Ser.*, Vol. 127, No. 1, pp. 012001, ISSN 1742-6596.

Onoda, M., Okamoto, T., Tada, K. & Nakayama, H. (1999a). Polypyrrole films with anisotropy for artificial muscles and examination of bending behavior. *Jpn. J. Appl. Phys. 2*, Vol. 38, No. 9AB, pp. L1070-L1072, ISSN 0021-4922.

Onoda, M., Tada, K. & Nakayama, H. (1999b). Polypyrrole films with anisotropy. *Synth. Met.*, Vol. 102, No. 1-3, pp. 1321-1322, ISSN 0379-6779.

Onoda, M. & Tada, K. (2004). Anisotropic bending machine using conducting polypyrrole. *IEICE Trans. Electron.*, Vol. E87C, No. 2, pp. 128-135, ISSN 0916-8524.

Onoda, M., Shonaka, H. & Tada, K. (2005). A self-organized bending-beam electrochemical actuator. *Curr. Appl. Phys.*, Vol. 5, No. 2, pp. 194-201, ISSN 1567-1739.

Orata, D. & Buttry, D. A. (1987). Determination of ion populations and solvent content as functions of redox state and pH in polyaniline. *J. Am. Chem. Soc.*, Vol. 109, No. 12, pp. 3574-3581, ISSN 0002-786.

Otero, T., Cascales, J. & Fernandez-Romero, A. (2007a). *Attempting a classification for electrical polymeric actuators*. Proceedings of Electroactive Polymer Actuators and Devices (EAPAD), San Diego.

Otero, T. F., Angulo, E., Rodriguez, J. & Santamaria, C. (1992a). Electrochemomechanical Properties from A Bilayer - Polypyrrole Nonconducting and Flexible Material Artificial Muscle. *J. Electroanal. Chem.*, Vol. 341, No. 1-2, pp. 369-375, ISSN 0022-0728.

Otero, T. F., Angulo, E., Rodriguez, J. & Santamaria, C. (1992b). Dispositivos laminares que emplean polímeros conductores capaces de provocar movimientos mecánicos, ES Patent 2 048 086.

Otero, T. F., Rodriguez, J. & Santamaria, C. (1992c). Músculos artificiales formados por multicapas: polímeros conductores-no conductores, ES Patent 2 062 930.

Otero, T. F. (1999). Conducting polymers, electrochemistry, and biomimicking processes. In: *Modern Aspects of Electrochemistry*, Bockris, J.O.M., White, R.E. & Conway, B.E. pp. 307-434, Kluwer Academic/Plenum Publ, ISBN 0076-9924, New York.

Otero, T. F. (2000). Biomimicking materials with smart polymers. In: *Structural biological materials. Design and structure-properties relationships*, Elices, M. & Cahn, R.W. pp. 187-220, Pergamon Materials Series, ISBN 0080434169, Amsterdam (The Netherlands).

Otero, T. F. & Cortes, M. T. (2003a). Artificial muscles with tactile sensitivity. *Adv. Mater.*, Vol. 15, No. 4, pp. 279-282, ISSN 0935-9648.

Otero, T. F. & Cortes, M. T. (2003b). A sensing muscle. *Sens. Actuators, B*, Vol. 96, No. 1-2, pp. 152-156, ISSN 0925-4005.

Otero, T. F. & Cortes, M. T. (2004). Artificial muscle: movement and position control. *Chem. Commun.*, Vol. No. 3, pp. 284-285, ISSN 1359-7345.

Otero, T. F., Arenas, G. V. & Cascales, J. J. L. (2006). Effect of the doping ion on the electrical response of a free-standing polypyrrole strip subjected to different preloads: Perspectives and limitations associated with the use of these devices as actuators. *Macromolecules*, Vol. 39, No. 26, pp. 9551-9556, ISSN 0024-9297.

Otero, T. F. & Broschart, M. (2006). Polypyrrole artificial muscles: a new rhombic element. Construction and electrochemomechanical characterization. *J. Appl. Electrochem.*, Vol. 36, No. 2, pp. 205-214, ISSN 0021-891X.

Otero, T. F. (2007). Artificial muscles. In: *Handbook of Conducting Polymers*, Skotheim, T.A., Elsenbaumer, R.L. & Reynolds, J.R. pp. 591-623, CRC Press, ISBN Boca Raton.

Otero, T. F., Cortes, M. T. & Arenas, G. V. (2007b). Linear movements from two bending triple-layers. *Electrochim. Acta*, Vol. 53, No. 3, pp. 1252-1258, ISSN 0013-4686.

Otero, T. F., Lopez Cascales, J. J. & Vazquez Arenas, G. (2007c). Mechanical characterization of free-standing polypyrrole film. *Mater. Sci. Eng., C*, Vol. 27, No. 1, pp. 18-22, ISSN 0928-4931.

Otero, T. F. (2008). Artificial Muscles, Sensing and Multifunctionality. In: *Intelligent Materials*, Shahinpoor, M. & Schenider, H.-J. pp. 142-190, Royal Society of Chemistry, ISBN 9780854043354, Cambridge (U.K.).

Otero, T. F. (2009). Soft, wet, and reactive polymers. Sensing artificial muscles and conformational energy. *J. Mater. Chem.*, Vol. 19, No. 681-689, ISSN 0959-9428.

Otero, T. F. (2011). From Electrochemically-Driven Conformational Polymeric States to Macroscopic Sensing and Tactile Muscles. In: *From Non-Covalent Assemblies to Molecular Machines*, Sauvage, J.-P. & Gaspard, P. pp. Wiley-VCH Verlag & Co., ISBN Weinheim, Germany.

Otero, T. F., Arias-Pardilla, J., Herrera, H., Segura, J. L. & Seoane, C. (2011). Electropolymerization of naphthaleneamidinemonoimide-modified poly(thiophene). *Phys. Chem. Chem. Phys.*, Vol. 13, No. 37, pp. 16513-16516, ISSN 1463-9076.

Otero, T. F., Sánchez, J. J. & Martinez, J. G. (2012) Biomimetic Dual Sensing-Actuators based on Conducting Polymers. Galvanostatic Theoretical Model for Actuators Sensing Temperature. J. Phys. Chem. B, Vol 116, No. 17, pp. 5279-5290, ISSN 1089-5647.

Ouyang, P. R., Tjiptoprodjo, R. C., Zhang, W. J. & Yang, G. S. (2008). Micro-motion devices technology: The state of arts review. *Int. J. Adv. Manuf. Technol.*, Vol. 38, No. 5-6, pp. 463-478, ISSN 0268-3768.

Paasch, G. (2000). The transmission line equivalent circuit model in solid-state electrochemistry. *Electrochem. Commun.*, Vol. 2, No. 5, pp. 371-375, ISSN 1388-2481.

Park, S. J., Cho, M. S., Nam, J. D., Kim, I. H., Choi, H. R., Koo, J. C. & Lee, Y. (2009). The linear stretching actuation behavior of polypyrrole nanorod in AAO template. *Sens. Actuat. B-Chem*, Vol. 135, No. 2, pp. 592-596, ISSN 0925-4005.

Pede, D., Smela, E., Johansson, T., Johansson, M. & Inganäs, O. (1998). A general-purpose conjugated-polymer device array for imaging. *Adv. Mater.*, Vol. 10, No. 3, pp. 233-237, ISSN 0935-9648.

Pei, Q. & Inganäs, O. (1993a). Electrochemical applications of the bending beam method; a novel way to study ion transport in electroactive polymers. *Solid State Ionics*, Vol. 60, No. 1-3, pp. 161-166, ISSN 0167-2738.

Pei, Q. & Inganäs, O. (1993b). Electrochemical applications of the bending beam method. 2. Electroshrinking and slow relaxation in polypyrrole. *J. Phys. Chem.*, Vol. 97, No. 22, pp. 6034-6041, ISSN 0022-3654.

Pei, Q. B. & Inganäs, O. (1992a). Conjugated Polymers and the Bending Cantilever Method - Electrical Muscles and Smart Devices. *Adv. Mater.*, Vol. 4, No. 4, pp. 277-278, ISSN 0935-9648.

Pei, Q. B. & Inganäs, O. (1992b). Electrochemical Applications of the Bending Beam Method .1. Mass-Transport and Volume Changes in Polypyrrole During Redox. *J. Phys. Chem.*, Vol. 96, No. 25, pp. 10507-10514, ISSN 0022-3654.

Perepichka, I. F. & Perepichka, D. F., Eds. (2009). *Handbook of Thiophene-Based Materials: Applications in Organic Electronics and Photonics.* John Wiley & Sons Ltd., ISBN 978-0-470-05732-2, Chichester (UK).

Plesse, C., Vidal, F., Randriamahazaka, H., Teyssie, D. & Chevrot, C. (2005). Synthesis and characterization of conducting interpenetrating polymer networks for new actuators. *Polymer*, Vol. 46, No. 18, pp. 7771-7778, ISSN 0032-3861.

Plesse, C., Vidal, F., Teyssie, D. & Chevrot, C. (2009). Conducting IPN Fibers: a new design for linear actuation in open air. In: *Artificial Muscle Actuators using Electroactive Polymers* Vincenzini, P., Bar-Cohen, Y. & Carpi, F. pp. 53-58, Trans Tech Publicactions Ltd., ISBN 1662-0356, Stafa-Zurich (Switzerland).

Plesse, C., Vidal, F., Teyssie, D. & Chevrot, C. (2010). Conducting polymer artificial muscle fibres: toward an open air linear actuation. *Chem. Commun.*, Vol. 46, No. 17, pp. 2910-2912, ISSN 1359-7345.

Ramírez-García, S. & Diamond, D. (2007). Biomimetic, low power pumps based on soft actuators. *Sens. Actuat. A-Phys*, Vol. 135, No. 1, pp. 229-235, ISSN 0924-4247.

Ranger, M. & Leclerc, M. (1998). Optical and electrical properties of fluorene-based pi-conjugated polymers. *Can. J. Chem.*, Vol. 76, No. 11, pp. 1571-1577, ISSN 0008-4042.

Ren, X. & Pickup, P. G. (1995). Impedance measurements of ionic conductivity as a probe of structure in electrochemically deposited polypyrrole films. *J. Electroanal. Chem.*, Vol. 396, No. 1–2, pp. 359-364, ISSN 1572-6657.

Roemer, M., Kurzenknabe, T., Oesterschulze, E. & Nicoloso, N. (2002). Microactuators based on conducting polymers. *Anal. Bioanal. Chem.*, Vol. 373, No. 8, pp. 754-757, ISSN 1618-2642.

Sansinena, J. M., Olazabal, V., Otero, T. F., daFonseca, C. N. P. & DePaoli, M. A. (1997). A solid state artificial muscle based on polypyrrole and a solid polymeric electrolyte working in air. *Chem. Commun.*, Vol. No. 22, pp. 2217-2218, ISSN 1359-7345.

Sansiñena, J. M., Gao, J. B. & Wang, H. L. (2003). High-performance, monolithic polyaniline electrochemical actuators. *Adv. Funct. Mater.*, Vol. 13, No. 9, pp. 703-709, ISSN 1616-301X.

Shakuda, S., Morita, S., Kawai, T. & Yoshino, K. (1993). Dynamic Characteristics of Bimorph with Conducting Polymer Gel. *Jpn. J. Appl. Phys. 1*, Vol. 32, No. 11A, pp. 5143-5146, ISSN 0021-4922.

Shapiro, B. & Smela, E. (2007). Bending Actuators with Maximum Curvature and Force and Zero Interfacial Stress. *J. Intel. Mat. Syst. Str.*, Vol. 18, No. 2, pp. 181-186, ISSN 1045-389X.

Shirakawa, H., Louis, E. J., Macdiarmid, A. G., Chiang, C. K. & Heeger, A. J. (1977). Synthesis of Electrically Conducting Organic Polymers - Halogen Derivatives of Polyacetylene, (Ch)X. *J. Chem. Soc., Chem. Commun.*, Vol. No. 16, pp. 578-580, ISSN 0022-4936.

Shoa, T., Madden, J. D., Fekri, N., Munce, N. R. & Yang, V. X. D. (2008). *Conducting polymer based active catheter for minimally invasive interventions inside arteries*. Proceedings of 30th Annual International Conference of the IEEE Engineering in Medicine and Biology Society, 2008.

Shoa, T., Madden, J. D. W., Munce, N. R. & Yang, V. (2010). Analytical modeling of a conducting polymer-driven catheter. *Polym.Int.*, Vol. 59, No. 3, pp. 343-351, ISSN 1097-0126.

Shoa, T., Yoo, D. S., Walus, K. & Madden, J. D. W. (2011). A Dynamic Electromechanical Model for Electrochemically Driven Conducting Polymer Actuators. *IEEE ASME Trans. Mechatronics*, Vol. 16, No. 1, pp. 42-49, ISSN 1083-4435.

Smela, E., Inganäs, O., Pei, Q. B. & Lundstrom, I. (1993). Electrochemical Muscles - Micromachining Fingers and Corkscrews. *Adv. Mater.*, Vol. 5, No. 9, pp. 630-632, ISSN 0935-9648.

Smela, E., Inganäs, O. & Lundstrom, I. (1995). Controlled Folding of Micrometer-Size Structures. *Science*, Vol. 268, No. 5218, pp. 1735-1738, ISSN 0036-8075.

Smela, E. (1999). A microfabricated movable electrochromic "pixel" based on polypyrrole. *Adv. Mater.*, Vol. 11, No. 16, pp. 1343-1345, ISSN 0935-9648.

Smela, E. & Gadegaard, N. (1999). Surprising volume change in PPy(DBS): An atomic force microscopy study. *Adv. Mater.*, Vol. 11, No. 11, pp. 953-957, ISSN 0935-9648.

Smela, E. & Gadegaard, N. (2001). Volume change in polypyrrole studied by atomic force microscopy. *J. Phys. Chem. B*, Vol. 105, No. 39, pp. 9395-9405, ISSN 1089-5647.

Smela, E., Christophersen, M., Prakash, S. B., Urdaneta, M., Dandin, M. & Abshire, P. (2007). *Integrated cell-based sensors and "cell clinics" utilizing conjugated polymer actuators.* Proceedings of SPIE - The International Society for Optical Engineering.

Smith, E. B. (2004). *Basic Chemical Thermodynamics*, Imperial College Press., ISBN 1-86094-445-0, London.

So, F., Ed. (2010). *Organic electronics : materials, processing, devices and applications.* CRC Press, ISBN 978-1-4200-7290-7, Boca Raton (USA).

Song, M. K., Cho, J. Y., Cho, B. W. & Rhee, H. W. (2002). Characterization of UV-cured gel polymer electrolytes for rechargeable lithium batteries. *J. Power Sources*, Vol. 110, No. 1, pp. 209-215, ISSN 0378-7753.

Spinks, G. M., Liu, L., Wallace, G. G. & Zhou, D. (2002). Strain Response from Polypyrrole Actuators under Load. *Adv. Funct. Mater.*, Vol. 12, No. 6-7, pp. 437-440, ISSN 1616-3028.

Spinks, G. M., Wallace, G. G., Ding, J., Zhou, D., Xi, B. & Gillespie, J. (2003a). *Ionic liquids and polypyrrole helix tubes: bringing the electronic Braille screen closer to reality.* Proceedings of Smart Structures and Materials 2003: Electroactive Polymer Actuators and Devices (EAPAD), San Diego, USA.

Spinks, G. M., Zhou, D. Z., Liu, L. & Wallace, G. G. (2003b). The amounts per cycle of polypyrrole electromechanical actuators. *Smart. Mater. Struct.*, Vol. 12, No. 3, pp. 468-472, ISSN 0964-1726.

Takashima, W., Uesugi, T., Fukui, M., Kaneko, M. & Kaneto, K. (1997). Mechanochemoelectrical effect of polyaniline film. *Synth. Met.*, Vol. 85, No. 1-3, pp. 1395-1396, ISSN 0379-6779.

Takashima, W., Pandey, S. S. & Kaneto, K. (2003). Bi-ionic actuator by polypyrrole films. *Synth. Met.*, Vol. 135, No. 1-3, pp. 61-62, ISSN 0379-6779.

Timoshenko, S. (1925). Analysis of bi-metal thermostats. *J. Opt. Soc. Am.*, Vol. 11, No. 3, pp. 233-255, ISSN 0093-4119.

Torresi, R. M. & Maranhao, S. L. D. (1999). Anion and solvent exchange as a function of the redox states in polyaniline films. *J. Electrochem. Soc.*, Vol. 146, No. 11, pp. 4179-4182, ISSN 0013-4651.

Tsai, E. W., Pajkossy, T., Rajeshwar, K. & Reynolds, J. R. (1988). Anion-Exchange Behavior of Polypyrrole Membranes. *J. Phys. Chem.*, Vol. 92, No. 12, pp. 3560-3565, ISSN 0022-3654.

Valero Conzuelo, L., Arias-Pardilla, J., Cauigh-Rodríguez, J. V., Smit, M. A. & Otero, T. F. (2010). Sensing and Tactile Artificial Muscles from Reactive Materials. *Sensors*, Vol. 10, No. 2638-2674, ISSN 1424-8220.

Valero, L., Arias-Pardilla, J., Smit, M., Cauich-Rodríguez, J. & Otero, T. F. (2010). Polypyrrole Free-Standing electrodes sense temperature or current during reaction. *Polym. Int.*, Vol. 59, No. 337-342, ISSN 1097-0126.

Valero, L., Arias-Pardilla, J., Cauich-Rodríguez, J., Smit, M. A. & Otero, T. F. (2011). Characterization of the movement of polypyrrole-dodecylbenzenesulfonate-

perchlorate/tape artificial muscles. Faradaic control of reactive artificial molecular motors and muscles. *Electrochim. Acta*, Vol. 56, No. 10, pp. 3721-3726, ISSN 0013-4686.

Vidal, F., Popp, J. F., Plesse, C., Chevrot, C. & Teyssie, D. (2003). Feasibility of conducting semi-interpenetrating networks based on a poly(ethylene oxide) network and poly(3,4-ethylenedioxythiophene) in actuator design. *J. Appl. Polym. Sci.*, Vol. 90, No. 13, pp. 3569-3577, ISSN 0021-8995.

Vidal, F., Plesse, C., Palaprat, G., Juger, J., Citerin, J., Kheddar, A., Chevrot, C. & Teyssie, D. (2009). Synthesis and Characterization of IPNs for Electrochemical Actuators. In: *Artificial Muscle Actuators Using Electroactive Polymers*, Vincenzini, P., BarCohen, Y. & Carpi, F. pp. 8-17, Trans Tech Publications Ltd, ISBN 978-3-908158-27-1, Stafa-Zurich.

Vidal, F., Plesse, C., Aubert, P. H., Beouch, L., Tran-Van, F. o., Palaprat, G., Verge, P., Yammine, P., Citerin, J., Kheddar, A., Sauques, L., Chevrot, C. & Teyssie, D. (2010). Poly(3,4-ethylenedioxythiophene)-containing semi-interpenetrating polymer networks: a versatile concept for the design of optical or mechanical electroactive devices. *Polym. Int.*, Vol. 59, No. 3, pp. 313-320, ISSN 1097-0126.

Vlad, A., Dutu, C. A., Jedrasik, P., Sodervall, U., Gohy, J. F. & Melinte, S. (2012). Vertical single nanowire devices based on conducting polymers. *Nanotechnology*, Vol. 23, No. 2, pp. 5, ISSN 0957-4484.

Wallace, G., Campbell, T. & Innis, P. (2007). Putting function into fashion: Organic conducting polymer fibres and textiles. *Fiber. Polym.*, Vol. 8, No. 2, pp. 135-142, ISSN 1229-9197.

Wang, H. L., Gao, J. B., Sansiñena, J. M. & McCarthy, P. (2002). Fabrication and characterization of polyaniline monolithic actuators based on a novel configuration: Integrally skinned asymmetric membrane. *Chem. Mater.*, Vol. 14, No. 6, pp. 2546-2552, ISSN 0897-4756.

Wilson, S. A., Jourdain, R. P. J., Zhang, Q., Dorey, R. A., Bowen, C. R., Willander, M., Wahab, Q. U., Al-hilli, S. M., Nur, O., Quandt, E., Johansson, C., Pagounis, E., Kohl, M., Matovic, J., Samel, B., van der Wijngaart, W., Jager, E. W. H., Carlsson, D., Djinovic, Z., Wegener, M., Moldovan, C., Iosub, R., Abad, E., Wendlandt, M., Rusu, C. & Persson, K. (2007). New materials for micro-scale sensors and actuators: An engineering review. *Mat. Sci. Eng. R*, Vol. 56, No. 1-6, pp. 1-129, ISSN 0927-796X.

Woosoon, Y., Joonsoo, L. & Kwang, J. K. (2007). An artificial muscle actuator for biomimetic underwater propulsors. *Bioinspir. Biomim.*, Vol. 2, No. 2, pp. S31, ISSN 1748-3190.

Wu, Y., Zhou, D., Spinks, G. M., Innis, P. C., Megill, W. M. & Wallace, G. G. (2005). TITAN: a conducting polymer based microfluidic pump. *Smart. Mater. Struct.*, Vol. 14, No. 6, pp. 1511-1516, ISSN 0964-1726.

Yang, F., Xiaobo, T. & Alici, G. (2008). Robust Adaptive Control of Conjugated Polymer Actuators. *IEEE Trans. Control Syst. Technol.*, Vol. 16, No. 4, pp. 600-612, ISSN 1063-6536.

Yao, Q., Alici, G. & Spinks, G. A. (2008). Feedback control of tri-layer polymer actuators to improve their positioning ability and speed of response. *Sens. Actuat. A-Phys*, Vol. 144, No. 1, pp. 176-184, ISSN 0924-4247.

Zainudeen, U. L., Careem, M. A. & Skaarup, S. (2008). PEDOT and PPy conducting polymer bilayer and trilayer actuators. *Sens. Actuat. B-Chem*, Vol. 134, No. 2, pp. 467-470, ISSN 0925-4005.

Chapter 13

Novel Self-Oscillating Polymer Actuators for Soft Robot

Yusuke Hara, Shingo Maeda,Takashi Mikanohara, Hiroki Nakagawa, Satoshi Nakamaru and Shuji Hashimoto

Additional information is available at the end of the chapter

http://dx.doi.org/10.5772/52474

1. Introduction

Intelligent materials with changing properties and functions have been studied in many fields [1-9]. In particular, stimuli-responsive polymer systems have been considerably investigated for the purpose of the many types of possible applications, such as soft actuators, microfluidics and medical devices, *etc.* due to their light weight, flexibility and low noise, *etc.* [10-18]. Polymer gels have the flexible bodies allowing volume changes and sensors properties for chemical substances at the same time. By utilizing these properties, the gel actuators have been considered to be applied for the biomimetic robots. In particular, a thermoresponsive poly(N-isopropylacrylamide) (PNIPAAm) gel have been significantly investigated especially for medical devices. [19, 20] However, in order to drive these polymer materials, external device for controlling the external stimulus must be needed. On the other hand, organic systems can generate autonomous motion without external stimuli. In recent years, self-oscillating gels and polymer chains were developed and researched in order to understand and mimic the spontaneous mechanical works. [21-27] The energy source of self-oscillating polymer systems is chemical energy like a living organism, that is, the polymer system cause the self-oscillation induced by the Belouzov-Zhabotinsky (BZ) reaction or the pH oscillating reaction. [28-43] The Self-oscillating reactions have been widely studied for their relevance to nonlinear, chaotic dynamics and biological signals. In previous investigations, the self-oscillating behaviors of polymer systems induced by the Belousov-Zhabotinsky (BZ) or the pH oscillating reactions were realized.[44-60] The BZ reaction is well known as an oscillating reaction accompanying spontaneous redox oscillations to generate a wide variety of nonlinear phenomena, e.g., a target or spiral pattern in an unstirred solution, and periodicity, multi-periodicity, or chaos in a stirred solution. The overall process of the BZ reaction is the oxidation of an organic substrate by an oxidizing agent in the presence of the catalyst under strong acidic conditions. In the BZ

reaction, the metal catalyst undergoes spontaneous redox self-oscillation, and it changes the solubility from hydrophilic to hydrophobic at the same time. In order to cause the self-oscillation under temperature constant condition, the different solubility of the Ru catalyst in the reduced and oxidized state is utilized. In previous investigations, poly(Nisopropylacrylamide) (poly(NIPAAm)) was covalently bonded to (ruthenium (4-vinyl-4′-methyl-2,2-bipyridine) bis(2,2′-bipyridine)bis(hexafluorophosphate)) ($Ru(bpy)_3$) and a negatively charged acrylamide-2-methylpropanesulfonic acid (AMPS) as a pH and solubility control site [47-50]. The AMPS-containing polymer systems caused self-oscillations originating from the different solubility of the $Ru(bpy)_3$ moiety in the oxidized and reduced state. However, the conventional-type self-oscillating polymer system has the temperature limitation for causing the self-oscillation. That is because the conventional-type self-oscillating polymer system causes the aggregation or the contraction above the LCST. In this chapter, we introduce novel type self-oscillating polymers systems that are constituted the non-thermoresponsive and biocompatible poly-vinylpyrrolidone (PVP) as a polymer main-chain in order to expand the application fields. In our previous investigations, we first succeeded in causing the aggregation-disaggregation self-oscillation of the novel polymer chain and polymer gel under the constant condition induced by the BZ reaction. As for the VP-based self-oscillating polymer chain, the influence of the concentration of the three BZ substrates (sodium bromate, malonic acid and nitric acid) other than the metal catalyst on the waveform and period of the self-oscillation was investigated. As a result, it was demonstrated that the amplitude of the self-oscillation is hardly affected by the initial concentration of the BZ substrates. Furthermore, we firstly succeeded in causing the swelling-deswelling self-oscillation at high temperature condition by adapting the VP-main chain. We studied the influence of the initial concentration of the three BZ substrates other than the metal catalyst and the temperature on the period of the self-oscillation. As a result, it was clarified that the period of the self-oscillation for the VP-based polymer gel can be controllable by the selection of the initial concentration of the three BZ substrates (malonic acid (MA), sodium bromated ($NaBrO_3$) and nitric acid) and the temperature. In addition, by optimizing the initial concentration of the BZ substrates and the temperature, it was demonstrated that the maximam frequency of the swelling-deswlling self-oscillation is 0.5 Hz. This value was 20 times as large as that of the conventional-type self-oscillating polymer gel (poly(NIPAAm-co-$Ru(bpy)_3$ gel) with the thermoresponsive nature.

Next, in this chapter, we introduce the challenge of mimicking the peculiar locomotion of living organism, such as mollusks and apods, by utilizing polymer gel that cause the swelling-deswelling induced by the pH oscillating chemical reaction. Gastropods, like snails and slugs, can obtain the peristaltic movement by forming contraction waves, which is the propagation of the shrinking part of the body [61]. In this research, we focused on producing a functional gels obtaining peristaltic movements by the chemical energy directly.

In addition, we introduce a novel-tpye nanofiber gel actuator that was manufactured by the electrospining method. The nanofiber gel actuator can cause bending-streching motion

synchronized with the pH oscilltion. Polymer nanofibers have been much studied for the application to various fields, such as chemical detector, scaffolds, wound dressing, and multifunctional membranes, sensors, filtration media and biomaterials, *etc.* [62-65]. That is because the nanofiber gel membranes have a high surface-to-volume ratio and a relatively small pore size compared to normal-type polymer hydrogel. The merit of the electrospining method to manufacutre nanofiber gel actutors is low cost, relatively high production rate, and applicability to many types of polymers. Moreover, by utilizing the electrospinning method, we can construct various shapes of nanofiber gel actuators because it does not require a mold to synthesize the gel. By controlling the external conditions, we can construct more complex structures such as porous fiber, core sheathe fiber, *etc.* [66-69]. Recently, some researchers succeeded in manufacturing functional nanofiber gels by utilizing electrospining method. For example, Hsieh and coworkers have reported ultrafine gel fibrous membranes that consist of poly(acrylic acid) (PAAc), poly(N-isopropylacrylamide-*co*-acrylic acid) and poly(vinyl alcohol)(PVA) and PVA/PAAc mixtures [70-72]. However, there are a few challenges to realize the nanofiber gel actuator [73,74]. In this study, in order to construct an autonomous nanofiber gel actuator, we utilized Landolt pH-oscillator based on a bromated/ sulfite/ferrocyanide reaction. By coupling with this pH-oscillator, we realized a nanofiber gel actuator that shows the bending and stretching motions with a constant period and displacement.

2. Experimental section

2.1.1. Synthesis of the poly(VP-co-Ru(bpy)3)

The polymer chain was prepared as follows. 0.5g of ruthenium(4-vinyl-4′-methyl-2,2′-bipylridine)bis(2,2′-bipyridine)bis(hexafluorophosphate) (Ru(bpy)3) as a metal catalyst for the BZ reaction, 9.5 g of vinylpyrrolidone (VP) and 0.35 g of 2,2′-azobis(isobutyronitrile) (AIBN) as an initiator were dissolved in the methanol solution (31 g) under total monomer concentration of 20 wt%. These polymerizations were carried out at 60 °C for 24 h *in vacuo*. These resulting reaction mixtures were dialyzed against graded series of water/methanol mixtures, for 1 day each in 0, 25, 50, 75, and 100 wt% of water, and then freeze-dried.

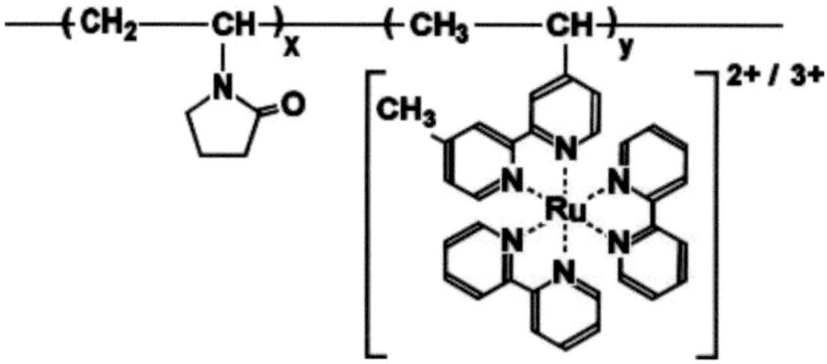

Figure 1. Chemical structure of poly(VP-*co*-Ru(bpy)3).

2.1.2. Measurement of Lower Critical Solution Temperature (LCST) for the Poly(VP-co-Ru(bpy)$_3$) solution

The lower critical solution temperature (LCST) of the polymer solution was measured under the reduced and oxidized states, by using $Ce(SO_4)_2$ as an oxidizing agent and $Ce_2(SO_4)_3$ as a reducing agent, respectively. The polymer solutions (0.5 wt%) of poly(VP-*co*-Ru(bpy)$_3$) were prepared by dissolving the polymer in a 0.3 M HNO_3 aqueous solution and adding 5 mM $Ce(SO_4)_2$ or 5 mM $Ce_2(SO_4)_3$, respectively. The LCST measurements were carried out with a spectrophotometer (JASCO, Model V-630) equipped with magnetic stirrers and a thermostatic controller. In this measurement, the 570 nm wavelength was used because it is the isosbestic point for the reduced and oxidized states of Ru(bpy)$_3$. The transmittance (%) of the polymer solution at 570 nm was then recorded by raising the temperature at a rate of 0.5 °C/min.

2.1.3. Measurement of optical oscillations for the poly(VP-co-Ru(bpy)$_3$) solution

The poly(VP-*co*-Ru(bpy)$_3$) solutions were prepared by dissolving the polymer (0.5 wt%) into an aqueous solution containing the three BZ substrates (malonic acid (MA) and sodium bromate ($NaBrO_3$), nitric acid (HNO_3)). The transmittance self-oscillations of the polymer solutions were measured under constant temperature and stirring. In order to detect the transmittance change which is based on the autonomous transmittance change, 570-nm wavelength was used. The time course of the transmittance at 570 nm was monitored by a spectrophotometer.

2.2.1. Synthesis of the poly(VP-co-Ru(bpy)$_3$) gel

The gel was prepared as follows. 0.110g of Ru(bpy)$_3$ as a metal catalyst for the BZ reaction was dissolved in 0.877g of vinylpyrrolidone (VP). 0.012g of N,N′-methylenebisacrylamide (MBAAm) as a cross-linker, and 0.020g of 2,2′-azobis(isobutyronitrile) (AIBN) as an initiator were dissolved in the methanol solution (3ml) (Figure 1). The two solutions were mixed together well, and then the mixed solution purged with dry nitrogen gas. The monomer solution was injected between Teflon plates separated by silicone rubber as a spacer (thickness: 0.5mm), and then polymerized at 60°C for 18 hours. After gelation, the gel strip was soaked in pure methanol for a week to remove unreacted monomers. The gel was carefully hydrated through dipping it in a graded series of methanol-water mixtures, for 1 day each in 75, 50, 25 and 0%.

2.2.2. Measurements of the equilibrium swelling ratio of poly(VP-co-Ru(bpy)$_3$) gels

The equilibrium swelling ratio of the gels was measured under the reduced and the oxidized state, by using the oxidizing and the reducing agents. Gels were put into two solutions of Ce(III) and Ce(IV) under the same acidity, $[Ce_2(SO_4)_3]$=0.001M and $[HNO_3]$=0.3M; $[Ce_3(SO_4)_2]$=0.001M and $[HNO_3]$=0.3M, respectively. The equilibrium swelling ratio of the gels was observed and recorded by using a microscope (Fortissimo Corp.WAT-

250D), a LED light (LEDR-74/40 W), and a video recorder (Victor corp. SR-DVM700). The analysis was conducted by using the image processing software (Image J 1.38x). Measurements of the equilibrium swelling ratio were performed in a water-jacketed cell made of acrylic plates.

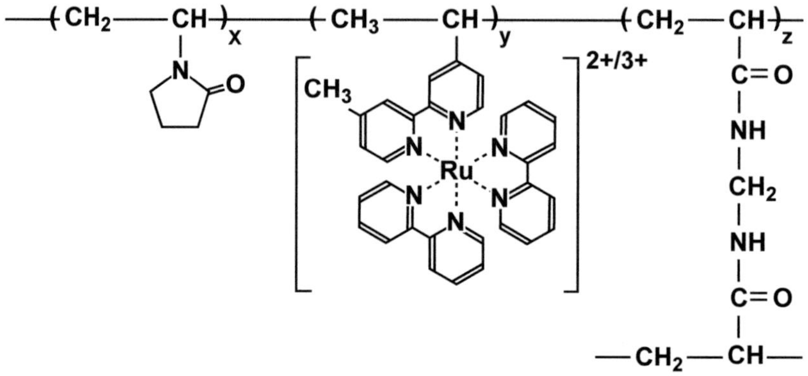

Figure 2. Chemical structure of poly(VP-co-Ru(bpy)3) gel.

2.2.3. *Measurements of the oscillating behavior for poly(Vp-co-Ru(bpy)3) gels*

The gel membrane was cut into rectangles(side length, about 2 × 20 mm) and immersed into 8mL of an aqueous solution containing malonic acid (MA), sodium bromate ($NaBrO_3$), and nitric acid (HNO_3). Shape changes of the gel strip were observed and recorded using a microscope (OLYMPUS.IX71), and a video recorder (Victor Corp. SR-DVM700). The analysis was conducted by using the image processing software (Image J 1.38x). A one-pixel line along the length of recorded gel image was stored at regular time intervals (0.05 s). The stored pixel line images were sequentially lined up as a function of time by the computer. This image processing procedure constructs a spatio-temporal diagram. From the obtained diagram, the time-dependent change in the gel edge was traced for observing the behavior of the volume change.

2.3.1. *Synthesis of the poly(AAm-co-AAC) gel*

The pH-responsive cylindrical hydrogels were prepared as follows. 0.533 g of acrylamide (AAm), 54.0 mg of acrylic acid (AAc), 12.8mg of N,N′-methylenebisacrylamide (MBAAm) as a cross-linker, and 22.6 mg of 2,2'-Azobis (2-methylpropionamidine) dihydrochloride as an initiator were dissolved in the O_2 free pure water (3.0 mL). The gelation was carried out in glass tubes (internal diameters from 1.0 mm to 2.0 mm) at 50 °C for 21 h. After removing from their molds, the cylindrical gels were cut in pieces 3.0 mm long and hydrated in pure water.

The pH-responsive cylindrical microphase-separated gels were prepared as follows. The composition of monomers is same as the cylindrical hydrogels, and they were dissolved in the solutions of O_2 free water/acetone mixtures (80/20, 70/30, 60/40, and 50/50 wt %). The polymerization was carried out in glass tubes (the size is same as above) at 50 °C for 21 h.

After removing from their molds, the cylindrical gels were also cut in pieces 3.0 mm long and hydrated in a graded series of methanol-water mixtures for 1 day each in 75, 50, 25, and 0 vol % methanol.

The pH-responsive microphase-separated tubular gel was prepared as follows. The composition of monomers is same as the cylindrical microphase-separated gels, and they were dissolved in the water/acetone mixture solution (70/30 wt %). The mold was made by coupling different size of glass tubes (external/internal diameters are 2.6/2.0, 1.0/0.6 mm), and the monomer solution was flowed into the tube-shaped space (external/internal diameters are 2.0/1.0 mm). The tubular gel was cut in pieces 4.0 cm long and hydrated in a graded series of methanol-water mixtures for 1 day each in 75, 50, 25, and 0 vol % methanol.

2.3.2. SEM Observation of the gels

Two samples were prepared in the following way. The microphase-separated gel cylinders were cut into pieces and soaked in acid (pH= 2) and base (pH = 11) solutions for 1 day until reaching the equilibrium. Then the gels were quickly frozen in liquid nitrogen and freeze-dried under vacuum for 1 day. The freeze-dried samples were fixed on the aluminum stubs and coated with the gold for 30 s under vacuum. The gel network structures were observed by the SEM (scanning electron microscope, HITACHI S2500CX).

2.3.3. Measurement of phase transition kinetics

In our measurement of the gels kinetics, two acid-base solutions with low and high pH (pH = 2 and pH = 11) were used. A piece of cylindrical gel under the equilibrium swelling in one solution was transferred quickly into the other. The changes of the cylindrical gel diameters were monitored by using a microscope equipped with a CCD camera controlled by a computer. The obtained images were analyzed by image processing software.

2.3.4. Experimental apparatus for causing the CT reaction in the pH-responsive tubular gel

The feeding solutions were stored in three separated reservoirs containing an alkaline sodium chlorite solution ([$NaClO_2$] = 1.2×10^{-1} M, [NaOH] = 1.5×10^{-4} M), an alkaline potassium tetrathionate solution ([$K_2S_4O_6$] = 3.0×10^{-2} M, [NaOH] = 1.5×10^{-4} M), and a sodium hydroxide solution ([NaOH] = 1.3×10^{-2} M). The solutions were pumped by peristaltic pumps (ATTO SJ-1211L) and premixed by a magnetic stirrer (AS-ONE CT-3A) before entering the tubular gel. The flows of chlorite and tetrathionate solutions were maintained at 30 mL/h, and the mixed solution flowed into the tubular gel by a peristaltic pump (ATTO SJ-1211H). The tubular gel was soaked in the mixed CT solution for 30 min in advance. After entering the flow of the mixed solution, an acid perturbation was applied by touching the gel at a particular spot with a paper soaked in a 1 M H_2SO_4 solution. The behavior of the tubular gel was observed by using a microscope equipped with a CCD camera controlled by a computer. Methyl red, a color indicator that changes from yellow

(basic) to red (acidic), was added to the solutions to directly visualize the propagation of acid regions. Fig. 3 shows the schematic illustration of experimental apparatus.

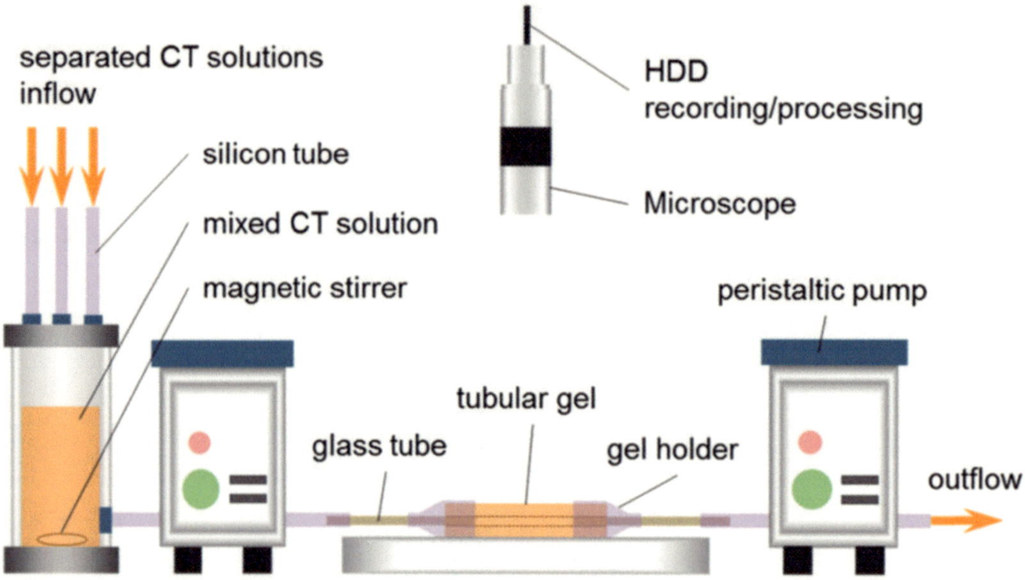

Figure 3. Schematic illustration of experimental apparatus for causing the CT reaction in the pH-responsive tubular gel. (Reprinted ref. 58, Copyright *IEEE*. Reproduced with permission.)

2.4.1. *Preperation of Poly(AAc-co-tBMA)*

Using AAc (25.9 g), tert- butyl methacrylate (tBMA) (34.1 g), and 2,2′-azobisiso-butyronitrile (AIBN) (0.49 g) as an initiator, *poly(AAc-co-tBMA)* (Figure 4) was synthesized by the radical polymerization in ethanol (139.5 g) using a total monomer concentration of 30 wt%. The molar ratio of tBMA incorporated into the copolymer was 30 mol%. The polymerization was carried out at 60 °C under nitrogen flow for 24 h. The resulting reaction mixture was dialyzed against ethanol for seven days.

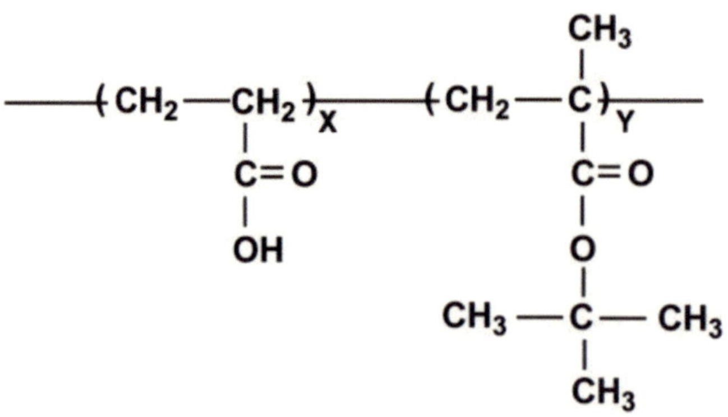

Figure 4. The chemical structure of poly(AAc-*co*-tBMA) polymer.

2.4.2. Electrospinning

The polymer solution (18 wt%) was poured into a 2.5 mL syringe. A potential of 10 kV was applied by connecting the power supply (GT80 GREEN TECHNO) to the syringe tip (Figure 5). In order to introduce anisotropic structure into the nanofiber gel, the 1.0 mL of the polymer solution in the syringe was sprayed at a flow rate of 2.0 mL/hour (sprayed for 30 minutes), and then the flow rate was changed to 1.0 mL/hour (sprayed for 60 minutes). The fibers were collected on the grounded glass substrate as a collector. The distance between the collector and the syringe tip was 15 cm. The temperature and humidity were 25 °C and 70%, respectively. After the electrospinning, the obtained sheet, with a thickness of about 200 μm, was dried overnight at 50 °C.

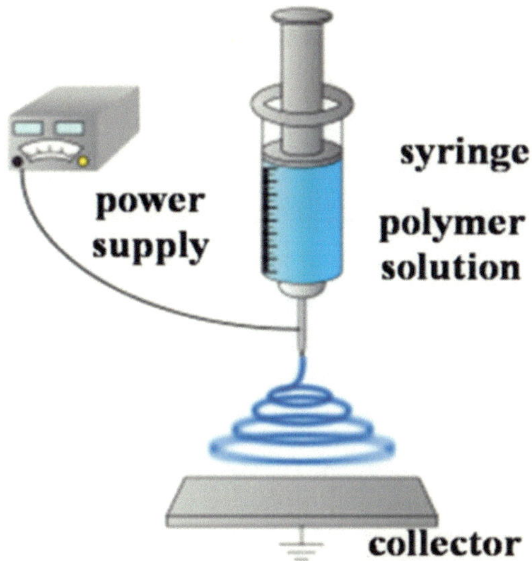

Figure 5. Schematic illustration of electrospinning set-up.

2.4.2. Measurement of motion of the nanofiber gel actuator

The open continuously stirred tank reactor (CSTR) (40 mL) was designed using an acrylic cell with a water jacket in order to control the solution temperature in the cell. Potassium bromated (0.26 mol/L), sodium sulfite (0.3 mol/L), potassium ferrocyanide (0.08 mol/L), and sulfuric acid (0.04 mol/L), solutions were pumped into the reactor at a flow rate of 50 mL/hour. The pH changes in the reactor were monitored continuously by utilizing a pH meter (F-55 HORIBA) held in the reactor, and its electronic output was directly recorded by a computer. The nanofiber gel (length 15 mm, width 3 mm) was set at the bottom of the water jacket. One end of the gel strip was sandwiched in the incision of the silicone rubber. Shape changes of the gel strip were recorded by a fixed microscope (Fortissimo Corp. WAT-250D) and a video recorder (Victor Corp. SR-DVM700). The temperature in the reactor was controlled at 25 °C by utilizing the water bath equipment.

3. Results and discussion

3.1. Self-oscillating behavior of poly(VP-*co*-Ru(bpy)$_3$) solution

The measurement of the solubility for the poly(VP-*co*-Ru(bpy)$_3$) solution in the reduced and oxidized state was conducted by changing the temperature from 10 to 50 °C. As shown in Figure 6, there are no the lower critical solution temperature (LCST) for the polymer solutions in the reduced and oxidized state. The polymer solutions in the reduced and oxidized state have different transmittance values. That is, this result indicates that the polymer solution has different solubility in the reduced and oxidized state, respectively. The solubility of the polymer solution in the reduced state is higher than that in the oxidized state. Figure 7 shows self-oscillating behaviors of the poly(VP-*co*-Ru(bpy)$_3$) solution in the different concentrations of malonic acid ([MA] = 0.04, 0.05, 0.06, 0.07, 0.08 and 0.09 M) under the fixed concentration of sodium bromate and nitric acid ([$NaBrO_3$] = 0.3 M and [HNO_3] = 0.3 M). As shown in Figure 7, the base line of the transmittance self-oscillation gradually decreased with time in all malonic acid concentrations. The damping behavior originates from the change in the ionic strength of the polymer solution when the transmittance measurement starts [47-49]. In order to cause the BZ reaction in the polymer solution, the self-oscillating polymer solution and the other solution of the BZ substrates are mixed just before the transmittance measurements. In general, the solubility of the polymer chain is significantly affected by the ionic strength of the solution. Therefore, when the ionic strength increased at the start point of the self-oscillation, the solubility of the polymer chain decreased. In the solution condition of this study, the ionic strength of the polymer solution was very high because the BZ reaction required a significant high concentration of the BZ substrates. Therefore, the damping behaviors occurred from the start point of the self-oscillation. In addition, as shown in Figure 7, the width of the waveform increased with decreasing the concentration of malonic acid. Basically, the width of the waveform of the transmittance self-oscillation depends on the rate of the BZ reaction because the self-oscillation was induced by the BZ reaction. As the concentration of the BZ substrates decreased, the rate of the BZ reaction decreased due to decrease in the collision rate among the BZ substrates. Therefore, the width of the waveform increased with decreasing the concentration of the BZ substrates. This tendency was observed in the transmittance self-oscillation of the AMPS-containing polymer solutions [48].

Figure 8 shows the transmittance self-oscillations of the novel polymer solution in the different concentration of sodium bromate ([$NaBrO_3$] = 0.1, 0.2, 0.3 and 0.4 M) at 14 °C under the fixed concentration of malonic acid and nitric acid ([MA] = 0.1M and [HNO_3] = 0.3 M). As shown in Figure 8, the amplitude of the self-oscillation gradually decreased with time in the same manner as in Figure 7. Moreover, the width of the waveform decreased with the increase in the concentration of sodium bromate due to the increase in the reaction rate of the BZ reaction. In addition, as shown in Figure 8, the amplitudes of the transmittance self-oscillations were hardly affected by the initial concentration of sodium bromate. In additon, Figure 9 showed the amplitude of the transmittance self-oscillation for the polymer solution under the different concentrations of the BZ substrates. As shown in Figure 9, all the BZ

substrate concentrations hardly influence the amplitude of the transmittance self-oscillation. That is, the amplitude values were almost the same in all BZ substrate conditions. In our previous investigations, we studied the effect of the concentration of the BZ substrates on the waveform of the transmittance self-oscillation for the AMPS-containing polymer solution. As a result, we clarified that the amplitude of the self-oscillation is significantly affected by the initial concentration of the BZ substrates [47-49]. This is because the AMPS-containing polymer chain caused damping, that is, the amplitude of the self-oscillation gradually decreased with time. The damping behavior of the polymer solution originates from the change in the size of the polymer aggregation with time. In the case of the NIPAAm-based polymer chains, the reduced Ru moiety in the polymer chain strongly interacts with the other reduced Ru one. Once the reduced Ru moiety strongly interacts with the other Ru one, the interaction hardly dissociates [48]. In the BZ reaction, the time in the reduced state is much longer than in the oxidized state. Therefore, the hydrophobic $Ru(bpy)_3^{2+}$ moiety in the polymer chain dominantly behaved for the determination of the polymer aggregation state in the self-oscillating behavior. For this reason, the mole fraction of the $Ru(bpy)_3^{2+}$ moiety in the polymer chain significantly affect the waveform of the transmittance self-oscillation. This influence can be explained by the overall process of the BZ reaction based on the Field-koros-Noyes (FKN) mechanism [33-36, 48]. On the other hand, in the case of the VP-based polymer chain, the bipyridine ligands interacted with the VP based main-chain. The strength of the interaction of the bipyridine ligands in the oxidized state is higher than in the reduced state. That is, the polymer aggregation increased in the oxidized state. However, the time in the oxidized state is much shorter than in the reduced state. Therefore, the size of the polymer aggregation changed very slowly compared to the AMPS-containing polymer solution. Consequently, the degree of the damping for the Vp-based polymer solution is considerably small. Therefore, the amplitude is hardly affected by the initial concentration of the BZ substrates.

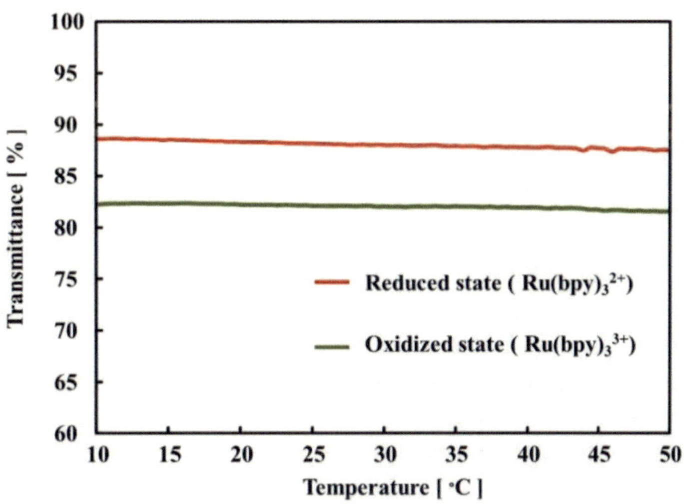

Figure 6. Temperature dependence of transmittance for poly(VP-*co*-$Ru(bpy)_3$) solutions under the different conditions of reduced Ru(II) (in Ce(III) solution) and Ru(III) (in Ce(IV) solution) states.

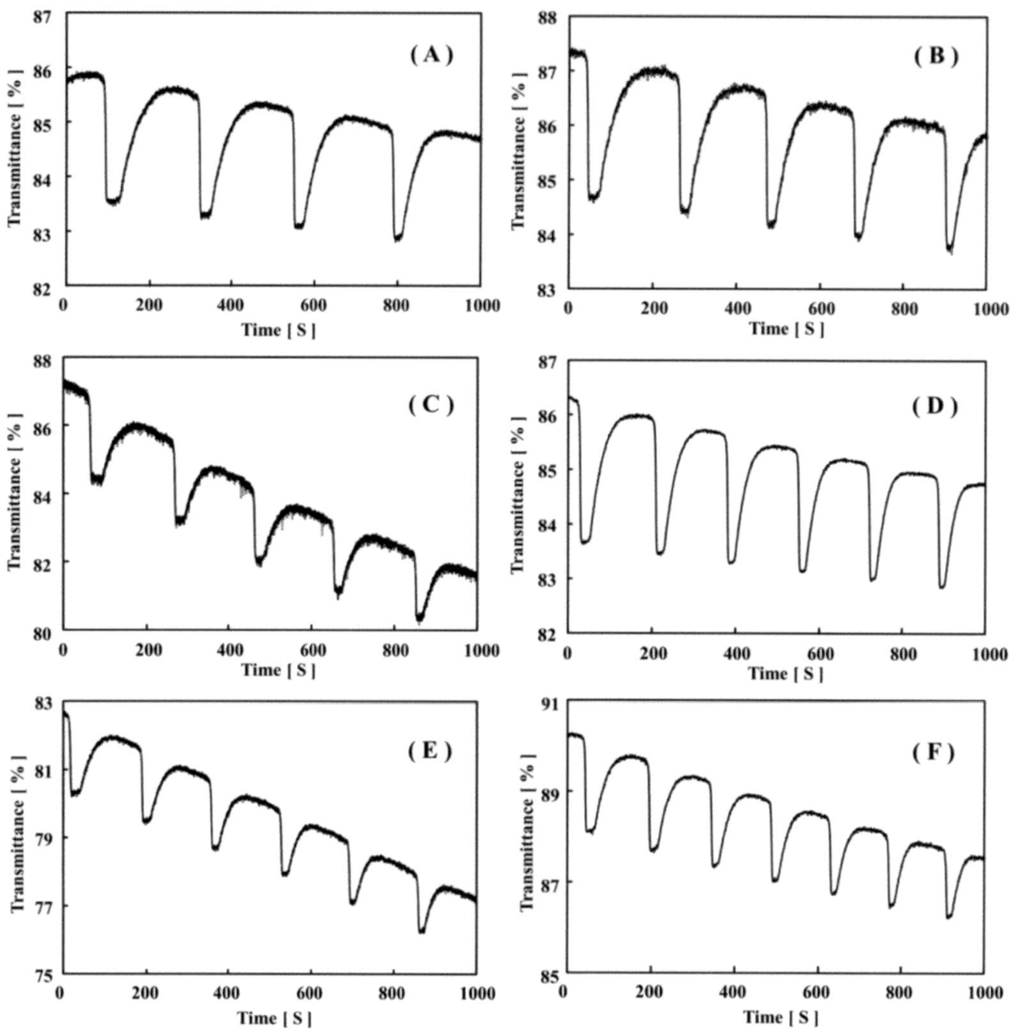

Figure 7. Oscillating profiles of transmittance at 14 °C for 0.5 wt% poly(VP-*co*-Ru(bpy)$_3$) solution in the fixed nitric acid and sodium bromate conditions ([HNO_3] = 0.3 M and [$NaBrO_3$] = 0.3 M): (**A**) [MA] = 0.04 M, (**B**) [MA] = 0.05 M, (**C**) [MA] = 0.06 M, (**D**) [MA] = 0.07 M, (**E**) [MA] = 0.08 M, (**F**) [MA] = 0.09 M.

3.2. Self-oscillating behavior of poly(VP-*co*-Ru(bpy)$_3$) gel

Figure 10 shows the equilibrium swelling behaviors of the poly(VP-*co*-Ru(bpy)$_3$) gels in the Ce(III) and Ce(IV) solutions under the same acidic condition. In the Ce(III) solution, the gel kept a tinge of orange, which indicated that the copolymerized Ru(bpy)$_3$ moiety in the gel was in the reduced state. On the other hand, in the Ce(IV) solution, the gel quickly turned from orange to green, which showed the Ru(bpy)$_3$ moiety in the gel had changed the reduced state to the oxidized state. In the oxidized state, the equilibrium volume of the gel was larger than that in the reduced state in all temperature condition. This is because the solubility of the Ru(bpy)$_3$ moiety has significantly different properties in the oxidized and

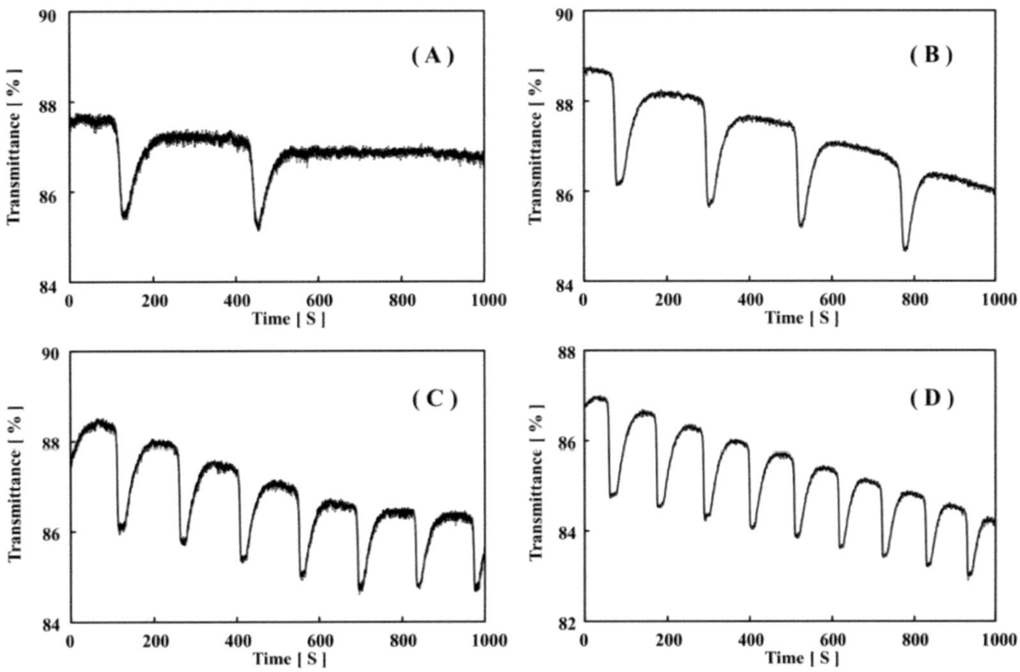

Figure 8. Oscillating profiles of transmittance at 14 °C for 0.5 wt% poly(VP-*co*-Ru(bpy)3) solution in fixed nitric acid and malonic acid conditions ([HNO_3] = 0.3 M, [MA] = 0.1 M) (**A**) [$NaBrO_3$] = 0.1 M, (**B**) [$NaBrO_3$] = 0.2 M, (**C**) [$NaBrO_3$] = 0.3 M, (**D**) [$NaBrO_3$] = 0.4 M.

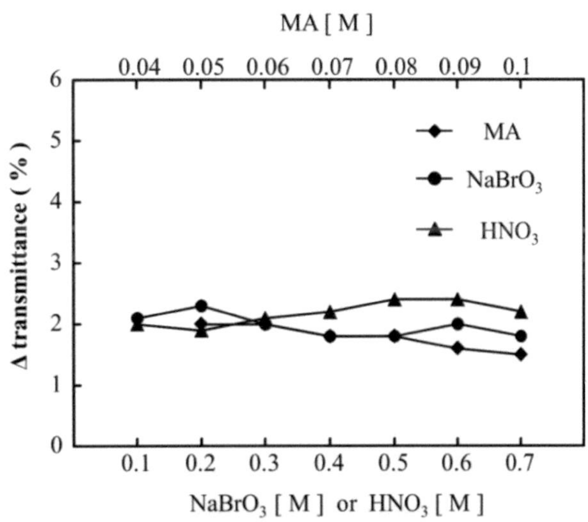

Figure 9. Dependence of amplitude of transmittance self-oscillation for polymer solution at 14 °C in the change in one BZ substrate under fixed concentrations of the other two BZ substrates: **MA** ([MA] = 0.04, 0.05, 0.06, 0.07, 0.08 and 0.09 M, fixed [$NaBrO_3$] = 0.3 M and [HNO_3] = 0.3 M); **$NaBrO_3$** ([$NaBrO_3$] = 0.1, 0.2, 0.3, 0.4, 0.5 and 0.6 M, fixed [MA] = 0.1 M and [HNO_3] = 0.3 M); **HNO_3** ([HNO_3] = 0.1, 0.2, 0.3, 0.4, 0.5 and 0.6 M, fixed [$NaBrO_3$] = 0.1 M and [HNO_3] = 0.3 M).

reduced state. The reduced Ru(bpy)3 moiety in the gel has an extreme hydrophobic property in the VP-based polymer gel. This property is attributed to the conformation of the bipyridine ligands surrounding the Ru ion, which induces the deswelling behavior. That is, in the VP-based gel, the bipyridine ligands surrounding the Ru ion exert a greater influence on the solubility of the polymer chain in the reduced state as compared with the ionization effect of the Ru ion.[47-52] On the contrary, the oxidized Ru(bpy)3 parts in the gel has a great hydrophilic property. The driving force of the swelling-deswelling self-oscillation is originated in the different solubility of the Ru(bpy)3 moiety in the reduced and oxidized states as shown in Figure 10. In the reduced and oxidized state, there is no observation of the volume phase transition because of the PVP main chain of the gel without LCST.

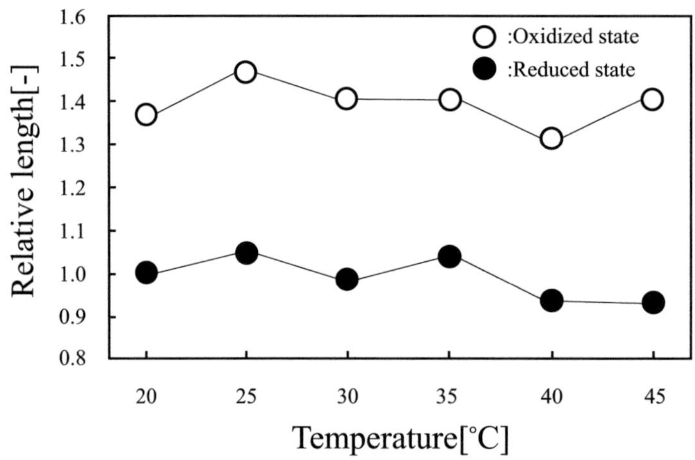

Figure 10. Equilibrium swelling ratio of poly(VP-*co*-Ru(bpy)3) gel in cerium sulfate solutions as a function of temperature. (●) $[Ce_2(SO_4)_3]$ = 0.001M and $[HNO_3]$ = 0.3M; (○) $[Ce(SO_4)_2]$ = 0.001M and $[HNO_3]$ = 0.3M. The relative length is defined as the ratio of characteristic diameter at the initial state at 20°C. (Reprinted ref. 57, Copyright American Chemical Society. Reproduced with permission.)

Figure 11 showed the logarithmic plots of the period against the initial concentration of one BZ substrate under fixed concentration of the other two BZ substrates at a constant temperature (T=20 °C). As shown in Figure 11, all the logarithmic plots had a good linear relationship. Therefore, the period [T(s)] of the swelling-deswelling self-oscillation can be expressed as $a[\text{substrate}]^b$ where a and b are the experimental constants, and the brackets indicate the initial concentration. Moreover, as shown in Figure 11, the period of the self-oscillation have the saturation point at the following initial concentration: [MA] = 0.07M (Figure 11(a)), $[NaBrO_3]$ = 0.5M (Figure 11(b)) and $[HNO_3]$ = 0.7M (Figure 11(c)). The period at the saturated point in Figure 11(a) was significantly higher than that in Figure 11(b) and Figure 11(c). This tendency can be explained by considering the mole fraction of the reduced Ru(bpy)3 moiety in the gel. This is because the reduced Ru(bpy)3 moiety in the gel has the significantly high hydrophobic property. Therefore, the number of the hydrophobic reduced Ru(bpy)3 moiety in the gel exert influence of the self-oscillating behavior. The Field-koros-Noyes (FKN) mechanism explained the overall process of the BZ reaction. [33-36, 48] According to the FKN mechanism, the overall reaction is divided into the following three

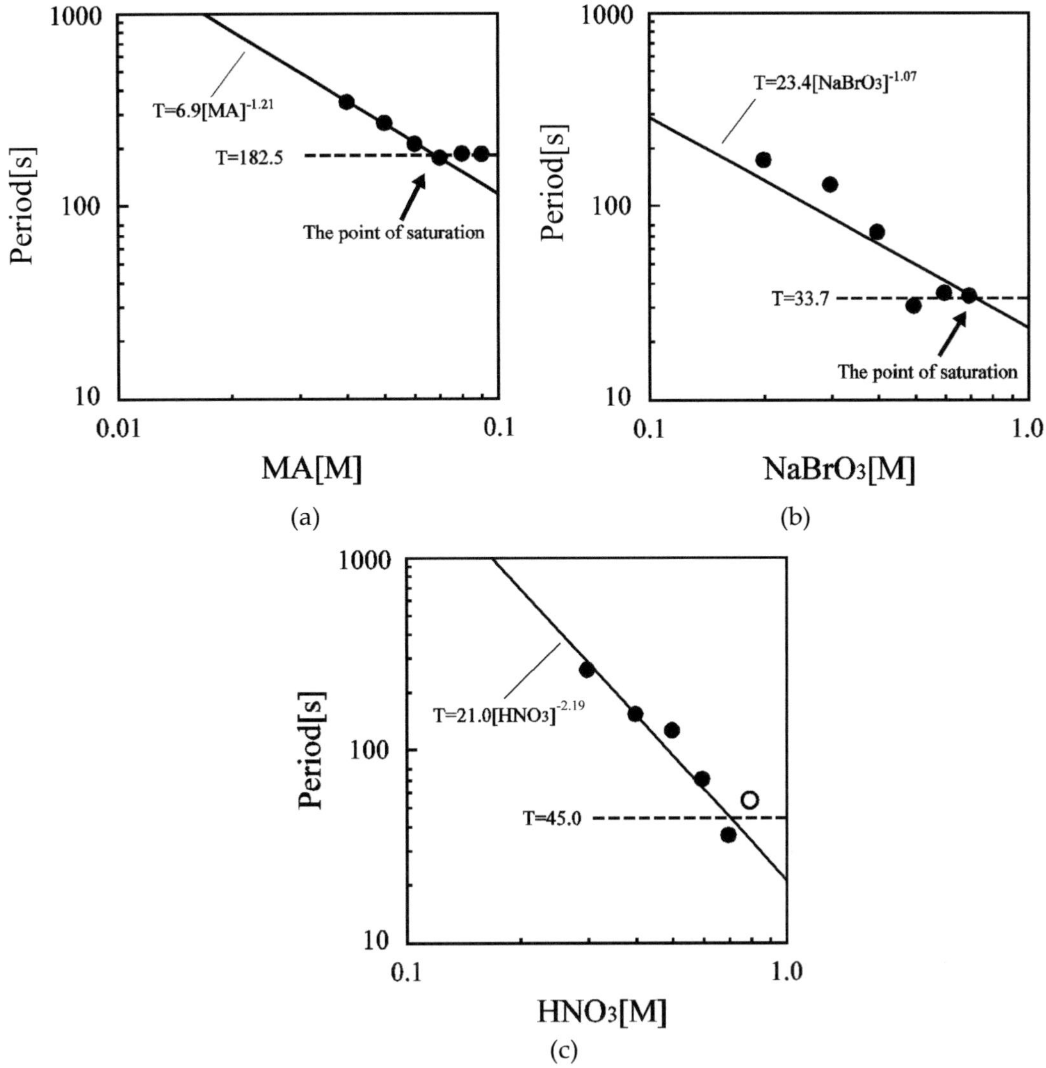

Figure 11. Logarithmic plots of period (T in s) vs initial molar concentration of one BZ substrate at a constant temperature (T=20 °C) under fixed concentrations of the other two BZ substrates; (a) [$NaBrO_3$] = 0.084M and [HNO_3] = 0.3M, (b) [MA] = 0.0625M and [HNO_3] = 0.3M, (c) [MA] = 0.0625M and [$NaBrO_3$] = 0.084M. (●) plots and (○) plots show the linear relation and the saturated line vs initial concentration of one BZ substrate, respectively. (Reprinted ref. 57, Copyright American Chemical Society. Reproduced with permission.)

main processes: consumption of Br^- ions (process A), autocatalytic formation of $HBrO_2$ (process B), and formation of Br^- ions (process C).

A: $BrO_3^- + 2Br^- + 3H^+ \rightarrow 3HOBr$

B: $BrO_3^- + HBrO_2 + 2Mred + 3H^+ \rightarrow 2HBrO_2 + 2Mox + H_2O$

C: $2Mox + MA + BrMA \rightarrow fBr^- + 2Mred + \text{other products}$

In the process of B and C, the $Ru(bpy)_3$ moiety in the gel works as the catalyst: the reduced $Ru(bpy)_3$ moiety is oxidized (process B), and the oxidized one is reduced (process C). Therefore, as the initial concentration of the MA increased, the mole fraction of the reduced $Ru(bpy)_3$ moiety in the gel increased in accordance with the FKN mechanism. With increasing in the mole fraction of the reduced $Ru(bpy)_3$ in the gel, the shrinking force originating in the hydrophobic reduced $Ru(bpy)_3$ greatly increased as well. Generally, as for a polymer gel, deswelling speed is faster than the swelling one. Once the gel collapsed, it takes a lot of time for the aggregated polymer domain in the gel to recover the elongated state. This is because the polymer aggregation state is thermodynamically more stable in the polymer gel. Therefore, as the shrinking force increased, the swelling speed of the poly(VP-*co*-$Ru(bpy)_3$) gel significantly decreased. As a result, in the higher MA condition, the period at the saturated point was long (T=182.5) compared with the other condition (T=33.7 (Figure 11(b)) and T=45.0 (Figure 11(c)). On the other hand, in the case of Figure 11(b), the period at the saturated point was greatly shorter than that in Figure 11(a). In the condition of Figure 11(b), the swelling force originating in the hydrophilic oxidized $Ru(bpy)_3$ moiety increased due to the increase in the mole fraction of the oxidized $Ru(bpy)_3$ moiety in the gel in accordance with the FKN mechanism. Therefore, the gel can cause the swelling-deswelling self-oscillation at the high speed due to the strong recovering force originating in the higher mole fraction of the hydrophilic oxidized $Ru(bpy)_3$ moiety in the gel. Moreover, in the condition of Figures 11(c), the period for the poly(VP-*co*-$Ru(bpy)_3$) gel had the different aspect from that of the conventional-type poly(NIPAAm-*co*-$Ru(bpy)_3$) gel.[75] The period of the self-oscillation decreased with increasing the initial concentration of the BZ substrates because of the increase in the collision frequency among the BZ substrates. Therefore, we considered that the relationship between the period and the [HNO_3] for the poly(VP-*co*-$Ru(bpy)_3$) gel is of a more natural tendency. In addition, in the condition of Figure 11(c), the control range of the period by changing the initial concentration of the HNO_3 for the gel was much wider than that for the poly(NIPAAm-*co*-$Ru(bpy)_3$ gel. [75]

Moreover, as shown in Figure 12, the period of the swelling-deswelling self-oscillation decreased with increasing the temperature because the temperature affects the BZ reaction rate in accordance with the Arrenius equation.[30] The period of the swelling-deswelling self-oscillation to the temperature for the poly(VP-*co*-$Ru(bpy)_3$) gel has the linear relationship. The period (2 second) reached the saturation at 46 °C in the BZ condition ([MA] = 0.08M, [$NaBrO_3$] = 0.48M and [HNO_3] = 0.48M). This is because the swelling-deswelling speed of the gel is slower than the changing rate of the redox states of the $Ru(bpy)_3$ in the gel. That is, the self-oscillating behavior of the gel cannot follow the changing the redox state of the $Ru(bpy)_3$ moiety. The maximum frequency (0.5Hz) of the poly(VP-*co*-$Ru(bpy)_3$) gel was 20 times as large as that of poly(NIPAAm-*co*-$Ru(bpy)_3$ gel. [75] The self-oscillating behaviors of the poly(VP-*co*-$Ru(bpy)_3$) gel at 20°C and 50°C were shown in the Figure 12(b) and 12(c), respectively. The displace of the volume change self-oscillation at 20°C and 50°C were about 10 μm and 4 μm, respectively. These results clarified that the displacement of the swelling-deswelling self-oscillation for the gel has the trade-off relationship against the period of the self-oscillation, that is, the length of the volume change decreased with increase in the period.

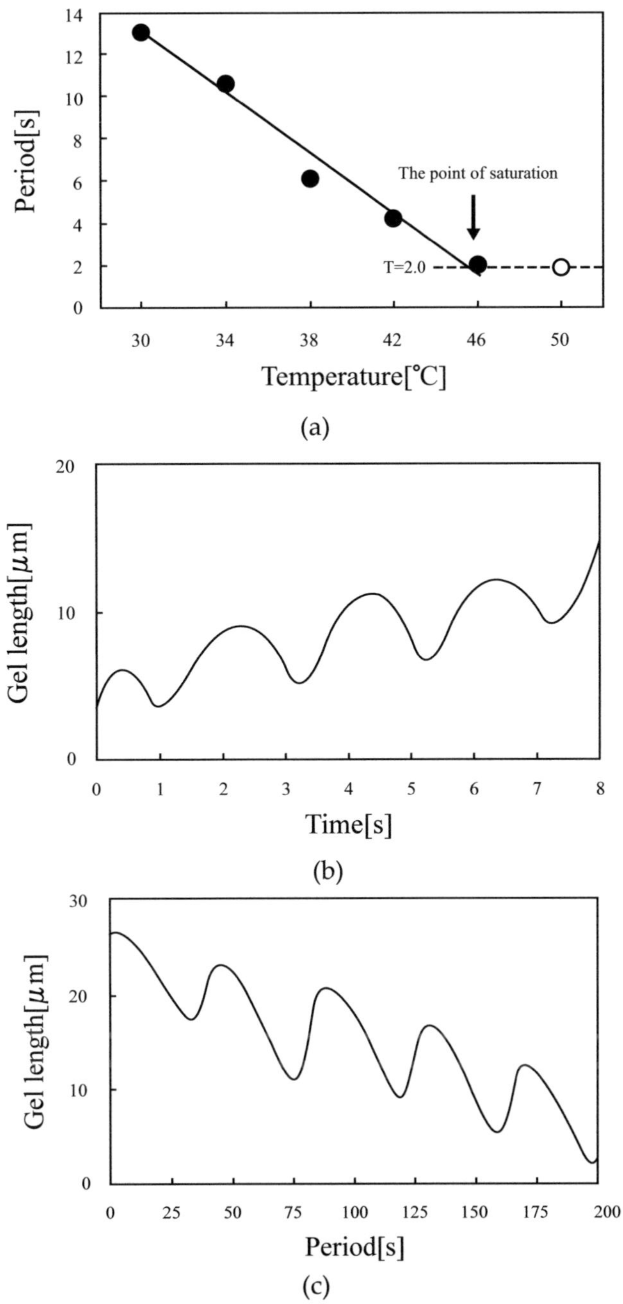

Figure 12. (a) Dependence of the self-oscillation period on the temperature. (●) plots and (○) plots show the linear relation and the saturated line vs temperature, respectively. (b) Self-oscillating profile of cubic poly(VP-*co*-Ru(bpy)3) gel at 50°C (MA = 0.08M, $NaBrO_3$ = 0.48M and HNO_3 = 0.48M). (c) Self-oscillating profile of cubic poly(VP-*co*-Ru(bpy)3) gel at 20°C (MA = 0.08M, $NaBrO_3$ = 0.48M and HNO_3 = 0.48M). Cubic gel (each side length is about 2mm and 20mm) was immersed in 8ml of the mixture solution of the BZ substrates. (Reprinted ref. 57, Copyright American Chemical Society. Reproduced with permission.)

3.3. Tubular gel motility driven by chemical reaction networks

In previous studies, De Kepper *et al.* showed a method to form the contraction waves by coupling the pH-responsive hydrogel and the acid autocatalytic chlorite tetrathionate (CT) reaction [42, 43]. They presented a system specially designed to show the chemomechanical instabilities, in other words, dynamical deformations of the functional gels, such as contraction waves and complex spatio-temporal volume oscillations. However, in this system, the pH responsive cylindrical gel is fixed to continuous stirred tank reactors (CSTR) because it must be permanently fed with constant flows of fresh CT reactants. Therefore, this system is unsuitable for the development of a locomotive gel robot. It is necessary to develop a novel concentration tuning mechanism without the stirred reactor for the coupled system of the functional gel and the CT reaction.

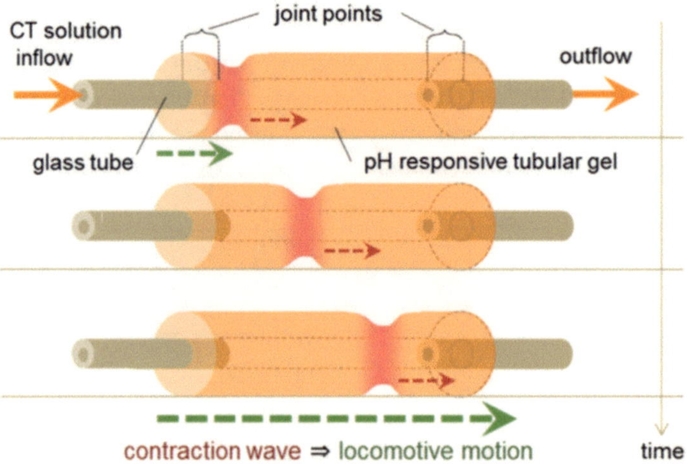

Figure 13. Schematic representation of the principles of the tubular gel feeding system (Reprinted ref. 58, Copyright *IEEE*. Reproduced with permission.)

In this chapter, we introduce a novel pH-responsive tubular gel and propose a new method of generating contraction waves on the functional gel. Figure 13 shows the schematic representation of the principle of the feeding system of the tubular gel. We controlled the reaction diffusion system by pouring the reactant solution into the hollow of the tubular gel, and attempted to achieve the peristaltic movement of the tubular gel by formulating contraction wave. This system enables us to feed with constant flow of fresh reactants, and also to be free from tank reactors.

The CT reaction is known as the acid autocatalytic reaction with the concentration changes of various sorts of reactive substrates, including the chlorite and tetrathionate ion, and this reaction exhibits a bi-stability of an acid steady state and an alkaline steady state. The CT reaction shows very complex kinetics which is still not completely understood [76]. However, it can be described by the following chemical reaction,

$$7ClO_2^- + 2S_4O_6^{2-} + 6H_2O \rightarrow 7Cl^- + 8SO_4^{2-} + 12H^+.$$

It runs in slight chlorite excess, and is associated with the following autocatalytic empirical rate law,

$$r = -\frac{1}{7}\frac{d[ClO_2^-]}{dt} = k[ClO_2^-][S_4O_6^-][H^+]^2,$$

r indicates the concentration per unit mol of the reactive substrates, and k is a proportionally coefficient. The changes between the steady states of the CT reaction take place depending on the feeding condition of the proton. When an alkali is constantly supplied, the CT reaction solution can maintain an alkaline steady state. As an acid perturbation is added or an alkali supply decreases below a certain concentration, the acid autocatalytic reaction develops and acid region spreads by the reaction-diffusion process, finally the solution becomes acid steady state. When autocatalytic reactions are operated in a pH responsive spherical gel, the gel shrinks following the one-way reaction, regularly. However, in some conditions, the relatively faster diffusivity of the proton can lead to the oscillatory reaction diffusion instability [42,43,77,78]. The volume oscillation of spherical gel is generated by coupling the CT reaction with the volume phase transition of the pH-responsive gel, and we call the concatenation of the two phenomena "chemical reaction networks."

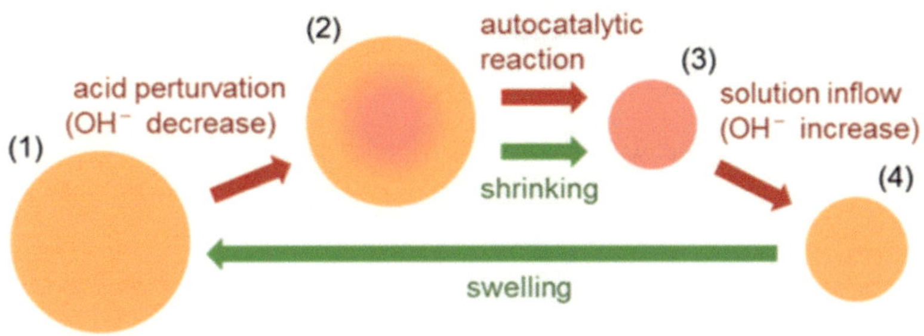

Figure 14. Flowchart of the chemical reaction networks. Each disk shows the condition of a spherical gel. The red arrows indicate changes of the interior CT solution and the green arrows indicate volume phase transition of the pH-responsive gel, that is, the gel causes shrinking or swelling. (Reprinted ref. 58, Copyright *IEEE*. Reproduced with permission.)

Figure 14 shows the flowchart of the chemical reaction networks. The red areas indicate the acid steady state and the yellow areas indicate the alkaline steady state. There are four steps in the networks:

1. The spherical gel is in the uniform alkaline steady state.
2. When the gel radius grows and exceeds a certain threshold value, an acid perturbation rises in the core of the gel.
3. The acid region fills the gel entirely, and the gel shrinks.
4. When the gel radius shrinks below a certain threshold value, it swells again by inflowing of alkali solution.

As described above, the stabilities of such steady states depend on the gel radius. It is shown theoretically and experimentally that the switching between the chemical reaction networks

occurs with hysteresis as a function of such geometric parameters. In this case, the reaction diffuses symmetrically from the center of the gel. When the reaction is applied to the cylindrical gels or tubular gels, it diffuses symmetrically to the diameter direction and also diffuses to the length direction. The disks shown in Figure 14 indicate the changes of the cross-section surface at one location of the cylindrical gel.

For coupling pH-responsive gels and the CT reaction, it is quite important to achieve large volume changes of the gels, because the stabilities of the steady states depend on the gel sizes. Also, the speed of volume change has a high correlation with the contraction wave shape. However, in general, the reactivities of hydrogels composed of chemically cross-linked polymer networks are low because the polymer chains are molecularly restricted by a large number of cross-links. An *N*-isopropylacrylamide (NIPAAm) gel with a microscale phase separation that underwent a quick response has previously been reported [79]. By preparing a NIPAAm gel above the LCST, the NIPAAm gel forms a porous structure with polymer-rich domains and aggregations in the matrix of loosely bound network structures [80]. Consequently, polymer-rich domains inside the gel aggregate or disperse rapidly through the porous structure within the gel by an effluent pathway of the solvent. Therefore, in this investigation, we prepared the microphase-separated AAm-based gels. The microphase-separation in the gel depends strongly on the methods of gel preparation. Meanwhile, a method for preparing the microphase-separated gels by altering the composition of the solvent has been reported [55]. Therefore, we synthesized the poly (AAm-*co*-AAc) gels in water/acetone solvents. Also, we investigated the gels kinetics at the different mixture proportions of the solvents.

(a) (b)

Figure 15. SEM images of the interior morphologies of poly (AAm-*co*-AAc) microphase-separated gel with the mixing ratio of water/acetone solvents at 50/50 ((a) swelling state and (b) shrinking state at equal magnification). (Reprinted ref. 58, Copyright *IEEE*. Reproduced with permission.)

First, we observed the structures of the gels. Figure 15 shows the SEM images of the interior morphologies of the poly(AAm-*co*-AAc) microphase-separated gels with the mixing ratio of water/acetone solvents at 50/50. There are polymer rich domains inside the both gels (swelling state and shrinking state) and their sizes change in proportion as the volume

phase transitions of the gels. This microscale structures are quite different from those of ordinary poly(AAm-*co*-AAc) gels with water solvents.

Next, we investigated the phase transition kinetics of the poly(AAm-*co*-AAc) cylindrical gels. When the external pH was rapidly changed, the gel diameters gradually changed to approach the equilibrium state. Figure 16 shows the plots of the gel diameters as a function of time on the shrinking process. Here, let us define L_f, $L(t)$ and L_i, as the final gel diameter, the gel diameter at t = t and the initial gel diameter (t = 0), respectively. The time evolution was found to be well described by a single exponential:

$$L(t) = L_i - \Delta L(1 - e^{-(t/\tau)}) \tag{1}$$

where ΔL and τ represents the total gel length change (=$L_i - L_f$) and the characteristic time of shrinking. When τ and L_f are determined, we can estimate the diffusion constant D by using the following relation for cylindrical gels [81]:

$$D = \frac{L_f^{\ 2}}{24\tau} \tag{2}$$

From the results, the corrective diffusion constant of the poly(AAm-*co*-AAc) cylindrical gels can be estimated as follows; (A) 4.39×10^{-5} mm²/s, (B) 5.38×10^{-5} mm²/s, (C) 1.75×10^{-3} mm²/s, (D) 1.99×10^{-3} mm²/s, (E) 2.07×10^{-3} mm²/s. In the case of sample (A) and (B), the corrective diffusion constants are almost the same as those of the normal type hydrogels. On the other hand, in the case of sample (C), (D) and (E), they are two orders of magnitudes larger than those of sample (A) and (B). These results indicate that the mixture fractions of acetone to the solvents have the threshold values for the microphase-separation. Also, these results confirm that the microphase-separation plays an important role for increasing speed of the volume phase transition.

We also investigated the detailed pH-responsiveness of the gels. Figure 17 shows the equilibrium swelling of the poly (AAm-*co*-AAc) cylindrical gels at various pH values. From this results, all samples caused volume phase transition between the pH variation range of the CT reaction (from pH = 2 to pH = 11). Also, the diameter change rates of sample (A) and (C) are larger than that of sample (E). It is notable that the stiffness of the gels depends on the mixture proportions of water/acetone solvents. The stiffness of the gels tends to get lower as the mixture fractions of acetone increase. In this experiment, sample (D) and (E) do not stand up under their own weight and became easily deformed in air. Therefore, for this reason, we chose sample (C) as the material of the tubular gel because of its stiffness and phase transition kinetics.

We succeeded in coupling the pH-responsive hydrogel and the CT reaction. Figure 18 shows the propagation of the acid region in the swollen part of the tubular gel. The mixture proportion of the solvent of the tubular gel was same as sample (C). As shown in Figure 18, the CT solution was colored with methyl red. The yellow area corresponds to alkaline composition and the red area corresponds to acid composition. The red allows indicate the

forefront of the acid region and the green allows indicate the forefront of the gel contraction. When the acid perturbation was applied to the left part of the tubular gel, the red area propagated to the right, and finally covered the entire gel. Also, the gel diameter shrank following the propagation of the acid region, as shown in Figure 18. These results suggest that the poly(AAm-*co*-AAc) tubular gel can be coupled with the CT solution, in terms of the acid propagation and the volume change of the gel.

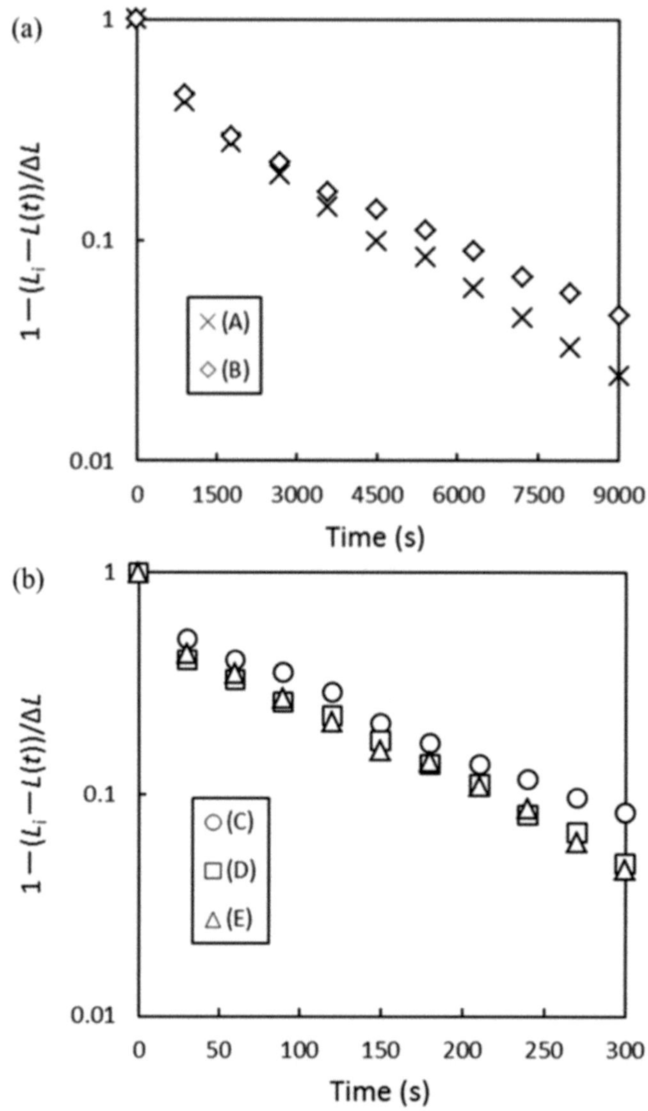

Figure 16. Shrinking behaviors of poly(AAm-*co*-AAc) cylindrical gels at the different mixture proportions of water/acetone solvents ((a): (A) 100/0, (B) 80/20, (b): (C) 70/30, (D) 60/40, and (E) 50/50 wt %). (Reprinted ref. 58, Copyright *IEEE*. Reproduced with permission.)

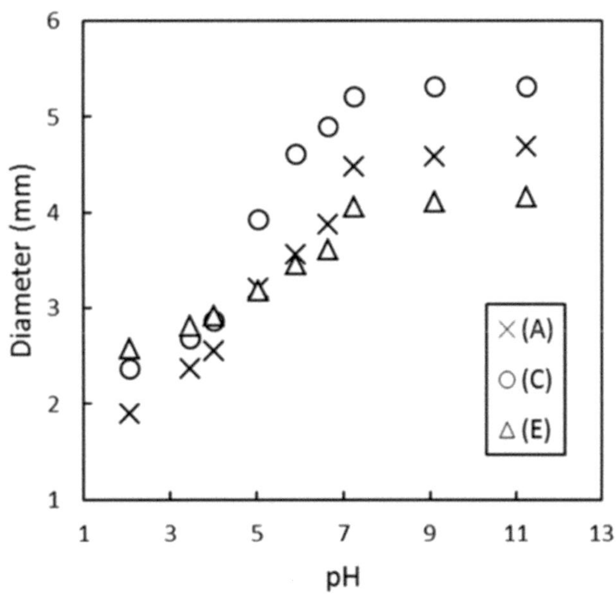

Figure 17. Equilibrium swelling of poly(AAm-*co*-AAc) cylindrical gels at various pH values adjusted using HCl and NaOH solutions, at the different mixture proportions of water/acetone solvents ((A) 100/0, (C) 70/30, and (E) 50/50 wt %). (Reprinted ref. 58, Copyright *IEEE*. Reproduced with permission.)

Fig. 19 shows the spatiotemporal diagram constructed from sequential images of the acid propagation. The extractive line is the black bar in Figure 18 (a). As shown in Figure 19, the acid region propagates at the constant speed, and the velocity of the acid propagation was 16 μm/s. It is 10 times faster than that of the previous research [42]. We also measured the velocity of the acid propagation in the mixed CT solution in the glass tube, and it was 200 μm/s. This is much larger value than that in the case of Figure 18. This result indicates that the propagation occurred inside the gel. These results confirm that a part of the chemical reaction networks, from step (1) to step (3) in Figure 14, was realized experimentally. In order to achieve autonomous swelling of the gel, from step (3) to step (4) in Figure 14, the gel diameter needs to shrink below a certain threshold value. It means that the minimum and maximum sizes of the tubular gels should be regulated for realizing oscillatory volume changes by inflowing of alkali in the CT solutions. The concentrations or mixing ratios of the CT solutions also affect the forming of the chemical reaction networks. The regulation of the gel sizes and the CT solutions will take an important role in the generation of the contraction waves in the tubular gels. This research will be the first step for realizing the biomimetic chemical robot which causes peristaltic locomotion.

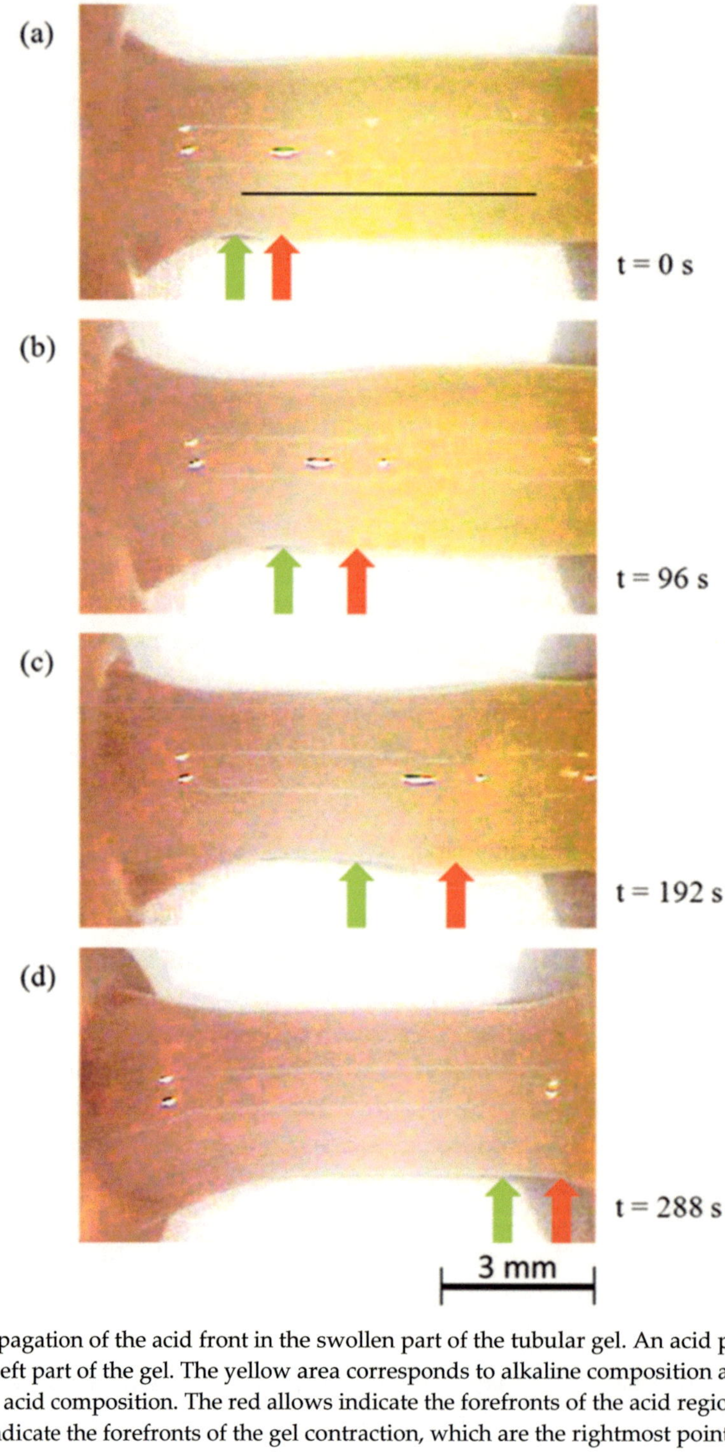

Figure 18. Propagation of the acid front in the swollen part of the tubular gel. An acid perturbation was applied to the left part of the gel. The yellow area corresponds to alkaline composition and the red area corresponds to acid composition. The red allows indicate the forefronts of the acid regions and the green allows indicate the forefronts of the gel contraction, which are the rightmost points obtained locally minimum diameters. (a) t = 0 indicates the criterion time after applying the acid perturbation. (Reprinted ref. 58, Copyright *IEEE*. Reproduced with permission.)

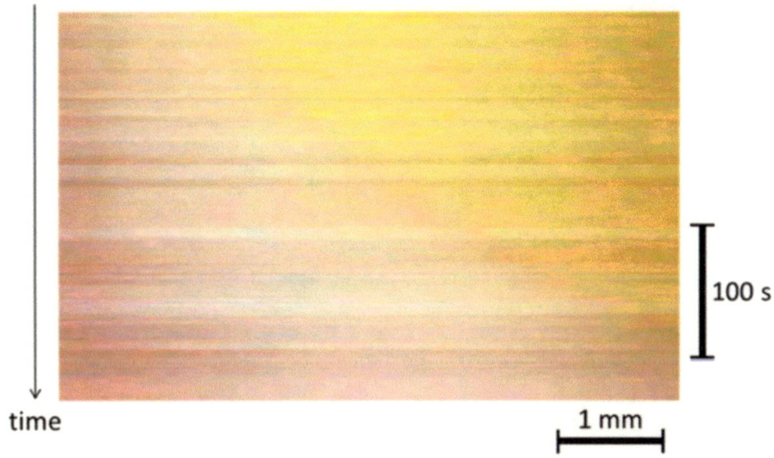

Figure 19. Spatiotemporal diagram constructed from sequential images of the acid propagation. The extracted horizontal lines correspond to the black bar in Fig. 18 (a). (Reprinted ref. 58, Copyright *IEEE*. Reproduced with permission.)

3.4. A pendulum-like motion of nanofiber gel actuator synchronized with external periodic pH oscillation

In this study, in order to drive the nanofiber gel actuator in response to the external pH changes, we selected the pH responsive poly(AAc) (PAAc) as a main polymer chain. The PAAc is protonated when the pH is below the pKa. When the pH of the solution is below pKa, the nanofiber gel collapses due to hydrogen bonding among the polymer chains. On the other hand, when the pH is above the pKa, the PAAc polymer chain is ionized, that is, the solubility of the polymer chain changes from a hydrophobic to a hydrophilic nature. As a result, the nanofiber gel expands because of the electrostatic repulsion force among the charged PAAc polymer chains. However, the fiber gel that consists of the only PAAc polymer chain, finally dissolves into the aqueous solution because the gel does not have the cross-linkage among the polymer chains into the nanofiber. In order to avoid the nanofiber gel dissolving, especially when it is above the pKa, we adopted the tBMA domain into the PAAc as a cross-linkage and a solubility control site, due to the hydrophobic interaction among the tBMA in the nanofiber. As a result, the poly(AAc-*co*-nBMA) nanofiber gel does not dissolve in the aqueous solution.

Figure 20 shows distributions of diameter of the poly(AAc-*co*-nBMA) nanofibers electrospun at two flow rates (2.0 mL/h and 1.0 mL/h) (See Figure 21). In general, the fiber diameters depend on the flow rate. In our experiment, the average diameter at the flow rate 2.0 mL/h (302 nm) was thicker than at 1.0 mL/h (233 nm).

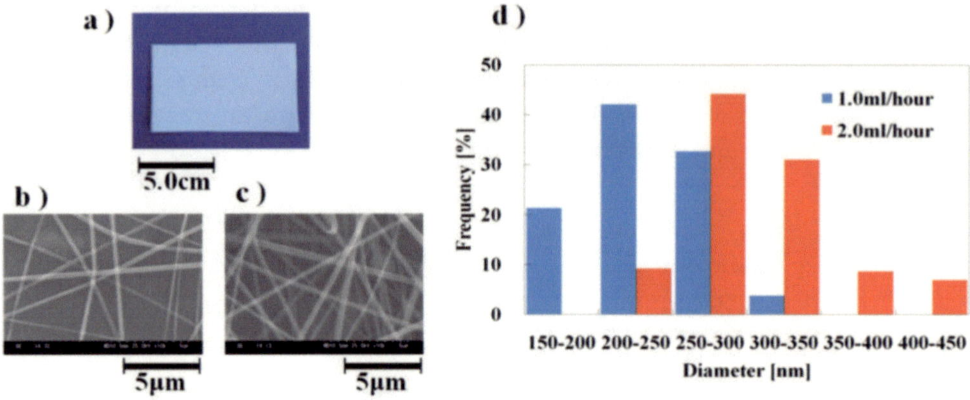

Figure 20. (**a**) An electrospun fiber sheet. (**b**) The SEM image of nanofibers at flow rate 1.0 mL/h. (**c**) The SEM image of nanofibers at flow rate 2.0 mL/h. (**d**) Distribution of fiber diameters of poly(AAc-*co*-nBMA) nanofibers electrospun at different flow rates.

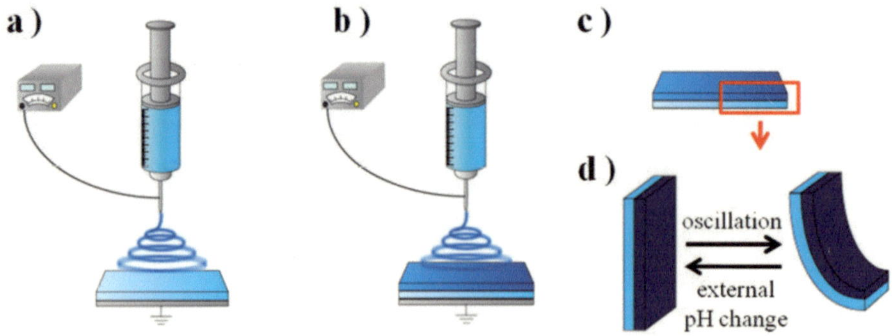

Figure 21. The method of introducing the anisotropic structure into the nanofiber gel. (**a**) Electrospinning at a flow rate of 2.0 mL/hour (sprayed for 30 minutes). (**b**) Electrospinning at a flow rate of 1.0 mL/hour (sprayed for 60 minutes). (**c**) Drying in 50 °C over night. (**d**) Cutting into 15 mm × 3 mm × 200 μm.

In order to drive the nanofibrous gel actuator synchronized with autonomous pH oscillation, we focused on the Landolt pH-oscillator, based on a bromated/ sulfite/ ferrocyanide reaction discovered by Edblom *et al.* [39,40]. This reaction causes the autonomous cyclic pH changes with a wide range at room temperature. The reaction has many reaction steps, so we estimated the main reactions as follow [41].

$$BrO_3^- + 3HSO_3^- + H^+ \rightarrow Br^- + 3SO_4^{2-} + 4H^+ \quad (3)$$

$$BrO_3^- + 6Fe(CN)_6^{4-} + 6H^+ \rightarrow Br^- + 6Fe(CN)_6^{3-} + 3H_2O \quad (4)$$

In Process (1), H_2SO_3 is oxidized by bromate, and ferrocyanide is oxidized by bromate in Process (2). In the above two processes, the hydrogen ions produced and consumed at comparable rates. Therefore, in this reaction, the pH oscillation takes place in the CSTR. Figure 22 shows the experimental set up of the CSTR. The CSTR was constructed by using

four peristaltic pumps in order to feed four solutions of potassium bromated, sodium sulfite, potassium ferrocyanide and sulfuric acid. Moreover, this system had one more peristaltic pump to drain the excess solution. The degree of changing the pH range (amplitude) and period of the oscillating reaction can be controlled by changing the feed concentration, flow rate and solution temperature.

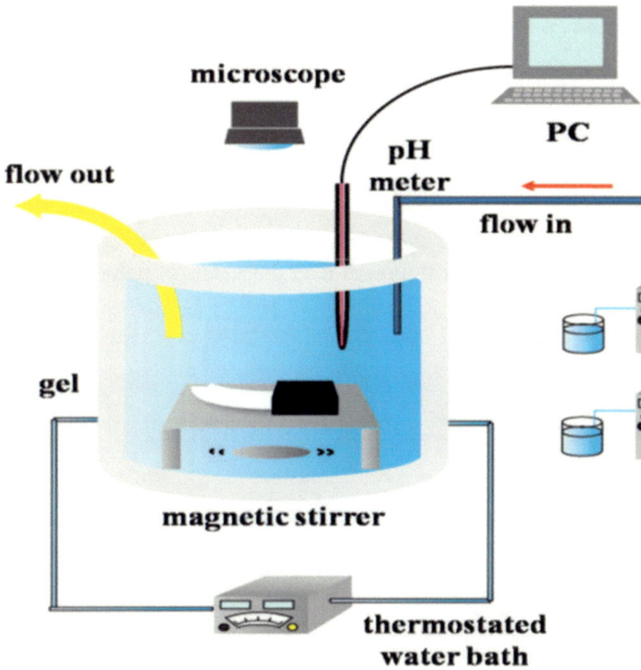

Figure 22. Continuous monitoring system for oscillation: monitoring both the medium pH and the gel motion in a CSTR.

Figure 23 shows a motion of the nanofiber gel actuator. The bending and stretching motions of the gel actuator synchronized with the pH oscillating reaction. As shown in Figure 23, we defined R as the length between two edges of the gel. Figure 24 shows the trajectory of the nanofiber gel strip. As shown in Figure 24, the gel strip caused the pendulum-like motion. As the external pH is below the pKa, the nanofiber gel stretches because of the deswelling originating from the hydrogen bonding (1→3). Next, when the pH is above the pKa, the gel bends because of the swelling originating from the repulsive force among the anionic polymer chains (4→6).

Figure 25 shows the temporal changes of R of the gel strip and the external pH, respectively. The range of the pH oscillation based on a bromate/sulfite/ferrocyanide reaction was $3.1 < pH < 7.2$, and the period was about 20 min. When the external pH changes periodically, the R of the gel strip cyclic changes synchronized with the external pH change. As shown in Figure 25, when the pH sharply decreased, the R of the gel strip starts to increase because the gel collapsed. Next, when the pH increased rapidly, R gradually decreased, due to the

gel actuator swelling originating from the repulsive force of AAc domain in the polymer chain. That is because the gel has different rates at swelling and deswelling. In general, the swelling motion of the gel is slower than the deswelling motion. Therefore, when the gel actuator bended, the R value gradually decreased.

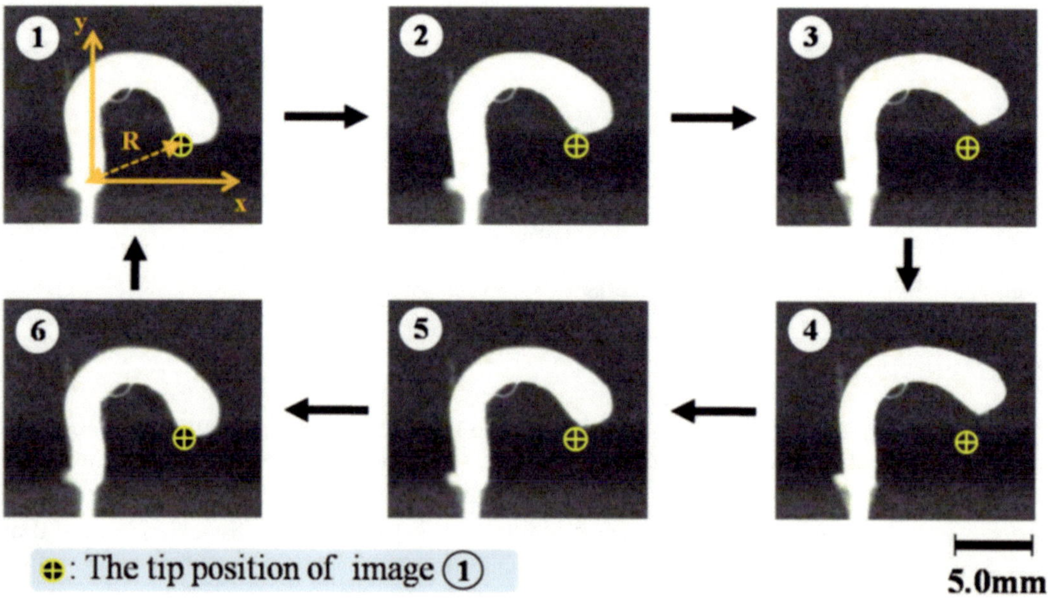

Figure 23. Periodical pendulum motion of poly(AAc-*co*-nBMA) nanofiber gel.

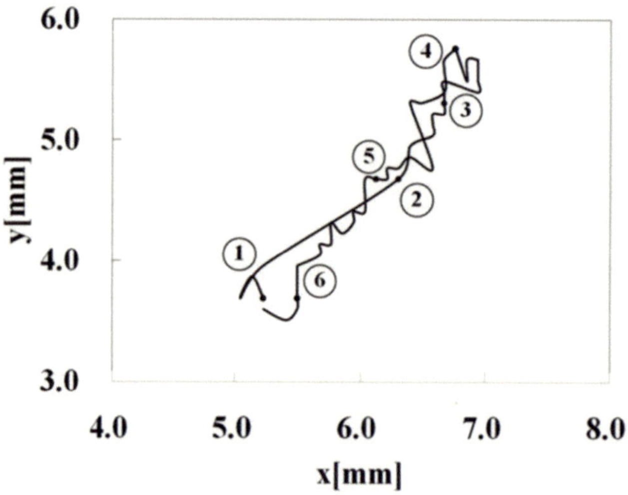

Figure 24. Trajectory of the tip of the gel relative to its attachment position during pH oscillation.

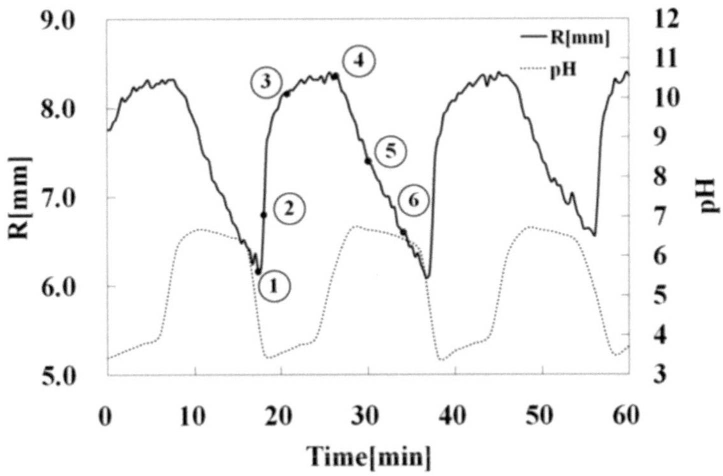

Figure 25. The time series of pH (discontinuous line) and R (solid line).

4. Conclusion

As for the VP-based self-oscillating polymer chain, we examined the influence of initial substrate concentrations of the BZ reaction on transmittance self-oscillation for the novel poly(VP-*co*-Ru(bpy)$_3$) solution. As a result we noted that the width of the self-oscillation is much affected by the initial concentration of the BZ substrates. In addition, we clarified that the amplitudes of the transmittance self-oscillation is hardly affected by the initial concentrations of the BZ substrates. This tendency was not observed in the case of the AMPS-containing polymer solution. Furthermore, we demonstrated that the period of the transmittance self-oscillation can be controlled by the selection of the initial concentration of the BZ substrates.

Furethermore we succeeded in clarifying the influence of the initial concentration of the BZ substrates and the temperature on the period of the swelling-deswelling self-oscillation for the novel poly(VP-*co*-Ru(bpy)$_3$) gel. The logarithmic plots of the period against the initial concentration of one BZ substrate under fixed the other two BZ substrates showed the good linear relationships. Moreover, the period of the self-oscillation increased with the increasing the temperature in accordance with Arrenius equation. The maximum frequency (0.5Hz) of the poly(VP-*co*-Ru(bpy)$_3$) gel was 20 times as large as that of poly(NIPAAm-*co*-Ru(bpy)$_3$ gel. Therefore, the period of the swelling-deswelling self-oscillation for the gel can be controllable in wide range by optimizing the initial concentration of the three BZ substrates and the temperature. In addition, we showed that the displacement of the self-oscillation for the gel has the trade-off relationship against the period of the self-oscillation.

Next, we introduced the new method of generating contraction waves in the functional gel, which controls the reaction diffusion system by flowing the CT solution into the hollow tubular gel. In order to realize this system, the poly(AAm-*co*-AAc) microphase-separated

gels were synthesized with various mixture proportions of the water/acetone solvents. From the view of the swelling kinetics, we experimentally clarified that the suitable mixture proportion for the fast volume changes and the adequate robustness for solution sending. Also, we coupled the poly (AAm-*co*-AAc) microphase-separated tubular gel and the CT solution. Moreover, we successfully demonstrated the propagation of the acidic contraction region of the gel. In the next step, we will accomplish the entire chemical reaction networks for realizing a swelling from the acid contraction region, which leads to a contraction wave in the tubular gel. This research will be the first step for realizing the biomimetic chemical robot which causes a peristaltic locomotion.

Finally, we introduced the fabrication of nanofiber gel actuator with anisotropic internal structure by changing the flow rate in electrospinning. The developed gel generates bending and stretching motion according to the external pH. In order to drive the nanofiber gel actuator automatically, we focused on the pH oscillating reaction based on a bromate/sulfite/ferrocyanide reaction. As a result, we succeeded in causing the bending-stretching motion of the nanofiber gel actuator synchronized with the external pH oscillation. By analyzing the motion of the gel, we found that the gel actuator caused the pendulum-like motion. Moreover, we clarified that the displacement and period of the nanofiber gel were stable, which makes it promising as a molecular device for potential applications.

Author details

Yusuke Hara*
Nanosystem Research Institute (NRI), National Institute of Advanced Industrial Science and Technology (AIST), Japan

Shingo Maeda
Shibaura Institute of Technology, Japan

Takashi Mikanohara, Hiroki Nakagawa, Satoshi Nakamaru and Shuji Hashimoto
Department of Applied Physics, Graduate School of Science and Engineering, Waseda University, Japan

5. References

[1] Fukushima, T.; Asaka, K.; Kosaka, A.; Aida, T. (2005). Fully plastic actuator through layer-by-layer casting with ionic-liquid-based bucky gel, *Angewandte Chemie International Edition*, 44, pp.2410-2413, ISSN 1433-7851

[2] Oguro, K.; Kawami, Y.; Takenaka, H. (1992). An actuator element of polyelectrotyte gel membrane-electrode composite, *Bulletin of the Government Industrial Research Institute, Osaka*, 43, pp. 21–24.

* Corresponding Author

[3] Otake, M.; Kagami, Y.; Inaba, M.; Kim, B. & Inoue, H. (2002). Motion design of a starfishshaped gel robot made of electro-active polymer gel. *Robotics and Autonomous Systems*, Vol.40, No.2-3, (August 2002), pp. 185-191, ISSN 0921-8890

[4] Steinberg, I, Z.; Oplatka, A.; Kachalsky, A. (1996). Mechanochemnical engines, *Nature*, 210, pp. 568–571, ISSN 0028-0836

[5] Feinberg, A. W. Feigel, A. Shevkoplyas, S. S. Sheehy, S. Whitesides, G, M. & Parker, K. K (2007). Muscular thin films for building actuators and powering devices, *Science*, Vol.317, No.5843, (September 2007), pp.1366-1370, ISSN 0036-8075

[6] Harada, A. ; Kataoka, K. (1999). Chain length recognition: core-shell supramolecular assembly from oppositely charged block copolymers. *Science*, 283, pp. 65-67 ISSN 1095-9203

[7] Kim, J.; Nayak, S.; Lyon, L. A. (2005). Bioresponsive hydrogel microlenses, *Journal of the American Chemical Society*, 127, pp. 9588–9592 ISSN 0002-7863

[8] Kwon, G.H.; Park, J.Y.; Kim, J.Y.; Frisk, M.L.; Beebe, D.J.; Lee, S.H. Biomimetic Soft Multifunctional Miniature Aquabots. *Small* 2008, 4, 2148-2153.

[9] Hoffman, A. S. (2002) Hydrogels for biomedical applications, *Advanced Drug Delivery Reviews*, 43, pp. 3-12, ISSN 0169-409X

[10] Tanaka, T. (1981). Gels, Scientific American.,244, pp.110-116, ISSN 0036-8733

[11] Hirokawa, Y.; Tanaka, T. (1984). Volume phase transition in a non ionic gel, *Journal of chemical Physics*, 81, pp.6379-6380, ISSN *0021-9606*

[12] Ilmain, F.; Tanaka, T., Kokufuta, E. (1991). Volume transition in a gel by hydrogen bonding, *Nature*, 349, pp.400-401, ISSN 0028-0836

[13] Tanaka, T. (1978). Collapse of gels and critical endpoint, *Physical Review Letters*, 40, pp.820-823, ISSN 0031-9007

[14] Suzuki, A.; Tanaka, T. (1990). Phase transition in a polymer gels induced by visible-light, *Nature*, 346, pp. 345-347, ISSN 0028-0836

[15] Kuhn, W. ; Hargity, B. ; Katchalsky, A. & Eisenberg, H. (1950). Reversible Dilation and Contraction by Changing the State of Ionization of High-Polymer Acid Networks, *Nature*, Vol.165, No.4196, (April 1950), pp. 514-516, ISSN 0028-0836

[16] Hu, Z.; Zhang, X. & Li, Y. (1995). Synthesis and Application of Modulated polymer gels. *Science*, Vol.269, No.5223, (April 1995), pp. 525-527, ISSN 0036-8075

[17] Yeghiazarian, L. Mahajan, S. Montemagno, C. Cohen, C. & Wiesner, U. (2005). Directed Motion and Cargo Transport Through Propagation of Polymer-Gel Volume Phase Transitions, *Advanced Materials*, Vol.17, No.15, (August 2005), pp. 1869-1873, ISSN 0935-9648

[18] Huber, D. L.; Manginell, R. P.; Samara, M. A.; Kim, B. & Bunker, B. C. (2003). Programmed Adsorption and Released of proteins in a Microfluidic device, *Science*, Vol.301, No.5631, (July 2003), pp. 352-354, ISSN 0036-8075

[19] Beebe, D. J.; Moore, J. S.; Bauer, J. M.;Yu, Q.;Liu, R. H.;Devadoss, C.; Jo, B. H. (2000). Functional hydrogel structures for autonomous flow control inside microfluidic channels, *Nature*, 404, pp. 588-590, ISSN 0028-0836

[20] Asoh, T.; Matsusaki, M.; Kaneko, T.; Akashi, M. (2008). Fabrication of Temperature-Responsive Bending Hydrogels with a Nanostructured Gradient, *Advanced Materials*, Vol.20, No.11, pp. 2080-2083, ISSN 0935-9648.

[21] Ross, J.; Muller, S. C.; Vidal, C. (1988). Chemical Waves, Science, Vol. 240, pp.460-465, ISSN 1095-9203.

[22] Epstein, I. R.; Showalter, K.(1996). Nonlinear Chemical Dynamics: Oscillations, Patterns, and Chaos, *The Journal of Chemcal Chemistry*, Vol. 100, No. 31, pp.13132-13147, ISSN 1932-7447.

[23] Epstein, I. R. (1995). The Consequences of Imperfect Mixing in Autocatalytic Chemical and Biological Systems, Nature Vol. 374, pp. 321-327, ISSN 0028-0836

[24] Lee, K. J.; Cox, E. C.; Goldstein, R. E. (1996). Competing Patterns of Signaling Activity in Dictyostelium Discoideum. *Physical Review Letters*, Vol. 76, pp. 1174-1177, ISSN 0031-9007.

[25] Shibayama, M.; Tanaka, T. (1993). Volume Phase Transition and Related Phenomena of Polymer Gels, *Advances in Polymer Science*, Vol. 109, pp. 1-62, ISSN 0065-3195.

[26] Landrot, V.; De Kepper, P.; Boissonade, J.; Szalai, I.; Gauffre, F. (2005). Wave Patterns driven bu chemomechanical instabilities in responsive gels, *The Journal of Physical Chemistry B*, Vol. 109, pp. 21476-21480, ISSN 1520-6106.

[27] Boissonade, J. (2003). Efficient Method for Predicting Crystal Structures at Finite Temperature: Variable Box Shape Simulations, *Physical Review Letters*, Vol. 90, pp. 188302 (4pages), ISSN 0031-9007.

[28] Yashin, V, V. & Balazs, A, C. (2006). Pattern formation and shape changes in self-oscillating polymer gels. *Science*, Vol. 314, No.5800, (September 2006) pp.798-801, ISSN 0036-8075

[29] Yashin, V, V. & Balazs, A, C. (2007). Theoretical and computational modeling of self-oscillating polymer gels, The *Journal of Chemical Physics*, Vol.126, No.12, (March2007) pp.124707-1-124707-17 ISSN 0021-9606

[30] Zaikin, A.N.; Zhabotinsky, A.M. (1970). Concentration Wave propagation in two-dimensional liquid-phase self-oscillating system, *Nature*, 225, pp. 535–537, ISSN 0028-0836

[31] Reusser, E.J.; Field, R.J. (1979). The transition from phase waves to trigger waves in a model of the Zhabotinskii reaction, *Journal of the American Chemical Society*, 101, pp. 1063–1071, ISSN 0002-7863

[32] Nicolis, G.; Prigogine, I. (1997). Self Orgainization in Nonequilibrium Systems; *Wiley: New York, NY, USA.*

[33] Field, R.J.; Burger, M. (1985). Oscillations and Traveling Waves in Chemical Systems; *John Wiley & Sons: New York, NY, USA.*

[34] Field, R.J.; Noyes, R. M. (1974). Oscillations in chemical systems. IV. Limit cycle behavior in a model of a real chemical reaction, *Journal of Chemical Physics*, 60, pp. 1877–1884, ISSN 0021-9606

[35] Gyorgyi, L.; Turanyi, T.; Field, R J. (1990). Mechanistic details of the oscillatory Belousov-Zhabotinskii reaction, *Journal of Chemical Physics*, 94, pp. 7162–7170, ISSN 0021-9606

[36] Turanyi, T.; Gyorgyi, L.; Field, R. J. (1993), Analysis and simplification of the GTF model of the Belousov-Zhabotinsky reaction, *Journal of Chemical Physics*, 97, pp. 1931–1941, ISSN 0021-9606.

[37] Scott, S. K. (1991), Chemical Chaos, *1st ed.; Oxford University Press: Oxford, UK.*

[38] Mori, H. ; Kuramoto, Y. (1997). Dissipative Structures and Chaos, *Springer-Verlag, Berlin.*

[39] Edblom, E.C.; Orban, M.; Epstein, I.R. (1986). A New Iodate Oscillator: The Landolt Reaction with Ferrocyanide in a CSTR. *Journal of the American Chemical Society*, Vol. 108, pp. 2826-2830, ISSN 0002-7863.

[40] Edblom, E.C.; Luo, Y.; Orban, M.; Kustin, K.; Epstein, I.R. (1989) Kinetics and Mechanism of the Oscillatory Bromate-Sulfite-Ferrocyanide Reaction. *The Journal of Physical Chemistry*, Vol. 93, pp. 2722-2727, ISSN 1089-5639.

[41] Crook, J.; Smith, A.; Jones, A.; Ryan, J. (2002). Chemically Induced Oscillations in a pH-Responsive Hydrogel. *Physical Chemistry Chemical Physics*, Vol. 4, pp. 1367-1369, ISSN 1463-9076.

[42] V. Labrot, P. De Kepper, J. Boissonade, I. Szalai, and F. Gauffre, (2005). Wave Patterns Driven by Chemomechanical Instabilities in Responsive Gels, *Journal of Chemical Physics*, vol. 109, No. 46, pp. 21476-21480, ISSN 0021-9606.

[43] J. Boissonade, P. De Kepper, F. Gauffre, and I. Szalai, (2006) Spatial bistability: A source of complex dynamics. From spatiotemporal reaction-diffusion patterns to chemomechanical structures, *Chaos*, Vol. 16, No. 3, pp. 037109 (8pages), ISSN 1054-1500.

[44] Ishiwatari, T.; Kawaguchi, M.; Mitsuishi, M. (1984). Oscillatry reactions in polymer systems, *Journal of Polymer Science Part A: Polymer Chemistry*, 22, pp. 2699-2704 ISSN 0887-624X

[45] Yoshida, R.; Takahashi, T.; Yamaguchi, T.; Ichijo, H. (1996). Self-oscillating gel, *Journal of the American Chemical Society*, 118, pp. 5134-5135, ISSN 0002-7863

[46] Yoshida, R.; Sakai, T.; Ito, S.; Yamaguchi, T. (2002). Self-oscillation of polymer chains with rhythmical soluble-insoluble changes, *Journal of the American Chemical Society*, 124, pp. 8095-8098, ISSN 0002-7863

[47] Hara, Y.; Yoshida, R. Self-oscillation of polymer chains induced by the Belousov-Zhabotinsky reaction under acid-free conditions. *Journal of Physical Chemistry B* 2005, 109, pp. 9451–9454, ISSN 1089-5647

[48] Hara, Y.; Yoshida, R. (2009). Damping behavior of aggregation-disaggregation self-oscillation for a polymer chain, *Macromolecular Rapid Communications*, 30, pp. 1656–1662, ISSN 1521-3927

[49] Hara, Y.; Yoshida, R. (2005). Control of oscillating behavior for the self-oscillating polymer with pH-control site, *Langmuir*, 21, pp. 9773–9776, ISSN 0743-7463

[50] Hara, Y.; Yoshida, R. (2008). A viscosity self-oscillation of polymer solution induced by the BZ reaction under acid-free condition, *Journal of Chemical Physics*, 128, 224904, ISSN 0021-9606

[51] Hara, Y.; Sakai, T.; Maeda, S.; Hashimoto, S.; Yoshida, R. (2005).b Self-oscillating soluble-insoluble changes of polymer chain including an oxidizing agent induced by the Belousov-Zhabotinsky reaction, *Journal of Physical Chemistry B*, 109, pp. 23316–23319, ISSN 1089-5647

[52] Hara, Y.; Yoshida, R. (2008). Self-oscillating polymer fueled by organic acid. *Journal of Physical Chemistry B,* 112, pp. 8427–8429, ISSN 1089-5647

[53] Maeda, S.; Hara, Y.; Yoshida, R.; Hashimoto, S. (2008). Control of the dynamic motion of a gel actuator driven by the Belousov-Zhabotinsky reaction, *Macromolecular Rapid Communications*, 29, pp. 401–405, ISSN 1521-3927

[54] Maeda, S.; Hara, Y.; Sakai. T.; Yoshida, R.; Hashimoto, S. (2007). Self-walking gel, *Advanced Materials*, 19, pp. 3480–3484, ISSN 1521-4095

[55] Maeda, S.; Hara, Y.; Yoshida, R.; Hashimoto, S. (2008). Peristaltic motion of polymer gels, *Angewandte Chemie International Edition*, 47, pp. 6690-6693, ISSN 1433-7851

[56] Hara, Y.; Yoshida, R. (2009). Influence of a positively charged moiety on aggregation-disaggregation self-oscillation induced by the BZ reaction, *Macromolecular Chemistry and Physics*, 210, pp. 2160–2166, ISSN 1022-1352

[57] Nakamaru, S,; Maeda, S. ; Hara, Y.; Hashimoto, S. (2009). Control of Autonomous Swelling-Deswelling Behavior for a Polymer gel. *Journal of Physical Chemistry B,* 2009, 113, pp. 4609-4613, ISSN 1089-5647

[58] T. Mikanohara, S. Maeda, Y. Hara, S. Hashimoto (2011). Tubular Gel Motility Driven by Chemical Reaction Networks", *Proceeding of 2011 IEEE International Conference on Robotics and Biomimetics*, pp.2008-2013, ISBN 978-1-4577-2136-6.

[59] Nakagawa, H. Hara, Y. Maeda, S. & Hashimoto, S. (2010). A Novel Design of Nanofibrous Gel Actuator by Electrospinning, *Proceedings of IEEE Nano* 2010. ISBN 978-4244-7031-0

[60] Nakagawa, H.; Hara, Y.; Maeda, S. & Hashimoto, S (2010). A Pendulum-Like Motion o Nanofiber Gel Actuator Synchronized with External Periodic pH Oscillation, *Polymers,* Vol.3, No.1, (December 2010) pp 405-412, ISSN 2073-4360

[61] R. McNeill Alexander, (2006). Principles of Animal Locomotion, *Princeton University Press*, ISBN: 9780691126340

[62] Stevens, M.M.; George, J.H. Exploring and Engineering the Cell Surface Interface. *Science* 2005, 310, 1135-1138.

[63] Cui, Y.; Wei, Q.; Park, H.; Lieber, C.M. (2001). Nanowire Nanosensors for Highly Sensitive and Selective Detection of Biological and Chemical Species. *Science*, Vol. 293, pp. 1289-1292, ISSN 0036-8075

[64] Gu, S.Y.; Wang, Z.M.; Li, J.B.; Ren, J. (2010). Switchable Wettability of Thermo-Responsive Biocompatible Nanofibrous Films Created by Electrospinning. *Macromolecular Materials and Engineering,* Vo. 295, pp. 32-36 ISSN 1438-7492.

[65] Adam, D. (2001). Fine Set of Threads, *Nature,* Vol. 411, No. 236, pp. 236-236, ISSN 0028-0836

[66] Yu, J.H.; Sergey, V.F.; Gregory, C.R. (2004). Production of Submicrometer Diameter Fibers by Two-Fluid Electrospinning. *Advanced Materials* Vol. 16, pp. 1562-1566, ISSN 1521-4095.

[67] Dan, L.; Gong, O.; Jesse, T.M.; Younan, X. (2005). Collecting Electrospun Nanofibers with Patterned Electrodes, *Nano Letter,* Vol. 5, pp. 913-916, ISSN 1530-6984.

[68] Donzhi, Y.; Jianfeng, Z.; Jing, Z.; Jun, N. Aligned (2008). Electrospun Nanofibers Induced by Magnetic Field, *Journal of Applied Polymer Science*, Vol. 110, pp. 3368-3372, ISSN 1097-4628.

[69] Bogntizki, M.; Frese, T.; Steinhart, M.; Greiner, A.; Wendoreff, J.H. (2001). Preparation of Fibers with Nanoscaled Morphologies: Electrospinning of Polymer Blends, *Polymer Engineering & Science*, Vol. 41, pp. 982-989, ISSN 1548-2634.

[70] Li, L.; Hsieh, Y. (2005). Ultra-fine Polyelectrolyte Fibers from Electrospinning of Poly(acrylic acid), *Polymer*, Vol. 46, pp. 5133-5139, ISSN 0032-3861.

[71] Chen, H.; Hsieh, Y. (2004). Ultrafine Hydrogel Fibers with Dual Temperatureand pH-Responsive Swelling Behaviors. *Journal of Polymer Science Part A: Polymer Chemistry* ,Vol. 42, pp. 6331-6339, ISSN 1099-0518.

[72] Jin, X.; Hsieh, Y. (2005). pH-Responsive swelling behavior of poly(vinyl alcohol)/poly(acrylic acid) bi-component fibrous hydrogel membranes, *Polymer*, Vol. 46, pp. 5149-5160, ISSN 0032-3861.

[73] Okuzaki, H.; Kobayashi, K.; Yan, H. (2009). Thermo-Responsive Nanofiber Mats, *Macromolecules* Vol. 42, pp. 5916-5918, ISSN 0024-9297.

[74] Jin, X.; Hsieh, Y. Electrospinning (2007). pH-Responsive Block Copolymer Nanofibers. *Advanced Materials*, Vol. 19, pp. 3544-3548, ISSN 1521-4095.

[75] Yoshida, R.; Tanaka, M.; Onodera, S.; Yamaguchi, T.; Kokufuda, E. (2000). In-phase synchronization of chemical and mechanical oscillations in self-oscillating gels, *The Journal of Physical Chemistry A*, Vol. 104, pp. 7549–7555, ISSN 1089-5639.

[76] AK. Horvath, I. Nagypal, G. Peintler, and IR. Epstein (2004). Autocatalysis and self-inhibition: Coupled kinetic phenomena in the chlorite-tetrathionate reaction, *Journal of the American Chemical Society*, vol. 126, No. 20, pp. 6246-6247, ISSN 0002-7863.

[77] J. Boissonade (2005). Self-oscillations in chemoresponsive gels: A theoretical approach, *Chaos*, vol. 15, No. 2, pp. 023703, ISSN 1054-1500.

[78] J. Boissonade (2003) Simple chemomechanical process for self-generation of rhythms and forms, *Physical Review Letters*, Vol. 90, No. 18, pp. 188302, ISSN 0031-9007.

[79] B. G. Kabra, and S. H. Gehrke (1991). Synthesis of fast response, temperature-sensitive poly(N-isopropylacrylamide) gel, *Polymer Communications*, vol. 32, no. 11, pp. 322-323, ISSN 0263-6476.

[80] T. Norisuye, Y. Kida, N. Masui, and Q. Tran-Cong-Miyata (2003). Studies on two types of built-in inhomogeneities for polymer gels: Frozen segmental concentration fluctuations and spatial distribution of cross-links, *Macromolecules*, Vol. 36, No. 16, pp. 6202-6212, ISSN 0024-9297.

[81] A. Onuki (1993). Theory of phase-transition in polymer gels, *Advances in Polymer Science*, Vol. 109, pp. 63-121, ISSN 0065-3195.

Section 3

Smart Fluids

Chapter 14

Optimal Design Methodology of Magnetorheological Fluid Based Mechanisms

Quoc-Hung Nguyen and Seung-Bok Choi

Additional information is available at the end of the chapter

http://dx.doi.org/10.5772/51078

1. Introduction

Magnetorheological fluid (MRF) is a non-colloidal suspension of magnetizable particles that are on the order of tens of microns (20-50 microns) in diameter. Generally, MRF is composed of oil, usually mineral or silicone based, and varying percentages of ferrous particles that have been coated with an anti-coagulant material. When inactivated, MRF displays Newtonian-like behavior. When exposed to a magnetic field, the ferrous particles that are dispersed throughout the fluid form magnetic dipoles. These magnetic dipoles align themselves along lines of magnetic flux. The fluid was developed by Jacob Rabinow at the US National Bureau of Standards in the late 1940's. For the first few years, there was a flurry of interest in MRF but this interest quickly waned. In the early 1990's there was resurgence in MRF research that was primarily due to Lord Corporation's research and development. Although similar in operation to electro-rheological fluids (ERF) and Ferro-fluids, MR devices are capable of much higher yield strengths when activated. For this advantage, many MRF-based mechanisms have been developed such as MR dampers, MR brake, MR clutch, MR valve... and some of them are now commercial. As well-known that performance of MRF based systems significantly depends on the activating magnetic circuit, therefore, by optimal design of the activating magnetic circuit, the performance of MRF-based systems can be optimized. Recently, there have been various researches on optimal design of MRF-based devices such as MR damper, MR valve, MR brake. The results from these studies showed that performance of MRF-based systems can be significantly improved via optimal design of the magnetic circuit of the systems.

Consequently, this chapter focuses on the methodology of optimal design of MRF-based devices. The chapter is organized as follows; in section 2, fundamentals and the theory behind MRF are overviewed. Section 3 deals with the modes used in MRF based devices and several different devices featuring MRF are discussed. In the fourth section, optimal

design methodology of MRF-based mechanisms is considered. In this section, firstly the necessity of optimal design and the state of the art are discussed. Then the magnetic circuit analysis and the modeling of MRF devices are considered. In addition, the optimization problem of MRF devices is figured out and the methods to solve the problem are investigated. Section 5 is devoted to deal with a case study of MR valve optimal design. In this case study, several valve configurations such as single-coil, multiple-coil and annular-radial MR valves are considered. The chapter is then closed by the conclusion

2. Fundamentals of MRF and applications

2.1. Composition of MRF

Generally, MRF consists of non-colloidal suspensions, magnetically soft ferromagnetic, ferrimagnetic or paramagnetic elements and compounds in a non-magnetic medium. In practice, MRF usually consists of suitable magnetizable particles such as iron, iron alloys, iron oxides, iron nitride, iron carbide, carbonyl iron, nickel and cobalt [1, 2]. Among these, a preferred magnetic responsive particle that is commonly used to prepare MRF is carbonyl iron. The possible maximum yield stress induced by MR effect is mainly determined by the lowest coercivity and the highest magnitude of saturation magnetization of the dispersed particles. Therefore, soft magnetic material with high purity such as carbonyl iron powder appears to be the main magnetic phase for most of the practical MRF composition [3]. Other than carbonyl iron, Fe-Co alloys and Fe-Ni alloys can also be used as MR materials, whereby, Fe contributes to the high saturation magnetization. In contrary, some of the ferrimagnetic materials such as Mn-Zn ferrite, Ni-Zn ferrite and ceramic ferrites have low saturation magnetizations and are therefore suitable to be applied in low yield stress applications [1]. MR particles are typically in the range of 0.1 to 10μm [4, 5], which are about 1000 times bigger than those particles in the ferrofluids [6]. In the MRF, magnetic particles within a certain size distribution can give a maximum volume fraction without causing unacceptable increasing in zero-field viscosity. For instance, fluid composition that consists of 50% volume of carbonyl iron powder was used in the application of electromechanically controllable torque-applying device.

The carrier liquid forms the continuous phase of the MRF. Examples of appropriate fluids include silicone oils, mineral oils, paraffin oils, silicone copolymers, white oils, hydraulic oils, transformer oils, halogenated organic liquids, diesters, polyoxyalkylenes, fluorinated silicones, glycols, water and synthetic hydrocarbon oils [7, 2]. A combination of these fluids may also be used as the carrier component of the MRF. In the earlier patents, inventors were using magnetizable particles dispersed in a light weight hydrocarbon oil [8], either a liquid, coolant, antioxidant gas or a semi-solid grease [9] and either a silicone oil or a chlorinated or fluorinated suspension fluid [10]. However, when the particles settled down, the field-induced particle chains formed incompletely at best in which MR response was critically degraded. Later, in order to prevent further sedimentation, new compositions of MRF with consideration on viscoplastic [11] and viscoelastic continuous phases [12] were formulated, so that the stability could be improved immensely. In addition, a composite MRF has been

prepared by Pan et al. [13] with a combination of iron particles powder, gelatine and carrier fluids. They showed that the MR effects were superior under low magnetic field strength, and had a better stability compared to pure iron carbonyl powder alone.

Surfactants, nanoparticles, nanomagnetizable or coating magnetizable particles can be added to reduce the sedimentation of the heavy particles in the liquid phase [14, 13]. The sedimentation phenomenon can cause a shear-thinning behaviour of the suspension [15]. With further sedimentation, with MRF under the influence of high stress and high shear rate over a long period of time, the fluid will thicken (in-use-thickening) [16, 17]. Sedimentation phenomenon will reduce the MR effect where the particles in the MRF are settled down and form a hard "cake" that consists of firmly bound primary particles due to incomplete chain formation [18]. Anti-settling agent such as organoclay can provide soft sedimentation. When the composition of MRF has relatively low viscosity, it does not settle hard and can easily re-disperse [2]. Coating of the polymer layer also influences magnetic properties of the particles and cause them to easily re-disperse after the magnetic field is removed [19]. However, specific properties of MRF such as shear and yield stresses under the same conditions were enormously degraded inevitably by addition of the coating layer. This is due to the shielding of the polymer layer that affects the magnetic properties of the particles [19, 20]. In addition, some additives can improve the secondary properties like oxidation stability or abrasion resistance.

2.2. Magnetic properties of MRF

The static magnetic properties of MRF are important to design any MRF-based devices and generally can be characterized by B-H and M-H hysteresis. Through the magnetic properties, the dependence of the MRF response on the applied current in the device can be predicted. Under the influence of the magnetic field, a standard model for the structure is used to predict the behaviour of the particle of MRF [21]. The model is based on a cubic network of infinite chains of the particles arranged in a line with respect to the direction of the magnetic field as shown in Figure 1.

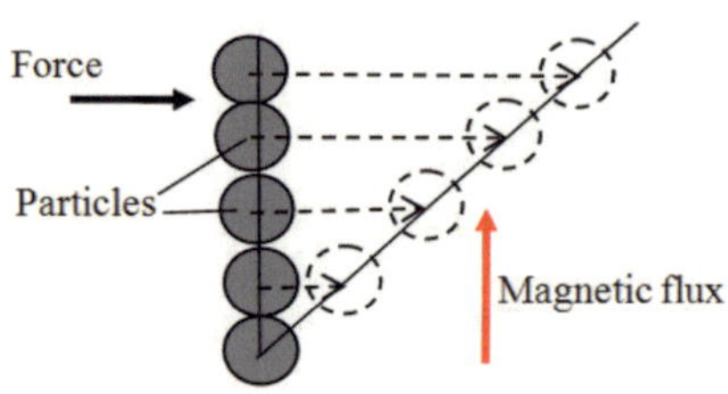

Figure 1. Schematic presentation of the affine deformation of a chain of spherical particles

The chains are considered to deform with the same distance between any pair of neighbours in the chains and increase at the same rate with the strain when the MRF is strained. This model is quite simple since the chains, in actual case, are formed into some more compact aggregates of spheres in which can be constituted in the form of cylinders. Under shear stress, these aggregates might deform and eventually break. Even though the particles

develop into different complicated structures under different conditions [22], the standard model still can be used in order to give a valid prediction of the yield stress [21]. The equation of motion of each particle under a magnetic field is required in order to evaluate the bulk property of MRF. At a very low magnetic field, the magnetic force tensor F_{ij} is obtained as point-dipole similar to the pair interaction, the magnetic dipole moment induced by other particles and surrounding walls for an unmagnetized and isolated sphere under a uniform magnetic field is given by [23]:

$$F_{ij} = \frac{3}{4\pi\mu_p\mu_0}\left[m^2 \frac{r_{ij}}{r_{ij}^5} - 5(mr_{ij})^2 \frac{r_{ij}}{r_{ij}^7} + 2(mr_{ij})m\frac{1}{r_{ij}^5} \right] \tag{1}$$

where F_{ij} is the magnetic force tensor acts on particle i from j, μ_p is the specific permeability of particles, μ_0 is the vacuum permeability, r_{ij} is position from particle j to i and m is magnetic dipole moment induced in particles within MRF given by [24],

$$m = 4\pi\mu_f\mu_0\beta a^3 H \tag{2}$$

where H is the uniform magnetic field, a is the diameter of the particles and β is given by,

$$\beta = \frac{\mu_f - \mu_p}{\mu_f - 2\mu_p} \tag{3}$$

where μ_f is the specific permeability of carrier liquid.

At high magnetic fields, the magnitude of the moment can be considered as independent point dipoles as magnetization of particles reaches saturation. In this case, the magnetic moment is given by [25].

$$m = \frac{4}{3}\pi a^3 \mu_s M_s \tag{4}$$

where $\mu_s M_s$ is the saturation magnetization of the particle, which is about 1.7 x $10^6 A/m$ for bulk iron and $0.48\text{x}10^6 A/m$ for the magnetite.

2.3. Fundamentals of rheological properties

Rheology is the response of materials to an applied stress [26]. Rheology is an interdisciplinary field and is used to describe the properties of a wide variety of materials such as oil, food, ink, polymers, clay, concrete, asphalt and others. Rheology measurements and parameters can be used to determine the processing behaviour of non-Newtonian materials, viscoelastic behaviour as a function of time, the degree of stability of a formulation at rest condition or during transport, and zero shear viscosity or the maximum viscosity of the fluid phase to prevent sedimentation [27]. The viscosity equation on the basis of a hydrodynamic theory for dilute dispersions of spherical particles has been developed by Einstein about 100 years ago [28]. The equation has been derived as

$$\eta_r = 1 + 2.5\phi \tag{5}$$

where η_r is the relative viscosity of the suspension and ϕ is the volume fraction of the suspended solutes or particles assumed to be spherical. The addition of the solid particles to a liquid will increase the amount of particles and consequently increases the volume fraction of the particles. Therefore, as the volume fraction of particles increases, there will be an increase in the fluid's viscosity. Shook [29] has suggested that the maximum concentration of the particles ϕ_{max} should be incorporated in the relationship between viscosity and concentration as

$$\eta_r = \frac{\phi}{(1-\phi)^{2.5\phi_{max}}} \tag{6}$$

However, these equations do not depend on the particle size but instead depend on the particle shape and solid concentration. Thus, Toda and Furuse [30] extended the equation in order to satisfy the viscosity behaviour of concentrated dispersion for small and large particles, respectively given by,

$$\eta_r = \frac{1-0.5\phi}{(1-\phi)^3} \tag{7}$$

$$\eta_r = \frac{1+0.5\kappa\phi-\phi}{(1-\kappa\phi)^3(1-\phi)} \tag{8}$$

where κ is the correction factor that may depend on the size and concentration of the particles. The viscosity of the fluid can be increased with additional amounts of the solid particles. However, at the same time, the fluid behaviour will change and diverge from a Newtonian fluid. Generally, shear stress τ increases with the shear rate du/dy which often can be represented by the relationship

$$\tau = \tau_y + \eta(\frac{du}{dy})^n \tag{9}$$

where τ_y, η and n are constants, τ_y is the yield stress and η is the dynamic viscosity. Newtonian fluids occur when the fluids show no yield stress or τ_y is equal to zero and n is equal to one. The viscosity of a Newtonian fluid is independent of time and shear rate. Figure 2 show the classification of fluids based on rheological properties. As shown in the figure, the behaviour of the fluids can be classified into Newtonian fluids and non-Newtonian fluids such as plastic, Bingham plastic, pseudo-plastic and dilatant fluids [31]. Fluids are said to be plastic when the shear stress must reach a certain minimum value before it begins to flow. If n, in Eq. (9), is equal to one, the material is known as a Bingham plastic. For the pseudo-plastic or shear-thinning fluid, the dynamic viscosity decreases as the shear rate increases. On the other hand, a shear-thickening or dilatant fluid exhibits the converse property of pseudo-plastic for which the dynamic viscosity increases as the shear rate increases. The shear thickening fluid is represented by $n > 1$ and shear thinning fluid by $n < 1$.

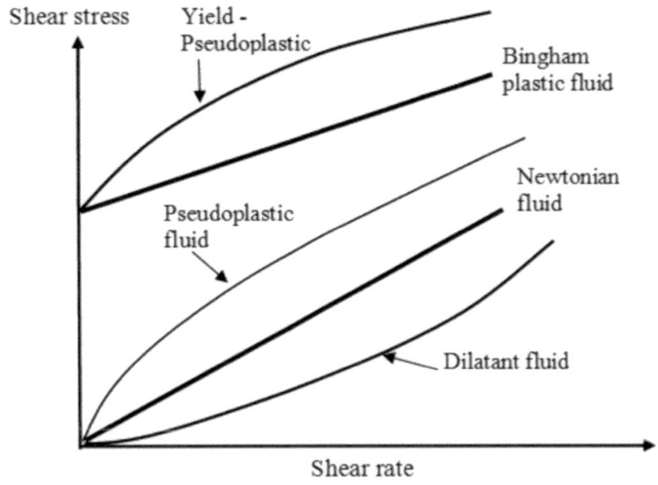

Figure 2. Classification of time-independent non-Newtonian fluid

2.4. Rheology of MRF

MRF responds to the external field, where the particles are held together to form chains parallel to the applied field. The interaction between the particles impedes to a certain level of the shear stress without breaking and simultaneously increases the viscosity of the fluids [32]. In many cases, the effect of MRF is described by Bingham Plastic model [33]. A modified or extended Bingham model, or a combination of Bingham model with other models such as viscous and coulomb friction have also been used to describe the behaviour of MRF [34]. In the absence of an external field, MRF behave like a normal fluid which is known as Newtonian fluid. There are many factors that influence the rheological properties of controllable MRF such as concentration and density of particles, particle size and shape distribution, properties of the carrier fluid [35], additional additives, applied field and temperature. The relationships of all these factors are very complex and are important in establishing methodology to improve efficiency of these fluids for suitable applications. Excellent MRF must have low viscosity and coercivity of particles without the influence of an external magnetic field and can achieve maximum yield stress in the presence of the external magnetic field. Gross [8] in his invention related to the valve for magnetic fluids, found that the advantage of large particle sizes or heavy suspensions can increase the size of the gap which also increases the flow of the fluid. Conversely, the large particles of the magnetically active phase of MRF lead to a strong tendency for particles to settle out of the liquid phase [19].

Some of the techniques are typically necessary in order to increase the yield stress; either by increasing the volume fraction of MR particles or by increasing the strength of the applied magnetic field. However, neither of these techniques is desirable since a higher volume fraction of the MR particles can add significant weight to the MR devices as well as increases the overall off-state viscosity of the material. In that connection, restricting the size and

geometry of the MR device capable of utilizing that material, and a higher magnetic field significantly increases the power requirement of the device. To overcome this difficulty, Carlson [36] in his patent introduced alloy-particles material that was used as a solid particle instead of the common carbonyl iron. This MRF independently increases the yield stress without requiring increment of either the volume fraction of particles or magnetic field strength.

2.5. MRF models

MRF models play an important role in the development of MRF based devices. Moreover, accurate models that can predict the performance of these MRF devices are an important part of implementation of such devices. MRF demonstrates nonlinear behavior when subjected to external magnetic fields. The rheological behavior of these materials can be separated into distinct preyield and post-yield regimes. A wide variety of nonlinear models have been used to characterize MRF, including the Bingham plastic model [37, 38], the biviscous model [39], the Herschel–Bulkley model [40, 41], and Eyring plastic model [42]. Although there have been several models have been developed and applied for MRF the two most popular models have been widely used with reasonable accuracy and computational cost are the Bingham plastic model and the Herschel–Bulkley plastic model. Therefore, in this chapter these two constitutive models are used.

i. Bingham plastic model

The so-called Bingham plastic model includes a variable rigid perfectly plastic element connected in parallel to a Newtonian viscosity element. This model assumes that the fluid exhibits shear stress proportional to shear rate in the post-yield region and can be expressed as [37, 38]

$$\tau = \tau_y(H)\operatorname{sgn}(\dot{\gamma}) + \eta\dot{\gamma} \tag{10}$$

where τ is the shear stress in the fluid, τ_y is the yielding shear stress controlled by the applied field H, η is the post-yield viscosity independent of the applied magnetic field, $\dot{\gamma}$ is the shear strain rate and sgn(·) is the signum function. That is, the fluid is in a state of rest and behaves viscoelastically until the shear stress is greater than the critical value τ_y, whereas it moves like a Newtonian fluid when such a critical value is exceeded. The Bingham plastic model is shown in Figure 3 to represent the field-dependent behaviour of the yield stress. The simplicity of this two-parameter model has led to its wide use for representation of field-controllable fluids, especially ER and MRF.

ii. Herschel-Bulkley plastic model

In cases where the fluid experiences post-yield shear thickening or shear thinning, especially when the MRF experiences high shear rate, this choice of constitutive equation can result in an overestimation. In this case, the Herschel-Bulkley plastic model is more suitable [43]. The Herschel-Bulkley model can be expressed by

$$\tau = (\tau_y(H)\mathrm{sgn}(\dot{\gamma}) + K|\dot{\gamma}|^{1/m})\mathrm{sgn}(\dot{\gamma}) \quad (11)$$

where K is the consistency parameter and m is fluid behavior index of the MRF. For $m > 1$, equation (4) represents a shear thinning fluid, while shear thickening fluids are described by $m < 1$. Note that for $m = 1$ the Herschel–Bulkley model reduces to the Bingham model.

It is noteworthy that, in the above, the post-yield parameters of the MRF such as the post-yield viscosity, the consistency parameter and the fluid behavior index are assumed to be independent on the applied magnetic field. In practice, these parameters are slightly affected by the applied magnetic field. Zubieta et al. [44] have proposed field-dependent plastic models for MRF based on the original Bingham plastic and Herschel-Bulkley plastic models. The models were then applied in several researches [45, 46] with experimental agreement. In the field dependent Bingham and Herschel-Bulkley model, the rheological properties of MRF depend applied magnetic field and can be estimated by the following equation

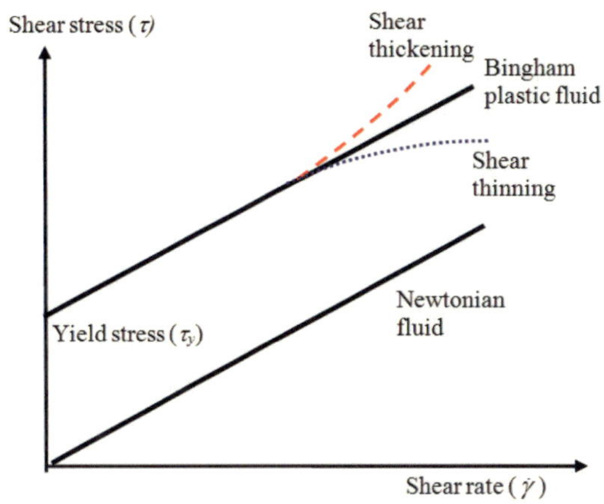

Figure 3. Viscoplastic models often used to describe MR fluids

$$Y = Y_\infty + (Y_0 - Y_\infty)(2e^{-B\alpha_{SY}} - e^{-2B\alpha_{SY}}) \quad (12)$$

where Y stands for a rheological parameters of MRF such as yield stress, post yield viscosity, fluid consistency and flow behavior index. The value of parameter Y tends from the zero applied field value Y_0 to the saturation value Y_∞ α_{YS} is the saturation moment index of the Y parameter. B is the applied magnetic density. The values of Y_0, Y_∞ α_{YS} are determined from experimental results using curve fitting method.

3. MRF mode of operation and its application

3.1. Valve mode

Figure 4a schematically show the valve mode which have been used in many MR devices where the flow of the MRF between motionless plates or a duct is created by a pressure drop. The magnetic field, which is applied perpendicular to the direction of the flow, is used

to change the rheological properties of the MRF in order to control the flow. Therefore, the increase in yield stress or viscosity alters the velocity profile of the fluid in the gap between two plates. A typical velocity profile for Bingham-plastic of the valve mode is illustrated in Figure 4b. The velocity profile contains a pre-yield region, where the velocity gradient is zero across the plug region. The velocity profile of MRF between two parallel plates can be represented by the following relation [47]

$$u_1(y)=\frac{n}{n+1}(\frac{\Delta P}{KL})^{1/n}\left[(\frac{d-\delta}{2})^{1+1/n}+(\frac{2y+\delta}{2})^{1+1/n};\ u_2(y)\right]=\frac{n}{n+1}(\frac{\Delta P}{KL})^{1/n}(\frac{d-\delta}{2})^{1+1/n}$$
$$u_3(y)=\frac{n}{n+1}(\frac{\Delta P}{KL})^{1/n}\left[(\frac{d-\delta}{2})^{1+1/n}-(\frac{2y-\delta}{2})^{1+1/n}\right] \tag{13}$$

Here, $n=1/m$, u_1 and u_3 are velocity profiles of the post-yield flow regions adjacent to the walls of the rectangular duct, and u_2 is the velocity profile across the central pre-yield or plug region. δ is the plug region thickness, which is a key parameter of the flow. As the field increases, so does the pre-yield thickness, thereby, constricting the flow through the duct, increasing the pressure drop. A high resistance produced by the valve mode can be used in many applications such as dampers, valves and actuators [48-51].

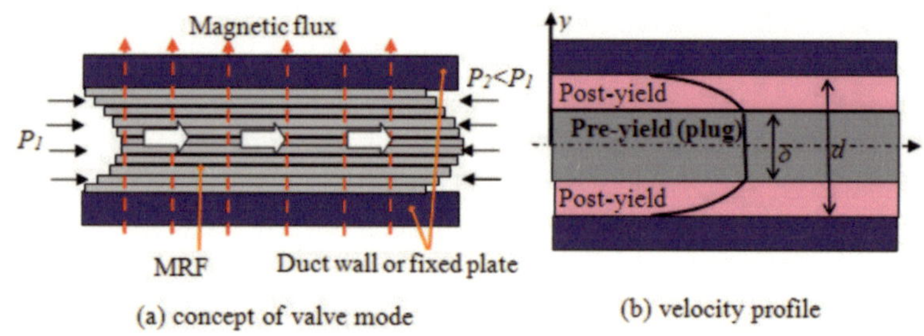

Figure 4. Valve mode in the MR application

3.2. Shear mode

The second working mode for controllable fluid devices is the direct shear mode. An MRF is situated between two surfaces, whereby one surface slides or rotates in relation to the other, with a magnetic field applied perpendicularly to the direction of motion of these shear surfaces. Figure 5 shows the concept of shear mode in MRF application

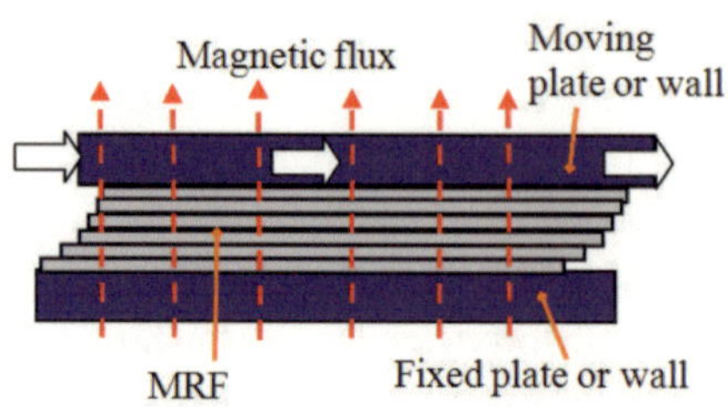

Figure 5. The concept of shear mode

The direct shear mode has been studied thoroughly especially in the MR damper technology. Masri et al. [52] proposed a curve fitting technique for representing the nonlinear restoring force of an ER device in order to characterize the ER material behaviour under static and dynamic loading over a wide range of electric fields. Spencer et al. [53] developed a phenomenological model which is based on the improved Bouc-Wen hysteresis model to represent MR dampers. Moreover, Wereley et al. have proposed a non-dimensional approach to model different types of shear damper (linear shear mode, rotary drum and rotary disc damper) [54]. In the research, the Bingham–plastic, biviscous, and Herschel–Bulkley models are considered. In terms of the behaviour of the damper under conditions of high-velocity and high field input, Lee et al. [55] recommended the Herschel-Bulkley shear model to analyze the performance of impact damper systems. Furthermore, Neelakantan et al. [56] incorporated a volume fraction profile of particles with an analytical technique for calculating the torque transmitted in clutches experiencing particle centrifuging. The effect of centrifuging at high rotational speeds and the subsequent sealing problems associated with it can be mitigated by the proposed model. Extraordinary features of the direct shear mode like simplicity, fast response, simple interface between electrical power input and mechanical power output using magnetic fields, and controllability are features that make MRF technology suitable for many applications such as dampers, brakes, clutches and polishing devices [56-59].

3.3. Squeeze mode

The third working mode of MRF is the squeeze mode shown in Figure 6. This mode has not been widely investigated. Squeeze mode operates when a force is applied to the plates in the same direction of a magnetic field to reduce or expand the distance between the parallel plates causing a squeeze flow. In squeeze mode, the MRF is subjected to dynamic (alternate between tension and compression) or static (individual tension or compression) loadings. As the magnetic field charges the particles, the particle chains formed between the walls become rigid with rapid changes in viscosity. The displacements engaged in squeeze mode are relatively very small (few millimeters) but require large forces. The squeeze mode was disclosed by Stanway et al. [60] in 1992. They studied the usage of ER fluids in squeeze mode and found that the yield stress produced under DC excitation could be several times greater than the shear mode. The same outcome was later confirmed by Monkman [61] for fluids under compressive stress. Consequently, systematic investigations have been carried out by many researchers to evaluate the mechanical and electrical properties of ER and MRF in squeeze flow. Despite the fact that the Bingham plastic model was used to describe the behaviour of ER fluid in shear mode, Nilsson and Ohlson [62] have not recommended to utilize that model in squeeze mode. Bingham parameters tested from shear mode are not well-founded for the calculation of the squeeze mode behaviour. Sproston et al. [63] characterized the performance of ER fluids in dynamic squeeze mode using a bi-viscous model under a constant potential difference or by a constant field. Later on, Sproston and El Wahed [64] utilized the model to assess the fluid's response to a step-change in the applied field and the influence of the size of solid phase. Even though the model was useful to

predict the peak values of the input and transmitted forces [65], a more refined model is needed, according to authors, to predict the detailed temporal variations. Therefore, a new approach of modified bi-viscous model was developed by El Wahed et al. in order to model the behaviour of an ER squeeze mode cell under dynamic conditions [66, 67]. Furthermore, Yang and Zhu [68] extended this latter model by incorporating Navier slip condition to obtain the radial velocity, pressure gradient, pressure and squeeze force.

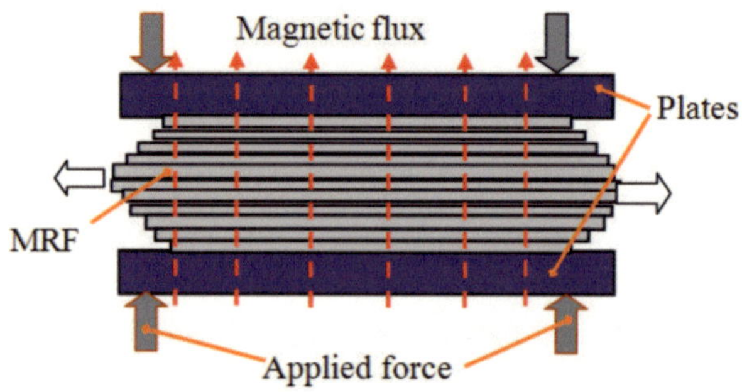

Figure 6. The concept of squeeze mode

The stress produced by the squeeze mode is the highest stress among other modes and can be used in damping vibrations with low amplitudes and high dynamic forces [69, 70]. In vibration isolation of structural system, the unwanted vibration in a relatively high frequency range can be attenuated by activating the MR mount. Examples for vibration control are isolation engine mount [71], turbo-machinery [72] and squeeze film damper [73]. Another interesting application on the squeeze mode is related to haptic devices where the user can feel the resistance forces by touching and moving a tool [74].

3.4. Combination of modes

Some of the applications of field responsive fluids take advantage of the combination of two modes for a greater strength and functionality. For instance, dampers can be constructed in three different modes [75]. In a general manner, shear mode exhibits Couette flow through the annular bypass, while a valve flow is characterized by Poiseiulle flow through the annular bypass. The combination of them often gives higher yield stress as compared to stress produced by individual operation modes. Kamath et al. [76] have shown in their analysis and testing of Bingham plastic behaviour that mixed (valve and shear) mode dashpot dampers have higher passive damping than flow mode dampers. The mixed mode damper has a secondary effect of viscous drag as a result of the motion of piston head, instead of relying on the pressure gradient developed by the piston head to push the fluid through the gap created by the fixed electrodes. Wereley and Pang [75] have developed nonlinear quasi-steady ER and MR damper models using idealized Bingham plastic shear flow mechanism to characterize the equivalent viscous damping constant of the dampers. Plug thickness is the strongest variable that contributed to the damper behaviour for both flow and mixed modes.

In another experimental study done by Kulkarni et al. [62], the performance ofhe combination of squeeze and shear modes of MRF in dynamic loading was investigated. Even though squeeze mode can produce the highest strength among all modes, the addition of squeeze mode to shear mode did not always give a better strength than the shear mode alone. However, Tang et al. [77] demonstrated that the yield shear tress can be significantly improved by compressing the MRF along the magnetic field direction before the shear process is performed.

4. Optimal design methodology of MRF-based mechanisms

4.1. Modeling of MRF based mechanisms

It is well-known that modeling of the MRF based systems is a coupled analysis problem: electromagnetic analysis and fluid system analysis. The purpose of the modeling of an MRF based device is to find the relation between the applied electric power (usually the current applied to the coils) and the output mechanical power such as pressure drops for MR valves, damping force for MR damper, braking torque for MR brakes and transmitted torque for MR clutches. In order to deal with modeling of MRF based devices, firstly the magnetic circuit of the MRF based devices should be solved. In general, the magnetic circuit can be analyzed using the magnetic Kirchoff's law as follows:

$$\sum H_k l_k = N_{turns} I \tag{14}$$

where H_k is the magnetic field intensity in the kth link of the circuit and l_k is the effective length of that link. N_{turns} is the number of turns of the valve coil and I is the applied current. The magnetic flux conservation rule of the circuit is given by

$$\Phi = B_k A_k \tag{15}$$

where Φ is the magnetic flux of the circuit, A_k and B_k are the cross-sectional area and magnetic flux density of the kth link, respectively. It is noteworthy that the more links are used the more exact solution can be obtained. However, this increases computation load. At low magnetic field, the magnetic flux density, B_k, increases in proportion to the magnetic intensity H_k as follows:

$$B_k = \mu_0 \mu_k H_k \tag{16}$$

where μ_0 is the magnetic permeability of free space (μ_0= $4\pi 10^{-7} Tm/A$) and μ_k is the relative permeability of the kth link material. As the magnetic field becomes large, its ability to polarize the magnetic material diminishes and the material is almost magnetically saturated. Generally, a nonlinear *B-H* curve is used to express the magnetic property of material. At low magnetic field, taking the linear relation (16) into consideration, the magnetic flux density and the field intensity of the kth link of magnetic circuit can be approximately calculated as follows:

$$B_k = \frac{\mu_0 N_{turns} I}{\frac{l_k}{\mu_k} + \sum_{i=1, i \neq k}^{n} \frac{l_i A_k}{\mu_i A_i}} \tag{17}$$

$$H_k = \frac{N_{turns} I}{l_k + \sum_{i=1, i \neq k}^{n} \frac{\mu_k A_k}{\mu_i A_i} l_i} \tag{18}$$

By assuming magnetic property of the structural materials of the MR devices is similar ($\mu_1 = \mu_2 = \ldots \mu_n = \mu$), the magnetic flux density and the field intensity across the active MRF volume can be approximately calculated as follows:

$$B_{mr} = \frac{\mu_0 N_{turns} I}{\frac{l_{mr}}{\mu_{mr}} + \frac{1}{\mu} \sum_i \frac{l_i A_{MR}}{A_i}} \tag{19}$$

$$H_{mr} = \frac{N_{turns} I}{l_{mr} + \frac{\mu_{mr} A_{mr}}{\mu} \sum_i \frac{l_i}{A_i}} \tag{20}$$

where μ_{mr} and μ are the relative permeability of MRF and the structural materials of the MR devices, respectively. It is noted that the permeability of the MRF is much smaller than that of the valve core material, therefore from Eq. (20) the magnetic field intensity of the MRF link can be approximated by

$$H_{mr} = N_{turns} I / l_{mr} \tag{21}$$

The inductive time constant (T_{in}) and the power consumption (N) of the MRF based devices can be calculated as follows:

$$T_{in} = L_{in} / R_w \tag{22}$$

$$N = I^2 R_w \tag{23}$$

where L_{in} is the inductance of the coil given by $L_{in} = N_{turns} \Phi / I$, R_w is the resistance of the coil wire which can be approximately calculated by

$$R_w = L_w r_w = N_{turns} \pi \bar{d}_c \frac{r}{A_w} \tag{24}$$

In the above, L_w is the length of the coil wire, r_w is the resistance per unit length of the coil wire, $\bar{d}_c$ is the average diameter of the coil, A_w is the cross sectional area of the coil wire, r is the resistivity of the coil wire, r = 0.01726E-6Ωm for copper wire, N_{turns} is the number of coil turns which can be approximated by $N_{turns} = A_c / A_w$, and A_c is the cross-sectional area of the coil.

In some applications, an electromagnet is used in combination with a permanent magnet to control the rheological properties of the MRF as shown in Figure 7. In this case, the permanent magnet is used to shift the off-state (no current in the coil) viscosity of the MRF to a selected value and the electromagnet is used to control the viscosity variations around this value. A frequent situation is that where the magnetic circuit is designed in such a way that the MRF viscosity is maximum when no current flows through the coil. This is particularly useful when the device based on such magnetic circuit has to be blocked the major part of its operation time (such as in release mechanisms, for instance). The magnetic intensity across the active volume of the MRF is determined by [78]

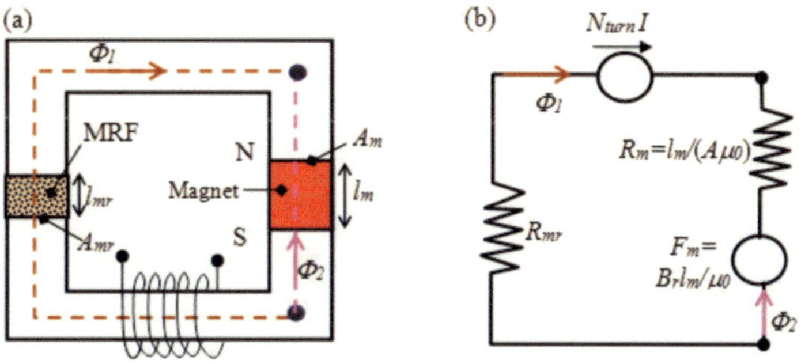

Figure 7. Magnetic coil with MRF filled gap and permanent magnet: a) magnetic circuit b) electric equivalence

$$H_{mr} = \frac{(\mu_r/l_m)N_{turn}I - B_r / \mu_0}{\mu_r l_{mr} / l_m + \mu_{mr} A_{mr} / A_m} \tag{25}$$

where l_m and A_m are the length and cross-sectional area of the permanent magnet, B_r and μ_r are remanent flux density and relative permeability of the magnet ($\mu_r \cong 1$). In order to cancel the flux inside the MRF, *NI* has to be equal to the magnetomotive force, $F_m = B_r l_m / \mu_0$, which is highly influenced by the magnet length. It should however be noticed that such a magnetic circuit will not be used in practice since it might lead to demagnetization of the permanent magnet. A solution to this problem is to include a secondary path inside the magnetic circuit as shown in Figure 8 [78]

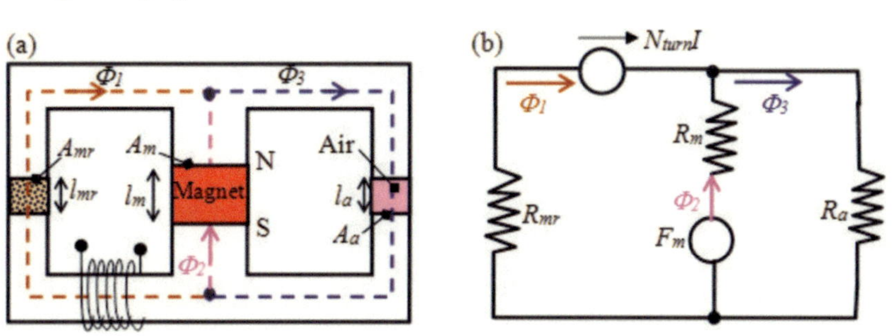

Figure 8. Magnetic coil with MRF filled gap and permanent magnet with secondary path: a) magnetic circuit b) electric equivalence

The electromagnet will thus not be used to completely cancel the flux produced by the permanent magnet but will only redirect it to the secondary path. This secondary path comprises a higher reluctance air gap in order to concentrate the major part of the flux generated by the permanent magnet in the primary path (comprising the MRF gap) when no current is flowing through the coil. In this case, the magnetic intensity across the active volume of the MRF is determined by [78]

$$H_{mr} = \frac{(\mu_r/l_m + A_a\mu_a / A_m l_a)N_{turn}I - B_r / \mu_0}{\mu_r l_{mr} / l_m + A_a \mu_a l_{mr} / A_m l_a + \mu_{mr} A_{mr} / A_m} \tag{26}$$

It is interesting to note that, if $g_a \rightarrow \infty$, we come back to Eq. (25). In order to cancel the magnetic flux inside the MRF gap, we need:

$$N_{turn}I = \frac{B_r l_m}{\mu_r \mu_{m0}\left[(\frac{A_a}{A_m})(\frac{l_m}{l_a})(\frac{\mu_a}{A_r})+1\right]} \tag{27}$$

This value may seem smaller than what was obtained in the previous case; however, to obtain the same magnetic field inside the MRF gap, the magnet has to be more powerful since it has to compensate for the loss of magnetic flux in the secondary circuit.

In the above, magnetic circuit of the MRF based devices is solved based on the approximation of the analytical analysis. This approach can only used in case of simple geometry. In case of complex geometry or several coils are used, the approach becomes very complicated. Therefore, practically, the magnetic circuit of the MRF based devices is solved by finite element method (FEM). Once the magnetic solution is obtained, the magnetic intensity and magnetic flux density across the active MRF volume can be calculated. The rheological properties of MRF in the active volume are then determined based on the behaviour characteristics of the employed MRF. The behavior characteristics of MRF are usually obtained from experimental results with a curve-fitting algorithm. The most important parameter of MRF is the field-dependent yield stress. There have been several approximate functions have been used to express the dependence of the induced yield stress of MRF on the applied magnetic field. The two most widely used functions are the exponential function and the polynomial function. The former can well expressed the saturation of MRF yield stress as a function of the applied magnetic intensity. However, it exhibits large error at the small value of the applied magnetic intensity. In general, the approximate exponential function of induced yield stress is expressed as following

$$\tau_y(H) = \tau_0 + \alpha H^{\beta} \tag{28}$$

where $\tau_y(H)$ is the induced yield stress of MRF as a function of the applied magnetic intensity (H), α and β are the curve parameters determined from experimental results using a curve-fitting algorithm, and τ_0 is the zero-field yield stress of the MRF.

The latter, the approximate polynomial function, can well predict the MRF yield stress at small value of the applied magnetic intensity. The higher order of the polynomial is the more accurate value of the yield stress can be predicted. In practice, the third order polynomial is often used. However, the polynomial function can not express the saturation of the induced yield stress. Therefore, a saturation condition should be added. The 3rd order approximate polynomial function of MRF yield stress can be expressed by

$$\tau_y(H) = \tau_0 + c_1 H + c_2 H^2 + c_3 H^3 \tag{29}$$

where c_1, c_2, and c_3 are the curve parameters determined from experimental results using a curve-fitting algorithm, and τ_0 is the zero-field yield stress of the MRF.

In many researches, other characteristics of MRF such as the post yield viscosity (η) in Bingham model, the consistency parameter (K) and the fluid behavior index (m) are determined from experimental results on rheological properties of MRF and assumed to be independent of the applied magnetic. However, in practice, these parameters are slightly affected by the applied magnetic field. In order to take this into account, Zubieta et al. [44] have proposed a field-dependent plastic model for MRF based on original Bingham plastic and Herschel-Bulkley plastic models as mention in Section 2.5. Once obtaining the yield stress and other rheological parameters of the MRF, the output mechanical power such as pressure drops, damping force for MR damper, braking torque and transmitted torque can be determined on governing equations of the MRF based devices.

4.2. Optimization problems in design of MRF based devices

As aforementioned modeling of MRF based systems is a coupled analysis problem. Therefore, output mechanical power of these systems depends not only on their mechanics behaviors but also on their magnetic circuits. It is obvious that in order to improve performance of the MRF based systems, the optimal design should be taken into account. Generally, the objective of the optimal design is to find significant geometric dimensions of the MRF based devices that maximize an objective function considering typical characteristics such as pressure drop, damping force, dynamic range, braking torque, transmitted force, mass, time response constant and power consumption. Some constraints such as available space, allowable operating temperature, uncontrollable torque etc. may be also considered in the optimal design. There have been several researches focusing on optimal design of MRF devices. Rosenfield and Wereley [48] proposed analytical optimization design method for MR valves and dampers based on the assumption of constant magnetic flux density throughout the magnetic circuit to ensure that one region of the magnetic circuit does not saturate prematurely and cause a bottleneck problem. Nguyen et al. [79] proposed a FEM based optimal design of MR valves (single-coil, two-coil, three-coil and radial-annular types) constrained in a specified volume. This work considered the effects of all geometric variables of MR valves by minimizing the valve ratio calculated from the FE analysis. Later on Nguyen el al. [80] have developed an optimization procedure based on the finite element method in order to find the optimal geometry of MR valves

constrained to a specific volume, satisfying a required pressure drop with minimal power consumption. The time response of the valves was also taken into account by considering the inductive time constant as a state variable. The optimization results showed the significance of the optimal design of the MR valves in order to minimize the power consumption. It was also shown that the wire diameter does not significantly affect the optimization solution and can be neglected. The optimal design of MR damper was also performed by Nguyen et al [81], in which the objective function was proposed by a linear combination of the ratios of the damping force, dynamic range and the inductive time constant and their reference values using corresponding weighting factors. Recently, there have been several researches on the optimal design of MR brakes and clutches. Park et al. [82] have performed multidisciplinary design optimization of an automotive MR brake, in which a multi-objective function considering both braking torque and mass of the brake was considered. Nguyen et al. [45] have performed a thorough research on optimal design of MR brake for middle-sized vehicle considering the available space, mass, braking torque and steady heat generated by a zero-field friction torque of the MR brake on cruising at a speed of 100*km/h*. Furthermore, different configurations of MR brake and different types of MRF are taken into account in that research. More recently, Nguyen et al. [83] have performed the optimal design of common types of MR brakes such as disc-type, drum-type, inverted drum-type, single-coil hybrid-types, inverted single-coil hybrid-types, two-coil hybrid-types, inverted two-coil hybrid-types and T-type. The objective of the optimization was to maximize the braking torque while torque ratio (the ratio of maximum braking torque and the zero field friction torque) is constrained not to exceed a certain value. Based on the optimal solutions, the advices on optimal selection of MR brakes type were addressed. It was showed that the guide on optimal selection of MR brake types can be applied for different types of MRF and different constrains of torque ratio.

4.3. Optimal design of MRF devices based on finite element analysis

As abovementioned, the magnetic circuit of the MRF based devices can be solved by an approximation of analytical solution or by FEM. Therefore, the optimal design of these devices can be performed based on either the analytical analysis or finite element analysis (FEA). The former is used only for simple devices such as single coil MR damper [84]. In this section, the optimal design of MRF devices based on FEA is introduced. First of all, an objective function should be proposed depending on the purpose of the optimal design and the application of the devices. It is noted that in the optimization problem the objective function is always minimized. Therefore, if the purpose of the optimization is to maximize a performance function of the devices, that function should be transform to an equivalent objective function. The equivalent objective function is the function that when it is minimized, the corresponding performance function is maximized. After the objective function is constructed, the design parameters of the optimization problem should be identified. In addition, the constraints of the optimization problem should be determined if there any. In the next step, an algorithm to obtain the optimal solution should be chosen. It is well-known that there have been numerous methods to find the optimal solution of an

optimization problem. They may be non-derivative, first-derivative or second-derivative methods. The non-derivative methods that do not require any derivative of the function are not usually applied to MRF based systems. Although they are generally easy to implement, their convergence properties are rather poor. They may work well in special cases when the function is quite random in character or the variables are essentially uncorrelated. Some typical non-derivative algorithms are the Simplex, Genetic Algorithms and Neural Networks. The second-derivative optimization methods are characterized by fast convergence and affine invariance. However, they require second derivatives and the solution of linear equation can be too expensive for large scale applications. The most popular optimization method, which is widely used in optimal design of MRF based devices, is the first order (derivative) method. Although the convergence rate of the first derivative method is somewhat slower than that of the second-derivative one, the first derivative method is still preferred in many applications because of its inexpensive cost for computation and programming. A typical first derivative optimization algorithm is the conjugate gradient method. The flow chart in Figure 9 shows how to find the optimal solution of MRF based devices based on ANSYS finite element software using the first order method. The procedures from the flow chart can be easily extended to other finite element software.

First of all, initial value of the design variables (DV) should be decided. Computation time of the optimization process significantly depends on the initial value of the DVs. Therefore, the initial value of design variables should be calculated based on a draft calculation or based on practical experience. Then, an analysis file for solving the magnetic circuit and calculating performance characteristics of the devices such as control energy, the inductive time constant, pressure drops, damping force, braking torque and transmitted torque is built. In ANSYS, the analysis file is built using parametric design language (APDL). It is noted that this analysis file can be created from a graphic user interface (GUI) model of ANSYS by using the *list>log file* submenu from the *file* menu of the ANSYS software. In the analysis file, the DVs must be used as symbolic variables and initial value is assigned to them. Generally, in order to calculate performance characteristics of the devices, the magnetic flux density across the active volume of MRF should be calculated. The magnetic flux density (*B*) and magnetic intensity (*H*) are not constant along the MRF duct, so an average should be used. The average magnetic flux density and intensity across the MR ducts was calculated by integrating flux density along a predefined path, then divided by the path length [79, 80]. In order to calculate the inductive time constant, firstly the magnetic flux is determined as follows:

$$\Phi = 2\pi R_d \int_{L_p} B(s)ds \tag{30}$$

where *B(s)* is the magnetic flux density at each nodal point on the path, *s* is a dummy variable for the integration. The integration was performed along the path length, L_p. It is noteworthy that geometric dimensions of the MRF devices change during the optimization process, so that the meshing size of FE model should be specified by the number of elements per line rather than element size.

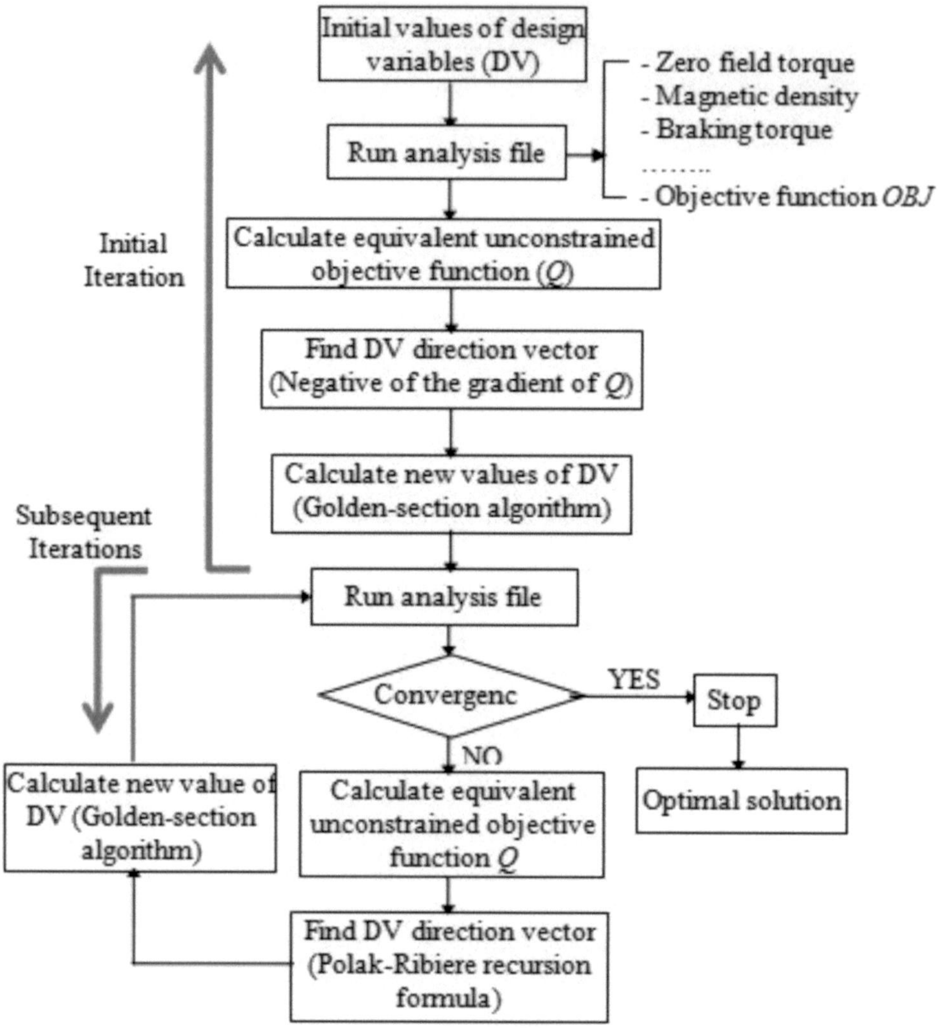

Figure 9. Flow chart for optimal design of MRF based devices using FEM.

After the analysis file is prepared, the procedures to achieve optimal design parameters of the MRF devices using the first order method of ANSYS optimization tool are performed as shown in Figure 9. Starting with initial value of DVs, by executing the analysis file, the initial value of the performance characteristics of the devices such as control energy, the inductive time constant, pressure drops, damping force, braking torque and transmitted torque are obtained. The ANSYS optimization tool then transforms the constrained optimization problem to an unconstrained one via penalty functions. The dimensionless, unconstrained objective function is formulated as follows:

$$Q(x,q) = \frac{OBJ}{OBJ_0} + \sum_{i=1}^{n} P_x(x_i) + q\sum_{i=1}^{m} P_g(g_i) \tag{31}$$

where OBJ_0 is the reference objective function value that is selected from the current group of design sets, q is the response surface parameter which controls constraint satisfaction. P_x is the exterior penalty function applied to the design variable x. P_g is extended-interior penalty function applied to state variable (the constraint) g. For the initial iteration ($j = 0$), the search direction of DVs is assumed to be the negative of the gradient of the unconstrained objective function. Thus, the direction vector is calculated by

$$d^{(0)} = -\nabla Q(x^{(0)},1) \tag{32}$$

The values of DVs in next iteration ($j+1$) is obtained from the following equation,

$$x^{(j+1)} = x^{(j)} + s_j d^{(j)} \tag{33}$$

where the line search parameter s_j is calculated by using a combination of a golden-section algorithm and a local quadratic fitting technique. The analysis file is then executed with the new values of DVs and the convergence of the objective function, OBJ, is checked. If the convergence occurs, the values of DVs at the jth iteration are the optimum. If not, the subsequent iterations will be performed. In the subsequent iterations, the procedures are similar to those of the initial iteration with the exception of the direction vectors which are calculated according to Polak-Ribiere recursion formula as follows:

$$d^{(j)} = -\nabla Q(x^{(j)},q_k) + r_{j-1} d^{(j-1)} \tag{34}$$

$$r_{j-1} = \frac{[\nabla Q(x^{(j)},q) - \nabla Q(x^{(j-1)},q)]^T \nabla Q(x^{(j)},q)}{\left|\nabla Q(x^{(j-1)},q)\right|^2} \tag{35}$$

Thus, each iteration is composed of a number of sub-iterations that include search direction and gradient computations.

It is noted that ANSYS software supports optimal design problems by integrating an optimization tool. Therefore, in most cases the optimal solution of the MRF based devices can be solved directly by the ANSYS software without interfacing with any programming software. In order to use the ANSYS optimization tool, it is necessary to set up optimization parameters. To do this, firstly the analysis file should be manually executed once to load all parameters in the analysis file into software buffer memory. After that, from the *Design Opt* menu, we specify the analysis file which will be used during optimization process, the DVs with their limits and tolerances, the state variables (if there are any) with limits and tolerances, the objective function with a convergence criteria, the method for solving the optimal solution, and the optimal output control option if necessary. In some cases, it is expected to employ some advanced optimization algorithms such as Genetic Algorithms, Neural Network, or user defined algorithms, the interfacing between the ANSYS and other software to perform the optimization such as Matlab, FORTRAN, C languages is required.

5. Case studies on optimal design of MR valves

5.1. Configuration and modeling of MR valve

Figure 10 shows the structural configurations of the two typical types of MR valves: the annular MR relief valve (Figure 10a) and MR valve with both annular and radial flow paths (Figure 10b). The valve in Figure 10a consists of valve coil, cores and covers. The MRF flows through annular ducts between core A and core B. When the power of the coil is turned on, a magnetic field is exerted on the MRF, which causes the MRF flowing through the ducts to change its state into semi-liquid or solid and stop the flow. Only when the supply pressure gets high enough to offset the yield stress, the fluid can flow through the valve again. The valve in Figure 10b consists of the valve core, magnetic disk and valve housing form a magnetic circuit of the valve. A non-magnetic washer is used to warrant the required thickness of the radial duct. When the magnetic disk is placed coaxially with the valve housing using the cone-shape cap, the annular and radial ducts are formed between the disk and the valve housing, and the disk and the valve core, respectively. MRF flows from the inlet through the first annular and radial duct, then flow along the hole at the center of the core and after that follows the second radial and annular duct to the outlet.

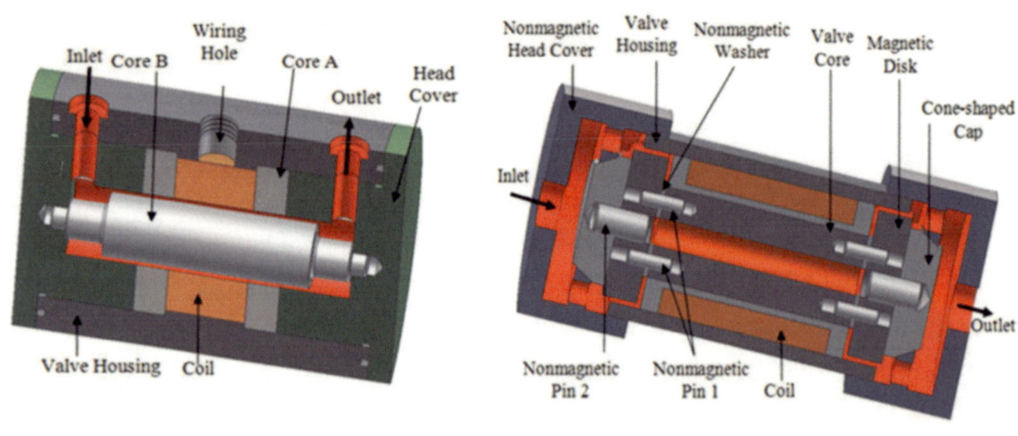

(a) annular MR valve (b) MR valve with both annular and radial flow paths

Figure 10. Schematic diagrams of MR valves.

Figure 11a shows a simplified structure and significant dimensions of a single-coil annular MR valve. The valve geometry is featured by the overall effective length L, the outside radius R, the valve housing thickness t_h, the MR duct gap t_g, the core flange (pole) thickness t_f, and the coil width w_c.

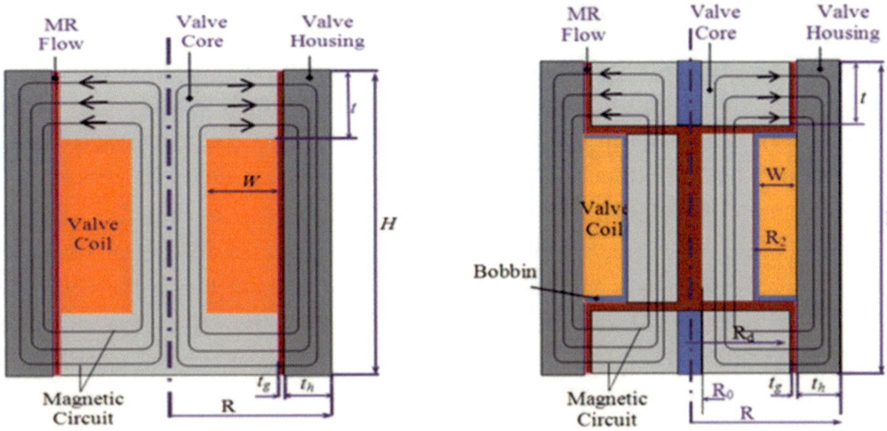

(a) single-coil annular MR valve (b) annular-radial flow MR valve

Figure 11. Simplified MR valve configurations

By using Bingham plastic model, the pressure drop of the valve is calculated by [80, 81]

$$\Delta P_A = \Delta P_{A,\eta} + \Delta P_{A,\tau} = \frac{6\eta L}{\pi t_g^3 R_d} Q + 2c \frac{t_f}{t_g} \tau_y \tag{36}$$

where $\Delta P_{A,\tau}$ and $\Delta P_{A,\eta}$ are the field-dependent and viscous pressure drop of the single annular MR valve, respectively, Q is the flow rate through the MR valve, R_d is the average radius of annular duct given by $R_d = R - t_h - 0.5t_g$, c is the coefficient which depends on flow velocity profile and has a value range from a minimum value of 2.0 (for $\Delta P_{A,\tau}/\Delta P_{A,\eta}$ greater than 100) to a maximum value of 3.0 (for $\Delta P_{A,\tau}/\Delta P_{A,\eta}$ less than 1).

The multi-coil MR valve, which was first employed by Spencer et al. to make a high damping force MR damper used in seismic protection system [85], is now widely used in many applications. For MR valve with two coils, the pressure drop is calculated by

$$\Delta P_{2A} = \Delta P_{2A,\eta} + \Delta P_{2A,\tau} = \frac{6\eta L}{\pi t_g^3 R_d} Q + 2c_1 \frac{t_{f1}}{t_g} \tau_{y1} + c_2 \frac{t_{f2}}{t_g} \tau_{y2} \tag{37}$$

where $\Delta P_{2A,\tau}$ and $\Delta P_{2A,\eta}$ are the field-dependent and viscous pressure drop of the two- coil annular MR valve respectively, τ_{y1} and τ_{y2} are the yield stresses of the MRF in the end ducts and the middle duct, respectively. c_1 and c_2 are coefficient which depends on flow velocity profile of MR flow in the end ducts and the middle duct, respectively.

Similarly, for the three-coil annular MR valve, the pressure drop is calculated by

$$\Delta P_{3A} = \Delta P_{3A,\eta} + \Delta P_{3A,\tau} = \frac{6\eta L}{\pi t_g^3 R_d} Q + 2(c_1 \frac{t_{f1}}{t_g}\tau_{y1} + c_2 \frac{t_{f2}}{t_g}\tau_{y2}) \quad (38)$$

For the MR valve with both annular and radial flow paths shown in Figure 11b, the pressure drop can be calculated by

$$\Delta P_{AR} = \Delta P_{AR,\eta} + \Delta P_{AR,\tau} \quad (39)$$

where $\Delta P_{AR,\tau}$ and $\Delta P_{AR,\eta}$ are determined by

$$\Delta P_{AR,\tau} = 2(c_a \frac{t_f}{t_g}\tau_{ya} + c_r \frac{R_2 - R_0}{t_g}\tau_{yr}) \quad (40)$$

$$\Delta P_{AR,\eta} = 2\left[\frac{6\eta t_f}{\pi t_g^3 R_d} Q + \frac{6\eta Q}{\pi t_g^3}\ln(\frac{R_d}{R_o})\right] + \frac{8\eta(L - 2t_f)Q}{\pi R_0^4} \quad (41)$$

In the above, τ_{ya} and τ_{yr} are the induced yield stresses of the MRF in the annular duct and the radial duct, respectively. R_o is the radius of the hole at the center of the valve core and R_2 is the outer radius of the radial duct. Here, c_a and c_r are coefficients that depend on the velocity profile of MRF flowing through the annular and radial ducts, respectively.

5.2. Optimization of MR valves considering pressure drop and dynamic range

The optimal objective is to minimize the valve ratio defined by the ratio of the viscous pressure drop to the field-dependent pressure drop of the MR valve. This ratio has great effect on the characteristics of the MR valve. It is desirable that the valve ratio takes a small value. The valves are constrained in a cylinder of the radius R=30*mm* and the height H=50*mm*. Magnetic properties of valve components are given in Table 1. The post-yield viscosity of the MRF is assumed to be constant, η=0.092*Pa.s* and the flow rate of the MR valves is Q=3E-4 m^3/s. The commercial MR fluid (MRF132-DG) from Lord Corporation is used. The induced yield stress of the MR fluid as a function of the applied magnetic field intensity (H_{mr}) can be approximately expressed by

$$\tau_y = p(H_{mr}) = C_0 + C_1 H_{mr} + C_2 H_{mr}^2 + C_3 H_{mr}^3 \quad (42)$$

In Eq. (42), the unit of the yield stress is *kPa* while that of the magnetic field intensity is *kA/m*. The coefficients C_0, C_1, C_2, and C_3, determined from experimental results by applying the least square curve fitting method, are respectively identified as 0.3, 0.42, -0.00116 and 1.05E-6.

It is noted that, a small change in the valve gap t_g would drastically alter the performance of the MR valve. Therefore, in MR valve design, a fixed gap is chosen according to each application. In this study, the valve gap is chosen as 1*mm*. From Eqs. (36)-(41), the valve ratios of the single-coil, two-coil and radial-annular MR valve are respectively calculated by

$$\lambda_A = \frac{\Delta P_{A,\eta}}{\Delta P_{A,\tau}} = \frac{3\eta HQ}{\pi t_g{}^2 R_d c t \tau_y} \tag{43}$$

$$\lambda_{2A} = \frac{\Delta P_{2A,\eta}}{\Delta P_{2A,\tau}} = \frac{3\eta HQ}{\pi t_g{}^2 R_d c(t\tau_{yt} + 0.5a\tau_{ya})} \tag{44}$$

$$\lambda_{AR} = \frac{\frac{2\eta Q}{\pi}\{[\frac{3t}{t_g{}^2 R_d} + \frac{3}{d^2}\ln(\frac{R_d}{R_o})] + \frac{2t_g(H-2t)}{R_0^4})}{c(t\tau_{ya} + (R_2 - R_0)\tau_{yr})} \tag{45}$$

The ANSYS APDL program is the analysis ANSYS APDL code used in optimal design of the annular single MR valve. The analysis ANSYS APDL code for other types of MR valve can be prepared in the same maner.

Valve Components	Material	Relative Permeability	Saturation Flux Density
Valve Core	Silicon Steel	B-H curve (Fig. 12a)	1.5 Tesla
Valve Housing	Silicon Steel	B-H curve (Fig. 12a)	1.5 Tesla
Coil	Copper	1	x
MR Fluid	MRF132-DG	B-H curve (Fig. 12b)	1.6 Tesla
Nonmagnetic Cap/Bobbin	Nonmagnetic Steel	1	x

Table 1. Magnetic properties of the valve components

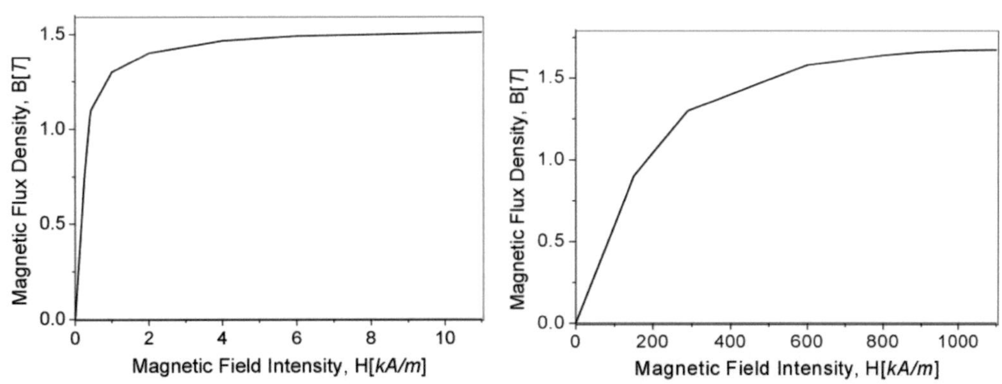

(a) B-H curve of silicon steel (b) B-H curve of MR fluid

Figure 12. Magnetic properties of silicon steel and MR fluid

ANSYS APDL program 1	
/COM, Magnetic-Nodal	CM,_Y1,AREA
/COM, Magnetic-Edge	CMSEL,S,_Y
/PREP7	CMSEL,S,_Y1
!Element definition	AATT, 3, , 1, 0,
ET,1,PLANE13	CMSEL,S,_Y
KEYOPT,1,1,0	CMDELE,_Y
KEYOPT,1,2,0	CMDELE,_Y1
KEYOPT,1,3,1	!* Coil
KEYOPT,1,4,0	CM,_Y,AREA
KEYOPT,1,5,0	ASEL, , , , 1
!Material definition	CM,_Y1,AREA
MAT,1, ! Silicon Steel (B-H curve)	CMSEL,S,_Y
TBDEL,ALL,_MATL	CMSEL,S,_Y1
MPDEL,ALL,_MATL	AATT, 4, , 1, 0,
TB,BH ,_MATL , 1, 9	CMSEL,S,_Y
TBPT,, 130.000000 , 0.380000000	CMDELE,_Y
TBPT,, 260.000000 , 0.750000000	CMDELE,_Y1
TBPT,, 400.000000 , 1.05000000	
TBPT,, 430.000000 , 1.10000000	!Mesh size per definition
TBPT,, 1000.00000 , 1.30000000	FLST,5,19,4,ORDE,9
TBPT,, 2000.00000 , 1.40000000	FITEM,5,3
TBPT,, 4000.00000 , 1.46500000	FITEM,5,-9
TBPT,, 6000.00000 , 1.49000000	FITEM,5,11
TBPT,, 11000.0000 , 1.51000000	FITEM,5,-12
!*	FITEM,5,14
MAT,3, ! MR Fluid (B-H curve)	FITEM,5,-17
TBDEL,ALL,_MATL	FITEM,5,21
MPDEL,ALL,_MATL	FITEM,5,23
TB,BH ,_MATL , 1, 8	FITEM,5,-27
TBPT,, 50000.0000 , 0.300000000	CM,_Y,LINE
TBPT,, 100000.000 , 0.600000000	LSEL, , , ,P51X
TBPT,, 150000.000 , 0.900000000	CM,_Y1,LINE
TBPT,, 290000.000 , 1.30000000	CMSEL,,_Y
TBPT,, 600000.000 , 1.58000000	LESIZE,_Y1,, , msize*1.5, , , , ,1
TBPT,, 800000.000 , 1.64000000	!*
TBPT,, 900000.000 , 1.66000000	FLST,5,8,4,ORDE,5
TBPT,, 1000000.00 , 1.67000000	FITEM,5,8
!*	FITEM,5,12
MPTEMP,,,,,,,,	FITEM,5,21
MPTEMP,1,0	FITEM,5,23
MPDATA,MURX,3,,1 ! Valve coil	FITEM,5,-27
MPTEMP,,,,,,,,	CM,_Y,LINE

```
MPTEMP,1,0
MPDATA,MURX,4,,1        ! Bobbin
!**********************
pi=3.1416
msize=12        !Basic No. of elements/line
!Geometric definition
H=0.05                  ! Height
R=0.03  ! Outer Radius
w=0.010  ! Coil width
t=0.017  ! Pole length
th=0.0065  ! Housing thickness
d=0.001  ! MRF duct gap
Rw=R-th-d
dc=0.00052  ! Wire Radius
res=0.0172e-6       ! Wire Resistivity
Ac=pi*dc**2/4
rrc=res/Ac
Nturn=w*(H-2*t)/Ac          ! No of turns
Rc=rrc*Nturn*pi*2*(Rw-0.5*w) ! Wire Resistance
I=2.5  ! Applied current
J=I*4/dc/dc/pi       ! Current density
PP=I**2*Rc           ! Consumption Power
! geometric model
RECTNG,0,R,0,H,
RECTNG,Rw,R-Th,0,H,
RECTNG,Rw-W,Rw,t,H-t,
RECTNG,0,Rw,0.0,t,
RECTNG,0,Rw,H-t,H,
RECTNG,Rw,R,t,H-t,
!*
FLST,2,6,5,ORDE,2
FITEM,2,1
FITEM,2,-6
AOVLAP,P51X
NUMCMP,LINE
NUMCMP,AREA
! Material assignment
! Housing
FLST,5,3,5,ORDE,3
FITEM,5,5
FITEM,5,9
FITEM,5,-10
LSEL, , , ,P51X
CM,_Y1,LINE
CMSEL,,_Y
LESIZE,_Y1, , ,msize*2, , , , ,1
!*
FLST,5,8,4,ORDE,7
FITEM,5,1
FITEM,5,-2
FITEM,5,10
FITEM,5,13
FITEM,5,18
FITEM,5,-20
FITEM,5,22
CM,_Y,LINE
LSEL, , , ,P51X
CM,_Y1,LINE
CMSEL,,_Y
LESIZE,_Y1, , ,msize, , , , ,1
! Meshing
MSHAPE,0,2D
MSHKEY,0
!*
AMESH,all
!*
FINISH
! Solving magnetic circuit
/SOL
! Boundary condition
FLST,2,12,4,ORDE,6
FITEM,2,1
FITEM,2,-2
FITEM,2,7
FITEM,2,17
FITEM,2,20
FITEM,2,-27
DL,P51X, ,ASYM
! Apply current density to the coil area
FLST,2,1,5,ORDE,1
FITEM,2,1
BFA,P51X,JS, , ,J,0
! Solving
NCNV,0,0,0,0,0,
SOLVE
```

CM,_Y,AREA	FINISH
ASEL, , , ,P51X	/POST1 ! Post processing
CM,_Y1,AREA	SET,LAST ! Choose the last set of solution
CMSEL,S,_Y	! Define path
!*	PATH,p1,2,30,20,
CMSEL,S,_Y1	PPATH,1,0,R-th-d/2,0,0,0,
AATT, 1, , 1, 0,	PPATH,2,0,R-th-d/2,t,0,0,
CMSEL,S,_Y	! Calculate the magnetic intensity along the path
CMDELE,_Y	PDEF,H1,H,sum,AVG
CMDELE,_Y1	/PBC,PATH, ,0
! Core	PCALC,INTG,H1in,H1,S,1,
FLST,5,3,5,ORDE,2	*GET,H1in,PATH, ,last,H1IN
FITEM,5,6	Hmr=1e-3*abs(H1IN)/t
FITEM,5,-8	!PADEL,P1 ! Delete path
CM,_Y,AREA	! Calculation of parameter
ASEL, , , ,P51X	neta=0.092
CM,_Y1,AREA	Q=0.0003
CMSEL,S,_Y	c=2. 5
CMSEL,S,_Y1	C0=0.3
AATT, 1, , 1, 0,	C1=0.42
CMSEL,S,_Y	C2=-0.00116
CMDELE,_Y	C3=1.0513e-6
CMDELE,_Y1	ty=(C0+C1*Hmr+C2*Hmr**2+C3*Hmr**3) ! yield st.
!* MR Fluid	R1=(R-Th-d/2)
FLST,5,3,5,ORDE,2	del_P1=(6*neta*H*Q/(3.14*d**3*R1))*0.00001
FITEM,5,2	del_P2=delta_P+2*C*t*ty*0.01/d
FITEM,5,-4	del_P=del_P1+del_P2 ! Pressure drop
CM,_Y,AREA	OBJ=del_P1/del_P2 ! Objective
ASEL, , , ,P51X	

Table 2. ANSYS APDL program

Figure 13 shows the optimal solution of a single-coil annular MR valve constrained in the specific volume when a current of 2.5*A* is applied to the valve coil. Initial values of t, w_c and t_h are 17*mm*, 10*mm* and 6.5*mm*, respectively. The valve ratio, pressure drop and power consumption of the valve at these initial values are λ_0=0.08274, ΔP_0=15*bar* and N_0=38.83*W*, respectively. From the figure, it is observed that the solution is convergent after 13 iterations and the minimum value of valve ratio (objective function) is λ_{opt} =0.033. The corresponding pressure drop isΔP_{opt}=37.32*bar*, which is also the maximum. At the optimum, the power consumption is N_{opt}=7.92*W* which is much smaller than at the initial. The DVs at the optimum are t_{opt}=7.23 *mm*, $w_{c,opt}$=1.78 *mm*, $t_{h,opt}$=7.43*mm*.

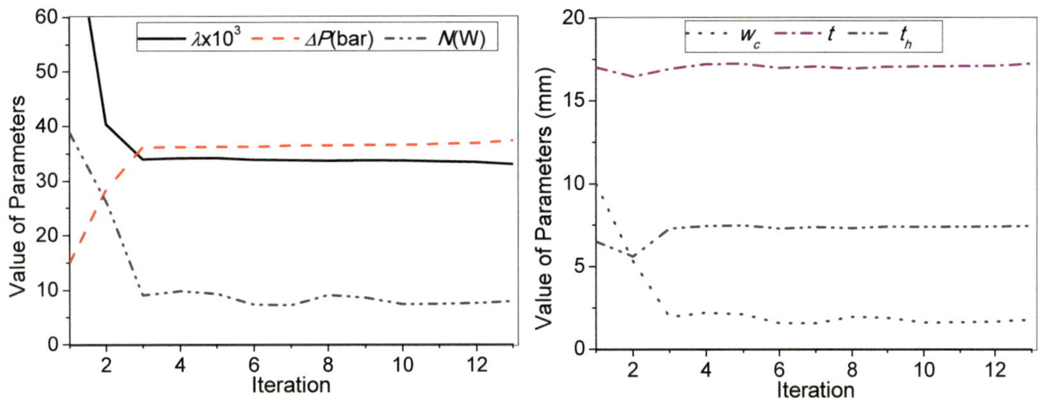

Figure 13. Optimal solution of single-coil MR valves considering the valve ratio and pressure drop

Figure 14 shows the optimal solution of the two-coil annular MR valve. Initial values of a, t, w_c and t_h are 10mm, 5mm, 10mm and 4mm, respectively. The valve ratio, pressure drop and power consumption at these initial values are λ_0=0.0381, ΔP_0=28.2bar and N_0=83.2W, respectively. The solution is convergent after 11 iterations and the minimum value of valve ratio isλ_{opt} =0.023. The corresponding pressure drop is ΔP_{opt}=48.6bar, which is also the maximum. The optimal DVs are a_{opt}=19.7 mm, t_{opt}=10.6 mm, $w_{c,opt}$=6.38 mm, $t_{h,opt}$=5.33mm.

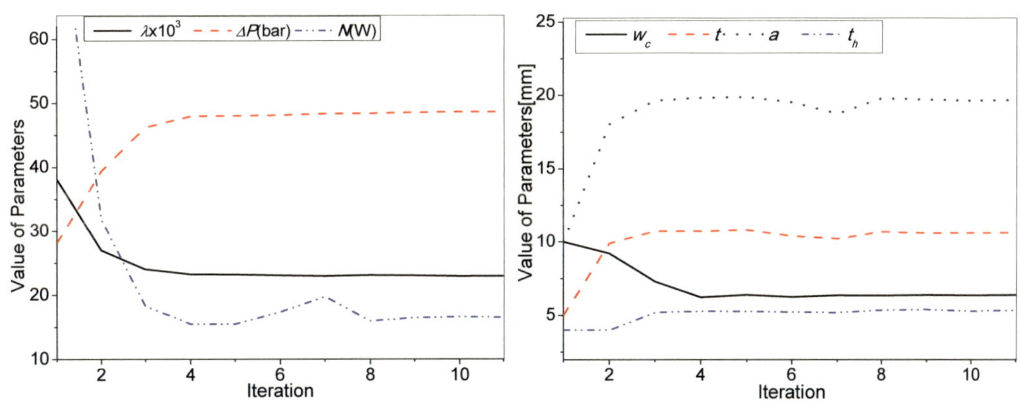

Figure 14. Optimal solution of two-coil MR valves considering the valve ratio and pressure drop

Figure 15 shows the optimization solution of the annular-radial MR valve. Initial values of R_0, t, w_c and t_h are 6mm, 10mm, 6mm and 8mm, respectively. The valve ratio, pressure drop and power consumption at these initial values are λ_0=0.041, ΔP_0=47bar and N_0=44.3W, respectively. The convergence occurs at 10 iteration, at which the minimum value of valve ratio is λ_{opt}=0.023 and the optimal design parameters are $R_{0,opt}$=14.41 mm, t_{opt}=6.47 mm,

$w_{c,opt}$=2.32 *mm*, $t_{h,opt}$=4.81*mm*. The corresponding pressure drop is ΔP_{opt} =37.2*bar*, which is **not** the maximum pressure drop. The reason for this is that the uncontrolled pressure drop (viscous pressure drop) of the valve significantly depends on the valve core radius. An increase of the valve core radius results in a decrease of the viscous pressure drop by which reduces the valve ratio. However, the increase of the valve core radius causes a decrease of the magnetic flux density, and by which reduces the pressure drop of the valve. In order to improve the valve performance, the valve core radius should be fixed at an appropriate. In case the valve core radius is fixed at 6*mm*, it was found that the optimal value of valve ratio is λ_{opt}=0.0293 and the corresponding pressure drop is ΔP_{opt} =64.4*bar* , which is also the maximum. The optimal DVs are t_{opt}=8.6*mm*, $w_{c,opt}$=3.1 *mm* and $t_{h,opt}$=6.36 *mm*. At these optimal DVs the power consumption is N_o=29.1*W*.

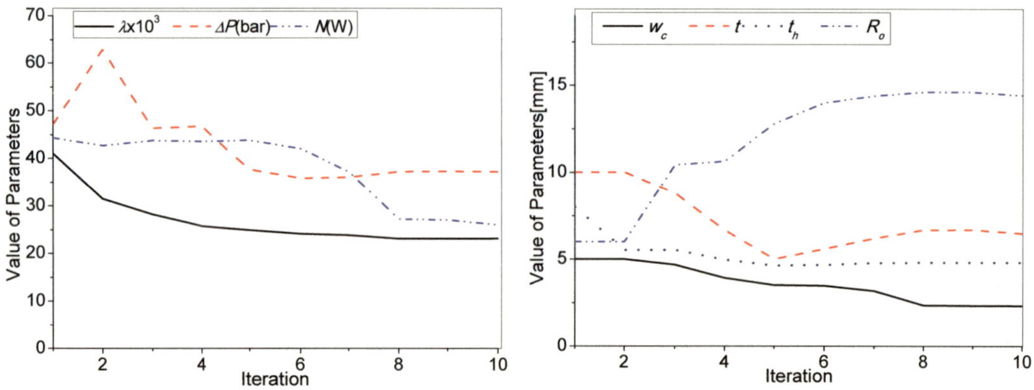

Figure 15. Optimal solution of annular-radial MR valves considering the valve ratio and pressure drop

Table 3 summarizes the optimization results for MR valve design abovementioned. The results show that the geometry of MR valve has a great effect on the valve performance. By choosing an optimal geometry, the valve performance such as pressure drop can be much improved and the power consumption can be significantly reduced. Among the MR valves constrained in the same volume, the two-coil annular MR valve provides the best value of valve ratio while the annular-radial can provide the best pressure drop at the optimal design parameters. For MR valves with three coils or more, it was shown that the performance of these valves is not better than that of the two-coil MR valve at optimal design parameters.

It was also shown by Nguyen et al. [79] that the optimal solution is affected by the applied current. The higher value of the applied current is the better performance of the valve is. However, when the applied current increases to a certain value the optimal solutions tends to be saturation. Therefore, it is advised that the applied current should be set by it maximum allowable value in the optimization problem of the MR valve.

Valve Type	Design Parameter [mm]		Valve Ratio, Pressure Drop & Power Consumption	
	Initial	Optimal	Initial	Optimal
Single-coil annular valve	$t = 17$ $w_c = 10$ $t_h = 6.5$	$t = 17.23$ $w_c = 1.78$ $t_h = 7.43$	λ_0=0.08274 ΔP_0=15bar N_0=38.83W	λ_{opt}=0.033 ΔP_{opt}=37.3bar N_{opt}=7.92W
Two-coil annular valve	a=10 $t = 5$ $w_c = 10$ $t_h = 4$	a=19.7 $t = 10.6$ $w_c = 6.38$ $t_h = 5.33$	λ_0=0.0381 ΔP_0=28.2bar N_0=83.2W	λ_{opt}=0.02296 ΔP_{opt}=48.62bar N_{opt}=16.5W
Three-coil annular valve	$b = 6$ $a = 7$ $t = 7$ $w_c = 10$ $t_h = 6$	$b = 6.36$ $a = 10.94$ $t = 5.84$ $w_c = 7.48$ $t_h = 3.35$	λ_0=0.0355 ΔP_0=32.72bar N_0=55W	λ_{opt}=0.024 ΔP_{opt}=42.95bar N_{opt}=37.4W
Annular-radial valve	$t = 10$ $w_c = 6$ $t_h = 8$	$t = 8.6$ $w_c = 3.1$ $t_h = 6.36$	λ_0=0.041 ΔP_0=47bar N_0=44.3W	λ_{opt}=0.0293 ΔP_{opt}=64.4bar N_{opt}=29.1W

Table 3. Optimization results for MR valve design

6. Conclusion

In this chapter, the methodology of optimal design of MRF-based devices was introduced. The chapter was started by a review of fundamentals and the theory behind MRF in section 2. In section 3, the operating modes used in MRF based devices were considered and several different devices featuring MRF are discussed. In the fourth section, optimal design methodology of MRF-based mechanisms was considered. In this section, firstly the necessity of optimal design and the state of the art were discussed. The magnetic circuit analysis and the modeling of MRF devices were then considered. In addition, the optimization problem of MRF devices was figured out and the methods to solve the problem were investigated. Section 5 dealt with a case study of MR valve optimal design. In this case study, several valve configurations such as single-coil MR valve, multiple-coil MR valve and annular-radial MR valve were considered. The optimization problem is to minimize the valve ratio by which maximized the control range and pressure drop of the MR valves. The results have shown the significance and the effectiveness of the proposed optimization methodology. Base on this study case, the optimal design of other MRF-based devices such as MR dampers, MR brakes can be performed.

Author details

Quoc-Hung Nguyen and Seung-Bok Choi
Mechanical Engineering, Inha University, Korea

7. References

[1] J. H. Park and O. O. Park, Electrorheology and Magnetorheology, *Korea-Aust. Rheol. J.* 13(1), 2001, 13-17.

[2] B. C. Munoz, G. W. Adams, V. T. Ngo and J. R. Kitchin, Stable Magnetorheological Fluids, *US Patent 6203717*, 2001.

[3] C. Fang, B. Y. Zhao, L. S. Chen, Q. Wu, N. Liu and K. A. Hu, The Effect of the Green Additive Guar Gum on the Properties of Magnetorheological Fluid, *Smart Mater. Struct.*14 (1), 2005

[4] S. Genc and P. Phu le, Rheological Properties of Magnetorheological Fluids, *Smart Mater. Struct.* 11, 2002, 140-146.

[5] M. R. Jolly, J. D. Carlson and B. C. Muñoz, A Model of the Behaviour of Magnetorheological Materials, *Smart Mater. Struct.* 5, 1996, 607-614.

[6] J. D. Carlson, D. M. Catanzarite and K. A. St. Clair, Commercial Magnetorheological Fluid Devices, *5th International Conference on ER Fluids, MR Suspensions and Associated Technology*, Singapore, 1996, 20-28.

[7] P. P. Phulé, Magnetorheological (MR) Fluids: Principles and Applications, Smart Materials Bulletin 2001(2), 2001, 7-10.

[8] W. A. Gross, Valve for Magnetic Fluids, *US Patent 3010471*, 1961.

[9] J. Rabinow, Magnetic Fluid Shock Absorber, *US Patent 2667237*, 1954.

[10] E. Germer, Magnetic Valve, *US Patent 2670749*, 1954.

[11] P. J. Rankin, A. T. Horvath and D. J. Klingenberg, Magnetorheology in Viscoplastic Media, *Rheological Acta* 38, 1999, 471-477.

[12] J. H. Park et al., Rheological Properties and Stability of Magnetorheological Fluids using Viscoelastic Medium and Nanoadditives, *Korean J. Chem. Eng.* 18(5), 2001, 580-585.

[13] H. J. Pan et al., Rheological Properties of Magnetorheological Fluid Prepared by Gelatin-Carbonyl Iron Composite Particles, *J.Cent. South Univ. T.*12(4), 2005, 411-415.

[14] S. T. Lim et al., Magnetorheology of Carbonyl-Iron Suspensions with Submicron-Sized Filler, *IEEE Transactions on Magnetics* 40(40), 2004, 3033-3035.

[15] H. Bose and A. Trendler, Smart Fluids - Properties and Benefit for New Electromechanical Devices, *AMAS Workshop on Smart Mater. Struct.*, SMART'03, 2003, 329-336.

[16] D. J. Klingenberg, Magnetorheology: Applications and Challenges, *AIChE Journal* 47(2), 2001, 246-249.

[17] J. D. Carlson, What Makes a Good MR Fluid? *J. Intell. Mater. Syst. Struct.*13, 2002, 431-435.

[18] J. Claracq, J. Sarrazin and J. P. Montfort, Viscoelastic Properties of Magneto-rheological Fluids, *Rheologica Acta* 43, 2004, 38-43.

[19] P. P. Phulé, Magnetorheological Fluid, *US Patent 5985168*, 1999.

[20] J. H. Park, B. D. Chin and O. O. Park, Rheological Properties and Stabilization of Magnetorheological Fluids in a Water-in-Oil Emulsion, *J. Colloid Interf . Sci* 240,2001, 349-54.

[21] G. Bossis, S. Lacis, A. Meunier and O. Volkova, Magnetorheological Fluids, *J. Magn. Mag. Mater.* 252, 2002, 224-228.

[22] G. L. Gulley and R. Tao, Structures of an Electrorheological Fluid, *Physical Review E* 56(4), 1997, 4328-4336.

[23] T. Shiraishi et al., Estimation of Equivalent Permeability in Magnetorheological Fluid Considering Cluster Formation of Particles, *J. Appl.Mech.*71(2), 2004, 201-207.

[24] B. J. de Gans, H. Hoekstra and J. Mellema, Non-Linear Magnetorheological Behaviour of an Inverse Ferrofluid, *Faraday Discussions* 112, 1999, 209-224.

[25] K. Butter et al., Direct Observation of Dipolar Chains in Ferrofluids in Zero Field Using Cryogenic Electron Microscopy, *J. Phys. Condens. Matter.* 15(15), 2003, 1451-1470.

[26] I. H. Shames, Mechanics of Fluids, 3rd Ed., McGraw-Hill Inc., New York, 1992.

[27] A. B. M. Saiful, The Behaviour of Magnetorheological Fluids in Squeeze Mode, *PhD thesis*, Dublin City University, 2008

[28] L. E. Nielson, Polymer Rheology, Marcel Dekker Inc., New York, 1977.

[29] C. A. Shook, Slurry Pipeline Flow, *in Proc. of Solid-Liquid Suspensions*, Butterworth-Heinemann, 1993, 287-309.

[30] K. Toda and H. Furuse, Extension of Einstein's Viscosity Equation to That for Concentrated Dispersions of Solutes and Particles, *J Biosci. Bioeng.* 102(6), 2006, 524-528.

[31] J. F. Douglas, J. M. Gasiorek and J. A. Swaffield, Fluid Mechanics, 3rd Ed., Longman Scientific & Technical, Harlow Essex, UK, 1995.

[32] Y. Grasselli, G. Bossis and E. Lemaire, Structure Induced in Suspensions by a Magnetic Field, *Journal De Physique II* 4, 1994, 253-263.

[33] N. M. Wereley et al., Bidisperse Magnetorheological Fluids using Fe Particles at Nanometer and Micron Scale, *J. Intell. Mater. Syst. Struct.*17, 2006, 393-401.

[34] H. Nishiyama, S. Fushimi and M. Nakano, Numerical Simulation of MR Fluid Damping Characteristics Using a Modified Bingham Model, *J Intell. Mater. Syst. Struct*13, 2002, 647-653.

[35] K. Goldsmith, Note on the Rheological Properties of Elasto-Plastic Materials, *Br. J. Appl. Phys.* 1(4), 1950, 107-109.

[36] J. D. Carlson and K. D. Weiss, Magnetorheological Materials Based on Alloy Particles, *US Patent 5382373*, 1995.

[37] Phillips, R., Engineering Applications of Fluids with a Variable Yield Stress, PhD Thesis, Mechanical Engineering, U.C. Berkeley, 1969.

[38] J. D. Carlson and M. R. Jolly, MR fluid, foam and elastomer devices, *Mechatronics* 10 , 2000, 555–69

[39] R. Stanway, J. Sproston and A. El Wahed, Application of Electrorheological Fluids in Vibration Control: A Survey, *Smart Mater. Struct.*, 5(4), 1996, 464–482.

[40] X. Wang and F. Gordaninejad, Flow analysis of field-controllable, electro- and magneto-rheological fluids using Herschel–Bulkley model, *J Intell. Mater. Syst. Struct.*10, 1999, 601–608.

[41] Y. T. Choi, J. U. Cho, S. B. Choi and N. M. Wereley, Constitutive models of electrorheological and magnetorheological fluids using viscometers, *Smart Mater. Struct.* 14 , 2005, 1025

[42] L. Bitman, Y. T. Choi, S. B. Choi and N. M. Wereley Electrorheological Damper Analysis Using an Eyring-Plastic Model, *Smart Mater. Struct.*14 (1), 2005, 237-246.

[43] O. Ashour, A. Craig, Magneto-Rheological Fluids: Materials, Characterization, and Devices, *J. Intell. Mater. Syst. Struct.* 7, 1996, 123–130.

[44] M. Zubieta, S. Eceolaza, M. J. Elejabarrieta, M. Bou-Ali, Magnetorheological fluids: characterization and modeling of magnetization, Smart Mater. Struct. 18, 2009.

[45] Q. H. Nguyen, S. B. Choi, Optimal design of an automotive magnetorheological brake considering geometric dimensions and zero-field friction heat, *Smart Mater. Struct.* 19, 2010.

[46] Q. H. Nguyen, S. B. Choi, Optimal design of a novel hybrid MR brake for motorcycles considering axial and radial magnetic flux, *Smart Mater. Struct.*21(5) 2012.

[47] N. M. Wereley, Nondimensional Herschel-Bulkley Analysis of Magnetorheological and Electrorheological Dampers, *J. Intell. Mater. Syst. Struct.* 19(3), 2008, 257-268.

[48] N. C. Rosenfeld and N. M. Wereley, Volume Constrained Optimization of Magnetorheological Valves and Dampers, *Smart Mater. Struct.* 13, 2004, 1303-1313.

[49] A. Milecki, Investigation and Control of Magneto-Rheological Fluid Dampers, *Int. J. Mach. Tool. Manu.* 41(3), 2001, 379-391.

[50] J. H. Yoo, J. Sirohi and N. M. Wereley, A Magnetorheological Piezohydraulic Actuator, *J. Intell. Mater. Syst. Struct.*16, 2005, 645-953.

[51] H. S. Lee and S. B. Choi, Control and response characteristics of a magneto-rheological fluid damper for passenger vehicles, *J. Intell. Mater. Syst. Struct.*11(1) 80-87.

[52] S. F. Masri, R. Kumar and R. C. Ehrgott, Modeling and Control of an Electrorheological Device for Structural Control Applications, *Smart Mater. Struct.* 4, 1995, 121-131.

[53] B. F. Spencer, S. J. Dyke, M. K. Sain and J. D. Carlson, Phenomenological Model of a Magnetorheological Damper, *J. Eng. Mech. ASCE* 123, 1997, 230-138.

[54] N. .M. Wereley, J. U. Cho, Y. T. Choi and S. B. Choi, Magnetorheological dampers in shear mode, *Smart Mater. Struct.* 17 , 2008.

[55] D. Y. Lee, Y. T. Choi and N. M. Wereley, Performance Analysis of ER/MR Impact Damper Systems Using Herschel-Bulkley Model, *J. Intell. Mater. Syst. Struct.*13, 2002, 525-531.

[56] V. A. Neelakantan and G. N. Washington, Modeling and Reduction of Centrifuging in Magnetorheological (MR) Transmission Clutches for Automotive Applications, *J. Intell. Mater. Syst. Struct.*16, 2005, 703-711.

[57] I. R. Park, Automobile Brake Device, *US Patent 2002/0074196 A1*, 2002.

[58] P. L. Wong, W. A. Bullough, C. Feng and S. Lingard, Tribological Performance of a Magneto-Rheological Suspension, *Wear* 247(1), 2001, 33-40.

[59] J. Huang, J. Q. Zhang, Y. Yang and Y. Q. Wei, Analysis and Design of a Cylindrical Magneto-Rheological Fluid Brake, *J. Mater. Process. Tech.* 129(1-3), 2002, 559-562.

[60] R. Stanway et al., ER Fluids in the Squeeze-Flow Mode: An Application to Vibration Isolation, *J. Electrostat.*28(1), 1992, 89-94.

[61] G. J. Monkman, The Electrorheological Effect under Compressive Stress, *J. Phys. D* 28, 1995, 588-593.

[62] P. Kulkarni et al., Study of the Behavior of MR Fluids in Squeeze, Torsional and Valve Modes, *J. Intell. Mater. Syst. Struct.*14, 2003, 99-104.

[63] J. L. Sproston et al., A Numerical Simulation of Electrorheological Fluids in Oscillatory Compressive Squeeze-Flow, *J. Phys. D* 27, 1994, 338-343.

[64] J. L. Sproston, A. K. El Wahed and R. Stanway, Electrorheological Fluids in Dynamic Squeeze Flow, *IEEE ICDL'96*, Roma, Italy, 1996, 515-519.

[65] A. K. El Wahed et al., Performance of an Electrorheological Fluid in Dynamic Squeeze Flow Under Constant Voltage and Constant Field, *J. Phys. D* 31(20), 1998, 2964-2974.

[66] A. K. El Wahed, J. L. Sproston and R. Stanway, The Rheological Characteristics of Electrorheological Fluids in Dynamic Squeeze, *J. Intell. Mater. Syst. Struct.*13, 2002, 655-660.

[67] A. K. El Wahed et al., An Improved Model of ER Fluids in Squeeze-Flow Through Model Updating of the Estimated Yield Stress, *J. Sound Vibrat.*268(3), 2003, 581-599.

[68] . P. Yang and K. Q. Zhu, Analytical Solutions for Squeeze Flow of Bingham Fluid with Navier Slip Condition, *J. Non-Newton.Fluid* 138, 2006, 173-180.

[69] J. Wang, G. Meng, N. Feng, and E. J. Hahn, Dynamic Performance and Control of Squeeze Mode MR Fluid Damper-Rotor System, *Smart Mater. Struct.* 14(4), 2005, 529-539.

[70] R. Stanway, J. L. Sproston and A. K. El-Wahed, Applications of Electro-Rheological Fluids in Vibration Control: A Survey, *Smart Mater. Struct.* 5, 1996, 464-482.

[71] S. R. Hong, S. B. Choi, W. J. Jung, Vibration Isolation of Structural Systems Using Squeeze Mode ER Mounts, *J. Intell. Mater. Syst. Struct.*13, 2002, 421-424.

[72] V. Hartel, Spring Element with Hydraulic Damping, *US Patent 4773632*, 1988.

[73] C. R. Burrows et al., Linearized Squeeze-Film Dynamics: A Model Structure and the Interpretation of Experimentally Derived Parameters, *J Mech Eng Sci* 204(4), 1990, 263-272.

[74] K. J. Kim and C. W. Lee, Design and Modeling of Semi-Active Squeeze Film Damper Using Magnetorheological Fluid, *MOVIC 2004*, Saint Louis, USA, 2004.

[75] N. M. Wereley and L. Pang, Nondimensional Analysis of Semi-Active Electrorheological and Magnetorheological Dampers Using Approximate Parallel Plate Models, *Smart Mater. Struct.* 7(5), 1998, 732-743.

[76] G. M. Kamath, M. K. Hurt and N. M. Wereley, Analysis and Testing of Bingham Plastic Behavior in Semi-Active Electrorheological Fluid Dampers, *Smart Mater. Struct.* 5(5), 1996, 576-590.

[77] X. Tang, X. Zhang, R. Tao and Y. Rong, Structure-Enhanced Yield Stress of Magnetorheological Fluids, *J. Appl. Phys.* 87, 2000, 2634-2638.

[78] M. T. Avraam, MR-fluid brake design and its application to a portable muscular rehabilitation device, PhD Thesis in Engineering Sciences, 2009.

[79] Q. H. Nguyen, Y. M. Han, S. B. Choi and N. M. Wereley, Geometry optimization of MR valves constrained in a specific volume using the finite element method *Smart Mater. Struct.* 16, 2007, 2242-2252.

[80] Q. H. Nguyen, S. B. Choi and N. M. Wereley, Optimal design of magnetorheological valves via a finite element method considering control energy and a time constant *Smart Mater. Struct.* 17, 2008,1-12.

[81] Q. H. Nguyen, S. B. Choi, Optimal Design of Vehicle MR Damper Considering Damping Force and Dynamic Range, *Smart Mater. Struct.* 18 (2008), 1-10.

[82] J. Park, L. F. Luz, A. Suleman, Multidisciplinary design optimization of an automotive magnetorheological brake design, *Computers and Structures* 86 (2008), 207-216.

[83] Q. H. Nguyen and S. B. Choi, Selection of magnetorheological brake types via optimal design considering maximum torque and constrained volume, *Smart Mater. Struct.* 21(1), 2012

[84] Q. H. Nguyen, S. B. Choi, Y. S. Lee and M. S. Han, An analytical method for optimal design of MR valve structures, *Smart Mater. Struct.* 18(9), 2009.

[85] J. B. F. Spencer et al., Smart dampers for seismic protection of structures: a full-scale study, *2nd World Conference on Structural Control*, Kyoto, Japan, 1998, 417–426.

Chapter 15

MR Fluid Damper and Its Application to Force Sensorless Damping Control System

D. Q. Truong and K. K. Ahn

Additional information is available at the end of the chapter

http://dx.doi.org/10.5772/51391

1. Introduction

Vibration suppression is considered as a key research field in civil engineering to ensure the safety and comfort of their occupants and users of mechanical structures. To reduce the system vibration, an effective vibration control with isolation is necessary. Vibration control techniques have classically been categorized into two areas, passive and active controls. For a long time, efforts were made to make the suspension system work optimally by optimizing its parameters, but due to the intrinsic limitations of a passive suspension system, improvements were effective only in a certain frequency range. Compared with passive suspensions, active suspensions can improve the performance of the suspension system over a wide range of frequencies. Semi-active suspensions were proposed in the early 1970s [1], and can be nearly as effective as active suspensions. When the control system fails, the semi-active suspension can still work under passive conditions. Compared with active and passive suspension systems, the semi-active suspension system combines the advantages of both active and passive suspensions because it provides better performance when compared with passive suspensions and is economical, safe and does not require either higher-power actuators or a large power supply as active suspensions do [2].

In early semi-active suspension, many researches on variable orifice dampers had been done ([3-4]). With these damper types, regulation on of the damping force can be achieved by adjusting the orifice area in the oil-filled damper, thus changing the resistance to fluid flow, but adjusting the speed is slow because of mechanical motion limitations. Another class of semi-active suspension uses controllable fluids. Two fluids that are viable contenders for development of controllable dampers are: electrorheological (ER) fluids and magnetorheological (MR) fluids. Although the discovery of both ER and MR fluids dates back to the late 1940's, researchers have primarily concentrated on ER fluids for civil engineering applications ([5-8]). Recently developed MR fluids appear to be an attractive alternative to ER fluids for use in controllable fluid dampers [9-15].

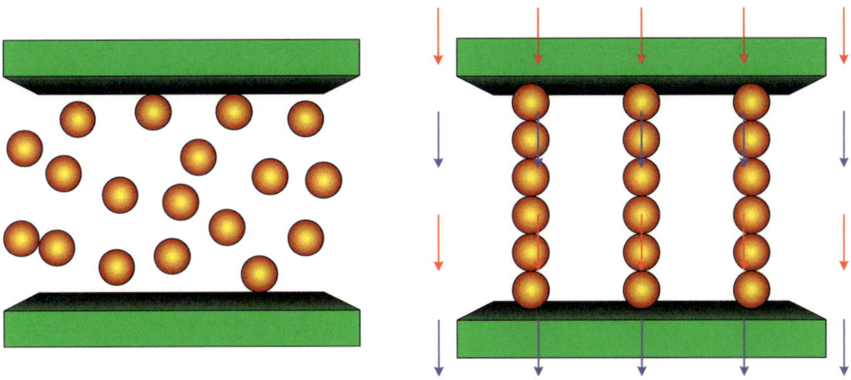

(a) Without Magnetic Field (b)With Magnetic Field

Figure 1. MR fluid – Working principle.

The initial discovery and development of MR fluid can be credited to Jacob Rainbow at the US National Bureau of Standards in the late 1940s [9,10]. These fluids are suspensions of micron-sized, magnetizable particles in an appropriate carrier liquid [11-15]. Normally, MR fluids are free flowing liquids having consistency similar to that of motor oil. However, in the presence of an applied magnetic field, the iron particles acquire a dipole moment aligned with the external field which causes particles to form linear chains parallel to the field, as shown in Fig. 1. This phenomenon can solidify the suspended iron particles and restrict the fluid movement. Consequently, yield strength is developed within the fluid. The degree of change is related to the magnitude of the applied magnetic field, and can occur only in a few milliseconds. A typical MR fluid contains 20-40% by volume of relatively pure, soft iron particles, e.g., carbonyl iron. These particles are suspended in mineral oil, synthetic oil, water or glycol. A variety of proprietary additives similar to those found in commercial lubricant are commonly added to discourage gravitational settling and promote suspension, enhance lubricity, modify viscosity, and inhibit wear. The ultimate strength of an MR fluid depends on the square of the saturation magnetization of the suspended particles. The key to a strong MR fluid is to choose a particle with a large saturation magnetization. The best available particles are alloys of iron and cobalt that have saturation magnetization of about 2.4 Tesla. Unfortunately, such alloys are prohibitively expensive for most practical applications. The best practical particles are simply pure iron, as they have saturation magnetization of 2.15 Tesla. Virtually all other metals, alloys and oxides have saturation magnetization significantly lower than that iron, resulting in substantially weaker MR fluids. Typically, the diameter of the magnetizable particles is 3 to 5 microns. Functional MR fluids may be made with larger particles; however, particle suspension becomes increasingly more difficult as the size increases. Smaller particles that are easier to suspend could be used, but the manufacture of such particles is difficult. Commercial quantities of relatively inexpensive carbonyl iron are generally limited to sizes greater than 1 or 2 microns.

Due to the special behavior of MR fluid, it has been used for a vast of applications such as: dampers, shock absorbers, rotary brakes, clutches, prosthetic devices, polishing and

grinding devices, etc. Among them, MR fluid dampers, which utilize the advantages of MR fluids, are semi-active control devices that are widely used in the modern industry nowadays. A typical MR damper includes MR fluid, a pair of wires, a housing, a piston, a magnetic coil, and an accumulator as displayed in Fig. 2a. Here, the MR fluid is housed within the cylinder and flows through a small orifice. The magnetic coil is built in the piston or on the housing. When a current is supplied to the coil, the particles are aligned and the fluid changes from the liquid state to the semi-solid state within milliseconds. Consequently, a controllable damping force is produced. The force procedured by a MR damper depends on magnetic field induced by the current in the damper coil and the piston velocity as in Fig. 2b. It is capable of generating a force with magnitude sufficient for rapid response in large-scale applications [16-18], while requiring only a battery for power [13]. Additionally, these devices offer highly reliable operations and their performance is relatively insensitive to temperature fluctuations or impurities in the fluid [12]. As a result, there has been active research and development of MR fluid dampers and their applications [9-19,21-29,32,35,36].

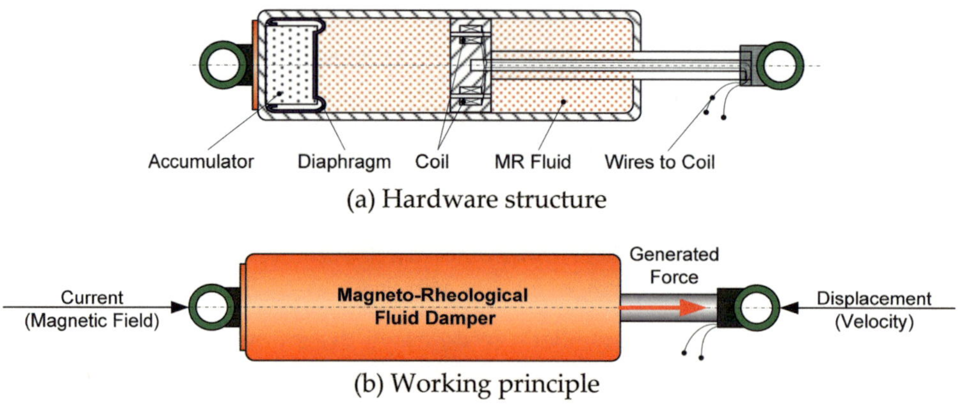

Figure 2. General configuration of a MR fluid damper.

However, major drawbacks that hinder MR fluid damper applications are their nonlinear force/displacement and hysteretic force/velocity characteristics. Therefore, one of the challenges involved in creating high efficiencies for MR fluid damper applications, especially in damping control field is to develop an accurate model that can take full advantages of the unique features of this device and to design proper control algorithms in order to improve the system working performances.

With MR fluid dampers modeling technologies, both parametric and non-parametric models have been built by researchers to describe the MR fluid damper behaviors. Savaresi *et al* [19] made a comparison of both the parametric and non-parametric methods and then developed a complete framework for the development of an accurate model of MR-dampers. Parametric models based on mechanical idealizations have been proposed such as Bingham model, Bouc-Wen model, phenomenological model, and others [20-25]. The Bingham model [20] represents the dry-friction as a signum function on the damper velocity and may be considered as a simple model for describing the hysteresis characteristic. The Bouc-Wen model uses a differential equation to depict the non-linear hysteresis with

moderate complexity and is widely applied in building controls. Once the characteristic parameters of the Bouc-Wen model are determined, the model can obtain the linearity and the smoothness of the transition from the pre-yield to the post-yield region. One of the major problems in the Bouc-Wen model is the accurate determination of its characteristic parameters which is obtained by using optimization or trial error techniques. Consequently, these techniques demand high computational cost to generate the model parameters. Moreover, the fact that each set of constant parameters is valid only for single vibration conditions makes the Bouc-Wen model inappropriate for varying excitation environments. Therefore, many researches on how to develop a MR fluid damper model for higher accuracy and higher adaptability in estimating the behavior of the damper have been done. Spencer *et al* [21] successfully developed a phenomenological model to improve the model accuracy with an additional internal dynamical variable. Choi and Lee [22] designed a hysteresis damper model based on a polynomial and a curve fitting to predict better the damping force when compared with conventional models. Dominguez *et al* [23] proposed a methodology to find out the characteristic parameter of Bouc-Wen model and then designed a new non-linear model to simulate the behavior of the MR fluid dampers. Kwok *et al* designed a hysteretic model based on a particle swarm optimization [24] or using GA technique [25] to modify the Bonc-Wen model and identify the characteristic parameters of the models. The effectiveness of these models with their identification process was proved through the experimental test data. However, the parametric modeling methods require assumptions regarding the structure of the mechanical model that simulates the system's behavior. These approaches could be divergent if the initial assumptions for the model structures are flawed or if the proper constraints are not applied to the parameters [24,25]. Unrealistic parameters such as negative mass or stiffness may be obtained [29].

On the contrary, non-parametric methods could avoid these drawbacks of the parametric approaches for modeling both the linear, nonlinear, and hysteretic systems with high adaptability. For modeling MR fluid dampers, some researches have been done. Chang and Roschke [26] proposed a non-parametric model using multilayer perceptron neural network with optimization method for a satisfactory representation of a damper behavior. Schurter and Roschke [27] investigated the modeling of MR fluid dampers with an adaptive neuro-fuzzy inference system. The fuzzy structure was simple for modeling; nevertheless, the training model process relied on input and output information on MR fluid dampers and took much computation time. Wang and Liao [28,29] explored the modeling of MR fluid dampers by using a trained direct identification based on recurrent neural network. Although, the designed models could predict the dynamic responses of the dampers with high precision, the model architectures and the training methods were complex.

Once an accurate model for the MR fluid damper is built, it is very useful to investigate the damper characteristics before applying to suspensions. In addition, the well-done model can effectively function as a virtual sensor to estimate the damping force which is used for closed-loop damping control systems with a self-sensing behavior. Self-sensing describes the technique of using a transducer to both actuate and sense concurrently [30,31]. Compared to typical self-sensing damping control systems using separated or integrated actuators and sensors [32], this technique can offer several advantages. A reduction in the

number of sensing and actuation devices, and associated power, wiring and interfacing, immediately reduces cost and complexity. A sensorless damping control system can also offer higher robustness than the corresponding conventional system in which the failure occurs due to faults in sensor hardware, reading/wiring signal, or measurement noises.

For these reasons and the current demands in MR fluid damper applications, this chapter includes mainly two contents: modeling and damping control technologies. The first one is to revise several typical MR fluid damper modeling methodologies as well as to develop an effective direct modeling method based on a so-called black-box model (BBM) [35,36]. This BBM using a self-tuning fuzzy system based on neural technique is designed to model simply MR fluid dampers and then can be apply to damping control systems as a virtual force sensor. The BBM built in the form of simple fuzzy mapping laws is considered to estimate directly the MR damping force with respect to the MR damper characteristics. In order to improve the accuracy of the suggested model, the back propagation learning rules based on gradient descent method was used to train the fuzzy parameters to minimize the modeling error function. Input information for the model training process is the current supplied for the MR fluid damper and its dynamic responses. The effectiveness of the BBM method as well as the self-sensing ability of a damping system using this model was clearly verified in a comparison with the other methods through modeling and experimental investigations on two damper test rigs. The comparison results show that the BBM has satisfactorily representative ability for the behavior of MR fluid damper with small computational requirement and it can be successfully used as the virtual force sensor for damping systems. The second content is to present a novel damping control methodology which is called force-sensorless damping control. This control technique is based on the developed BBM, and its inverse back-box model, IBBM, which were designed for any given MR fluid damper, to apply to general systems using this damper for damping control with force self-sensing behavior. The IBBM was derived from the optimized BBM and suggested for usage as an effective force controller. In addition, the IBBM structure is online adjusted with respect to the control error to improve the system performance. Consequently, the closed-loop force controller, based on the 'virtual' force sensor - BBM and the adaptive force controller – IBBM, was built for the force-sensorless damping system. Simulations and real-time experiments have been finally carried out to verify that the designed models have satisfactorily representative ability for the behavior of MR fluid damper with small computational requirement and they were successfully applied for force-sensorless control of the damping systems.

2. MR fluid damper analysis

2.1. MR fluid damper

For later discussions on modeling and control of MR fluid dampers, a MR fluid damper of the small size damper series, RD-1005-3, manufactured by LORD Corporation was used. The damper RD-1005-3 can be adapted to a wide variety of applications because of its simple design, quiet operation, and compact shock absorption with low voltage and current demands that allow for damping control. Specifications of this damper are displayed in Table 1.

The purpose of this research is to investigate the characteristics of the damper RD-1005-3 which are then used to derive the accurate damper models as well as to design the force-sensorless controller for damping systems using this damper. Therefore, a test rig employing the damper RD-1005-3, named as TR01, is needed for following tasks:

- To perform a series of experiments on this rig for a full investigation of the damper working performance.
- To acquire a set of experimental input/output data which is used to analyze, design and optimize the damper models in a comparison with some typical models.
- Based on the optimized BBM model, the inverse model – IBBM is designed for damping control purpose.

The optimized BBM model as well as the proposed force-sensorless control system based on BBM and IBBM models might be applied not only on this system but also on other damping systems using the same dampers. Consequently, the ability of using these models for modeling and damping control needs to be verified. Therefore, a second damping system, which is named TR02, is indispensable for following purposes:

- This system functions as another typical damping system using the damper RD-1005-3 as same as the one used in the first damping system.
- The optimized BBM model is applied to this system for evaluating the effectiveness of the proposed self-sensing methodology.
- The suggested force-sensorless control method is applied to this rig to evaluate the damping force control performance.

LORD MR fluid damper – RD-1005-3 Series	
Parameter	Value
Compressed length (m)	0.155
Extended length (m)	0.208
Weight (kg)	0.800
Magneto-Rheological fluid Viscosity (N-s/m) at 40°C Density (kg/m^3) Solids content by weight, %	MRF-132DG 0.092 ± 0.015 2980-3180 80.98
Electrical characteristics: Maximum input current (A) Input voltage (VDC)	 2 12
Mechanical characteristics: Maximum extension force (N) Maximum operating temperature (deg)	 4448 71
Response time (s) (amplifier & power supply dependent)	<0.025 (time to reach 90% of max level during 0-1A step input)

Table 1. Technical data for the MR fluid damper RD-1005-3

Based on the dimensions and characteristics of the damper RD-1005-3 as well as the design purposes, the two testing systems were designed and set up as described below.

2.2. Test rig setup

2.2.1. Test rig 01 (TR01) to build the BMM and IBBM models

The schematic diagram of the test rig TR01 for the damper RD-1005-3 is depicted in Fig. 3a. In this system, a hydraulic actuator and a driving controller (VibMaster) manufactured by Park Electronics were employed to drive the damper. The end-effector of the actuator is a hydraulic cylinder with 0.035m diameter which is driven by a servo valve. The servo valve with a nominal operational frequency range of 0-50Hz, made by Moog Inc., was used as the final control target to adjust the actuator motion. A linear variable differential transformer (LVDT) was set up to measure the piston-rod displacement of the MR fluid damper. In addition, a compatible load cell with 5000N capacity made by Bongshin Corp. was attached in series with the damper rod to measure the damping force. A PC installed with the VibMaster control program was used to generate system vibrations, while the PC with a

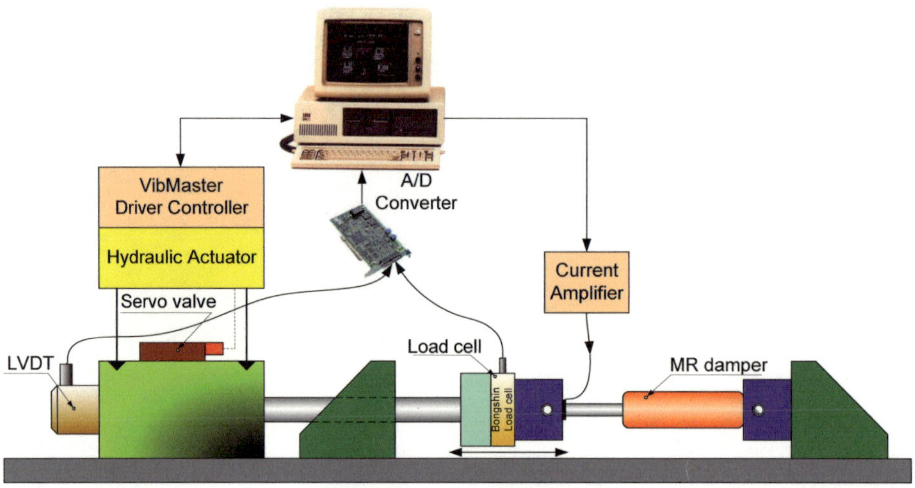

(a) Diagram of TR01

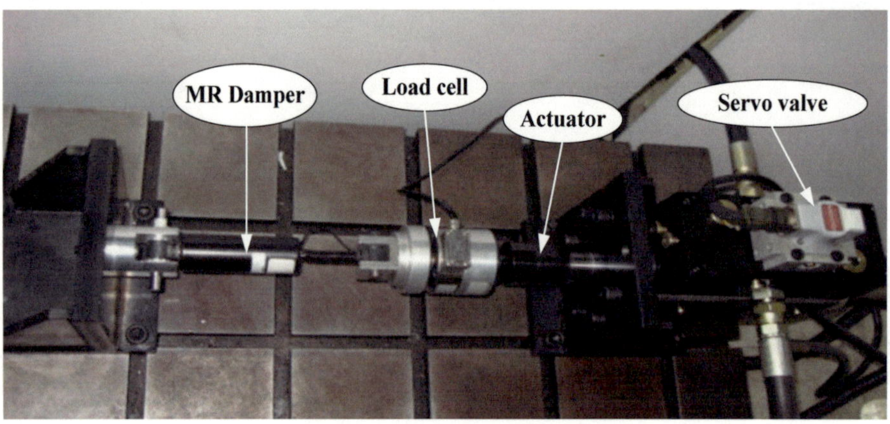

(b) Photograph of TR01

Figure 3. Test rig 01 – TR01 using damper RD-1005-3.

current amplified circuit sent the current signal to adjust the damper characteristics. Consequently, the feedback signals measured by the LVDT and load cell were sent back to the PC through an Advantech A/D PCI card 1711 to perform a full input-output data acquisition. Finally, the load frame shown in Fig. 3a was designed and fabricated as shown in Fig. 3b for the purpose of obtaining the MR fluid damper responses.

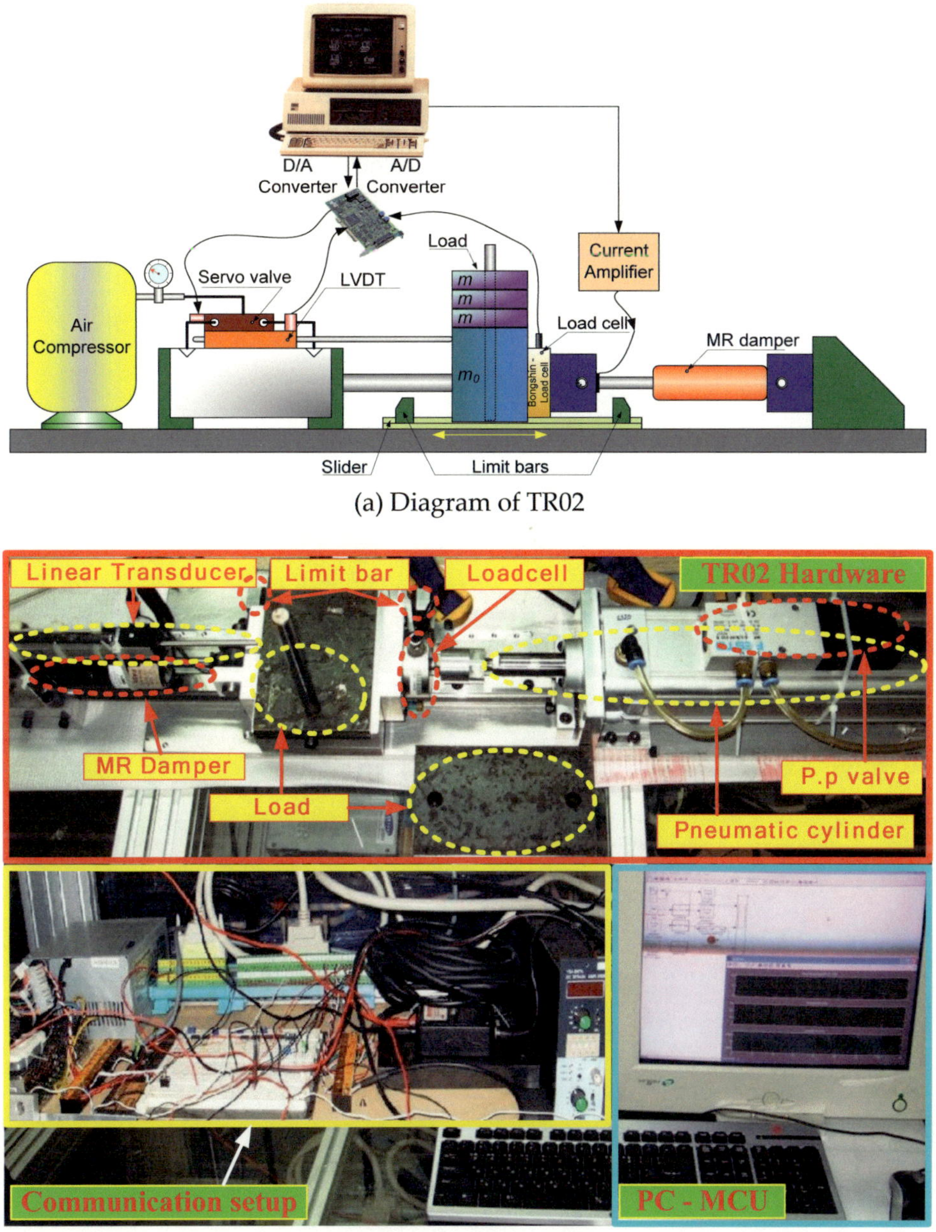

(a) Diagram of TR02

(b) Photograph of TR02

Figure 4. Test rig 02 – TR02 using damper RD-1005-3.

2.2.2. Test rig 02 (TR02) to verify the self-sensing behavior and force-sensorless controller

In order to evaluate the ability of using the proposed self-sensing method and force-sensorless controller for any damping system using the same MR fluid damper, the second test rig (TR02) was set up as depicted in Fig. 4a. The system consists of the MR fluid damper investigated in the TR01 and a pneumatic actuator. The end-effector of this actuator is a pneumatic cylinder with 0.063m diameter which is driven by a 5/3-way proportional valve manufactured by Festo Corp. to generate the vibration for the damping system. The spool motion of this servo valve is proportional to its control signal sent from the PC through a D/A converter of the Advantech card 1711. A linear transducer (Novotechnik TR 100) was fixed on the TR02 base and this sensor slider was in contact with the cylinder end-position to feed back the vibration information to the PC. In addition, a loading system was installed in series with the damper rod to create the working environment. Here, the load can be manually varied while the compatible load cell with 5000N capacity made by Bongshin Corp. was chosen to measure and feedback the actual damping force in order to compare with the force predicted by the BBM and, consequently, to evaluate the self-sensing method. In addition, the proposed force-sensorless controller is applied to the TR02 to drive the MR fluid damper to follow desired force targets for controller verification.

For safety when doing experiments on the TR02, two limit bars were positioned at two sides of the loading system to restrict the piston movement, consequently, protecting the load cell and MR fluid damper from damages. A photograph of the TR02 configuration is described in Fig. 4b.

2.3. Experiments on the TR01and data analysis

2.3.1. Experimental data obtained from the TR01

To obtain the data used to characterize the RD-1005-3 MR fluid damper behavior, a series of experiments on the rig TR01 was conducted under various sinusoidal displacement excitations while simultaneously altering the magnetic coil in a varying current range. The output of each test was the force generated by the damper. The setting parameters for experiments are listed in Table 2. During all the experiments, the damping force response was measured together with the variation of piston displacement and supplied current for the damper at each step of time, 0.002 second. Fig. 5 depicts an example of relationship between the piston velocity, applied current and dynamic response of the damper corresponding to a sinusoidal excitation with 1Hz of frequency and 0.005m of amplitude applied to the damper.

Test No.	Displacement – Sine wave		MR fluid damper current (A)
	Amplitude (m)	Frequency (Hz)	
01 to 06	0.005	1.0	(0,0.5, 0.75, 1.0, 1.25, 1.5)
07 to 12	0.005	1.5	(0,0.5, 0.75, 1.0, 1.25, 1.5)
13 to 18	0.005	2.0	(0,0.5, 0.75, 1.0, 1.25, 1.5)
19 to 24	0.005	2.5	(0,0.5, 0.75, 1.0, 1.25, 1.5)

Table 2. Setting parameters for experiments on the test rig TR01

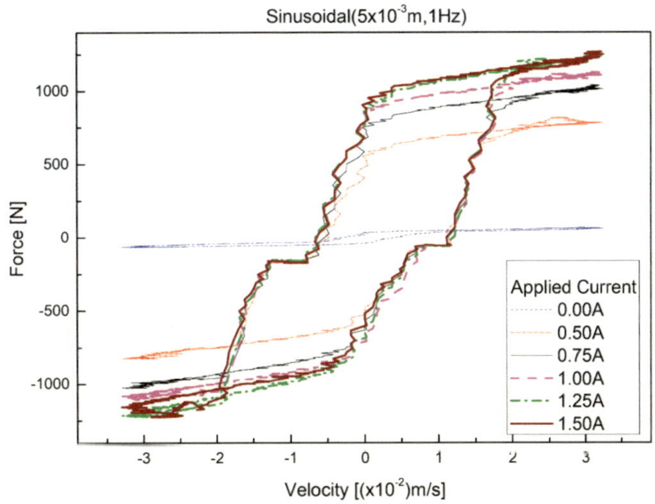

Figure 5. Performance curves for the RD-1005-3 MR fluid damper for a sinusoidal excitation at frequency 1Hz and amplitude 0.005m, and supplied current in range (0, 1.5A)

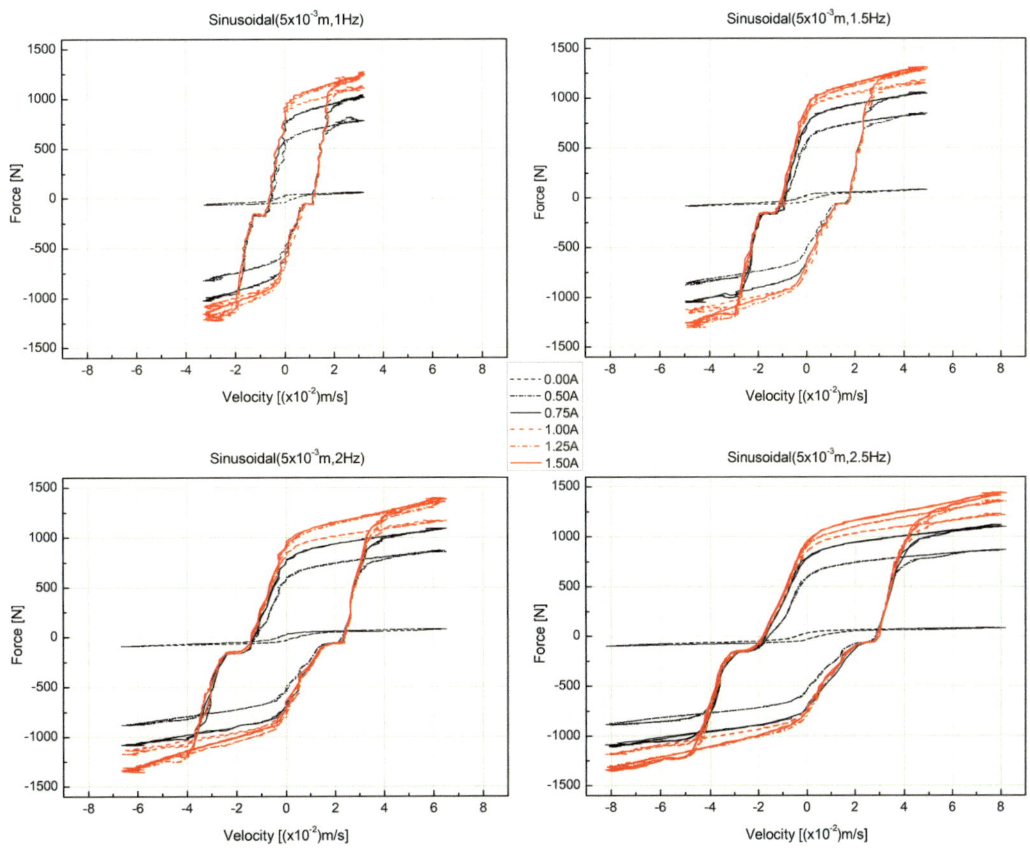

Figure 6. Experimental data measured at sinusoidal excitations (frequency range (1, 2.5)Hz, and amplitude 0.005m), and supplied current in range (0, 1.5)A.

2.3.2. *MR fluid damper characteristic analysis*

Remark 1 (*affecting factors*). In order to design the MR fluid damper models, an investigation into factors which affect the dynamic responses of the damping system has been done. The first affecting factor is the applied displacement/velocity on the piston rod of the damper. Fig. 6 displays a comparison between damping results under various sine excitations with 0.005m amplitude and frequency range from 1Hz to 2.5Hz while the supplied current level was in range from 0 to 1.5A. The results show that at fixed current level applied to the damper, the damping force varies due to the piston rod velocity which is caused by the simultaneous change of frequency and/or amplitude of the applied excitation. The second factor affecting the damper behavior is the change in current applied to the damper coil. Fig. 7 shows an example of measurement results in plots of force-time, force-displacement, and force-velocity relations with respect to a 2.5Hz sinusoidal excitation and 0.005m of amplitude while the current supplied to the damper was in range between 0 and 1.5A. From these figures, it is readily apparent that:

- The force produced by the damper is not centered at zero. This effect is due to the effect of an accumulator containing high pressure nitrogen gas in the damper. The accumulator helps to prevent cavitations in the fluid during normal operation and accounts for the volume of fluid displaced by the piston rod as well as thermal expansion of the fluid.

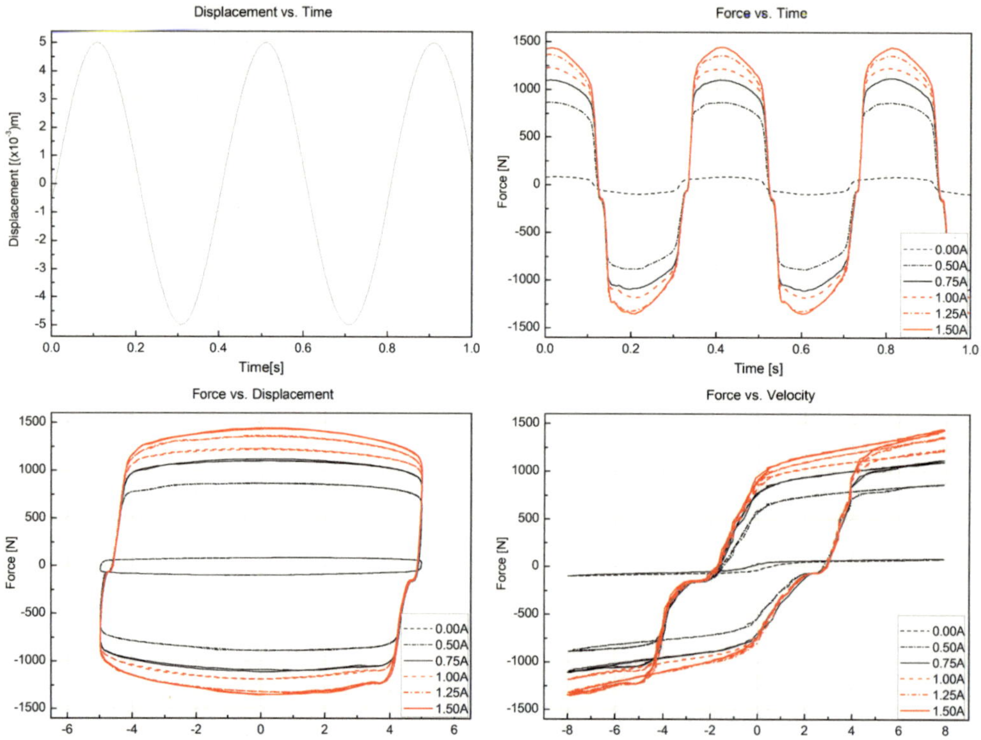

Figure 7. Experimental data measured at sinusoidal excitation (frequency 2.5Hz, and amplitude 0.005m), and supplied current in range (0.5, 1.5)A.

- The greater current level, the greater damping force.
- The change rate of force is faster at lower current levels because of the effect of magnetic field saturation.

Based on the above analyses, it is clear that the damping force of the MR fluid damper depends on the displacement/velocity of the damper rod and the current supplied for the coil inside the damper.

3. MR damper modeling technologies

3.1. Typical parametric models

3.1.1. Bingham model

The stress-strain behavior of the Bingham visco-plastic model [37] is often used to describe the behavior of MR fluid. In this model, the plastic viscosity is defined as slop of the measured shear stress versus shear strain rate data. For positive values of the shear rate, $\dot{\gamma}$, the total stress is given:

$$\tau = \tau_{y(field)} + \eta\dot{\gamma} \tag{1}$$

where $\tau_{y(field)}$ is the yield stress induced by the magnetic field and η is the viscosity of the fluid.

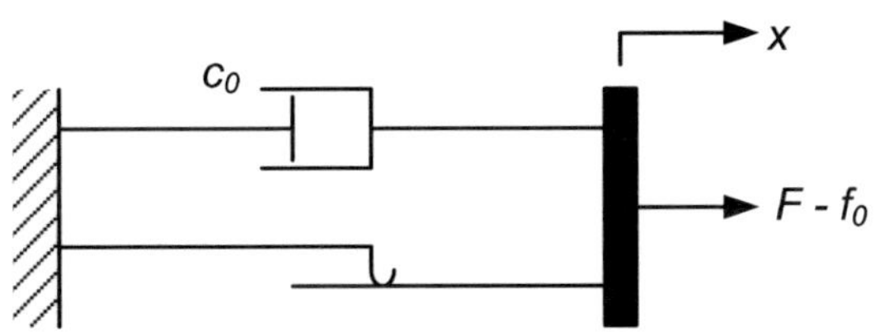

Figure 8. Bingham model of a MR fluid damper.

Based on this model, an idealized mechanical model referred to as the Bingham model was proposed to estimate the behavior of an MR fluid damper by Standway *et al* [20]. This model consists of a Coulomb friction element placed in parallel with a viscous damper as depicted in Fig. 8. Here, for nonzero piston velocities, $\dot{x}$, the force F generated by the device is given by:

$$F = f_c \text{sign}(\dot{x}) + c_0\dot{x} + f_0 \tag{2}$$

where c_0 is the damping coefficient; f_c is the frictional force related to the fluid yield stress; and an offset in the force f_0 is included to account for the nonzero mean observed in the measured force due to the presence of the accumulator. Note that if at any point the velocity of the piston is zero, the force generated in the frictional element is equal to the applied force.

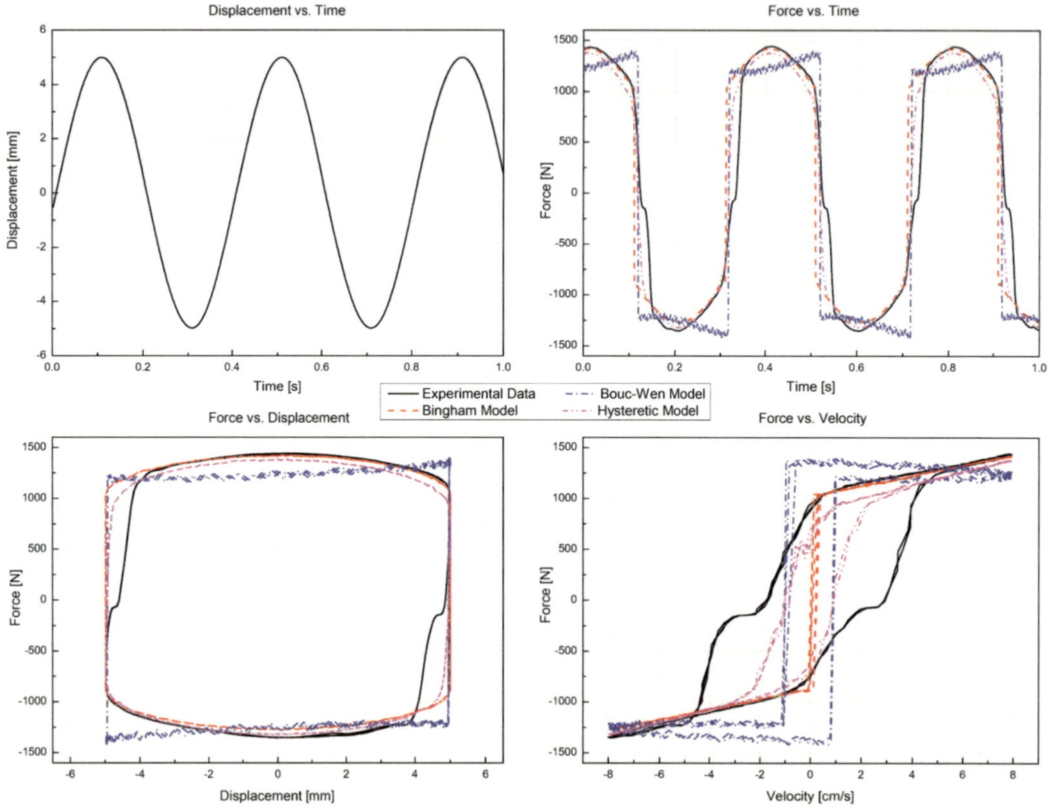

Figure 9. Comparison between experimental data and the predicted damping forces for a 2.5Hz sinusoidal excitation with amplitude 5mm while current supplied to the damper is 1.5A.

To present the damper behavior, the characteristic parameters of the Bingham model in equation (2) need to be chosen to fit with the experimental data of the damping system. For example, those parameters were chosen as c_0 = 50Ns/cm; f_c = 950N and f_0 = 75N for a 2.5Hz sinusoidal excitation with amplitude 5mm while the current supplied to the damper was 1.5A. Consequently, the predicted damping force by using the Bingham model was compared with the experimental response as plotted in Fig. 9.

From the results, although the force-time and force-displacement behavior were reasonably modeled, the predicted force-velocity relation was not captured, especially for velocities that were near zero. By using this model, the relationship between the force and velocity was one-to-one, but the experimentally obtained data was not one-to-one. Furthermore, at zero velocity, the measured force had a positive value when the acceleration was negative (for positive displacements), and a negative value when the acceleration was positive (for negative displacements). This behavior must be captured in a mathematical model to adequately characterize the device. Hence, Gamota and Filisko [38] developed an extension of the Bingham model, which is given by the viscoelastic-plastic model shown in Fig. 10.

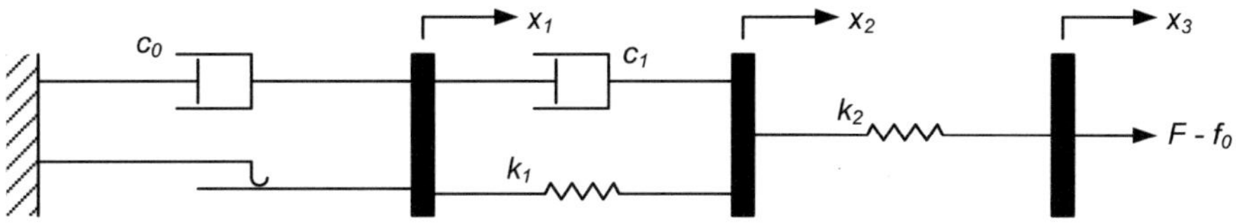

Figure 10. Extened Bingham model of a MR fluid damper.

The model consists of the Bingham model in series with a standard model. The governing equations for this model are given as followings

$$\left.\begin{aligned} F &= k_1(x_2 - x_1) + c_1(\dot{x}_2 - \dot{x}_1) + f_0 \\ &= c_0\dot{x}_1 + f_c \operatorname{sign}(\dot{x}_1) + f_0 \\ &= k_2(x_3 - x_2) + f_0 \end{aligned}\right\}, \quad |F| > f_c \tag{3}$$

$$\left.\begin{aligned} F &= k_1(x_2 - x_1) + c_1\dot{x}_2 + f_0 \\ &= k_2(x_3 - x_2) + f_0 \end{aligned}\right\} \quad , \quad |F| \le f_c \tag{4}$$

where c_0 is the damping coefficient associated with the Bingham model; k_1, k_2 and c_1 are associated with the linear solid material.

This model can present the force-displacement behavior of the damper better the Bingham model. However, the governing equations (3), (4) are extremely stiff, making them difficult to deal with numerically [21]. Therefore, the Bingham model or extended Bingham model are normally employed in case there is a significant need for a simple model.

3.1.2. Bouc-Wen model

One model that is numerically tractable and has been extensively used for modeling hysteretic systems is Bouc-Wen model. This model contains components from a viscous damper, a spring and a hysteretic component. The model can be described by the force equation and the associated hysteretic variable as given

$$F = c\dot{x} + kx + \alpha z + f_0 \tag{5}$$

$$\dot{z} = -\gamma|\dot{x}|z|z|^{n-1} - \beta\dot{x}|z|^n + \delta\dot{x} \tag{6}$$

where: F is the damping force; f_0 is the offset force; c is the viscous coefficient; k is the stiffness, $\dot{x}$ and x are the damper velocity and displacement; α is a scaling factor; z is the hysteretic variable; and γ, β, δ, n are the model parameters to be identified. Note that when $\alpha = 0$, the model represents a conventional damper.

In order to determine the Bouc-Wen characteristic parameters predicting the MR fluid damper hysteretic response, Kwok *et al* [25] proposed the non-symmetrical Bouc-Wen model with following modifications

$$F = c\left(\dot{x} - \mu \operatorname{sign}(z)\right) + kx + \alpha z + f_0 \tag{7}$$

$$\dot{z} = \left\{ -\left[\gamma \operatorname{sign}(z\dot{x}) + \beta \right] |z|^n + \delta \right\} \dot{x} \tag{8}$$

where μ is the scale factor for the adjustment of the velocity.

As the optimization results for the test rig applied the damper RD-1005-3 by using GA in [25], the relationships between the Bouc-wen parameters and the supplied magnetization current, i, were given as

$$\begin{array}{l} c = 2.65\times10^3 i + 2.05\times10^3; k = 1.99\times10^3 i + 5.57\times10^3; \alpha = 2.11\times10^3 i + 1.68\times10^3 \\ f_0 = 0.6i - 12.43; \mu = -0.02i + 1.25; n = 0.12i + 1.58 \\ \delta = 0.5\times10^5 i + 2.5\times10^5; \beta = -0.45\times10^6 i + 3.18\times10^6; \gamma = 0.39\times10^6 i + 3.6\times10^6 \end{array} \tag{9}$$

Because of using the same researched damper, the Bouc-Wen model built from equations from (7) to (9) was tested for modeling the damping force in this study. As a result, the predicted force was plotted as the 'dash-dot' line in Fig. 9 for a 2.5Hz sinusoidal excitation with amplitude 5mm while the current supplied to the damper was 1.5A. The estimated damping performance when compared with the real damping performance shows that the proposed Bouc-wen model in [25] could not represent for the damping behavior in the TR01. It is because that the model parameters in equation (9) were only optimized for the damping system using the damper RD-1005-3 in [25]. From the result, it is clearly that to obtain good predicted behavior of a MR fluid damper in a specific system, the Bouc-Wen parameters must be tuned again by using optimization or trial error techniques which causes high computational cost to obtain the optimal parameters.

Furthermore, to obtain better modeling performance, some modified Bouc-Wen models have been proposed. The research results in [21] show that the modified Bouc-Wen model improves the modeling accuracy. However, the model complexity is unavoidably increased with an extended number of model parameters (14 parameters need to be identified in [21]) which may impose difficulties in their identification and take much time for optimization process [27].

3.1.3. A hysteretic model

For a simple model, Kwok *et at* [24] proposed a hysteretic model to predict the damping force of the MR fluid damper RD-1005-3 as illustrated in Fig. 11. The model can be expressed as following equations

$$F = c\dot{x} + kx + \alpha z + f_0 \tag{10}$$

$$z = \tanh\left(\beta \dot{x} + \delta \operatorname{sign}(x)\right) \tag{11}$$

where: c and k are the viscous and stiffness coefficients; α is the scale factor of the hysteresis; z is the hysteretic variable given by the hyperbolic tangent function; f_0 is the damper force offset; and β, δ are the model parameters to be identified.

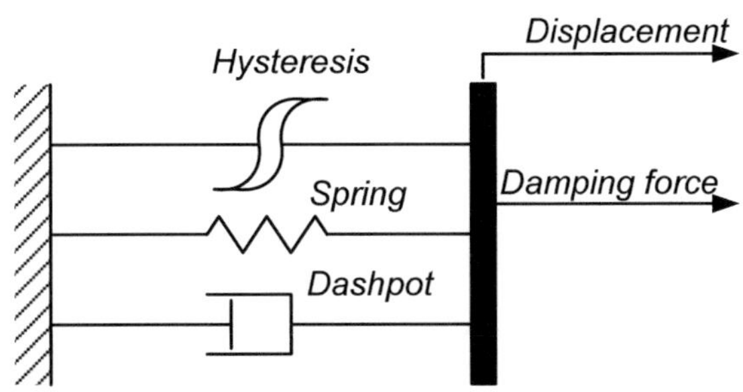

Figure 11. Hysteretic model of a MR fluid damper.

As the results in [24], the parameters in equations (9) and (10) were given:

$$\begin{aligned} &c = 1929i + 1232; k = -1700i + 5100; \alpha = -244i^2 + 918i + 32; f_0 = -18i + 57 \\ &\beta = 100; \delta = 0.3i + 0.58 \end{aligned} \tag{12}$$

However, to obtain the parameters as in equation (12), a swam optimization [24] must be used to select the most suitable values with respect to each specific system using the damper RD-1005-3. Hence, when using the set of resulting parameters in [24] to apply to the test system of the MR fluid damper RD-1005-3 in this study, the hysteretic model cannot present well the damper behavior. For example, the modeling result by using the hysteretic model, for a 2.5Hz sinusoidal excitation with amplitude 5mm while the current supplied to the damper was 1.5A, is depicted in Fig. 9 as the 'short dash' line. The result proves that although the estimated force in this case was better than in case of using Bingham or Bouc-Wen model, the nonlinear characteristic of the damper could not be described well. Moreover, the swam optimization also requires training time to generate the parameters of the hysteretic model.

3.2. Proposed non-parametric model

It is known that the typical parametric models show their possibility to be applied for the MR fluid damper identification. However, the decisive parameters of parametric models need to be tuned by using optimization or trial error techniques which causes high computational cost to generate their suitable values. In addition, those models only adapt with specific damping systems. For a new system using the same MR fluid damper series, the optimization process must be done again for a full prediction of the damper behavior [35,36]. Therefore, a non-parametric method based on intelligent techniques, for example, is an effective solution to estimate directly the MR fluid damper behavior with high precision.

Fuzzy system is an intelligent tool imitating the logical thinking of human and then is capable of approximating any continuous function. However, there is no systematic method

to design and examine the number of rules, input space partitions and membership functions (MFs). Meanwhile, neural network mimics the biological information processing mechanisms. This technique modifies its behavior in response to the environment, and is ideal in case that the expected mapping algorithm is un-known and the tolerance to faulty input information is required. Hence an identification system using fuzzy and neural network theories can be easily selected as an effective method for directly modeling MR fluid dampers purpose.

3.2.1. BBM model design

As mentioned in section 2.3, the MR fluid damper force is affected by the rod displacement/velocity and supplied current. Therefore, the designed BBM contains two parts: one is the neural-fuzzy inference (NFI) that is used to estimate the damping force caused by the displacement of the damper rod, and the other is the scheduling gain fuzzy inference (SGFI) which is used to switch between the damping force levels with respect to the current levels supplied for the damper coil. Consequently, the estimated damping force (f_{MR_est}) is computed as a multiplication of the NFI estimated force and the SGFI gain as:

$$f_{MR\_est} = K \times U \tag{13}$$

where: K is the damping force level corresponding to the current level supplied for the damper coil; U is the damping force caused by the displacement applied to the damper rod.

$$\begin{cases} K = \alpha_{SGFI} k \\ U = \alpha_{NFI} u \end{cases}; \begin{cases} k \in [-1;1] \\ u \in [0;1] \end{cases} \tag{14}$$

where: k and α_{SGFI} are the SGFI output and a scale factor chosen from the current range for the MR fluid damper coil, respectively; u and α_{NFI} are the NFI output and a scale factor chosen from the MR fluid damper specifications, respectively.

To evaluate the accuracy of the BBM model, an error function (E) was derived from the difference between the damping force (f_{MR_est}) estimated from the MR model and the actual damping force (f_{MR}) when the input conditions (current and displacement/velocity) for both the model and real MR fluid damper system are the same. Therefore, the error function is defined as following equation:

$$E = 0.5(f_{MR\_est} - f_{MR})^2 \tag{15}$$

Based on the Remark 2, the overall structure of the proposed BBM to model the MR fluid damper is shown in Fig. 12a while the internal structure of the NFI system is described in Fig. 12b. For all of the fuzzy designs, triangle membership functions are used to represent for partitions of fuzzy inputs and outputs. Fuzzy control is applied using local inferences. That means each rule is inferred and the inferring results of individual rules are then aggregated. The most common inference method (aggregation-fuzzy implication operators)

- max-min method, which offers a computationally nice and expressive setting for constraint propagation, is used. Finally, a defuzzification method is needed to obtain a crisp output from the aggregated fuzzy result. Popular defuzzification methods include maximum matching and centroid defuzzification. The centroid defuzzification is widely used for fuzzy control problems where a crisp output is needed, and maximum matching is often used for pattern matching problems where the output class needs to be known. Hence in this research, the fuzzy reasoning results of outputs are gained by aggregation operation of fuzzy sets of inputs and designed fuzzy rules, where max-min aggregation method and centroid defuzzification method are used.

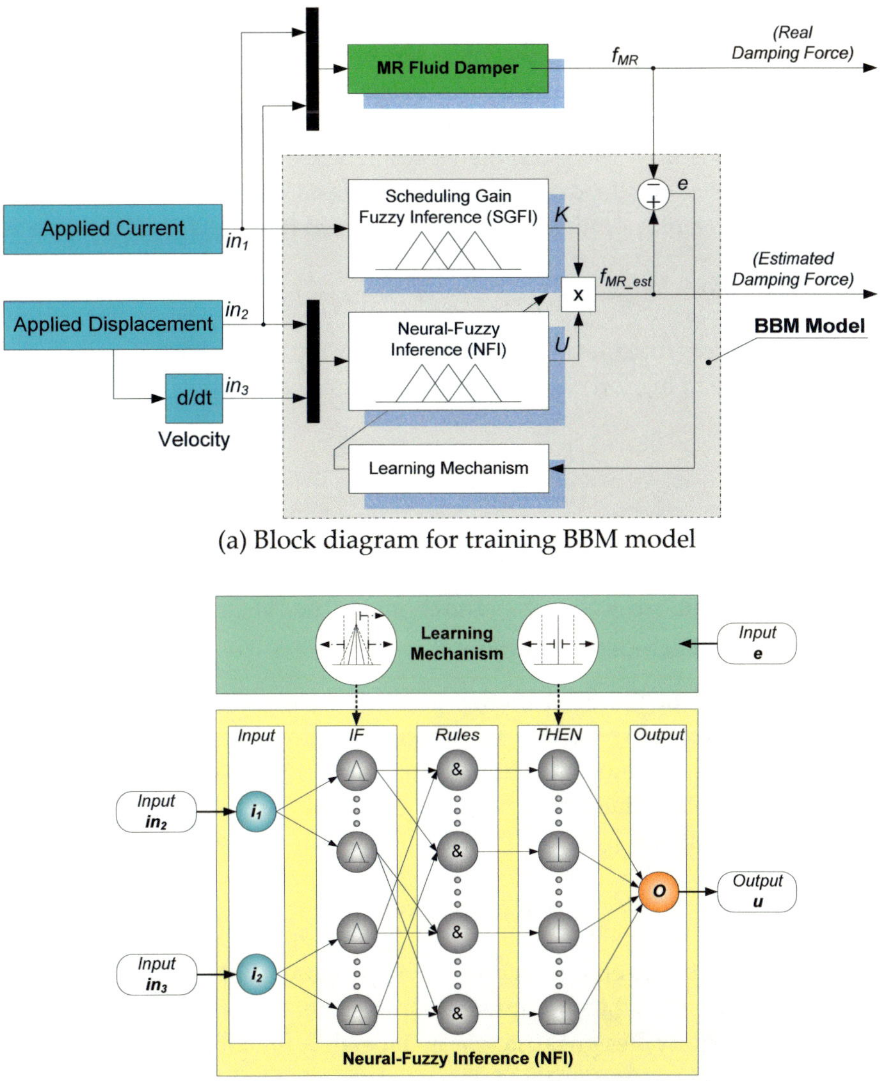

(a) Block diagram for training BBM model

(b) Internal structure of NFI system

Figure 12. Structure of identification for a MR fluid damper using proposed BBM.

3.2.1.1. Neural-Fuzzy inference (NFI)

The NFI system takes part in estimating the damping force caused by the applied displacement/velocity to the damper. As seen in Fig. 12, the NFI fuzzy set was therefore designed with two inputs (in_2, and in_3) and one output (u). The ranges of these inputs were from -1 to 1, which were obtained from the applied displacement, and its derivative (velocity) through scale factors chosen from the range of displacement and specifications of the MR fluid damper. The fuzzy output range was also set from -1 to 1. Consequently, the estimated force can be obtained by multiplying the NFI output and the suitable scale factor α_{NFI} (see equation (2)). The input/output ranges need to be divided into several partitions in order to construct the fuzzy rule map. Based on design experience obtained from the previous researches [33,34], five triangle MFs were used for each the NFI fuzzy input/output for smooth estimating the damping force while it does not require much calculating time consumption.

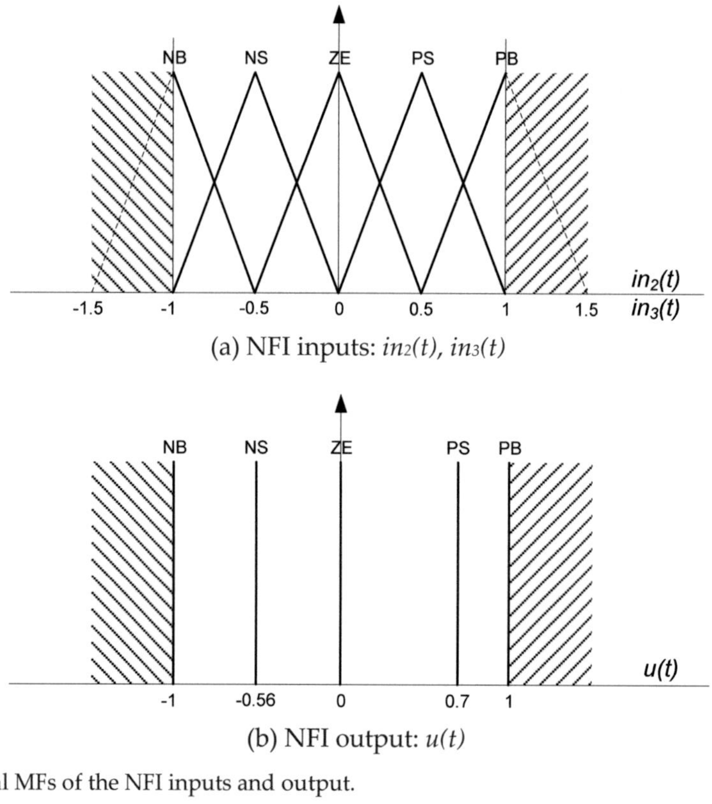

Figure 13. Initial MFs of the NFI inputs and output.

For each input variable, five triangle MFs $\left(\mu(.)\in\left[0,1\right]\right)$ were designed and named as "NB", "NS", "ZE", "PS" and "PB" which mean "Negative Big", "Negative Small", "Zero", "Positive Small" and "Positive Big", respectively. These MFs and their centroids were initially set with a same shape size and at same intervals, respectively, in Fig. 13a. Because all of the MFs are triangle shapes, so we can express these MFs as follows:

$$\mu_j(x_i) = \frac{1-2\left|x_i - a_{ji}\right|}{b_{ji}}, \ j = 1,2.........N, \ i = 1 \text{or} 2, \begin{cases} x_1 \equiv in_2 \\ x_2 \equiv in_3 \end{cases} \tag{16}$$

where a_j and b_j are the centre and width of the jth triangle MF; N is the number of triangles ($N = 5$).

The fuzzy reasoning result of the NFI output is gained using an aggregation operation of fuzzy sets of the inputs and designed fuzzy rules, where the max-min aggregation and centroid defuzzification methods are used. For a pair of inputs (in_2, in_3), the NFI output can be computed as:

$$u = \frac{\sum_{k=1}^{M} \mu(w_k) w_k}{\sum_{k=1}^{M} \mu(w_k)} \tag{17}$$

where: w_k and $\mu(w_k)$ are the weight and its height of the NFI output, respectively; M is the number of fuzzy output sets (M = 5). The height $\mu(w_k)$ is computed by using the fuzzy output function:

$$\mu(w_k) = \sum_{i,j} \mu_{ij}(w_k) \tag{18}$$

where $\mu_{ij}(w_k)$is defined as the consequent fuzzy output function when the first and second NFI input are in the i and j class, respectively:

$$\mu_{ij}(w_k) = \delta_{ij}\mu_{ij} \tag{19}$$

where δ_{ij} is an activated factor which is active when the input in_2 is in class i, and the input in_3 is in class j; μ_{ij} is the height of the consequent fuzzy function obtained from the input class i and j:

$$\mu_{ij} = \min\left[\mu_i(x_1), \mu_j(x_2)\right] \equiv \min\left[\mu_i(in_2), \mu_j(in_3)\right] \tag{20}$$

where $\mu_i(in_2)$and$\mu_j(in_3)$ are obtained from equation (16).

The output u of the NFI system contains five single output values: "NB", "NS", "ZE", "PS", and "PB", within the range from -1 to 1, with the same meaning as the MFs of the inputs. The initial output weights were decided from the experimental results with constant supplied current where the damping force values were caused by the corresponding points of input displacement and velocity [35,36]. Consequently, the output weights were initially set at the different intervals as in Fig. 13b.

By using the above fuzzy sets of input/output variables, experimental data, damper behavior, and experience, the fuzzy rules for the NFI part of the MR model are established

in Table 3. Five MFs for the each input were used to decide the total twenty five rules by using an IF-THEN structure. Here, one fuzzy rule is composed as follows:

RULE *i*: IF displacement (in_2) is A_i and velocity (in_3) is B_i THEN MR force (u) is C_i (i=1,2, .., 25)

where A_i, B_i, and C_i are the ith fuzzy sets of the input and output variables used in the fuzzy rules. A_i, B_i, and C_i are the linguistic variable values in_2, in_3, and u, respectively.

NFI output u		NFI input 02 - Scaled velocity - in_3				
		NB	NS	ZE	PS	PB
NFI input 01 - Scaled displacement - in_2	NB	NB	NB	NS	ZE	PB
	NS	NB	NS	NS	PS	PB
	ZE	NB	NS	ZE	PS	PB
	PS	NB	NS	PS	PS	PB
	PB	NB	ZE	PS	PB	PB

Table 3. Rules table for the neural-fuzzy inference of the black box model

As the above description, the fuzzy MFs and rules were dependent on the characteristics of the damper which were investigated from the experimental data. These MFs and rules were then determined from both the intuition and practical experience. There is no systematic method for designing and examining the input space partitions, MFs, and rules which affect directly the modeling quality. As a result, an optimization methodology was indispensible to be used for tuning the NFI structure to fit with the damping behaviors.

Remark 2 (*model optimization*). In order to improve the identification quality of the proposed models, a learning mechanism using neural network methodology, including the back-propagation algorithm and gradient descent method, has been used to adjust the fuzzy structures of the BBM and IBBM models. The back-propagation algorithm is a first order approximation of the steepest descent technique in the sense that it depends on the gradient of the instantaneous error surface. The algorithm is therefore stochastic in nature which means that it has a tendency to zigzag its way about the true direction to a minimum on the error surface. The basic idea of using the proposed method to optimize the fuzzy designs is to use the back-propagation to tune the input as well as output MF shapes of the models during the system operation process in order to minimize a defined error cost function.

The ability of using the training mechanism based on the back-propagation algorithm and gradient descent method for optimizing the fuzzy scheduling systems were clearly proved in previous researches [33,34]. As Remark 2, the proposed optimization method is used to tune the input MF shapes and output weights of the NFI system during the system operation process in order to minimize the modeling error function (15). Consequently, the decisive factors in the inputs MFs a_{ji}, b_{ji}, and the output weights w_k were automatically adjusted by computing efficiently partial derivatives of the error function E realized by the model network with respect to all these decisive elements. A following set of equations shows the back-propagation algorithm based on the gradient descent method for updating the decisive factors at a step of time $(t+1)^{th}$:

$$\left.\begin{aligned} a_{ji}\big|_{t+1} &= a_{ji}\big|_{t} - \eta_a \frac{\partial E}{\partial a_{ji}}\bigg|_{t} \\ b_{ji}\big|_{t+1} &= b_{ji}\big|_{t} - \eta_b \frac{\partial E}{\partial b_{ji}}\bigg|_{t} \\ w_{k}\big|_{t+1} &= w_{k}\big|_{t} - \eta_w \frac{\partial E}{\partial w_{k}}\bigg|_{t} \end{aligned}\right\} \tag{21}$$

where and are the learning rate which determine the speed of learning; E is the error function defined by (15).

With the self learning of neural network technique and the decreasing of the modeling error, the optimized NFI system works more effectively with high accuracy when compared to the real damping response.

3.2.1.2. Scheduling gain fuzzy inference (SGFI)

This section provides a description of the scheduling gain fuzzy inference which works as an intelligent switch to tune the damping force levels (k) with respect to the current levels supplied for the MR fluid damper. The SGFI system was then designed with a single input (in_1) and a single output (k) (see Fig. 12).

The range of the input was from 0 to 1, which was obtained from the supplied current through a scale factor chosen from the current range for the MR fluid damper coil. Five triangle MFs, "Z"(Zero), "VS"(Very Small), "S"(Small), "M"(Medium), and "B"(Big), were used for this input variable. These MFs and their centroids were initially set with a same shape size and at same intervals, respectively, in Fig. 14a. These MFs can be then expressed in the same form as in (16). By using the same fuzzy design method as that of the NFI system in section 3.2.1a to design the SGFI, the output gain (k) corresponding to an input value (in_1) can be computed as

$$k = \frac{\sum_{q=1}^{Q} \mu\left(w_q\right) w_q}{\sum_{q=1}^{Q} \mu\left(w_q\right)} \tag{22}$$

where: w_q and $\mu(w_q)$ are the weight and its height of the SGFI output, respectively. Q is the number of fuzzy output sets ($Q = 5$).

For the output k of the SGFI system, five MFs were used. Here, "VS", "S", "M", "B", and "VB" are "Very Small", "Small", "Medium", "Big", and "Very Big", respectively. The output range was set from 0 to 1. The estimated damping force level is then obtained by multiplying the SGFI output and the suitable scale factor α_{SGFI} (see equation (14)). The output weights were decided based on the experimental results and characteristics of the

MR fluid damper [35,36]. Therefore, the output weights were set as in Fig. 14b. By using the above fuzzy sets of input and output variables, the fuzzy rules for the SGFI part in the BBM model are established in Table 4 by using the IF-THEN structure.

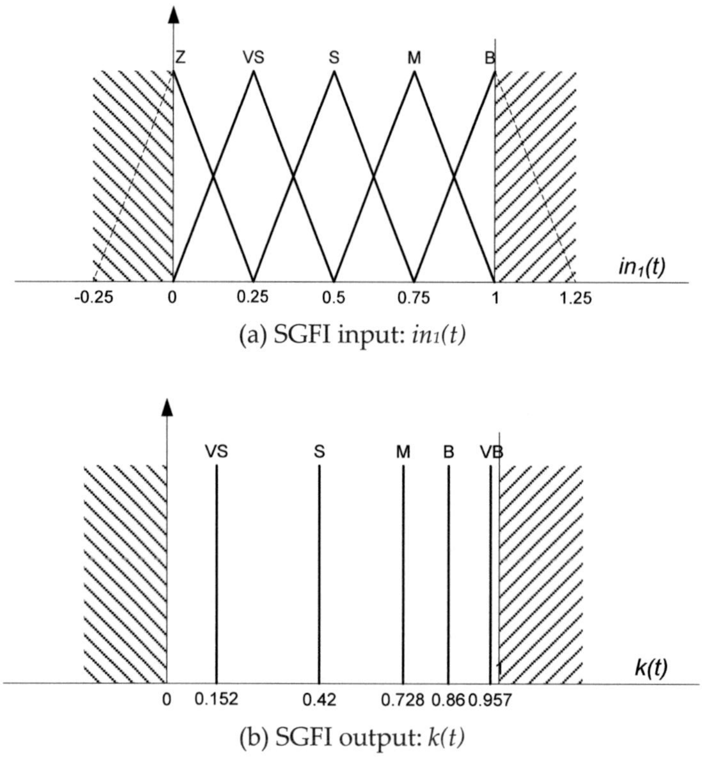

Figure 14. MFs of the SGFI inputs and output.

SGFI input – Scaled supplied current (in_1)	Z	VS	S	M	B
SGFI output (k)	VS	S	M	B	VB

Table 4. Rules table for the scheduling gain inference of the black box model

Finally, the output of the proposed black-box model (f_{MR_est}) can be computed from the NFI output (u) and SGFI output (k) using equations (13) and (14).

3.2.2. BBM model verification

3.2.2.1. Comparison between modeling results and experimental data on the TR01

From [35,36], the model training process and simulations have been carried out to find out the BBM model with optimized parameters and to evaluate the ability of the optimized BBM model when comparing with the measured dynamic responses of the damper, respectively.

As the result, the BBM with parameters optimized by the leaning mechanism with respect to the modeling error cost function was found. Fig. 15 shows the MFs of the BBM system after training to obtain high accuracy in estimating force of the MR fluid damper. The optimized BBM model was then evaluated in the comparison with the actual measurement data.

Fig. 16 displays the modeling results of the proposed BBM model in a comparison with the real damping behavior for a 2.5Hz sinusoidal displacement. The results show that with the designed modeling method, the nonlinear characteristic of the MR fluid damper can be directly estimated with high accuracy for both the force/time, force/displacement, and force/velocity relations despite the variation in applied current for the damper.

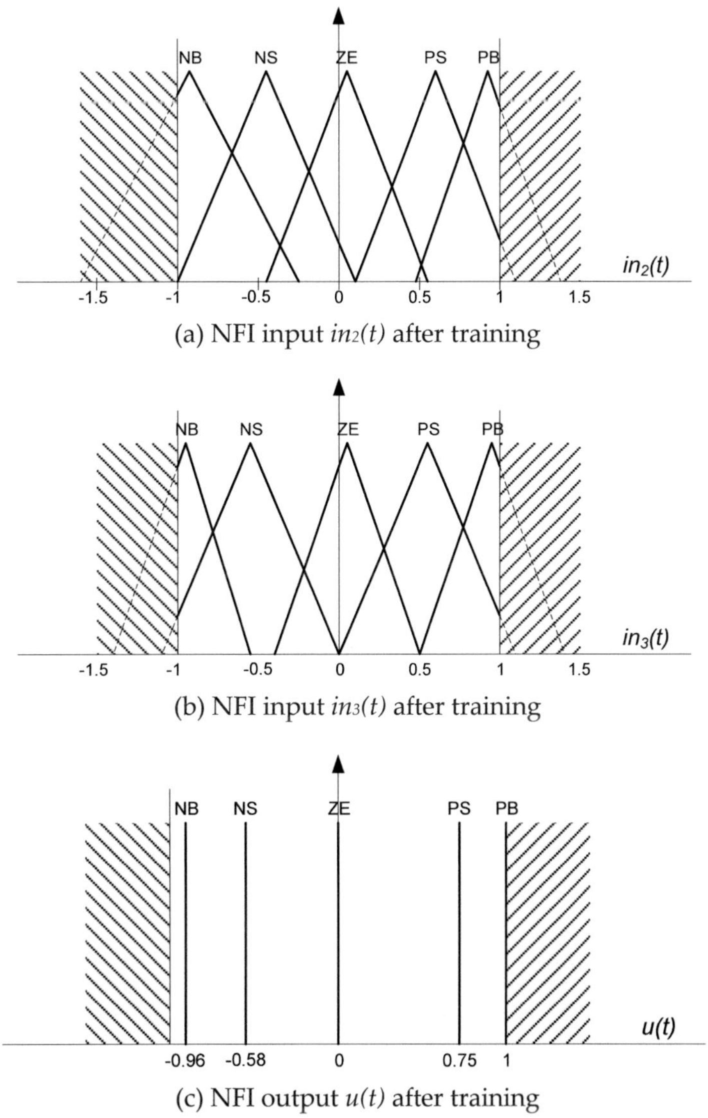

Figure 15. MFs of the NFI inputs and output after training.

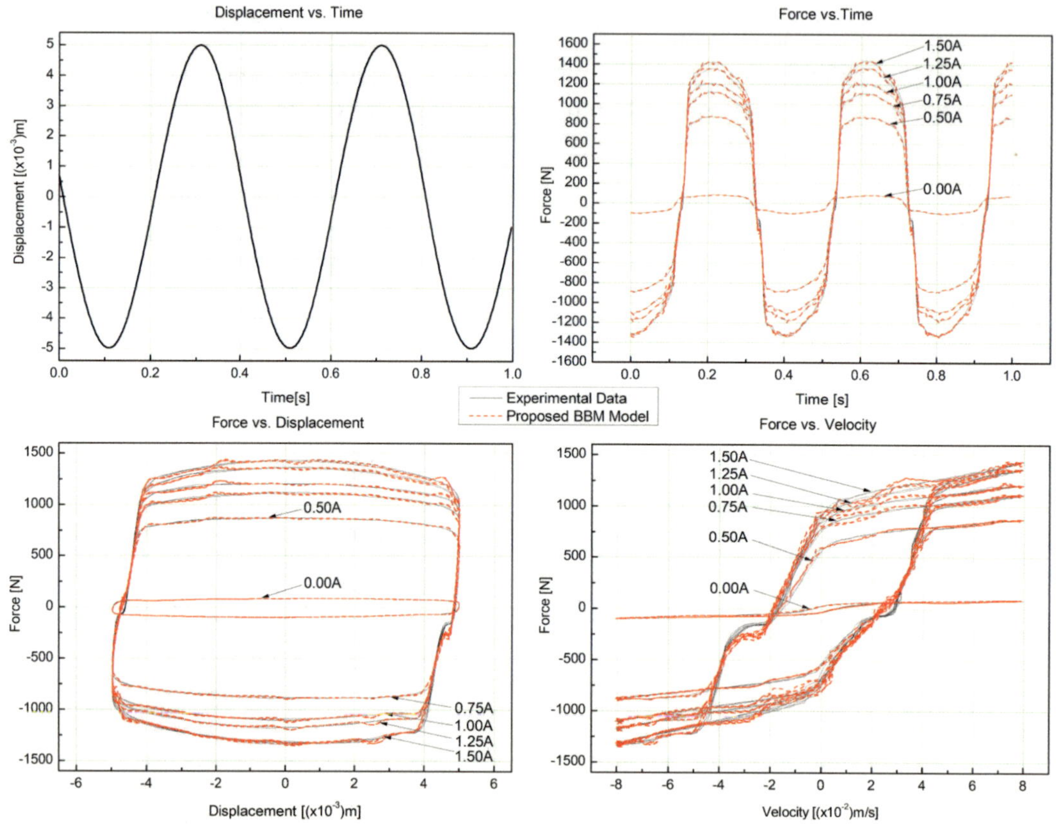

Figure 16. Comparison between the estimated force and actual damping force for an applied current range (0, 1.5)A at a sinusoidal excitation (frequency 2.5Hz and amplitude 0.005m).

Secondly, displacement excitations with a continuous variation of the frequency were generated to fully check the ability of the designed modeling method in case of varying excitation environments. In addition, experimental data were measured from the damping system, rig TR01, with the chirp displacement excitations of which the frequencies were varied from 1Hz to 2.5Hz. Figures 17 and 18 depict the comparisons of the real damping responses and the estimated forces using the different models in cases: 0A and 1.5A of the applied current for the MR fluid damper coil. As seen in figures 17a and 18a, the damping behavior could not be modeled by using the Bingham model, Bouc-Wen model, or Hysteretic model. Here, the Bingham model could only predict the relation between damping force and time/ or displacement/or velocity as the one-to-one relation. Meanwhile, the Bouc-wen and Hysteretic models can only predict force for a particular damping system as the TR01 when their parameters are optimized with respect to this system. In contrast to the unfavorable modeling results in figures 17a and 18a, figures 17b and 18b show a good damping force prediction using the designed BBM model. From these results, it is clearly that with the self-tuning ability, the BBM has enough strength to describe well the nonlinear behavior of the damper under various excitation environments, especially in case of low supplied current levels.

3.2.2.2. Investigation of self-sensing behavior of the TR02 using the optimized BBM

In this section, the ability of optimized BBM model is investigated when it is applied as a virtual force sensor to any damping system using the same MR fluid damper, such as the rig TR02, for the self-sensing behavior.

In the test rig TR02, the vibration was generated by the pneumatic cylinder and proportional valve of which the control signal was a voltage signal. This signal was a sinusoidal of which the frequency was in a range from 1 to 2.0 Hz while the amplitude was 3V as:

$$ValveControlSignal = A\sin(2\pi ft); A = 3V; f \in [1, 2.0]Hz \tag{23}$$

There were two cases of working load: 3kg and 9kg while the supplied current for the MR fluid damper was changed from 0 to 2A in order to create different test conditions. Consequently, setting parameters for experiments on the TR02 system are shown in Table 5.

Test No.	Working load (kg)	Displacement – Sine wave		MR fluid damper current (A)
		Amplitude (V)	Frequency (Hz)	
Case 01	9	3.0	1.0	0.0
Case 02	3	3.0	1.5	1.0
Case 03	3	3.0	2.0	2.0

Table 5. Test conditions on the TR02 for verification of the BBM with self-sensing behavior

A testing program for the BBM model verification using the rig TR02 was built in Simulink with the real-time toolbox of MATLAB as shown in Fig. 19. Experiments had been done on the TR02 in which the vibration was created as in Table 5 by using the program in Fig. 19. During the system operations, the real damping force caused by the vibration was measured by the load cell in order to make the comparison with the damping force 'measured' by the virtual force sensor - BBM. Consequently, the evaluation results of self-sensed force with respect to the test conditions are depicted on figures from 20a to 20c.

As seen in these figures, the predicted performances using the BBM model were mostly close to the real damping performances. However, there were some predicting errors at the limits of damper compressions which can be seen in Fig. 20 as the tips of excitations. There are some reasons for these errors. The first reason is that the BBM model was optimized using experimental investigations on the rig TR01 of which the hardware included compliances and the control system included measuring noises. In addition, the rig TR02 is a damping system activated by using the pressurized pneumatic cylinder. Consequently, the changing of the generated excitations was fast, especially when the cylinder was extracted and then retracted which caused the damper compression limits. Therefore, there were some small errors in the estimated damping force near the damper compression limits.

From the results in Fig. 20, it is clear that the proposed self-sensing methodology based on the BBM model has strong ability to apply to damping control systems without using the force sensor.

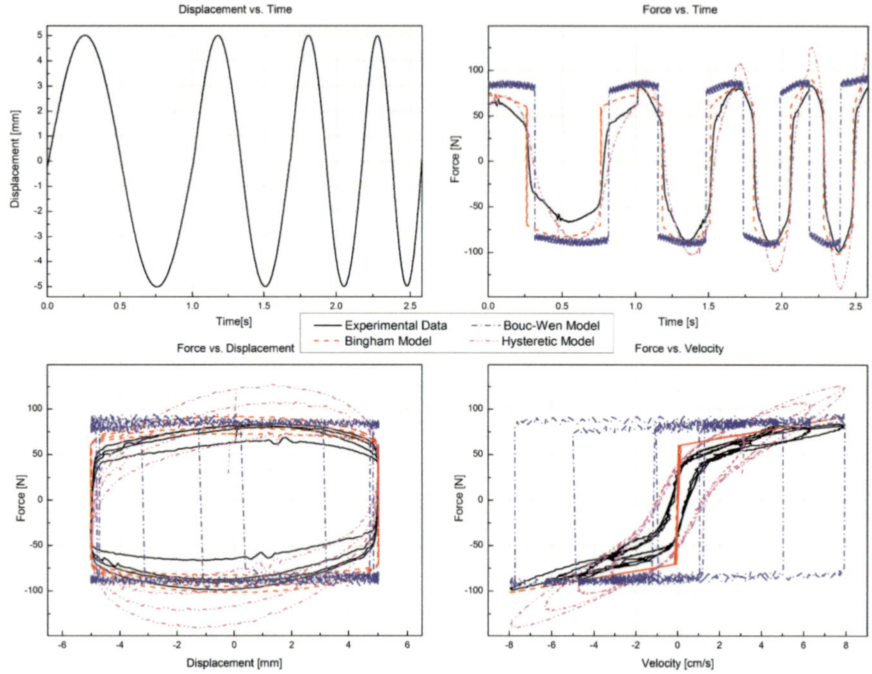

(a) Modeling results using the conventional models

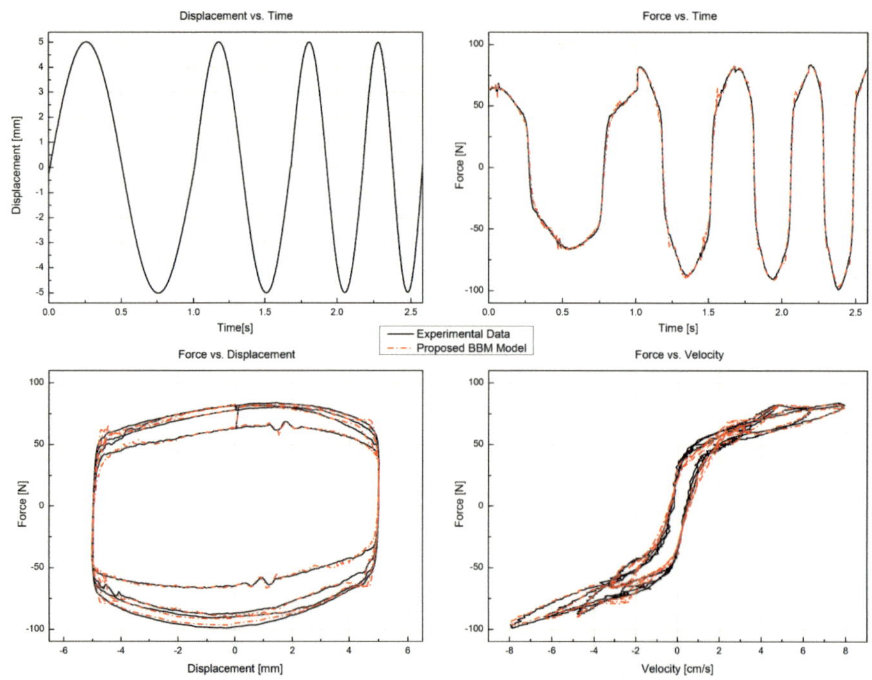

(b) Modeling results using the proposed BBM Model

Figure 17. Comparison between estimated forces and actual damping force for an applied current 0A at a chirp excitation (frequency range (1, 2.5)Hz and amplitude 5mm).

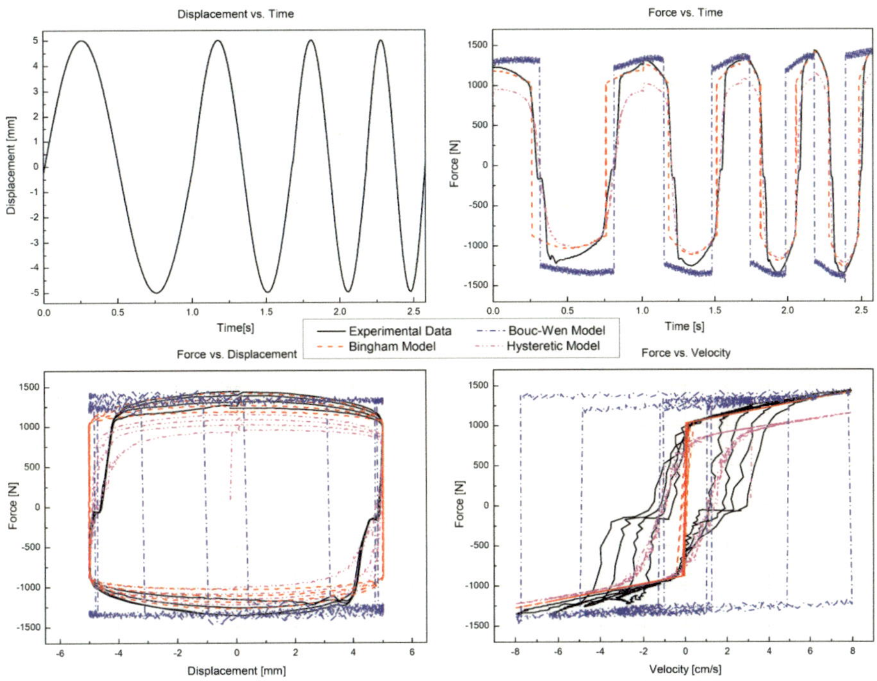

(a) Modeling results using the conventional models

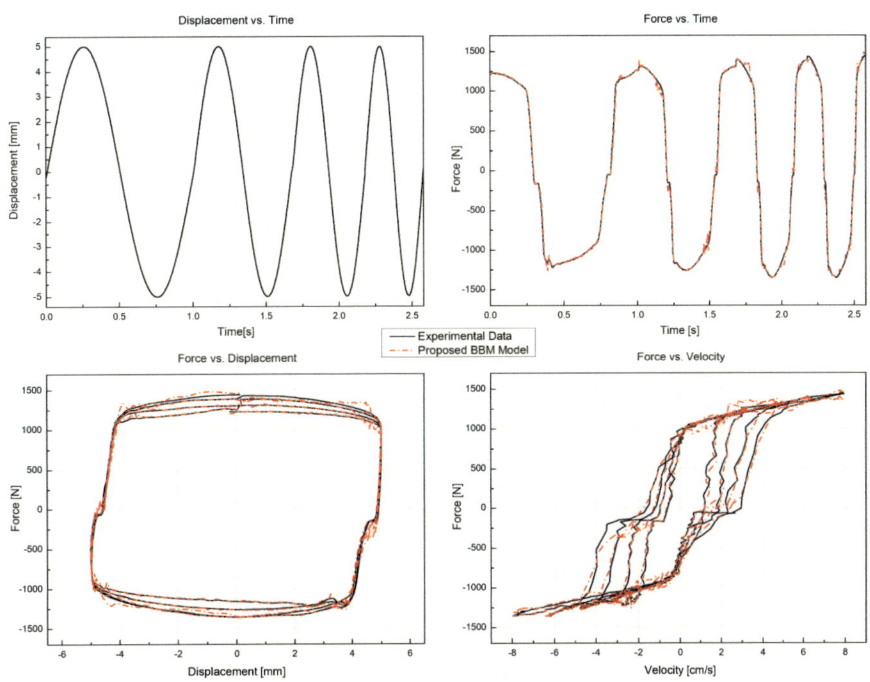

(b) Modeling results using the proposed BBM Model

Figure 18. Comparison between estimated forces and actual damping force for an applied current 1.5A at a chirp excitation (frequency range (1, 2.5)Hz and amplitude 5mm).

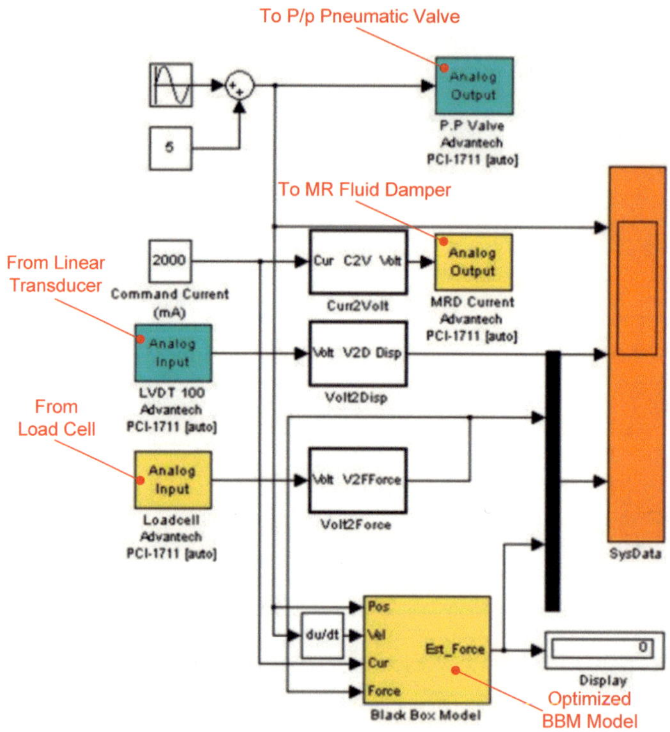

Figure 19. Simulink program for verification of the BBM sensor with self-sensing behavior.

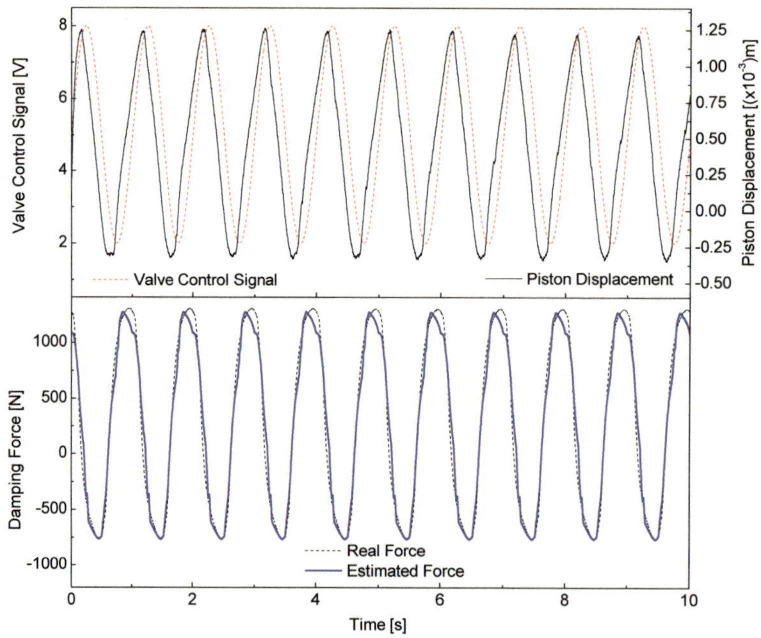

(a) Case 01 (Working load: 9kg; Vibrating control signal: (3V, 1Hz); Damper applied current: 0.0A)

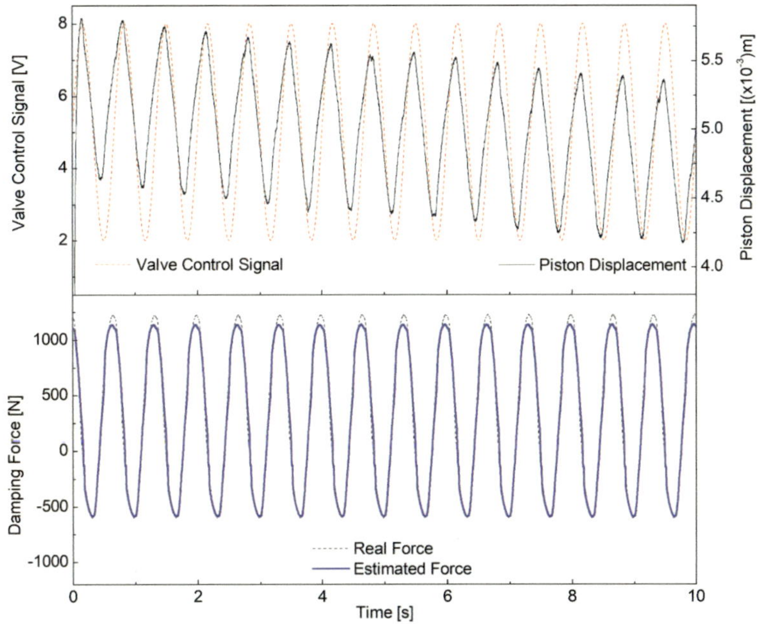

(b) Case 02 (Working load: 3kg; Vibrating control signal: (3V, 1.5Hz); Damper applied current: 1.0A)

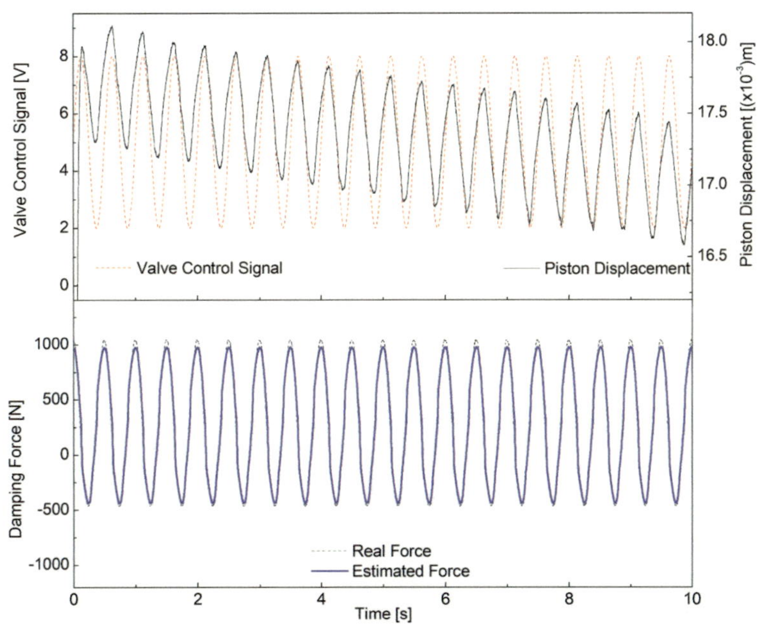

(c) Case 03 (Working load: 3kg; Vibrating control signal: (3V, 2.0Hz); Damper applied current: 2.0A)

Figure 20. Comparisons between the real and estimated damping forces of the TR02 system using the self-sensing method based on the optimized BBM.

4. Force sensorless damping control system design

In this section, the force sensorless control technique, combined from the BBM and IBBM models, is developed to be applied to systems using this damper for damping control with force self-sensing behavior. The BBM model, which uses a self-tuning fuzzy system optimized by neural network technique, was designed to overcome the disadvantages of conventional models. The BBM built in the form of the simple fuzzy mapping laws is considered to estimate directly the damping force output with respect to the characteristics and inputs of the MR damper. In order to improve the accuracy of the suggested model, the back-propagation learning rule based on the gradient descent method has been used to train the fuzzy parameters to minimize a defined modeling error function. Next, the IBBM model is derived from the optimized BBM and suggested to use as an effective force controller. In addition, the IBBM structure is online adjusted with respect to the control error to improve the system performance.

Finally, the proposed controller based on the BBM and IBBM is built for the force sensorless damping system as shown in Fig. 21. Here, the BBM model works as a virtual force sensor while the IBBM takes part in direct damping control without using other control techniques.

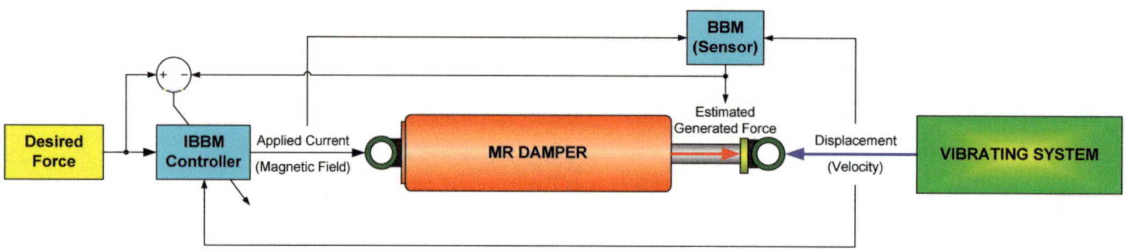

Figure 21. Structure of the proposed force sensorless damping control system.

4.1. Design of IBBM model for the proposed force-sensorless damping control method

Remark 3 (*inverse model selection*). Based on the advantages of the direct modeling method for the MR fluid damper using the black-box model, the inverse black-box model has been derived as the damping force controller. The BBM optimized by using neural network technique has been used to set initial parameters for the IBBM controller. The proposed inverse model estimates current levels should be applied to the MR fluid damper to create desired damping forces.

Remark 4 (*force-sensorless control solution and model optimization*). The proposed control method is designed for damping systems without using any force sensor. Hence, the optimized BBM is integrated in the control system as a virtual force sensor to estimate the generated damping force. This estimated force with the desired force are fed back to the IBBM controller to compute a control signal (current signal) to supply the MR fluid damper and, subsequently, to perform a closed-loop force-sensorless controller. In order to improve the control performance using the inverse model, the self-learning mechanism based on

neural network technique is also integrated into the IBBM of which the parameters are continuously adjusted during the damping control process with respect to the control error minimization.

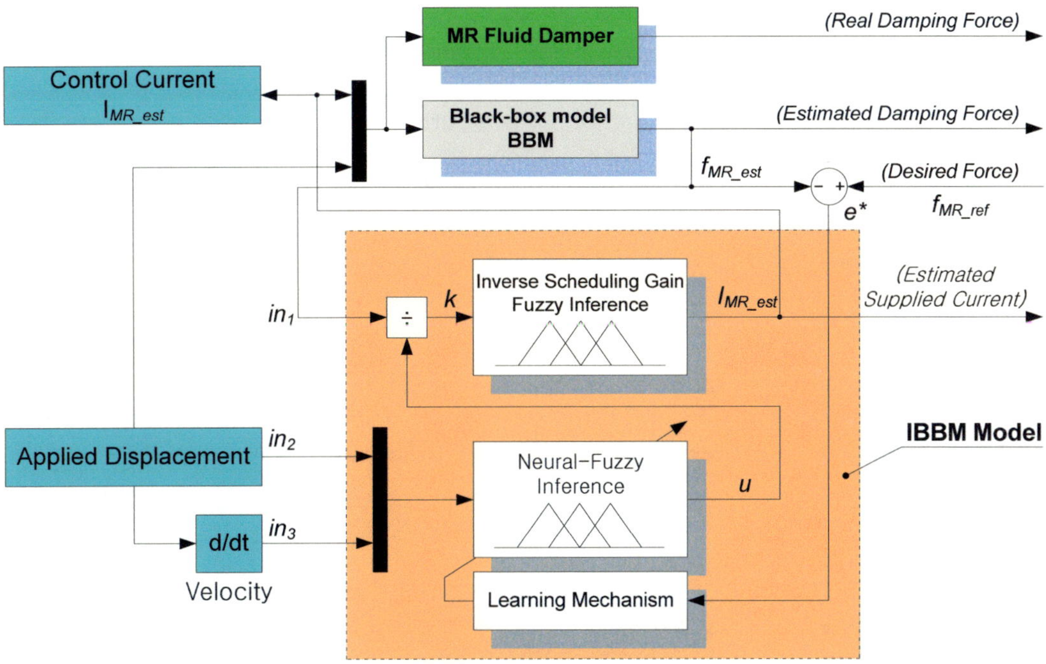

Figure 22. Structure of proposed force-sensorless damping control system based on BBM and IBBM.

Consequently, the force-sensorless control system is suggested as in Fig. 22. As seen in this figure, the proposed IBBM model contains two parts. The first part is a neural-fuzzy inference (NFI*), which was derived from NFI system of the optimized BBM, to estimate the damping force (u) caused by the damper rod displacement/velocity. The second part is an inverse scheduling gain fuzzy inference (ISGFI) which was developed from SGFI system of the optimized BBM. The ISGFI selects the current (I_{MR_est}) level needed to supply for the MR fluid damper for obtaining the damping force level (k). This damping force level can be computed from the damping force (f_{MR_est}) estimated by using the BBM and the estimated damping force (u) caused by the damper rod displacement/velocity (see Fig. 22) as:

$$k = \begin{cases} f_{MR\_est} / u & if\, u \neq 0 \\ 0 & if\, u = 0 \end{cases} \tag{24}$$

For improving the IBBM control accuracy, an error function (E^*) was derived from the difference between the damping force (f_{MR_est}) measured by the BBM sensor and the desired force (f_{MR_ref}). Therefore, the error function is defined as following equation:

$$E^* = 0.5(f_{MR\_est} - f_{MR\_ref})^2 \tag{25}$$

4.1.1. Neural-Fuzzy inference (NFI)*

The NFI in the optimized BBM was used to construct the inverse model. Hence, the structure of the NFI* in the IBBM is the same as the descriptions in section 3.2.1a except the initial MFs of the fuzzy inputs/output (Fig. 13). In the IBBM, these initial MFs of the NFI* are set as the optimized MFs of the NFI in the BBM which were shown in Fig. 15 of section 3.2.2.

Furthermore in the IBBM, the decisive factors of the NFI* input MFs, a_j, b_j, and the weights of NFI* output, w_j, are automatically online-trained by using neural network technique during the damping force control process. Therefore, these affecting factors are updated as follows

$$\left.\begin{aligned} a_{ji}\Big|_{t+1} &= a_{ji}\Big|_{t} - \eta_a^* \frac{\partial E^*}{\partial a_{ji}}\Big|_{t} \\ b_{ji}\Big|_{t+1} &= b_{ji}\Big|_{t} - \eta_b^* \frac{\partial E^*}{\partial b_{ji}}\Big|_{t} \\ w_{k}\Big|_{t+1} &= w_{k}\Big|_{t} - \eta_w^* \frac{\partial E^*}{\partial w_{k}}\Big|_{t} \end{aligned}\right\} \tag{26}$$

where and are the learning rate which determine the speed of learning;

E^* is the error function defined by (25).

The factor $\frac{\partial E^*}{\partial w_k}$ in equation (25) can be calculated as

$$\frac{\partial E^*}{\partial w_k} = \frac{\partial E^*}{\partial f_{MR\_est}} \frac{\partial f_{MR\_est}}{\partial I_{MR\_est}} \frac{\partial I_{MR\_est}}{\partial k} \frac{\partial k}{\partial u} \frac{\partial u}{\partial w_k} \tag{27}$$

where:

$$\frac{\partial E^*}{\partial f_{MR\_est}} = e^*(t) = f_{MR\_est}(t) - f_{MR\_ref}(t) \tag{28}$$

$$\frac{\partial f_{MR\_est}}{\partial I_{MR\_est}} \approx \frac{\delta f_{MR\_est}}{\delta I_{MR\_est}}\Big|_t \tag{29}$$

$$\frac{\partial I_{MR\_est}}{\partial k} \approx \frac{\delta I_{MR\_est}}{\delta k}\Big|_t \tag{30}$$

$$\frac{\partial k(t)}{\partial u(t)} = -\frac{f_{MR\_est}}{u^2} \tag{31}$$

$$\frac{\partial u}{\partial w_k}=\frac{\mu(w_k)}{\left(\sum_{l=1}^{M}\mu(w_l)\right)} \tag{32}$$

The next factors $\frac{\partial E^*}{\partial a_{ji}}$ in (26) can be computed by:

$$\frac{\partial E^*}{\partial a_{ji}}=\frac{\partial E^*}{\partial u}\frac{\partial u}{\partial \mu_j(x_i)}\frac{\partial \mu_j(x_i)}{\partial a_{ji}} \tag{33}$$

where:

$\frac{\partial E^*}{\partial u}$ is calculated by using equations from (28) to (31).

$$\frac{\partial u}{\partial \mu_j(x_i)}=\frac{\partial u}{\partial \mu(w_k)}=\frac{\sum_{l=1}^{M}\mu(w_l)(w_k-w_l)}{\left(\sum_{l=1}^{M}\mu(w_l)\right)^2} \tag{34}$$

$$\frac{\partial \mu_j(x_i)}{\partial a_{ji}}=sign(x_i-a_{ji})\frac{2}{b_{ji}} \tag{35}$$

The final factor $\frac{\partial E^*}{\partial b_{ji}}$ in (26) can be found by:

$$\frac{\partial E^*}{\partial b_{ji}}=\frac{\partial E^*}{\partial u}\frac{\partial u}{\partial \mu_j(x_i)}\frac{\partial \mu_j(x_i)}{\partial b_{ji}} \tag{36}$$

where:

$\frac{\partial E^*}{\partial u}$ and $\frac{\partial u}{\partial \mu_j(x_i)}$ are calculated by using equations from (28) to (31), and (34), respectively.

$$\frac{\partial \mu_j(x_i)}{\partial b_{ji}}=\frac{2\left|x_i-a_{ji}\right|}{b_{ji}^{\ 2}} \tag{37}$$

By using the above self-learning algorithm (from (26) to (37)), the NFI* can work more precisely in estimating the damping force (u) with respect to the applied displacement/velocity.

4.1.2. Inverse scheduling gain fuzzy inference (ISGFI)

The ISGFI system, designed with a single input (k) and a single output (I_{MR_est}), works as an intelligent switch to select the current (I_{MR_est}) level needed to supply for the MR fluid damper with respect to the damping force level (k). This fuzzy system was obtained from the SGFI mechanism of the BBM design (section 3.2.1) in which the SGFI input/output became the ISGFI output/input. Consequently, the input/output MFs and rule table of the ISGFI system were set as in Fig. 23 and Table 6, respectively.

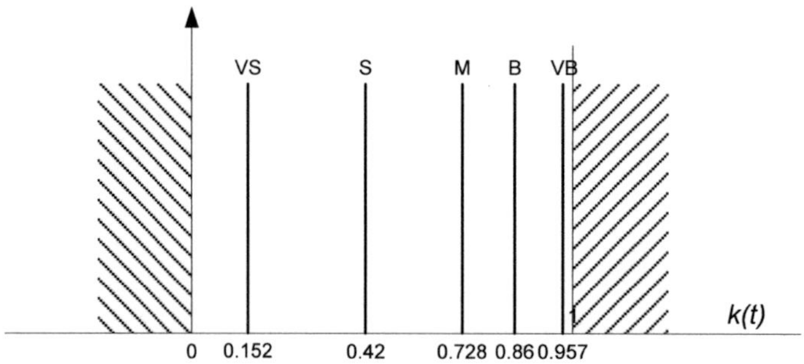

(a) ISGFI input: $k(t)$

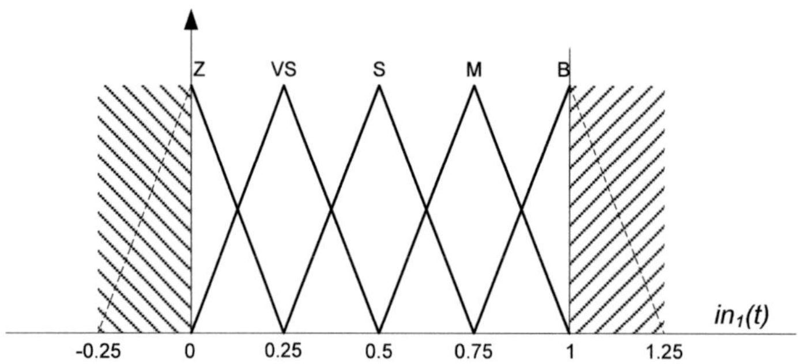

(b) ISGFI output: $I_{MR\_est}(t)$

Figure 23. MFs of the ISGFI inputs and output.

Damping Force Gain (k)	VS	S	M	B	VB
Estimated Supplied Current (I_{MR_est})	Z	VS	S	M	B

Table 6. Rules table for the inverse scheduling gain inference of the inverse black box model

4.2. Verifications

4.2.1. Checking the control ability of a damping system using the designed IBBM

As the descriptions for the proposed force-sensorless control method, the designed IBBM controller was firstly examined by simulations before its application to the real-time damping control. Therefore, a simulating scheme for the IBBM controller validation was built in Simulink/MATLAB as in Fig. 24. As seen in this figure, the validating damping

control system contains three main blocks. The two blocks labeled as 'BBM1' and 'BBM2' are similar and represent for the optimized BBM model which was obtained in section 3.2. These blocks then function as the actual MR fluid dampers. The remained block is the IBBM controller which was designed from section 4.1.

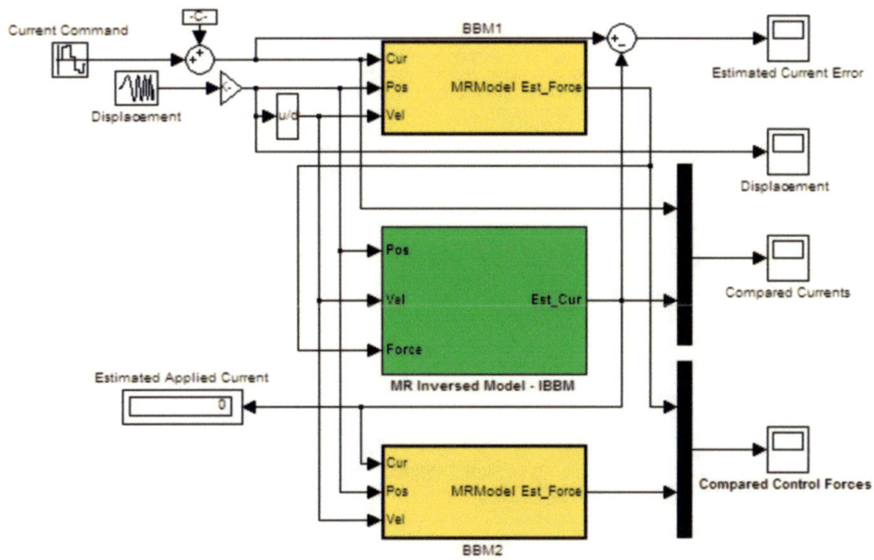

Figure 24. Simulation program for testing the IBBM controller.

The goal of the validation process is using the IBBM controller to control the second damper model, BBM2, to create the damping force to follow the reference force given from the first damper model, BBM1. Therefore, a displacement/velocity signal was generated and input into both the two damper models and the force controller. A current command signal was applied to the first damper model, BBM1. The output of this model, called the first simulated damping force, was used as a reference force signal for the damping system based on the second damper model, BBM2, and the IBBM controller. Hence, corresponding to a force command sent from the BBM1, the IBBM generated a simulated current command to control the damper model BBM2. This simulated current was then fed into the BBM2 together with the applied displacement/velocity to produce the second simulated damping force. As a result, the validation process carried out the comparison between the reference current command and simulated current command obtained from the IBBM, and the comparison between the first and second simulated damping forces.

The simulation results are displayed in Fig. 25. The results show that almost not only the simulated current command coincided with the reference current command, but also the simulated damping force of the BBM2 coincided with the simulated damping force obtained from the BBM1. As the result, it points out that the damping system using the self-learning IBBM controller with the optimized BBM virtual sensor can control accurately the MR fluid damper for a good tracking force performance. In the next section, experiments on the test rig TR02 have been carried out in order to verify the real-time control ability of the proposed force-sensorless damping control method.

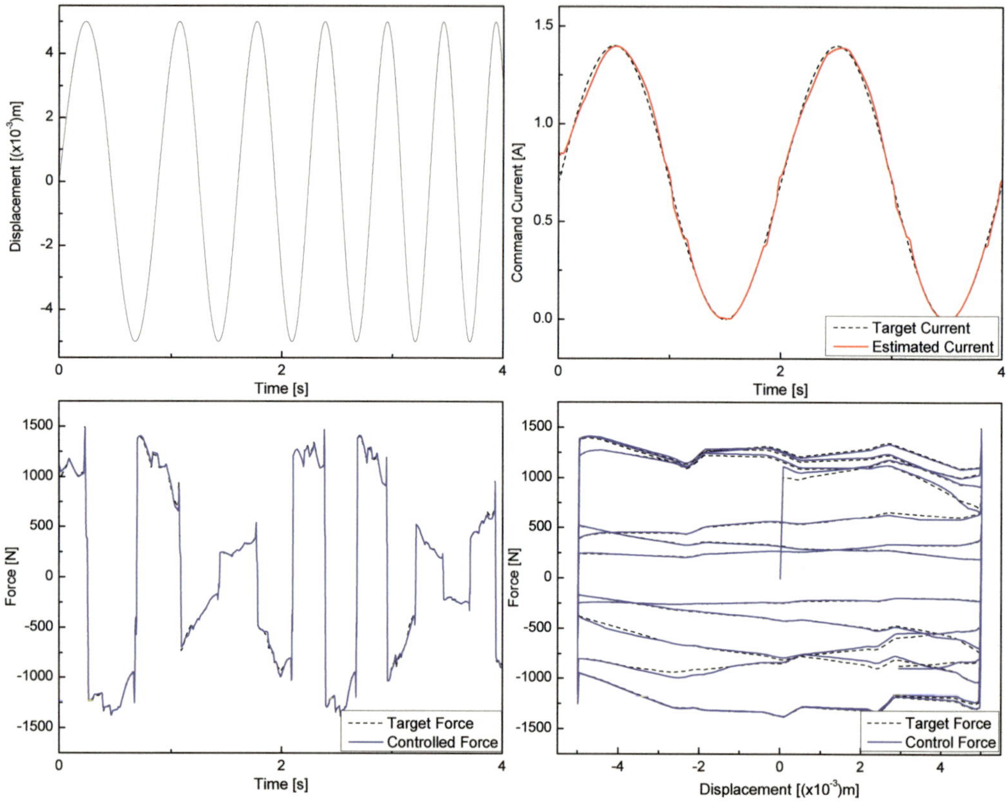

Figure 25. Comparison of simulation results between estimated current command with given current command corresponding to the damping force output.

4.2.2. Verification of the force-sensorless control system based on the BBM and IBBM

Based on the advanced characteristics of the designed BBM and IBBM models which were proven to be effective in the previous sections, the proposed force-sensorless control methodology was applied to the system TR02 for the real-time damping force control. Here, a harmonic excitation was applied to the damping system through the pneumatic cylinder controlled by the proportional valve. Meanwhile, a desired force performance was given in order to validate the damping control ability. Therefore, the control signal (voltage) for the pneumatic valve and the reference damping force were given as equations (38) and (39), respectively.

$$\text{ValveControlSignal} = 5\sin\left(\pi t + \pi / 2\right)\left(V\right) \tag{38}$$

$$\text{ReferenceForce} = A\sin\left(\pi t\right)\left(N\right) \tag{39}$$

here: the amplitude A of the reference force signal was set at two values: 500N and 1000N to perform two testing cases. The working load condition was set with a 9kg of load.

The control program for the damping system TR02 was built in Simulink with the real-time toolbox of MATLAB as shown in Fig. 26. The program contained two main blocks in which the first one is the motion generating block to generate the system displacement while the second one is the damping control system to ensure that the system could track the desired damping force by only using the IBBM controller and BBM sensor, without using the mechanical force sensor (load cell). In this control system, the optimized BBM estimated the damping force and fed back to the self-learning IBBM controller to create the current control signal for the MR fluid damper. In order to evaluate the control performance, the load cell was used separately with the controller to measure the actual damping force.

Experiments on the TR02 controlled by the system in Fig. 26 had been done under the testing conditions as shown in (38) and (39). Consequently, the validating results for the force-sensorless damping control system based on the BBM and IBBM corresponding to the two testing cases, 500N and 1000N of the desire force amplitude, are shown in figures 27a and 27b, respectively. The figures show that the system using the IBBM controller with the virtual sensor BBM tracked the desired damping force well. From the results, it strongly indicates that the damping force can be completely controllable by this proposed control methodology.

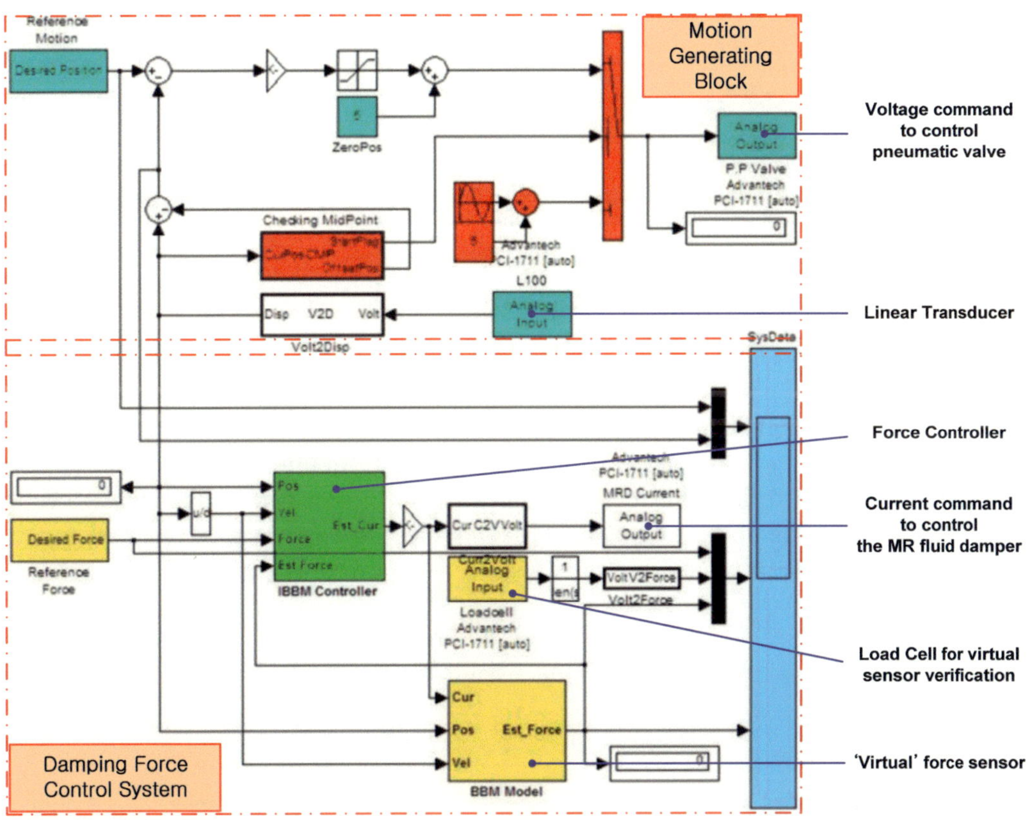

Figure 26. Simulink program for damping force control based on IBBM controller and BBM sensor.

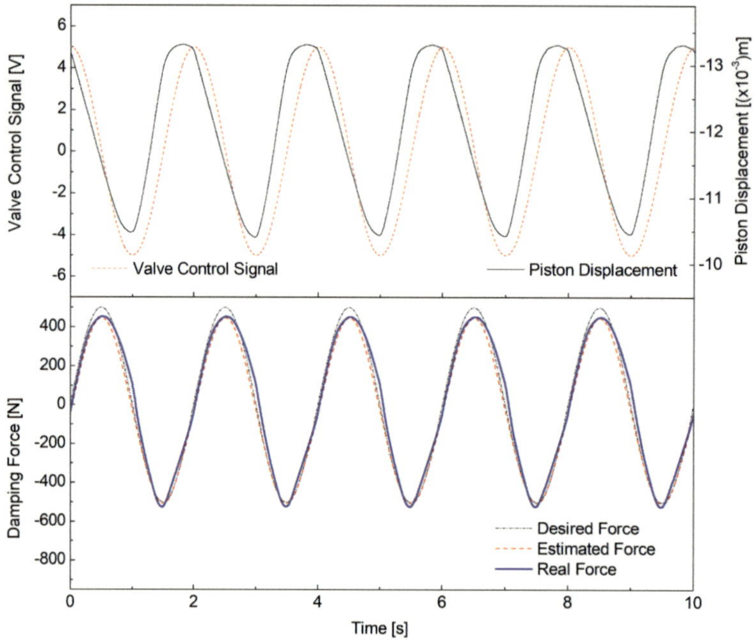

(a) Case 1 (Working load: 9kg; Vibrating control signal: (5V, 0.5Hz, 90°); Reference damping force: (500N, 0.5Hz, 0°))

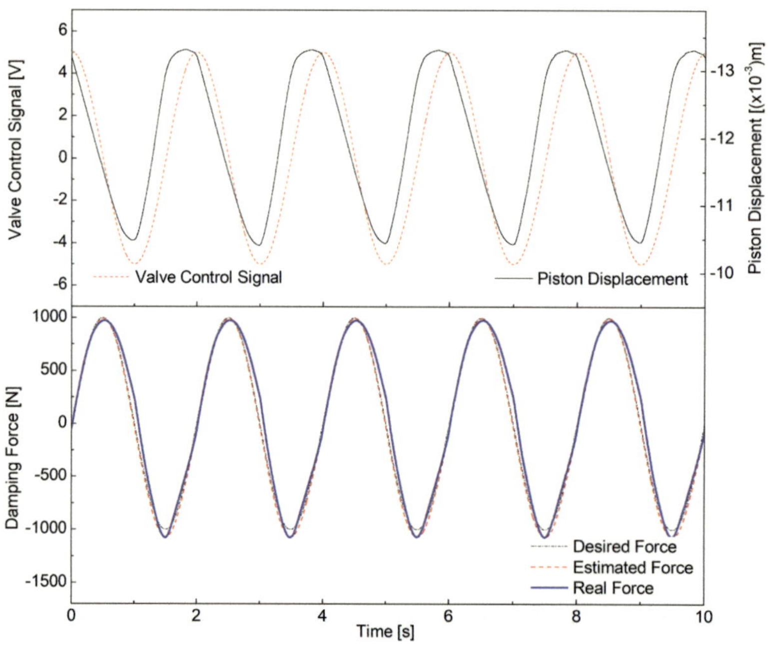

(b) Case 2 (Working load: 9kg; Vibrating control signal: (5V, 0.5Hz, 90°); Reference damping force: (1000N, 0.5Hz, 0°))

Figure 27. Damping force control performance using IBBM controller and BBM sensor.

5. Conclusions

In this chapter, physical phenomena of a MR fluid damper have been carefully investigated through both experimental data and modeling methodologies. Furthermore, a novel force-sensorless control method was proposed and successfully applied to the damping system using the MR fluid damper and without using any mechanical force sensor. The designed force control system is based on the optimized BBM model, functioning as the virtual force sensor, and the IBBM model, functioning as the self-learning force controller.

The two test rigs using the MR fluid damper have been fabricated in order to design the models as well as to evaluate the proposed controller. The BBM model is built as the simple direct modeling method and optimized by using neural network technique. The IBBM controller is the direct inverse model of the optimized BBM. In addition, the IBBM has the online self-learning ability by using neural network, consequently, improving the system performance. As the result, the optimized BBM and IBBM models have a strong ability to be applied to damping systems using the MR fluid damper as a virtual force sensor and a simple direct controller, respectively.

The modeling as well as control validation process has been done by both the simulations and real-time experiments. Based on the simulation and experimental results, it is clear that the optimized BBM can predict the force-displacement behavior of the MR damper with high precision while the IBBM can control the damping system to follow the desired damping force well. As a result, the proposed force-sensorless damping control methodology can become an effective and economical solution for practical MR fluid damper applications in the near future.

Author details

D. Q. Truong and K. K. Ahn*
School of Mechanical Engineering, University of Ulsan, Korea

6. References

[1] Karnopp D, Crosby M.J, Farwood R.A (1974) Vibration control using semi-active force generators. ASME j. eng. ind. 96(2): 619-626.
[2] Yi K, Song B.S (1999) A new adaptive sky-hook control of vehicle semi-active suspensions. Proc. of IMechE, Part D: j. auto. eng. 213(3): 293-303.
[3] Kawashima K, Unjoh S, Shimizu K (1992) Experiments on Dynamics Characteristics of Variable Damper. Proc. of the japan national symp. on structural response control, Tokyo, Japan: 121.
[4] Mizuno T, Kobori T, Hirai J, Matsunaga Y, Niwa N (1992) Development of Adjustable Hydraulic Dampers for Seismic Response Control of Large Structure. ASME PVP conf.: 163-170.

* Corresponding Author

[5] Gavin H.P, Hose Y.D, Hanson R.D (1994) Design and Control of Electrorheological Dampers. Proc. of the first world conf. on structural control, Pasadena, CA: 83-92.

[6] Park W.C, Choi S.B, Suh M.S (1999) Material characteristics of an ER fluid and its influence on dampingforces of an ER damper. Part I: material characteristics. Materials & design 20(6): 317-323.

[7] Park W.C, Choi S.B, Suh M.S (1999) Material characteristics of an ER fluid and its influence on dampingforces of an ER damper. Part II: damping forces. Materials & design 20(6): 325-330.

[8] Choi S.B, Choi Y.T (1948) Sliding Model Control of a Shear-Mode Type ER Engine Mount. KSME Int. j. 13(1): 26-33.

[9] Rabinow J (1948) Proceedings of the AIEE trans. 67: 1308-1315.

[10] Rabinow J (1951) US Patent 2,575,360.

[11] Carlson J.D, Chrzan M.J (1994) Magnetorheological Fluid Dampers. U.S. Patent 5277281.

[12] Carlson J.D, Weiss K.D (1994) A growing attraction to magnetic fluids. J. machine design 66(15): 61-64.

[13] S.J Dyke, Spencer B.F. Jr, Sain M.K, Carlson J.D (1996) Modelling and control of magneto-rheological fluid dampers for seismic response reduction. Smart material and structures 5: 565-575.

[14] Boelter R., Janocha H (1998) Performance of long-stroke and low-stroke MR fluid damper. Proc. of SPIE, smart structures and materials: passive damping and isolation, San Diego, CA: 303-313.

[15] Carlson J.D, Jolly M.R (2000) MR fluid, foam and elastomer devices. Mechatronics 10: 555-569.

[16] Hong S.R, Choi S.B, Choi Y.T, Wereley N.M (2005) Non-dimensional analysis and design of a magnetorheological damper. J. of sound vib. 288(4): 847-863.

[17] Choi K.M, Jung H.J, Cho S.W, Lee I.W (2007) Application of smart passive damping system using MR damper to highway bridge structure. KSME Int. j. 21(6): 870-874.

[18] Spelta C, Previdi F, Savaresi S.M, Fraternale G, Gaudiano N (2009) Control of magnetorheological dampers for vibration reduction in a washing machine. Mechatronics 19(3): 410-421.

[19] Savaresi S.M, Bittanti S, Montiglio M (2005) Identification of semi-physical and black-box non-linear models: the case of MR-dampers for vehicles control. Automatica 41(1): 113-127.

[20] Stanway R, Sproston J.L, Stevens N.G (1987) Non-linear Modelling of an Electro-rheological Vibration Damper. J. of electrostatics 20(2): 167-184.

[21] Spencer B.F, Dyke S.J, Sain M.K, Carlson J.D (1996) Phenomenological Model of a Magneto-Rheological Damper. ASCE J. of eng. mech. 123(3): 230-238.

[22] Choi S.B, Lee S.K (2001) A Hysteresis Model for the Field- dependent Damping Force of a Magneto-rheological Damper. J. of sound vib. 245(2): 375-383.

[23] Dominguez A, Sedaghati R, Stiharu I (2004) Modelling the hysteresis phenomenon of magnetorheological dampers. Smart mat. struc. 13(6): 1351–1361.

[24] Kwok N.M, Ha Q.P, Nguyen T.H, Li J, Samali B (2006) A Novel Hysteretic Model for Magneto-Rheological Fluid Damper and Parameter Identification Using Particle Swarm Optimization. Sens. actuators A 132(2): 441-451.

[25] Kwok N.M, Ha Q.P, Nguyen M.T, Li J, Samali B (2007) Bouc-Wen Model Parameter Identification for a MR Fluid Damper Using Computationally Efficient GA. ISA Trans. 46(2): 167-179.

[26] Chang C.C, Roschke P (1998) Neural network modeling of a magnetorheological damper. J. intell. mater. syst. struct. 9(9): 755-764.

[27] Schurter K.C, Roschke P.N (2000) Fuzzy Modeling of A Magneto-Rheological Damper Using Anfis. Proc. of IEEE int. conf. on fuzzy sys.: 122-127.

[28] Wang D.H, Liao W.H (2001) Neural Network Modeling and Controllers for Magneto-rheological Fluid Dampers. Proc. of 10th IEEE int. conf. on fuzzy sys.: 1323-1326.

[29] Wang D.H, Liao W.H (2005) Modeling and Control of Magneto-Rheological Fluid Damper Using Neural Networks. Smart mat. struc. 14(1): 111-126.

[30] Hanson B, Levesley M (2004) Self-sensing applications for electromagnetic actuators. Sens. actuators A. 116(2): 345-351.

[31] Mizuno T, Ishii T, Araki K (1998) Self-sensing magnetic suspension using hysteresis amplifiers. Con. eng. prac. 6(9): 1133-1140.

[32] Lama K.H, Chena Z.H, Nia Y.Q, Chanb H.L.W (2010) A magnetorheological damper capable of force and displacement sensing. Sens. actuators A. 158(1): 51-59.

[33] Ahn K.K, Truong D.Q, Thanh T.Q, Lee B.R (2008) Online Self Tuning Fuzzy PID Control for Hydraulic Load Simulator. Proc. of the IMechE - part I: j. sys. and con. eng. 222(2): 81-95.

[34] Truong D.Q, Ahn K.K (2011) Parallel Control for Electro-Hydraulic Load Simulator Using Online Self Tuning Fuzzy PID Technique. Asian j. con. 13(4): 522-541.

[35] Ahn K.K, Truong D.Q, Islam M.A (2009) Modeling of a Magneto-Rheological (MR) Fluid Damper using a Self Tuning Fuzzy Mechanism. J. of mech. sci. and tech. 23(5): 1485-1499.

[36] Truong D.Q, Ahn K.K (2010) Identification and application of black-box model for a self-sensing damping system using a magneto-rheological fluid damper. Sens. actuators A. 161(1-2): 305-321.

[37] Shames I.H, Cozzarelli F.A (1992) Elastic and Inelastic Stress Analysis. Prentice Hall, Englewood Cliffs, New Jersey 120-122 p.

[38] Gamota D.R, Filisko F.E (1991) Dynamic Mechanical Studies of Electrorheological Materials: Moderate Frequencies. J. of rheology 35(3): 399-425.

Chapter 16

New Magnetic Translation/Rotation Drive by Use of Magnetic Particles with Specific Gravity Smaller than a Liquid

Makoto Nokata

Additional information is available at the end of the chapter

http://dx.doi.org/10.5772/50999

1. Introduction

Magnetic fluid has unique properties[1], by which it can, not only be attracted to, but also solidified by magnets. Many researchers have proposed useful applications for these properties, such as in dampers, sensors, actuators [2][3] and medical applications[4]-[6]. J.Furusho et al. [7] developed MR-Fluid Actuator, controlled it by use of torque Feedback. He also developed 6-DOF Rehabilitation System driven by ER-Actuator[8]. The functional fluid has performed well in clinical use. However, as all Magnetic fluids have a greater specific gravity than liquids like water, the applications are restricted.

This chapter presents a new type of magnetic fluid and the rotational drive with simple magnetic field. This chapter consists of two sections.

The first section describes the development of a unique magnetic fluid and the proposal of a driving principle by use of the magnetic fluid. We have developed a lighter fluid than liquid, mixed it with liquid, and created a new impetus in this research. We have proposed a driving principle and developed the verification models and the generator of magnetic field. It became clear that our proposed driving principle results from collision of the magnetic particles to the outer cover, deformation of the whole form of magnetic particles and movement of the center of gravity.

The second section describes an advanced translation drive that produces non-slipping motion in soft and closed environment by use of the magnetic fluid. New movement principle of the robot, which has a soft and deformable body that can move through a confined space is proposed. The mechanism of a toy water snake is applied to this principle. Magnetic particles inside the water balloon are affected by the external magnetic field and push the inner side of a balloon to the direction of a magnetic field. We construct an experimental model to verify the proposed principle, the sliding movement is measured using the model.

2. A unique magnetic fluid and driving principle

This section describes the development of a unique magnetic fluid and the proposal of a driving principle by use of the magnetic fluid.

2.1. Principle of magnetic drive

The explanation shows the magnetic drive principle by use of floating magnetic fluid and a simple magnetic field.

Fig.1 shows the schematic diagram of the rolling object that is driven by the proposed magnetic drive principle. The functional section(1) and the liquid(2) are sealed in the interior by the outer cover(3).

The inner wall(4) and outer wall(5) of the outer cover are covered with the small projections or a textured material. The projections play the role of creating friction. Floating magnetic particles(6) are employed in a magnetic field. The specific gravity of these magnetic particles is smaller than the liquid, so they gather in the upper part of the rolling object when no magnetic field is applied.

When a magnetic field is applied, the magnetic particles move in the direction of the magnetic field quickly. The path they take is just inside the outer cover, because both the magnetism and buoyancy are working on the magnetic particles. The particles collide with the projections on the inner wall of the outer cover (Fig. 1(b)).

The external side of the outer cover does not slip on the ground, because there is friction between the ground and the outer wall. Consequently, the rolling object moves in the direction of the magnetic field (Fig. 1(c)). After the magnetic field is removed, the magnetic particles will return to the upper part of the rolling objet slowly (Fig. 1(d)), the rolling object stayed at the same position. By repeating the above, the rolling object can move a long distance.

If the specific gravity of a magnetic particle is larger than the liquid, the object rotates in the opposite direction (Fig.2). The movement becomes unstable because it may move in the opposite direction to the magnetic field.

2.2. Verification experiment and discussion

2.2.1. Float magnetic particle

In order to realize the above-mentioned principle of operation, magnetic particles with a specific gravity smaller than the liquid are required. Fig.3 shows a photograph and schematic diagram of the particles. The diameter of each particle is 1-2 mm. The particles are composed of magnetic iron oxide 4.5%, Kerosene 95.2% and Additives 0.3%.

2.2.2. Verification models

Fig.4 shows the photograph of two types of the model used for the verification of the above-mentioned principle. Fig.4(a) is a hard model for multi-directional drive on a solid surface, and Fig.4(b) is a soft model for crawling on uneven ground.

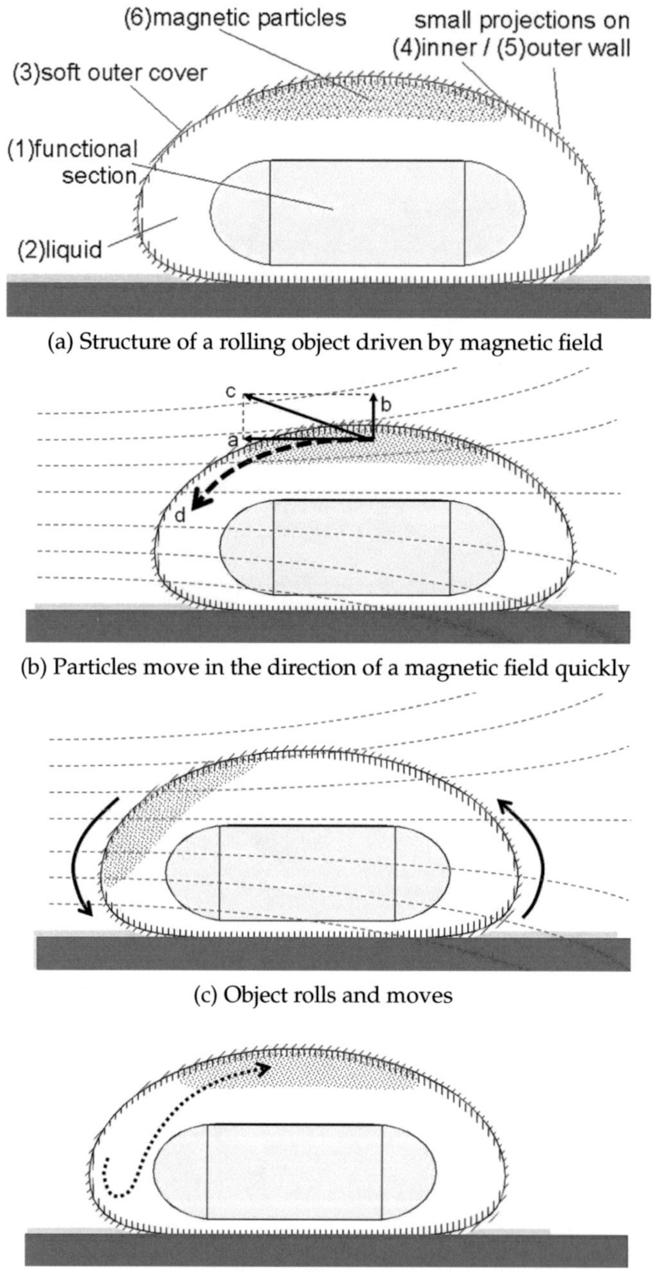

(a) Structure of a rolling object driven by magnetic field

(b) Particles move in the direction of a magnetic field quickly

(c) Object rolls and moves

(d) Magnetic particles return to the upper part slowly

Figure 1. Magnetic drive principle with simple magnetic field

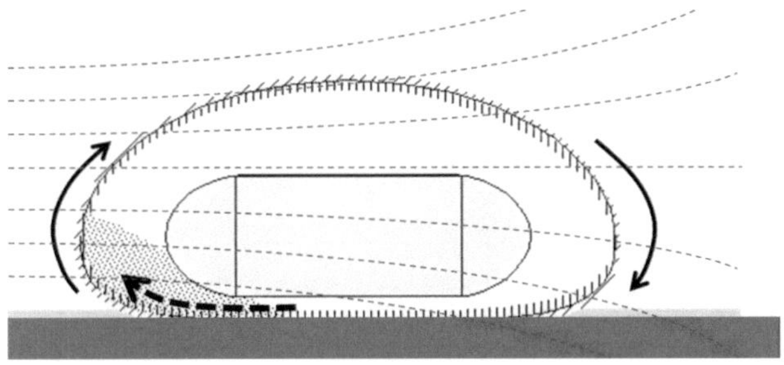

Figure 2. Reverse drive. If the specific gravity of a magnetic particle is larger than a liquid, the rolling object rotates conversely.

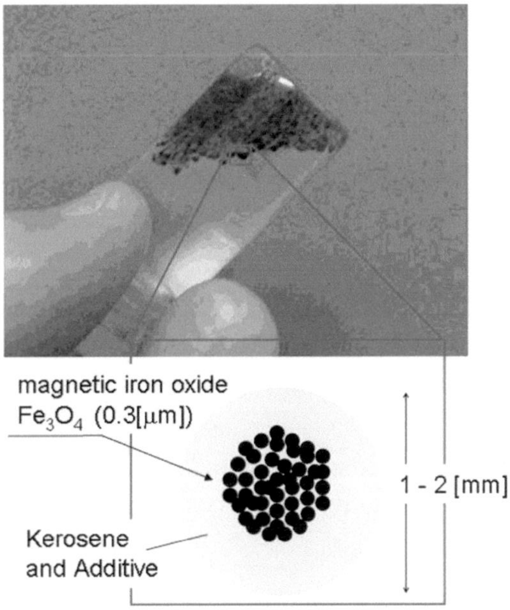

Figure 3. Photograph and schematic diagram of a magnetic particle with specific gravity smaller than a liquid

(a)Hard model for omni directional drive

(b)Soft model for crawling on an uneven surface

Figure 4. Two types of the model for the verification of the above-mentioned principle

2.2.3. *Generator of magnetic field*

Fig.5 shows the system that generates the magnetic field necessary to move the model, it consists of a solenoid (outer diameter: 100 mm, inner diameter: 50 mm, length: 100 mm, cross-sectional dimension of coil line: 2 × 5mm, 163 turns); a power supply (applied voltage: 50 - 900 V, capacitance of capacitor: 10 mF, maximum permissible current: 10000 A); and a sensor probe that measures the flux density of three axes. The charging time of the power supply is 0 - 5 s, the solenoid generates a magnetic field with a peak because the stored charge is discharged all at once. Fig.6 shows a graph of the magnetic field generated by the system; about 2 T can be generated. Fig.7 shows the wave shape of the magnetic flux density at a point on the center axis that is 7 cm from the edge of the solenoid.

2.2.4. *Results and discussions*

In order to verify the drive principle, an experiment to drive a solid model as shown in Fig.4 (a) was conducted. At first, The magnetic flux density:150 mT was generated horizontally. The model motions captured by camera are shown in Fig.8. The monitor in the captured image shows the value of the flux density. The magnetic particles were shifted to the upper part of the solid model by magnetic force, contacting the inside of the outer cover. All the particles were moved a distance of 2 cm by one 150 mT magnetic field impulse. As a result, this experiment demonstrated the practicality of the magnetic drive principle using a simple magnetic field. Next, a magnetic flux density of 150 mT was generated obliquely downward. The magnetic particles were shifted obliquely downward and didn't contact the inside the outer cover, so the model didn't move.

Fig.9 and Fig.11 show the model motions captured by camera, when the magnetic flux density was generated horizontally. Magnetic particles clumped together and adhered to the inside wall. The form of the clump did not disintegrate when shaken and the center of gravity of the clump was maintained (Fig.9(1)). When a magnetic field was applied, the model began to rotate and roll in the direction of rotation, with the clump maintaining its form (Fig.9(2)(3)). When the inclination of the clump in the model became large, each particle began to move

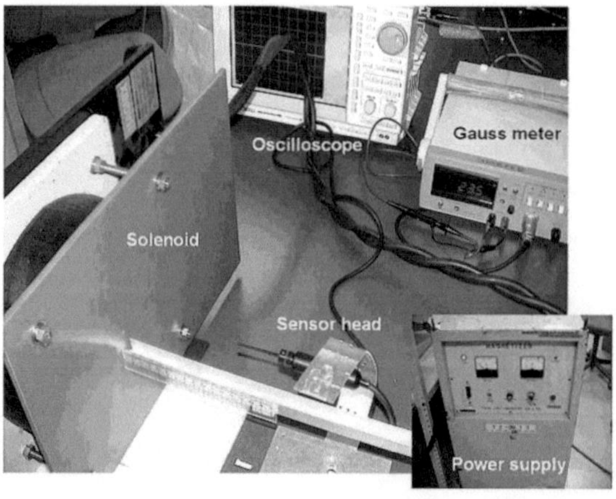

Figure 5. Total system of generating the impulse to the magnetic field for driving the models, it consists of a solenoid, a power supply, and a sensor probe that measures the flux density of three axes.

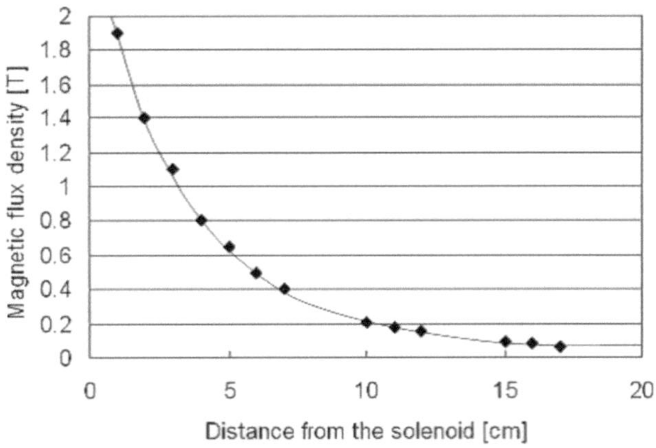

Figure 6. Graph of the magnetic field produced by the system, about 2 Tesla (T) can be generated.

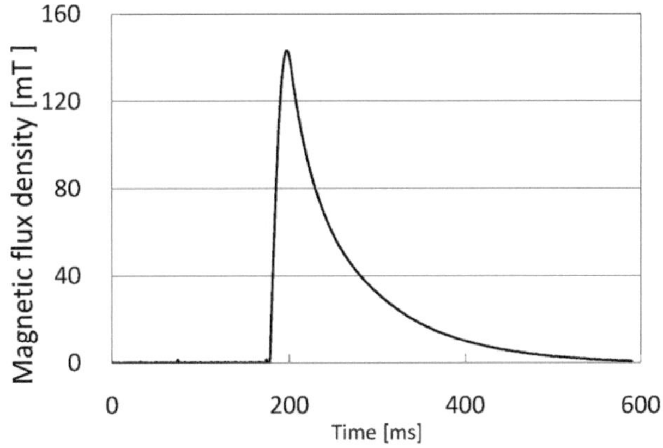

Figure 7. Wave shape of the magnetic flux density at the point of 12 cm on a center axis from the edge of the solenoid.

Figure 8. Captured motions of the rolling model. The magnetic flux density:150[mT] was generated horizontally. The monitor in back shows the value of flux density.

Figure 9. Captured motion which there is slide in the particle

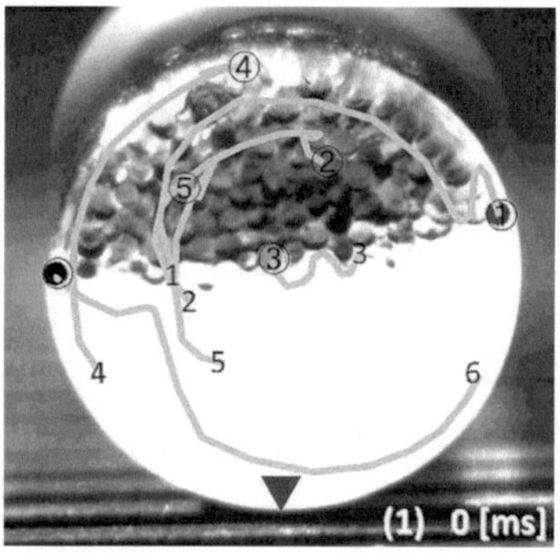

Figure 10. Move Path of each particle

separately due to its buoyancy, and the shape of the clump changed (Fig.9(4)(5)). The position of the center of gravity became stable, so the rotation stopped (Fig.9(6)). If the form did not change, the model turned back to its original position (Fig.11, left-side figures).

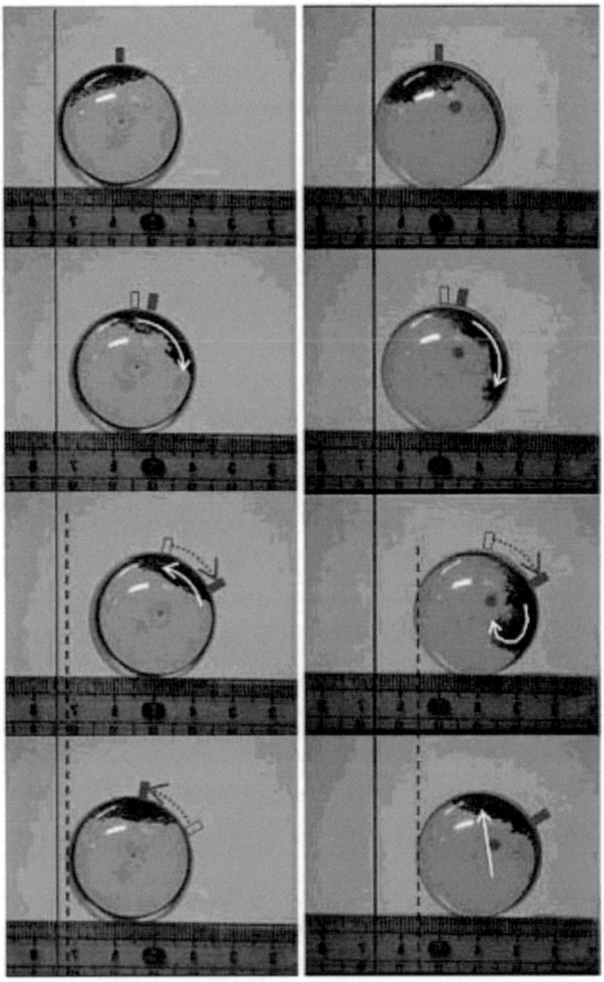

Figure 11. Captured motion which there is slide in the particle and the inner wall.

Fig.10 shows the path of each particle. The particles were circulated by convection, that is, a sequential convection current can cause sequential rolling. This result says that there is another magnetic wave pattern in addition to the impulse one shown in Fig.7. As a result, it became clear that our proposed driving principle results from not only collision of the magnetic particles with the outer cover but also the flow resistance between the particles and the cover. Equation 1 shows the force F produced on the magnetic particles by the external magnetic field.

$$F = M\frac{dH}{dr} = \frac{[wbm][N/wb]}{[m]} = [N] \tag{1}$$

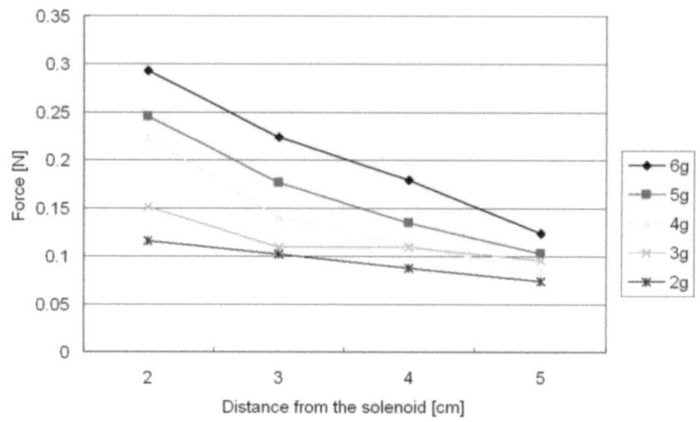

Figure 12. The graph of the measured magnetic field produced by the equipment of Fig.5. The values are measured by the gauss meter for every mass of a magnetic particle.

M is the magnetic moment, *H* is the magnetic field strength and *r* is position of the magnetic particle. Fig.12 shows the graph of the measured magnetic field produced by the equipment in Fig.5. The mass, of each magnetic particle is measured using the gauss meter.

2.3. Applications

This chapter introduces some mechanisms that make use of the magnetic drive principle.

(1)Ring type (Fig.13(a)) This mechanism consists of an outer cover in ring form and an axis of rotation. The torque can be taken from the main axis.

(2) Propeller type (Fig.13(b)) After a magnetic field is applied and the magnetic particles flow downward, the particles then float back up due to buoyancy. When a particle contacts a propeller, torque is generated.

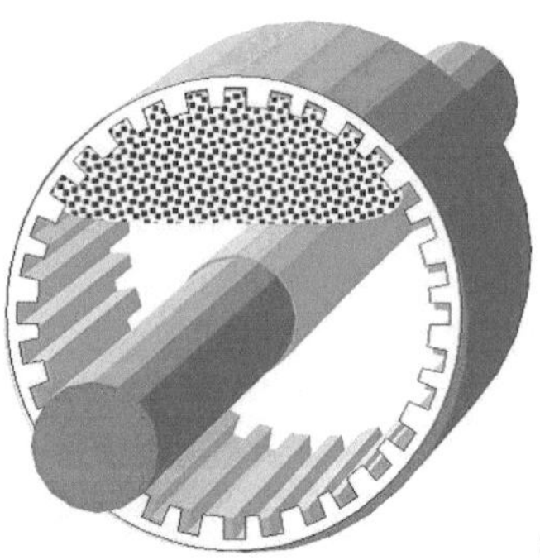

(a) Ring type

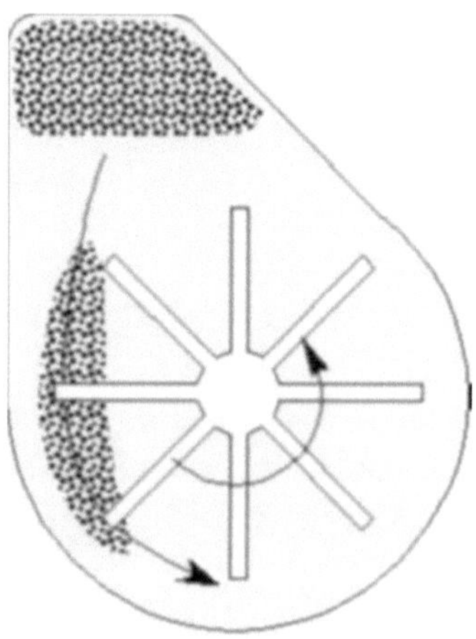

(b) Propeller type

Figure 13. Mechanisms which applied the magnetic drive principle

Fig.14 shows the prototype of Ring type (Fig.13(a)), Fig.15 explains the principle of driving. In order to verify the drive principle, an experiment to drive a simple model was developed. Fig.16 shows the model motions captured by camera, when the magnetic flux density was

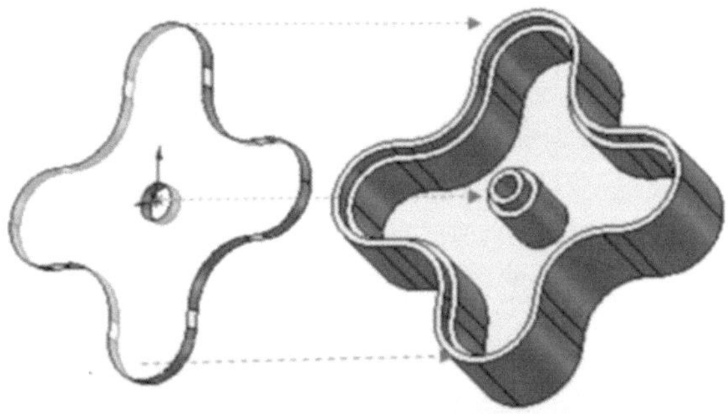

(a)Design of cross wheel

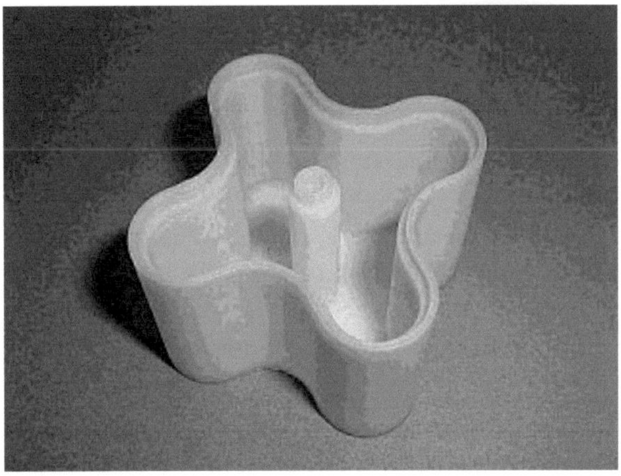

(b)Prototype fabricated by 3D printer

Figure 14. Prototype of ring type (cross wheel)

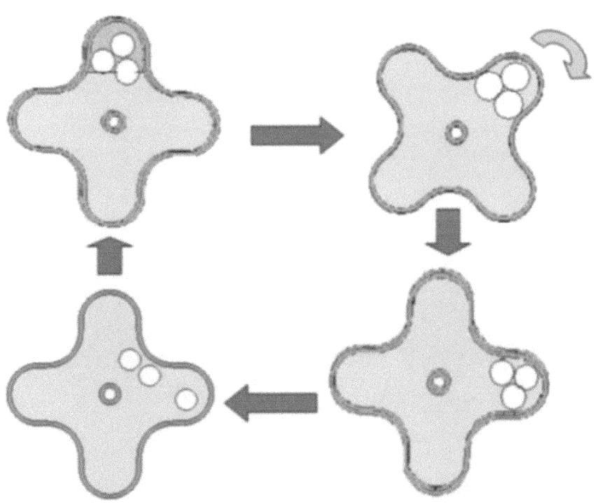

Figure 15. Driving Principle of cross wheel

generated horizontally. Magnetic particles pushed the inside wall of the cross wheel, rotated the wheel, then the model drove itself forward.

This driving principle could be applied to medical robots shown in Fig.17. One impulse of magnetic force produces 90 degree rotation of cross wheel. This mechanism is able to control the travel distance, little influenced by friction and viscosity of organs[9]. In addition, No actuator and no battery are required for devices inserted in the body. The external magnetic field is suitable for supplying the rotating or driving force to the device. This application has the advantage over a capsule robot that has a permanent magnet mounted in its interior.

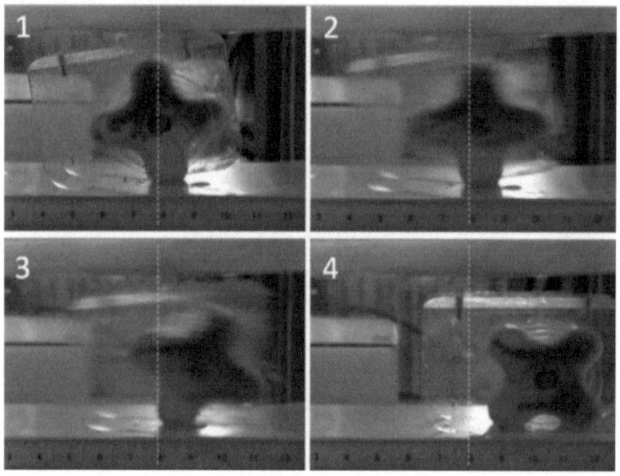

Figure 16. Motions of cross wheel captured by camera

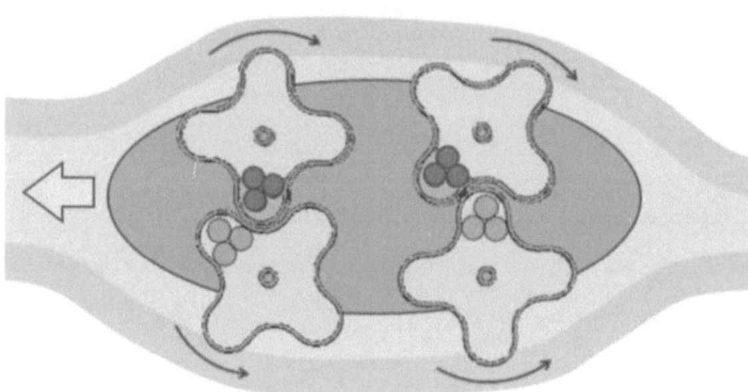

Figure 17. Schematic diagram of internal medical robot in bowel

3. Deformable robot actuation by magnetic particle

The endoscope, which is a flexible tube, is used to examine the interior of the human gastrointestinal tract. The standard endoscope is inserted via the mouth or the anus and manually pushed and pulled into the organ to be examined. In contrast, a capsule endoscope is swallowed by a patient and naturally exits the body within eight hours. [10] Another capsule endoscope mounts a permanent magnet inside the body, rotates and moves actively by applying a rotational magnetic field. [11][12] M. Shikanai et al. developed a robotic endoscope that consists of front and rear bodies with bidirectional rotational helical fins. [13] The front and rear bodies are connected by a DC motor. The robot advances through the

intestines by clockwise rotation of the front body and anticlockwise rotation of the rear body. These capsule-type robots require power supply wires and a permanent magnet. A potential problem with this design is that the rotational drive could cause engulfment. P. Glass et al. developed new controllable capsule endoscopes, the prototypes have used shape memory alloy or micro motors to press elastomer micropillar legs into the gastrointestinal lining to anchor the capsule in place.[14][15] Bio-inspired adhesives by the elastomer micropillar legs to stick the capsule to the intestinal lining allow the clinician to control the progress of the capsule along the patient's gastrointestinal tract. Another type of capsule robot mounts a ferromagnetic material inside the body and is moved by applying a gradient magnetic field. [16] The limitation of this design is that slipping locomotion is affected by the surface friction of organs. A magnetic particle has been reportedly developed that is lighter than liquid, and on mixing with liquid creates a new impetus. [17]

This section presents an advanced locomotion method that produces non-slipping motion in digestive organs and the abdominal cavity by developing an endoscopic robot. The developed endoscopic robot is soft, deformable, and can move through a confined environment of internal organs.

3.1. Proposal and features of movement principle

This section describes the proposed movement principle of the robot, which has a soft and deformable body that can move through a confined space. The mechanism of a toy water

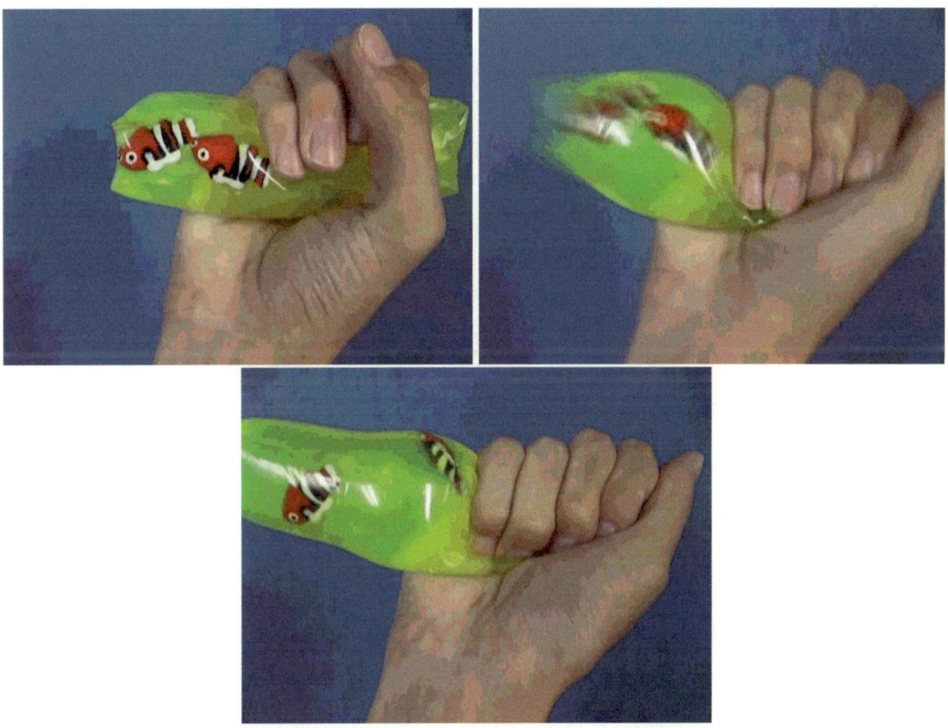

Figure 18. A toy WATER SNAKE, a cylindrical water balloon which will jump out if it grasps tightly

snake is applied to this principle (Fig.18). The water snake is similar to a cylindrical water balloon. The internal pressure of water inside the balloon is increased by externally grasping and pressing the balloon. Not grasped part is swelled by water pressure and pushed by movement. In the water snake, the external skin rotates and moves inward simultaneously with the outward pushing of the internal skin. Due to this buildup in pressure, no slippage in rotation occurs and the water snake slips from the examiner's grasping hand.

Fig.19 shows the proposed movement principle of the cylindrical soft robot. Magnetic particles inside the water balloon are affected by the external magnetic field and push the inner side of a balloon to the direction of a magnetic field in a manner similar to the examiner's grasping hand, as explained for the case of the toy water snake (Fig.19(2)). This causes the water balloon to rotate (Fig.19(3)(4)).

The key features of the proposed movement principle are

1. The movement does not cause slippage between the contiguous environment and the outside of the cylindrical balloon.
2. The actuator and the battery are not required.
3. The flexible and deformable structure allows the proposed robot to pass through narrow tubes and spaces.
4. The robot can be miniaturized because the structure is simple.

This section also explains the condition that realizes the movement principle. Each parameter is shown in Fig.20. The parameters of the condition are

- F: The force that the external magnetic field imparts to the magnetic particles.
- a, b: The maximum frictional force generated by contact with the contiguous environment.
- c: The force necessary to rotate the flexible skin of the cylindrical balloon and that required to deform it.

The condition required for movement without slippage is given by Eq. 2. This equation determines the value of the external magnetic field to be input.

$$c < F < a + b \tag{2}$$

3.2. Prototype construction for principle verification

Experimental prototype was constructed an experimental model to verify the proposed principle, as shown in Fig.21. A commercially available toy water snake was used as a cylindrical balloon, which is made of a thermoplastic elastomer. Two types of magnetic particles (TEC-BALL, JFE Techno-Research Corporation)[18], one with specific gravity less than that of a liquid, and the other with specific gravity higher than that of a liquid were placed inside the cylindrical balloon, and the balloon was filled with water. These magnetic particles push the top and bottom parts of the cylindrical balloon simultaneously and generate a well-balanced rotation. The dimensions of the cylindrical balloon became 115 mm in length, 35 mm in height, and 42 mm in width. There is no balloon of dimensions other than this, the size of the abdominal wall model was united with that of the balloon. The black band, which is a material joint, in the center of the cylindrical balloon is used to verify the sliding movement.

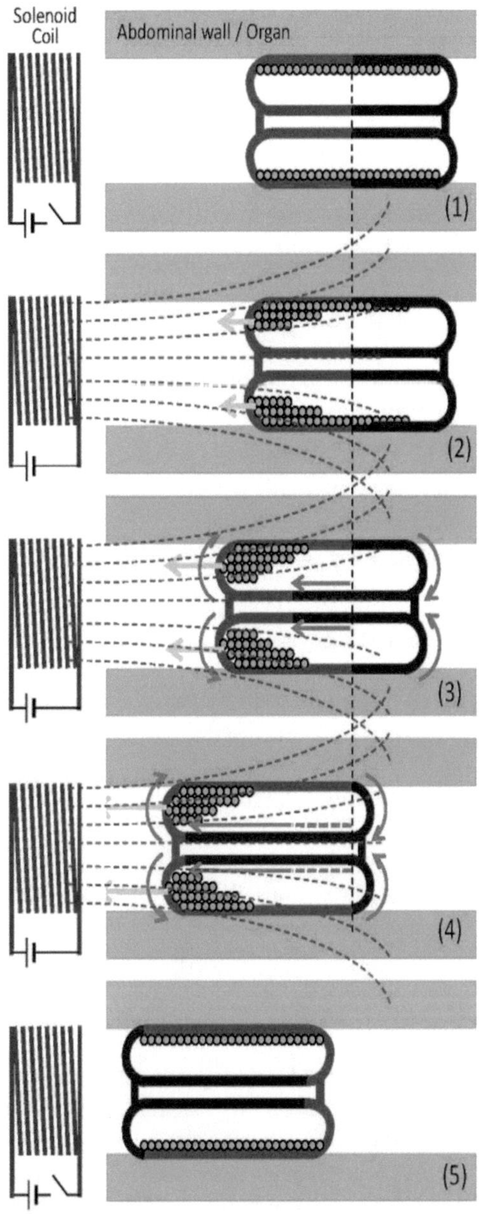

Figure 19. Schematic diagrams of the proposed movement principle of the cylindrical soft robot. Initial state (1), Magnetic particles inside the water balloon are affected by the external magnetic field (2), push the inner side of a balloon, rotate the water balloon to the direction of a magnetic field (3)(4), Magnetic field disappears and rotational movement stops (5).

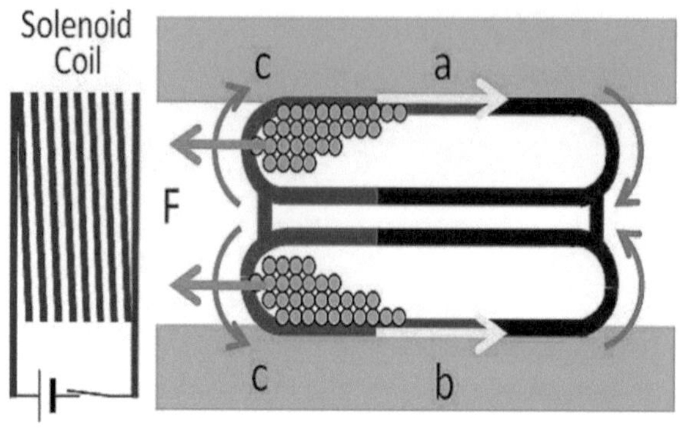

Figure 20. Parameters, F, a, b and c for analyzing the condition that the proposal principle enables the movement.

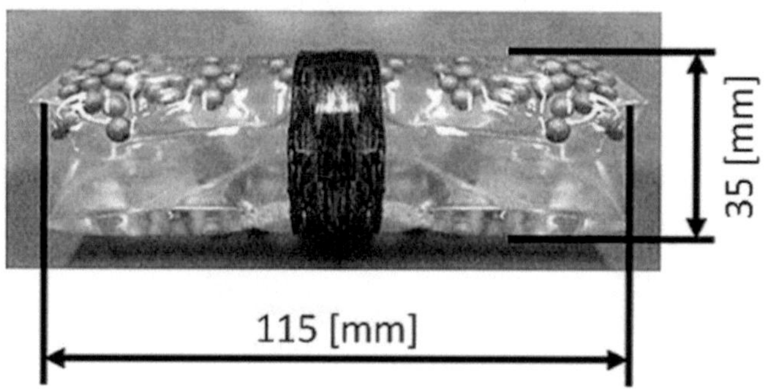

Figure 21. Experimental prototype for the principle verification, it consists of a commercially available toy water snake, two types of magnetic particles and water.

3.3. Experiment on principle verification

The sliding movement is measured using an experimental model (Fig.21). The model is pulled with a wire, and it passed between two sheets that simulated internal organs (SEIWA DENTAL Ltd., Japan; Polyvinyl alcohol, 20 × 10 × 1cm). The sheet imitates the softness of an abdominal wall but not the colonic peristalsis and the organs' shape. The gap between the two sheets is gradually increased. The force required to slide the model without rotation is measured as frictional force a, b, and the force required to rotate and deform the model without slippage is measured as the force necessary for movement c in Eq.(2). The measuring system is shown in Fig.22, it consists of a load cell (KYOWA ELECTRONIC INSTRUMENTS CO., LTD.; LUX-A-1KN), a stage (SIGUMA KOKI CO.,LTD; SGSP 20-30), and a PC. The measured values are shown in Table 1. The values of the magnetic force necessary for movement were determined at intervals from 28 to 33 mm.

Verification experiment of the proposed movement principle were conducted by use of the experimental prototype(Fig.21), simulated internal organs' sheets and magnetic field generator(Fig.5).

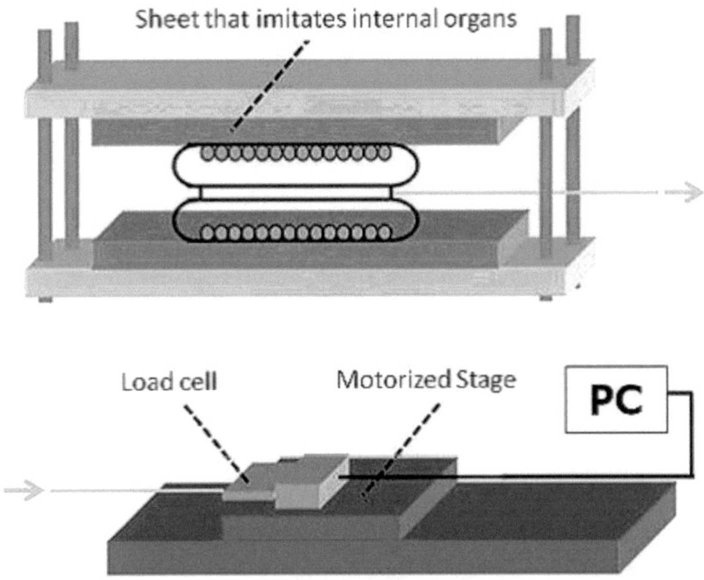

Figure 22. Measuring setup of movement condition. The model is pulled with a wire, and it passed between two sheets that simulated internal organs. The gap between the two sheets is gradually increased.

interval[mm]	friction(a+b)[N]	deformation(c)[N]
28	2.31	0.378
29	2.15	0.346
30	1.99	0.328
31	1.81	0.304
32	1.73	0.284
33	1.69	0.263

Table 1. Measured value of movement condition

Fig.23 shows a graph of the measured magnetic force exerted by magnetic particles inside the water ballon. The magnetic force of 0 - 1.5 N can be applied to the model by the magnetic field generator.

Eq.(3) shows theoretical formulas of the magnetic force F, measured values are consistent with theoretical ones.

$$F = m\frac{dH}{dx} \tag{3}$$

$$m = \chi_{eff}HV \tag{4}$$

m: magnetic moment, H:magnetic field,

x: distance from coil, V:volume of material,

χ_{eff}: magnetic susceptibility

In the experiment, the model was positioned at this point and placed between the two sheets that simulate internal organs. The gap between the two sheets was 30 mm. The magnetic force of 0.52 N was momentarily applied to the model by the magnetic field generator.

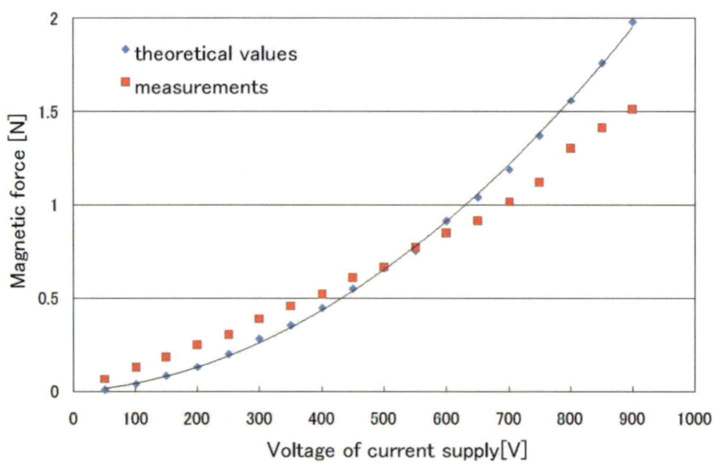

Figure 23. Graph of the measured magnetic force exerted by magnetic particles inside the water balloon.

3.4. Results and discussions

The movement of the model is shown in Fig.24. The movement was filmed by a high-speed camera, and the image was captured at intervals of 50 ms. The value written on each figure is the movement distance. The head point on which the triangle is marked moved, while the position of the black band did not move between the two sheets, confirming movement without slippage. The seam part of the cylindrical balloon skin stopped moving, and the total moving distance was 6.2 cm. The moving distance increases by reducing the width of the seam; the method of manufacturing the cylindrical balloon will be improved in the future. Periodic expansion and contraction were observed in the entire film. To improve the efficiency of movement, appropriate material of the cylindrical balloon and sufficient amount of water inside the balloon should be used.

The method of applying the magnetic field to the tube and the patient's body is described. A method using a midair solenoid, as shown in Fig.25, is examined here. The tube or the specimen material is placed in the solenoid, and the part to which the magnetic field is applied is set on the edge of the coil. The magnetic gradient on the edge of the coil generates the magnetic field necessary for movement.

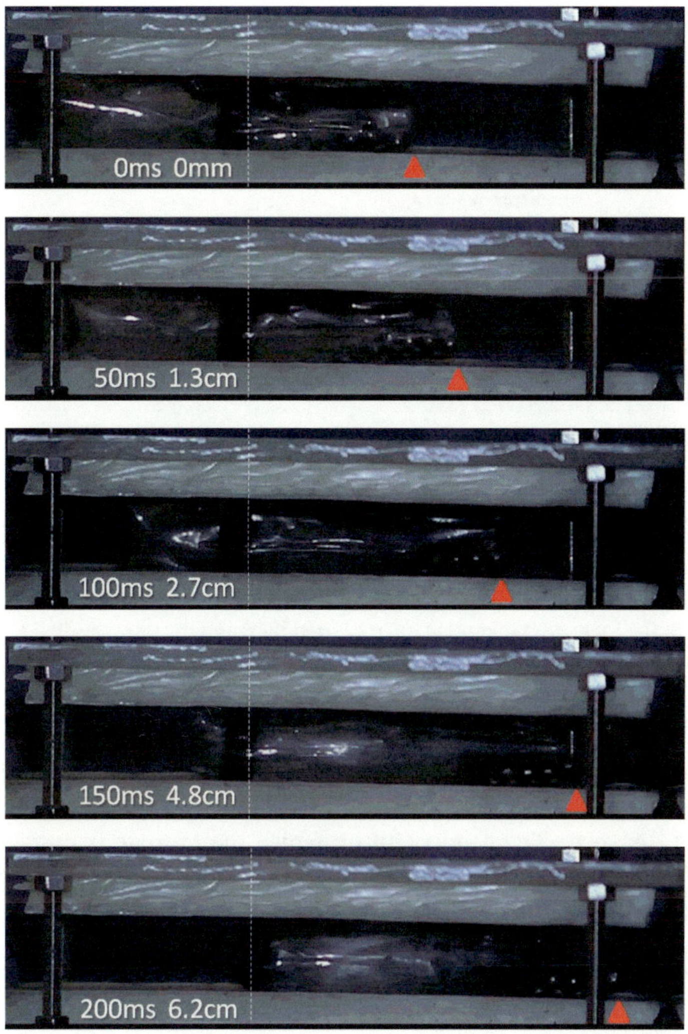

Figure 24. Movement of the model filmed by a high-speed camera, and the image was captured at intervals of 50 ms. The total moving distance was 6.2 cm.

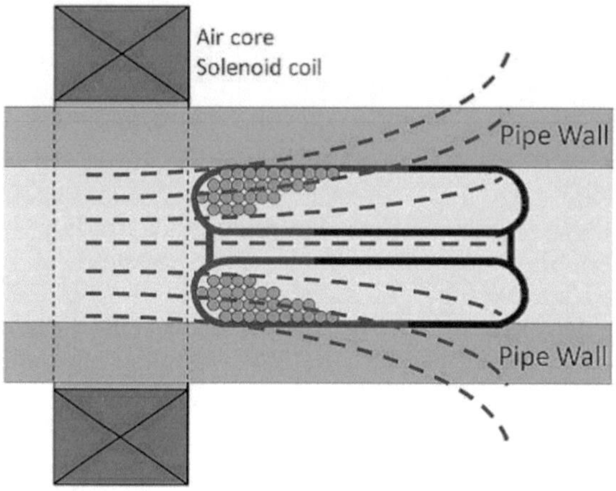

Figure 25. Method of applying the magnetic field to the tube and the patient's body using a midair solenoid.

4. Conclusions

This chapter presents a new type of magnetic fluid, a rotational drive and an advanced locomotion method with simple magnetic field.

First, a new type of magnetic fluid and a rotational drive using a simple magnetic field by use of magnetic fluid were presented. It is clear that our proposed driving motions are created by collision of the magnetic particles with the outer cover and the flow resistance between the particles and the cover. The drive for the rolling model was achieved from a small-scale external magnetic field generator and a simple magnetic field control. The device in this study could be applied to new mechanical elements and micro internal rolling objects as a useful medical tool.

Second, an advanced locomotion method that produced non-slipping motion in soft and closed environment by use of magnetic fluid were presented. New movement principle of the robot, which had a soft and deformable body that could move through a confined space was proposed. The mechanism of a toy water snake was applied to this principle. Magnetic particles inside the water balloon were affected by the external magnetic field and push the inner side of a balloon to the direction of a magnetic field. Experimental prototype to verify the proposed principle were conducted, the sliding movement was measured using the model. Confirmatory experiments of movement were conducted in the two sheets that imitated internal organs, the total moving distance was 6.2 cm. Drive a rolling model by simple magnetic field control were succeeded.

Author details

Makoto Nokata
Department of Robotics, Ritsumeikan University, Japan

5. References

[1] http://www.sigma-hc.co.jp/english/index.html

[2] N. Takesue, H. Asaoka, J. Lin, M. Sakaguchi, G. Zhang and J. Furusho, Development and Experiments of Actuator Using MR Fluid, Proceedings of 2000 IEEE International Conference on Industrial Electronics, Control and Instrumentation, pp.1838-1843, 2000.

[3] T. Kiyota, N. Sugimoto and M. Someya, 3D-Free Rescue Robot System, Proceedings of the 2006 IEEE International Conference on Robotics and Automation (ICRA'06), pp.3983-3988, 2006.

[4] K. Berk Yesin, Philipp Exner, Karl Vollmers and Bradley J. Nelson, Design and Control of In-Vivo Magnetic Microrobots, Proc. of Medical Image Computing and Computer-Assisted Intervention (MICCAI2005), pp.819-826, 2005.

[5] Mathieu, Jean-Baptiste Martel, Sylvain, Magnetic Steering of Iron Oxide Microparticles Using Propulsion Gradient Coils in MRI, 28th Annual International Conference of the IEEE Engineering in Medicine and Biology Society (EMBS'06), pp.472-475, 2006.

[6] CHIBA ATSUSHI, SENDO MASAHIKO, ISHIYAMA KAZUSHI, ARAI KEN'ICHI, Moving of a Magnetic Actuator for a Capsule Endoscope in the Intestine of a Pig, Journal of Magnetics Society of Japan, Vol.29, No.3 pp. 343-346, 2005.

[7] N.TAKESUE, J.FURUSHO, M.SAKAGUCHI, Improvement of Response Properties of MR-Fluid Actuator by Torque Feedback Control, Proceedings of the 2001 IEEE International Conference on Robotics & Automation, pp.3825-3830, 2001.

[8] J.Furusho and T.Kikuchi, A 3-D rehabilitation System for Upper limbs EMUL and a 6-DOF Rehabilitation System Robotherapist and Other Rehabilitation System with High Safety, Rehabilitation Robotics, I-Tech Education and Publishing, pp.115-136, 2007.

[9] Nokata,M. Kitamura,S. Nakagi,T. Inubushi,T. Morikawa,S., Capsule type medical robot with magnetic drive in abdominal cavity, Proc. of Biomedical Robotics and Biomechatronics, 2008 (BioRob 2008). 2nd IEEE RAS & EMBS International Conference, pp. 348-353, 2008.

[10] CapsuleEndoscopy, Given Imaging Ltd., http://www.capsuleendoscopy.org/

[11] K.Ishiyama, M.Sendoha, A.Yamazakia, K.I Arai: Swimming micro-machine driven by magnetic torque, Sensors and Actuators A: Physical, Vol. 91, Issues 1-2, pp.141-144, 2001.

[12] A.Chiba, M.Sendoh, K.Ishiyama, K.Arai: Magnertic Actuator for capusle endoscope navigation system INTERMAG Asia 2005: Digest of the IEEE International Magnetics Conference.

[13] M. Shikanai et al.: Development of robotic endoscope that locomotes in the colon with flexible helical fin, The 31st Annual International Conference of the IEEE Engineering in Medicine and Biology Society (EMBC2009).

[14] P. Glass, E. Cheung, and M. Sitti: A legged Anchoring Mechanism for Capsule Endoscopes Using Micropatterned Adhesives, IEEE Transactions on Biomedical Engineering, Vol.55, No.12, pp. 2759-2767, 2008.

[15] Eugene Cheung, Mustafa Emre Karagozler, Sukho Park, Byungkyu Kim, and Metin Sitti: A New Endoscopic Microcapsule Robot using Beetle Inspired Microfibrillar Adhesives, Proceedings of the 2005 IEEE/ASME International Conference on Advanced Intelligent Mechatronics, pp.551-557,2005.

[16] M.Nokata et al.: Capsule Type Medical Robot with Magnetic Drive in Abdominal Cavity, IEEE RAS / EMBS International Conference on Biomedical Robotics and Biomechatronics (BioRob 2008), pp.348-353, 2008.

[17] M.Nokata et al.: New magnetic rotational drive by use of magnetic particles with specific gravity smaller than a liquid, 2010 IEEE International Conference on Robotics and Automation (ICRA2010).

[18] T.Fukuda: Characteristic and application development of TECBALL, KINZOKU, vol.79, No.11, 2009 (in Japanese).

Chapter 17

New Perspectives for Magnetic Fluid-Based Devices Using Novel Ionic Liquids as Carriers

Laura Rodríguez-Arco, Ana Gómez-Ramírez, Juan D.G. Durán and Modesto T. López-López

Additional information is available at the end of the chapter

http://dx.doi.org/10.5772/51398

1. Introduction

Magnetic materials, that is to say, materials which respond to external magnetic fields, have attracted great interest since around 5th century BC. The ability of materials such as iron or magnetite to be attracted by permanent magnets has always involved some kind of mystery, although magnetism principles are well understood nowadays, especially after J.C. Maxwell stated the basis of electromagnetism in 1865. Magnetism plays a crucial role in our lives these days, being present in many technological applications surrounding us. Nevertheless, only solid magnetic materials can be found in nature. In the search for new ones that could be used in novel technological applications, scientists in the 18th century tried to prepare field-responsive liquids by dispersing particles of magnetic materials in liquid carriers. Actually, the first attempt can be attributed to G. Knight (1779), who suspended iron filings in water that quickly settled (Popplewell, 1984). The preparation of magnetic fluids has undergone great development since then, and magnetic fluids with particles of different nature, size or shape and a wide range of liquid carriers have been reported in the literature.

In this chapter an overview of magnetic fluids and their applications is made, focusing on the latest developments in the field. More specifically, the use of novel ionic liquids as carrier fluids is described. The interest in doing so lies in the fact that ionic liquids may give rise to a new generation of magnetic fluids with promising technological applications.

2. Magnetic fluids

As it has been pointed out above, the term "magnetic fluids" (MFs) is used to describe a group of smart materials whose properties can be controlled by means of external magnetic fields. They all are suspensions of magnetisable particles dispersed in a liquid carrier.

Basically, two kinds of MFs can be defined according to the size of the dispersed phase: magnetorheological fluids (MRFs) and ferrofluids (FFs).

MRFs consist of micron-sized ferro- or ferrimagnetic particles dispersed in a liquid carrier. Traditionally, organic solvents such as kerosene or mineral oil are used as carriers. With regards to the dispersed particles, they can be considered as multi-domain from the magnetic viewpoint, since their size is much higher than the one of a single magnetic domain for the constituent material in question (Bossis et al., 2002). For example, typical MRFs can be composed of iron particles with diameters around 1 μm, while the size of one magnetic domain in this material is approximately 30 nm (Herpin, 1968). As a result, MRFs strongly respond to external magnetic fields, giving rise to considerable changes in their flow (rheological) behaviour. More specifically, they behave as Newtonian fluids in the absence of magnetic field application, while their behaviour becomes that of a non-Newtonian plastic fluid when a magnetic field is applied. In this latter case, the MRF only flows when the shear stress applied to the suspension overcomes a finite value, the so-called yield stress. Therefore, it could be said that their rheological behaviour changes from a *liquid-like* to a *solid-like* one due to the application of external fields. This change, known as magnetorheological (MR) effect is a consequence of the formation of magnetically-induced structures by the dispersed particles, that have to be broken by the hydrodynamic forces to make the suspension flow (Bossis et al., 2002). Figure 1 shows typical MRF rheograms (shear stress vs. shear rate curves) both in the absence of and upon magnetic field application; the appearance of the yield stress (σ_y) can be clearly observed. The field-induced structures and their evolution as the shear rate is increased are schematized too.

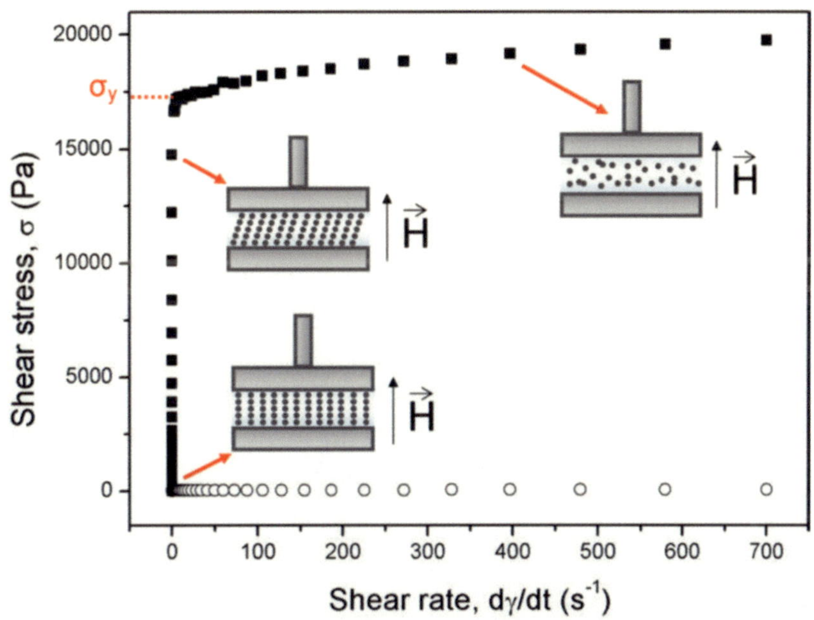

Figure 1. Rheograms for a MRF consisting of 30 vol % of iron microparticles in mineral oil at magnetic field strengths (H) of 0 and 250 kA/m (circles and squares respectively). The evolution of the field-induced structures is showed.

The strong magnetic interaction between the dispersed particles in MRFs, responsible for the MR effect, has its disadvantages too. Actually, it makes them agglomerate and settle down if no stabilizing additives are used. Since complete avoidance of sedimentation in MRFs is almost impossible, the efforts in formulation have focused on trying to reduce it and making particle redispersion easy. Different additives such as surfactants (oleic acid, lecithin, aluminum stearate) or thixotropic agents (organoclay particles) have been used for this purpose (de Vicente et al., 2003; López-López et al., 2005a, 2008).

FFs, on the other hand, are ultrastable suspensions of magnetic nanoparticles (size around 10 nm) dispersed in a liquid carrier. From the magnetic point of view, these particles are single-domain and therefore, FFs behave as superparamagnetic systems (Charles, 2002; Rosensweig, 1985). The term superparamagnetism is used to describe the magnetic behaviour of systems that exhibit high saturation magnetization values (i.e. typical of ferromagnetic materials), and no magnetic remanence or hysteresis as it happens in paramagnetic materials. FFs only undergo slight changes of viscosity in the presence of external magnetic fields. This phenomenon, known as magnetoviscous (MV) effect is very interesting from the technological viewpoint (Odenbach et al., 2002).

Nevertheless, the most important feature related to FFs is their so-called ultrastability: ideally, true FFs should not settle either when subjected to strong magnetic field gradients or in the presence of gravitational forces during their lifetime (Rosensweig, 1985). As a result, they really behave as magnetic liquids, since no phase separation appears. Figure 2 shows how a FF climbs the tube walls in order to move towards the zones where the magnetic field provided by the magnet is higher.

Figure 2. FF (5 vol % of magnetite) subjected to magnetic field application by a powerful rare-earth magnet. It can be seen that the liquid moves as a whole towards the zone where the field is more intense.

Ultrastability in FFs is usually achieved as follows (Charles, 2002; Rosensweig, 1985): (i) particle size is low enough so that Brownian motion prevents from possible particle aggregation induced by the magnetic interaction between the particles; (ii) particle

aggregation by means of van der Waals attraction (short range forces) must also be avoided by imposing some kind of repulsion between the particles. Such repulsion is usually obtained by adsorbing surfactants or polymers on the surface of the particles, and thus providing with a strong enough barrier (steric repulsion) to overcome van der Waals interactions. In polar media (i.e. water-based FFs), this can be alternatively achieved by electrostatic repulsion between the particles derived from their surface charge (Charles, 2002; Rosensweig, 1985).

The efforts made in order to improve MFs in terms of stability, durability, MR and MV effects have led to the development of a new generation of MFs. This group would include inverse FFs, bimodal suspensions, and MRFs consisting of anisotropic particles. In the first case, non-magnetic microparticles are dispersed in a concentrated enough FF. These particles behave as non-magnetic holes in a practically continuous magnetic medium, and, as a result, are able to orientate and form magnetic structures when a magnetic field is applied (figure 3). This sort of MR effect is stronger than in the case of FFs, but it is still far from that of MRFs (de Gans et al., 1999; Ramos et al., 2011).

Figure 3. Microscopic picture upon magnetic field application of an inverse FF consisting of PMMA microparticles dispersed in a FF (5 vol % of magnetite in mineral oil). It can be seen that the non-magnetic particles (white spots) form chains in the direction of the magnetic field. Bar length: 100 μm.

In the case of bimodal suspensions, micron-sized particles are again dispersed in a FF, but this time, the dispersed particles are magnetic. These MFs have proved to be more stable against sedimentation than conventional MRFs due to the stabilization achieved by the formation of nanoparticle clouds around each micron-sized particle. The formation of such clouds can be observed in figure 4 (López-López et al., 2005b, 2006, 2010).

Finally, the anisotropic nature of fibrillar particles leads to an increase of the MR effect in MRFs consisting of them (Kuzhir et al., 2009; López-López et al., 2007, 2009, 2012). Figure 5 shows magnetic fibrillar particles, with a longitudinal axis of about 50 nm, which have been used in the preparation of stable FFs.

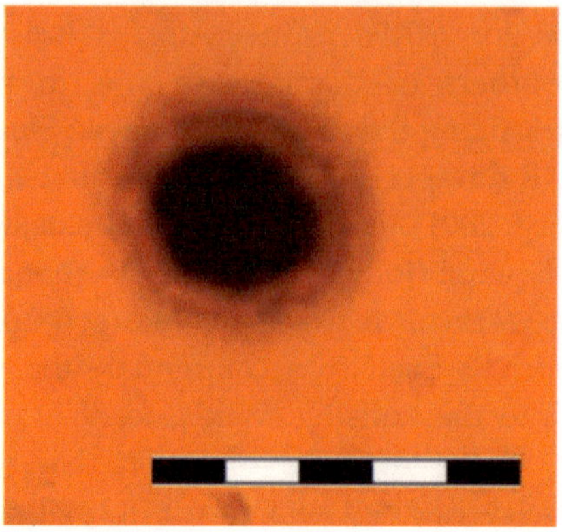

Figure 4. Optical microscope picture of a diluted bimodal suspension consisting of ferromagnetic microparticles dispersed in a FF. Note the halo around each microparticle formed by the magnetite nanoparticles of the FF carrier. Bar length: 20 μm.

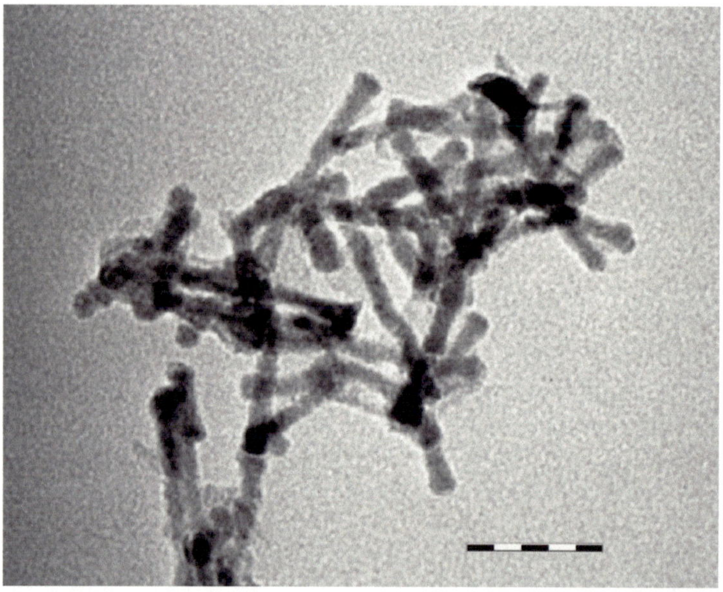

Figure 5. High resolution TEM image of CoNi fibrillar particles for the preparation of MFs. Bar length: 50 nm.

3. Magnetic fluid-based applications and devices

The possibility of controlling the physical properties of MFs by external magnetic fields makes them versatile smart materials that can be employed in numerous technological applications, ranging from fields as different as civil and mechanical engineering, aerospace, biomedicine or optics (Carlson et al., 1996; Durán et al., 2007; Jeong et al., 2007; Jolly et al., 1999; Klingenberg, 2001; Park et al., 2010; Popplewell, 1984; Raj et al., 1990; Tran et al., 2010).

The best known application for MRFs is their use as lubricants with field-dependent viscoelasticity in shock absorbers or dampers. It was in the early 1990s when *Lord Corporation*© started commercializing the first MR dampers. Since then, further progress has followed, and commercial MR dampers have become available for large trucks or race and high-quality cars (Klingenberg, 2001). A typical monotube design of a MR damper used in vehicles was reported by Jolly et al. (Jolly et al., 1999). As shown in this previous work, this kind of MR damper has one reservoir for the MR fluid and an accumulator receptacle full with a compressed gas, which is used to accommodate the volume changes due to the entrance of the piston rod into the housing. When needed, an electric current in the coil generates a magnetic field in the required direction that makes the MRF change its rheological behaviour to a *solid-like* one; as a result, the vibrations provoked by external forces during driving are absorbed. Therefore, this kind of damping function can be labeled as semi active (Jolly et al., 1999).

The technological applications of MR dampers are not only restricted to vehicles. For example they are also used as shock/vibration absorbers in structures (i.e. seismic control of buildings or bridges). In this particular case, the stability against sedimentation plays a very important role, since the damper is expected to remain inactive most of its lifetime (Jolly et al., 1999; Klingenberg, 2001; Park et al., 2010). In addition, MR shock absorbers can also be used in semi-active human prosthetic legs.

In the case of FFs, many biomedical applications have been described, which take advantage not only of their superparamagnetic behaviour, but also of their high surface-volume ratio, high reactivity, etc. Medical applications include cell labeling and targeting, separation and purification of cell populations, tissue repair, targeted drug delivery, magnetic resonance imaging or hyperthermia for cancer treatment (Durán et al., 2007; Tran et al., 2010). In all these applications biocompatibility and non-toxicity are of crucial importance. For this reason, iron oxides are preferred as the material for the dispersed phase and, in addition, they are often made biocompatible by means of surface coating by polymers (PEG, dextran, polyvinyl alcohol) or functional groups (thiols, amines, and carboxyls). All these additives prevent from particle aggregation too, which should be almost completely avoided so that particle size does not increase (Tran et al., 2010).

From a more engineering point of view, the control of both the position and the physicochemical properties of FFs by using magnets or solenoids makes them very interesting too. In fact, FFs have been used as lubricants, heat transfer agents or integrated in devices such as dynamic seals, dampers, magnetic inkjets or optic devices. As an example, companies like *Ferrotec*© have successfully commercialised FF-based dynamic seals for 40 years. Dynamic seals are used when two different environments have to be isolated one from each other but energy has to be carried between them, by a shaft for example (see figure 6). Here, FFs can be used as filling materials, since they hermetically isolate both environments due to their fixed position if the hole for the shaft is made into a magnet. An important advantage of using a seal like this, apart from being almost completely hermetic, is a prolongation of the seal lifetime, since friction between its rotating and stationary components is almost negligible (i.e. the FF acts as a lubricant agent too) (Raj et al., 1990).

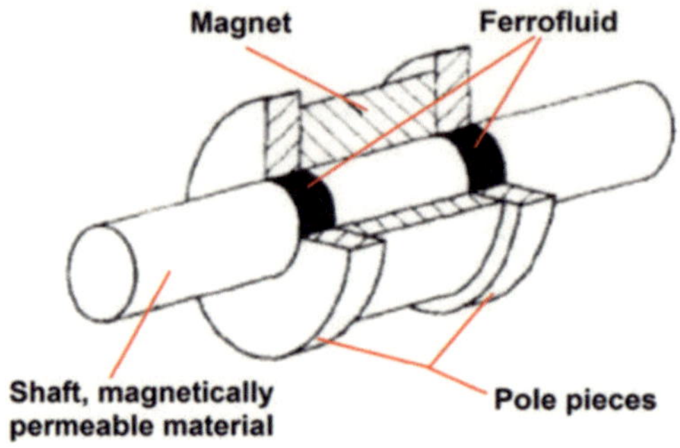

Figure 6. Sketch of a FF-based seal. Reprinted from Journal of Magnetism and Magnetic Materials, 85/1-3, K. Raj and R. Moskowitz, Commercial applications of ferrofluids, 233-245, Copyright (1990), with permission from Elsevier.

Finally, as it has been pointed out above, FFs can also be used as heat transfer materials, provided that their thermal conductivity increases when applying a magnetic field in a direction parallel to the temperature gradient. In this way, magnetically-induced particle chains, able to transfer heat by conduction, are formed in the field direction. When the field is removed, thermal conductivity reaches its original value (Shima et al., 2011). This working principle is used to dissipate heat from devices such as high power loudspeakers. In this application, the use of FFs has additional advantages like the absorption of undesirable vibrations and avoiding resonances without disturbing the quality of the sound (due to their liquid condition), or the possibility of fixing their position by magnets (as it happens in dynamic seals) (Popplewell, 1984; Raj et al., 1990).

All the applications mentioned in this section are quite well known, and some of them have already been developed at the industrial scale. However, it is quite clear that the range of technological applications of MFs will expand as progress in their dynamics and preparation is made.

4. Ionic liquids as carriers for magnetic fluids

Traditionally, organic carriers (mineral oil, kerosene, etc.) have been used in the preparation of MFs, although in some cases (i.e. medical applications) water-based MFs are needed too. These conventional MFs have proved to be useful for many applications as pointed out above. However, the use of novel ionic liquids (ILs) as carriers may promote further improvements in the performance of MFs, especially in ultra-high vacuum applications, in which organic solvents would easily evaporate. But the preparation of IL-based MFs is not only interesting from an applied, technological approach, but also from the fundamental one. For example, their use as carriers allows studying phenomena involving MFs in environmental conditions that conventional carriers would not withstand. But, what are ILs and why do they appear as promising candidates in the preparation of MFs? These and other questions are discussed in next sections.

4.1. Ionic liquids

ILs are substances composed entirely of ions that are liquids at room temperature (Endes et al., 2008). While common salts, such as table salt (i.e. sodium chloride), have melting points around 800 °C, the upper limit for the melting point so that they can be classified as ILs is usually set at 100 °C. Therefore, they can be considered as room-temperature molten salts (Keskin et al., 2007). Such behaviour is obviously related to their chemical composition: ILs are usually composed of bulky and asymmetric organic cations and smaller inorganic or organic anions. This feature makes their lattice energy quite low and, therefore, so does their melting point (Keskin et al., 2007). There are many of them depending on their chemical composition, but almost every IL can be classified, according to its cationic structure, into one of the seven families reported by Torimoto et al. (Torimoto et al., 2010).

Connected to their composition, one of the most interesting properties concerning ILs is their tunability. As a matter of fact, by changing the nature of the constituent ions it is possible to obtain ILs with very different physical and chemical properties (Endes et al., 2008). Theoretically, the combination of cations and anions would lead to a number of ILs as high as 10^{18} (Chiappe et al., 2005). In practice, however, this number would be much smaller, but it gives an idea about the broad range of physicochemical properties that can be obtained. For instance, ILs with the same imidazolium cation, [BuMeIm][1], can be either hydrophilic (combined with [BF_4] or [$AlCl_4$]) or hydrophobic (if the anion is [PF_6] or [Tf_2N]) (Torimoto et al., 2010). Here it is worth mentioning the fact that some ILs with magnetic properties have also been prepared by using [$FeCl_4$] anions. Tunability is especially interesting in the case of MFs, since it would allow synthesizing tailor-made MFs specifically designed for every single application.

Nevertheless, there are two common features for many of them that are rather interesting from the technological viewpoint. These are their negligible vapour pressure and flammability, at least when compared to those of conventional solvents such as volatile organic compounds (VOCs). Both characteristics ensure, first of all, thermal stability and vacuum resistance and, what is more, an insignificant contribution to air pollution. As a matter of fact, they have been labelled as "green solvents" due to their negligible impact on atmospheric pollution (Keskin et al., 2007; Torimoto et al., 2010). Finally, it is possible to recycle them, which would contribute to reduce their release to the aquatic environment and lead to more efficient and economic industrial processes (Haerens et al., 2010; Wu et al., 2009).

4.2. Ionic liquid applications

Since the academic world began to become interested in ILs in the 1990s, the number of scientific articles and patents related to the topic has incredibly increased. For example, in 2000 about 100 patent applications had been reported, while in 2004 this number increased to 800 (Keskin et al., 2007). Nowadays, about 28200 patents related to ILs can be described.

---

[1] See apendix (section 7) for nomenclature abbreviations.

Some applications arise from the fact that many ILs are powerful solvents, able to dissolve salts, fats, proteins, amino acids, surfactants or sugars. For instance, [(MeOEt)MeIm][BF_4] can dissolve glucose 100 times better than acetone (Park et al., 2001). They can even dissolve crude oil, inks, plastics or DNA. For this reason, they can be used as reaction media for electrochemical and chemical syntheses (Torimoto et al., 2010), giving rise to reaction rates similar or even better than those obtained when using aqueous or organic solvents. Another advantage of using them with this aim is the ease in recovering the resulting products from the reaction medium, especially in those cases in which distillation is not practical, for example, because the reaction products are thermally sensible (Keskin et al., 2007). In addition, it has been reported that ILs can be used as solvents for insoluble bio-related materials like cellulose, which is easily dissolved by strong hydrogen bond-acceptor ILs (like chloride anion-based ones) (Keskin et al., 2007; Torimoto et al., 2010).

Another typical application is their use in gas purification because many gases are soluble in ILs (for example CO_2 is highly soluble in [BuMeIm][PF_6]) (Anthony et al., 2002). This feature also makes them excellent candidates to be employed in gas absorption applications, together with the fact that gas separation from the IL stream would be very easy (i.e. by simple flash distillation) (Keskin et al., 2007).

Related to materials science, ILs have been widely used in chemical synthesis of nanoparticles, with self-evident advantages like the fact that almost no stabilizing agents that prevent from particle coalescence are needed (requisite almost indispensable in the case of aqueous or organic solvents) (Torimoto et al., 2010). For example, it is possible to synthesize particles for the preparation of MFs such as magnetite (Wang et al., 2009) or cobalt (Behrens et al., 2012) nanoparticles. This would allow an *in situ* synthesis of the dispersed particles.

As a final point, the preparation of IL-based MFs would allow the use of some of the devices mentioned in section 3 under extreme environmental conditions (ultra-high vacuum or high temperature applications). This would be particularly interesting in the case of space technology such as dampers, dynamic seals or hydrodynamic bearings in gyroscopes for mini-satellites. As a matter of fact, the first MF patent was aimed to prepare controlling fluids for space applications (Papell, 1965).

5. Ionic liquid-based magnetic fluids

In this section the research highlights in the field of IL-based MFs are described both in the case of MRFs and FFs. Some details about our group contributions to the field are given too.

5.1. Ionic-liquid based magnetorheological fluids

Guerrero-Sánchez et al. (Guerrero-Sánchez et al., 2007) were pioneers in the preparation of IL-based MRFs. More specifically, they prepared several suspensions consisting of either micron- or nanosized magnetite particles dispersed in eight different ILs. The most stable MRF was obtained for a composition of 10 vol % of magnetite microparticles suspended in

[BuMeIm][PF_6]. As a matter of fact, this MRF only exhibited a sedimentation ratio of 0.95 after two months of preparation. Having prepared a highly stable IL-based MRF was important, but even more important was the fact that colloidal stability was achieved without the addition of any stabilizing agents. The enhancement of stability was attributed to the physical adsorption of the IL ions on magnetite surface (for which they had strong affinity) that gave rise to steric repulsion between the dispersed particles. In the same work, the rheological behaviour of the samples upon magnetic field application was also studied concluding that the MR effect of the IL-based MRFs was comparable to those obtained for conventional MRFs, and high enough for practical purposes.

In a different work, Guerrero-Sánchez et al. (Guerrero-Sánchez et al., 2009) also used IL-based MRFs to study the influence of temperature on their MR behaviour. Here, ILs allowed extending such study to a broader range of temperature than in the case of conventional MRFs, due to their thermal stability and low vapour pressure. Something similar was pointed out by Dodbiba et al. (Dodbiba et al., 2007) who also took advantage of the good performance of ILs at high temperature to study the rheological behaviour upon magnetic field application of a mixture of two types of carbonyl iron powders dispersed in an IL. Therefore, these works make evident that the preparation of IL-based MFs is important from the fundamental point of view too.

Our group has also studied IL-based MRFs, especially for what concerns their stability. With this aim, Gómez-Ramírez et al. (Gómez-Ramírez et al., 2011) prepared IL-based MRFs differing both in the disperse phase material (pure iron and silica-coated iron particles) and in the liquid carrier ([EtMeIm][$EtOSO_3$] and ([EtMeIm][Et_2OPO_3]). Additionally, conventional MRFs with mineral oil as liquid carrier were also synthesized. The particle volume fraction of all the samples was kept as 10 vol %. The most stable MRF was obtained by the combination of silica-coated iron particles and [EtMeIm][Et_2OPO_3]; for example, the viscosity of such MRF for shear rates over 100 s^{-1} almost coincided with the theoretical prediction of Batchelor equation, what means that only two-body hydrodynamic interactions existed between the particles, and, therefore, almost no magnetic aggregates were present. In fact, microscopic observations of the suspensions upon magnetic field application showed that field-induced particle chains had a more regular structure in the case of the most stable MRF (see figure 7). The worst results were obtained for mineral oil-based MRFs in which the particles were strongly aggregated and the resulting structures became more irregular. An intermediate behaviour was found for the suspensions with [EtMeIm][$EtOSO_3$] as liquid carrier.

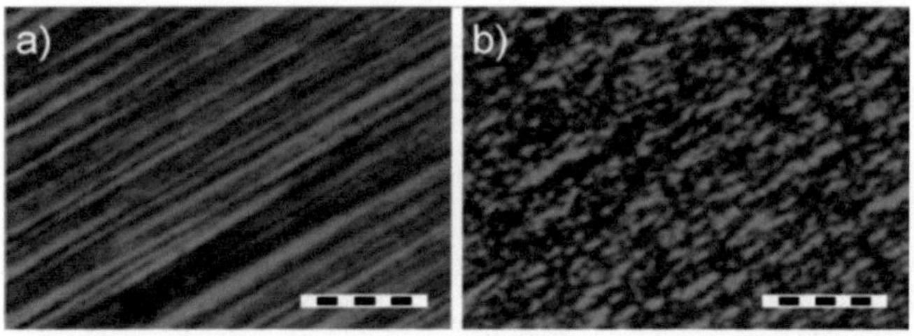

Figure 7. Field-induced particle structures in MRFs consisting of silica-coated iron microparticles dispersed in [EtMeIm][Et_2OPO_3] (a) and mineral oil (b). Bar length: 200 μm.

The improvement of MRF stability when using ILs as carriers, with respect to the ones prepared in mineral oil, was also explained by the adsorption of the IL ions on the surface of the dispersed particles. In a first approximation, such stability appears to be related to the electric conductivity of ILs, which, on its part, is also connected to the anion volume, being the first lower when the latter increases. Accordingly, if the adsorbed ions are larger, the steric repulsion between the particles is stronger too, leading to a decrease in particle aggregation. This is the reason for which a low conductivity-IL like [EtMeIm][Et_2OPO_3] gave rise to better results than the high conductivity-one [EtMeIm][$EtOSO_3$]. The explanation for the higher stability in the case of MRFs consisting of silica-coated iron particles was the interaction between silica and the constituent IL ions. In fact, the wettability of both ILs was better in the case of silica-coated iron than for bare iron (Gómez-Ramírez et al. 2011).

The direct consequence of the improved stability of IL-based MRFs is a simpler MRF formulation, since no stabilizing additives are needed. This makes easier their preparation at larger scale given that no additional mixing steps are needed, for example. Therefore, stable MRFs with adequate MR response that can be subjected to ultra-high vacuum or very high/low temperatures are obtained, just by using ILs as carrier fluids.

5.2. Ionic-liquid based ferrofluids

As it has been pointed out above, one of the advantages of using ILs as carriers is the stability that the physical adsorption of their constituent ions provides with. In the case of MRFs it has been seen that this improvement may be good enough for practical purposes. But what happens in the case of FFs that, by definition, need to be ultrastable? What are the mechanisms involved in the stabilization of IL-based FFs?

The first reference to IL-based FFs appeared in 2009, when Oliveira et al. prepared suspensions of bare maghemite (γ-Fe_2O_3) and cobalt ferrite ($CoFe_2O_4$) nanoparticles dispersed in [BuMeIm][BF_4] (Oliveira et al., 2009). Such suspensions were said to be stable in the presence of an external magnetic field, even without the addition of stabilizing agents. The authors attributed such stability to the formation of a semi-organized protective layer by the IL ions around the dispersed particles. Nevertheless, when water was added to the suspensions, they became completely unstable, apparently due to the destruction of such a protective layer. The authors also tried to prepare suspensions using hydrophobic ILs (namely [BuMeIm][PF_6] and [BuMeIm][Tf_2N]), but these suspensions were not stable at all.

Jain et al. (Jain et al., 2011) also attempted to synthesize IL-based FFs using maghemite nanoparticles, but with particle concentrations much higher than in the case of Oliveira et al. They succeed when using [EtMeIm][Ac] and [EtMeIm][SCN] as carriers, without needing any stabilizers, but they did not in the case of the protic IL [EtN][NO_3] and [BuMeIm][BF_4]. In the latter case, they failed even at lower concentrations in contrast with the results of Oliveira et al. As a matter of fact, there is still some controversy related to the stabilization of bare magnetic nanoparticles in IL media, although the stabilization provided by the IL ions has been pointed out to work quite well for metal nanoparticles in general (Pârvulescu et al., 2007). This is the case of positively charged Pt nanoparticles dispersed in [BuMeIm][BF_4] and

[BuMeIm][PF_6], for example (Scheeren et al., 2006). However, in some other cases it is not high enough, and additional stabilization is needed (Pârvulescu et al., 2007; Torimoto et al., 2010).

In most of the suspensions mentioned above, the material for the dispersed phase is maghemite. At this point it is important to mention that both Guerrero-Sánchez et al. (Guerrero-Sánchez et al., 2007) and Oliveira et al. (Oliveira et al., 2009) failed to synthesize ultrastable IL-based suspensions when using magnetite (Fe_3O_4) nanoparticles. [BuMeIm][BF_4] that, according to Oliveira et al., had given rise to good results for maghemite and cobalt ferrite, did not work at all in the case of magnetite. This could be due to the fact that magnetic interaction between magnetite nanoparticles is higher than in the case of maghemite ones, since the saturation magnetization of the first (90 Am^2kg^{-1}) is higher than the one of the latter (80 Am^2kg^{-1}). Given that both iron oxide surfaces must be composed of the same Fe-OH groups, the formation of the protective layer around magnetite nanoparticles could have taken place, but probably was not protective enough to overcome the magnetic attraction between the particles. An additional stabilization mechanism seemed to be needed.

In 2011 our group reported a true IL-based magnetite FF (Rodríguez-Arco et al., 2011a). In this work, emphasis was made in the fact that only strong steric repulsion was useful to obtain an ultrastable suspension of magnetite in [EtMeIm][$EtOSO_3$]. This was achieved by adsorbing surfactants with long tails on the surface of magnetite nanoparticles that gave rise to such strong steric repulsion (more details about this IL-based FF are given below). In fact, we also tried to prepare suspensions of bare magnetite nanoparticles in the same IL and failed. Later, something similar was pointed out by Jain et al., who observed that bare maghemite nanoparticles were unstable in the protic IL [EtN][NO_3], but they became highly stable (even at highly particle concentration) when coated by a layer of short acrylic-acid-*b*-acrylamide copolymer (Jain et al., 2012). As a result, steric repulsion seems to be the best stabilization mechanism for the preparation of IL-based FFs.

In view of the results presented above a question may arise: how can we explain that in some cases stabilization is ensured without the addition of stabilizing agents (i.e. surfactants) while in others it is strictly necessary? Here, it is particularly interesting to mention the work by Ueno et al. in which they prepared suspensions of bare and PMMA-grafted silica particles dispersed in imidazolium-based ILs (Ueno et al., 2008). Given that ILs are non-volatile, the *in situ* behaviour of these suspensions was studied by means of transmission electron microscopy (i.e. no escapes of the liquid to the vacuum system took place). They observed that strong aggregation, which appeared for bare silica particles, was almost absent in the case of PMMA-grafted ones (figure 8). They concluded that electrostatic repulsion, that was the only possible stabilization mechanism in the case of bare silica, did not work. They attributed this to the screening effect of the high ionic atmosphere surrounding the particles, since ILs are actually room-temperature molten salts. They also corroborated such affirmation by estimating the interaction potential energy between the dispersed particles and checking that the electrostatic repulsion component was negligible when compared with the steric repulsion one (Ueno et al., 2008).

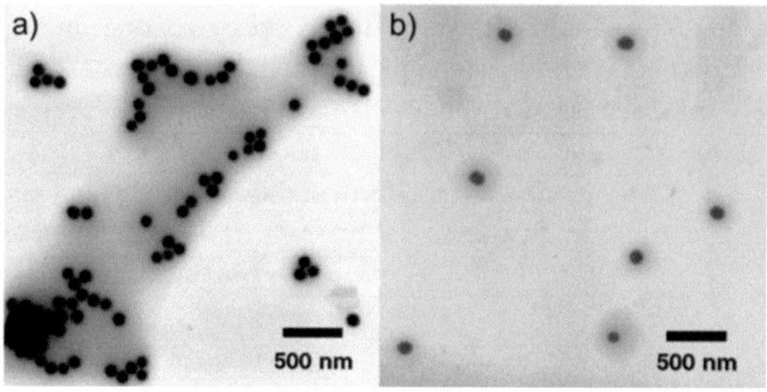

Figure 8. In situ TEM picture of dilute suspensions of bare (a) and PMMA-grafted (b) silica particles in [EtMeIm][Tf_2N]. Reprinted (adapted) with permission from (Ueno et al., 2008). Copyright (2008) American Chemical Society.

If the results of Ueno et al. (Ueno et al., 2008, 2009) are taken into account, there are two points that seem clear: (i) electrostatic repulsion does not work in IL media because of the highly ionic environment; (ii) only steric repulsion gives rise to long-term stability. On its part, the protective layers formed by the adsorption of the IL ions (i.e. physically or by coordination compounds) could provide with steric repulsion. However, in some cases (as for magnetite suspensions), it is not strong enough to ensure long-term stability and other additives (i.e. surfactants) have to be used.

As it has been pointed out above, our group succeeded in the preparation of true magnetite IL-based FFs. This was only achieved when the surface of the particles was coated with some stabilizing agents, but not all additives worked (Rodríguez-Arco et al., 2011b). For example, citric acid, that had been previously used to stabilize water-based FFs, did not work properly in [EtMeIm][$EtOSO_3$], since the samples consisting of citric acid-coated magnetite slightly settled when subjected to a magnetic field gradient of 10 $mTmm^{-1}$, and gave rise to strong phase separation after 5 min of centrifugation at 8000*g*. This was due to the fact that citric acid molecule is very short and, therefore, the resulting steric repulsion was quite weak. As a consequence, it seems that only long enough molecules should be used.

In addition, compatibility between the tails of the surfactant and every particular IL must exist. For example, if magnetite particles were just coated with a single layer of oleic acid, the resulting suspension was as unstable as those consisting of bare magnetite. Nevertheless, if a second layer of oleic acid molecules was formed when an excess of oleic acid was added, a true IL-based FF was obtained. The ultrastability of such FF was accurately proved, since it did not settle either upon magnetic field gradient application or after 30 min of centrifugation at 8000*g*. The differences between both particle coatings can be easily observed in figure 9.

It can be seen that the tails facing the IL in figure 9a are non-polar, while the situation is just the opposite when the oleic acid-double layer is formed (figure 9b). In the first case, there is not compatibility between the surfactant tails and the IL (highly polar), whereas in the second one there is. As it has been previously pointed out, Jain et al. (Jain et al., 2012) were

able to stabilize maghemite nanoparticles in [EtN][NO_3] by coating them by a layer of short acrylic-acid-*b*-acrylamide copolymer; however, they were not in the case of [BuMeIm][BF_4] and [EtMeIm][SCN] just because the polyacrylamide block of the acrylic-acid-*b*-acrylamide copolymer is not soluble in these ILs, and therefore, no compatibility exists. Therefore, as it has been said above, the best way to ensure long-term stability is the use of surfactants with long enough, carrier liquid-compatible tails adsorbed on the surface of magnetite nanoparticles.

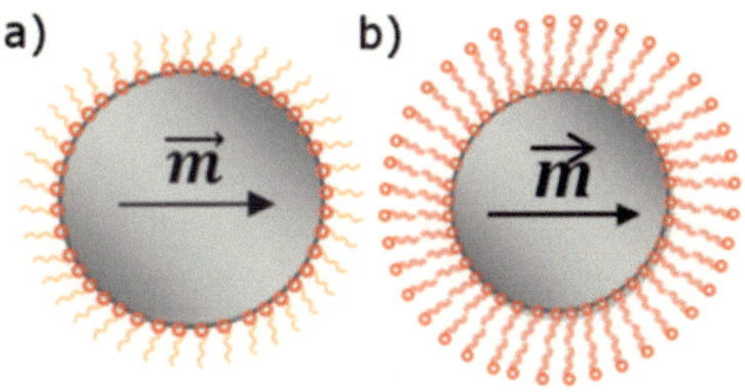

Figure 9. Monolayer of oleic acid molecules adsorbed on the surface of magnetite nanoparticles (a). Oleic acid-double layer formed when adding an excess of oleic acid (b).

Finally, some results about the rheological behaviour upon magnetic field of the samples mentioned above are shown (Rodríguez-Arco et al., 2011b). In particular, figure 10 shows the yield stress obtained for all of them.

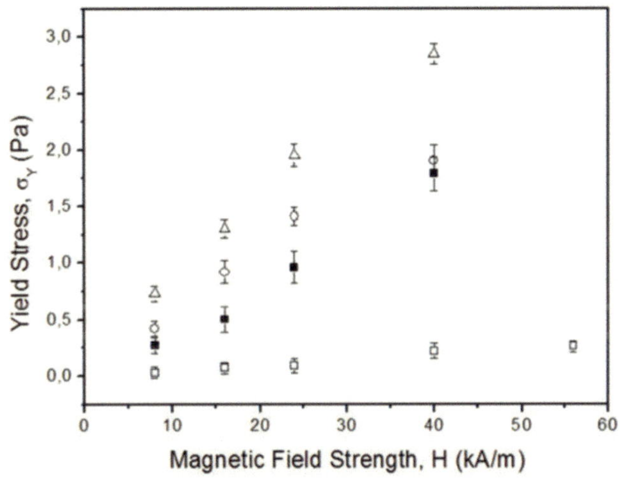

Figure 10. Yield stress for IL-based suspensions consisting of bare (triangles), citric acid-coated (circles) and oleic acid-double layer-coated (squares) magnetite. Full and open squares correspond to this latter sample before and after centrifugation respectively (see text).

With regards to the results presented in figure 10 it is important to remind that, theoretically, a true FF should not display considerable yield stresses, since its response to magnetic field is too weak for this to happen. However, in the case of the suspension of bare magnetite, the yield stress is quite high, likely due to the strong particle aggregation, and therefore, to the formation of field-induced structures by the aggregates. When the particles

are coated by citric acid, the aggregation degree decreases, and so does the yield stress. In the case of the oleic acid-double layer two different situations appear. The first one corresponds to the original sample, which displays similar values (a little bit lower) than the sample of citric acid-coated magnetite. However, when this sample is subjected to centrifugation, although the supernatant itself is still magnetic, the yield stress of the supernatant becomes negligible. This can be explained because particle aggregates that could be present in the original sample were removed by centrifugation. Similar results were found for the MV effect (Rodríguez-Arco et al., 2011b). In fact, the MV response of these new IL-based MFs was theoretically fitted by a model that was originally proposed by Zubarev et al. (Zubarev et al., 2005) for oil-based FFs, showing that the MV effect increases with both the volume of the largest particles and their concentration and when the distance between the magnetic cores of the particles decreases.

In conclusion, it could be said that much work is being done in the field of IL-based FFs. Thanks to such studies our understanding of the phenomena involving them (i.e. stability) is improving, and it is very likely that better IL-based FFs will be prepared in the future. However, more applied studies are needed, which, for example, analyze their real performance in technological devices like those of section 3.

6. Future challenges

Many of the future challenges in the field of IL-based MFs are related to ILs in general. One of the most important disadvantages of ILs, for instance, is their cost, that can be much higher than that of conventional organic solvents. However, in some specific applications, it is probably worth using them (i.e. space applications). Nevertheless, their price will decrease if they begin to be produced at a larger scale.

Another disadvantage of ILs when compared with traditional carriers is the lack of enough data about their physicochemical properties and toxicity (Keskin et al., 2007). In the same direction, a thorough analysis of the relationship (e.g. adsorption, wettability) between the particle surface and the constituent ions in MFs would be of the utmost importance, in order to gain a better understanding of the stability mechanisms.

Since ILs could be used to prepare tailor-made MFs for each specific application, it would be interesting to broaden their preparation by changing the IL carrier, the nature of the dispersed phase or its concentration. But before this is accomplished, it is necessary to determine which solid materials (or surfactants in the case of FFs) fit best each particular IL. As a result, deeper studies on the compatibility among all the MF constituents may be needed in the future. Additionally, further research has to be done, not only with regards to the formulation of IL-based MFs, but also to their performance in each particular application. As a consequence, detailed magnetization and magnetorheological studies should be faced too.

The final step for IL-based MFs would need a much more applied study which could result in a number of patents susceptible of being exploited. For example, they should be included

in prototypes before developing any industrial device. This would allow additional industrial implementation and commercialization of this new breed of MFs. Here it is worth mentioning the work by companies like *Ioniqa Technologies*© that are starting to commercialize different IL-based magnetic smart materials, including MFs, magneto-responsive elastomers, and magnetic ILs. Therefore, this new type of MFs will pose new questions not only to scientists and engineers, but also to businesspeople.

7. Conclusions

In this book chapter an overview of the latest advances and research highlights in the field of MFs has been done, especially concerning the use of room-temperature ILs as fluid carriers. As it has been described, there are numerous MF applications that could benefit from IL features such as low flammability, negligible vapor pressure and tunability. However, although a lot of work has been done, there are still problems that have to be overcome if a wider commercialization of IL-based MFs is desired. Therefore, these new horizons broadened by their potential uses are, at the same time, encouraging and challenging.

Appendix

Here we list the IL nomenclature abbreviations used in the chapter.

Cations	[EtMeIm]	1-ethyl-3-methylimidazolium
	[BuMeIm]	1-butyl-3-methylimidazolium
	[(MeOEt)MeIm]	1-methoxyethyl-3-methylimidazolium
	[EtN]	Ethylammonium
Anions	[Ac]	Acetate
	[Tf_2N]	Bis(trifluoromethylsulfonyl) imide
	[Et_2OPO_3]	Diethylphosphate
	[$EtOSO_3$]	Ethylsulphate
	[PF_6]	Hexafluoroborate
	[NO_3]	Nitrate
	[$AlCl_4$]	Tetrachloroaluminate
	[$FeCl_4$]	Tetrachloroferrate
	[BF_4]	Tetrafluoroborate

Table 1.

Author details

Laura Rodríguez-Arco, Ana Gómez-Ramírez,
Juan D.G. Durán and Modesto T. López-López
University of Granada, Spain

Acknowledgement

This work is supported by projects P08-FQM-3993 and P09-FQM-4787 (Junta de Andalucía, Spain) and FIS2009-07321 (MICINN, Spain). L.R.-A. and M.T.L.-L. also acknowledge financial support by the Secretaría de Estado de Educación, Formación Profesional y Universidades (MECD, Spain) through its FPU Program and University of Granada (Spain), respectively.

8. References

Anthony, J.L.; Maginn, E.J. & Brennecke, J.F. (2002). Solubilities and Thermodynamic Properties of Gases in the Ionic Liquid 1-n-butyl-3-methylimidazolium hexafluorophosphate. *The Journal of Physical Chemistry B,* Vol. 106, No. 29, (July 2002), pp. 7315–7320, ISSN 1520-6106

Behrens, S. & Essig, S. (2012). A Facile Procedure for Magnetic Fluids Using Room Temperature Ionic Liquids. *Journal of Materials Chemistry,* Vol. 22, No. 9, (March 2012), pp. 3811-3816, ISSN 0959-9428

Bossis G.; Volkova, O.; Lacis, S. & Meunier, A. (2002). Magnetorheology: Fluids, Structures and Rheology, In: *Ferrofluids. Magnetically Controllable Fluids and Their Applications*, S. Odenbach, pp. 186-230, Lecture Notes in Physics (Springer-Verlag), ISBN 3-540-43978-1, Berlin, Germany

Cao, L.; Park, H. S.; Dodbiba, G. & Fujita, T. (2010). Synthesis of an Ionic Liquid-based Magnetorheological Fluid Dispersing $Fe_{84}Nb_3V_4B_9$ Nanocrystalline Powders. *International Journal of Modern Physics B,* Vol. 24, No. 10, (April 2010), pp. 1227-1234, ISSN 0217-9792

Carlson, J.D.; Catanzarite, D.M. & St. Clair, K.A. (1996). Properties and Applications of Commercial Magnetorheological Fluids. *International Journal of Modern Physics B,* Vol. 10, No. 23-24, (October 1996), pp. 2857-2865, ISSN 0217-9792

Charles, S.W. (2002). Magnetic Fluids. Ferrofluids, In: *Ferrofluids. Magnetically Controllable Fluids and Their Applications*, S. Odenbach, pp. 3-18, Lecture Notes in Physics (Springer-Verlag), ISBN 3-540-43978-1, Berlin, Germany

Chiappe, C. & Pieraccini, D. (2005). Ionic liquids: Solvent Properties and Organic Reactivity. *Journal of Physical Organic Chemistry,* Vol. 18, No. 4, (April 2005), pp.275-297, ISSN 1099-1395

Dodbiba, G.; Park, H. S.; Okaya, K. & Fujita, T. (2008). Investigating Magnetorheological Properties of a Mixture of Two Types of Carbonyl Iron Powders Suspended in an Ionic Liquid. *Journal of Magnetism and Magnetic Materials,* Vol. 320, No. 7, (April 2008), pp. 1322-1327, ISSN 0304-8853

de Gans, B.J.; Blom, C.; Philipse, A.P. & Mellema, J. (1991). Linear Viscoelasticity of an Inverse Ferrofluid. *Physical Review E,* Vol. 60, No. 4, (October 1999), pp. 4518-4527, ISSN 1539-3755

Durán, J.D.G.; Arias, J.L.; Gallardo, V. & Delgado, A.V. (2007). Magnetic Colloids as Drug Vehicles. *Journal of Pharmaceutical Sciences,* Vol. 97, No. 8, (December 2007), pp. 2948-2983, ISSN 1520-6017

de Vicente, J.; López-López, M.T.; González-Caballero, F. & Durán, J.D.G. (2003). Rheological Study of the Stabilization of Magnetizable Colloidal Suspensions by Addition of Silica Nanoparticles. *Journal of Rheology*, Vol. 47, No. 5, (November 2003), pp. 1093-1109, ISSN 0148-6055

Endes, F.; Abbot, A.P. & MacFarlane, D.R. (2008). *Electrodeposition from Ionic Liquids,* Wiley-VCH, ISBN 978-3-527-31565-9, New York, USA

Guerrero-Sánchez, C.; Lara-Ceniceros, T.; Jiménez-Regalado, E., Raşa, M. & Schubert, U.S. (2007). Magnetorheological Fluids Based on Ionic Liquids. *Advanced Materials,* Vol. 19, No. 13, (July 2007), pp.1740-1747, ISSN 1521-4095

Guerrero-Sánchez, C.; Ortiz-Alvarado, A. & Schubert, U.S. (2009). Temperature Effect on the Magneto-Rheological Behavior of Magnetite Particles Dispersed in an Ionic Liquid. *Journal of Physics: Conference Series,* Vol. 149, pp. 012052-012056, ISSN 1742-6588

Haerens, K.; Van Deuren, S.; Matthijs, E. & Van der Bruggen, B. (2010). Challenges for recycling ionic liquids by using pressure driven membrane processes. *Green Chemistry,* Vol. 12, No. 12, (December 2010), pp. 2182-2188, ISSN 1463-9262

Herpin, A. (1968). *Théorie du Magnétisme,* Institut National des Sciences et Techniques Nucléaires, PUF, Saclay-Paris, France

Jain, N; Zhang, X.; Hawkett, B.S. & Warr, G.G. (2011). Stable and Water-Tolerant Ionic Liquid Ferrofluids. *Applied Materials & Interfaces,* Vol. 3, No. 3, (March 2011), pp. 662-667, ISSN 1944-8244

Jeong, U; Teng, X.W.; Wang, Y.; Yang, H. & Xia, Y.N. (2007). Superparamagnetic Colloids: Controlled Synthesis and Niche Applications. *Advanced Materials,* Vol. 19, No. 1, (January 2007), pp. 33-60, ISSN 1521-4095

Jolly, M.R.; Bender, J.W. & Carlson, J.D. (1999). Properties and Applications of Commercial Magnetorheological Fluids. *Journal of Intelligent Material Systems and Structures,* Vol. 10, No. 1., (January 1999), pp. 5-13, ISSN 1530-8138

Keskin, S.; Kayrak-Talay, D.; Akman, U. & Ortaçsu, O. (2007). A Review of Ionic Liquids Towards Supercritical Fluid Applications. *Journal of Supercritical Fluids,* Vol. 43, No. 1, (November 2007), pp. 150-180, ISSN 0896-8446

Klingenberg, D.J. (2001). Magnetorheology: Applications and Challenges. *AlChhE JournalK,* Vol. 47, No. 2, (February 2001), pp. 246-249, ISSN 1547-5905

Kuzhir, P.; López-López, M.T. & Bossis, G. Magnetorheology of Fiber Suspensions. II. Theory. (2009). *Journal of Rheology,* Vol. 53, No. 1, (January 2009), pp. 127-151, ISSN 0148-6055

López-López, M.T.; de Vicente, J.; González-Caballero, F. & Durán, J.D.G. (2005a). Stability of Magnetizable Colloidal Suspensions by Addition of Oleic Acid and Silica Nanoparticles. *Colloids and Surfaces A: Physicochemical and Engineering Aspects,* Vol. 264, No.1-3, (August 2005), pp. 75-81, ISSN 0927-7757

López-López, M.T.; de Vicente, J.; Bossis, G.; González-Caballero, F. & Durán, J.D.G. (2005b).

Preparation of Stable Magnetorheological Fluids Based on Extremely Bimodal Iron-magnetite Suspensions. *Journal of Materials Research,* Vol. 20, No. 4, (April 2005), pp. 874-881, ISSN 0884-2914

López-López, M.T.; Kuzhir, P.; Lacis, S.; González-Caballero, F.; Durán, J.D.G. & Bossis, G. (2006). Magnetorheology for Suspensions of Solid Particles Dispersed in Ferrofluids. *Journal of Physics: Condensed Matter,* Vol. 18, No. 38, (September 2006), pp. S2803-S2813, ISSN 0953-8984

López-López, M.T.; Vertelov, G.; Bossis, G.; Kuzhir, P. & Durán, J.D.G. (2007). New Magnetorheological Fluids Based on Magnetic Fibers. *Journal of Materials Chemistry,* Vol. 17, No. 36, (September 2007), pp. 3839-3844, ISSN 0959-9428

López-López, M.T.; Gómez-Ramírez, A.; Durán, J.D.G. & González-Caballero, F. (2008). Preparation and Characterization of Iron-Based Magnetorheological Fluids Stabilized by Addition of Organoclay Particles. *Langmuir,* Vol. 24, No. 14, (June 2008), pp. 7076-7084, ISSN 0743-7463

López-López, M.T.; Kuzhir, P. & Bossis, G. (2009). Magnetorheology of Fiber Suspensions. I. Experimental. *Journal of Rheology,* Vol. 53, No. 1, (January 2009), pp. 115-126, ISSN 0148-6055

López-López, M.T.; Zubarev, A. & Bossis, G. (2010). Repulsive Force Between Two Attractive Dipoles, Mediated by Nanoparticles Inside a Ferrofluid. *Soft Matter,* Vol. 6, No. 18, (September 2010), pp. 4346-4349, ISSN 1744-683X

López-López, M.T.; Gómez-Ramírez, A.; Rodríguez-Arco, L.; Durán, J.D.G., Iskakova, L. & Zubarev, A. (2012). Colloids on the Frontier of Ferrofluids. Rheological Properties. *Langmuir,* in press, DOI: 10.1021/la204112w, ISSN 0743-7463

Odenbach, S. & Thurm S. (2002). *Magnetoviscous Effects in Ferrofluids,* Lecture Notes in Physics (Springer-Verlag), ISBN 3-540-43068-1, Berlin Germany

Oliveira, F.C.C.; Rossi, L.M.; Jardim, R.F. & Rubim, J.C. (2009). Magnetic Fluids Based on γ-Fe_2O_3 and $CoFe_2O_4$ Nanoparticles Dispersed in Ionic Liquids. *The Journal of Physical Chemistry C,* Vol. 113, No. 20, (May 2009), pp. 8566–8572, ISSN 1932-7447

Papell, S.S. (1965). US Patent Specification 3215572.

Park, S. & Kazlauskas, R.J. (2001). Improved Preparation and Use of Room-temperature Ionic Liquids in Lipase-catalyzed Enantio- and Regioselective Acylations. *The Journal of Organic Chemistry,* Vol. 66, No. 25, (December 2001), pp. 8395-8401, ISSN 0022-3263

Park, B.J.; Fang F.F. & Choi, H.J. (2010). Magnetorheology: Materials and Application. *Soft Matter,* Vol. 6, No. 21, (November 2010), pp. 5246-5253, ISSN 1744-683X

Pârvulescu, V.I. & Hardacre, C. Catalysis in Ionic Liquids. (2007). *Chemical Reviews,* Vol. 107, No. 6, (June 2007), pp. 2615–2665, ISSN 0009-2665

Popplewell, J. (1984). Technological Applications of Ferrofluids. *Physics in Technology,* Vol. 15, No. 3, (May 1984), pp. 150—156, ISSN 0305-4624

Ramos, J.; Klingenberg, D.J., Hidalgo-Álvarez, R. & de Vicente, J. (2011). Steady Shear Magnetorheology of Inverse Ferrofluids. *Journal of Rheology,* Vol. 55, No. 1, (January 2011), pp. 127-152, ISSN 0148-6055

Raj, K. & Moskowitz, R. (1990). Commercial Applications of Ferrofluids. *Journal of Magnetism and Magnetic Materials,* Vol. 85, No. 1-3, (April 1990), pp. 233-245, ISSN 0304-8853

Rodríguez-Arco, L.; López-López, M.T.; González-Caballero, F. & Durán, J.D.G. (2011a). Steric Repulsion as a Way to Achieve the Required Stability for the Preparation of Ionic Liquid-based Ferrofluids. *Journal of Colloid and Interface Science,* Vol. 357, No. 1, (May 2011), pp. 252-254, ISSN 0021-9797

Rodríguez-Arco, L.; López-López, M.T.; Durán, J.D.G.; Zubarev, A. & Chirikov, D. (2011b) Stability and Magnetorheological Behaviour of Magnetic Fluids Based on Ionic Liquids. *Journal of Physics: Condensed Matter,* Vol. 23, No. 45, (November 2011), pp. 455101-455116, ISSN 0953-8984

Rosensweig, R.E. (1985). *Ferrohydrodynamics,* Cambridge University Press, ISBN 0-486-67834-2, New York, USA

Scheeren, C.W.; Machado, G.; Teixeira, S.R.; Morais, J.; Domingos, J.B. & Dupont, J. (2006). Synthesis and Characterization of Pt(0) Nanoparticles in Imidazolium Ionic Liquids. *The Journal of Physical Chemistry B,* Vol. 116, No. 26, (July 2006), pp. 13011-13020, ISSN 1520-6106

Shima, P.D. & Philip, J. (2011). Tuning of Thermal Conductivity and Rheology of Nanofluids Using an External Stimulus. *The Journal of Physical Chemistry C,* Vol. 115, No. 41, (October 2011), pp. 20097-20104, ISSN 1932-7447

Torimoto, T.; Tsuda, T.; Okazaki, K. & Kuwabata, S. (2010). New Frontiers in Materials Science Opened by Ionic Liquids. *Advanced Materials,* Vol. 22, No. 11, (March 2010), pp. 1196-1221, ISSN 1521-4095

Tran, N. & Webster T.J. (2010). Magnetic Nanoparticles: Biomedical Applications and Challenges. *Journal of Materials Chemistry,* Vol. 20, No. 40, (October 2010), pp. 8760-8767, ISSN 0959-9428

Ueno, K., Inaba, A., Kondoh, M. & Watanabe, M. (2008). Colloidal Stability of Bare and Polymer-Grafted Silica Nanoparticles in Ionic Liquids. *Langmuir,* Vol. 24, No. 10, (May 2008), pp. 5253-5259, ISSN 0743-7463

Ueno, K., Imizumi, S., Hata, K. & Watanabe, M. (2009). Colloidal Interaction in Ionic Liquids: Effects of Ionic Structures and Surface Chemistry on Rheology of Silica Colloidal Dispersions. *Langmuir,* Vol. 25, No. 2, (January 2009), pp. 825-831, ISSN 0743-7463

Wang, Y. & Yang, H. (2009). Synthesis of Iron Oxide Nanorods and Nanocubes in an Imidazolium Ionic Liquid. *Chemical Engineering Journal,* Vol. 147, No. 1, (April 2009), pp. 71-78, ISSN 1385-8947

Wu, B.; Liu, W.; Zhang, Y. & Wang, H. (2009). Do We Understand the Recyclability of Ionic Liquids? *Chemistry - A European Journal,* Vol. 15, No. 8, (February 2009), pp. 1804-1810, ISSN 1521-3765

Zubarev, A. Yu.; Fleisher, J. & Odenbach, S. (2005). Towards a theory of dynamical properties of polydisperse magnetic fluids: effect of chain-like aggregates. *Physica A,* Vol. 358, No. 2-4, (December 2005), pp. 475–91, ISSN: 0378-4371

Section 4

Smart Transducer Applications

Chapter 18

Trade-off Analysis and Design of a Hydraulic Energy Scavenger

Enrico Zenerino, Joaquim Girardello Detoni, Diego Boero,
Andrea Tonoli and Marcello Chiaberge

Additional information is available at the end of the chapter

http://dx.doi.org/10.5772/50719

1. Introduction

In the last years there has been a growing interest in intelligent, autonomous devices for household applications. In the near future this technology will be part of our society; sensing and actuating will be integrated in the environment of our houses by means of energy scavengers and wireless microsystems. These systems will be capable of monitoring the environment, communicating with people and among each other, actuating and supplying themselves independently. This concept is now possible thanks to the low power consumption of electronic devices and accurate design of energy scavengers to harvest energy from the surrounding environment.

In principle, an autonomous device comprises three main subsystems: an energy scavenger, an energy storage unit and an operational stage. The energy scavenger is capable of harvesting small amounts of energy from the surroundings and converting it into electrical energy. This energy can be stored in a small unit like a small battery or capacitor, thus being available as a power supply. The operational stage can perform a variety of tasks depending on the application.

Inside its application range, this kind of systems presents several advantages with respect to devices that exploit external energy supplies. They can be simpler to employ and install, as no external connections are needed; they are environmentally friendly and might be economically advantageous in the long term. Furthermore, their autonomous nature permits the use in locations where the local energy grid is not present and allows them to be 'hidden' in the environment, being independent from interaction with humans.

The idea is to make autonomous and more energy efficient processes in some very specific areas, particularly in the management of household heating/cooling systems, and in the

environmental monitoring. The basic concept is to convert a fraction of the energy that would be normally dissipated by the process into electrical energy. These "secondary" energy sources are then used as primary sources in micro generators whose electrical current will be used to power the devices distributed along the process. The physical phenomena involved in this energy conversion can be essentially: the piezoelectric, photovoltaic and thermoelectric effects and phenomena related to fluid dynamics. Significant examples of energy harvesting are evident in particular in the following scientific fields: the construction of electric generators coupled to microturbines (Chunyan et al., 2010), (Bansal et al, 2009), (Yan et al., 2011), (Zainuddin, H et al., 2009); the Stirling thermodynamic cycle (Valdes, 2004), the Seebeck effect in thermoelectric generators (Lineykin et al., 2007), (Lu et al., 2010) and an Helmholtz resonator based generator (Kim et al. 2009).

Within this context the objective of this work is to present a trade-off analysis between different types of hydraulic machines, electric generators and energy storage units to reach a good compromise in the design of harvesting devices to be integrated in fluid distribution systems.

2. System description

The present section describes the system configuration of the energy harvester (or scavenger) used to supply a thermostatic motorized valve of a heating system for residential applications.

The Scavenger of Fig. 1 is composed by three main subsystems:

- hydraulic machine
- electric generator
- energy storage unit

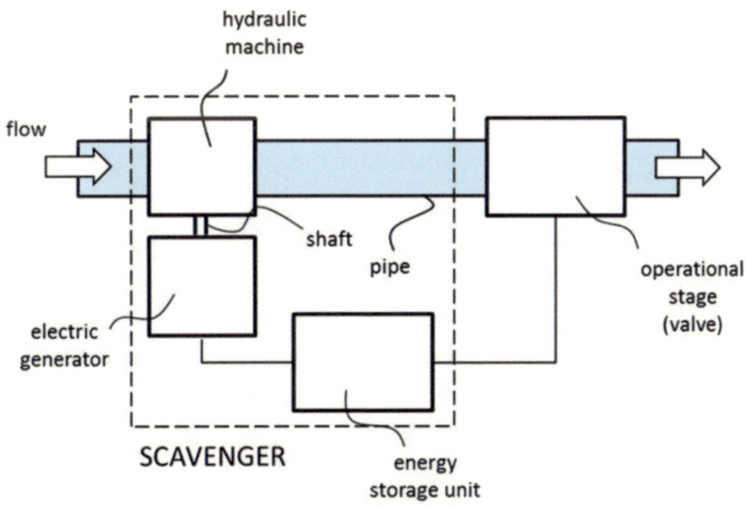

Figure 1. Hydraulic Energy Harvester system scheme

A cross flow turbine transforms the hydraulic power of the water flow into mechanical power used to drive a small electric generator. The device includes an energy storage unit to match the relatively constant power production profile of the generator unit with the more discontinuous one that characterizes the load. Additionally, since the generator and the valve are in series, the energy storage ensures the possibility to open the valve from the closed state (no flow). The energy is then stored during the peaks of production and then reuse it in a second time when prompted by the operational unit, the valve in our case. To this end the power management system is made to operate at its Maximum Power Point (MPPT) through closed loop control of recovered current from the scavenger unit.

2.1. Specifications and design choices

The collection of specifications starts from the hydraulic data available for the system from which we want to extract the energy. The typical flow rate of household heating system pipelines is between 1.3 and 4 l/min. The geometrical size of the device must be compliant with the available space at the interface between the heating element of conditioning systems and the pipeline. Other specifications are related to electrical power and voltage that must be generated. A campaign of experimental tests performed on motorized thermostatic valves shows that the average power is about 100mW with peaks that can reach 1 W for a short time during valve actuation. The minimum nominal voltage must be compatible to the voltage generated by a couple of AA-type batteries currently used to power the electronic thermostatic valves actuation units. Table 1 lists the above-mentioned specifications.

Parameter	Symbol	Value	Unit
water nominal flow rate	Q	2	l/min
radial size	r	<60	mm
axial size	l	<90	mm
min nominal voltage generated	V_n	>3	V
nominal power generated	P_n	100	mW

Table 1. System specification

In addition to specification some key choices are taken to proceed with the design, they are summarised in Tab. 2. One aspect which strongly influences the design of the system is the size of the inlet nozzle of the hydraulic machine. This choice is driven by the need to avoid chocking of the nozzle because of the large amount of dirt particles that characterize the fluid of heating systems. The design choice is to have a nozzle diameter d larger than 4 mm, as shown in the cross-section view of Fig. 17. Moreover a relatively large inlet nozzle diameter reduces the hydraulic losses due to the introduction of the scavenger into the heating system. The advantage is of reducing the need of increasing the size of the main pump that produces the hydraulic flow in the system. Another design choice about the realization of the turbine concerns the speed rotation of the runner. Here, a rated nominal speed of 1000 rpm has been chosen for the device. This rotation speed should be compatible

with the precision that can be reached with standard production process of some details of the generator unit as impeller and bushings. A limited angular speed also ensures an adequate degree of durability and strength. This speed value is precautionary as regards the possibility of creating vibrations, that may occur at higher speeds and lead to the failure of the rotating parts.

Parameter	Symbol	Value	Unit
inlet nozzle diameter	d	>4	mm
generator rated speed	n	1000	rpm

Table 2. Design choices

3. Trade off and design

This phase is carried on by splitting the system in its three subsystems. Following the transformations of energy that take place in the device the hydraulic machine is met first, then the electric machine and finally the energy storage unit, without neglecting their mutual interactions.

3.1. Hydraulic machine

We start from the conversion of the kinetic energy of the water into mechanical rotational energy of the hydraulic machine runner. Different typologies of machine are considered for the hydraulic machine: water turbines for mini and micro power plant and a gear pump. The water turbines for micro-hydropower generation can be classified into two main categories: impulse turbines and reaction turbines as described in (Inversin 1994) and listed in Fig. 2. In impulse turbines there is no expansion of the flow within the moving blades of the runner and, as such, the pressure remains constant while passing over the blades. In reaction turbines the stream expands as it flows over the blades, therefore producing a drop in pressure which gives a reaction and hence motion to the rotor.

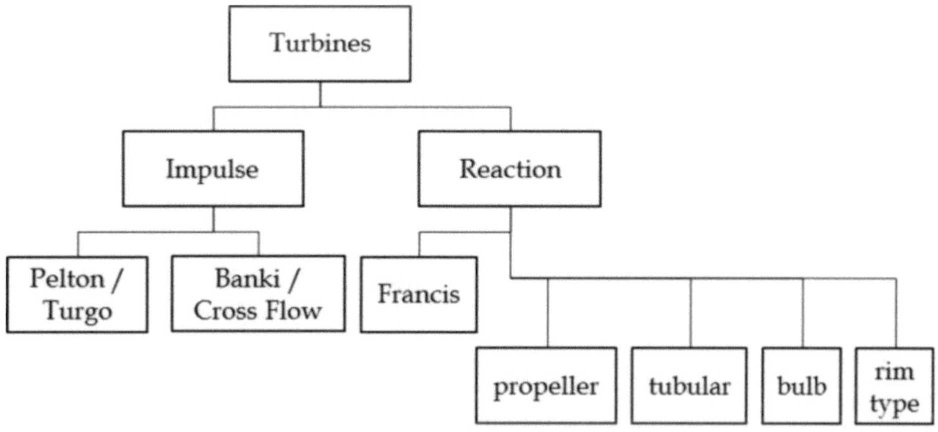

Figure 2. Classification of turbines for micro-hydropower generation

In general, impulse turbines are used for high head plants while reaction turbines for low head sites.

The turbines available on the market are usually very large compared to what needed for the present application, so a new turbine needs to be designed following the indications and design criteria of classical hydropower turbines of much larger dimensions.

For a given nominal output rated power and flow rate, equation (1) allows evaluating the pressure drop across the hydraulic machine:

$$P_n = Q \cdot \Delta p \tag{1}$$

Then, from the energy conservation principle of an ideal fluid inside a pipeline (i.e. the Bernoulli's principle) it is possible to determine the head of the water flow H.

$$\Delta p = \rho \cdot g \cdot H \tag{2}$$

From the head H, the absolute mean speed c of the water jet at the nozzle output can be found following the classical design indications of a Pelton turbine:

$$c = \varphi \cdot \sqrt{2 \cdot g \cdot H} \tag{3}$$

where φ is the nozzle flow factor, evaluated in the present case as equal to *0.97*. As reported in (Nechleba, 1957), the best efficiency is reached when the circumferential speed of the runner (also indicated as drag speed) is related to the water jet speed as follows:

$$u = 0.46 \cdot c \tag{4}$$

The circumferential speed u can then be related to the diameter A of the rotor and to the angular speed n of the runner:

$$A = \frac{60 \cdot u}{\pi \cdot n} \tag{5}$$

where u is expressed in meter per second and n in revolutions per minute.

A trade-off analysis has been performed between different kind of turbines by making the preliminary design of the runner. The case of hydraulic gear motor has been included in the trade off study along with the classical Francis, Pelton, and Banki turbines . The results of the preliminary design are presented in Tab. 3. All of them are compatible in size with the design specifications.

In addition to the preliminary design, other critical aspects related both to functionality and practical feasibility have been considered in the trade-off. In particular, the overall dimensions of the different layouts have been evaluated along with the possibility to integrate the rotor of the electric machine to that of the turbine, and with the need of avoiding chocking due to the debris transported by the water flow. Finally, the constructive complexity of the different parts has also been considered, in particular focusing on rotor

and stator. Each of these aspects have been ranked in a three step scale. The results of this trade-off are reported in Tab. 4.

Hydraulic machines	**Turbines**			
	Francis	**Pelton**	**Banki**	**Gear Pump**
pressure drop (Δp)	4200 Pa			
head (H)	0.43 m			
nozzle jet speed (c)	2.17 m/s	2.78 m/s	2.84 m/s	
circ. runner speed (u)	1,65 m/s	1.33 m/s	1.33 m/s	
Runner diameter (A)	31 mm	25 mm	24 mm	
Displacement (V)				2,5 cm^3/turn

Table 3. Preliminary design of hydraulic machines. Comparison of the main parameters of the different solutions.

Hydraulic machines	**Turbines**			**Volumetric**
runner	**Reaction**	**Action**		
	Axial	**Tangential**		
	Francis	**Pelton**	**Banki**	**Gear Pump**
Radial size	**	**	**	*
Axial size	**	**	***	**
Rotor complexity	*	*	***	***
Stator complexity	*	**	***	**
Electric machine interface	*	**	***	**
Choking risk	*	**	**	*

*** good, ** average, * bad

Table 4. Trade-off analysis between hydraulic machines.

3.1.1. Cross Flow or Banki turbine design

The Cross Flow or Banki turbine appears to be the best in almost all aspects examined. Its strengths are the simplicity of construction, the compact size, a good interfacing capability with the generator and a limited risk of choking.

Following the Banki water turbine theory reported in (Mockmore et al. 1959) the two main parts of the turbine, namely the nozzle and the runner, have been designed. The design drawings of the Banki turbine runner are shown in Fig. 3, and its characteristic parameters are given in Tab. 5.

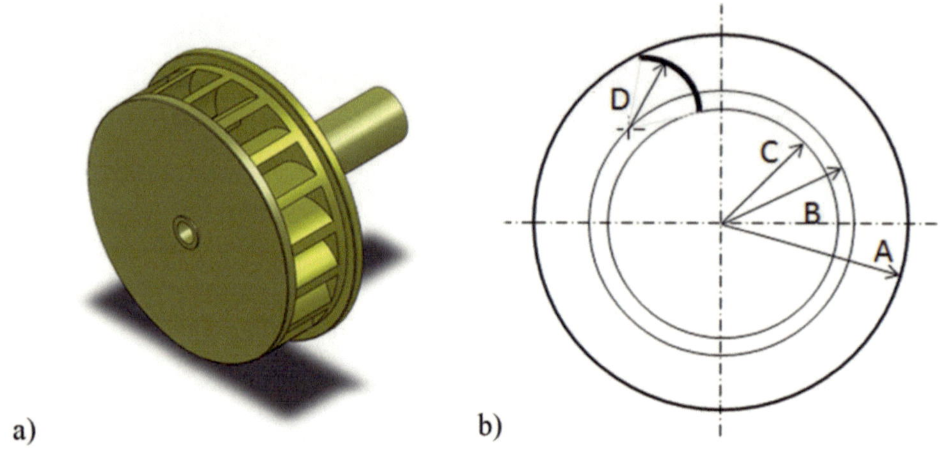

Figure 3. Banki turbine runner: a) 3D model. b) geometrical parameters.

parameter	symbol	value	unit
inlet nozzle diameter	*d*	4	mm
runner diameter	*A*	24	mm
center of blade curvature diameter	*B*	17.7	mm
blade root diameter	*C*	15.8	mm
blade curvature radius	*D*	7.8	mm
runner breadth	*L*	5	mm
number of blades	*N*	18	-

Table 5. Cross flow or Banki turbine construction parameters.

3.2. Electric generator

In order to convert the rotational mechanical energy from the turbine shaft into electrical energy which can be used to power the wireless spot and operate the valves of the system, a miniaturized electrical generator has been designed especially for this application. Two different configurations of generators have been investigated in order to obtain a clear perspective on the advantages and drawbacks of each one. The first configuration is based on a multiphase permanent magnet generator layout. The second is a single phase permanent magnet generator having claw pole structure. Fig. 4 shows the two different configurations describing the main components of the electrical machines, namely, rotating permanent magnet (1), generator's coil (2), and stator yoke (3).

Both cases consider permanent magnet excitation on the rotor. It is known that for reduced size applications such as the present one, it is better to use permanent magnet excitation instead of electrically excited magnetic systems. The electrical excitation is disadvantageous in these cases owing to unfavourable scaling of the currents (Arnold 2007).

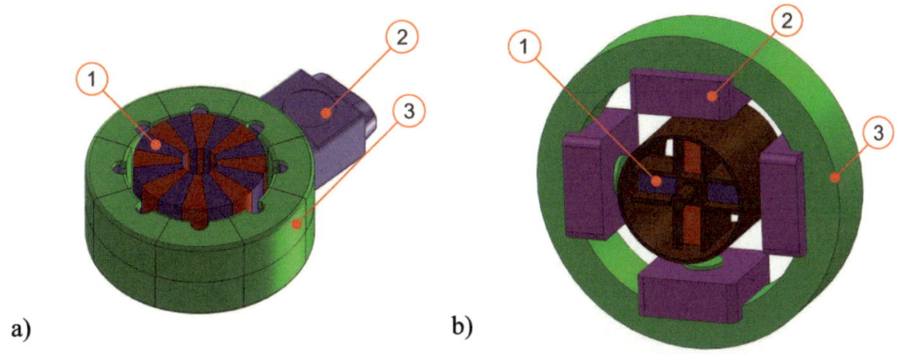

Figure 4. Configurations of electrical machines studied during the trade-off analysis. a) Two phase generator; b) Claw pole generator.

A trade-off analysis is performed using virtual prototyping tools. The difficulties related to the mechanical layouts were studied using CAD models while the electrical and magnetic properties were analyzed using analytical and finite element (FE) models.

	Configuration 1 Multiphase	Configuration 2 Claw pole
Rotor complexity	***	***
Stator complexity	*	**
Overall volume	*	**
Number of pole pairs	*	***
Winding complexity	**	***
Detent torque	*	**

*** good, ** average, * bad

Table 6. Comparison between multiphase and claw pole layouts.

From these models it is possible to obtain a relatively accurate perspective of the critical aspects related to the feasibility of each configuration of the generator. From the application point of view, the most important characteristics are compared in Tab. 6. Analyzing the table, it is easy to conclude that the claw pole configuration is more suitable for this application. Its layout makes it possible to obtain a simple and compact structure, and, since the output voltage must be rectified to supply the batteries, there is no advantage in having a multiphase winding, such as configuration 1. Furthermore, the single phase winding enables having a larger number of magnetic pole pairs, thus resulting in an increase of the frequency of the induced electromotive force (EMF), which is beneficial from the electronic point of view. Another aspect that cannot be neglected is the amplitude of the detent torque generated by the interaction between rotor's permanent magnets and stator's yoke (Lossec et al. 2010). A larger number of pole pairs tends to reduce the amplitude of the detent torque for the same rotor radius since the slot opening is reduced (Hendershot et al 1994). Moreover, the geometry of the teeth in the claw pole configuration can be adjusted in order

to further reduce the cogging torque. On the other hand, the multiphase configuration creates problems in this aspect due to the difficulty in realizing yoke and windings having such small dimensions.

3.2.1. Finite element modeling

The prediction of the generator's performance is developed by means of FE simulations. The simulations are performed using a stationary formulation without electric currents for one single pole pair of the electrical machine. Non-linear magnetic properties were considered in the iron parts of the structure. Fig. 5 shows the model used in the finite element modeling, evidencing the use of cyclic symmetry boundary conditions to improve the modeling quality with reduced computational cost. The analyses are conducted in order to calculate the flux linking the coil for different values of rotor angles. This information is then used to evaluate the induced voltage with respect to the rotor's spin speed. To this end, the problem is set to allow the rotor mesh to move with respect to the stator mesh, thus enabling the calculation of the magnetic quantities for different values of angular position between the two. Notice that the air surrounding the rotor and stator of the electrical machine is modeled, but is not shown in the illustration. Fig. 6 shows the results obtained from the FE model evidencing the path of flux lines inside the stator's yoke (Fig. 6b). Moreover, it can be noticed that the flux densities inside the iron are relatively low (Fig. 6a) resulting in very little or no saturation.

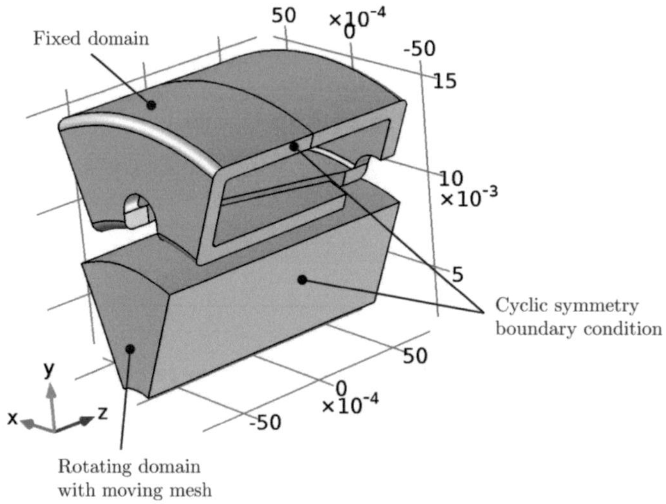

Figure 5. Settings of the FE model for the magnetic simulations.

The parameters characterizing the system described in the FE simulations are summarized in Tab. 7, and the flux linkage wave calculated with the FE model is illustrated in Fig. 7. Observing the graph it can be noticed that the flux linking the coil realizes one complete period every 45 mechanical degrees, evidencing the existence of eight magnetic pole pairs.

Furthermore, the flux linkage is a sinusoidal function of the rotor angle and, since the generator operates at constant rotating speed during most of its operative life, the analysis can be performed in terms of RMS quantities.

The velocity constant k_e can be obtained from the data plotted in Fig. 7 considering the flux λ linkage derivative with respect to the rotor angle θ as:

$$k_e = \frac{1}{\sqrt{2}}\frac{\partial \lambda}{\partial \theta} \tag{6}$$

The induced electromotive force e can then be obtained as:

$$e = k_e \omega \tag{7}$$

where ω is the generator rotational speed expressed in radians per second.The value of the velocity constant obtained for the generator is reported in Tab. 7.

Parameter	Symbol	Value	Unit
Stator's outer diameter	D_e	31.5	mm
Stator's inner diameter	D_i	20	mm
Rotor's outer diameter	d_e	19	mm
Active length	l_a	12.5	mm
Air gap length	t	0.5	mm
Number of pole pairs	p	8	-
Coil turns	N	1200	-
Permanent magnet induction	B_r	0.42	T
Velocity constant	k_e	9.41	V/krpm

Table 7. Claw pole generator nominal parameters.

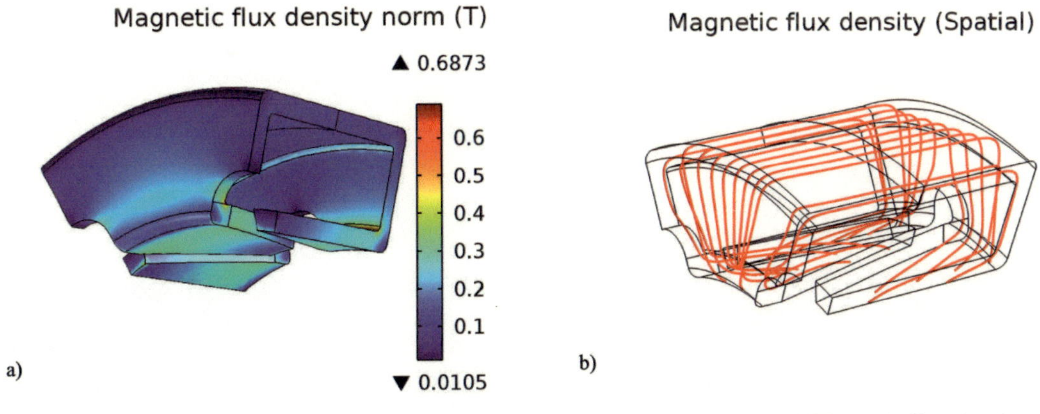

Figure 6. Results of the finite element simulations on the claw pole generator. a) Magnetic flux density in the stator in teslas (T); b) magnetic flux lines inside the iron.

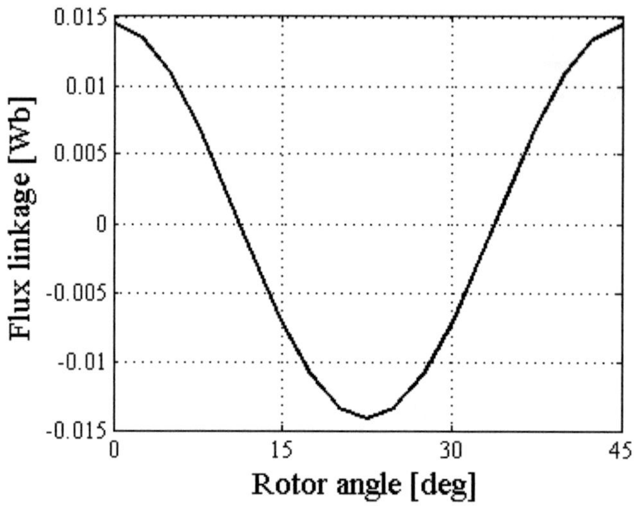

Figure 7. Flux linking the generator's windings for different values of rotor angle.

3.3. Energy storage unit

In order to properly manage the energy coming out from the generator, it is necessary to consider that the maximum amount of power is generated when the impedance of the load is nearly equal to the impedance of the generator. Due to this limitation it is important to design a control system shown in Fig.8 that can monitor, manage and store the energy in order to increase the efficiency of the whole system. For this reason the energy converter can be divided into several subsystem:

- rectifier
- DC/DC regulator
- storage system
- control system

The main idea was to develop each single subsystem in order to have more degree of freedom for each subsystem: active rectifiers to reduce energy losses and perform a Power Factor Corrector (PFC) regulation, independent DC/DC regulator with different Maximum Power Point Tracker (MPPT) according to the instantaneous situation, charge controller to ensure a good storage reduce as much as possible the memory effects in the battery, and a unique controller to manage properly all the interaction and the functions of these systems.

Due to the limited time for testing and to simplify the construction of the first prototype, an integrated solution shown in Fig. 9 is preferable to reduce cost and to obtain a suitable industrial solution. For these reason some solution from Linear Technologies (LT) turned out to be useful because they include into a single chip the DC/DC regulator, the charge controller and the main control system, therefore reducing cost and implementation time. Only few additional components have been selected and added to the integrated chip to obtain the final requested solution.

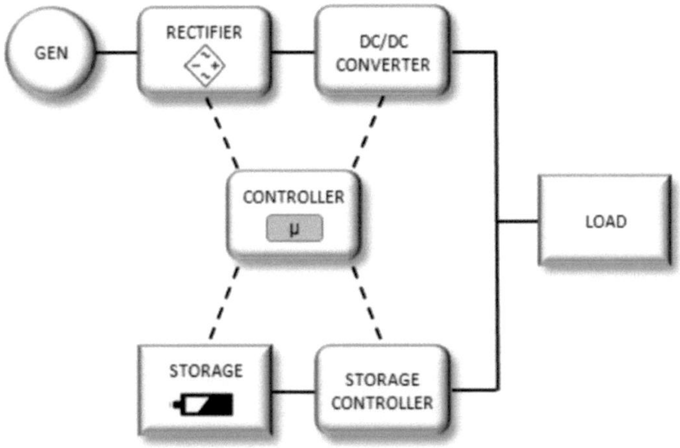

Figure 8. Architecture component interaction and interconnection.

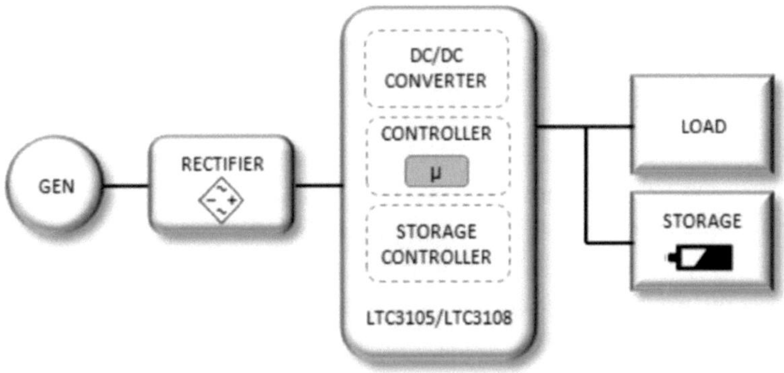

Figure 9. Block diagram of the integrated solution with LT components

The electric generator produces a sine wave with an electric frequency proportional to the mechanical velocity of the hydraulic turbine n expressed in rpm with the following relation:

$$f_e = \frac{n}{60} p \tag{8}$$

where p is the number of the pole pairs as reported in Tab. 7.

The easiest way to convert this sinusoidal voltage into continuous voltage is using a rectifier bridge; the ideal solution is based on active rectifier to obtain a voltage drop as lower as possible, but for this application, during the first tests, the results were not so different using a traditional passive rectifier instead. For this reason the first prototype was developed using a simple single phase rectifier composed by four Schottky diodes to reduce as much as possible the power losses. Using BAT54 diodes the voltage drop is about 250÷400 mV, equal to 500÷800 mV for each stage of conversion according to the current flow. A 10uF capacitor is enough to keep constant the voltage with a low output ripple (<1% of the maximum peak voltage).

The second stage is the regulator used to increase or reduce the input voltage to provide an output voltage around 3÷3.3 V to properly supply any kind of microprocessor or actuator. Using a buck-boost converter it is easy to satisfy this requirement but, due to the lower amount of energy generated by the turbine, it is necessary to implement also the MPPT algorithm. The Maximum Power Point Tracker is an algorithm normally used in photovoltaic cells; its role is to constantly check the input voltage and current to know exactly how much power is available and to limit the current absorption from the generator so as to keep always the condition of maximum production (see Fig. 10). To perform this operation it is necessary to use two different feedbacks: the first is used to check the output voltage to keep it constant with a low ripple; the second one checks the input current, limiting its absorption by varying the duty cycle of the DC/DC converter. In this way even if a very heavy load, alike a completely discharged battery, is connected, the converter can make the generator work in the most efficient condition. This algorithm needs to be included into the general control system alike an independent microprocessor or an integrated chip. The use of a microprocessor allows to modify these conversion algorithms without hardware changes, but only varying the internal control parameters.

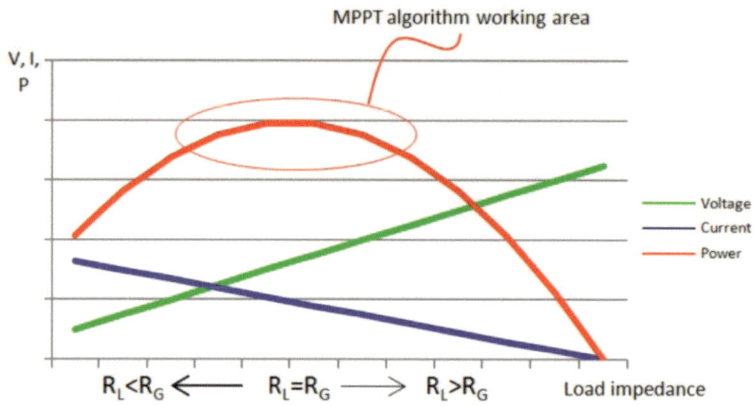

Figure 10. Current adjustment from MPPT to achieve maximum efficiency

In this system, the recovered energy can be used directly by the loads or can be partially stored for future use. The storage system can be divided in two main components based on different technologies:

- electrostatic storage (super capacitors): most suitable when the load requires small quantity of energy in short interval (less than 5÷15 minutes) and the generator can provide always the average energy required. The auto-discharge factor is higher than in the chemical solution but in the short period is a value than can be neglected.
- chemical storage (NiMH o LiIon batteries): most suitable if the generator has an intermittent production or the load requires medium/high quantity of energy with long pause interval also when the generator is switched off. The auto-discharge factor is low but the storage efficiency is lower than in the electrostatic storage, causing efficiency reduction during charge and discharge.

If the load is composed by a combination of continuous small absorptions with occasional high requests of energy, it can be useful to combine the above storage technologies in order to reduce energy losses.

The first prototype was not developed using an independent control system and energy converter, but using an integrated solution were all the three elements are included into one single chip. Two different boards were developed using two chips from Linear Technologies. The first solution used the LTC3108, a buck-boost converter without MPPT algorithm but capable of converting input voltages lower than 200mV. This solution was adopted due to the extremely low power coming out from the generator in the first prototype developed. Increasing the generator production, it was possible to move to the LTC3105 regulator, that is capable of converting input voltages higher than 500mV till to 5V, including an internal MPPT control in order to adapt the load absorption according to the generator production.

Figure 11. DC/DC converter: LTC3108 controller (1), output voltage selectors (2), input terminal, switching inductance and filters (3), regulated output 2.3V÷5V (4), 5.25V storage battery or capacitor (5)

In the test board two different topologies were tested. The first one used an output voltage set to 2.2÷2.3V connected to a 1F 2.3V super-capacitor (see Fig. 11). This solution was adopted to supply a very light load like a microcontroller (PIC16F886) that run a simple code to only switch on and off one led. The second solution used an output voltage of 4V to supply directly the load and to recharge the 3.6V 220mA battery; specifically, when the converter was on, the remaining current that was not used by the load was employed to recharge the battery, whereas when the input generator was off, the OUT pin was disconnected from the converter so that all the energy required was provided by the battery. A custom external controller is necessary to ensure battery protection from deep discharges (< 3V) during switch-off period to prevent irreversible damage to the cells.

4. Experimental validation

The system described in the previous sections was implemented and experimental tests were carried out to verify the correspondence between the design model and the real

system. The experimental tests were first addressed to investigate the performance of each subsystem and in the second instance of the whole system. For this purpose several test rigs have been built in order to characterize the single components and the interactions between them. Experimental tests verified the nominal design expectations and have been used also to perform sensitivity analysis on different parameters of some components.

4.1. Dry test rig

To characterize the electric generator, a "dry" test rig was developed as shown in Fig. 12. It is constituted of a DC electric motor (2), which is used to drive the generator's rotor, and the generator's stator (1). A DC power supply is used to energize the drive motor and to put the generator rotor into rotation at different rotational speeds. The values of induced EMF measured during the tests are compared to the FE model in Fig. 13.

Figure 12. Dry Test Rig realization

Different configurations of the generator have been implemented and investigated with several values of the air gap *t*: 0.5 mm for the nominal configuration as reported in Tab. 7, 1 mm for the preliminary supply of claw poles generators tested in the dry test rig, and 1.25 mm for claw poles generators compliant with the wet test rig. The experimental values represent the RMS value of the voltage measured between the two ends of the stator coil, whereas the FE model results were obtained by using Eq. 6 and Eq. 7 with the data corresponding to the above mentioned configuration. The obtained experimental results show good correlations with the numerical values resulting from the model simulation.

In Fig. 14 the generated power is plotted as function of the rotating speed of the generator for the different air gaps. It is clear that the rated generated power is significantly less than the 100 mW required. This is due to the fact that it was not possible to find a ferrite with a value of the permanent magnet induction equal to 0.42T but only 0.27T, with a consequent loss in performance. It must also be emphasized the drastic lack of performance in terms of

power generated by the increasing of the air gap from the nominal value of 0.5 mm up to the value of the wet test rig equal to 1.25 mm.

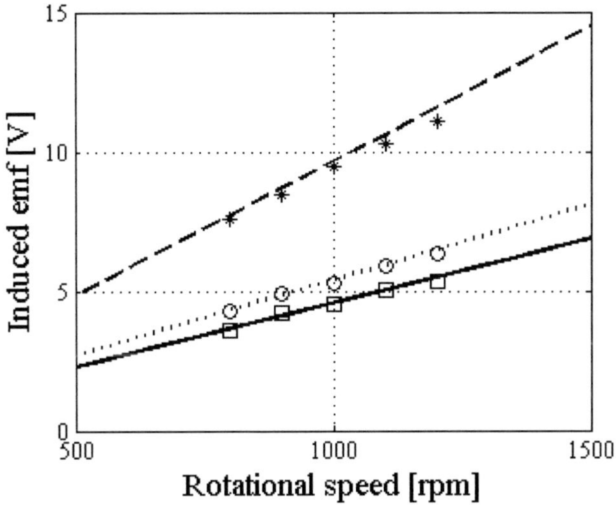

Figure 13. Induced emf for different values of the air gap : *t*=0.5mm (dashed line) vs exp. (star markers); *t*=1mm (dotted line) vs exp. (circular markers); *t*=1.25mm (solid line) vs exp. (square markers).

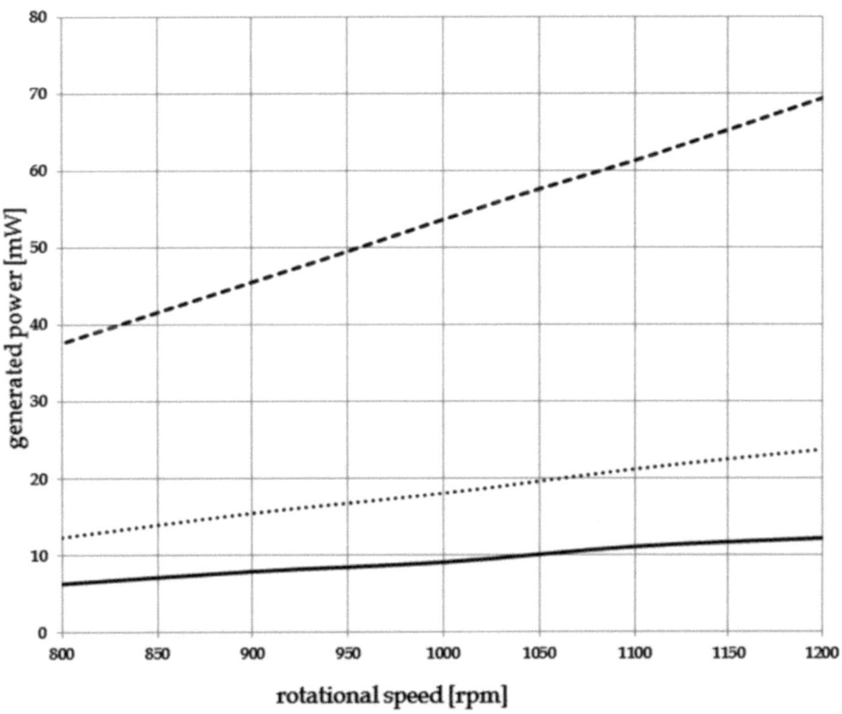

Figure 14. Air gap Sensitivity Analysis: generated power with external load equal to generator resistance. *t*=0.5 mm (dashed line), *t*=1 mm (dotted line), *t*=1.25mm (solid line).

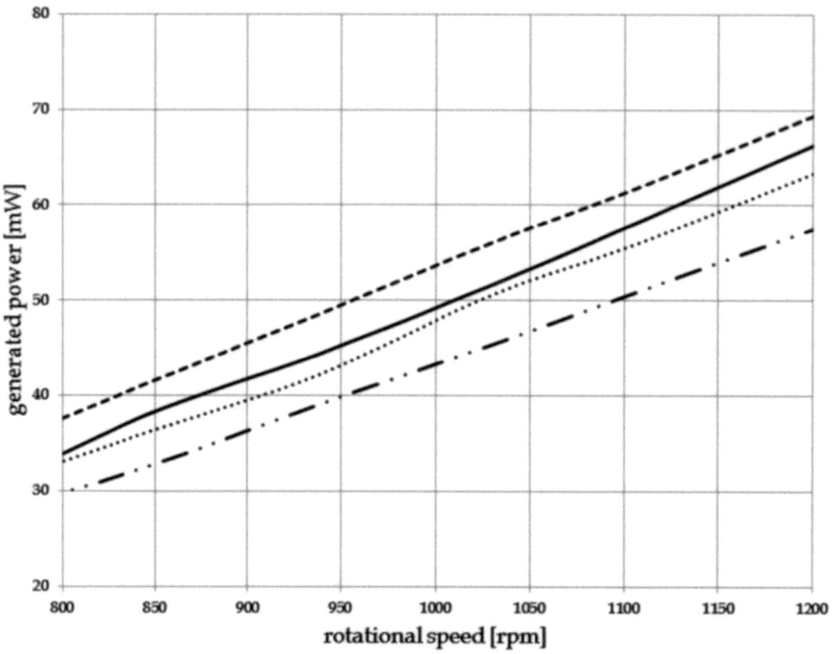

Figure 15. Number of coils Sensitivity Analysis: generated power with external load equal to generator resistance. Nominal *N*=1200 (dashed line), *N*=800 (dash dotted line), *N*=620 (dotted line) and *N*=490 (solid line).

A lower influence has been found with the variation of the number of coils of the generator stator *N* as shown in Fig. 15, leaving a certain degree of freedom from this point of view.

4.2. Wet test rig

A second test rig has been developed in order to test the electrical generator coupled with the hydraulic turbine, which makes it possible to characterize both the generator and the hydraulic machine as shown in Fig. 16. This wet test rig is composed by the scavenger (1), connected in series to a flow meter (2), which are both supplied by a domestic water pipeline, where the water flow is adjustable by a tap (3). The measuring system consists of a flow meter (2), a multimeter (4) and an oscilloscope (5) in order to correlate the generated power with the available flow rate. In the lower left box of Fig. 16 an enlarged view of the rapid prototyping realization of the scavenger (1) is also shown.

In Fig. 17 a cross-section of the scavenger is reported; the device incorporates a eighteen blades Banki turbine (1) (see Tab. 5 for other specifications) and a claw-poles voltage generator (2) (see Tab. 7 for nominal specifications). The design of the integration between the two parts required special attention, in particular to ensure the sealing between rotating and fixed parts and to prevent the direct contact between the main water flow and the electrical generator. The presence of the magnet in the generator rotor and ferromagnetic

residues in the water may lead to choking risk. To solve the problem of sealing an O-ring (4) has been introduced; to limit as much as possible the choking risk labyrinth seals have been used among the rotor housing and the runner.

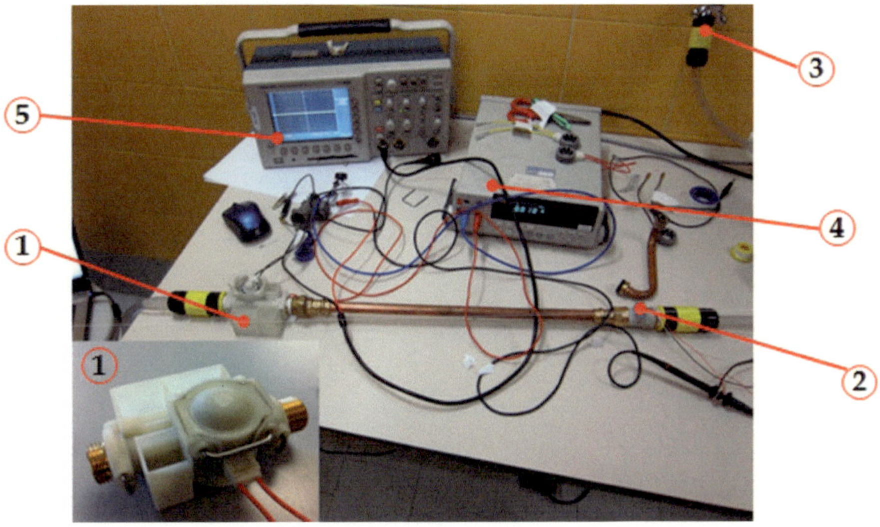

Figure 16. Wet Test Rig: scavenger (1), flow meter (2), domestic water supply and tap (3), multimeter (4) and oscilloscope (5).

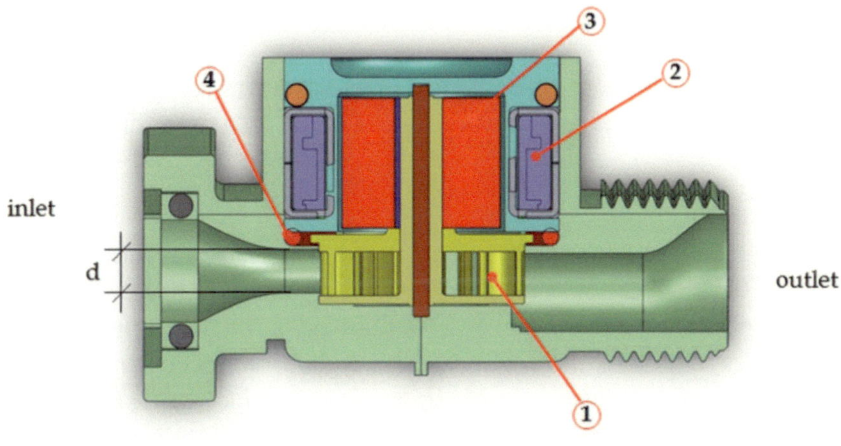

Figure 17. Cross-section view of the scavenger: Banki turbine (1), claw pole generator (2), permanent magnet rotor (3) and O-ring (4).

Owing to manufacturing problems related to the choice of the rapid prototyping process, the implementation of the first scavenger prototype presented an air gap equal to 1.25 mm which is greater than the nominal value. This fact has a significant impact on the system performances as shown in Fig. 18. Comparing the data of the power generated by the wet test rig with those of the dry test rig, it is noticed a further drop of performance of the generated power. It goes from 9.1 mW produced at 1000 rpm on the dry test rig to 5.2 mW

generated by the wet test rig. This decay can be explained with stick-slip phenomena present among rotating and non rotating parts of the turbine, and it will be addressed and fixed in future studies.

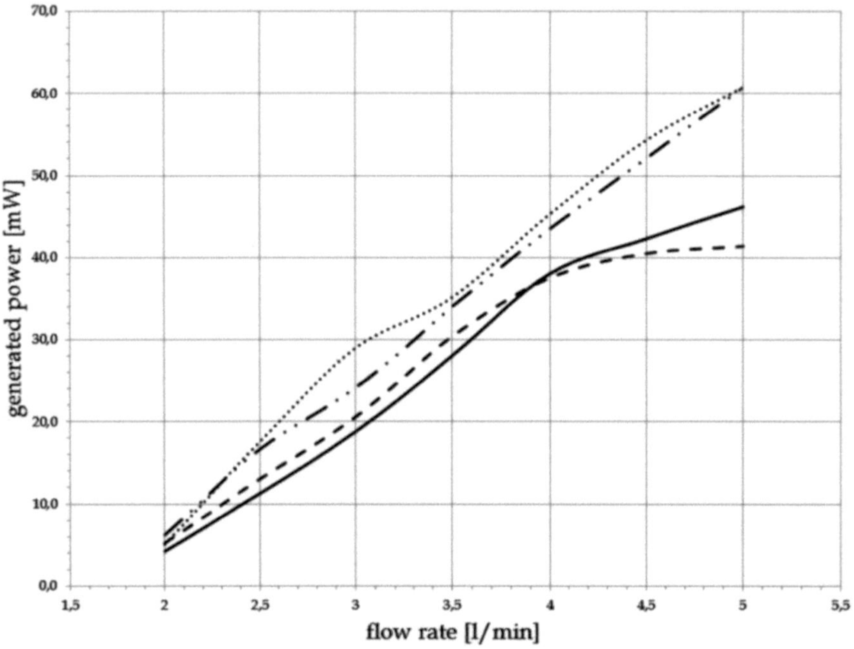

Figure 18. Wet Test Rig: generated power with external load equal to generator resistance. Nominal *N*=1200 (dashed line), *N*=800 (dash dotted line), *N*=620 (dotted line) and *N*=490 (solid line).

5. Conclusions

This study presents the trade-off analysis, the design, and the experimental validation of an Hydraulic Energy Scavenger applied to a motorized valve for domestic heating systems. The trade-off analysis conducted on the hydraulic and electric machines has identified the Banki turbine coupled to a claw poles generator as the solution to investigate and design. In this configuration the axis of rotation of the machine results to be perpendicular to the flow of the water thereby limiting problems of choking. The models underlying the design are validated from the electrical perspective by the dry test rig, and for the whole system by the wet test rig. The investigation was performed for different values of the air gap t, of the number of coils N, and of the resistive load of the device. The comparison between model and experiments show a good correlation, even though the power generated by the device resulted to be lower than the desired design value. This lower power production is essentially related to manufacturing issues, namely an higher value of the air gap and a lower value of the permanent magnet induction. However, the good correlation between the experimental and theoretical data makes it possible to predict the achievement of the desired performance in case the indicated designed parameters are respected.

Author details

Enrico Zenerino, Joaquim Girardello Detoni and Andrea Tonoli
Department of Mechanical and Aerospace Engineering, Politecnico di Torino, Italy
Mechatronics Lab, Politecnico di Torino, Italy

Marcello Chiaberge
Department of Electronics and Telecommunications, Politecnico di Torino, Italy
Mechatronics Lab, Politecnico di Torino, Italy

Diego Boero
Mechatronics Lab, Politecnico di Torino, Italy

6. References

Arnold, D.P.; (2007). Review of Microscale Magnetic Power Generation, *IEEE Transactions on Magnetics*, Vol. 43, No. 11, pp. (3940-3950).

Bansal, A.; Howey, D.A.; Holmes, A.S. (2009) - *CM-Scale Air Turbine and Generator for Energy Harvesting from Low-Speed Flows*, IEEE, ISBN 978-1-4244-4193-8

Chunyan, M.; Gengxin, L. (2010). *Research on a self powered wireless ultrasonic flow sensor system*, IEEE, ISBN 978-1-4244-6789-1

Hendershot, J. R.; Miller T. J. E. (1994). *Design of brushless permanent-magnet motors*, Magna Physics Publishing.

Inversin, A.(1994). *Micro-Hydropowe Soucebook*, A Pratical Guide to Design and Implementation in in Developing Cuntries, Third Printing

Kim, S.; Ji, C.; Galle, P.; Herrault, F.; Wu, X.; Lee, J.; Choi, C and Allen, M. G. (2010). An electromagnetic energy scavenger. *Journal of Micromechanics and Microengineering*, Vol.19, (August 2009), pp. (1-8), ISSN 0960-1317/09/094010

Lineykin, S.; Ben-Yaakov, S. (2007). Modeling and Analysis of Thermoelectric Modules, *IEEE Transactions On Industry Application*, Vol.43, No.2, pp. (505-512)

Lossec, M.; Multon, B and Ahmed, H.B. (2010). Micro-kinetic Generator: Modeling, Energy Conversion Optimization and Design Considerations Proceedings of the IEEE Mediterranean Electrotechnical Conference, pp. (1516-1521).

Lu, X.; Shuang-Hua, Y. (2010). *Thermal energy harvesting for WSNs*

Nechleba, M (1957). *Hydraulic Turbines. Their Design and Equipment.* ATIA ,Prague

Valdes, L.C. (2004). Competitive solar heat engines, *Renewable energy*, Vol. 29, pp. 1825-1842

Yan, T.C. ; Ibrahim, T. ; Nor, N. M. (2011). *Micro Hydro Generator Applied on Domestic Pipeline*, IEEE, ISBN 978-4577-0752-0

Zainuddin, H. ; Yahaya, M. S. ; Lazi, J. M. ; Basar, F. M. ; Ibrahim, Z. (2009). Design and Development of Pico-hydro Generator System for Energy Storage Using Consuming Water Distributed to Houses, *World Academy of Science, Engineering and Technology*, Vol.59, pp. (154-159)

Chapter 19

Magnetoelastic Energy Harvesting: Modeling and Experiments

Daniele Davino, Alessandro Giustiniani and Ciro Visone

Additional information is available at the end of the chapter

http://dx.doi.org/10.5772/50892

1. Introduction

Magnetoelastic materials belong to the wide category of smart materials because of their capability of coupling mechanical quantities (force, strain) to magnetic ones (field, induction) and viceversa. Recently, they have received a lot of interest for actuating and sensing purposes. Moreover, in the general framework of recovering some environmental energy, this kind of smart materials have been considered to recover the mechanical energy of vibrations [51].

Indeed, by employing the inverse magnetostriction or *Villari Effect* [26], it is possible to scavenge the vibration energy by means of the induced magnetization change in the material to generate electrical power. This opens the possibility to have a regenerative source of electrical power, especially useful in harsh environments. For this reason, this kind of technology can result of great interest in several application fields, such as health monitoring of civil infrastructure (bridges, buildings), automotive and biomedical tasks. Magnetostrictive alloys (Terfenol, Galfenol, Metglas) are actually the most known and employed magneto-elastic materials in this kind of applications. Nevertheless, they have interesting properties like high energy densities, high bandwidth, absence of depolarization phenomena that make them complementary to the piezoelectrics.

Energy harvesting techniques from vibrations have a promising future in civil engineering, where a strong need of structural monitoring of the health of ageing bridges and structures is occurring [17]. Indeed, for all civil infrastructure in general, it is possible to infer the structural health by measuring accelerations and resonant frequencies [21, 31]. Usually, the resonant frequencies are measured in suitable places along the structure by using the vibrations induced by wind or traffic [4, 27, 30]. Moreover, the sensors can detect other local environment parameters as well, namely temperature, wind speed, humidity, etc. Once the data are measured, among all the possible transmission techniques, the wireless method is

undoubtedly the more effective because, for example, it reduces costs with respect to periodic human intervention and it improves reliability with respect to wired solutions. The use of sensors together with wireless transmission results in the so called wireless sensors networks (WSN) [7]. Of course, in the same line of reasoning, the source of electrical power for the WSN should be self-contained too. The easiest choice could be then the use of batteries but, due to the limited life-span of them, an increment of the sensors maintenance whole costs should be considered, with the aggravation of workers safety concerns because the sensors can be located in inconvenient places along the bridge. Then, smart renewable energy methods should be used instead. Solar and thermal harvesting have been proposed [43] but these solutions can be both costly and bulky. Nevertheless, bridges as many other civil infrastructure vibrate because of the wind action and of traffic loadings or, when presents, for trains traffic loadings. The possibility to convert this ambient mechanical energy, otherwise wasted, into electrical energy is very attractive in those applications [36, 39]. So, together with measurements purposes, vibrations can be harvested to feed the sensors.

It is worth that this type of conversion could be performed by means of linear electromagnetic generators too [20, 47]. In those devices, a proof mass oscillates with the structure, making a permanent magnet move linearly in a coil. This solution is undoubtedly well assessed and reliable. But, usually, this type of device has a narrow bandwidth that can be broaden at the cost of a very sophisticate mechanical construction. Another conversion technique, as already introduced, can make use of smart materials instead. These devices can be in principle less bulky and with higher reliability, because of their simpler mechanical design and construction.

Another field of application for energy harvesting from vibration is the automotive one. In fact, on one hand, the society is looking forward for vehicles more and more efficient. On the other hand, moving vehicle are site of vibrations of two different kinds. The first is due to the *internal* combustion engine operations and so, whenever the engine is on, a source of vibration is present. These vibrations are usually damped by means of engine rubber supports or even more complicated ways, while they could be used to recover some more energy. Obviously, the vibrations level is higher in big vehicles like trucks or tractors. A second source of vibrations can be referred as *external* and it is intrinsically related to the vehicle movement and interaction with the roads irregularities. It is apparent that both of them can be exploited to scavenge electrical power, instead of wasting the energy in rubber damping or shock absorbing. Also in this case, the harvested power can be used to recharge the vehicle battery or for feeding monitoring sensor nodes. A typical example is a tyre pressure sensor, [46]. In this case, the knowledge of the tyre health and pressure status allows to reduce the fuel consumption and to limit accidents. Another example of application on vehicles is the mechanical energy harvesting on the suspension system. Noting that the electrical energy transfer on the load (*e.g.* the battery) is strictly related to the mechanical source damping, this kind of solution can be conceived also for obtaining shock absorption, [54]. This additional feature is very attractive for enhancing the comfort of the passengers, particularly for workers (*e.g.* on trucks or tractors).

The harvesting from vibrations can be also considered in flying vehicles. In this case the vibrations are induced on the wings by the movement in the air flow. In [2], a sheet of

piezoelectric material on the wings of an unmanned plane is applied making a harvester in a sort of unimorph cantilever arrangement.

Another potential huge field of application for vibrations harvesting is the biomedical one [32]. In this case the source of vibrations is the human gait (walking and running) and it could be used to power devices aimed to monitor human health conditions, [18], out of personal multimedia readers, smartphone, etc. In this framework of human activities induced vibrations, even the movement of the heart muscle has been conceived as a possible energy source to feed internal biomedical devices as pacemakers, [23].

Aim of this chapter is to discuss the main achievements and the open challenges in the field of vibrations energy harvesters based on magnetostrictive materials. This is a very attractive field involving the modeling of active materials that, with their complex behavior, are the link between the mechanical and the electrical *worlds* and so represent the path by which a *smart* electromechanical conversion can take place.

Different modeling strategies will be considered, ranging from the basic linear one [22, 45] allowing to understand the device working principles to the more accurate nonlinear approaches [10, 50], outlining their impact on the practical design of the harvester. To this aim, some detail on the experimental setup for material characterization, [1, 11], the modeling of the mechanical source and its coupling to the active material will be presented.

Finally, many open problems will be also reviewed, such as the power conversion stage requirements, the main arrangements (bulk or cantilevers) of magnetostrictive harvesters in connection to the different fields of applications.

The chapter is structured as follows:

- Section 2 focuses the modeling of a magnetostrictive energy harvester discussing the basic components of the device, with particular emphasis on the active material. To this aim both linear and nonlinear modeling approaches are considered, highlighting their impact on the harvester performances prediction.
- Section 3 presents some experimental results necessary to the characterization of the material parameters and to evidence the basic phenomena involved.
- Section 4 is focused on some open problems related to the magnetostrictive harvesting, on new materials and on modeling challenges.

2. Magnetoelastic energy harvester modeling

This section aims to introduce different basic approaches that can be considered when a magneto-elastic energy harvester is modeled. Particular emphasis will be devoted to the material modeling, starting with the classic description under the general hypothesis of linearity (in analogy with the typical modeling of piezoelectric materials), as well as with more sophisticated approaches, taking into account the material nonlinearities and rate-independent hysteresis. Then, the whole device is modeled from the magneto-mechanical point of view. In the scientific literature, usually, a *semicoupled* modeling approach is adopted: the mechanical stimulus is *ideal* and only its effect on the magnetic

characteristic is considered. On the other hand, a *fully coupled* approach can be considered and, in this case, the influence of the material response on the mechanical side can be taken into account too, if the vibration source is not ideal.

This analysis will be done employing an analogy with a circuital description by means of the two-port circuits formalism. As interesting application, the capabilities of the harvesting process to damp the vibrations of the mechanical source will be discussed.

All the analysis will be performed under the following fundamental assumptions:

(a) All the field are coaxial and directed along the magneto-elastic material axis,

(b) the length of the structure along the field axis is much larger than the other dimensions,

(c) the vibrations frequencies are much lower than the mechanical resonance of the structure, i.e. sound propagation can be neglected,

(d) the electric load is a lumped element, i.e. electromagnetic field propagation is neglected.

The first one is necessary to have, together with an isotropic material behavior, a scalar description of the constitutive relationships. The second one allows, for instance, to have the *long solenoid* hypothesis for the coil and to treat the mechanical stress as uniform along the material. The third and the fourth assumptions allow to neglect any propagation effect into the device.

2.1. Material modeling - constitutive relationships

In the hypothesis of monodimensional operating regime (all the fields and the mechanical input along the same direction), the magneto-elastic characteristics can be written as:

$$\begin{cases} S = S(H,T) \\ B = B(H,T) \end{cases} \tag{1}$$

where S is the strain, H is the applied magnetic field, T is the compressive stress and B is the magnetic induction. Typical behaviors of those characteristics are shown in Fig. 1 and the following general properties can be inferred:

- $S(H)$ and $B(H)$ are non-linear with non-local memory, i.e. hysteresis,
- $S(H)$ and $B(H)$ show saturation when $|H| \to \infty$,
- S and B are even and odd function of H, respectively,
- $S(H)$: there is an optimum T_0 that makes ΔS largest at $H_{\max}$,
- $S(H)$: if $T > T_0$ then the $S(H)$ cycles drop down,
- $S(H)$: the $S(H)$ cycles are self-similar with respect to the stress,
- $B(H)$: if the compressive T increases then the $B(H)$ cycles drop down (the material becomes magnetically harder under increasing constant stress).

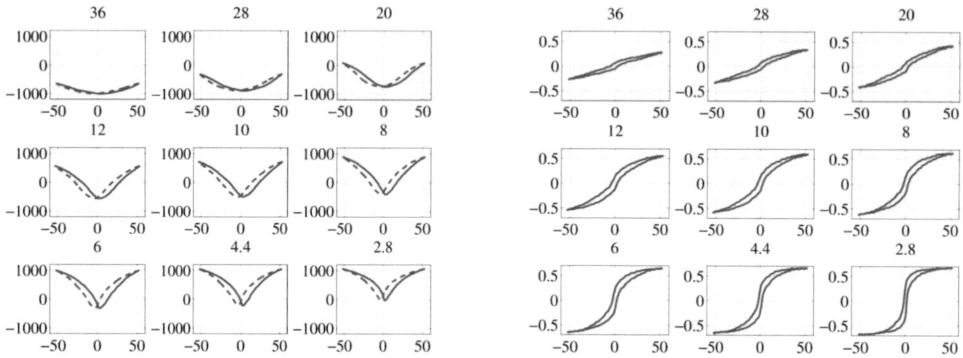

(a) Magnetostrictive cycles (S in ppm vs H in kA/m).

(b) Magnetic cycles (B in Tesla vs H in kA/m).

Figure 1. Typical magnetostrictive characteristics (Terfenol-D) at different constant stresses (the titles are $|T|$ [MPa]).

The eqs. (1) can be obtained by suitable derivatives of the Gibbs free energy expression:

$$G(T,H) = \frac{T^2}{2E} + \frac{\mu_0}{2}H^2 + \Psi(T,H) \rightarrow \begin{cases} S = \left.\dfrac{\partial G}{\partial T}\right|_H \\ B = \left.\dfrac{\partial G}{\partial H}\right|_T \end{cases} \tag{2}$$

where the first and second terms are pure linear elastic and magnetic energies, respectively, and the third one is the magneto-elastic energy. E is the Young modulus and μ_0 is the vacuum magnetic permeability. It is worth noting that $|_{H,T}$ means that the derivatives are made at constant H and T, respectively.

The simplest way to model the magnetoelastic materials behaviour is the linear one. In that case the Gibbs free energy expression reads:

$$G(T,H) = \frac{T^2}{2E} + \frac{\mu_0}{2}H^2 + \frac{\mu_0}{2}\chi H^2 + dTH \rightarrow \begin{cases} S = dH + \dfrac{T}{E} \\ B = \mu H + dT \end{cases} \tag{3}$$

where $\mu = \mu_0(1+\chi) = \mu_0\mu_r$ and d is the, so called, piezo-magnetic coefficient. It is worth noting that in this case it is assumed that any transformation taking place in the material is *lossless*. This assumption, from the thermodynamic viewpoint, leads to the following constraint [48]:

$$\left.\frac{\partial S}{\partial H}\right|_T = \left.\frac{\partial B}{\partial T}\right|_H. \tag{4}$$

It can be noted that the eqs. (3) are the magnetic counterpart of the piezoelectric ones. Obviously, in a B-H plane (or in a S-H) at a fixed stresses T (or at a fixed magnetic fields H), the linear modeling predicts parallel lines. This behavior is far from the actual material response because magnetoelastic materials show saturation phenomena and also a magnetic hardening, as stated above. So, this approach can fail in predictions when large variations of

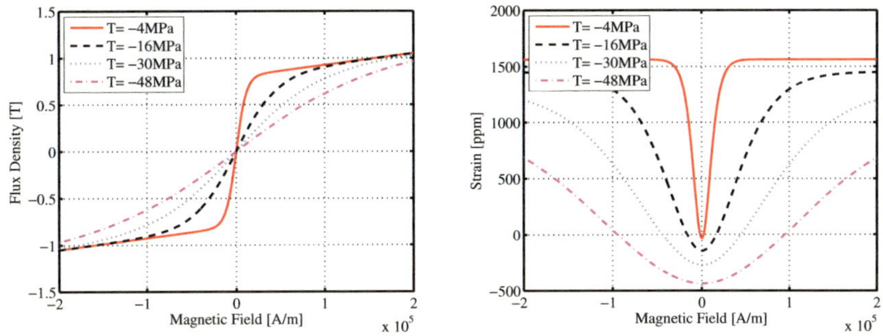

Figure 2. Nonlinear magnetostrictive characteristics obtained from eqs. (6). The values $M_s = 0.8$ T, $\gamma = -347$ T, $E = 110$ GPa match the typical behavior of Terfenol.

the inputs are considered but can be fruitfully used if the inputs have small variations around a magneto-mechanical working point.

A more advanced model, with the only further hypothesis to neglect the hysteresis, is represented by the following magnetic characteristic:

$$G(T,H) = \frac{T^2}{2E} + \frac{\mu_0}{2}H^2 + f(T)\chi\left(\frac{H}{f(T)}\right) \rightarrow \begin{cases} S = \left(\frac{\partial G}{\partial T}\right)_H = \frac{T}{E} - f'(T)\left[z\chi'(z) - \chi(z)\right] \\ B = \left(\frac{\partial G}{\partial H}\right)_T = \mu_0 H + \chi'(z) \end{cases} \tag{5}$$

where $z = H/f(T)$ and $f(\cdot)$ and $\chi(\cdot)$ are suitable one-variable functions and can be chosen in order to model the physical behavior of the material, as saturation effect and stress dependence [19]. Indeed, if $\chi'(z) = M_s \tanh(z)$ [3] and $f(T) = \frac{T}{\gamma}$ then the magnetic characteristic can be well approximated. In this case, $\chi(z) = M_s \ln(\cosh z)$ and eqs. (5) become:

$$\begin{cases} S = \frac{T}{E} - \frac{M_s}{\gamma}\left[z\tanh(z) - \ln(\cosh z)\right] \\ B = \mu_0 H + M_s \tanh\left(\gamma\frac{H}{T}\right) \end{cases} \tag{6}$$

where M_s is the magnetic polarization saturation and γ is a parameter modifying the approach to saturation which should be properly identified, as reported in [10]. As can be noted, it shows saturation by means of hyperbolic tangent and the magnetic hardening by means of the z dependence for compressive stresses.

Finally, considering that also hysteresis is shown by magnetoelastic materials another generalization can be made. A phenomenological approach for including hysteresis is to introduce a Prandtl-Ishilinskii operator [24]:

$$\pi[x] = \int_0^{+\infty} \xi(r)\mathcal{P}_r[x]dr, \tag{7}$$

where $\mathcal{P}_r$ is a Play operator of threshold r and $\xi(r)$ is a weight function that can be identified from experimental data. It is constructed by a linear superposition of Play operators, but it

cannot model saturation phenomena. In order to circumvent this limitation and to model the magneto-mechanical coupling taking place in the magnetoelastic material, the above operator can be generalized. In particular, it can be written:

$$B = G\left(\pi[H], T\right), \tag{8}$$

with

$$G(x, T) = \mu_0 x + M_s \tanh\left(\gamma \frac{x}{T}\right). \tag{9}$$

The function G is the same considered in the memoryless modeling approach and allows to take into account saturation.

However, it is worth to note that when the hysteresis is introduced the method adopted for the memoryless case basing on the thermodynamic constraint (eq. (4)) cannot be considered because the process is not lossless. Moreover, this approach models only one between magnetostrictive and magnetic characteristics. This can be still valid if the vibration source can be considered ideal (*semi-coupled* approach). On the other hand, the problem of a fully coupled approach with hysteresis is still open.

2.2. Device modeling

A magnetoelastic harvester can be arranged in different configurations. They can be summarized in two main categories:

- *Direct force* harvesters also known as force-driven.
- *Inertial* harvesters also known as velocity-driven.

All the harvesters where the force source is in mechanical direct contact to the active material, in the so called longitudinal mode, belong to the first group. A representative device of such a harvester is shown in Fig. 3 (a). In this case, usually, the material is fabricated in the form of rods, disks or cymbals.

Instead, the second group exploits the inertial forces of a proof mass oscillations induced by a vibrating bond. Typical example is a cantilever, where a thin sheet of the active material is bonded on a elastic structure and bounded on one side, while on the other there is a mass free to vibrate, as sketched in Fig. 3 (b). The thin sheet of active material undergoes heavy stress variations when the mass vibrate. Conceptually, the energy conversion follows the same path of the force driven harvester. But, this configuration has a resonant-like mechanical behavior with a relatively small bandwidth and a resonant frequency related to the mass (the larger is the mass, the lower is the frequency) [44]. The first group has, instead, a larger bandwidth from the vibration frequencies point of view but a lower peak specific power. In the following we will refer to a generic harvester with particular reference to the force-driven.

Despite of the arrangements, the main elements of a magnetostrictive harvester are:

- the active material,
- the magnetic circuit,
- the coil.

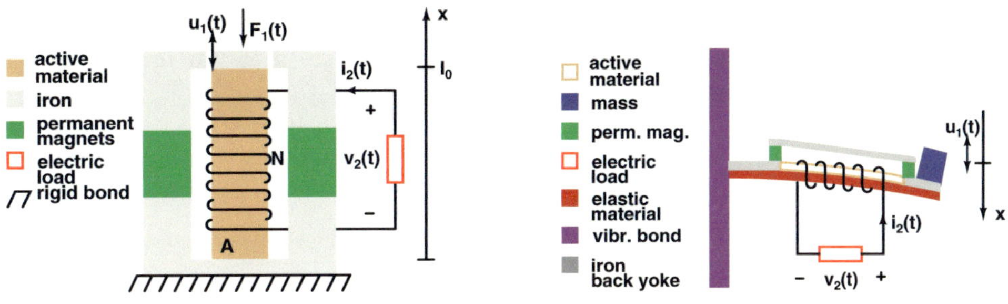

(a) Force-driven. (b) Velocity-driven.

Figure 3. Magnetostrictive energy harvester concept devices.

Let us consider the sketch of a magnetostrictive harvester under an external ideal time-variable force $F_1(t)$, as sketched in Fig. 3 (a). The active material of such a force-driven harvesters has a rod shape. In order to achieve an easier modeling and design of the device, the rod should be a cylinder with the length l_0 far larger than the section diameter. Indeed, in this case, two hypothesis can be made: the stress and the magnetic field lines are uniform along the rod leading to, eventually to 1-D problems along the radial direction [15]. The section area A should be chosen such that the available force variation gives the best stress variation for harvesting (see section 3.3). The magnetic circuit is needed to provide the magnetic bias to the active material and suitable permanent magnets can be used then. But, it is worth noting that the magnets cannot be coaxial to the active material because, usually, they are sintered and cannot stand the time-variable mechanical stresses. The N-turns coil is wound around the active material where N is chosen in order to achieve the best compromise between available voltage and current over the electric load. Finally, it should be noted that the modern electronic allows the construction of suitable energy conversion circuits that helps to match the electric impedance seen by the harvester itself to the best value, improving conversion efficiency then [8, 35].

Let us recall the eqs. (3):

$$\begin{cases} S = dH + \dfrac{T}{E} \\ B = \mu H + dT \end{cases} \tag{10}$$

Starting from these relationships, a general model of the energy harvester of Fig. 3 can be obtained. The procedure is based on the analogy of the whole system with a two-port: the first port with *mechanical* variables and the second with *electrical* variables. Then, a two-port circuit model of the magneto-elastic device can be obtained by exploiting the following analogy: the input force F_1 corresponds to a *primary two-port voltage* ($F_1 \Leftrightarrow v_1$); while the rod tip velocity $u_1 = dx/dt$ corresponds to a *primary current* ($u_1 = l_0(dS/dt) \Leftrightarrow i_1$). Instead, the current i_2 and voltage v_2 on the actual magnetostrictive rod coil represent the current and voltage on the second port.

It is worth noting that such an approach exploits all the power of the lumped circuit theory, making very easy to treat different cases. Moreover, the two-port equivalent model can be solved by means of standard circuit simulators [10].

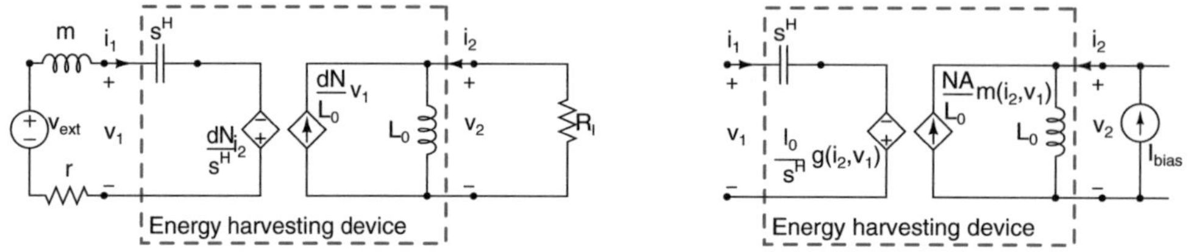

(a) Linear modeling and external elements.

(b) Non-linear modeling.

Figure 4. Two-port equivalent circuits. (a) the part inside the dashed box implements the eqs. (12). (b) The part inside the dashed box implements the eqs. (15).

In order to do so, the local quantities of eqs. (10) have to be related to the measurables one, i.e. F_1, u_1, v_2 and i_2, by using the hypothesis already made: $H = Ni_2/l_0$ (long solenoid), $\Phi_2 = NAB$, $x = l_0 S$, and $F_1 = AT$ (an applied uniform in the space force on the material cross section A is assumed), where Φ_2 is the magnetic flux and x is deformation.

So the eqs (10) can be recasted as follows:

$$\begin{cases} x & = \dfrac{l_0}{EA}F_1 + dNi_2 \\ \dfrac{\Phi_2}{NA} & = \dfrac{d}{A}F_1 + \mu\dfrac{Ni_2}{l_0}. \end{cases} \rightarrow \begin{cases} x & = s^H F_1 + dNi_2 \\ \Phi_2 & = dNF_1 + L_0 i_2. \end{cases} \tag{11}$$

where $L_0 = \mu\frac{N^2 A}{l_0}$ and $s^H = \frac{l_0}{EA}$ have been defined, the latter being the mechanical compliance at constant magnetic field H.

By exploiting the analogies $F_1 \Leftrightarrow v_1$ e $dx/dt \Leftrightarrow i_1$, the following relationships can be derived then:

$$\begin{cases} \dfrac{1}{s^H}\displaystyle\int_0^t i_1 d\tau & = v_1 + \dfrac{dN}{s^H}i_2 \\ \dfrac{1}{L_0}\displaystyle\int_0^t v_2 d\tau & = \dfrac{dN}{L_0}v_1 + i_2. \end{cases} \tag{12}$$

These equations are implemented in the equivalent circuit of Fig. 4 (a). The first one can be interpreted as a voltage balance at the primary port: the primary voltage v_1 is equal to the voltage on a capacitor of value s^H minus a voltage generator controlled by the secondary current i_2. Furthermore, the second of eqs. (12) can be interpreted as a current balance at the secondary port: the secondary current i_2 is equal to the current in an inductor of value L_0 minus a current generator controlled by the primary voltage v_1. For example, if the harvester undergoes a vibrating force generator with a certain mass m and it is connected to a resistor R_l then, as represented in Fig. 4 (a), this can be simply solved by connecting the series of a voltage generator and a inductor[1] to the first port and, of course, the resistor to the second port.

The proposed methodology can be also employed to derive a two-port representation in the memoryless non-linear case, (5). By The eqs. (6) can be recasted as follows:

[1] The equation of a rigid mass is $F = md^2x/dt^2 \leftrightarrow v = mdi/dt$ that is the inductor characteristic. A linear elastic effect is equivalent to a capacitor: $F = kx \leftrightarrow v = k\int idt$ (with the capacitance equal to $1/k$) and, finally, a viscous friction is represented by a resistor: $F = rdx/dt \leftrightarrow v = ri$.

$$\begin{cases} x = \dfrac{l_0}{SA} F_1 + g(i_2, F_1)\, l \\ \Phi_2 = \mu_0 \dfrac{N^2 A}{l} i_2 + m(i_2, F_1)\, NA. \end{cases} \tag{13}$$

where

$$\begin{cases} g(i_2, F_1) = -\dfrac{M_s}{\gamma} \left[(\dfrac{\gamma NA}{l} \dfrac{i_2}{F_1}) \tanh(\dfrac{\gamma NA}{l} \dfrac{i_2}{F_1}) - \ln(\cosh(\dfrac{\gamma NA}{l} \dfrac{i_2}{F_1})) \right] \\ m(i_2, F_1) = M_s \tanh(\dfrac{\gamma NA}{l} \dfrac{i_2}{F_1}). \end{cases} \tag{14}$$

Assuming the same previous analogies $i_1 = dx/dt$ and $v_2 = d\Phi_2/dt$, the following system describing the nonlinear dynamic two-port can be derived:

$$\begin{cases} \dfrac{1}{s^H} \displaystyle\int_0^t i_1 d\tau = v_1 + \dfrac{l_0}{s^H} g(i_2, v_1) \\ \dfrac{1}{L_0} \displaystyle\int_0^t v_2 d\tau = i_2 + \dfrac{NA}{L_0} m(i_2, v_1), \end{cases} \tag{15}$$

and the two-port is formally unchanged, apart from the controlled generators and a DC current generator, as shown in Fig. 4 (b). The generator implements the effect of a magnetic bias over the magnetostrictive characteristics. It does not produce any dissipated power into the electric load and, therefore, does not contribute to the energy balance of the two-port.

Finally, the two-port modeling approach makes easier to consider different types of vibration sources, either force or velocity driven [8]. Indeed, the simplest model of *non ideal* mechanical source can be defined as:

$$\begin{cases} l_0\, dS/dt = u_1(t) - \dfrac{A}{\eta} T(t) \quad \text{for } \eta > 0; \\ AT(t) = F_1(t) - \eta l_0 \dot{S} \quad \text{for } \eta < +\infty. \end{cases} \tag{16}$$

In the former case, u_1 is an impressed velocity (velocity-driven source), while in the latter F_1 is an impressed force (force-driven source). These two elements can be added to the two-port circuit as the classical current and voltage *non ideal* generators, respectively. The first equation leads to the ideal velocity-driven source when the mechanical resistance $\eta \to \infty$, while $\eta = 0$ describes a ideal stress-driven mechanical source in the latter.

2.3. Numerical results

In this section the different aforementioned approaches for describing the material behavior are compared, highlighting the main features and limitations. To this aim, it is useful considering a semi-coupled problem where the mechanical stress is imposed. In this case, in fact, only the effect of the mechanical source on the harvesting performances can be considered and evaluated. In the following examples, a cylindrical active material sample with length $l_0 = 0.1$ m and area $A = \pi$ cm^2 wounded by a coil with $N = 100$ turns is considered. In Fig. 5, the harvested power on a resistor $R = 10\Omega$ is reported, for a mechanical sinusoidal stress $T = T_0 + T_m \cos(2\pi f t)$ with $T_0 = -27.4$ MPa and $T_m = 13.9$ MPa, at different frequencies.

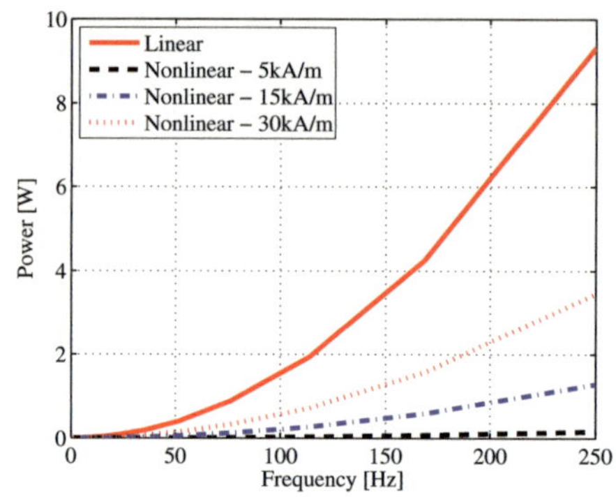

Figure 5. Harvested Power at different magnetic biases. It has been considered $\mu_r = 12$, $E = 30$ GPa, $d = 2 \cdot 10^{-8}$ m/A in eqs. (12), while $M_s = 0.8$ T, $\gamma = -347$ T, $E = 110$ GPa in eqs. (6).

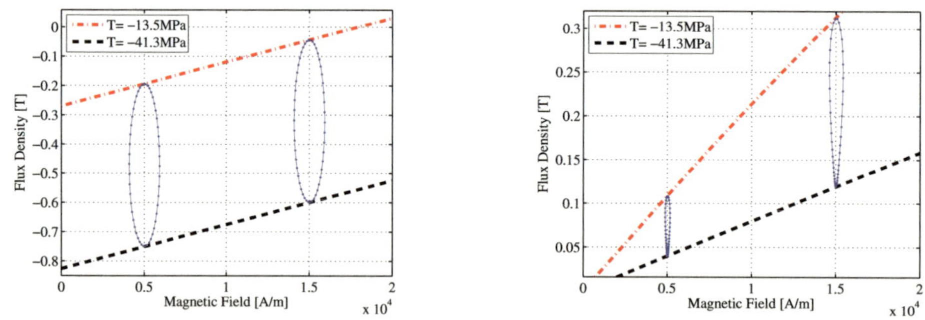

(a) Linear Modeling.

(b) Nonlinear Memoryless Modeling.

Figure 6. Harvesting periodic trajectories - varying external magnetic bias.

A first important difference between the linear and the nonlinear memoryless approach can be noted. In fact, in the linear case the harvested power is independent on the magnetic bias, while in the nonlinear case this strongly influences the harvester performances. This phenomenon can be easily understood referring to Fig. 6 where the periodic working trajectories are reported in the B-H plane, for a fixed frequency $f = 168.5$ Hz. It is worth to note that the area of this loops is directly proportional to the harvested energy, i.e. the averaged harvested power [9]. In the linear case (Fig. 6 (a)), the described loops are equal independently on the magnetic bias. Instead, in the non-linear case (Fig. 6 (b)), the loops area changes on this parameter. In the same figures the static characteristics of the two modeling approaches are reported. As it can be seen, these loops are within the static characteristics: in the linear case the distance is equal (parallel lines), in the nonlinear one this relative distance depends on the magnetic bias level. To support this interpretation, another test has been performed comparing again the linear and the nonlinear memoryless models, but varying the mechanical prestress level, keeping the magnetic bias constant, see Fig. 7. In this case $T_0 = -20.45$ MPa or $T_0 = -34.35$ MPa have been considered, while the stress amplitude $T_m = 6.95$ MPa is fixed. Again the harvested energies in the linear case are the same, while in the nonlinear case are strongly influenced by the prestress level. To explain this other

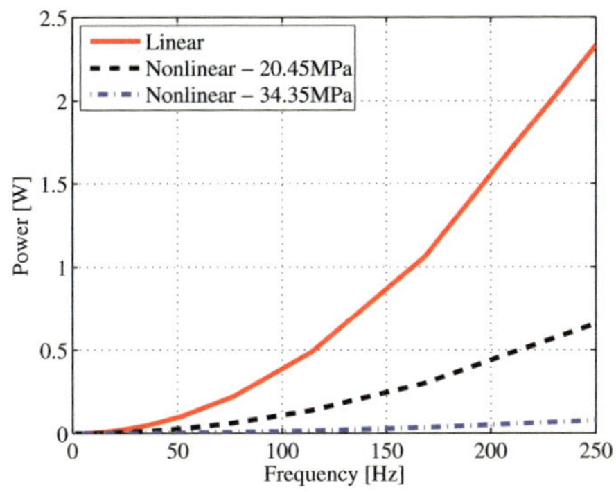

Figure 7. Harvested Power at different mechanical prestresses. The models parameters are the same of Fig. 5.

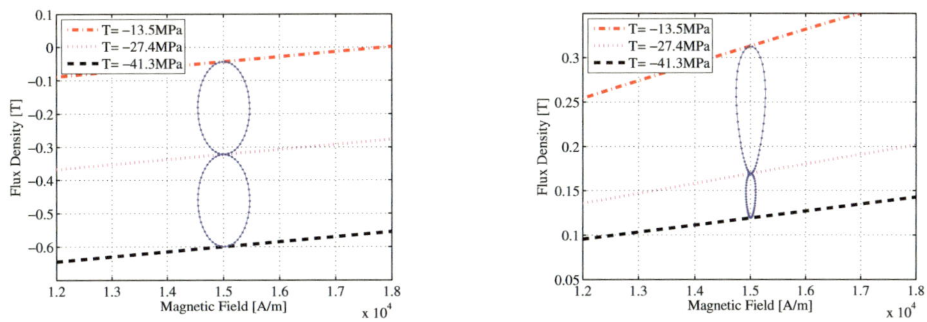

(a) Linear Modeling.

(b) Nonlinear Memoryless Modeling.

Figure 8. Harvesting periodic trajectories - varying mechanical prestress. The models parameters are the same of Fig. 5.

behavior, let us consider the Figs. 8 with loops in the B-H plane. In Fig. 8 (a), the linear static characteristics are also reported and as it can be noted the relative distance is still constant while in the nonlinear case it is not. So the described loops by the linear model are equal (same power level) independently on the prestress level. So summarizing, the linear model cannot take into account the magnetic bias and the mechanical prestress in the description of the harvester performances.

In Fig. 9, a comparison between the memoryless non-linear and hysteretic models is reported. In Fig. 9 (a), the static characteristics of the two models for two different mechanical stresses are drawn. For a mechanical input with a prestress $T_0 = -27.4$ MPa and amplitude $T_m = 13.9$ MPa, the harvested powers on a resistor $R = 10\Omega$ are reported in Fig. 9 (b). In this case, as additional phenomenon related to the hysteresis, it can be noted that the harvested power levels changes with the initial state (in this case, the demagnetized initial state and the saturation). This behavior can be also analyzed on the B-H plane, as reported in Figs. 10 (a) and (b). While in the memoryless nonlinear case the periodic loops are within the static characteristics, when the hysteresis is considered the periodic trajectories are within the

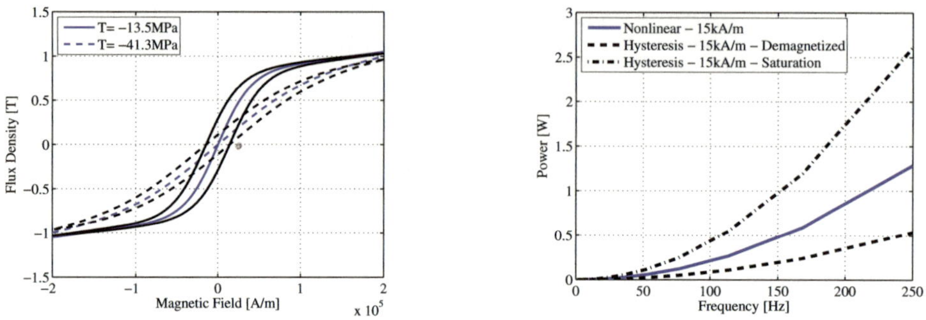

(a) Static Characteristics. (b) Average Power.

Figure 9. Comparison between Nonlinear Memoryless and Hysteretic modeling.

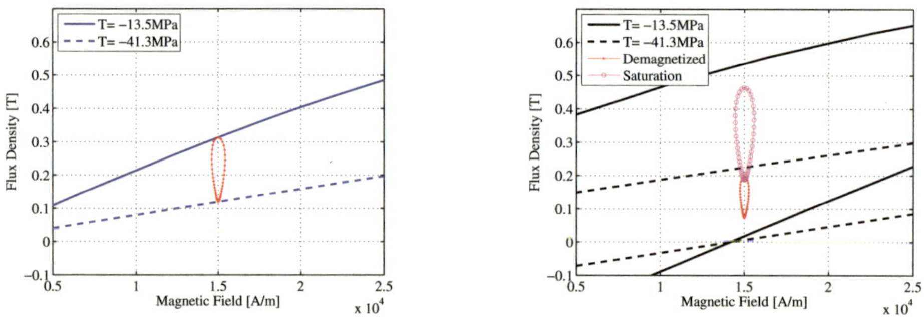

(a) Non-linear Memoryless model. (b) Hysteretic model.

Figure 10. Harvesting periodic trajectories: comparison between non-linear and hysteretic models, starting from different initial internal states.

hysteresis loops but their position and area are strongly influenced by the initial state of the hysteresis model.

In order to analyze the effect of the harvesting process on the mechanical side, a fully coupled problem is also addressed. In this case, the first port of the equivalent two-port model is closed with a lumped mass $m = 0.5$ kg and a viscous friction $r = 500$ Ns/m. In Fig. 11, the free response of the model with an initial velocity $v = 0.1$ m/s is shown, for the linear and the nonlinear memoryless models. As it can be noted, the external magnetic bias strongly influences also the strain time evolution (see Fig. 11 (a)) as done on the electrical side (see Fig. 11 (b)). The linear modeling approach cannot take into account this dependence, however it can be considered for a preliminary analysis of the magneto-elastic harvesting device.

3. Experimental characterization and results

The good design of a magnetoelastic energy harvester needs the characterization of the raw active materials, by knowing their magnetomechanical characteristics. In order to do so, a rather complex experimental setup is needed because both magnetic and mechanical

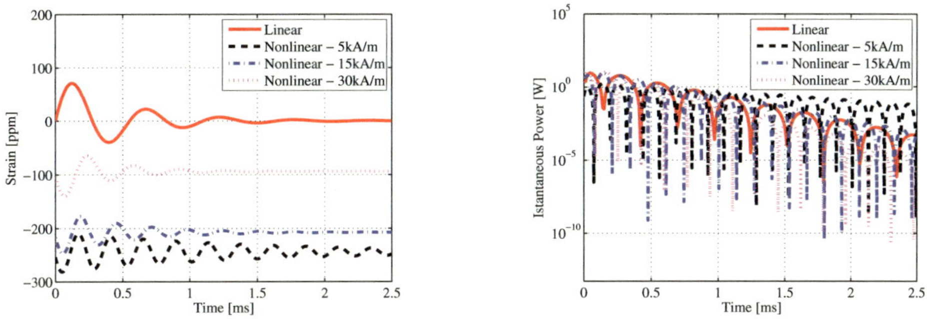

(a) Strain. (b) Instantaneous power.

Figure 11. Impulse response - Comparison

variables must be measured and/or enforced at the same time. Morever, the constitutive relationships of magnetostrictive materials, like Terfenol-D or Galfenol, show nonlinear and rate-independent memory effect (hysteresis) and the mechanical (S, T) and magnetic variables (H, B) are cross-coupled [19]. The cross-coupling terms are usually named piezo-magnetic coefficients and, in the energy harvesting framework, a deep knowledge of them is crucial because they are related to the material ability to convert energy [19]. For example, the accurate knowledge or modeling of their behavior, with respect to the magnetic field and prestress biases, can be useful to choose the working point (H_0, T_0) that guarantees the best performance in a energy harvesting device [10] or in other applications like a sensor. Now, in the case of characteristics without hysteresis, the piezo-magnetic parameters have to satisfy the thermodynamic constraint of eq. (4).

Several papers have tackled the experimental problem of measuring the piezo-magnetic coefficients of magnetostrictive material, as well other parameters as mechanical compliance, Young modulus, etc. This study is mandatory to explore the performance of new magnetostrictive compounds [5, 6, 37]. In a early paper, the problem of hysteresis of those coefficients started to be considered [34]. Also the variability of the piezo-magnetic parameter with the applied prestress has been considered [40], leading to the conclusion that magnetic and mechanical bias must be chosen with attention in order to get the best performance [49]. This result have been confirmed by a later paper on non-linear modeling [10].

The magnetostriction and the magnetic characteristics have to be measured in different stress and magnetic field conditions in order to explore the previous conjecture.

In order to measure the magnetostrictive characteristics, a combination of different standard and custom instruments have to be used, as sketched in Fig. 12. The mechanical compressive load is applied through a computer-controlled test-machine while the magnetic field is imposed through an electromagnet and measured by using a digital gaussmeter. The material magnetic induction is obtained after a numerical time-integration of a voltage, measured over a pick-up coil with 500 turns. The demagnetizing effect must be taken into account. The strain is directly measured with a strain gauges bridge, configured to have temperature self-compensation, while the applied stress is measured by a load cell.

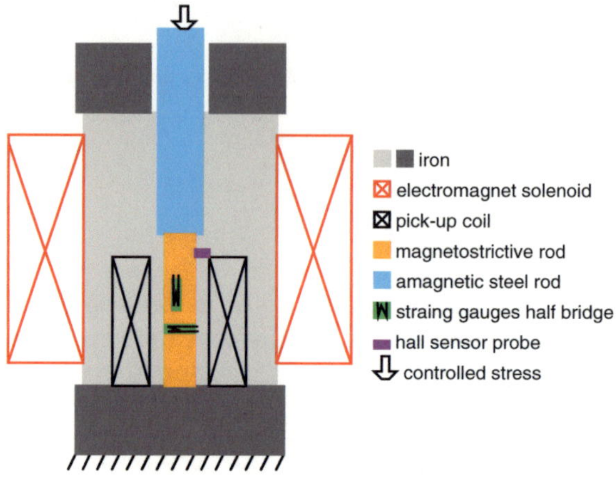

Figure 12. Sketch of a possible setup to measure the magnetostrictive characteristics.

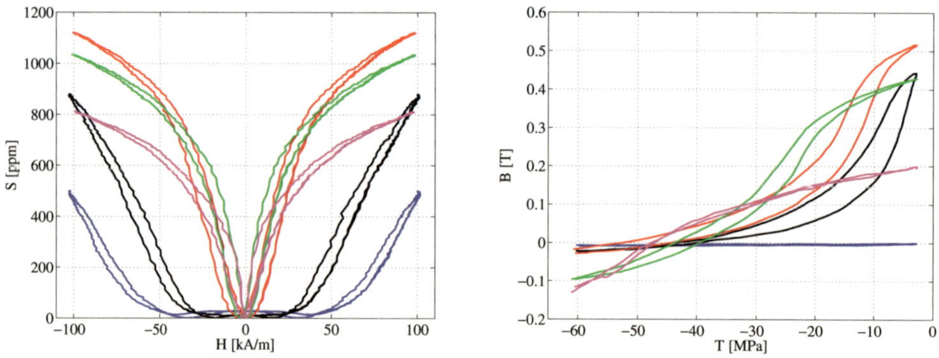

(a) Strain vs magnetic field.

(b) Magnetic induction vs stress.

Figure 13. Magnetostrictive cycles. (a) legend: 4 MPa–fuchsia, 6 MPa–green, 12 MPa–red, 40 MPa–black, 60 MPa–blue. (b) legend: 0 kA/m–blue, 10 kA/m–black, 20 kA/m–red, 40 kA/m–green, 100 kA/m–fuchsia.

3.1. Magnetostrictive characteristics

The magnetostrictive characteristics $S(H)$ under variable magnetic field and constant compressive stress are shown in Fig. 13 (a). The magnetostrictive characteristics does not include the elastic effect, i.e. the mechanical strain induced by the constant stress. It is worth noting that the dependence from the applied constant stress is somewhat complex. Indeed, the cycles change in shape and maximum amplitude with the stress.

Finally, the Fig. 13 (b) shows the magnetic induction induced by a cycled stress. These measurements are very important to infer the harvesting potentialities of the material. Indeed, the larger is the ΔB at a certain ΔT the better would be the energy conversion. The measurements show wider cycles, i.e. more hysteresis, at low-intermediate magnetic field. Moreover, the maximum ΔB is reached at intermediate field too.

3.2. Piezo-magnetic coefficients

The definition of the piezo-magnetic parameters considered here are the following [12, 34]:

$$\begin{cases} d_{33} = \left.\dfrac{\partial S}{\partial H}\right|_T \\ d^*_{33} = \left.\dfrac{\partial B}{\partial T}\right|_H \end{cases} \tag{17}$$

As stated in eq. (4), they are equal if the hypothesis of lossless material is considered while, in general, they are not. These parameters are usually considered constant in datasheets and in linear models of the material [34]. Such an approximation holds if the material is employed in actuators where the mechanical prestress is higher than the external stress applied to the device. On the other hand, the approximation cannot be applied if general dynamic conditions are considered, as the case of energy harvesting applications.

Indeed, the piezo-magnetic parameters are nonlinear function of the considered (H, T) couple. Moreover, they show hysteresis. Nevertheless, the piezo-magnetic parameters can be obtained by a numerical derivative of the experimental hysteresis loops. The two branches of the loops are considered as *up* and *down* curves in the following.

In order to compare the two parameters, they have been sampled in a same matrix of (H, T) points. The corresponding surfaces are shown in Figs. 14,15. It is apparent that the orders of magnitude and the shapes are the same, but the parameters are not exactly equal for each (H, T) couple. The maximum for each set of curves (up and down) moves at higher H field for higher T and this confirms nonlinear models results in literature (see [8], Fig. 3 or [48]). Nevertheless, for all of them, the maximum values are achieved at relatively low stress and magnetic field, that are the ranges $T \in (-10, 0)$ MPa and $H \in (10, 20)$ kA/m. Those values are useful to choose the best magnetic bias and mechanical prestress.

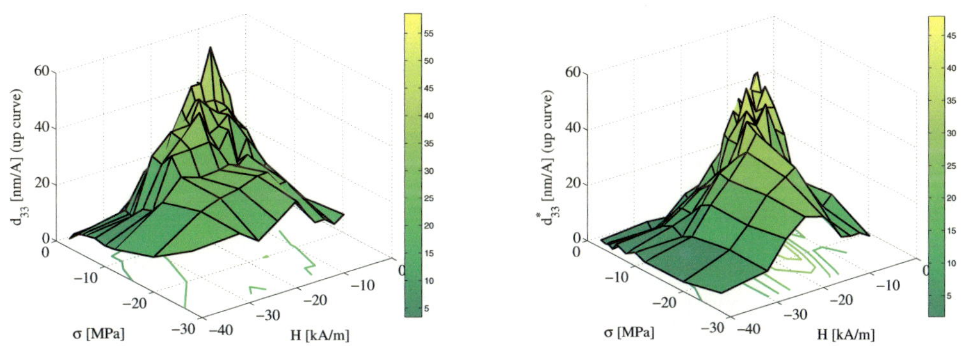

(a) d_{33}. (b) d^*_{33}.

Figure 14. The piezo-magnetic parameters (up curves) sampled at different magnetic field and stress.

3.3. Harvesting results - Terfenol

Let us consider now the specific powers generated by a laboratory harvesting device with a Terfenol-D rod 18 mm long and with 9 mm^2 square section. The harvester has a pick-up coil

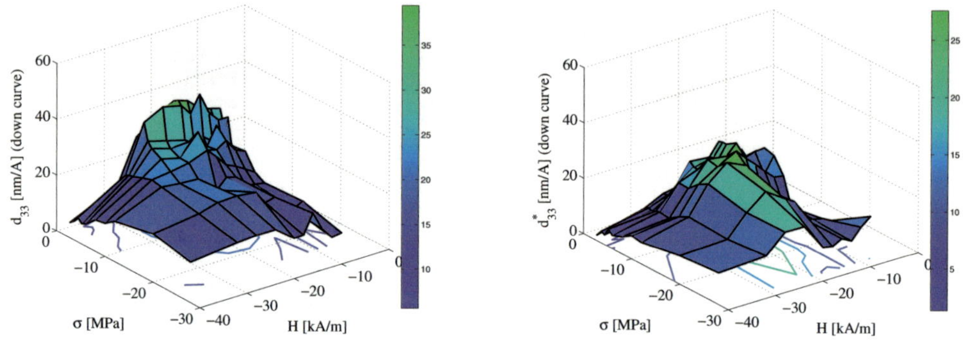

(a) d_{33}.

(b) d^*_{33}.

Figure 15. The piezo-magnetic parameters parameters (down curves) sampled at different magnetic field and stress.

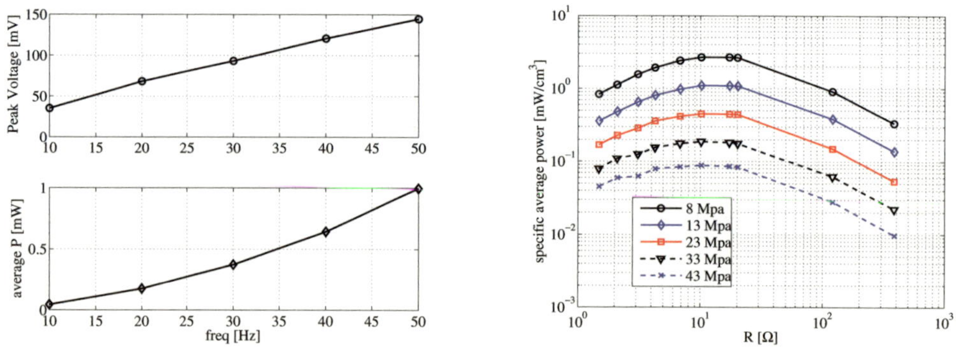

(a) Peak voltage and average power with H_0 = 18 kA/m, ΔF= 300 N.

(b) Specific average power at f = 50 Hz, H_0 = 4.8 kA/m, ΔF= 310 N at different pre-stresses.

Figure 16. Harvesting results of a force-driven device with a terfenol rod.

with 800 turns. The Fig. 16 (a) shows the peak voltage and average power with vibration frequency and H_0 = 18 kA/m, ΔF= 300 N, corresponding to ΔT = 33.3 MPa. It is worth noting that, at 50Hz, a relevant specific power of 6.17 mW/cm^3 is reached, confirming the potentialities of vibration harvesting when high frequencies are concerned. The Fig. 16 (b) shows the average power at 50Hz by varying the resistance load, at different prestresses. It is apparent that the choices of the correct prestress and resistance are crucial to harvest the maximum power.

3.4. Harvesting results - Galfenol

Let us consider now the specific powers generated by a laboratory harvesting device with a Galfenol rod 30 mm long and with 5 mm diameter. The Fig. 17 shows the average power densities (left) and the peak to peak voltage with a 3000 turns pick-up coil (parasitic resistance of 1 kΩ) and 1 kΩ termination resistor. The compressive stress variation is 30 MPa and

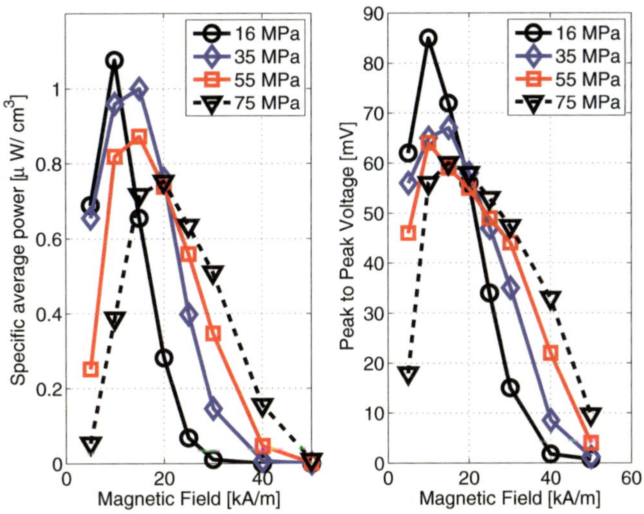

Figure 17. Average power densities (left) and peak to peak voltage (right) at $f = 0.35$ Hz, with a $\Delta T = 30$ MPa stress variation.

$f = 0.35$ Hz, a very low frequency that could be found in structural vibrations of civil constructions. It is worth noting that the peaks are reached at relatively low magnetic field bias, confirming one of the advantages to use Galfenol for harvesting applications. Indeed, those magnetic biases could be easily obtained with a permanent magnet, in a engineered harvesting device. The peak specific power is about 1 μW/cm^3, reached at 10 kA/m magnetic bias and 16 MPa prestress. Those values are reasonable and in agreement with literature values [33]. The Fig. 18 (a) shows the average power densities with respect to different termination resistors and with a 800 turns coil (parasitic resistance of 19Ω). Moreover, the resistor value near the parasitic coil resistance gives the maximum power. The Fig. 18 (b) shows the average power density behavior with respect to the stress variation frequency with 20Ω resistance, 15kA/m magnetic bias and with a $1 - 80$ MPa stress variation. As expected, the power increases with frequency.

3.5. Harvesting loops behavior in the B-H plane

It is a fundamental result the fact that magnetic loops in the B-H plane represent losses if the loops are passed through counterclockwise. On the other hand, if passed clockwise, those loops area represent *generated* energy. Then, it is apparent that a good design of the device should aim to enlarge as much as possible that area. Now, as it was shown in the previous sections, magnetic bias, prestress and other parameters can be used to do so because if in steady-state vibrations the power is increased then the loop area in the B-H plane is increased too. The Fig. 19 (a) shows the magnetic characteristics at 1.58 and 48.2 MPa constant stresses. The blue lines within the cycles represent the loops due to harvesting tests at different magnetic bias (8.7, 14.8 and 23 kA/m), 1 MΩ resistance and $f = 0.8$ Hz. It is noticeable that the loops are contained within the magnetic characteristics, as it was theoretically foreseen in [9]. Moreover, this is still true if different resistors are employed, as shown in Fig. 19 (b). The loops are wider for smaller resistor because a larger current circulates in the pick-up

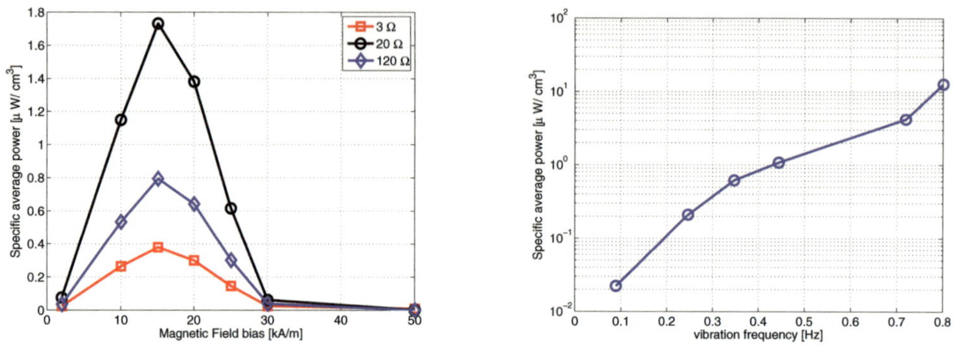

(a) Average power densities at $f = 0.35$ Hz, with a $1 - 80$ MPa stress variation.

(b) Average power densities at 15 kA/m magnetic bias, with a $1 - 80$ MPa stress variation, 20Ω resistance.

Figure 18. Harvesting results of a force-driven device with a a galfenol rod.

coil and a magnetic field variation is introduced then. This result is important because it is evident that the effort to improve the material harvesting behavior should concentrate on the *static* magnetic characteristics at different stresses, trying to improve as much as possible the *available area* among limit cycles at different stresses.

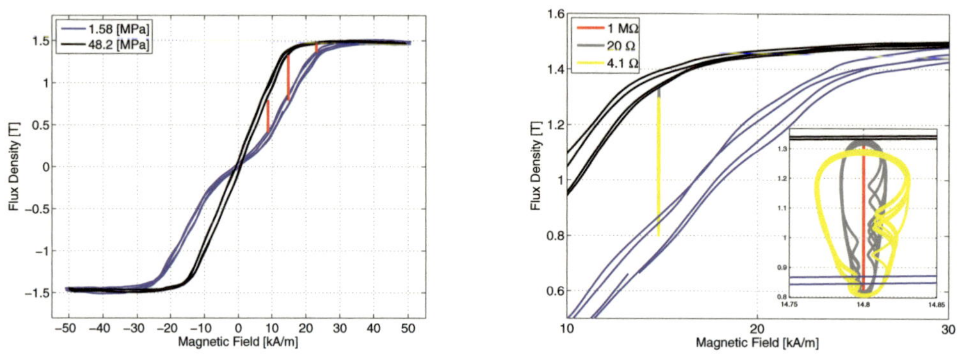

(a)

(b)

Figure 19. (a) Magnetic characteristics with converted energy loops at different magnetic bias (8.7, 14.8 and 23kA/m), 1 MΩ resistance and $f = 0.8$ Hz. (b) Converted energy loops zoom with different resistance termination at 14.8kA/m magnetic bias.

4. Conclusions and open problems

Materials

It is evident that the research on magnetoelastic material synthesis can give a strong advancement to the energy harvesting applications. Indeed, the research should focus to maximize the piezo-magnetic coefficients. This, as explained in section 3.5, would means to have more space for harvesting loops within static limit cycles. In other words, to have a larger ΔB for a certain ΔT.

Another topic about materials synthesis concerns the *mechanical impedance* matching of the harvesters. Indeed, materials like terfenol or galfenol are quite *rigid* with a mechanical behavior near the one of bulk iron. In that case, the ideal vibrations have high stresses and low strains, in the $0 - 1000$ppm range. If a softer magnetoelastic materials would exist, it would be possible to have vibrations harvesting with lower stresses and higher strains, in the $0.1 - 1\%$ range, with a rubber-like behavior.

Recently the use of another magnetoelastic material (Metglass) has been proposed for energy harvesting [45]. The main advantage respect to the others magnetostrictive material is that it can be laminated achieving a higher harvester compactness. The material is a Fe-based amorphous ribbons with excellent magnetic softness and elastic response and it is cheaper than Fe-Ga, Fe-Tb-Dy alloys. A recent new application design reached 20 $\mu W/cm^3$ at 100 Hz [53].

Models

The further steps in the material modeling should aim to include the hysteresis in both characteristics in a thermodynamic compatible way. Indeed, up to now, this step is made only on one of the two characteristics when the other one is not relevant for the specific applications. This is valid when the vibration source can be considered almost *ideal*.

Another topic in the analysis of a harvester, based on magnetoelastic materials, is the eddy currents effect due to the mechanical stress variations, that can have a detrimental effect on the harvester performances [51]. This phenomenon is present when the material has a finite electric resistivity, that is the case of magnetostrictive materials as Terfenol and Galfenol. The effect can be more harmful on the latter because of the high relative permeability. From the modeling point of view, the eddy currents problem can be formulated, as for standard magnetic materials, starting from Maxwell's equations in the magneto-quasistatic limit. The main difference is that the magnetic characteristics depends on the applied stress. Then, the stress acts as the *forcing term* in the magneto-mechanical problem [14, 15].

Damping

An effective energy harvesting from vibrations results in an amplitude damping on the mechanical source, no matter what the harvesting method is. In other terms, though the principal aim of an harvesting device is to convert the available mechanical energy into electrical energy, at the same time a kind of *passive* mechanical damping occurs. The passive damping by smart materials has been generally addressed in [41]. Recent contributions deepen this damping effect for Piezo [28, 29] and Magnetostrictive materials [10, 13]. The latter presents a preliminary analysis showing the effects of the energy harvesting on mechanical damping which requires further effort. As already mentioned, the passive damping could be exploited in several application fields, for example, in automotive applications where the vibrations reduction is a concern and energy harvesting can help to improve the overall efficiency. Moreover, this concept is even more attractive if compared to the classical *active* mechanical damping, where an additional properly controlled actuator is needed with a consequent increase of cost, system complexity and use of energy.

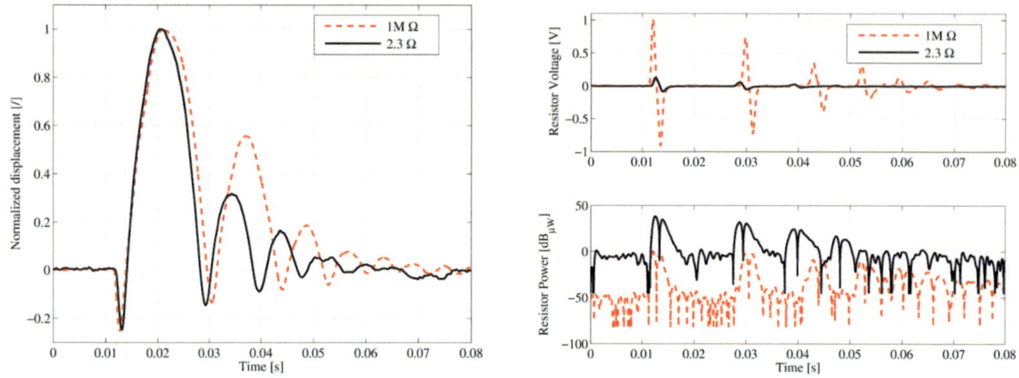

(a) Displacement of a 7.5Kg mass.

(b) Measured resistors voltage (up) and instantaneous dissipated power (down).

Figure 20. Damping effect of the vibration harvesting. Two different resistors case are reported with a $7kA/m$ magnetic bias. The power $p(t) = v^2(t)/R$ is represented with $10\log_{10}(\frac{p(t)}{1\mu W})$.

In Fig. 20 (a) the normalized displacement comparison between two different resistive loads is reported (2.3Ω- solid line, 1MΩ-dashed line). The Fig. 20 (b) shows the corresponding resistor voltage (up) and the instantaneous dissipated electric power (down). The mass is in mechanical contact with the active material and a controlled initial velocity is applied to them. The active material has suitable bias conditions (magnetic bias of $7kA/m$ and a mechanical prestress of 2.9MPa). The damping with the 1MΩ resistor is due only to mechanical friction, while with the 2.3Ω resistor there is a faster damping of the oscillation, both in amplitude and in time, because of the energy harvesting. Indeed, in this case a far larger power is extracted. Finally, the voltage is higher with the 1MΩ resistor as expected (a sort of open circuit condition).

Power conversion circuits

A magnetoelastic energy harvester is an unregulated AC power source. So it can not be directly interfaced to common electronic loads that need a regulated DC supply. Then, a power conversion stage must be a part of the harvester to overcome this issue. In particular, this stage should accomplish to two main functions:

- an efficient rectification;
- the regulation and level shifting of the output voltage.

These tasks can be afforded in different ways that strongly depend on the available AC power from the harvester and on the specific field of application. In fact, magnetoelastic energy harvesters output powers can range from milliwatt to watt levels depending on the mechanical source characteristics and its coupling with the active material. Moreover, also the dimensions of the magnetoelastic material influence the obtainable power, but in many applications they are limited by compactness requirements.
The choice of the topology and circuit implementation of the power conversion stage is based also on the following criteria, [42]:

- *efficiency*. The power stage for its operation must consume less power as possible. This is a strict requirement in energy havesting application due to the limited generated power.
- *stand-alone operation*. The mechanical source is strongly time varying: if there are vibrations there is electrical power output. The power stage should also contain an independent start-up circuit.
- *circuit complexity*. This influences the choice of the control strategies and their implementation for the power stage.
- *adaptivity*. The power stage must work with a wide swing of electric inputs (due to the mechanical source) and of electric output (due to the load requests), guaranteeing the maximum power transfer to the electrical load.

The possible solutions to the above requirements and criteria can be identified in two different approaches, according to what has been done for piezoelectrics or other mature harvesting technologies, like electromagnetic and electrostatic generators:

- *Single stage power conversion*. In this case the rectification and the DC voltage regulation are made in a single step. To this approach belong solutions like the Direct AC-DC Switch-Mode Converters (*e. g.* single inductor with split capacitor, single inductor with secondary-side switches, dual-boost converters), [16]. Also specific solutions for a particular harvesting technology are possible. For the piezoelectrics, due to their intrinsic capacitive behavior, switched inductor converters (*e.g.* the synchronized switch harvesting on inductor (SSHI) and its generalizations) have been proposed [25, 52]. In the case of magnetoleastic harvester similar solutions could be conceived by considering their inductive nature, by duality.
- *Double stage power conversion*. In this case the rectification and the voltage regulation are separated. The rectification can be made with classical solutions as diodes full bridges or with the so called *active diodes*. The second stage instead should regulate the output voltage and its shifting. Due to the relatively low voltages obtainable from a magnetoelastic harvester a DC-DC boost based topology with high boosting gain can be considered, [38].

Finally, another feature of a magnetoelastic harvester that challenges the definition and the modeling of the power conversion stage is its strong nonlinearities. For example, these can create on the AC side an additional harmonic content that is not present in the mechanical stimulus and it depends heavily on the harvester operating conditions (*e.g.* mechanical prestresses and magnetic biases). This pushes to the definition of new circuital multidomain modeling approaches for analyzing the coupling among the mechanical, magnetic and electronic worlds.

Author details

Daniele Davino, Alessandro Giustiniani, Ciro Visone
University of Sannio - Engineering Department, 82100 Benevento (BN), Italy

5. References

[1] Adly, A.A., Davino, D., Giustiniani, A. & Visone, C. (2010). Experimental tests of a magnetostrictive energy harvesting device towards its modelling, *AIP Journal of Applied Physics*, Vol. 107, No. 09A935, May 2010, pp. 1-3.

[2] Anton, S.R. & Inman D.J. (2008). Vibration energy harvesting for unmanned aerial vehicles, *Proceedings of SPIE*, Vol. 6928, No. 692824.

[3] Basso, V., Bertotti, G., Serpico, C. & Visone, C. (1998). Application of an exactly inverse hysteresis model to magnetic field computations, *Journal de Physique IV*, Vol.8, pp. 639-642.

[4] Bewes, O.G., Thompson, D.J., Jones, C.J.C. & Wang, A. (2006). Calculation of noise from railway bridges and viaducts Experimental validation of a rapid calculation model, *ELSEVIER Journal of Sound and Vibration*, Vol. 293, pp. 933-943.

[5] Clark, A.E., Teter, J.P. & McMasters, O.D. (1988). Magnetostriction "jumps" in twinned $Tb_{0.3}$ $Dy_{O.7}Fe_{1.9}$, *AIP Journal of Applied Physics*, Vol. 63, No. 8, pp. 3910-3912.

[6] Clark, A.E., Wun-Fogle, M., Restorff, J.B. & Lindberg, J.F. (1992). Magnetomechanical properties of single crystal Tb_xDy_{1-x} under compressive stress, *IEEE Transactions on Magnetics*, Vol. 28, No.5, pp. 3156-3158.

[7] Culler, D. E. & Hong, W. (2004). Wireless sensor networks, *Commun. ACM*, Vol.47, pp.30-33.

[8] Davino, D., Giustiniani, A. & Visone, C. (2009). Capacitive load effects on a magnetostrictive fully coupled energy harvesting device, *IEEE Transactions on Magnetics*, Vol. 45, No. 10, Oct 2009, pp. 4108-4111.

[9] Davino, D., Giustiniani, A. & Visone, C. (2009). Analysis of a magnetostrictive power harvesting device with hysteretic characteristics, *AIP Journal of Applied Physics*, Vol. 105, 07A939.

[10] Davino, D., Giustiniani, A. & Visone, C. (2011). A two-port nonlinear model for magneto-elastic energy harvesting devices, *IEEE Transactions on Industrial Electronics*, Vol. 58, No. 6, Jun 2011, pp. 2556-2564.

[11] Davino, D., Giustiniani, A., Visone, C. & Adly, A.A. (2011). Experimental analysis of vibrations damping due to magnetostrictive based energy harvesting, *AIP Journal of Applied Physics*, Vol. 109, No. 07E509, Mar 2011, pp. 1-3.

[12] Davino, D., Giustiniani, A. & Visone, C. (2011). The piezo-magnetic parameters of Terfenol-D: An experimental viewpoint, *ELSEVIER Physica B: Condensed Matter*, Vol. 407, No. 9, pp. 1427-1432.

[13] Davino, D., Giustiniani, A., Visone, C. & Adly, A.A. (2011). Experimental analysis of vibrations damping due to magnetostrictive based energy harvesting, *AIP Journal of Applied Physics*, Vol.109, 07E509.

[14] Davino, D., Giustiniani, A. & Visone, C. (2011). Effects of hysteresis and eddy currents in magnetostrictive harvesting devices, *ELSEVIER Physica B: Condensed Matter*, Vol.407, No.9, pp. 1433-1437.

[15] Davino, D., Giustiniani, A., Visone, C. & Zamboni, W. (2012). Stress-Induced Eddy Currents in Magnetostrictive Energy Harvesting Devices, *IEEE Transactions on Magnetics*, Vol. 48, No.1, pp. 18-25.

[16] Dayal, R., Dwari, S. & Parsa, L. (2011). Design and Implementation of a Direct AC-DC Boost Converter for Low-Voltage Energy Harvesting, *IEEE Transactions on Industrial Electronics*, Vol. 58, No. 6, pp. 2387-2396.

[17] Del Grosso, A., Inaudi, D. & Pardi, L. (2002). Overview of european activities in the health monitoring of bridges, *Proceedings 1st International Conference on Bridge Maintenance, Safety Management (IABMAS)*, Barcelona, Spain, pp. 14-17.

[18] Elvin, N.G. & Elvin A.A., 2011 Vibrational Energy Harvesting From Human Gait, *IEEE/ASME Transactions on Mechatronics*, DOI: 10.1109/TMECH.2011.2181954, forthcoming.

[19] Ed. Engdahl, G. 2000, Handbook of Giant Magnetostrictive Materials, (Academic, New York).

[20] Galchev, T.V., McCullagh, J., Peterson, R.L. & Najafi, K. (2011). Harvesting traffic-induced vibrations for structural health monitoring of bridges, *IOP Journal of Micromechanics and Microengineering*, Vol. 21, No. 104005.

[21] Garibaldi, L., Giorcelli, E. & Piombo, B.A.D. (1998). ARMAV techniques for traffic excited bridges, *ASME Journal of Vibration and Acoustics*, Vol. 120, No 3, pp. 713-718.

[22] Hu, J., Xu, F., Huang, A.Q. & Yuan, F.G. (2011). Optimal design of a vibration-based energy harvester using magnetostrictive material (MsM), *IOP Smart Materials and Structures*, Vol. 20, No. 015021, Dec 2011, pp. 1-12.

[23] Karami, M.A. & Inman D.J. (2012). Powering pacemakers from heartbeat vibrations using linear and nonlinear energy harvesters, *AIP Applied Physics Letters*, Vol. 100, No. 042901, pp. 1-4.

[24] Krejčí, P. (1996). Hysteresis, convexity and dissipation in hyperbolic equations, *Gakuto International Series: mathematical sciences and applications*, 8, Gakkōtosho, Tokyo.

[25] Lallart, M. & Guyomar, D. (2011). Nonlinear energy harvesting, *IOP Conference Series: Materials Science and Engineering*, Vol. 18, 092006, pp. 1-6.

[26] Lee, E.W. (1955). Magnetostriction and magnetomechanical effects, *IOP Reports on Progress in Physics*, vol. 18, pp. 184-227.

[27] Lee, J.W., Kim, J.D., Yun, C.B., Yi, J.H. & Shim, J.M. (2002). Health-monitoring method for bridges under ordinary traffic loadings, *ELSEVIER Journal of Sound and Vibration*, Vol. 257, No. 2, pp. 247-264.

[28] Lesieutre, G.A., Ottman, G.K. & Hofmann, H.F. (2004). Damping as a Result of Piezoelectric Energy Harvesting, *ELSEVIER Journal of Sound and Vibration*, Vol. 269, pp. 991-1001.

[29] Liang, J.R. & Liao W.H. (2009). Piezoelectric Energy Harvesting and Dissipation on Structural Damping, *SAGE Journal of Intelligent Material Systems and Structures*, Vol. 20, No. 5, pp. 515-527.

[30] Lin, C.W. & Yang, Y.B. (2005). Use of a passing vehicle to scan the fundamental bridge frequencies: An experimental verification, *ELSEVIER Engineering Structures*, Vol. 27, pp. 1865-1878.

[31] Lynch, J.P. (2007). An overview of wireless structural health monitoring for civil structures, *Philosophical Transactions of the Royal Society A*, Vol. 365, pp. 345-372.

[32] Mhetre, M.R., Nagdeo, N.S. & Abhyankar, H.K. (2011). Micro Energy Harvesting for Biomedical Applications: A Review, *Proceedings of 3rd International Conference on Electronics Computer Technology (ICECT)*, pp. 1-5.

[33] Mitcheson, P.D., Yeatman, E.M., Rao, G.K., Holmes, A.S. & Green, T.C. (2008). Energy Harvesting From Human and Machine Motion for Wireless Electronic Devices, *Proceedings of IEEE*, 96(9), pp. 1457-1486.

[34] Moffett, M.B., Clark, A.E., Wun-Fogle, M., Linberg, J., Teter, J.P. & McLaughlin, E.A.(1991). Characterization of Terfenol-D for magnetostrictive transducers, *Journal of the American Acoustical Society*, Vol. 89, No.3, pp. 1448-1455.

[35] Ottman, G.K., Hofmann, H.F., Bhatt, A.C. & Lesieutre, G.A. (2002). Adaptive piezoelectric energy harvesting circuit for wireless remote power supply, *IEEE Transactions on Power Electronics*, Vol. 17, No.5, pp. 669-676.

[36] Pereyma, M. (2007). Overview of the Modern State of the Vibration Energy Harvesting Devices, *Proceedings of International Conference on Perspective Technologies and Methods in MEMS Design*, pp. 107-112.

[37] Restorff, J.B., Wun-Fogle, M. & Summers, E. (2011). Hysteresis, d_{33}^* and d_{33} of $Fe_{81.6}Ga_{18.4}$ textured polycrystals, *AIP Journal of Applied Physics*, Vol. 109, 07A922.

[38] Richelli, A., Comensoli, S. & Kovacs-Vajna, Z.M. (2012). A DC/DC Boosting Technique and Power Management for Ultralow-Voltage Energy Harvesting Applications, *IEEE Transactions on Industrial Electronics*, Vol. 59, No. 6, pp. 2701-2708.

[39] Roundy, S. (2005). On the Effectiveness of Vibration-based Energy Harvesting, *SAGE Journal of Intelligent Material Systems and Structures*, Vol. 16, Oct 2005, pp. 809-823.

[40] Schulze, M.P., Greenough, R.D., & Galloway, N. (1992). The Stress Dependence of k_{33}, d_{33}, λ and μ in $Tb_{0.3}Dy_{0.7}Fe_{1.95}$, *IEEE Transactions on Magnetics*, Vol. 28, No.5, pp. 3159-3161.

[41] Smith, C.A. & Anderson, E.H. (1995). Passive damping by smart materials: analysis and practical limitations, *proceedings of SPIE*, **2445**, Vol.136, pp. 136-148.

[42] Szarka, G.D., Stark, B.H. & Burrow, S.G. (2012). Review of Power Conditioning for Kinetic Energy Harvesting Systems, *IEEE Transactions on Power Electronics*, Vol. 27, No. 2, pp. 803-815.

[43] Turner, J., Cartwright, J., Ha, D.S., Zhang, D. & Banerjee, S. (2011). On Energy Harvesting Module for Scalable Cognitive Autonomous Nondestructive Sensing Network (SCANSn) System for Bridge Health Monitoring, *Proc. of SPIE*, 79831T.

[44] Ueno, T. & Yamada, S. (2011). Performance of Energy Harvester Using Iron-Gallium Alloy in Free Vibration, *IEEE Transactions on Magnetics*, Vol. 47, No.10, pp. 2407-2409.

[45] Wang, L. & Yuan, F.G. (2008). Vibration energy harvesting by magnetostrictive material, *IOP Smart Materials and Structures*, Vol. 17, No. 045009, Jun 2008, pp. 1-14.

[46] Westby, E.R. & Halvorsen E. (2011). Design and modeling of a patterned-electret-based energy harvester for tire pressure monitoring systems, *IEEE/ASME Transactions On Mechatronics*, DOI: 10.1109/TMECH.2011.2151203, forthcoming.

[47] Williams, C.B. (1997). Feasibility Study of Vibration-Electric generator for Bridge Vibration sensors, *Proceedings Of IMAC-XVI.*

[48] Yongping, W., Daining, F., Soh, A.K. & Hwang, K.C.(2003). Experimental and theoretical study of the nonlinear response of a giant magnetostrictive rod, *SPRINGER Acta Mechanica Sinica*, Vol. 19, No.4, pp. 324-329.

[49] Yoo, J.H., Pelligrini, G., Datta, S. & Flatau, A.B. (2011). An examination of Galfenol mechanical magnetic coupling coefficients, *IOP Smart Materials and Structures*, Vol. 20, No. 075008, Jun 2011, pp. 1-8.

[50] Zhang, H. (2011). Power generation transducer from magnetostrictive materials, *AIP Applied Physics Letters*, Vol. 98, No. 232505, Jun 2011, pp. 1-3.

[51] Zhao, X. & Lord, D.G. (2006). Application of the Villari effect to electric power harvesting, *AIP Journal of Applied Physics*, Vol. 99, No. 08M703, Apr 2006, pp. 1-3.

[52] Zhu, L. & Chen, R. (2012). A new synchronized switching harvesting scheme employing current doubler rectifier, *ELSEVIER Sensors and Actuators A: Physical*, Vol. 174, No. 6, pp. 107-114.

[53] Zucca, M., Bottauscio, O., Beatrice, C. & Fiorillo, F. (2011). Modeling Amorphous Ribbons in Energy Harvesting Applications, *IEEE Transactions on Magnetics*, Vol. 47, No.10, pp. 4421-4424.

[54] Zuo, L., Scully, B., Shestani, J. & Zhou, Y. (2010). Design and characterization of an electromagnetic energy harvester for vehicle suspensions, *IOP Smart Materials and Structures*, Vol. 19, No. 045003, pp. 1-10.

Chapter 20

Feedforward and Modal Control for a Multi Degree of Freedom High Precision Machine

Andrea Tonoli, Angelo Bonfitto, Marcello Chiaberge,
Mario Silvagni, Lester D. Suarez and Enrico Zenerino

Additional information is available at the end of the chapter

http://dx.doi.org/10.5772/52213

1. Introduction

High precision industrial machines suffer the presence of vibrations mostly due to two noise sources: ground vibration and direct force disturbances. They can generate several problems at different levels and of different natures, causing performance losses on sensitive systems (Crede, 1951), (Rivin, 1979).

In the last years the growing processing quality level and the need to increase throughput resulted in a continuing demand for higher accuracy. Therefore active isolation and vibration damping systems became mandatory to satisfy these requests (Pneumont, 2002), (Hyde, 1997).

In general, machine supports are designed for high stiffness to obtain a robust machine alignment with respect to its surroundings. However, when significant ground vibration levels occur, the support stiffness is commonly sacrificed to reduce vibration transmission to the payload stage. Efforts to go towards these issues are recorded in several applications and the solutions are different for any particular situation, depending on the nature of vibration sources, the amount of disturbances and the machine environment.

Several actuation technologies are used to face this kind of problem: shape memory alloys, electromagnetic, piezoelectric, magnetostrictive and magneto-rheological fluids actuators (Thayer, 1998). Among them, electromagnetic actuators revealed themselves as effective and performing. Methods for vibration suppression can be classified in a rough approach in three families: passive, active and semi-active actuators. Completely passive solutions have almost reached their maximum potential which is still not sufficient to satisfy stringent requirements. On the opposite, the exponential growth in electronics and actuators fields made the use of active and semi-active isolation more feasible. In particular, active control

architectures allow to perform an effective isolation at low frequencies, which is a common requirement for very demanding applications like micrometer motion control, defect inspections, critical dimensions measurement and overlay metrology.

In general, active control arrangements are provided with sensors, actuators and controllers (Watters, 1988). Each of them can be classified depending on their technology and physical working principle. The choice of sensors and actuators is strictly related to the type of application and requirements and has also influence on the selection of the control strategies to be employed. Depending on the type of controller, the system model can be used only to support the control design or can play itself a fundamental role on the control action (model based strategies) (Beadle et al, 2002), (Sullivan, 1997). Typically the main control approaches are feedback, classical or model based, and feed-forward technique, mostly with adaptive reference filtering (Anderson, 1996).

This chapter focuses on the evaluation of an active isolation and vibration damping device mounted in the working cell of a micro-mechanical laser center, which is based on active electromagnetic actuators. Two different models and three control strategies are developed and illustrated.

To clarify the goal of this study it is important to point out that: a) the vibration damping is defined as the reduction of the response amplitude of the system within a limited bandwidth near the natural frequencies of the system; b) vibration isolation is defined as the attenuation of the response of the system after its corner frequency to cut-off all the disturbances after that frequency, while allowing all the signals below it to pass with no alterations.

The machine object of study is composed by two main parts: a frame support and a payload stage where the laser cutting operation is performed. The system performance in terms of accuracy and precision is reduced by the presence of two main vibration sources: the ground and the stage itself. The active device should meet two goals: the payload vibrations damping and the reduction of the transmissibility of ground disturbances.

In this work, after a review of the major actuators families usually employed to damp and isolate high precision machines, the phases followed to design, implement and validate the proposed device are illustrated with a particular emphasis on the mechatronics aspects of the project.

A detailed analysis of the plant components is reported along with an exhaustive explanation of the design criteria followed for the choice of supports, actuation and sensing subsystems. The actuation block consists in four electromagnetic Lorentz type actuators (two per axis).

The absolute velocities of the frame support and of the stage are measured by means of eight geophone sensors to determine the amount of disturbances (Huan, 1985), (Riedesel, 1990). The considerations leading to the choice of this sensing system are reported along with the description of the related signal conditioning stage. The design of the supports between the ground and the frame and of the connections between the frame and the stage is also explained.

Furthermore, all the subsystems described in the first part of the chapter are modeled along with their interactions. The Lagrange equations approach is used to represent the system behavior and in particular the links between the mechanical and electrical subsystems are illustrated.

Two models are developed: a) four degrees of freedom model and b) six degrees of freedom model. Both of them include the plant, the sensing, the control and the actuation blocks. Time and frequency domain computations are carried out from the models to evaluate vibration levels and displacements and to identify which control parameters need to be carefully designed to satisfy the requirements.

The last section exposes in detail the proposed control strategies along with the modeling approach validation. Three different control strategies are developed:

a. Feedback control: the control law consists in a couple of decentralized actions exerted along X and Y-axis allowing to minimize the ground vibrations transmission and damp the payload vibrations. Specifically, a Lead-Lag control strategy, performed with a digital platform based on DSP and FPGA, is used to compensate the high-pass band dynamic of the geophone sensors and to damp the vibrations (Kuo, 1996), (Elliott, 2001). The payload isolation is achieved by feeding the control block with the difference of frame and stage velocities and giving the proper current command to the actuators. The four degrees of freedom model is used to design this control law. In the section describing the control strategy the comparisons between simulation and experimental tests is presented, which illustrates the validity of the model and the effectiveness of the proposed approach. In particular, the performance of the vibration damping has been evaluated by using the frequency responses between the actuators force and the payload velocities, whereas the performance of the active isolation is evaluated by simulating numerically the disturbances coming from the ground and evaluating their transmission through all the system till the payload in closed loop configuration.
b. Feedforward control: this action is focused on the rejection of the direct disturbance coming from the payload. The command is not generated on-line as in classical feedforward applications, but it is computed in advance from the data obtained from a direct disturbance from the payload to the machine. That is, here the compensation is computed numerically in the case of known disturbances profiles. The design of this strategy is based on the four degrees of freedom model, as in the case of the feedback control technique.
c. Modal control: the approach allows the controller to focus on the rotational and translationalal modes of the machine. Results show that the performance of this strategy are comparable to those of the standard feedback control (a), though significant advantages exist in the design procedure where the control effects can be evaluated directly on the motion modes. This technique makes use of the six degrees of freedom model.

2. Actuators technology for damping and active isolation: An overview

Undesired noise and vibrations are since ever a major problem in many human activities and domains. Airplanes, space trusses and satellites, cars, machine tools and large bridges,

all can be disturbed in their normal functions by vibrations and noise. Actuators play a critical role in the active control of vibration and different technologies must be considered in order to obtain compact and efficient smart structures.

Selection and use of these technologies is greatly influenced by the user's technical knowledge, the project's budget, available energy sources, and performance tradeoffs. For example, pneumatic actuators don't deliver high force output, but are well suited when a cost-effective, easy start-up solution is required. Hydraulic actuators generate a lot of noise and can leak nasty fluid, but are ideal for high force applications that require precise control. Electromechanical actuators have high energy requirements and are more difficult to install and maintain, but are preferred for complex, multi-axis, motion control applications.

Pneumatics: pneumatic actuation is the conversion of compressed air into, typically, linear force. Typical applications involve extreme temperature and magnetic systems because pneumatic actuators don't have the magnetic field issues of electric motors. Position feedback with proximity sensors is used in modern control-loop systems, bringing pneumatics beyond simple bang-bang applications.

Pressure losses and the compressibility of air make pneumatics less efficient than other actuator technologies. In addition compressor and delivery system limitations dictate that pneumatic systems operate at lower pressures, providing lower forces and lower bandwidths than other systems. Pneumatic cylinders typically operate with compressed air at 100 psi or less, in contrast with hydraulic cylinders, which operate on pressurized hydraulic fluids at over 500 psi. Speed, force and bandwidth are directly connected with these characteristics.

Hydraulics: hydraulic actuators are suitable for rugged applications that require high force output. However, hydraulic systems generate noise and, without proper maintenance, they can leak. More equipment is needed as well: hydraulic systems require a fluid reservoir, motors and pumps, release valves, and equipment to reduce noise and heat levels. Moreover external sensors are needed to determine piston velocity, acceleration and position in a closed-loop system. Hydraulic systems can deliver much tighter control than pneumatic systems and higher force density than any other actuator technologies. Bandwidth is better than pneumatic actuators but still under hundreds of Hertz.

Electromechanical: electromechanical actuators can be based on rotatory motors (using ball screw, roller screw or belt drive), linear motors or moving coils. This type of actuator have high dynamic performance, with accelerations greater than 20 g and velocities of 10 m/sec and eventually higher. Sub-micron resolution and repeatability are commonplace. Because the actuator is directly coupled to the load, there are fewer components with the chance of failure, which adds long term value.

Piezoelectric: piezomotors and piezoactuators rely on the electromechanical response of crystals. Electrical excitation causes the crystals to slightly change shape and distort, therefore generating large forces and small displacements. Exciting the crystals at a high frequency generates smooth, precise motion, making piezoelectric actuators suitable for applications with very fine positioning and high bandwidth requirements.

Actuator Technology	Advantages	Drawbacks
Pneumatic	Strong, light, simple, fast.	Precise position control impossible except at full stops.
Hydraulic	Very high forces possible.	Can leak. Requires position feedback for repeatability. External hydraulic pump required. Some designs good in compression only.
Electro-mechanical rotary motor	Cheap. Repeatable. Operation can be automated. Self-contained. Identical behaviour extending or retracting. DC or stepping motors. Position feedback possible.	Many moving parts prone to wear.
Electro-mechanical Linear motor	Simple design. Minimum of moving parts. High speeds possible. Self-contained. Identical behaviour in extending or retracting.	Low force.
Moving coil	Force, position and speed are controllable and repeatable. Capable of high speeds and precise positioning. Linear, rotary, and linear + rotary actions possible.	Requires position feedback to be repeatable.
Piezoelectric	Very small motions possible.	Requires position feedback to be repeatable. Short travel. Low speed. High voltages required. Expensive. Good in compression only, not in tension.

Table 1. Actuators technology comparison

3. System architecture

In this section of the chapter a full description of machine subsystems is provided. The mechanical, electrical, electronic, and control parts are identified and fully described separately in the first part. Furthermore, since the project can be assumed as a classical mechatronics application, the different blocks are analyzed with their interactions in order to provide an overall view of the system.

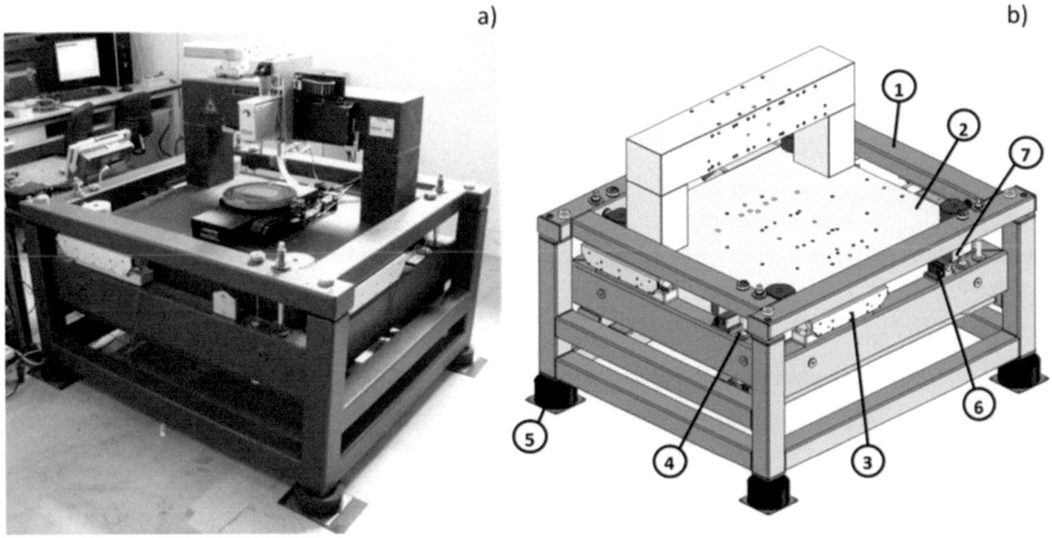

Figure 1. a) Picture of the machine. b) Sketch of the system. 1: Frame; 2: Stage; 3: Actuators; 4: Frame–Stage Springs; 5: Air springs; 6: Frame sensors; 7: Stage sensors.

Figure 1.a shows a picture of the laser cutting machine while in the sketch of Figure 1.b all the components of the system are highlighted. The stage (2) consists in a granitic base that can move freely within the work volume and is surrounded by four electromechanical actuators (3) acting between the frame (1) and the stage. The machine is partially isolated from the ground by means of four air springs (5). Four mechanical springs (rods) (4) are placed between the frame and the stage. The vibrations due to the machine process and coming from the ground are measured on the stage and on the frame by means of eight velocity inertial sensors (6, 7). A schematic representation of the actuators, sensors, and springs position is reported in Figure 2, where c_{GF} and k_{GF} represent the damping and the stiffness, respectively introduced by the supports, whereas c_{FS} and k_{FS} are the damping and the stiffness, respectively of the springs acting as connections between frame and stage. Actuators and sensors positions can be considered collocated, in order to minimize the couplings between the axes actions by keeping the proper alternation between resonances and anti-resonances in the system dynamics. The main machine parameters and specifications are listed in Table 2.

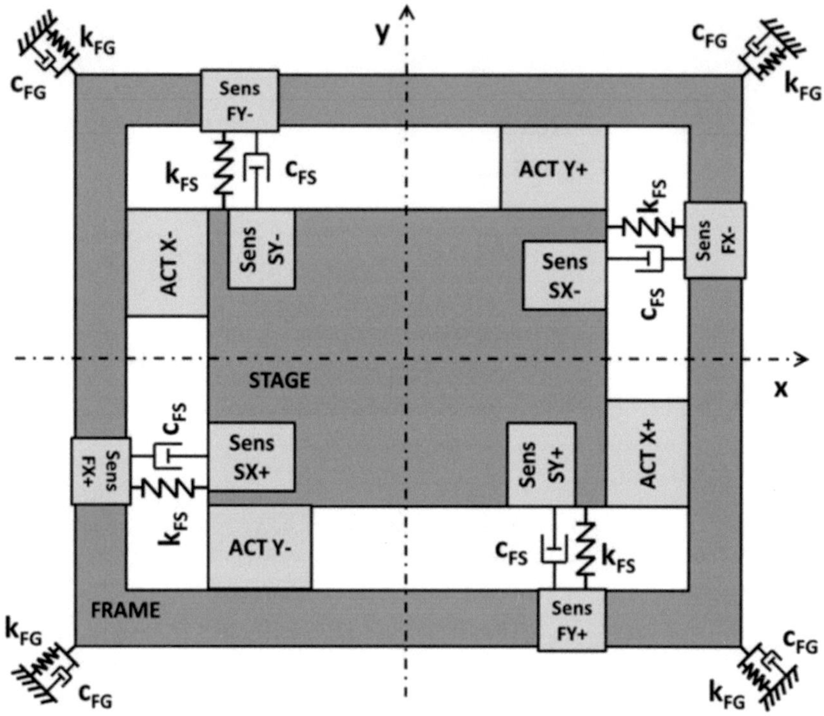

Figure 2. XY plane view of the system. Stage-Frame spring (k_{SF} , c_{SF}), electromagnetic actuator (ACT), velocity sensor (Sens.), Ground-Frame spring (k_{GF} , c_{GF}).

Stage mass	1450 kg
Frame mass	300 kg
Maximum displacement of the stage	2.5 mm
Inertia of the stage along X -axis in YZ -plane	200 kg m^2
Inertia of the frame along X -axis in YZ -plane	100 kg m^2

Table 2. Main parameters and specifications of the machine.

The design phases have been performed considering the mechatronics nature of the system and the interactions between the machine subsystems, illustrated in Figure 3. Regarding overall controller architecture, a classical feedback behavior is performed: eight velocities are acquired by the sensors measurements and elaborate with conditioning and filtering stages in order to feed the actuators with the proper commands by means of power electronics action. The filtering stage consists in the implementation of a Lead-Lag control strategy designed to fulfill the machine requirements in terms of: a) active isolation from the disturbances coming from the ground and b) damping of the vibrations generated by the machine processes. Feedforward action is also included which allows to reject the direct disturbances coming from the payload. These feedback and feedforward control actions are completely independent one from the other.

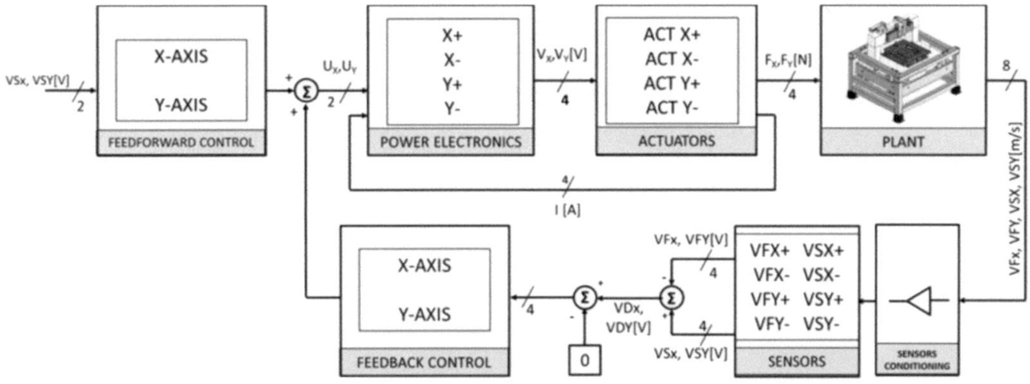

Figure 3. Block diagram of the system.

3.1. Actuators subsystem

The actuation on the system is realized by means of four electromagnetic Lorentz type actuators placed as illustrated in Figure 1 and Figure 2.

The picture and the section view of the actuator architecture are reported in Figure 4, being A and B permanent magnets, while C indicates the coil.

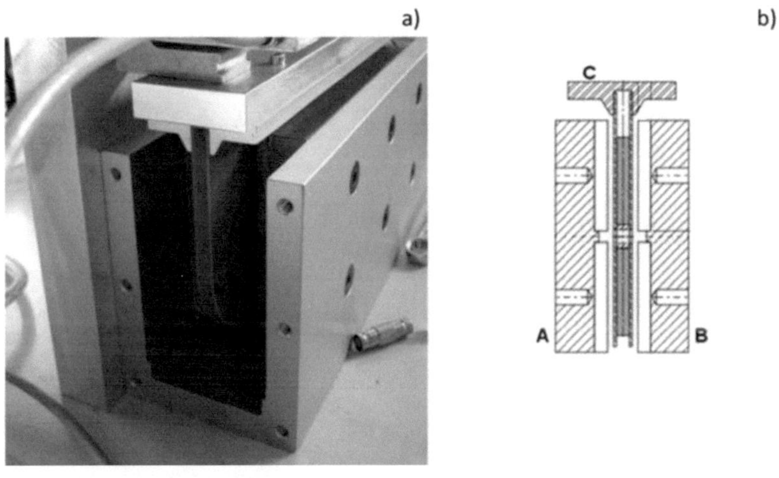

Figure 4. a) Picture of the Lorentz actuator. b) Section view (A and B: permanent magnets, C: coil).

The force F_{ACT} generated by each actuator is:

$$F_{ACT} = BNli \tag{1}$$

where B is the magnetic field, N is the number of turns of the coil, i is the current flowing in the coil, l is the coil length. The direction of the resulting force is illustrated in Figure 5. The amount of required force for each actuator is equal to 200 N while the main parameters of the designed actuator are reported in Table 3.

Coil thickness	6	mm
Coil length	3.3	mm
Coil active section	198	mm²
Copper current density	12	A/mm²
Coil length (l)	200	mm
Coil max actuation force (F_{ACT})	200	N
Number of turns (N)	263	-
Number of coils per axis	2	-

Table 3. Actuators main parameters.

The design of the actuators has been performed starting from the requirements of force and maximum displacement of the stage, then a current density and the wire section have been selected in order to perform a FEM analysis and to compute the magnetic field. Finally, once known all the electrical parameters, the coil length l has been computed.

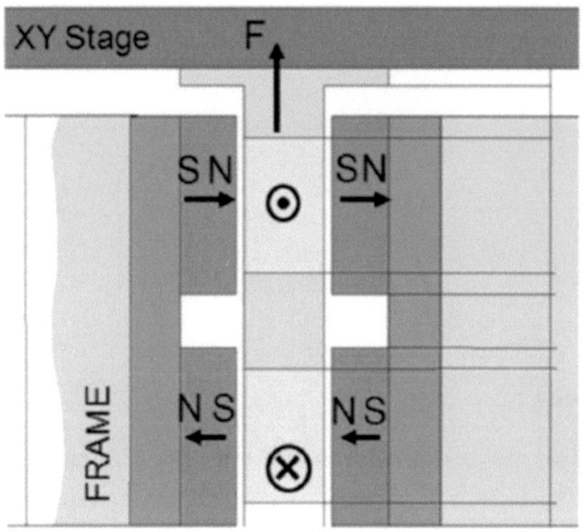

Figure 5. Actuator force generation.

The actuators parameters have been identified experimentally. The resulting values are: resistance $R = 4.33\,\Omega$, $L = 9.64\ mH$. The actuator electrical dynamics can be expressed as:

$$G_{ACT}(s) = \frac{1}{Z(s)} = \frac{1}{sL + R} = \frac{\frac{1}{L}}{s + \frac{R}{L}} \tag{2}$$

The stationary gain $G(s = 0)$ is:

$$G(s = 0) = 20\log_{10}\left(\frac{1}{R}\right) = -12.73\ dB \tag{3}$$

The electrical pole ω_e is:

$$\omega_e = \frac{R}{L} = 449\ rad/s = 72\ Hz \tag{4}$$

The resulting actuator trans-conductance (Current/Voltage) transfer function is reported in Figure 6.

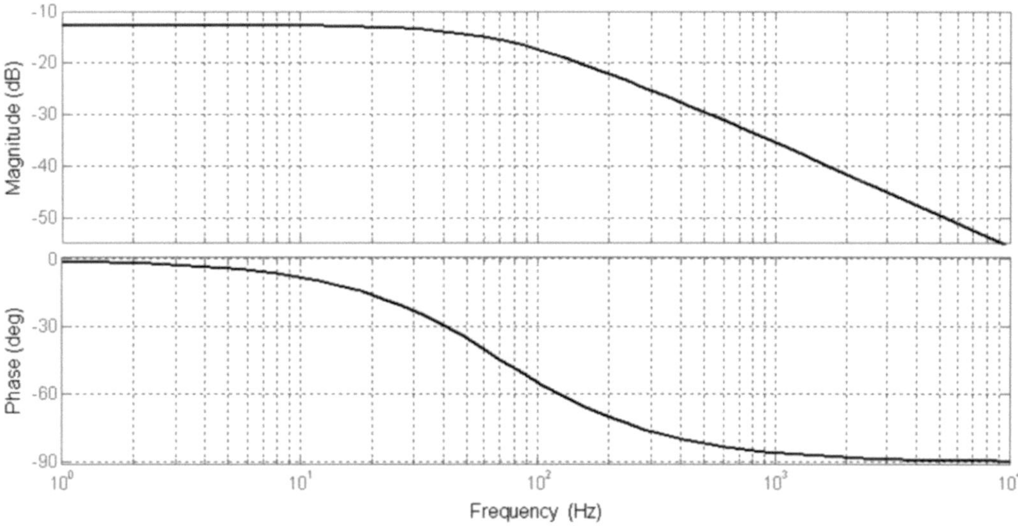

Figure 6. Actuator trans-conductance (Current/Voltage) transfer function (magnitude and phase).

3.2. Springs and supports

The frame and the stage are connected in the vertical direction by means of four linear springs indicated by 4 in Figure 1 as well as c_{SF} and k_{SF} in Figure 2. The design has been performed computing displacements and stresses with a FEM software, starting from the following requirements:

- infinite fatigue life;
- stiffness $\begin{cases} k_{SFx} = 40\,N/mm; \\ k_{SFy} = 40\,N/mm; \\ k_{SFz} = 32500\,N/mm; \end{cases}$
- damping $\begin{cases} c_{SFx} = 228\,Ns/m; \\ c_{SFy} = 228\,Ns/m; \\ c_{SFz} = 4313\,Ns/m; \end{cases}$
- maximum displacement $z_{MAX} = 2.5mm$;

The designed spring is made of harmonic steel and is characterized by:

- length $l_{SPRING} = 125mm;$
- diameter $d_{SPRING} = 5mm;$
- maximum value of stress $\sigma_{MAX} = 500MPa.$

Four air-springs (indicated by5 in Figure 1 as well as k_{GF} and c_{GF} in Figure 2) consisting in a resilient element air and neoprene diaphragm, have been chosen as supports to provide the system of a partial level of isolation from the ground. The springs are characterized by the following properties:

- Nominal natural frequency: $\begin{cases} f_{GFx} = 12.3Hz; \\ f_{GFy} = 12.3Hz; \\ f_{GFz} = 5.4Hz; \end{cases}$
- stiffness $\begin{cases} k_{GFx} = 450\,N/mm; \\ k_{GFy} = 450\,N/mm; \\ k_{GFz} = 500\,N/mm; \end{cases}$
- damping $\begin{cases} c_{GFx} = 575\,Ns/m; \\ c_{GFy} = 575\,Ns/m; \\ c_{GFz} = 1700\,Ns/m; \end{cases}$
- Transmissibility at resonance: 8:1;
- The maximum load is equal to 545 kg;
- The maximum air pressure is equal to 80 psi (5.5 bar).

3.3. Sensing subsystem

The disturbances on the plant are evaluated by measuring the velocities of the stage and of the frame along X -axis and Y –axis, by means of eight geophones placed as indicated in Figure 2. They are the most common inertial velocity sensors used to monitor seismic vibrations and can be classified as electromagnetic sensors that measure the velocity and produce a voltage signal thanks to the motion of a coil in a magnetic field (Hauge et al, 2002). One configuration of the conventional geophones consists of a cylindrical magnet coaxial with a cylindrical coil as shown in Figure 7. The coil is made up of a good conductor like copper and is wound around a nonconductive cylinder to avoid eddy currents effects, caused by the currents induced in the coil. The wire diameter and the dimensions of the holding cylinder are designed according to the application requirements.

The internal core is a permanent magnet selected to maximize the magnetic field density and consequently the induced voltage in the coil. The coil is fixed to the geophone housing by means of leaf springs (membranes). These springs are designed to ensure the alignment during the relative motion between coil and magnet, by keeping as low as possible the stiffness in order to minimize the geophone resonant frequency.

The reverse configuration shown in Figure 8 is realized using a coil fixed to the housing while the moving mass is the permanent magnet. Since the mass of the magnet is heavier than that of the coil, this configuration leads to a lower natural frequency, but the moving part is larger and heavier.

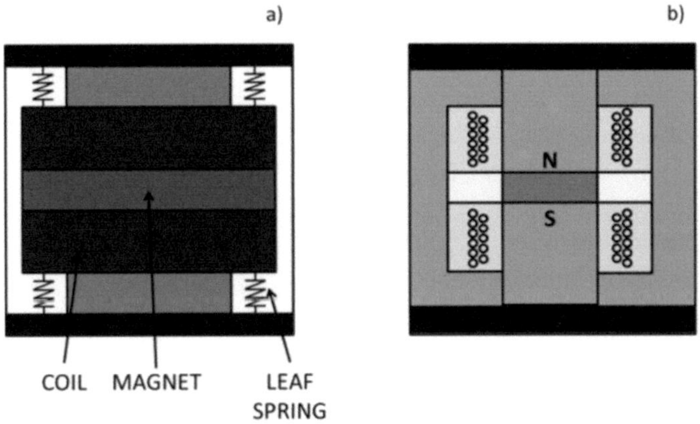

Figure 7. Geophone active configuration scheme. a) Coil and springs installation. b) Cross section.

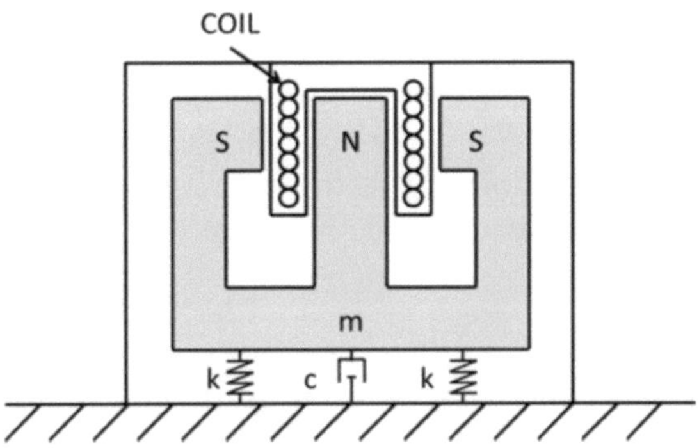

Figure 8. Geophone reverse configuration scheme.

Two different geophones of the Input-Output Inc. sensors have been tested: an active sensor model LF24 (configuration in Figure 7) and a passive sensor model SM6 (configuration in Figure 8). The LF-24 Low Frequency Geophone is characterized by the following parameters: natural frequency at 1Hz, distortion measurement frequency at 12Hz and sensitivity equal to 15V/(m/s).

The sensor chosen is the passive model SM6 because it allows to have an extreme low noise, though the output needs to be amplified by an active conditioning stage.

The sensor response transfer between the velocity of the housing and the induced voltage in the coil, can be written in the well known second order form:

$$TFG = -\frac{Gs^2}{s^2 + 2\xi\omega_n s + \omega_n^2} \tag{5}$$

where $\omega_n = \sqrt{K/m}$ is the natural frequency of the geophone, $\xi = C/2m\omega_n$ is the damping ratio including the eddy current effects and $G = Bl$ is the transduction constant, where B is the magnetic field generated by the permanent magnet and l is the length of the coil.

Considering that the first natural frequency of the system is at about 1.8 Hz, close to the geophone natural frequency, the sensor sensitivity cannot be simply modeled as a constant value. Thus the transfer function of the geophone response must be identified to make the result more reliable.

SM6 geophone is a passive velocity sensor with the following parameters: natural frequency 4.5Hz and sensitivity 28V/(m/s). The damping ratio coefficient has been experimentally identified for both sensors and is equal to 1 (model SM6 is represented in Figure 9.a and model LF24 in Figure 9.b).

Since the generated voltage is proportional to the crossing rate of the magnetic field, the output of the sensor will be proportional to the velocity of the vibrating body. A typical instrument of this kind may have a natural frequency between 1 Hz to 5 Hz. The sensitivity of this kind of sensor is in the range 2-3.5 V/ms^{-1} with the maximum peak to peak displacement limited to about 5 mm (Thomson, 1981). When a geophone is used to measure vibrations with a frequency below its natural frequency, the proof-mass tends to follow the motion of the vibrating body rather than staying stationary. This motion of the proof-mass reduces the relative motion between the same proof-mass and the housing decreasing the induced voltage. In these conditions the sensitivity of the sensor (ratio between the voltage and the casing velocity) becomes very small limiting its range of usage to frequencies above its corner frequency. It is important to underline that both displacement and acceleration can be obtained from the velocity by means integration and differentiation operations.

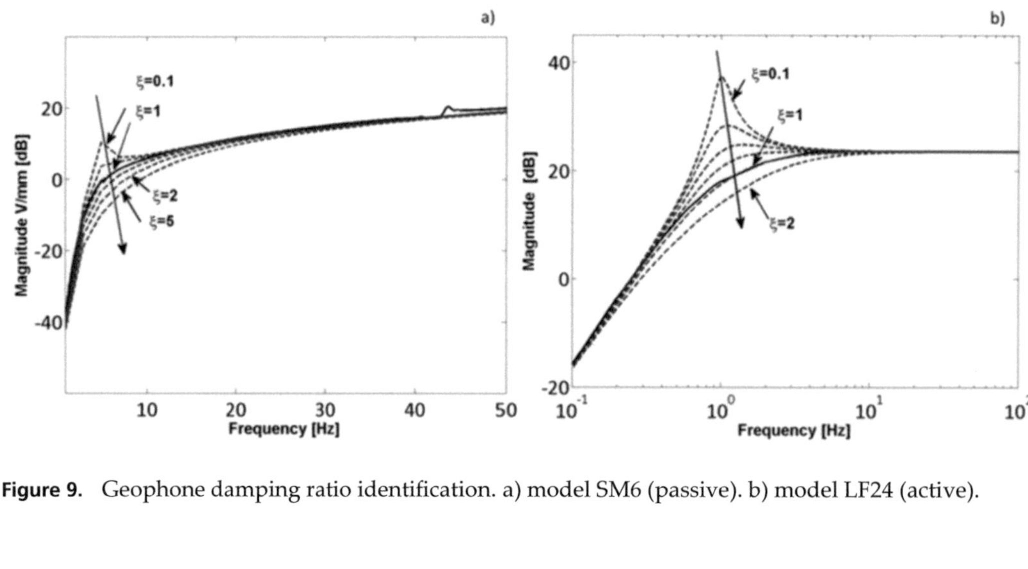

Figure 9. Geophone damping ratio identification. a) model SM6 (passive). b) model LF24 (active).

3.4. Electronics subsystem

In this section the subsystems related to sensor acquisition and conditioning, power electronics and control implementation (Sensor Conditioning, Power Electronics, Feedforward Control, and Feedback Control in Figure 3) are illustrated.

The electronics system architecture is shown in Figure 10. The main characteristic of this architecture is the serial communication input/output line that provides high noise immunity, which can be useful when signals must travel through a noisy environment, such as with remote sensors.

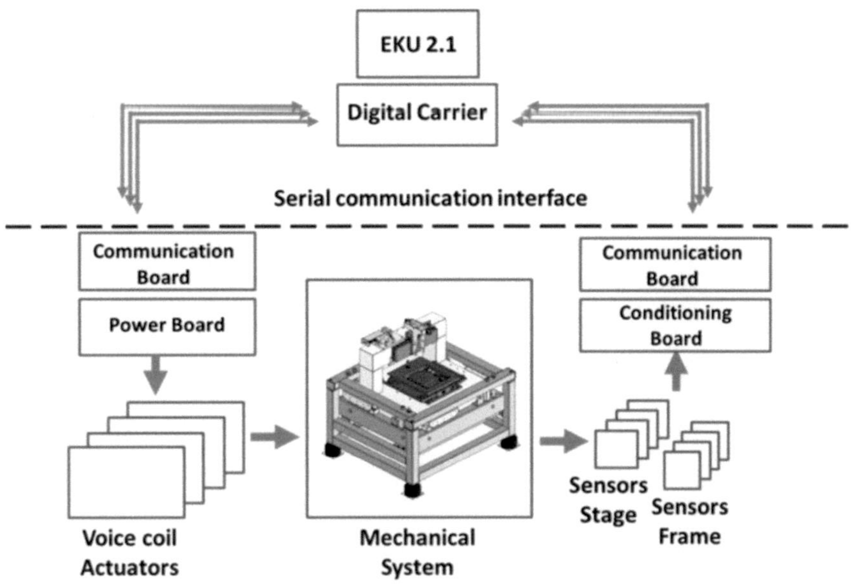

Figure 10. Electronics subsystem.

The digital carrier is used like a buffer to provide the proper current level for the serial communication. Here, multiples system buses manage data exchange between the main serial communication core (FPGA) and the communication boards placed on the plant.

The communication boards are provided with one digital-to-analog converter (DAC) and two analog-to-digital converters (ADC). The DAC is a 16-bit, high-speed, low-noise voltage-output DAC with 30-MHz serial interface that is capable of generating output signal frequencies up to 1 MHz. The ADC is a single channel 12-bit analog-to-digital converter with a high-speed serial interface and sample rate range of 50 ksps to 200 ksps.

Control Unit

The control modules are supported by a DSP/FPGA–based digital control unit. Hence the overall control implementation can be divided between the two digital devices in order to fulfill different requirements: control strategy realization on DSP and serial communication implementation on FPGA.

The overall control strategy is characterized with a nested and decentralized control structure, where only the outer loop is implemented on DSP while the inner current loop is realized on the power module directly. In particular, the outer loop computes the right reference for the inner one starting from required error compensation. The same strategy is applied for each axis.

Sensors Conditioning

The Sensors Conditioning Module provides the output signal from geophone by means of an instrumentation amplifier circuits. The component is configured for dual-channel operation, in order to connect two geophones together. Figure 11, shows the circuit layout for dual-channel. R1A and R1B are the gain setting resistors.

With the ADC input in the range [0-3] V and assuming the maximum magnitude of noise in geophone measurement nearly equal to 1000 m/s, the setting resistors are selected to achieve a gain of 100.

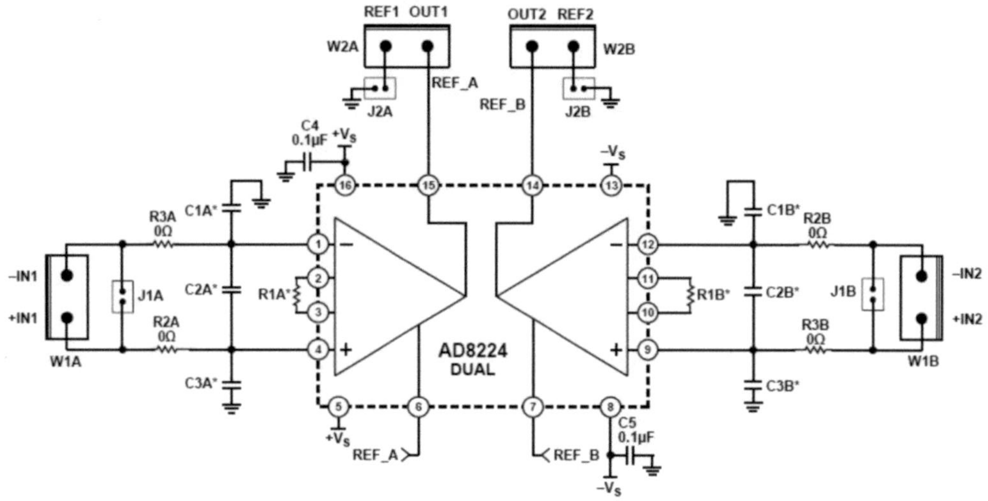

Figure 11. Instrumentation amplifier circuits AD8224. R1A and R1B are the gain setting resistors.

Power Electronics

The Power Electronics Module is based on a trans-conductance amplifier instead of a switching amplifier in order to avoid noise due to the switching frequency. This kind of amplifier operates as a voltage-to-current converter whit a differential input voltage (voltage controlled current source configuration).

The electronics layout that is divided in three main stages: a) the trans-conductance amplifier, b) the current amplifier and c) the feedback resistor.

The power module uses the voltages reference $\left(V_{in}\right)$ from the control unit to generate the proper current $\left(I_L\right)$ to the load (electromagnetic actuator assumed as a RL load). The first

stage performs the current control by means of an operational amplifier that is unity-gain stable with a bandwidth of 1.8MHz and it is internally protected against over-temperature conditions and current overloads. The second stage is a classical current amplifier with bipolar transistors in Darlington configuration to increase the current gain. The last stage provides the feedback signal to ensure the desired current in the load. The power supply is in the range of ±30V.

4. Modeling

Two different models have been developed to permit the design of the three proposed control strategies:

1. Four degrees of freedom model used for the design of: a) a feedback controller with a Lead Lag approach, b) Feedforward control strategy.
2. Six degrees of freedom model used for the design of c) Modal controller.

4.1. Four degrees of freedom model

The system has been modeled by using four degrees of freedom describing the dynamics in *YZ* plane. Four flexural steel springs have been used to link the stage to the frame, four air springs are placed at the bottom of the frame, two actuators are working in series between the stage, and the frame and two geophones are used to measure the velocities of stage and frame respectively. As the axial stiffness of the flexural springs is very high, it can be assumed that there is no relative displacement between stage and frame along the vertical direction, which means that the relative displacement along the z axis between stage and frame are the same. Both stage and frame are assumed as moving about the frame mass center with the same rotating speed. The model reference frames are defined in Figure 2 (XY -plane view) and in Figure 13 (YZ -plane view).

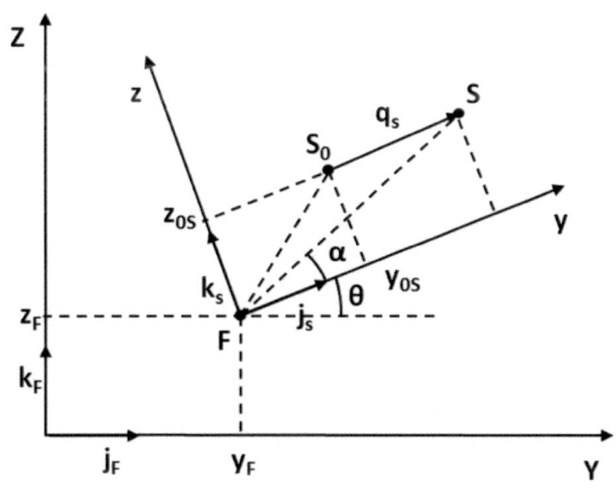

Figure 12. YZ plane 4 dof kinematic relationships scheme.

The degrees of freedom of the model are:

$$\bar{X} = \left[y_F; \quad z_F; \quad \theta; \quad q_S \right] \tag{6}$$

that indicate the displacement of the frame along Y-axis and Z-axis, the rotation of the frame (and stage) around the X-axis mass center and the stage displacement along its Y-axis.

Referring to Figure 12, it is possible to obtain the formulation of the velocity of a generic point S of the stage:

$$\begin{aligned} \vec{V}_S &= \vec{V}_F + \dot{q}_S \vec{j}_S + \dot{\theta} \overline{FS_0} \left[\cos\alpha \vec{k}_S - \sin\alpha \vec{j}_S \right] = \\ &= \dot{y}_F \vec{j}_F + \dot{z}_F \vec{k}_F + \left(\dot{q}_S - \dot{\theta} z_{0S} \right) \vec{j}_S + \dot{\theta} (y_{0S} + q_S) \vec{k}_S \end{aligned} \tag{7}$$

The kinetic energy T of the system can be expressed as:

$$T = \frac{1}{2} m_S \vec{V}_S^{\,2} + \frac{1}{2} J_S \dot{\theta}^2 + \frac{1}{2} m_F \vec{V}_F^{\,2} + \frac{1}{2} J_F \dot{\theta}^2 \tag{8}$$

Where m_S and J_S are the mass and the rotating inertia measured in the center of mass of the stage S, and m_F and J_F the mass and the rotating inertia measured in the center of mass of the frame F.

The potential energy U is obtained starting from the diagram reported in Figure 13.

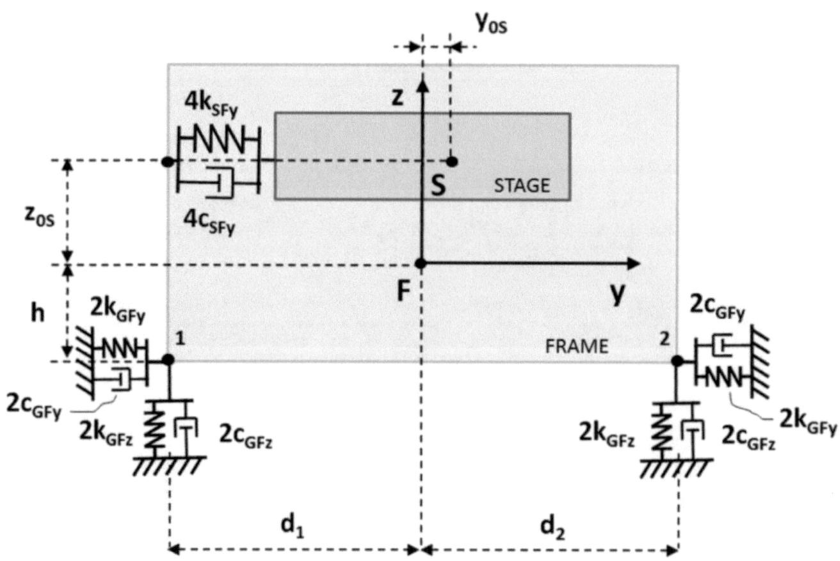

Figure 13. YZ plane 4 dof model scheme.

The potential energy U is:

$$U = k_{GFz}\left(z_F - \theta d_1 - z_G\right)^2 + k_{GFy}(y_F + \theta h - y_G)^2 + k_{GFz}\left(z_F + \theta d_2 - z_G\right)^2 + \ldots$$
$$+k_{GFy}(y_F + \theta h - y_G)^2 + k_{SFy}{q_S}^2 \tag{9}$$

where y_G and z_G are the displacement of the ground and d_1, d_2, and h the quantities reported in Figure 13.

Owing to the Rayleigh formulation, the damping of the system is governed by the following dissipation function:

$$\Re = c_{GFz}\left(\dot{z}_F - \dot{\theta} d_1 - \dot{z}_G\right)^2 + c_{GFy}(\dot{y}_F + \dot{\theta} h - \dot{y}_G)^2 + c_{GFz}\left(\dot{z}_F + \dot{\theta} d_2 - \dot{z}_G\right)^2 + \ldots$$
$$+c_{GFy}(\dot{y}_F + \dot{\theta} h - \dot{y}_G)^2 + c_{SFy}{\dot{q}_S}^2 \tag{10}$$

where each damping term c_i is obtained starting from the experimental identification of damping ratios ς_i:

$$c_i = 2\varsigma_i\sqrt{k_i m_i} \tag{11}$$

The inputs of the system are: the force of the electromagnetic actuators F_{act}, the force of the stage F_S and the velocities from the ground in y direction v_{Gy} and z direction v_{Gz}. The output are the velocities v_F of the frame and v_S of the stage measured with geophones sensors. Inputs and outputs are graphically represented in Figure 14.

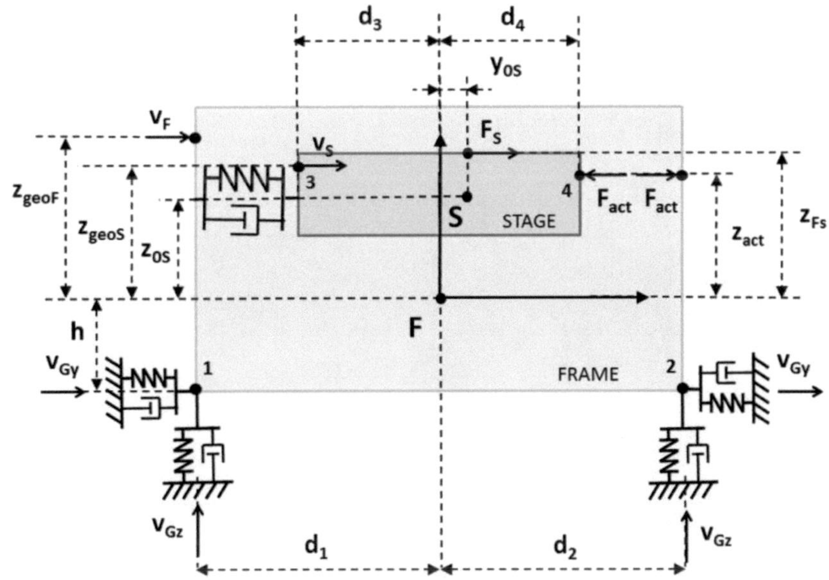

Figure 14. YZ plane 4dof model scheme – input and output.

Using the Lagrange formulation is possible to write the equations of motion in the form:

$$\mathrm{M}\ddot{q} + C\dot{q} + \mathrm{K}q + T_G q_G = \mathrm{T}\{F\} \tag{12}$$

where

$$q = \begin{pmatrix} y_F & z_F & \vartheta & q_S \end{pmatrix}^T \tag{13}$$

$$q_G = \begin{pmatrix} y_G & z_G \end{pmatrix}^T$$

are the vectors of the generalized coordinates,

$$F = \begin{pmatrix} v_{Gy} & v_{Gz} & F_S & F_{act} \end{pmatrix}^T \tag{14}$$

is the vector of the generalized forces and M is the mass matrix

$$M = \begin{bmatrix} m_{tot} & 0 & -m_S z_{0S} & m_S \\ 0 & m_{tot} & m_S y_{0S} & 0 \\ -m_S z_{0S} & m_S y_{0S} & J_{tot} & -m_S z_{0S} \\ m_S & 0 & -m_S z_{0S} & m_S \end{bmatrix} \tag{15}$$

with $m_{tot} = m_S + m_F$, $J_{tot} = J_S + J_F + m_S(y_{0S}{}^2 + z_{0S}{}^2)$. y_{0S}, z_{0S} are the initial position of the stage. The matrix is symmetric and not diagonal because it takes into account the coupling between the stage and the frame dynamics.

The stiffness matrix K is:

$$K = \begin{bmatrix} 4k_{GFy} & 0 & 4k_{GFy}h & 0 \\ 0 & 4k_{GFz} & -2k_{GFz}d_1 + 2k_{GFz}d_2 & 0 \\ 4k_{GFy}h & -2k_{GFz}d_1 + 2k_{GFz}d_2 & 2k_{GFz}d_1{}^2 + 4k_{GFy}h^2 + 2k_{GFz}d_2{}^2 & 0 \\ 0 & 0 & 0 & 4k_{SFy} \end{bmatrix} \tag{16}$$

The damping matrix C is:

$$C = \begin{bmatrix} 4c_{GFy} & 0 & 4c_{GFy}h & 0 \\ 0 & 4c_{GFz} & -2c_{GFz}d_1 + 2c_{GFz}d_2 & 0 \\ 4c_{GFy}h & -2c_{GFz}d_1 + 2c_{GFz}d_2 & 2c_{GFz}d_1{}^2 + 4c_{GFy}h^2 + 2c_{GFz}d_2{}^2 & 0 \\ 0 & 0 & 0 & 4c_{SFy} \end{bmatrix} \tag{17}$$

The selection matrix T of the generalized forces is:

$$T = \begin{bmatrix} -4c_{GFy} & 0 & 1 & 0 \\ 0 & -4c_{GFz} & 0 & 0 \\ 4c_{GFy}h & 2c_{GFz}d_1 - 2c_{GFz}d_2 & 0 & 0 \\ 0 & 0 & 1 & -1 \end{bmatrix} \quad (18)$$

$$T_G = \begin{bmatrix} -4k_{GFy} & 0 \\ 0 & 4k_{Gzy} \\ 4k_{GFy}h & 2k_{GFz}d_1 - 2k_{GFz}d_2 \\ 0 & 0 \end{bmatrix}$$

In the state space formulation the equations of motion of the system can rewritten as:

$$\dot{X} = AX + BU \quad (19)$$

where the state vector X and the input vector are:

$$X = \{q \quad \dot{q} \quad q_G\}^T, U = \{v_{Gy} \quad v_{Gz} \quad F_S \quad F_{act}\}^T \quad (20)$$

with A the state matrix, B the input matrix

$$A = \begin{bmatrix} 0 & I & 0 \\ -M^{-1}K & -M^{-1}C & M^{-1}T_G \\ 0 & 0 & 0 \end{bmatrix}, \qquad B = \begin{bmatrix} 0 \\ M^{-1}T \\ I \end{bmatrix} \quad (21)$$

The relationship between input and output can be represented as:

$$Y = CX + DU \quad (22)$$

where Y is the output vector, C the output matrix and D the feedthrough matrix

$$Y = \{v_S \quad v_F\}^T, C = \begin{bmatrix} 0 & 0 & 0 & 0 & 1 & 0 & -z_{geoS} & 1 & 0 & 0 \\ 0 & 0 & 0 & 0 & 1 & 0 & -z_{geoF} & 1 & 0 & 0 \end{bmatrix}, D = \begin{bmatrix} 0 & 0 & 0 & 0 \\ 0 & 0 & 0 & 0 \end{bmatrix} \quad (23)$$

4.2. Six degrees of freedom model

As well as the dynamics on the YZ plane described in the previous section, it has been developed a six degrees of freedom model of system dynamics on the XY plane. In this case, the degrees of freedom of the model are:

$$\overline{X} = [x_S; \; y_S; \; \theta_S; \; x_F; \; y_F; \; \theta_F] \quad (24)$$

indicating the stage displacements x_S along X-axis, y_S along Y-axis, the rotation θ_S about the axis passing through the mass center and oriented along the Z-axis, the frame displacements x_F along X-axis, y_F along Y-axis, and the rotation θ_F about the axis passing through the mass center oriented along the Z-axis. Stage and frame degrees of freedom, inputs, and geometric properties are illustrated in Figure 15 and 16.

Resorting to the Lagrange formulation as reported in (12), the q vector of the generalized coordinates is:

$$q = \begin{pmatrix} x_S & y_S & \theta_S & x_F & y_F & \theta_F \end{pmatrix}^T \tag{25}$$

and the F the vector of the generalized forces is

$$F = \begin{pmatrix} F_{X+} & F_{X-} & F_{Y+} & F_{Y-} \end{pmatrix}^T \tag{26}$$

it is possible to obtain the corresponding mass matrix M, stiffness matrix K and damping matrix C (not reported due to its excessive size).

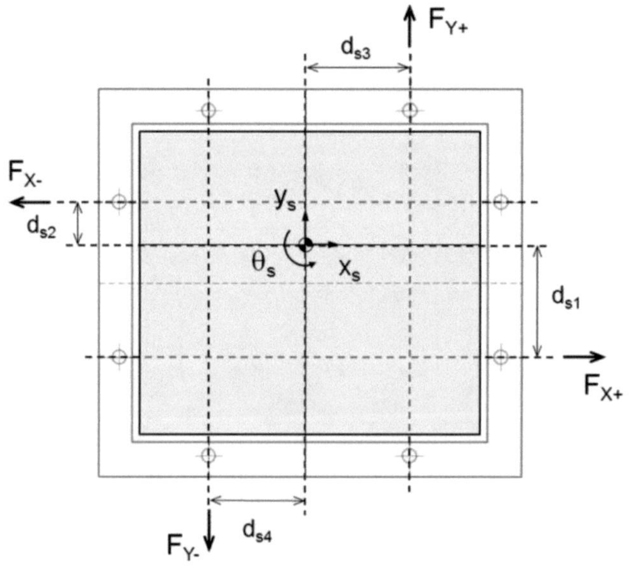

Figure 15. XY Plane 6 dof model scheme: stage degrees of freedom and inputs.

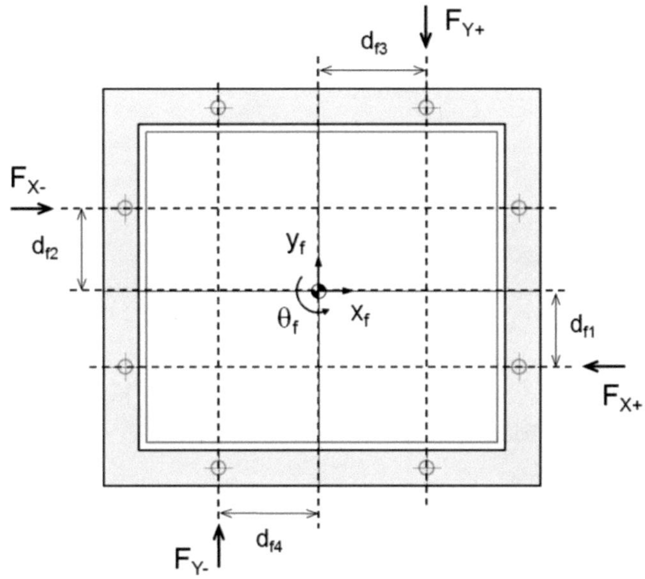

Figure 16. XY Plane 6 dof model scheme: stage degrees of freedom and inputs.

The selection matrix T of the generalized forces is:

$$T = \begin{bmatrix} 1 & -1 & 0 & 0 \\ 0 & 0 & 1 & -1 \\ d_{s1} & d_{s2} & d_{s3} & d_{s4} \\ -1 & 1 & 0 & 0 \\ 0 & 0 & -1 & 1 \\ -d_{f1} & -d_{f2} & -d_{f3} & -d_{f4} \end{bmatrix} \tag{27}$$

Similarly in the state space formulation the equations of motion of the system can rewritten as:

$$\dot{X} = AX + BU \tag{28}$$

where the state vector X and the input vector U are:

$$X = \{q \quad \dot{q}\}^T, U = \{F\}^T \tag{29}$$

with the following state and input matrix

$$A=\begin{bmatrix} 0 & I \\ -M^{-1}K & -M^{-1}C \end{bmatrix}, B=\begin{bmatrix} 0 \\ M^{-1}T \end{bmatrix} \tag{30}$$

The relationship between input and output can be represented as:

$$Y=CX+DU \tag{31}$$

where Y is the output vector that contains the derivative time of the generalized coordinates (25):

$$Y=\{\dot{q}\}^T \tag{32}$$

C is the output matrix and D the feedthrough matrix:

$$C=\begin{bmatrix} 0 & I \end{bmatrix}, D=\begin{bmatrix} 0 \end{bmatrix} \tag{33}$$

5. Control design & results

In this section three different control strategies to damp vibration and isolate the machine are proposed: a) Feedback control by the use of a Lead-Lag technique, b) Feedforward control and c) Modal control. The experimental validation has been carried out just for the first strategy as proof of the correctness of the modeling approach. Feedforward and modal controls are validated numerically.

5.1. Feedback control

The control action is designed to achieve two main goals: active isolation of the payload from the ground disturbances and vibration damping during the machine work processes. These two actions allow to operate on the stage without external disturbances. Dynamics on XZ and YZ -planes are considered the same and decoupled so the control laws along the two planes are equivalent.

Furthermore, from the control point of view, the adopted model is oversized with respect to the control requirements if the goal is the isolation of the stage. As a matter of fact, in this case a two degrees of freedom model is sufficient while if also the dynamics of the frame is required to be controlled, then a 4 dof model is necessary.

The considered system can be regarded as intrinsically stable due to the presence of mechanical stiffness between the stage and the frame, which allows to obtain a negative real part for all the eigenvalues of the system.

Root loci of the system in open and closed loop configurations are reported in Figure 17.

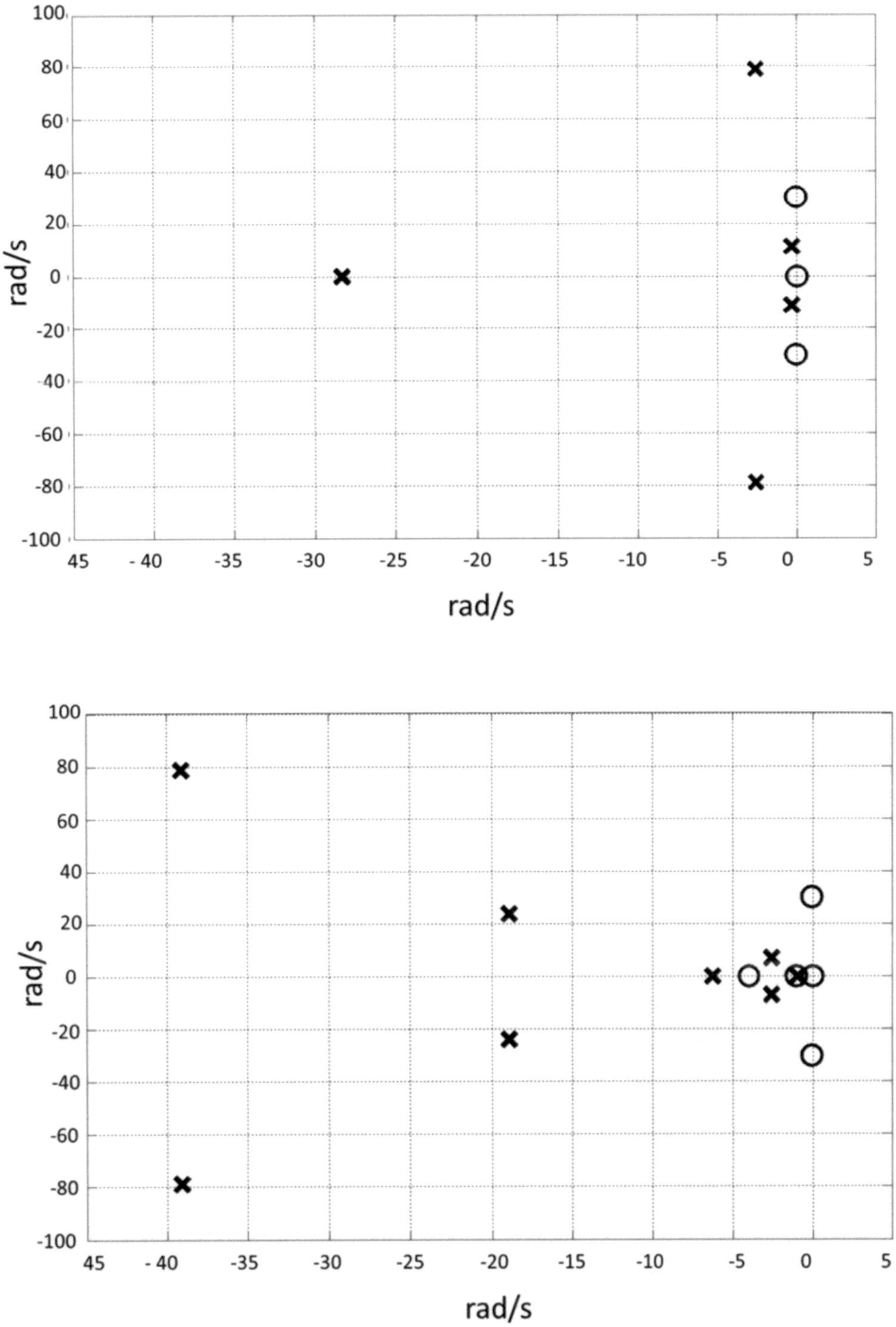

Figure 17. Root loci of open loop (a) and closed loop (b) configurations (Circles: zeros; Crosses: poles).

Poles and zeros of the system are reported in Table 4.

	Poles [rad/s]	Zeros [rad/s]
Geophone	-28.2743	-
	-28.2743	-
Feedback Controller	-1	-75
	-4	-12
Open Loop Plant	-6.4136 + 79.8805i	-5.8786 + 80.8581i
	-6.4136 - 79.8805i	-5.8786 - 80.8581i
	-3.7383 + 46.4898i	-2.0791 + 37.9468i
	-3.7383 - 46.4898i	-2.0791 - 37.9468i
	-0.2371 + 10.9133i	-1.9325 + 33.6912i
	-0.2371 - 10.9133i	-1.9325 - 33.6912i
	-1.9557 + 33.7951i	0
	-1.9557 - 33.7951i	0
	-28.2743	0
	-28.2743	-
Close Loop Plant	-6.3137 + 79.8109i	-5.8786 + 80.8581i
	-6.3137 - 79.8109i	-5.8786 - 80.8581i
	-5.3540 + 46.2140i	-2.0791 + 37.9468i
	-5.3540 - 46.2140i	-2.0791 - 37.9468i
	-1.9649 + 33.7933i	-1.9325 + 33.6912i
	-1.9649 - 33.7933i	-1.9325 - 33.6912i
	-27.7026 + 12.6862i	-1
	-27.7026 - 12.6862i	-4
	-1.8090 + 9.2149i	0
	-1.8090 - 9.2149i	0
	-4.7413	0
	-0.9885	-

Table 4. Poles and zeros of the system

Since the system along $YZ\,(XZ)$ presents one actuation point and a couple of sensors (frame and stage velocities), a solution with a SISO control strategy is not feasible. A simplest solution to this problem considers the difference between the measured velocities as the feedback signal, so the system can be assumed as SISO and the control design becomes simpler.

Figure 18 shows that the system dynamics has a peak at 1.8 Hz related to the stage and higher modes related to the interaction of the stage with the frame and the ground at 10 Hz and beyond.

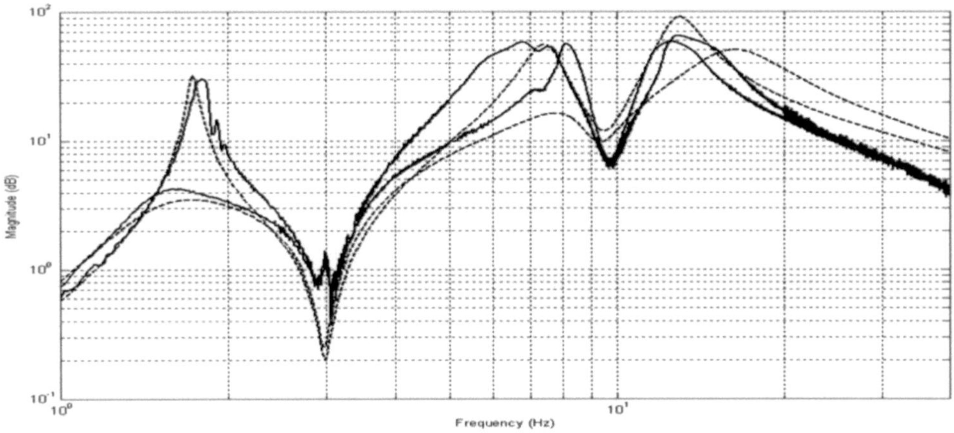

Figure 18. Vibration damping action. Transfer function from the actuator force to the difference of frame and stage velocities ($(\dot{q}_S - \dot{q}_F)/F_{ACT}$). Open-loop vs Closed-loop. Solid line: experimental; Dashed line: numerical.

The feedback controller is focused on damping the mode related to the stage by adding on the loop a lead-lag compensator.

The two actions can be expressed as:

$$C_{LAG} = \frac{s + z_{LAG}}{s + p_{LAG}} \qquad C_{LEAD} = \frac{s + z_{LEAD}}{s + p_{LEAD}} \tag{34}$$

The C_{LAG} action is used to improve the transient response at low frequency, while the C_{LEAD} is useful to increase the stability margin of the closed-loop system.

Therefore the resulting Lag-Lead action allows to compensate the critical phase behavior of the geophones and furthermore guarantees a quick damping action with good levels of stability margins.

The experimental tests have been performed to validate the two control actions. Figure 18 shows the numerical and experimnental frequency response function in open loop and closed loop, obtained from the actuator force to the velocity measured on the stage. The force acts both on the stage and the frame, the dynamics of both the subsystems are visible. The vibration damping effect of the control action is validated on the stage mode (1.8 Hz peak) and the good correspondence shown between the simulated and experimental response is useful to validate the modeling approach.

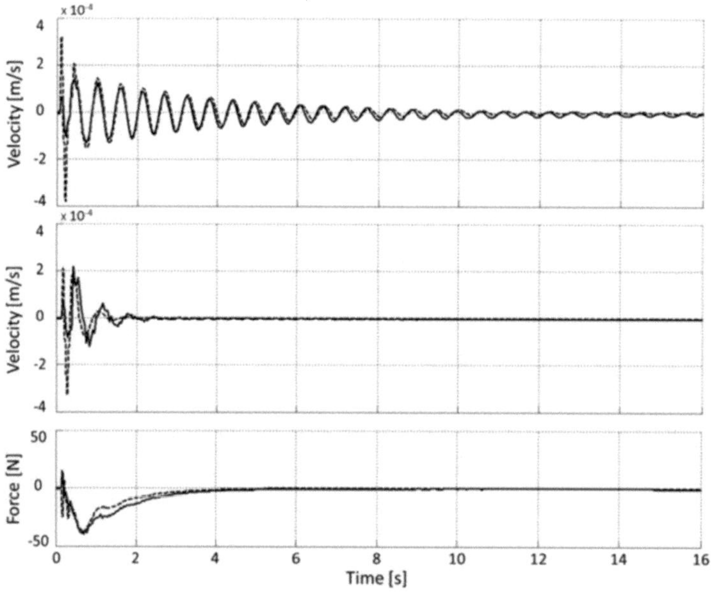

Figure 19. Impulse time response, force from the actuator and velocity measured on the stage. Open-loop (a), Closed-loop (b), Force exerted by the actuators. Solid line: experimental results. Dashed line: numerical results.

A further demonstration of the correctness of the damping action is the velocity time response reported in Figure 19. In this case the system is excited with an impulse from the actuator and the velocity is measured on the stage. Numerical and experimental responses are superimposed to provide a further validation of the model (the position time response is not reported since the machine is not provided with displacement sensors and hence this validation could not be possible to performed). Figure 19.a shows open loop response, Figure 19.b shows closed loop response while in Figure 19.c the force exerted by the actuators is reported.

The excitation coming from the laser-axis action on the stage is controlled in an effective way as shown in Figure 20 where the numerical transfer function between a force impulse on the stage and the related measured velocity is reported.

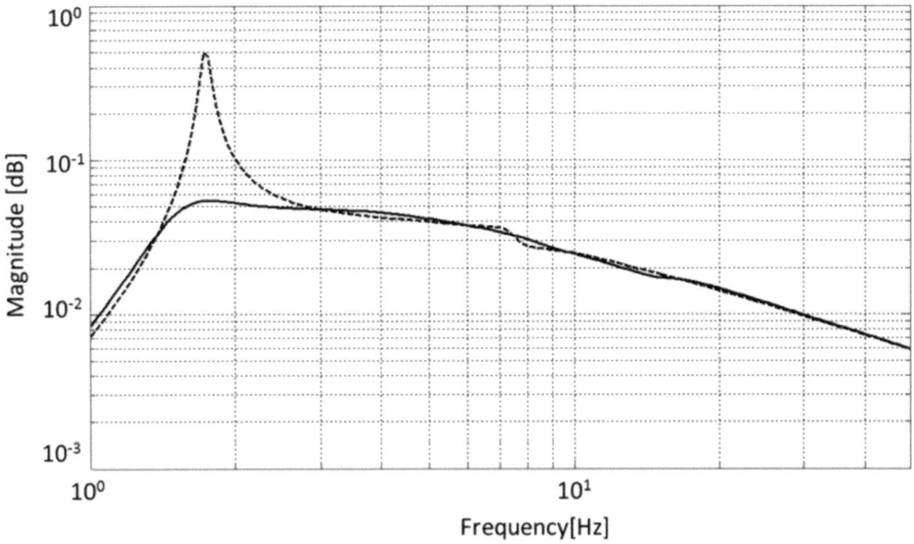

Figure 20. Vibration damping action. Transfer function from a force applied on the stage to the velocity measured on the stage $\left(\dot{q}_S/F_S\right)$. Numerical response. Solid line: closed-loop; Dashed line: Open-loop.

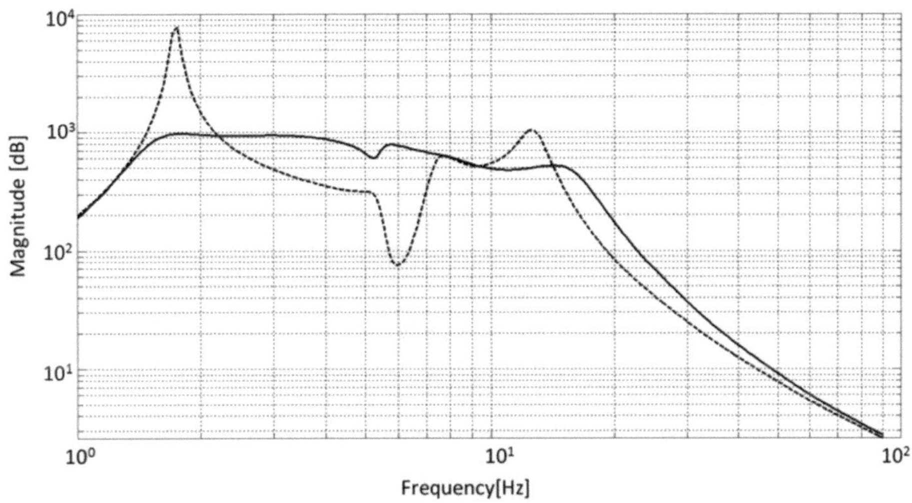

Figure 21. Active isolation action. Transfer function from a simulated ground velocity to the velocity measured on the stage $\left(\dot{q}_S/\dot{q}_G\right)$. Numerical response. Solid line: Open loop configuration. Dashed line: Closed loop configuration.

The active isolation action is verified by simulating the excitation coming from the ground. The experimental test in this case has not been performed since in reality it is difficult to excite the machine from the ground in a controlled and effective way. Nevertheless the model is reliable as proved in Figure 14 and the obtained results can be assumed as a good validation of the control action.

Figure 21 illustrates that the closed loop system is capable to reject the disturbances coming from the ground in an effective way.

5.2. Feedforward control

Although the feedback control explained in Section 5.1 is strongly effective for external disturbances coming from the ground, it could not be sufficient to make the machine completely isolated from the direct disturbance generated by the movement of the payload. It is indeed possible that in the case of high precision requests, feedback control approaches such as PID, Lead-Lag or LQR are not able to satisfy by themselves severe specifications. Hence different schemes, operating selectively on the stage direct disturbances, are required.

In this section an off-line feedforward scheme allowing to isolate the machine from the action of payload direct disturbance in operating condition is proposed. The scheme is not classical, i.e. the command is not generated on-line but it is computed in advance on the basis of the data response to the direct disturbance and the transfer function between the control command and the controlled output. As illustrated in Figure 3, the action of feedforward control is superimposed to the one of the Lead-Lag feedback control and acts exclusively on the disturbance acting from the payload.

The technique is based on the complete knowledge of the fixed pattern followed by the payload of the machine during operations. Since also the operation timing is known, it is possible to compute in advance a feedforward command, so as to be able to suppress the effects of the direct disturbance that are generated by the payload movements, and that cannot be measured. These commands are stored in the electronic control unit and are summed to the feedback control action at the appropriate time.

The model used to design the control law is the four degrees of freedom model exposed in Section 4.1. Being the XZ-plane and YZ-plane symmetric, just the latter is considered in the design phases.

The controlled output is the velocity measured on the stage $v_s(s)$ and it can be considered as the sum of two contributions: the effect of the direct disturbance on the output $v_{Ds}(s)$ and the effect of the feedforward action on the output $v_{FFs}(s)$. Then the total response is:

$$v_s(s) = v_{Ds}(s) + v_{FFs}(s) = v_{Ds}(s) + h(s)u_{FF}(s) \tag{35}$$

where $h(s)$ is the transfer function between the control command $u_{FF}(s)$ to the controlled output $v_{FF}(s)$.

The control signal is:

$$u_{FF}(s) = -h(s)^{-1}v_{Ds}(s) \tag{36}$$

Since the operation pattern and timing are known (Figure 23 (a)), the transfer function $h(s)$ can be obtained by using an FFT analyzer, the command signal $u_{FF}(s)$ (Figure 23 (b)) can be computed offline, stored in the control unit and applied to the system at the proper time when the payload is moving.

It is worthy to notice that the inversion of $h(s)$ leads to a non-causal function with a numbers of zeros equal or higher than the number of poles. This issue is overcome by adding the required number of poles at a frequency sufficiently high (more than 100 Hz), in order to make the feedforward filter proper and fit to be used in the control scheme.

Bode diagram of *h(s)* is reported in Figure 22 (feedback control is on, vibrations coming from the ground are damped).

Figure 23 (c) shows that the proposed technique is effective and allows to isolate the machine from the direct disturbance generated by the payload operations. The excitation signal reproduces a standard laser cut periodic profile.

The coupling of this action with the feedback control system permits to obtain a full vibration damping and active isolation from external disturbance coming from the ground and direct disturbance coming from the stage.

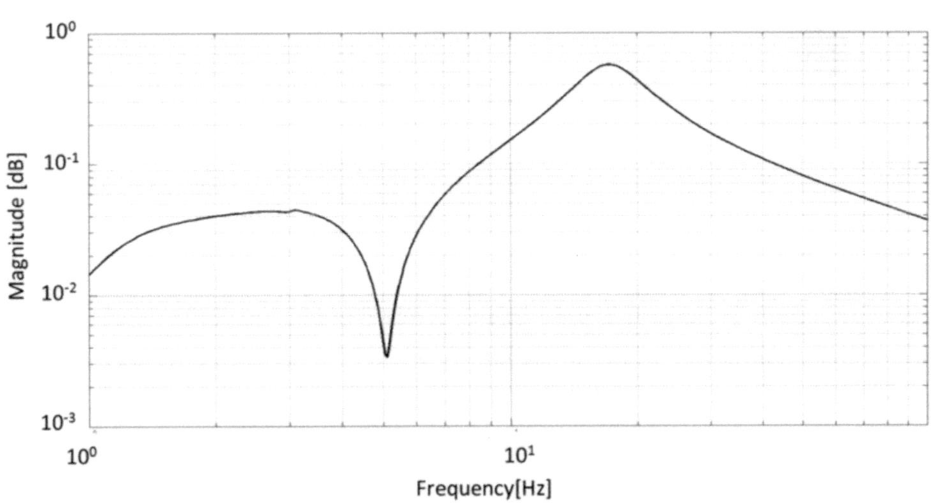

Figure 22. Control command to controlled output stage velocity transfer function (h(s)) Bode diagram.

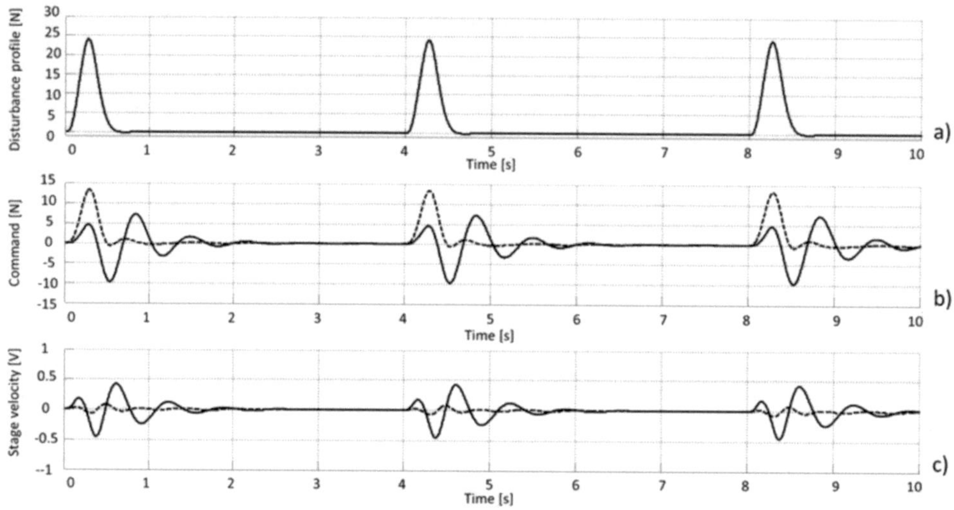

Figure 23. a) Feedforward control: disturbance profile; b) Control signal. Solid line: feedforward off, dashed line: feedforward on; c) Controlled output: stage velocity. Solid line: feedforward off, dashed line: feedforward on.

5.3. Modal control

The third and last control technique proposed in this chapter is a modal approach to perform a feedback control scheme. This strategy is similar in performance to the Lead–Lag strategy illustrated in Section 5.1, but it simplifies the control design procedure once it gives a direct feeling on actuators action on machine modes.

The method is based on the scheme reported in Figure 24. The goal of the technique is to decouple the rotational and translational motion modes of the machine to direct the action of the controller selectively on the dynamic of interest.

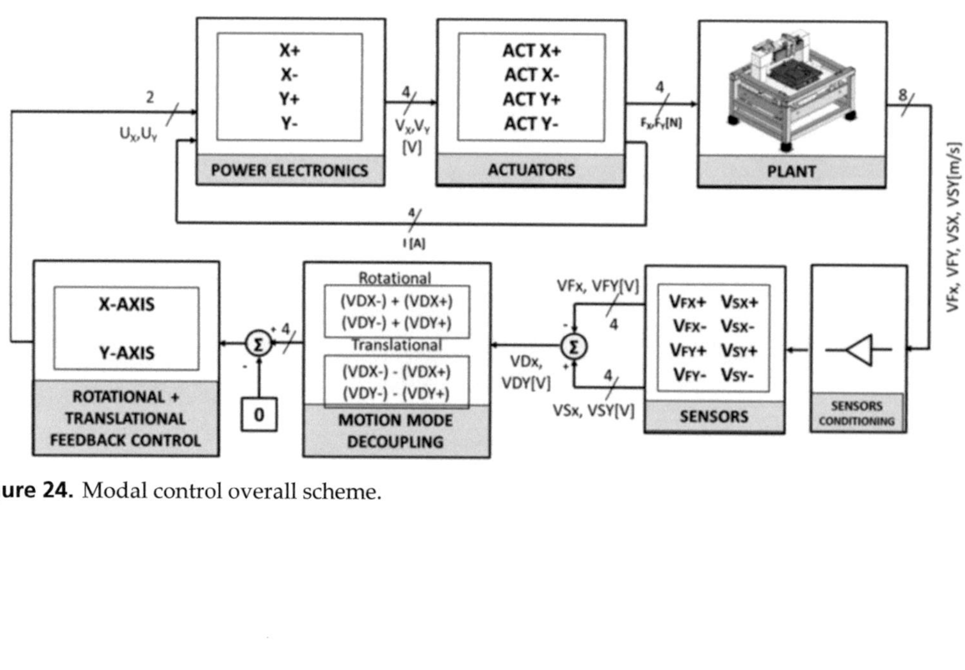

Figure 24. Modal control overall scheme.

The eight geophones measurements on stage and frame are elaborated to obtain four velocity differences:

$$\begin{aligned} V_{DX+} &= V_{SensSX+} - V_{SensFX+} \\ V_{DX-} &= V_{SensSX-} - V_{SensFX-} \\ V_{DY+} &= V_{SensSY+} - V_{SensFY+} \\ V_{DY-} &= V_{SensSY-} - V_{SensFY-} \end{aligned} \tag{37}$$

These values are then summed and subtracted in order to obtain the motion mode uncoupling.

Rotational mode:

$$\begin{aligned} V_{RX} &= V_{DX+} + V_{DX-} \\ V_{RY} &= V_{DY+} + V_{DY-} \end{aligned} \tag{38}$$

Translational mode

$$\begin{aligned} V_{TX} &= V_{DX+} - V_{DX-} \\ V_{TY} &= V_{DY+} - V_{DY-} \end{aligned} \tag{39}$$

The control dynamic is the same of Lead-Lag approach, the difference consisting in the error fed to the controller. The poles of the system in open and closed loop are reported in Table 4.

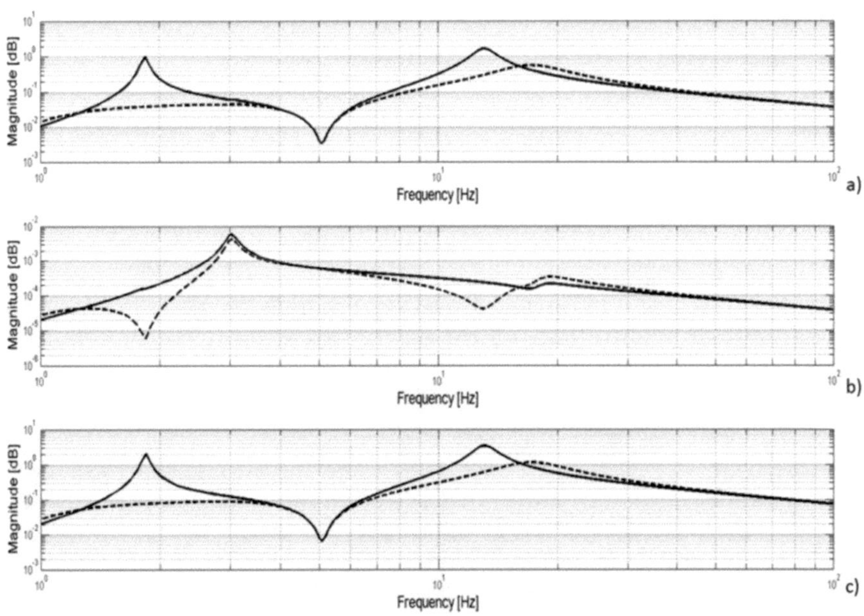

Figure 25. Modal control. a) Control command to stage-frame velocities difference transfer function. b) Control command to translational dynamics transfer function. c) Control command to rotational dynamics transfer function. Solid line: open loop. Dashed line: closed loop.

Figure 25 shows the motion modes uncoupling and system behaviour in open and closed loop. Figure 25.a illustrates control command to stage-frame velocities difference transfer function where translational and rotational modes are coupled. Figure 25.b and Figure 25.c report the translational ($V_{TX,TY}$) and rotational ($V_{RX,RY}$) dynamics respectively. It is worthy to notice that the influence of rotational dynamics is dominant, being its response amplitude higher than translational one. Due to this consideration it can be easily explained the low action of the feedback control on the translational dynamics (b)) is compared to the rotational one (c)).

6. Conclusions

In this chapter the design of three different control techniques for vibration damping and active isolation for high precision laser cutting machines has been illustrated. After an overview on the main actuation technologies in this field the work explains the advantages of electromechanical actuators and focuses on the mechatronics approach of the machine subsystem design. For controller implementation, two different models (four and six degrees of freedom) have been developed. The considered controllers are:

a. Feedback control with a Lead-Lag approach;
b. Off-line Feedforward scheme;
c. Modal control.

Experimental and simulation results used to check the effectiveness of the modeling approach and of the three proposed control techniques.

Author details

Andrea Tonoli[1,3], Angelo Bonfitto[3], Marcello Chiaberge[2,3],
Mario Silvagni[1,3], Lester D. Suarez[3] and Enrico Zenerino[1,3]
1Department of Mechanical and Aerospace Engineering/ Politecnico di Torino, Italy
2Department of Electronics and Telecommunications/ Politecnico di Torino, Italy
3Mechatronics Lab/Politecnico di Torino, Italy

7. References

Anderson E., Leo D.J., Holcomn M.D, (1996). UltraQuiet platform for active vibration isolation, Proceedings Smart Structures and Integrated Systems 2717, San Diego, CA, pp. 436–451.

Beadle B.M., Hurlebaus S., Gaul L., Stöbener U. (2005). Active control strategies for vibration isolation, in Proc. IUTAM Symp. Vibration Control of Nonlinear Mechanisms and Structures 2005, Munich, Germany, pp. 91-100.

Crede, C. (1951). Vibration and shock isolation, John Wiley and Sons Inc., New York, USA

Elliott S.J. (2001). Signal processing for active control, Academic Press.

Hauge, G.S. Campbell, M.E.. (2004). Sensors and control of a space-based six-axis vibration isolation system. Journal of Sound and Vibration 269. Pp 913–931.

Huan, S. L., & Pater, A. R. (1985). Analysis and prediction of geophone performance parameters. Geophysics, 50(8), 1221:1228.

Hyde T.T, (1997) An experimental study of active isolation. Proceedings of the 38th AIAA/ASME/ASCE/AHS/ASC, Structures, Structural Dynamics and Materials Conference, Orlando, FL, pp. 1763–1773.

Kuo S.M., Morgan D.R., (1996). Active noise control systems: algorithms and DSP implementations, John Wiley & Sons.

Preumont, A. (2002). Vibration control of Active Structures. Kluwer Academic Publishers, 2nd Edition, Netherlands

Riedesel, M., Moore, R. D., & Orcutt, J. A. 1990. Limits of Sensitivity of Seismometers with Velocity Transducers and Electronic Amplifiers. Bulletin of the Seismological Society of America, 80(6), 1725:1752.

Rivin, E. I. (1979). Principles and criteria of vibration isolation of machinery. Trans. of the ASME, Journal of Mechanical Engineering, 101:682–692.

Sullivan J., Rahman Z., Cobb R., Spanos J. (1997). Closed-loop performance of a vibration isolation and suppression system, Proceedings of the American Control Conference, Vol. 6, Albuquerque, NM, pp. 3974–3978.

Thayer, D., Vagners, J., von Flotow, A., Hardham, C. and Scribner, K., (1998). "Six Axis Vibration Isolation Using Soft Actuators and Multiple Sensors," Proceedings of the 21st Annual AAS Guidance and Control Conference, Feb 1998.

Thomson, W. (1971). Theory of vibration with applications. George Allen and Unwin.

Tonoli A., Bonfitto A., Silvagni M., Suarez L.D., Zenerino E. (2011) Active Isolation and Damping of Vibrations for High Precision Laser Cutting Machine. In: Vibration Control/Beltrán-Carbajal F. InTech, Rijeka. ISBN 9789533074337

Watters B.G., Coleman R.B., Duckworth G.L., Berkman E.F. (1988). A perspective on active machinery isolation, Proceedings of the 27th Conference on Design and Control, Vol. 3, Austin, TX, pp. 2033–2038.

Chapter 21

Segmented Foot with Compliant Actuators and Its Applications to Lower-Limb Prostheses and Exoskeletons

Qining Wang, Jinying Zhu, Yan Huang, Kebin Yuan and Long Wang

Additional information is available at the end of the chapter

http://dx.doi.org/10.5772/51495

1. Introduction

In recent years, there has been an increasing interest in the functionality of the foot in human normal walking. Different from the existing methods that represent the foot as a single rigid bar, several multi-segmented foot models have been studied to evaluate the effects of the segmented foot structures on human walking for clinical applications [1], adolescent gaits [2] and pediatric gaits [3]. The results show that the segmented foot with a toe joint has several advantages compared to the rigid foot in: walking step, walking speed, range of joint angle, change in angular velocity and joint energy-output. In addition, biomechanical studies conducted on ten donated limbs [4] indicate that the human foot can not be considered as a single rigid body with no intrinsic motion.

Inspired by biological investigations, several studies implemented segmented feet in robotic systems to improve walking performance. One of the main applications is in humanoid robots. Simulations and experiments on prototypes showed that adding toe joints could increase the walking speed of biped robots [5, 6]. These works were carried on humanoid robots based on the trajectory-control approach [7]. By controlling joint angles precisely, the robots can achieve static equilibrium postures at any time during motion. However, this kind of bipedal walking features low resemblance to human normal gaits and high energy consumption [8]. In contrast to the active-control bipedal walking mentioned above, passive dynamic walking [9] has been developed as a possible explanation for the efficiency of the human gait. Investigations on the effects of segmented foot, which are based on passivity-based model, may reveal more insights on real human walking. Though several efforts have been made in adding flat feet to passivity-based models [10–15], only a few studies have investigated passive dynamic bipedal walking model with segmented feet. Recently, [16] proposed a passive dynamic walking model with toed feet. Specifically, in the work, the authors contributed to the investigation of the passive bipedal walking behavior under toe joint rotation. The toe-rotation phase is initiated by ankle-strike. Simulation results showed that the advantages of the proposed walker come from its relation to arc-feet walker.

However, the effects of heel-strike and toe-strike during normal walking are ignored, which may influence the characteristics of bipedal walking [12]. In addition, the phase of rotation of the stance foot about the toe tip is ignored in this model, which makes the bipedal walking gait far from natural human-like gait.

Another area that can use segmented foot is rehabilitation robotics, e.g. lower-limb prostheses and exoskeletons. Although foot prosthesis was invented thousands of years ago, the development of foot prostheses is not as fast as people expect. Most of today's commercial foot prosthesis are passive and do not comprise segmented foot. In 1998, [17] first built a powered ankle-foot prosthesis which was powered by a pneumatic actuator. Then, several pneumatic actuated prostheses have been developed [18, 19]. Though the pneumatic actuator is lightweight, inherently compliant and capable of generating high forces, its control difficulties, large size and noise restrain the development of pneumatic driven prosthesis. Recently, [20] developed a powered ankle-foot prosthesis driven by a DC motor. The motor implemented compliant actuation and was placed on the ankle joint. The prosthesis can provide net positive work to propel the body upward and forward during the stance period. Experimental results show that the prosthesis can decrease the amputee's metabolic cost by 14% on average as compared to the conventional passive-elastic prosthesis. However, the prosthesis functionality is not comparable to that of the human foot because of the absence of segmented foot with toe joint. Similarly, although exoskeletons were invented several decades ago [21], there is no powered toe joint implemented in existing exoskeleton systems, e.g. [22, 23].

In this chapter, we discuss the effects and applications of adding segmented feet with compliant joints to lower-limb prostheses and exoskeletons. To analyze the effects of segmented foot and compliant joints on energetic efficiency and stability of bipedal walking, we first propose a passivity-based dynamic bipedal model which shows resemblance to human normal walking. Phase switching is determined by the direction of ground reaction force. The push-off phase includes rotation around toe joint and rotation around toe tip, which show a great resemblance to natural human gait. The effects of foot structure on motion characteristics including energetic efficiency and walking stability is investigated through simulation experiments. Starting from the theoretical analysis, we introduce segmented foot with toe joint in both ankle-foot prosthesis and exoskeleton prototypes. Both the ankle and toe joints are driven by two series-elastic actuators (SEA), which not only provide the required torque, but also shock tolerance and energy storage during walking. Preliminary studies on sensory based feedback control are carried out to improve the movement of the proposed systems. Experimental results validate the effectiveness of the proposed structure and actuation method.

The rest of the chapter is organized as follows. In section 2, we introduce the idea of adding segmented feet with compliant actuators placed on ankles and toes. Specifically, a theoretical model with segmented feet is proposed which is based on the simplest walking model. In Section 3, detailed investigations are presented to analyze the effects of segmented foot with joint compliance on dynamic walking. Then, starting from the theoretical analysis, the applications of segmented feet to lower-limb prostheses and exoskeletons are introduced in Section 4. After an overview of current compliant actuators in robotics, the development of a lower-limb prosthesis and an exoskeleton with powered compliant ankle and toe joints is presented. In section 5, the basic control method and related experiments on the prototypes are described. Experimental results validate the effectiveness of the proposed systems.

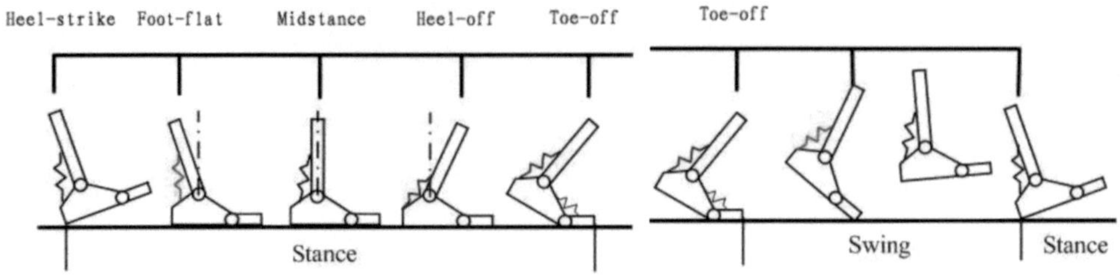

(a) Stance phase (b) Swing phase

Figure 1. One typical gait cycle. Each cycle includes two main subdivisions: (a) stance phase, (b) swing phase. The four important instants in every cycle are heel-strike (HS), foot-flat (FF), heel-off (HO) and toe-off (TO).

2. Dynamic walking model

2.1. Human normal walking gaits

Human walking is a cyclic pattern of bodily movements that is repeated over and over, step after step. Every gait cycle starts with heel-strike (HS) when the heel initially touches the ground and ends with the next HS of the same leg. As shown in Fig. 1, each cycle can be divided into two main phases: stance phase and swing phase [44]. The stance phase begins at the moment of HS and ends at the moment of toe-off (TO) when the forefoot pushes off the ground. The swing phase begins at the moment TO and ends at the next HS. The stance phase takes up 60% of the gait cycle and includes four subphases: 1) Hs to foot-flat (FF); 2) FF to midstance (MS); 3) MS to heel-off (HO); 4) HO to TO.

As introduced in [44], at the moment of HS, the ankle has to endure the impact force when the heel initially contacts the ground. From HS to FF, the ankle stores elastic energy within plantarflexor muscles. During FF to MS, the energy stored during the preceding period is released to help the body change the center of gravity from the support leg to the foreleg. At MS, the ankle begins to store elastic energy to dorsiflexor muscles and at HO the ankle reaches a state of maximum dorsiflexion. Then in the period HO to TO, ankle releases the energy stored in the last period to propel the body. However, the stored energy is much less than the energy needed. Then the ankle has to output much more net positive work. Meanwhile, the forefoot is bent to store energy. At the end of the period HO to TO, the toe joint releases the stored energy and supplements some net work to propel the body. During swing phase, the main function of the ankle and toe joints is to adjust the position of the foot to prepare for the next stance phase.

2.2. Bipedal model

To further investigate the effects of compliant joints and segmented feet on dynamic bipedal walking, in this section, we propose a passivity-based dynamic bipedal walking model that is more close to human beings. In the model, compliant ankle joints and flat segmented feet with compliant toe joints are included. As shown in Fig. 2, the two-dimensional model consists of two rigid legs interconnected individually through a hinge. Each leg contains segmented foot. The mass of the walker is divided into several point masses: hip mass, leg masses, masses of foot without toe, toe masses. Each point mass is placed at the center of corresponding stick. Torsional springs are mounted on both ankle joints and toe joints to represents joint stiffness.

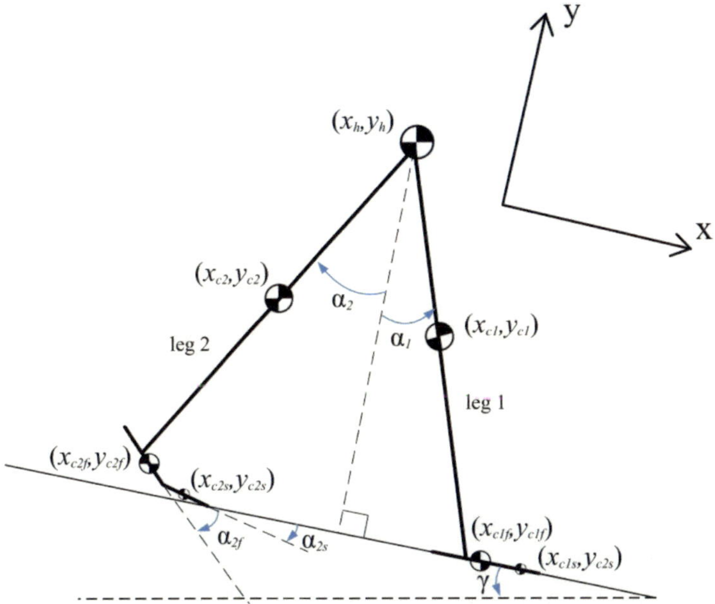

Figure 2. Passivity-based dynamic bipedal walking model with flat segmented feet and compliant ankle joints.

To simplify the motion, we have several assumptions, including that legs suffering no flexible deformation, hip joint with no damping or friction, the friction between walker and ground is enough, thus the flat feet do not deform or slip, and strike is modeled as an instantaneous, fully inelastic impact where no slip and no bounce occurs. The passive walker travels on a flat slope with a small downhill angle.

The process of push-off is dissipated into foot rotation around toe joint and around toe tip, which is the main difference between the passive walking models with rigid flat feet and with segmented flat feet. The toe and foot are restricted into a straight line during the swing phase. When the flat foot strikes the ground, there are two impulses, "heel-strike" and "foot-strike", representative of the initial impact of the heel and the following impact as the whole foot contacts the ground. After foot-strike, the stance leg and the swing leg will be swapped and another walking cycle will begin. The passive walking is restricted to stop in two cases, including falling down and running. We deem that the walker falls down if the angle of either leg exceeds the normal range. The model is considered to running when the stance leg lifts up while the swing foot has not yet contacted the ground. Foot-scuffing at mid-stance is neglected since the model does not include knee joints.

We suppose that the x-axis (hereafter called horizontal coordinate) is along the slope while the y-axis (hereafter called vertical coordinate) is orthogonal to the slope and pointing upwards. The configuration of the walker is defined by the coordinates of the point mass on hip joint and six angles (swing angles between vertical coordinates and each leg, foot angles between horizontal coordinates and each foot, toe angles between horizontal coordinates and each toe), which can be arranged in a generalized vector $q = (x_h, y_h, \alpha_1, \alpha_2, \alpha_{1f}, \alpha_{2f}, \alpha_{1t}, \alpha_{2t})^T$ (see Fig. 2). The positive direction of all the angles are counter-clockwise.

2.3. Walking dynamics

The model can be defined by the rectangular coordinates r, which can be described by the x-coordinate and y-coordinate of the mass points and the corresponding angles (suppose leg 1 is the stance leg):

$$r = [x_h, y_h, x_{c1}, y_{c1}, \alpha_1, x_{c2}, y_{c2}, \alpha_2, x_{c1f}, y_{c1f}, \alpha_{1f}, x_{c2f}, y_{c2f}, \alpha_{2f}, x_{c1s}, y_{c1s}, \alpha_{1s}, x_{c2s}, y_{c2s}, \alpha_{2s}]^T \quad (1)$$

The walker can also be described by the generalized coordinates q as mentioned before:

$$q = [x_h, y_h, \alpha_1, \alpha_2, \alpha_{1f}, \alpha_{2f}, \alpha_{1t}, \alpha_{2t}]^T \quad (2)$$

The definitions of variables mentioned above can be found in Fig. 2.

We defined matrix T as follows:

$$T = \frac{dr}{dq} \quad (3)$$

Thus T transfers the independent generalized coordinates $\dot{q}$ into the velocities of the rectangular coordinates $\dot{r}$. The mass matrix in rectangular coordinates r is defined as:

$$M = diag(m_h, m_h, m_l, m_l, I_l, m_l, m_l, I_l, m_f - m_s, m_f - m_s, I_f, m_f - m_s,$$

$$m_f - m_s, I_f, m_s, m_s, I_s, m_s, m_s, I_s) \quad (4)$$

Denote F as the active external force vector in rectangular coordinates. The constraint function, which is used to maintain foot contact with ground and detect impacts, is marked as $\xi(q)$. Note that $\xi(q)$ in different walking phases may be different since the contact conditions change.

We can obtain the Equation of Motion (EoM) by Lagrange's equation of the first kind:

$$M_q \ddot{q} = F_q + \Phi^T F_c \quad (5)$$

where F_c is the contact force acted on the walker by the ground to meet the constraint of the stance foot.

$$\xi(q) = 0 \quad (6)$$

where $\Phi = \frac{\partial \xi}{\partial q}$. M_q is the mass matrix in the generalized coordinates:

$$M_q = T^T M T \quad (7)$$

F_q is the active external force in the generalized coordinates:

$$F_q = T^T F - T^T M \frac{\partial T}{\partial q} \dot{q} \dot{q} \quad (8)$$

Equation (6) can be transformed to the following equation:

$$\Phi \ddot{q} = -\frac{\partial (\Phi \dot{q})}{\partial q} \dot{q} \quad (9)$$

Then the EoM in matrix format can be obtained from Equation (5) and Equation (9):

$$\begin{bmatrix} M_q & -\Phi^T \\ \Phi & 0 \end{bmatrix} \begin{bmatrix} \ddot{q} \\ F_c \end{bmatrix} = \begin{bmatrix} F_q \\ -\frac{\partial(\Phi\dot{q})}{\partial q}\dot{q} \end{bmatrix} \tag{10}$$

The equation of strike moment can be obtained by integration of Equation (5):

$$M_q\dot{q}^+ = M_q\dot{q}^- + \Phi^T I_c \tag{11}$$

where $\dot{q}^+$ and $\dot{q}^-$ are the velocities of generalized coordinates after and before the strike, respectively. Here, I_c is the impulse acted on the walker which is defined as follows:

$$I_c = \lim_{t^- \to t^+} \int_{t^-}^{t^+} F_c dt \tag{12}$$

where I_c is the impulse acted on the walker by ground. Since the strike is modeled as a fully inelastic impact, the walker satisfies the constraint function $\xi(q)$. Thus the motion is constrained by the following equation after the strike:

$$\frac{\partial \xi}{\partial q}\dot{q}^+ = 0 \tag{13}$$

Then the equation of strike in matrix format can be derived from Equation (11) and Equation (13):

$$\begin{bmatrix} M_q & -\Phi^T \\ \Phi & 0 \end{bmatrix} \begin{bmatrix} \dot{q}^+ \\ I_c \end{bmatrix} = \begin{bmatrix} M_q\dot{q}^- \\ 0 \end{bmatrix} \tag{14}$$

More detailed description of the bipedal model can be found in [24].

3. Effects of compliant joints and segmented feet

Based on the EoMs mentioned above, we analyze the effects of compliant ankles and toes on energetic efficiency and stability of dynamic bipedal walking. All masses and lengthes are normalized by the leg mass and leg length respectively. The spring constants (stiffness of ankle joint and toe joint) are normalized by both the mass and the length of leg.

3.1. Energetic efficiency

Energetic efficiency is an important gait characteristics. The energy consumption of passive dynamic based models is usually represented in the nondimensional form of "specific resistance": energy consumption per kilogram mass per distance traveled per gravity [8]. However, for passive walkers on a gentle slope, specific resistance is not a suitable measure of efficiency, since all walkers have the same specific resistance for a given slope [12]. Therefore, similar to [12], normalized walking velocity is used as the measure of efficiency, such that "most efficient" is synonymous of "fastest".

The walking velocity of the rigid foot model (without toe joints) decreases monotonically as foot length or foot ratio (the ratio of distance between heel and ankle joint to distance between ankle joint and toe tip) grows (see Fig. 3(a)). For the segmented foot model, the walker moves slower for longer foot according to the main tendency (see Fig. 3(b)). Walking velocity achieves the maximum value when foot ratio is near 0.3 for any fixed foot length.

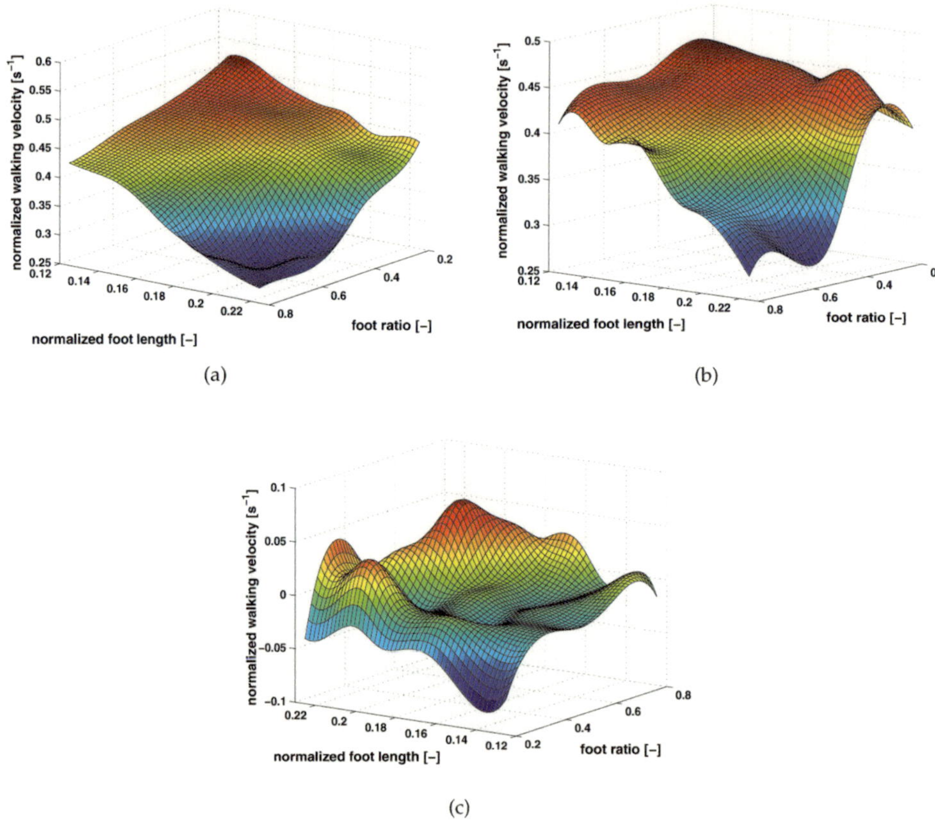

Figure 3. Comparison of walking velocities of rigid foot model and segmented foot model. The curved surfaces are smooth processed based on the sample data. (a) Average walking velocity versus foot length and foot ratio for rigid flat-foot passive walking model. (b) Average walking velocity versus foot length and foot ratio for segmented flat-foot passive walking model. (c) The difference of walking velocity of the two models, obtained from (b) subtracted by (a). Both walking velocity and foot length are normalized by leg length. Foot ratio is defined as the ratio of distance between heel and ankle joint to distance between ankle joint and toe tip.

A peak appears at relative large foot length (larger than 0.2) and foot ratio near 0.3, which is similar to the foot structure of human beings [25]. The comparison of the two models shows that the walker with segmented feet moves slower than rigid foot model with small foot length, however, the velocity of segmented foot model is larger when the foot is long enough, especially when foot ratio is near 0.3. In another word, if the segmented foot ratio is close real human foot, the segmented foot model is more efficient than the rigid foot model.

3.2. Walking stability

We evaluate adaptive walking of the model on uneven terrain to analyze the walking stability. Figure 4 shows the relationship between the maximal allowable ground disturbance (a step down) the walker can overcome and the foot ratio of rigid foot bipedal model. The maximal allowable ground disturbance decreases monotonically as the foot ratio grows, which is

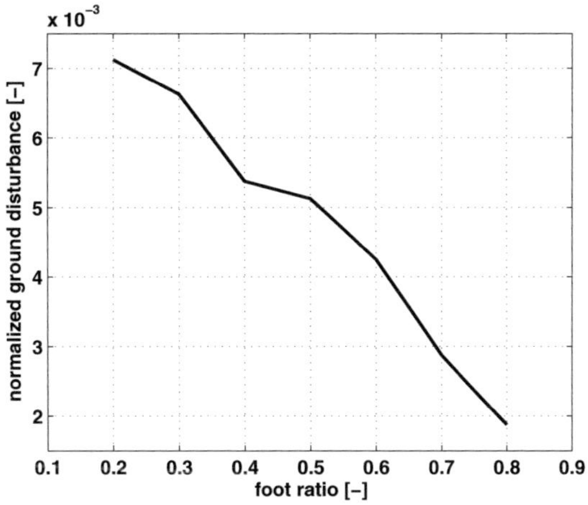

Figure 4. Adaptability of the rigid foot model with different foot ratios. The normalized foot length is 0.1875. The ground disturbance is also normalized by leg length.

similar to the trend of walking velocity. In case of short hindfoot and long forefoot (foot ratio is 0.2), the walker can return to stable motion cycle after a ground disturbance larger than 0.7 percent of leg length. However, the maximum disturbance the model can overcome decreases below 0.2 percent of leg length when the lengths of hindfoot and forefoot are comparable (foot ratio is 0.8). The relationship between the maximal allowable ground disturbance and foot ratio of segmented foot model also shows a great resemblance to the trend of walking velocity (Fig. 5 shows the results). The maximum value is obtained when the foot ratio is 0.3.

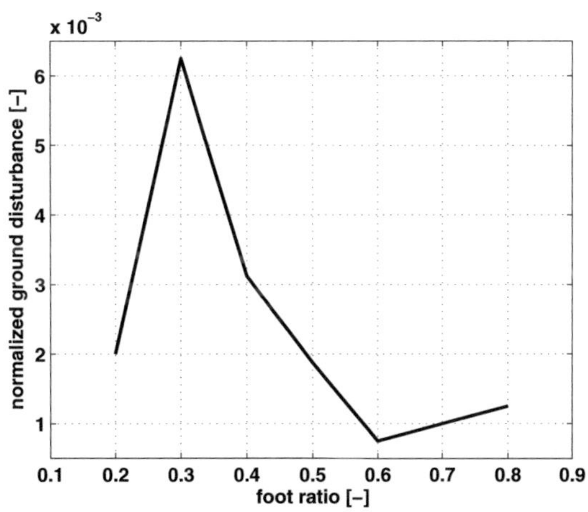

Figure 5. Adaptability of the segmented foot model with different foot ratios. The normalized foot length is 0.1875. The ground disturbance is also normalized by leg length.

In that case the model can overcome ground disturbance larger than 0.6 percent of leg length. The adaptability of the segmented foot model decreases significantly if the foot ratio changes to other values. The results indicate that there exists a best foot structure of the segmented foot model, which achieves both excellent adaptability and walking velocity.

From the analysis above, one can find that the segmented foot model has comparable walking adaptability with rigid foot models in the case of suitable foot ratios. However, the walkers with segmented feet perform worse in other cases.

4. Applications to prosthesis and exoskeleton

The theoretical analysis mentioned above indicates that the segmented foot model with compliant joints is more complicated than the rigid foot model in both foot structure and walking sequence. However, segmented foot models with foot structure close to human body show better walking performance (energetic efficiency and stability) compared with rigid-foot walking. Based on the results, here we introduce the design of the lower-limb prosthesis (PANTOE 1) [26] and exoskeleton prototype (EXO-PANTOE 1) [27] with compliant joints and segmented foot. At first, a brief overview of compliant actuators in robotics is presented. It shows the reasons of using spring based compliant actuators for joints in the proposed prosthesis and exoskeleton.

4.1. Compliant actuators in joints

In practice, a combination of sophisticated nonlinear control strategies and analytical methods is required to deal with the uncontrollable and underactuated degrees of freedom introduced by joint compliance. As a result, many robotic applications have used stiff actuation schemes and high-stiffness materials. However, mechanically compliant systems, such as elastic joints, may be used for shock absorbtion and be exploited to store energy and decrease control complexity. As a new trend in robotics, various kinds of compliant actuators and compliant joints have been introduced to real robot applications, for instance, hexapod robot [28], quadruped system [29] and bipedal walking [30].

According to the composition of the elastic elements, compliant actuators can be roughly divided into several main classes. The first category is the spring based compliance. The well-known examples are the Series Elastic Actuators (SEA) developed by Pratt and Williamson [31] and the two-legged robot Spring Flamingo [32]. Starting from the Series Elastic Actuators, various legged robot applications have been equipped with motor-spring systems. Yamaguchi *et al* [33] used a non-linear spring mechanism to make predefined changes in stiffness of biped robot possible. Meyer *et al* [34] described a simple and low-cost humanoid leg with compliant joints and springy feet, aiming for a repetitive jumping system. Hurst *et al* [35] designed the Actuator with Mechanically Adjustable Series Compliance (AMASC) which consists of a drive motor connected in series with a pair of large, variable stiffness springs. Van Ham *et al* [36] reported a controlled passive walker Veronica actuated with the Mechanically Adjustable Compliance and Controllable Equilibrium Position Actuator (MACCEPA) which uses a dedicated servo motor and solid springs to control the compliance and equilibrium position.

Another category is the pneumatic artificial muscles based compliance. The well-known artificial muscles in the field of dynamic walking are the pneumatic McKibben muscles [37]. In addition, the Pleated Pneumatic Artificial Muscle (PPAM) designed by [38], has been successfully implemented in the biped Lucy [30]. However, such air muscles based mechanisms may be the barrier that constrains the dynamic walking systems, especially the lower-limb prostheses and exoskeletons, in practical use. One of the disadvantages is that with air muscles, it is difficult to perform precise control. Other problems, for example, include the use of compressed air and the inefficient down-regulation of pressure of the onboard

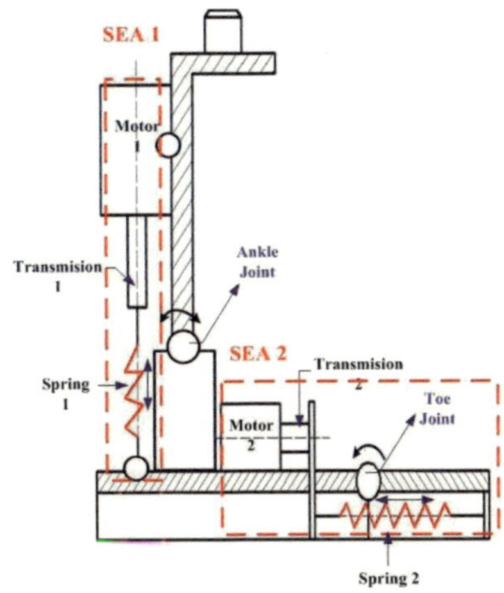

Figure 6. Schematics diagram of the segmented foot with compliant joints. The two main components of the prosthesis are two SEAs, which are used to drive the ankle and toe joints respectively.

air storage system. Thus, several recent studies try to change air muscles into motor-spring systems as compliant actuators for dynamic walking systems, e.g. [39].

In addition, there are other types of compliant actuators in robotics. For example, shape memory alloys show impressive actuation characteristics, while suffer from slow response and motion constraints [40]. Other interesting compliant actuators include artificial muscle actuator using fluid [41], polymer materials [42], dielectric elastomers [43], etc. Most of them are used in particular environments and difficult to implement in autonomous systems, especially in lower-limb prostheses and exoskeletons.

As mentioned above, currently, spring based compliance is the most promising compliant actuators in the field of dynamic walking. By using the proper spring based elastic mechanisms, lower-limb prostheses and exoskeletons may be capable of performing stable walking on different terrains and with controllable walking velocity. Applying compliant actuators to lower-limb rehabilitation systems provides not only new challenges for bipedal locomotion but also improvement of practical use.

4.2. Powered ankle and toe with compliant actuators

As discussed above, we finally used two series elastic actuators (SEA) to construct ankle and toe joints. Each SEA includes a DC motor, a transmission and a spring structure. Compared with other actuators, SEA has several benefits: 1) the actuator exhibits lower output impedance and back-driveability; 2) shock tolerance is greatly improved by the spring placed in series between the transmission and the load; 3) the motor's required force is reduced; 4) force control stability is improved, even in intermittent contact with hard surfaces; 5) energy can be stored and released in the elastic element, potentially improving efficiency [31, 45]. The design of the two SEAs will be introduced in details in the following part of the paper.

The design of the powered joints is based on the functionality of human toe and ankle joints. As shown in Fig. 6, the basic architecture of the segmented foot is an integration of two SEAs, which are used to drive the ankle and toe joints respectively. Each SEA comprises a DC motor, a ball screw transmission and a spring structure. Because human toe joint only outputs net positive work at the moment TO, the toe joint is designed to rotate counterclockwise passively,

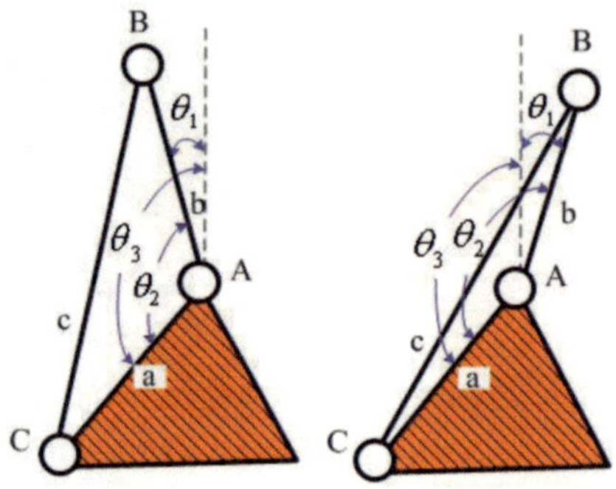

Figure 7. The ankle joint can be simplified as a special three-bar mechanism, of which the length of c can be modulated by SEA 1.

and to rotate clockwise actively. When toe joint is forced to rotate counterclockwise, Spring 2 will be extended to store energy. At the moment TO, Spring 2 will release the stored energy and Motor 2 will drive the toe joint to rotate clockwise via Transmission 2 and Spring 2. Spring 2 comprises four drawsprings set in parallel and the stiffness is $200N/cm$. Motor 2 used in the current design is a $30W$ DC motor with an angle encoder.

The functionality of the human ankle joint can be realized by a special three-bar mechanism, shown in Fig.10. The special three-bar mechanism comprises three bars (a, b and c) and three hinges (A, B and C). Hinge A is the ankle joint. Bars a and b are the foot and the shank respectively. c is a special bar, of which the length can be regulated by SEA 1. θ_1 is the angle of the ankle joint, which is used to control the movement of the ankle. θ_1 can be calculated by the following equation:

$$\theta_1 = \arccos \frac{L_a^2 + L_b^2 - L_c^2}{2L_a L_b} - \theta_3 \tag{15}$$

where L_a, L_b and L_c are the length of the bars a, b and c respectively. L_a and θ_3 can be calculated by L_3 and H_1:

$$\theta_3 = \pi - \arctan \frac{L_3}{H_1} \tag{16}$$

$$L_a = \sqrt{L_3^2 + H_1^2} \tag{17}$$

where L_3 is the level distance of the ankle joint from the end of the heel and H_1 is the height of the ankle joint from the ground. L_c is determined by SEA 1. When Spring 1 is compressed or pulled by Transmission 1 (the nut of the ball screw, Nut 1, move upward), the L_c becomes shorter. When Spring 1 is extended or pushed by Transmission 1 (Nut 1 move downward), the length of c becomes longer. In fact, the state of Spring 1 and the movement of Transmission 1 are not independent, namely, L_c is determined by a combined motion of Spring 1 and Transmission 1. L_c can be obtained by the following equation.

$$L_c = L_s + L_t + \frac{F}{K_1} + \Delta_1 \tag{18}$$

Where L_s is the length of Spring 1 with no load. L_t is the length of Transmission 1 with the nut at the initial position; F is the load on Spring 1, which is determined by the weight of the subject and the walking state; K_1 is the spring stiffness of Spring 1. Δ_1 is the displacement of the nut of Transmission 1.

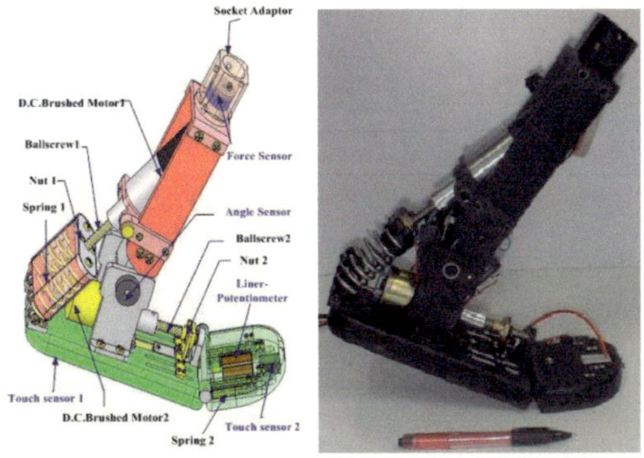

(a) (b)

Figure 8. The CAD model and the prototype of PANTOE 1 with compliant joints and segmented foot. The prototype is made of aluminium-alloy. The weight is 1.47 kg (not including the Li rechargeable battery about 1 kg), comparable to the weight of the subject's limb. The full angles of the ankle and the toe joints are 45° and 90° respectively.

4.3. Powered prosthesis with segmented foot

PANTOE 1 is short for foot-prosthesis with Powered Ankle and Toe Joints. Our aim is to build an ankle-foot prosthesis with compliant powered joints and segmented foot to replicate the functionality of human foot as closely as possible. SEA 1 was designed based on the angle range of θ_1 from -16° to 27°. Spring 1 was constructed with three springs placed together in parallel. The stiffness of Spring 1 is about $500N/cm$. The pitch of the ball screw Transmission 1 is $4mm$ per revolution, then the nut is self-locking. Because the ankle needs to provide high power output to propel the body, we chose a $83W$ FAULHABER brushed DC motor (with rotary encoder) as Motor 1. Figure 8 shows the CAD model and the prototype of PANTOE 1 in detail. In order to decrease the total weight, PANTOE 1 is made of aluminium-alloy. This design weighs $1.47kg$, not including the rechargeable Li battery and the molded socket.

4.4. Lower-limb exoskeleton with segmented foot

EXO-PANTOE 1 is short for below-knee exoskeleton with powered ankle and toe joints. It is designed for a subject suffering from some ankle pathology. The information of the subject is shown in Table 1. Figure 9 shows the CAD model and the prototype of EXO-PANTOE 1 in detail. The design concept of EXO-PANTOE 1 is similar to PANTOE 1. Two SEAs are used to drive the ankle and toe joints respectively.

Parameter	Value
length of the foot L_1	$265mm$
length of the forefoot L_2	$79mm$
level distance of the ankle joint from the end of the heel L_3	$68mm$
height of the ankle joint from the ground H_1	$83mm$
maximal plantar flexion angle of the ankle joint θ_p	27°
maximal dorsiflexion angle of the ankle joint θ_d	16°
maximal angle of the toe joint θ_t	90°

Table 1. Information of the subject.

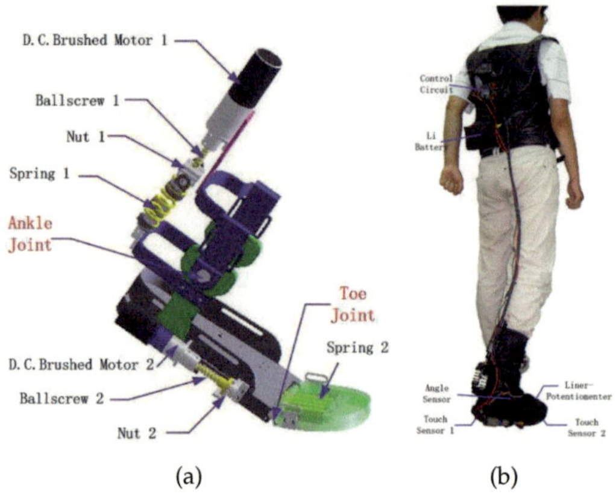

Figure 9. The CAD model and the prototype of EXO-PANTOE 1 with compliant joints and segmented foot. The prototype is made of aluminium-alloy. The weight is $1.2kg$ (not including the Li rechargeable battery about $0.5kg$), acceptable to the subject. The full angles of the ankle and the toe joints are 45° and 90° respectively.

4.5. Sensors and control

Sensors and control method used in PANTOE 1 and EXO-PANTOE 1 are similar. The sensor system includes two touch sensors, a angle sensors, a linear-potentiometer and a force sensor. The reactions between the ground and the foot are detected by touch sensors 1 and 2. The angle sensor is used to measure the rotation angle of the ankle joint. Assembled in parallel with series Spring 2 (see Fig. 9(a)), the linear-potentiometer is utilized to measure the rotation angle of the toe joint and the force of the Spring 2 at the same time. The force between the subject and the system is measured by the force sensor.

For gait identification, we divide the walking gait cycle into seven phases. Each phase represents a unique state where the system will be controlled by a specific control strategy and perform a specific behavior. Besides, a state can switch to one another if the triggering transition requirements are meet. This method can be called finite-state control and can be describe as follows [46]:

$$A_i = f_a(S_i) \tag{19}$$

where f_a is the action function indicating the specific output behavior A_i of the specific state S_i.

$$S_{i+1} = f_s(S_i, I_i) \tag{20}$$

where f_s is the transition function indicating the transition between two adjacent states and I_i represents the input triggering information.

As described, a level ground walking gait cycle begins with the heel strike of one foot and ends with the next heel strike of the same foot [20]. It is generally divided into the stance phase when the foot is on the ground and the swing phase when the foot is off the ground. To get a more accurate description of the prosthesis state, we divide each phase into different sub-phases:

1) Stance Phase:

Controlled plantar flexion of ankle (CPA): CPA begins at heel-strike when the ankle joint begins to plantar flex and ends at foot-flat when toe-strike occurs. The behavior of the ankle

joint at this state is consistent with a linear spring response since the output joint torque is proportional to the joint angle.

Controlled dorsiflexion of ankle (CDA): CDA begins at foot-flat when the ankle joint begins to dorsiflex and ends when the ankle reaches the maximum dorsiflexion angle. The behavior of the ankle joint at this state can be described as a nonlinear spring.

Powered plantar flexion of ankle (PPA): PPA begins after CDA when the ankle joint begins to plantar flex again and ends at the instant of toe-off. The function of the ankle joint at this state is the superposition of a nonlinear spring and a torque source. This phase is the main phase that can reflects the dynamical characteristics of the ankle joint.

Controlled dorsiflexion of toe (CDT): CDT begins after heel-off when the toe joint is compressed and begin to plantar flex. It ends when the toe joint is compressed to specific extent. The behavior of the toe joint at this state can be modeled as a linear spring.

Powered plantar flexion of toe (PPT): PPT begins when the toe joint angle reaches a specific value and the toe joint begins to dorsiflex to push the body forward and upward together with the ankle joint. It ends at toe-off. The toe joint's behavior at this state can be modeled as the superposition of a linear spring and a torque source.

2) Swing Phase:

Early swing (ESW): ESW begins at toe-off when the ankle joint and the toe joint begin to return to the equilibrium position and ends after a predefined time period when the two joints have reached back to the equilibrium position. Both of the two joints play the role of position source to reset the prosthesis to equilibrium position.

Late swing (LSW): LSW begins just after the ESW and ends at the next heel-strike. The toe joint and the ankle joint just keep the balanced state and get ready for the beginning of the next gait cycle. The function of the two joints can also be modeled as a position source.

Figure 10 shows the finite-state control scheme for level-ground walking with segmented foot. We define the ankle joint angle to be zero when the shank is perpendicular to the foot. From the zero position, the angle will be negative if the ankle plantar flexs and positive if the ankle dorsiflexs. The toe joint angle is defined to be zero when the joint is not compressed and to be positive we compressed to dorsiflex. To accurately identify each gait state and decide the transition between different states, we collect the information below: a) Heel contact (H), $H=0$ indicates that the heel is off the ground and vice versa; b) Heel pressure (F_H), F_H indicates the pressure that the ground exerts to the heel; c) Toe contact (T), $T=0$ indicates that the toe is off the ground and vice versa; d) Toe pressure (F_T), F_T indicates the pressure that the ground exerts to the toe; e)Ankle joint angle (θ_a); f) Ankle joint's angular velocity ($\dot{\theta}_a$), $\dot{\theta}_a$ indicates the rotatory direction of the ankle joint; g) Toe joint angle (θ_t); h) Ankle torque (T_a); i) Stance phase time period (Δt_{stance}), Δt_{stance} can be used as an indicator of the walking speed; j) Swing phase time period (Δt_{swing}), Δt_{swing} can be used as an indicator of the walking speed. Figure 11 shows the hardware of the proposed control platform.

5. Experimental results

5.1. Walking performance

For PANTOE 1, we have done several preliminary experiments to evaluate the functionality of the prototype. The amputee subject was 45 years of age, 1.70m in height and 71kg in weight.

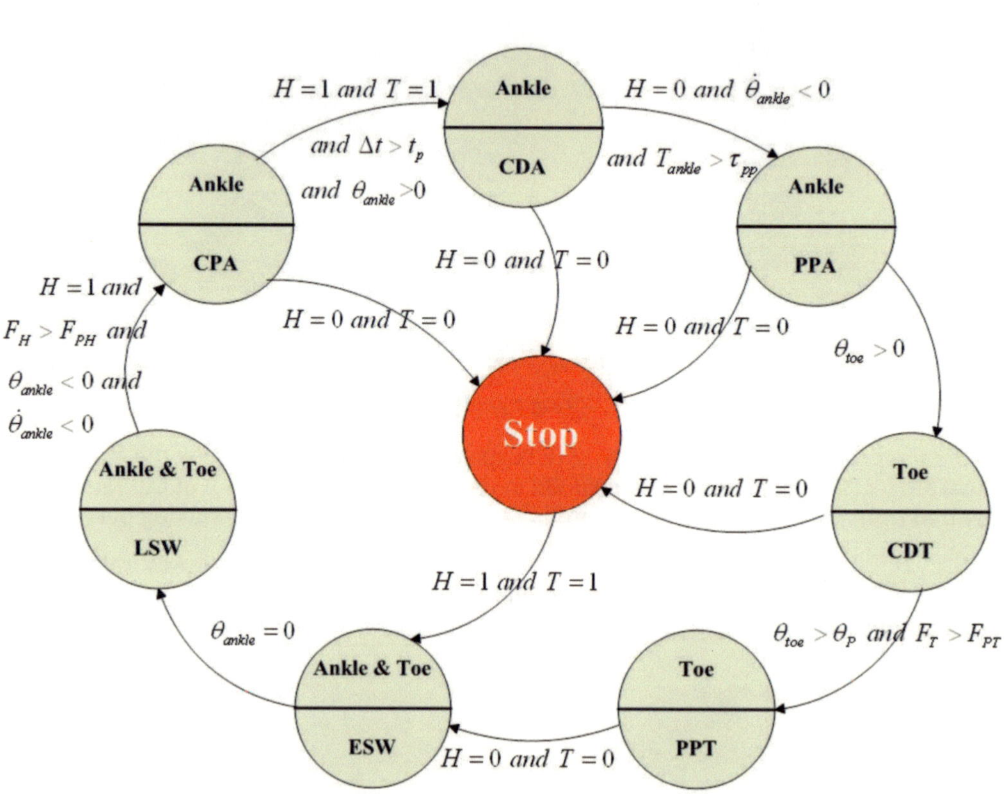

Figure 10. Finite-state control for level-ground walking with segmented foot.

Figure 11. The hardware of the control platform.

Figure 12. A sequence of photos captured during walking of the amputee subject.

He wore the proposed prosthesis during the experiment. The ratio between the length of the residual shank (the distance between patella to the amputated site) to that of the sound shank (measured from patella to malleolus lateralis) is 0.4m. Fig. 12 shows the walking performance.

EXO-PANTOE 1 is worn by a subject whose right ankle is injured and can not output sufficient power during walking. With the powered ankle and toe joints, EXO-PANTOE 1 is able to provide enough energy to the subject and help him relearn the normal walking gait (shown in Fig. 13).

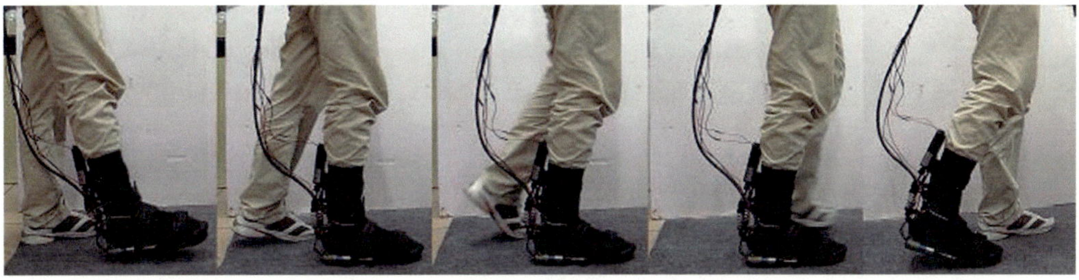

(a) Stance phase

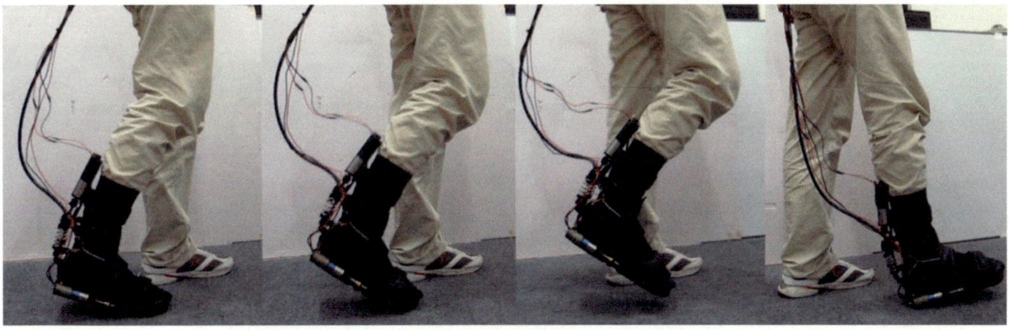

(b) Swing phase

Figure 13. Sequence pictures captured from the walking motion of the subject wearing EXO-PANTOE 1 in a walking gait cycle beginning with heel-strike. The result indicates that EXO-PANTOE 1 can assist the subject relearn the human walking gait.

5.2. Energy consumption

In order to analyze the effects of the segmented foot structure on the energetic efficiency during walking, we have measured the energy consumption of EXO-PANTOE 1 during the subject walking at his most comfortable speed ($1.1m/s$) in three cases. In the first case, the segmented foot is locked with a mechanical structure and the foot just acts as a single rigid plate. In the second case, Motor 2 does not work at all and the toe joint can only be bent passively. In the third case, the toe joint is active and it is able to output sufficient net positive work to the subject during the TO phase. In these three cases, the ankle joint always provides enough energy to the subject. Before the measurement of the energy consumption, in each case the subject is allowed to have a long enough training period to adapt to the exoskeleton.

As shown in Fig. 14, one can find that EXO-PANTOE 1 consumes the most energy in the first case (see Fig. 14(a)), which indicates that the segmented foot plays important role in improving energetic efficiency. The total energy consumed in the second case and the third case is close, where the segmented foot with active toe performs slightly better (see Fig. 14(b)

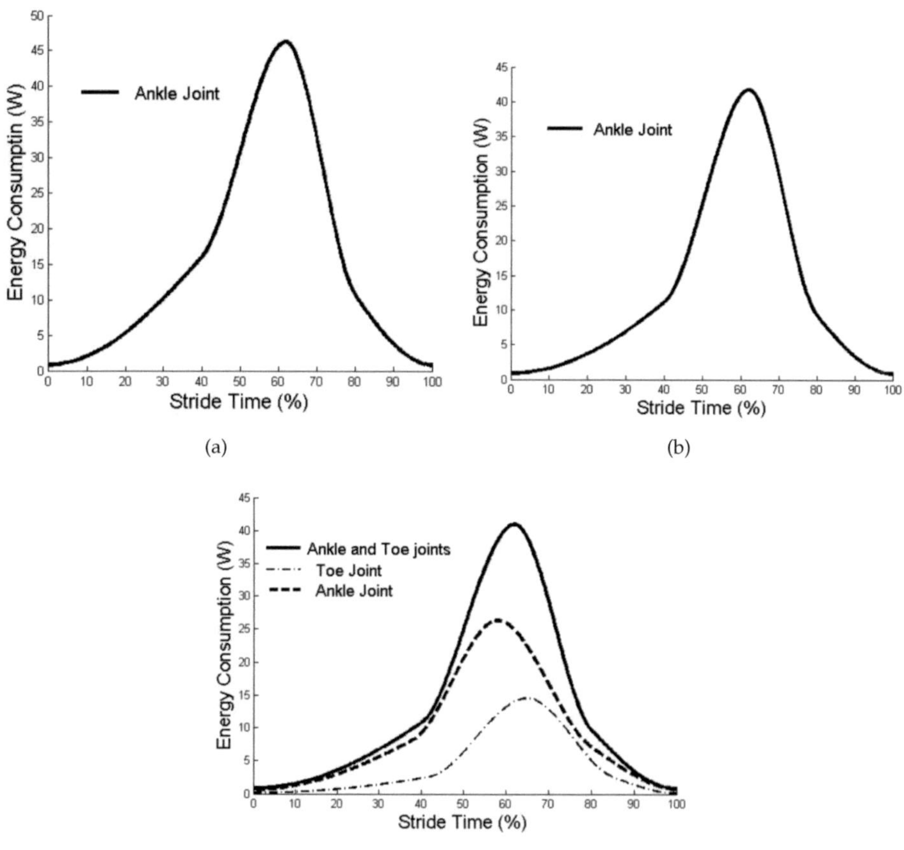

Figure 14. Energy consumption of EXO-PANTOE 1 in one stride cycle. (a)Energy consumption of EXO-PANTOE 1 when the foot is a single rigid plate. (b) Energy consumption of EXO-PANTOE 1 when the toe joint can only be bent passively. (c)Energy consumption of EXO-PANTOE 1 when the toe joint can output sufficient power.

and Fig. 14(c)). The ankle joint in the third case consumes much less energy than that in the second case. The result shows that powered toe joint can share the energy cost of the ankle joint, and enables the development of more efficient and effective powered lower-limb exoskeleton.

Acknowledgements

This work was supported by the National Natural Science Foundation of China (No. 61005082, 61020106005), Doctoral Fund of Ministry of Education of China (No. 20100001120005), PKU-Biomedical Engineering Joint Seed Grant 2012 and the 985 Project of Peking University (No. 3J0865600).

Author details

Qining Wang, Jinying Zhu, Yan Huang, Kebin Yuan and Long Wang
Intelligent Control Laboratory, College of Engineering, Peking University, Beijing 100871, China
Beijing Engineering Research Center of Intelligent Rehabilitation Engineering, Beijing 100871, China

6. References

[1] Carson M. C, Harrington M. E, Thompson N, O'Connor J. J, Theologis T. N (2001) Kinematic analysis of a multi-segment foot model for research and clinical applications: a repeatability analysis. J. Biomech. 34, 1299–1307.
[2] MacWilliams B. A, Cowley M, Nicholson D. E (2003) Foot kinematics and kinetics during adolescent gait. Gait Posture 17, 214–224.
[3] Myers K. A, Wang M, Marks R. M, Harris G. F (2004) Validation of a multisegment foot and ankle kinematic model for pediatric gait. IEEE T. Neur. Sys. Reh. 12(1), 122–130.
[4] Okita N, Meyers S. A, Challis J. H, sharkey N. A (2009) An objective evaluation of a segmented foot model. Gait Posture 30, 27–34.
[5] Nishiwakiz K, Kagamiy S, Kuniyoshiz Y, Inabaz M, Inouez H (2002) Toe joints that enhance bipedal and fullbody motion of humanoid robots. Proceedings of IEEE International Conference on Robotics and Automation, pp. 3105–3110.
[6] Sellaouti R, Stasse O, Kajita S, Yokoi K, Kheddar A (2006) Faster and smoother walking of humanoid hrp-2 with passive toe joints. Proceedings of IEEE/RSJ International Conference on Intelligent Robots and Systems, pp. 4909–4914.
[7] Hirai K, Hirose M, Haikawa Y, Takenaka T (1998) The development of the Honda humanoid robot. Proceedings of IEEE International Conference on Robotics and Automation, pp. 1321–1326.
[8] Collins S, Ruina A, Tedrake R, Wisse M (2005) Efficient bipedal robots based on passive-dynamic walkers. Science 307, 1082–1085.
[9] McGeer T (1990) Passive dynamic walking. Int. J. Robot. Res. 9, 68–82.
[10] Mochon S, McMahon T. A (1980) Ballistic walking. J. Biomech. 13(1), 49–57.
[11] Ruina A, Bertram J. E. A, Srinivasan M (2005) A collisional model of the energetic cost of support work qualitatively explains leg sequencing in walking and galloping, pseudo-elastic leg behavior in running and the walk-to-run transition. J. Theor. Biol. 237(2), 170–192.
[12] Kwan M, Hubbard M (2007) Optimal foot shape for a passive dynamic biped. J. Theor. Biol. 248, 331–339.
[13] Hobbelen D. G. E, Wisse M (2008) Ankle actuation for limit cycle walkers. Int. J. Robot. Res. 27, 709–735.

[14] Wang Q, Huang Y, Wang L (2010a) Passive dynamic walking with flat feet and ankle compliance. Robotica 28, 413–425.

[15] Wang Q, Huang Y, Zhu J, Wang L, Lv D (2010b) Effects of foot shape on energetic efficiency and dynamic stability of passive dynamic biped with upper body. Int. J. Hum. Robot. 7(2), 295–313.

[16] Kumar R. P, Yoon J, Christiand, Kim G (2009) The simplest passive dynamic walking model with toed feet: a parametric study. Robotica 27, 701–703.

[17] Klute G. K, Czerniecki J, Hannaford B (1998) Development of powered prosthetic lower limb. The 1st Natl. Meeting, Veterans Affairs Rehabil. R and D service.

[18] Versluys R, Peeraer L, Van der Perre G, Van Gheluwe B, Lefeber D (2007) Design of a powered below-knee prosthesis. The 12th World Congr. Int. Soc. Prosthet. Orthot., 2007.

[19] Versluys R, Desomer A, Lenaerts G, Van Damme M, Bey P, Van der Perre G, Peeraer L, Lefeber D (2008) A pneumatically powered below-knee prosthesis: design specifications and first experiments with an amputee. Proceedings of the 2nd IEEE/RAS-EMBS International Conference on Biomedical Robotics and Biomechatronics, pp. 372–377.

[20] Au S. K, Weber J, Herr H (2009) Powered ankle-foot prosthesis improves walking metabolic economy. IEEE Trans. Robot. 25(1), 51–66.

[21] Dollar A. M, Herr H (2008) Lower extremity exoskeletons and active orthoses: challenges and state-of-the-art. IEEE Trans Robot. 24(1), 144–158.

[22] Chu A, Kazerooni H, Zozz A (2005) On the biomimetic design of the Berkeley Lower Extremity Exoskeleton (BLEEX). Proceedings of the IEEE International Conference on Robotics and Automation, pp. 4345–4352.

[23] Hayashi T, Kawamoto H, Sankai Y (2005) Control method of robot suit HAL working as operator's muscle using biological and dynamical information. Proceedings of the IEEE/RSJ International Conference on Intelligent Robots and Systems, pp. 3063–3068.

[24] Huang Y, Wang Q, Gao Y, Xie G (2012) Modeling and analysis of passive dynamic bipedal walking with segmented feet and compliant joints. Acta Mechanica Sinica. (accepted)

[25] Wang W. J, Crompton R. H (2004) Analysis of the human and ape foot during bipedal standing with implications for the evolution of the foot. J. Biomech. 37, 1831–1836.

[26] Zhu J, Wang Q, Wang L (2010) PANTOE 1: Biomechanical design of powered ankle-foot prosthesis with compliant joints and segmented foot. Proceedings of the IEEE/ASME International Conference on Advanced Intelligent Mechatronics, pp. 31–36.

[27] Zhu J, Wang Q, Huang Y, Wang L (2011) Adding compliant joints and segmented foot to bio-inspired below-knee exoskeleton. Proceedings of the IEEE International Conference on Robotics and Automation, pp. 605–610.

[28] Saranli U, Buehler M, Koditschek D (2001) RHex: A simple and highly mobile hexapod robot. Int. J. Robot. Res. 20(7), 616–631.

[29] Poulakakis I, Papadopoulos E, Buehler M (2006) On the stability of the passive dynamics of quadrupedal running with a bounding gait. Int. J. Robot. Res. 25(7), 669–687.

[30] Vanderborght B, Verrelst B, Van Ham R, Lefeber D (2006) Controlling a bipedal walking robot actuated by pleated pneumatic artificial muscles. Robotica. 24, 401–410.

[31] Pratt G. A, Williamson M (1995) Series elastic actuators. Proceedings of the IEEE International Conference on Intelligent Robots and Systems, pp. 399–406.

[32] Pratt J (2000) Exploiting inherent robustness and natural dynamics in the control of bipedal walking robots, Ph.D. Thesis, Computer Science Department, MIT, Cambridge, MA.

[33] Yamaguchi J, Nishino D, Takanishi A (1998) Realization of dynamic biped walking varying joint stiffness using antagonistic driven joints. Proceedings of the IEEE International Conference on Robotics and Automation, pp. 2022–2029.

[34] Meyer F, Spröwitz A, Lungarella M, Bettbouze L (2004) Simple and low-cost compliant leg-foot system. Proceedings of the IEEE/RSJ International Conference on Intelligent Robots and systems, pp. 515–520.

[35] Hurst J. W, Chestnutt J. E, Rizzi A. A (2004) An actuator with physically variable stiffness for highly dynamic legged locomotion. Proceedings of the IEEE International Conference on Robotics and Automation, pp. 4662–4667.

[36] Van Ham R, Vanderborght B, Verrelst B, Van Damme M, Lefeber D (2006) Controlled passive walker Veronica powered by actuators with independent control of equilibrium position and compliance. Proceedings of the IEEE International Conference on Humanoids, pp. 234–239.

[37] Caldwell D, Medrano-Cerda G, Goodwin M (1995) Control of pneumatic muscle actuators. IEEE Control Syst. Mag. 15(1), 40–48.

[38] Daerden F, Lefeber D (2001) The concept and design of pleated pneumatic artificial muscles. Int. J. Fluid Power 2(3), 41–50.

[39] Wisse M, Schwab A. L, van der Linde R. Q, van der Helm F. C. T (2005) How to keep from falling forward: elementary swing leg action for passive dynamic walkers. IEEE Trans. Robot. 21(3), 393–401.

[40] Choi S. B, Han Y. M, Kim J. H, Cheong C. C (2001) Force tracking control of a flexible gripper featuring shape memory alloy actuators. Mechatronics 11, 677–690.

[41] Boblan I, Bannasch R, Schwenk H, Prietzel F, Miertsch L, Schulz A (2004) A human-Like robot hand and arm with fluidic muscles: biologically inspired construction and functionality. Lecture Notes in Artificial Intelligence 3139, 160–179.

[42] Kornbluh R, Pelrine R, Eckerle J, Joseph J (1998) Electrostrictive polymer artificial muscle actuators. Proceedings of the IEEE International Conference on Robotics and Automation, pp. 2147–2154.

[43] Wissler M, Mazza E (2007) Electromechanical coupling in dielectric elastomer actuators. Sensers and Actuators A - Physical 138(2), 384–393.

[44] Inman V. T, Ralston H. J, Todd F (1981) Human Walking, Baltimore, MD: Williams and Wilking.

[45] Robinson D (2000) Design and an analysis of series elasticity in closed-loop actuator force control. Ph.D. Thesis, MIT.

[46] Zlatnik D, Steiner B, Schweitzer G (2002) Finite-state control of a trans-femoral prosthesis. IEEE Trans. Control Syst. Technol. 10(3), 408–420.

Chapter 22

Compliant Robotics and Automation with Flexible Fluidic Actuators and Inflatable Structures

I. Gaiser, R. Wiegand, O. Ivlev, A. Andres, H. Breitwieser, S. Schulz and G. Bretthauer

Additional information is available at the end of the chapter

http://dx.doi.org/10.5772/51866

1. Introduction

The fields of Robotics and Automation have been experiencing a boom the last few years. According to the "International Federation of Robotics" (IFR) in 2011, 140,000 robot units have been sold. This made 2011 the year with the most robot sales ever [52, 55]. Experts predict a worldwide population of 1.3 million industrial robots until 2014. The prognoses for the field of service robotics are also very promising [53, 54, 56]. The growing number of service robots changes the general requirements for those robots. Whereas industrial robots are optimized regarding precision, repeatability, and reliability, requirements for service robots are different, due to more human-robot interaction (HRI) and operation in unstructured environments. This means these robots need to integrate new concepts in terms of adaptivity, safety, and universality, necessitating change in the characteristics of the actuation systems and the structures.

Safety strategies can be divided into pre-collision and post-collision strategies [75].

Post-collision strategies traditionally refer to the fields of measuring and control. The important questions in these fields are how human injury and robot damage can be minimized after a collision has occurred. Existing standards regarding robot safety [59, 60] are currently revised based on the most recent studies in the area of post-collision safety [12, 43–45].

Pre-Collision safety strategies have been discussed over the last 30 years [49, 67, 91, 93] including their limitations. More recent works discuss the fact that since the robots are operating in a human-centered environment, this must inform the design of the robots. Kathib et al. [68] describe how a robot should be designed to establish autonomous tasks as well as human guided tasks. Haegele et al. [46] determine that robots should either have a broad sensory infrastructure to limit forces and moments through measuring them, or should be inherently compliant. In 2006 Alami et al. [4] published a novel design paradigm for robots: **"design robots that are intrinsically safe and control them to deliver performance"**. In summary the requirements for a high degree of pre-collision safety are:

1. Lightweight
2. Inherent Compliance

These requirements are obvious in the field of service robotics since any interaction with animals, technology, or humans will be safer by implementing them. Operating industrial robots usually requires a strict division between the working area of the robot and the working area of the human, since industrial robots normally do not fulfill these requirements. However, there are many tasks that could be accomplished resulting in higher quality or more efficiency, with closer human-robot interaction.

A lower inertia of the robot allows faster operating speed. At the same time lower inertia reduces the impact in case of a collision. The application of composite materials can achieve this while maintaining stiffness and precision of the robot. An impressive example is the DLR "Lightweight Robotic Arm III" (LWR III) [50]. Another example presenting a robot for the manipulation of small masses is described in [75].

When looking at compliant actuation systems for robots it becomes clear that there is currently much effort to add compliance to conventional drives by adding elastic elements to the drive chain. This concept is referred to as series elastic actuation and has been carried out in many various forms [24, 121, 138, 140]. Other drives with more or less inherent compliance are piezo-drives, shape-memory-actuators (SMA), electrorheological drives, and polymeric actuators. Fluidic actuators are a well-suited actuation principle for compliant actuation. Whereas pneumatic actuators are already compliant because of the compressibility of gases, hydraulic actuators need the integration of compliant membrane structures in order to achieve compliance. This group of actuators is referred to as "Flexible Fluidic Actuators".

In addition to its drive elements a robot consists of structural elements connecting the drive elements. So independently from the drive system a robot can exhibit compliant characteristics via the integration of compliant structural elements.

2. Historical background

2.1. Flexible fluidic actuators

A Flexible Fluidic Actuator generally consists of a flexible shell that transmits potential energy, delivered by the pressurized fluid, into a mechanical force, which then can be used to create a motion.

The flexible fluidic operating principle has a strong background in biomimetics. Gutmann [41, 42] established the "Hydroskelett-Theorie" as an approach to explain evolutionary biology via the concepts of constructional morphology. He understood that the design-principles of hydrosceletts are responsible for the general designs of organisms. Gudo et al. [39, 40] developed this idea further by introducing the term "Engineering Morphology which describes how Gutmann's ideas can be applied to technical design. The biomimetic background of flexible fluidic actuation was specifically discussed in [11, 130]. These works discuss several examples on weevils and spiders and describe how crucial the membrane properties are regarding efficiency and durability of the whole actuation system. The transfer of the evolutionary optimized flexible fluidic drives developed by nature into powerful technical systems is one focus of this work.

A useful classification of Flexible Fluidic Actuators can be made according to their operating principle, or the kind of force they are creating. Here we differentiate between designs that use expansion, contraction or bending directions to drive a system.

2.1.1. Flexible fluidic "Expansion" actuators

The history of this group of actuators starts with simple designs for lifting applications [32, 90, 133, 136]. Current "lifting bags" are mainly used for rescue operations [134]. However, the biggest group of expansion actuators is air-springs or damping elements [17]. These systems are either bellow-type actuators or rolling lobe-type actuators. The first rolling lobe systems were developed in the 19th century [77] but the principle is still popular today. Some differences between types of air springs are depicted in figure 1.

Bellow-type actuators are suitable drives for application in harsh environments since there is no friction and they can compensate or create tilt motions up to 30° without any additional transmission elements. The smallest commercially available bellow-type actuators have a diameter of 160 *mm*. The following patents [10, 31, 37, 80, 96] give an idea as to how these air actuators have developed throughout the last century. However, these patents only describe the general design. Newer patents discuss problems concerning fatigue of the elastomer as well as the connectors [107].

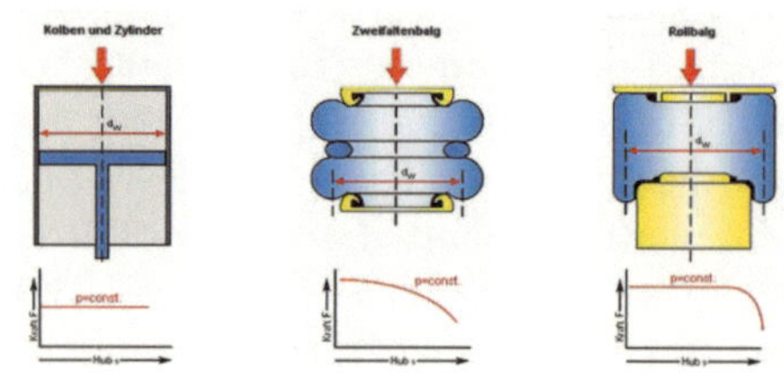

(a) Conventional Cylinder | (b) Bellow-Type Actuator | (c) Rolling Lobe-Type Actuator

Figure 1. Comparison of Different Kinds of Expansion Actuators [17]

Another big group is expansion actuators that work as rotary drive elements. [119] describes a solution where the structural integrity is created by the housing and the torque is created by an internal bellow-type actuator (figure 2(a))

The design proposed in [62] describes a bellow-type actuator suitable for linear or rotary motions as well as single chamber actuators (figure 2(b)-2(d)). The proposed materials for the bellow include rubbers as well as metals.

A very interesting design is introduced in [13]. The patent refers to fabrication techniques used in the tire industry and describes a layered membrane design with several reinforcement layers (figure 2(e), 2(f)).

[23] discusses the stress distribution in the membrane of a rotary actuator and requirements for achieving low bending stiffness and high tensile strength. Another approach is described in [34]. Here the drive element consists of a mono-material system. The whole drive is fabricated in one step (figure 2(g)).

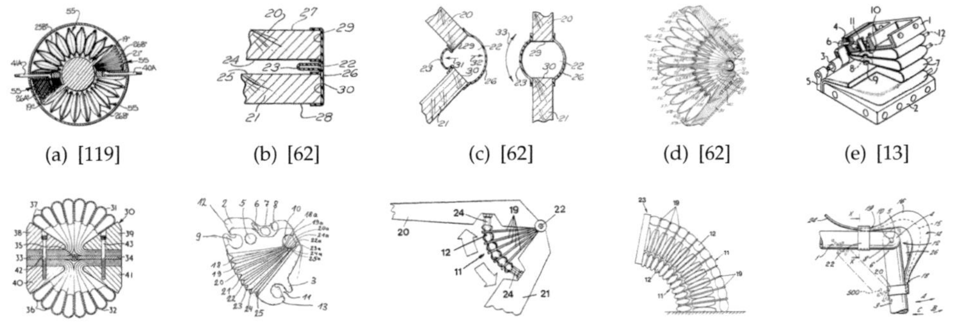

(a) [119] (b) [62] (c) [62] (d) [62] (e) [13]

(f) [13] (g) [119] (h) [74] (i) [74] (j) [47]

Figure 2. Radial Cross-Section View of Different Fluidic "Expansion" Drives

While the presented concepts are designed to operate with pressures in the range of $0 - 20\, bar$ the development in [74] is operated with pressures up to $200\, bar$. The actuator set-up allows both, linear and rotary actuation (figure 2(h), 2(i)). The focus in these works lays on heavy duty machinery but rotary drives or trunk-like structures are discussed for robotic applications as well. Detailed concepts regarding layered fiber reinforcements in the shell are introduced.

While the previous example requires complex knowledge and technology to produce an individually shaped membrane, other examples implement standard materials for flexible fluidic actuators. In [61, 81, 110] a design is proposed that uses bulky materials such as ordinary water hoses to form the actuator. Figure 2(j) shows one set-up of this FLEXATOR muscle. Subsequent developments applied the FLEXATOR technology to the fields of rehabilitation [103] and horticultural robotics [129]. The works of Prior et al. [103] introduced a unique approach that came to be known as "hybrid actuation" [117]. Here the powerful fluidic actuators are combined with precisely controllable electrical actuators in a parallel configuration.

2.1.2. Flexible fluidic "Contraction" actuators

This type of actuators generates a tensile force when pressure is applied. There is large variety of "contraction" actuators. They must be sorted into two groups. The first group includes actuators that generate a tensile force due to **"Anisotropic Membrane Stiffness"**. Daerden and Lefeber described some of those actuators in their review article [20]. These actuators increase in surface area when pressurized. The axial contraction is coupled to a radial expansion in which some of the energy is used for membrane deformation. Generally Joseph L. McKibben is said to be the inventor of the most popular design, often referred to as "McKibben Muscle". However, earlier patents describe the same design. In 1929 Dimitri Sensaud de Lavaud [22] introduced a fluidic muscle as shown in figure 3(a). This early work was later followed up by the patents of Morin [89] in 1947 and Woods [141] in 1953, where the design and characteristics of the fluidic muscle were described in detail. The actuators consist of a highly elastic inner membrane that is covered with a helically wound fiber reinforcement like a braided fiber hose (figure 3(b)). When pressurized the fiber angles change until the critical fiber angle of $\theta = 54,4°$ is reached (figure 3(c)) [141].

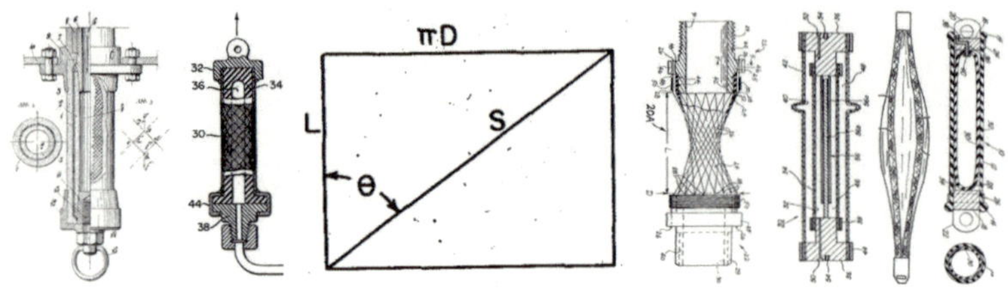

(a) [22] (b) [141] (c) geometric correlation between fiber angle θ and muscle diameter D [141] (d) [119] (e) [92] (f) [76] (g) [88]

Figure 3. Different Fluidic Muscles

In [101] Paynter describes a variation of this type of fluidic actuator. The "hyperboloid muscle" is equivalent to a prestretched fluidic muscle, which extends the range of motion (figure 3(d)). Other set-ups from Paynter are discussed in [98] and [100].

Commercially available fluidic actuators were introduced by Bridgestone Corporation, Japan, FESTO AG&Co. KG, Germany, and Shadow Robot Company, UK. Bridgestone introduced a single-acting [128] and a antagonistic [92] actuator design (figure 3(e)) but soon stopped their activities in the field. With operating pressures up to $2\,bar$ and a fatigue life of $67,000$ load cycles these actuators weren't really competitive.
Nowadays FESTO offers the biggest portfolio of fluidic muscles [29, 30]. Operating pressures are $0 - 8\,bar$ in connection with a fatigue life of $10,000 - 1\,Mio$ load cycles depending on the load case.

Lewis [76] and Monroe [88] proposed a design with only axially fiber reinforcements. Thus actuation is connected with a radial stretch of the pure rubber sections between the axial fiber strands (figure 3(f), 3(g)).

The second group of "Contraction" actuators generates the force due to **"Vectored Structural Degrees of Freedom"**. These actuator designs try to raise efficiency and to minimize the hysteresis compared to the first group of actuators. Ideally there is no strain of the membrane and almost no internal friction. When pressurized these actuators increase in volume while maintaining the same surface area. Yarlott [143] proposed a folded structure that unfolds when pressurized and thus contracts (figure 4(a))

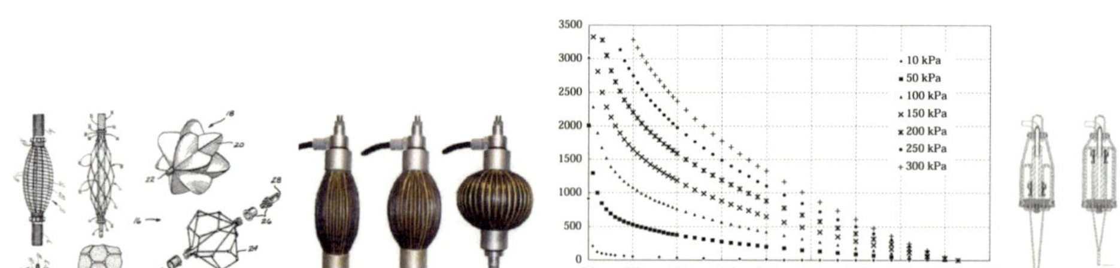

(a) [143] (b) [73] (c) [58] (d) [19] (e) [21] (f) [25]

Figure 4. Some FFAs based on "Vectored Structural Degrees of Freedom"

The work of Kukolj [73] shows an actuator with a net as the fiber reinforcement. This eliminates the friction between the fiber strands, but the friction between membrane and fiber reinforcement remains (figure 4(b)). Immega [58] enhanced this idea by implementing a stiff, folded membrane in between the fiber mesh (figure 4(c)).

A newer design known as "Pleated Pneumatic Artificial Muscles (PPAM)" was introduced by Daerden and Lefeber [19, 21]. The design is similar to the Yarlott muscle. Figure 4(d) and 4(e) show the design and the force-displacement characteristics of these artificial muscles.

Erickson [25] described a contraction actuator that can be considered an inverse rolling-lobe cylinder. This set-up has a large working range of 40-60% of the initial length (figure 4(f)).

2.1.3. Flexible fluidic "Bending" actuators

Bending actuators are generating a bending motion when pressurized, which is used to manipulate objects in an adaptive and compliant way. Staines [120] presented vacuum operated and Baer [7, 8] pressure operated conceptual solutions for this problem (figure 5(a) and 5(b)).

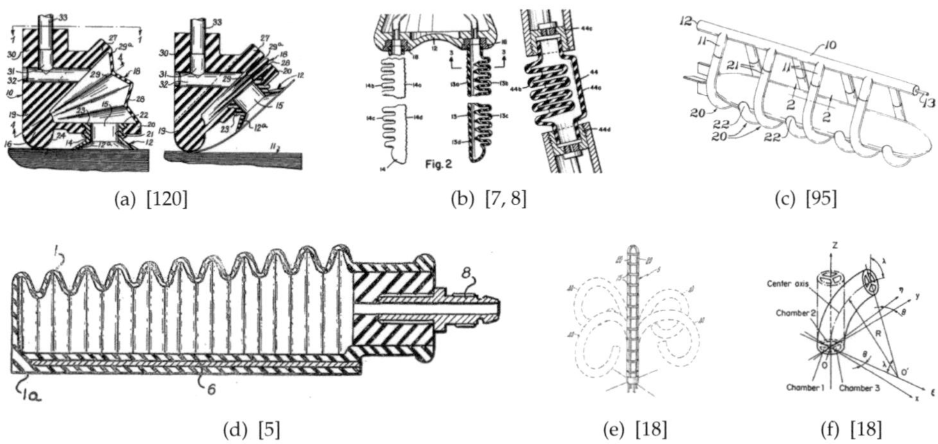

(a) [120] (b) [7, 8] (c) [95]

(d) [5] (e) [18] (f) [18]

Figure 5. Bending Actuators

The work by Orndorff and Ewing [27, 95] as well as Andorf et al. [5] introduce designs where bending occurs due to "anisotropic membrane stiffness". Craig et al. [18] point out that these types of actuators can be folded to reduce shipping volume specially for space applications. Figure 5(c)-5(e) show the different designs.

Bending actuators designed with multi-lumen hoses are represented by the work of Suzumori et al. [122–126] shown in figure 5(f). Radial reinforcements inhibit radial expansion so that the operating pressure is $1.4 - 4\,bar$.

There is a large variety of trunk-like bending actuators that create bending motion by adding structural constraints. A few examples are shown in figures 6(a)-6(c).

Monolithic bending actuators represent the last group in this section. Different research groups have been working on this topic during the last years [57, 71, 145, 146]. These actuators

(a) [15] (b) [84] (c) [6] (d) [145]

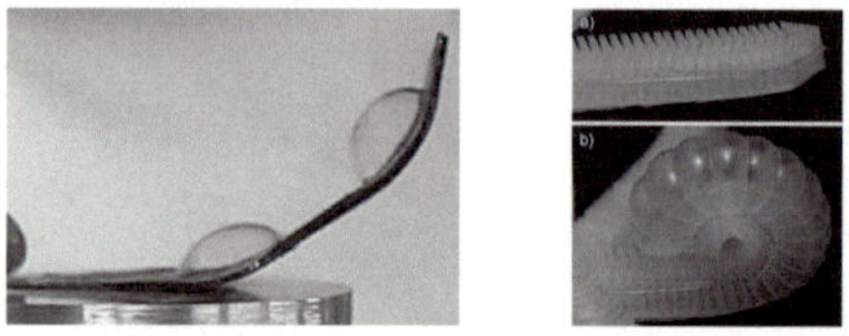

(e) [71] (f) [57]

Figure 6. Trunk-Like Bending Actuators and Monolithic Bending Actuators

are single material devices and mainly fabricated in on step. Operating pressures are mostly $< 1\,bar$. Some prototypes are shown in figure 6(d)-6(e).

2.1.4. Flexible fluidic actuators - combined motion

This group of actuators produces multiple directions of motion. Griebel et al. [38] developed a placement actuator for EEG-electrodes that conducts a linear expansion in combination with a coaxial rotation (figure 7(a)). Paynter [99, 102] on the other hand supresses the linear motion on purpose in order to design an pure rotary drive. As shown in figure 7(b) a pre-twisted membrane is straightened when pressurized and thus generates a rotary motion.

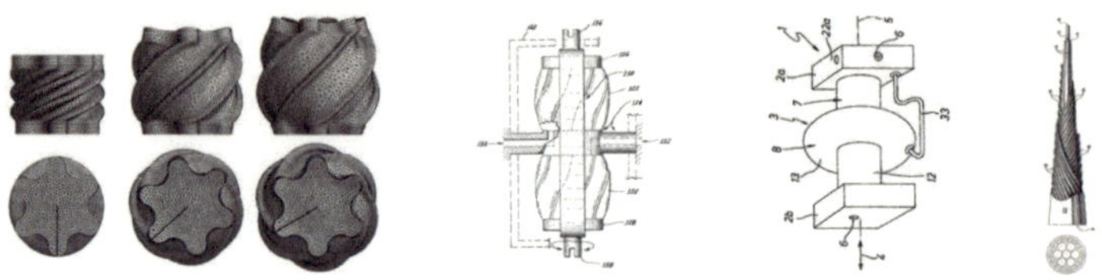

(a) [38] (b) [99] (c) [16] (d) [139]

Figure 7. Actuators with two directions of motion

Other concepts use several different cavities to create multi-motion actuators. Claus [16] presents a push-pull actuator while Wilson [139] combines expandable hoses to build a versatile robotic arm (figure 7(c) and 7(d)).

Alternative approaches are presented by Kimura and Brown. Kimura et al. [69] describe a principle that is called "whole skin locomotion". Here an elongated toroid turns itself inside out and hence can move over surfaces or through gaps (figure 8(a)). The concept by Brown et al. [14] is referred to as "jamming of granular material". The idea here is known from vacuum mattresses in ambulances. A flexible bag containing a granular material is shaped around an object and then evacuated. Friction, suction and mechanical interlocking connects gripper and object (figure 8(b)).

(a) Whole Skin Motion by Kimura et al. [69]

(b) Whole Skin Gripper by Brown et al. [14]

Figure 8. Actuators based on Whole Skin Effects

2.2. Flexible fluidic structural elements

Flexible fluidic structural elements complete the biomimetic approach. There are many examples of hydrostatic skeletons in nature [130].

Inflatable structures are well known in fields of crisis intervention [26] and exhibition stand construction [94]. Other applications include space structures like antennas [33, 79, 82, 132] and rovers [28, 48, 63, 72].

Most robotic designs with inflatable structures aim at space applications since they have a small shipping volume when deflated. Koren et al. [72] proposed a design for zero gravity applications and operating pressures of about 3.5 *bar* (figure 9(c)). Shoham et al. [104, 105, 118] developed a inflatable robot as shown in figures 9(a) and 9(b). They also characterized the robot regarding its stiffness at internal pressures up to 2 *bar*.

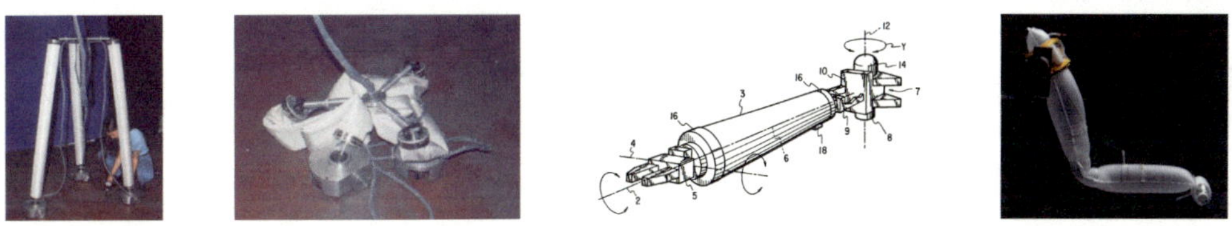

(a) inflated [118]

(b) deflated [118]

(c) [72]

(d) [106]

Figure 9. Different Robots with Inflatable Structures

The works of Sanan et al. [106], Maruyama et al. [83], and Voisemebert et al. [135] focus more on service and inspection robotics. Operating pressures are in the range of $0.4 - 0.6\,bar$. The 1-DOF arm developed by Sanan is shown in figure 9(d).

3. Flexible fluidic actuators - fabrication

3.1. FFA - Materials

As the name implies FFAs are made of flexible shell materials. This fact limits the number of relevant materials naturally. Normally laminated foils [112], vulcanized elastomers [36], coated fabrics [97], layered set-ups [1], or various combinations are used. A reasonable differentiation can be made between primary shaping processes like rubber molding and

[1] http://www.otherlab.com/

processes that use semifinished products like foils, in order to form the actuator's cavity. Both processes have their pros and cons.

3.1.1. Semifinished materials

Material selection cannot be considered without looking at the fabrication process. Of all welding technologies high-frequency (HF) welding can produce the most resilient seams [2]. However, HF-welding can only be achieved with materials that contain sufficiently strong dipoles. Thermoplastic Urethanes (TPU) have compared to PVC and PA the best material properties [144]. The latest material developments regarding composite sheet materials will be described in the following sections. The HF-welding technology is mostly used for more complex actuator technologies since the technology does not require very complicated molds.

3.1.2. Vulcanized elastomers

Vulcanized elastomers are often processed in compression molding processes. The uncured rubber monomer is put in the heated cavity of a compression mold. The press is closed and under the influence of heat and pressure the rubber is cross-linked. The variety of rubber compounds is infinite. The final material properties can be influenced by fillers, reinforcements and a large variety of chemical additives. How to achieve materials with tailored properties for flexible fluidic actuators will be presented. Compression molding requires complex molds, which limits variations in shape but provides actuators for high operating pressures.

3.2. FFA - fabrication processes

The two main processes for FFA fabrication are compression molding and HF-welding. A detailed description of both process developments is given here.

3.2.1. Compression molding

The requirements for vulcanized flexible fluidic actuators are simple:

- High pressure resistance
- High fatigue resistance
- Modular design
- Reproducible fabrication process

In order to achieve these properties the actuator shell is divided into two layers and fabricated in a two-step compression molding process. In the **first step** the inner shell is made. The inner shell (figure 10(b)) is responsible for tightness and fatigue strength of the actuator. It is vulcanized in a mold as shown in figure 10(a).

The **second step** needs some preparation. First the inner shell is covered with a braided aramid fiber sleeve, which is fixed around the rubber with a Vectran® yarn. This fiber reinforcement determines the pressure resistance of the actuator. After that the metal connectors are inserted and the mold adapters are added. Finally a thin layer of rubber is applied to the surface (figure 11(a)). The set-up is now ready for the second vulcanization

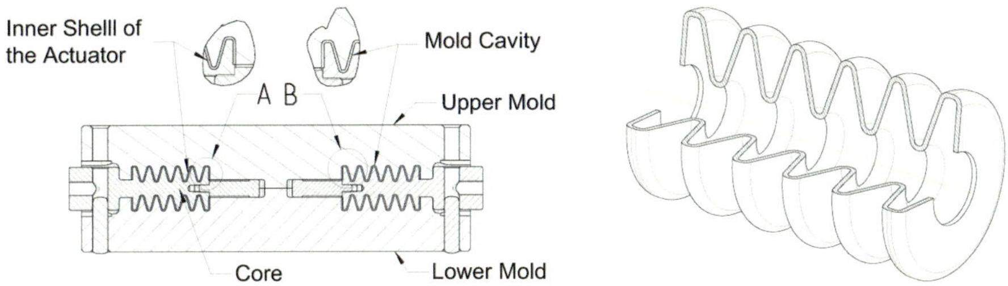

(a) Cross-section View of the first Mold

(b) Partial Section of the demolded Inner Shell

Figure 10. First Fabrication Step

step. The prepared actuator is inserted in the second mold and pressure is applied as depicted in figure 11(b). Fiber reinforcement and metal connectors are vulcanized to the inner shell.

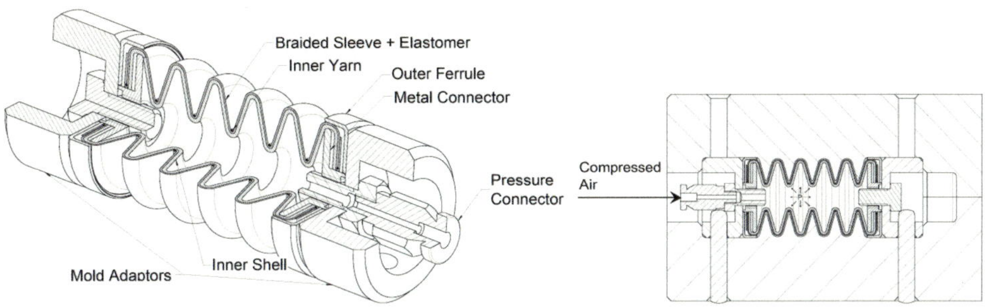

(a) Partial Section of the Actuator Prepared for the Second Fabrication Step

(b) Cross Section View of the second Mold

Figure 11. Second Fabrication Step

Operation shows that fatigue crack growth is the main failure mode of vulcanized flexible fluidic actuators. There are three main mechanisms to enhance crack growth resistance: Use of stress and strain crystallizing rubbers, use of fillers (mainly carbon black), and dispersion of pulp fibers. The first two mechanisms are implemented by the proper choice of the basis material (in our case a chloroprene rubber mixture (CR)). However, the particle morphology of carbon blacks limits their contribution to fatigue resistance [142]. Crack bridging effects can only be achieved by dispersing micro fibers in the basis elastomer. Aramid fiber pulps are well-suited for this purpose [3, 70, 131]. With a specific surface are of $5 - 15\frac{m^2}{g}$ and a minimal fiber diameter of $10\,\mu m$ they can enhance the overall material properties significantly (figure 14).

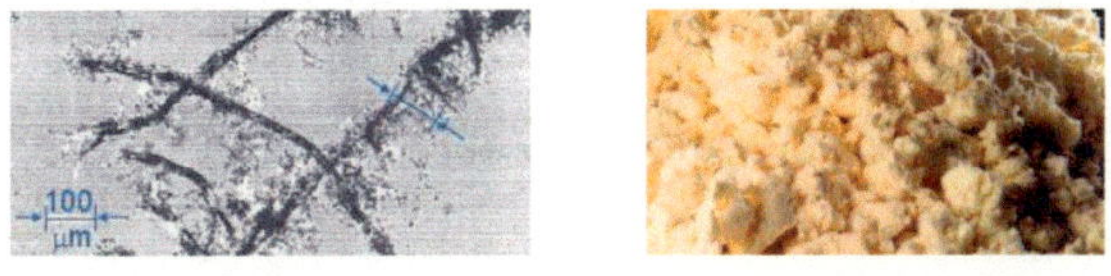

(a) Single Pulp Fiber

(b) Aramid Fiber Pulp

Figure 12. Morphology of Fiber Pulps

In order to determine the best concentration masterbatches with 0,5%, 1%, 2%, and 5%-mass have been tested according to DIN 53 504. Figure 13(a) shows how the pulp fibers get oriented in flow direction. Hence the samples with parallel and orthogonal pulp fiber orientation have been tested. The orientation has a significant influence on the force-displacement characteristics. The figures 13(c)-13(d) shows crack surfaces with different fiber orientations.

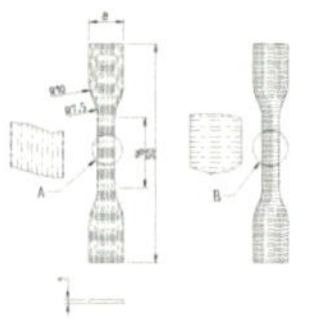

(a) Sample Geometry and Pulp Fiber Orientation

(b) Sample Plate

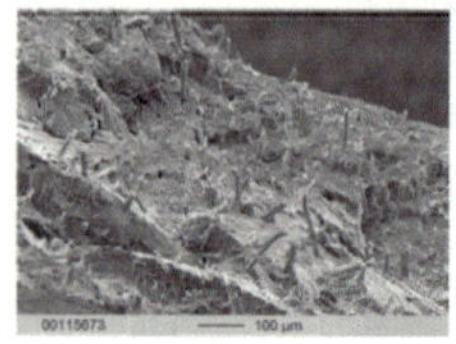

(c) Pulp Fiber Orientation Parallel to Crack Surface

(d) Pulp Fiber Orientation Orthogonal to Crack Surface

Figure 13. Material Characterization

Generally the goal is to define the maximum fiber concentration that does not lower the tensile strength significantly. The results show that a concentration of 1%-mass leads to the best material properties. The force-displacement plots show a clear influence of the fiber orientation, but in parallel configuration there is no significant difference compared to the pure rubber material (figure 14).

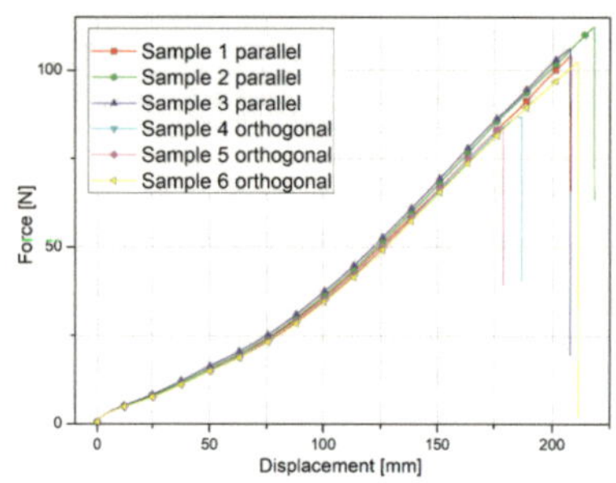

(a) 0% Pulp Fibers

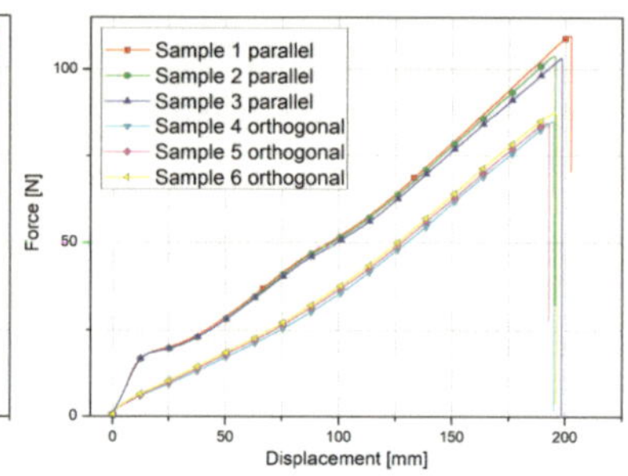

(b) 1% Pulp Fibers

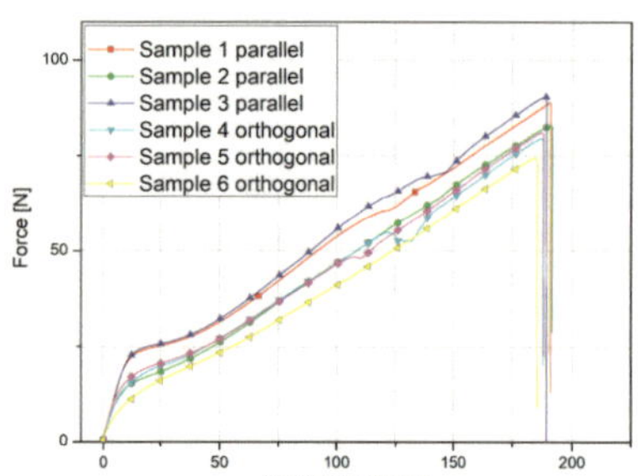

(c) 2% Pulp Fibers

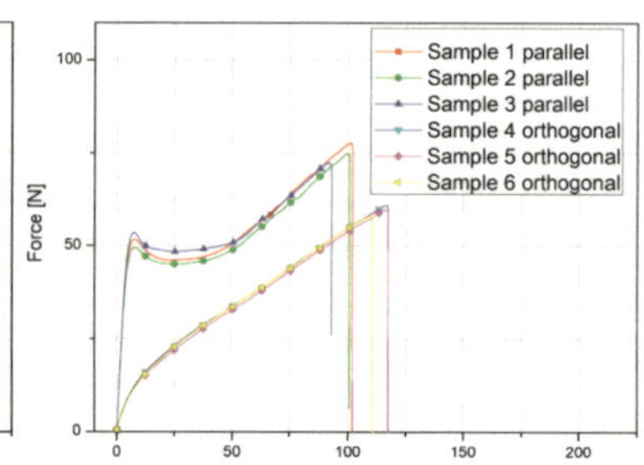

(d) 5% Pulp Fibers

Figure 14. Force-Displacement Plots for different Pulp Fiber Concentrations

3.2.2. High frequency welding

Commercially available fiber reinforced TPU foils show several drawbacks, such as low tensile strength, delamination between fibers and TPU-matrix, and axial fiber porosity. The general requirements for HF-welded FFAs are:

- A maximum thickness of 700 μm
- A tensile strength between 30 and 100 $\frac{N}{mm^2}$
- A modulus of 100 to 150 $\frac{N}{mm^2}$
- Gas-tight including no axial fiber porosity
- Odourless
- Processable with HF technology

Two material systems meet these requirements. **PEEK-monofilament reinforced TPU films** are composite sheets with two layers of monofilament PEEK-mesh between three layers of TPU-films. The overall material properties are: tensile strength 33 $\frac{N}{mm^2}$, Young's modulus of 168 $\frac{N}{mm^2}$, specific weight 269 $\frac{g}{m^2}$, thickness 450 μm.

Aramid-fiber reinforced TPU films are processed slightly different. Since the aramid fibers are spun to a yarn and are not available as monofilament material with sufficient strength, the yarn must be sealed prior to the laminating process. This is done by applying a TPU solution on the aramid fabric. After the evaporation of the solvent all yarn surfaces are covered with a thin layer of TPU. The pretreated aramid fabric is then laminated between two layers of TPU film. The resulting material properties are: thickness 580 μm, specific weight 365 $\frac{g}{m^2}$, tensile strength 60 $\frac{N}{mm^2}$.

A schematic view of the general production process is illustrated in figure 15. The inter-chamber connections of the pre-cut foil pieces are welded first. These pre-assembled pieces are then welded together around the outer contour in as many layers as desired. This second step closes the actual actuator chamber.

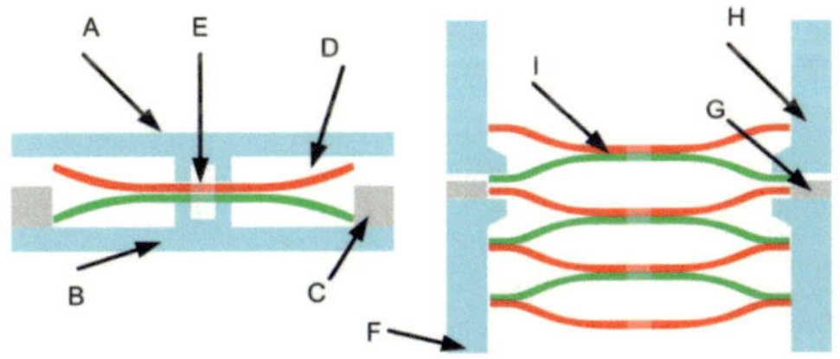

Figure 15. (A) upper welding electrode, (B) lower welding electrode, (C) centering coil, (D) two foil pieces, (E) hole for air passage, (F) lower welding electrode with undercut, (G) centering coil, (H) upper welding electrode with undercut, (I) foil segments from first step

3.3. FFA - evaluation

All dynamic testing was accomplished on a fatigue test rig consisting of a pressure-tight container, in which the actuators are mounted. The pressure in the container is monitored which allows detection of leaks in the actuators.

3.3.1. *Vulcanized actuators*

The relationship between fatigue resistance, operating pressure, and material combination is shown in figure 16. The choice of the base rubber compound influences the fatigue resistance significantly. The best fatigue resistance can be achieved with CR-rubber containing 1%-mass of aramid pulp fibers. At 6 *bar* the actuator withstands over 1,200,000 load cycles. In order to keep testing simple all fatigue evaluation was carried out with actuators 18 *mm* in diameter. All actuators failed due to fatigue cracks in the inner rubber shell. This results in excellent fail-safe characteristics, since the actuator never bursts and can be changed without any danger to the whole system.

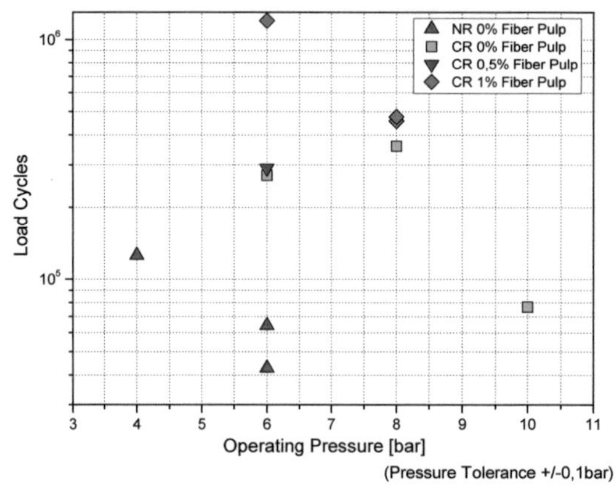

Figure 16. Fatigue Resistance of 18 *mm*-Actuators

3.3.2. *HF-welded actuators*

To evaluate the influence of the various production parameters during HF-welding on lifespan, several endurance tests were performed. In these tests, one parameter is varied and all others kept constant. Each experiment was performed with ten actuators. This was necessary because the deviation in lifespan was about 30% of the average lifespan. The influence of the following parameters on the lifespan has been determined:

- **Determination of the geometry of the welding tool:** In this test it was determined how the geometry of the welding tool influences the lifespan of the actuators. The first tool set-up consists of two electrodes (ground and high frequency) with the same geometry and mass. The second tool set-up consists of one electrode (high frequency) with the shape of the weld seam. The other electrode (ground electrode) is a flat plate.
- **Identification of the most durable seam thickness:** The seam thickness was varied between 95% and 15% of the total strength of the two films in order to determine which weld strength achieved the longest lifespan. In this experiment it was also determined how thick or thin a seam can be produced.
- **Variation of the welding force:** It was determined how the welding force, pressing the two electrodes together, affects the lifespan. This parameter indirectly sets the temperature at which the seam is formed.

- **Varying the time of the load cycle:** In addition to previous experiments where the load cycle time was six seconds (3 *s* pressure / 3 *s* no pressure) the cycle times were varied at 2, 4, 6, 10, and 20 seconds.
- **Evaluate the best pressure rise time:** In this experiment the actuators were filled via a restrictor. So the time to fill is adjustable. These different times were 40, 125 and 250 *ms*. A shorter time than 40 *ms* between the electrical signal to the switching valve and the reach of 95% of the desired pressure was technically not possible.
- **Evaluate the best release time:** Analogous to the determination of the pressure rise time, the pressure reduction period was varied. The restrictors were used with the same settings as the pressure rise test.
- **Limitation of the expansion:** This test was necessary to evaluate the effect on durability when the maximum deformation is limited. The expansion is limited to 100, 66, 50, and 33%.

It has been found that the lifespan increases when the deformation speed is low. When the pressure rise time is six times higher (40 $\rightarrow$ 250ms), the lifespan increases by a factor of 2.5. The pressure release time influences the lifespan similarly. The limitation of the maximum expansion to 50% of the nominal stroke increases the lifespan by a factor of 4. The use of two electrodes with the same mass and same shape also creates an increase by a factor of 2. The production of a seam with a higher initial welding force also has a positive effect on the lifespan of the actuators. This increase is only about 10%. The optimal seam thickness varies depending on the structure of the material used. The increases on the lifespan which can be achieved with this optimization are by a factor of two. Weld seams with a thickness less than 20% or greater than 80% of the nominal thickness of the two films are not useful. The variation of the load cycle time shows that the actuators fail very quickly at an operating frequency of 6 seconds. For longer load cycles, the composite material has more time to relax and reduce internal stresses. With shorter cycle times the internal stresses caused by the previous load cycle are not yet dissipated and a kind of solidification occurs.

Endurance tests of actuators with different film materials were conducted in order to compare the film properties.

- **Determination of the lifetime curve:** These tests were conducted with ten actuators for each pressure step. This pressure was applied cyclically to determine the durability of the actuator as a function of pressure. Pressure values of 3, 4, 6 and 8 *bar* were used. All tests were carried out using a commercial film material.
- **Continuous operation of the actuators of PEEK reinforced TPU film:** With ten actuators made of this material an endurance test at six bar and a cycle time of 6 seconds was performed. In this endurance test, the results of some of the previous test runs were considered. The thickness and the initial load were adjusted. The other parameters like pressure rise/release time, expansion ratio, and welding tools have not been altered to ensure the comparability of results.
- **Continuous operation of the actuators of aramid reinforced TPU film:** The parameters of this endurance test were identical to those of PEEK actuators.

Determining the lifespan trajectory has shown that the operating pressure has a big effect on the durability. If the average lifetime at 6 *bar* of about 500 cycles is set to one, the actuators

achieve at 8 *bar* only the value of 0.2. At a pressure of 4 *bar* they reach a 8 and at 3 *bar* they reach a value of 66. In comparison, the average lifetime of PEEK actuators at 6 *bar* is around 112 and 205 for the aramid-reinforced actuators. The failure mode mostly was breaking of the weld between the tube and the TPU.

4. Series of flexible fluidic actuators

Over the years, a large variety of flexible fluidic actuators have been developed. The following sections give an overview of the most recent models either fabricated by compression molding or HF-welding.

4.1. Vulcanized Actuators

All vulcanized actuators are fabricated using the process described in chapter 3. This process represents a scalable and reproducible fabrication process resulting in the following actuator geometries. Figure 17 gives an overview over all rubberbased FFAs. Each section includes the general properties and the actuator characteristics. The standard operating pressure is $0 - 10\,bar$ for all vulcanized FFAs.

Figure 17. Series of Vulcanized FFAs: (from left to right) 36mm, 18mm, 11mm, and single chamber FFA

4.2. HF-welded actuators

HF-welded actuators and the corresponding joint modules have been called GPA (Gelenkmodul fuer Pneumatische Aktoren). GPAs have already a fairly long and successful history as can be seen from the different generations shown in figure 18. The emphasis of continuing developments is improving the lifespan of actuators, the welding process, the maintainability, and some economic aspects.

GPA Type	Torque [Nm] (min. - middle - max.) at 4 bar	Range of Motion [°] at 4 bar	Dimensions [mm]	Weight [kg]	number of parts to be machined
T1-14T	3 - 10 - 14.5	±83	85x75x70	0.25	16
T2-13RVD	2.1 - 9.9 - 17.6	±85	110x95x80	0.26	6
T3-12RVD	1.7 - 7.1 - 13	±116	103x80x69	0.18	10

Table 1. Properties of different GPA versions

In table 1, the actuator designators define the number of chambers and the material. The range of motion depends on the applied pressure because the antagonistic actuator needs to

∅ 11mm Actuator:

- Effective Diameter of 11 *mm*
- 5 Pleats
- Range of Motion 90°
- Symmetric Shape

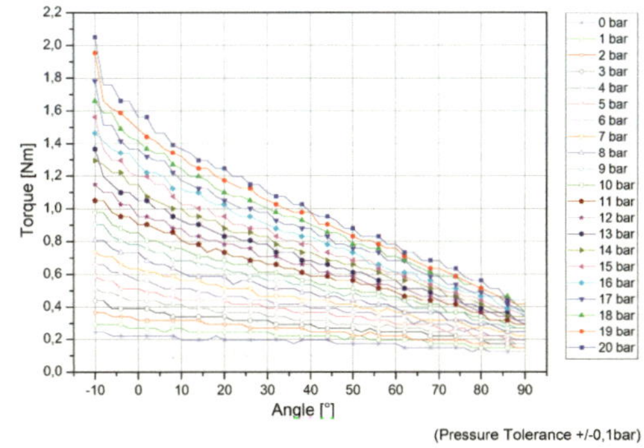

∅ 18mm Actuator:

- Effective Diameter of 18 *mm*
- 5 Pleats
- Range of Motion 90°
- Symmetric Shape
- Burst Pressure > 40 *bar*

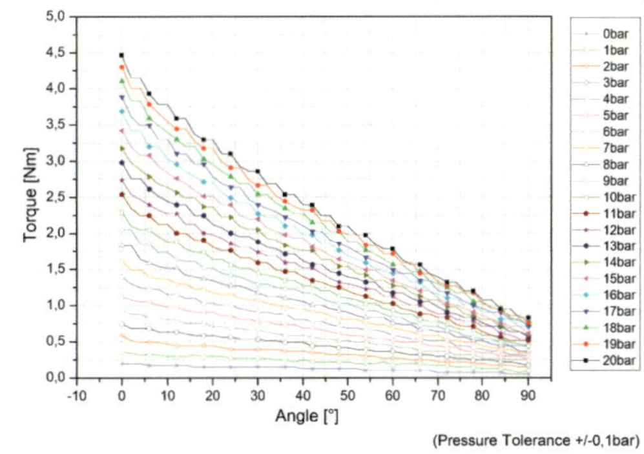

∅ 36mm Actuator:

- Effective Diameter of 36 *mm*
- 6 Pleats
- Range of Motion ±90°
- Asymmetric Shape

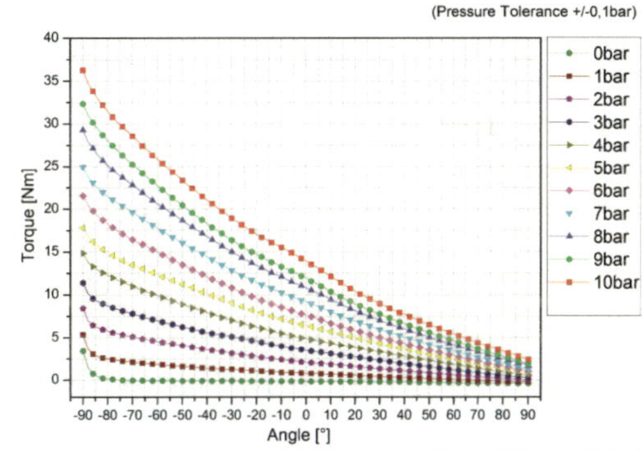

Single Chamber Actuator:

- Effective Diameter of 40 *mm*
- 1 Pleat
- Range of Motion ±12° or 19*mm* Linear Motion
- Symmetric Shape

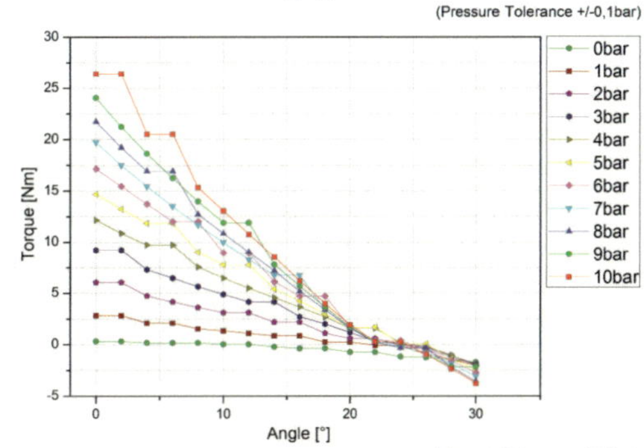

be compressed and thus the range of motion depends on the applied torque. Except for the extreme angular ranges, the torque over angle is fairly linear. Figure 19 shows the torque-angle plot for different pressure values.

(a) GPA 1 (b) GPA 2 (c) GPA 3

Figure 18. The three GPA generations

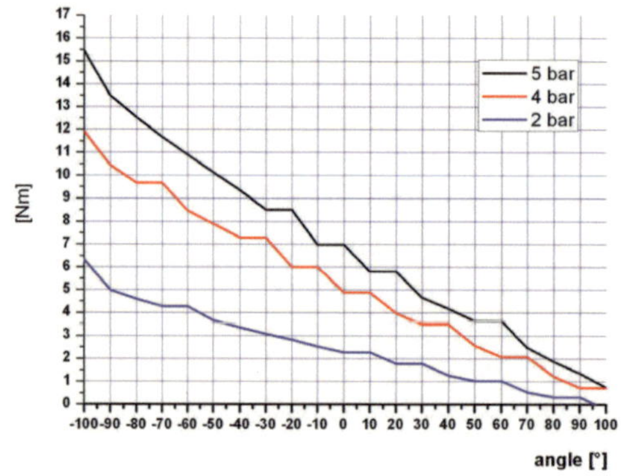

Figure 19. Torque-Angle Plot for GPA-3 module with a 10 chamber actuator

Of course table 1 does not reflect all the important parameters. For example, the lifespan of actuators has grown dramatically from a view thousand up to more than 350, 000 stress cycles (full range motion with 4 *bar* forth and back in 3.6 *s*) for GPA3.2 joint modules. Lifespan depends very much on the maximum applied pressure and the valves to be used. The value of 4 *bar* is a good compromise between maximal torque and lifespan. Most GPA types may be operated using higher pressure (up to 6 *bar*), but the lifespan of actuators will decrease dramatically at this pressure. In many cases repair welding is possible.

5. Designing with flexible fluidic actuators

5.1. Design principles

Using flexible fluidic actuators as drives makes the designing process comparatively easy. Since the compressive force of the actuator is directly used to generate torque, no additional transmission elements are necessary. However, the flexible properties of the actuators require guiding components along the the actuators track of motion. If those guiding elements are not included, the actuator will work, but will eventually have decreased range of motion and less torque. The guiding elements for vulcanized actuators are metal connectors at the end of the actuator and additional metal brackets in the midsection of the actuator. HF-welded actuators mainly use additional lugs around the chambers which are included in the seams or the end areas of the actuators. These lugs are then clamped in the structure of the corresponding joint. Position and type of the guiding element are additional parameters to determine the overall characteristics of the flexible fluidic drive.

5.2. Operating flexible fluidic actuators

The requirements to operate FFAs are pressure supply, control members, and sensory infrastructure. Depending on the complexity of control that is desired, control members and sensors can include valves, position sensors, pressure sensors, and micro controllers. A detailed overview for highly integrated FFAs is given in [1]. Using hydraulics for mobile applications instead of pneumatics can help to avoid bulky pressure supplies.

5.3. Structural modeling of flexible fluidic actuators

Structural and material nonlinearities as well as large strains make it very hard to predict the behavior and characteristics of flexible fluidic actuators. Thus implementing a structural model of FFAs is a challenging task. The FEM-model was developed with ANSYS in order to provide a design tool for FFAs. The configuration of the model is illustrated in figure 20. The different layers of the actuator shell are modeled using different elements. The inner and outer rubber shell is modeled with 941,902 SOLID285-elements using a hyperelastic material model according to YEOH [2]. The fiber reinforcement is represented by 13,428 SHELL181-elements which are connected to REINF265-elements respectively [3]. The structural integrity of the corresponding joint is modeled using 1,502 MPC184-Link/Beam-elements. The metal brackets are implemented using 14.042 SOLID185-elements [4]. Including the contact elements, the model consists of 1,235,804 elements.

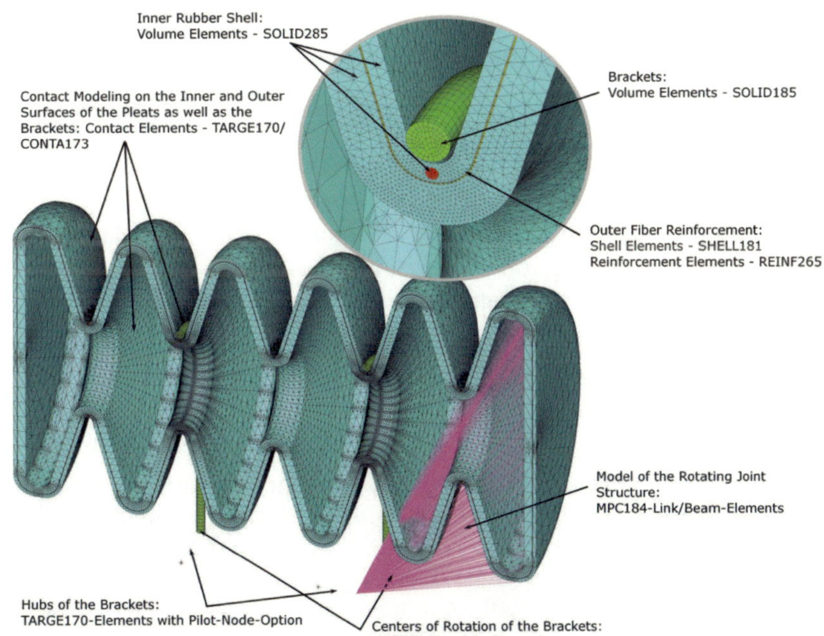

Figure 20. Set-up of the FEM-Model for a Flexible Fluidic Actuator

[2] $C_{10} = 0.477330421717; C_{20} = -0.148261658100; C_{30} = 0.225732915194$

[3] Young's Modulus $E_{\|} = 124\frac{kN}{mm^2}$; $E_{\perp} = 8\frac{kN}{mm^2}$; Poisson's Ratio $\nu = 0.32$

[4] Young's Modulus $E = 210\frac{kN}{mm^2}$; Poisson's Ratio $\nu = 0.32$

In order to compare the structural model with the behavior of the real actuator at first a strain analysis is conducted. Figure 21(a) compares the strain behavior of the model and the real actuator for joint motions of $0° - 90°$. The comparison shows good compliance of model and actuator.

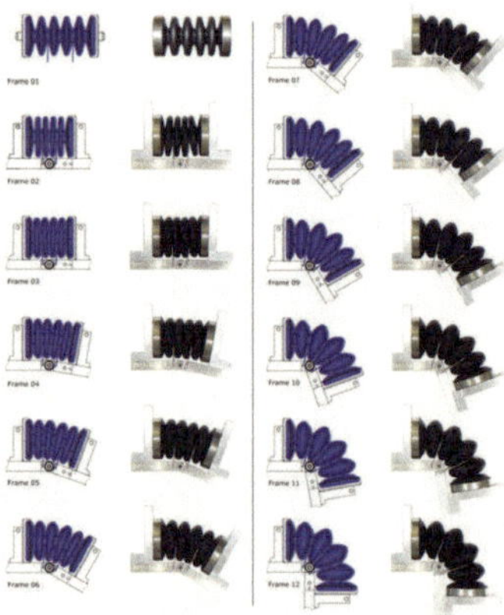

(a) Strain Analysis of a Flexible Fluidic Actuator ($\varnothing$ 18mm), FEM-Model vs. Real Actuator

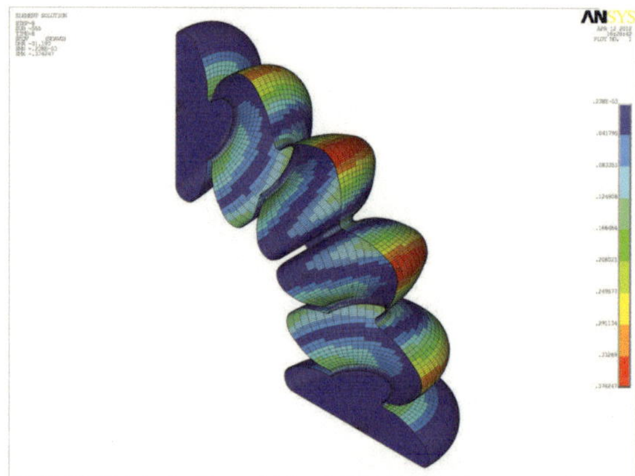

(b) Stress Analysis of the Fiber Reinforcement of a Flexible Fluidic Actuator ($\varnothing$18 mm)

Figure 21. Structural Analysis of the Actuator Model

The stress analysis of the fiber reinforcement in the actuator shell is shown in figure 21(b). The regions of high stress correspond with the areas where fatigue failure occurs.

6. Modeling and control of flexible fluidic actuators

From the control point of view, FFAs represent a complex dynamic system of high nonlinearity. In contrast to conventional fluidic actuators, FFAs posess some important properties, which are essential with respect to modeling and control problems:

- as a consequence of using viscoelastic materials for the chamber manufacturing, actuator torque and volume characteristics are complex non-linear functions not of pressure only, but also of angular displacements
- there are no "classical" sealing elements with static friction and stick-slip effects, but the viscose friction and damping effects can play an important part
- hysteresis effects by torque generating are indicated

Due to restricted volume of this section, the modeling and parameter identification approaches as well as different control concepts will be presented briefly. More details can be found in the related references.

6.1. Modeling

Generally, the model of FFA as a torque source consists of a model of actuator mechanics and pressure dynamic model. The modeling process includes the experimental investigation of the basic actuator characteristics with following data fitting.

6.1.1. *Mechanical model*

Using FFAs, a compliant robotic joint of rotary type can be realized either as an antagonistic set up with two working chambers (Fig. 22a) or as a unit with one working chamber and a retraction spring (Fig. 22a). Fig. 22 shows mechanical schematics of the both joint types.

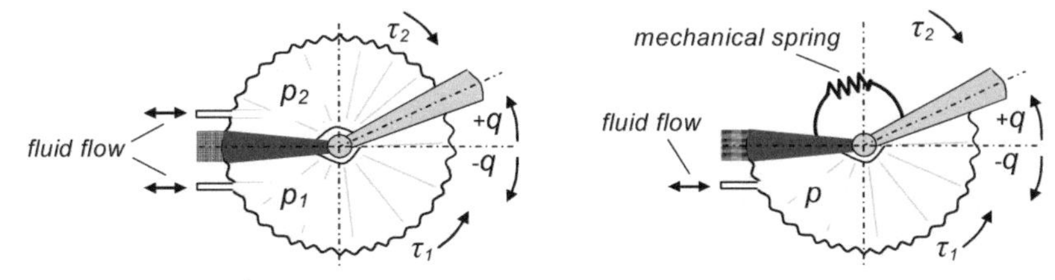

(a) with two antagonistic assembled flexible chambers

(b) with one flexible chamber and a retraction spring

Figure 22. Working principle of a fluidic actuated rotary robotic joint

As shown in previous sections by means of experimental investigations (static load tests), the static torque characteristics of the FFA are in general nonlinear functions not only of the operating pressure p only, but also of the working angle q. This is a basic feature of FFAs, in contrast to conventional fluidic actuators like pneumatic and/or hydraulic cylinders or vane actuators and the source of inherent compliancy of FFAs. Taking into account this basic feature, the total torque for the compliant robotic joint with antagonistic set-up can be determined as

$$\tau\left(q,\dot{q},p_1,p_2\right) = \tau_1\left(q,p_1\right) - \tau_2\left(q,p_2\right) - \tau_{loss}\left(q,\dot{q},p_1,p_2\right) \tag{1}$$

or as

$$\tau\left(q,\dot{q},p\right) = \tau_1\left(q,p\right) - \tau_2\left(q\right) - \tau_{loss}\left(q,\dot{q},p\right) \tag{2}$$

for the joint with one flexible chamber and a retraction spring. Here τ_1 and τ_2 represent static actuator torques acting in positive and negative directions respectively, τ_{loss} represents the torque losses like viscous friction (damping) and can include also tourque/angle hysteresis. These effects are typical for soft fluidic actuators (including linear pneumatic muscles) and are an inevitable consequence of using viscoelastic materials in the chamber. The results of experimental investigations of basic characteristics and dynamic modelling of compliant fluidic robotic joint operated by gaseous as well as liquid working media, was reported in [86, 87] and summarized in [85], whereby the torque losses τ_{loss} was measured in first approximation in the experiments at constant velocities. The hysteresis was measured, performing the torque measurements in opposite angle directions. The torque characteristic $\tau_i\left(q,p_i\right)$ of an individual chamber i, experimentally determined during static load tests, can be analytically described by means of a third-order polynomial with angle q as independent variable, as

$$\tau_i\left(q,p_i\right) = \kappa_3\left(p_i\right)\cdot q^3 + \kappa_2\left(p_i\right)\cdot q^2 + \kappa_1\left(p_i\right)\cdot q + \kappa_0\left(p_i\right), \tag{3}$$

where the polynomial coefficients $\kappa_3...\kappa_0$ are functions of pressure in the chamber. For control purposes, the torque characteristic 3 can be approximated well by using the polynomial fit with six constant polynomial coefficients $k_5...k_0$ [86]:

$$\tau_i\left(q,p_i\right) = k_5\cdot q^3 + k_4\cdot q^3 + \left(k_3\cdot p_i + k_2\right)\cdot + k_1\cdot p_i + k_0, \tag{4}$$

The mean relative error of this simplest satisfactory approximation is < 9% (see [85] for more details).
Assuming the identity of both actuator chambers, the torque τ_2 of the antagonist chamber in (1) can be obtained by mapping the torque τ_1 of the agonist symmetrically with respect to the ordinate as

$$\tau_2(p_2, q) = \tau_1(p_1, -q) \;\; for \;\; p_1 = p_2. \tag{5}$$

The second term in (2) when using a mechanical spring with a linear characteristic can be determined as $\tau_2(q) = c_S q$, where c_S is a spring constant. Despite obvious uncertainties, especially in FFA torque losses and fluctuations in model parameters due to behaviour of viscoelastic material, model (1) - (5) can be useful for approximate description of FFA torque characteristics.

6.1.2. Pressure dynamic model

In the case of pneumatics (i.e. for FFA, operated with a gas/air), the pressure dynamics in an actuator chamber i can be described by

$$\dot{p}_i = \frac{\chi}{V_i}\left(RT\dot{m}_i - p_i\dot{V}_i\right), \tag{6}$$

where R is the universal gas constant, T is the air temperature, V_i is the chamber volume, the mass flow rate $\dot{m}_i$ defines the amount of air, passing through a valve into or out of the chamber in a time unit. It is assumed that the heat transfers at chamber's charging and discharging are the same polytropic processes, characterized by the polytropic coefficient χ. This topic is discussed in more detail in [85]. The volume characteristic of an elastic chamber is a nonlinear function of current displacement angle q and pressure p_i. Based on the experimental data it can be approximated as

$$\dot{V}_i(q, p_i) = v_1(p_i)\, q^2 + v_2(p_i)\, q + v_3(p_i) \tag{7}$$

where values of polynomial coefficients v_1, v_2, v_2 are functions of pressure [64]. Similar to the torque characteristic, the volume characteristic of the FFA-based soft robotic with antagonistic set up is assumed to be symmetrical with respect to the joint angle:

$$\dot{V}_1(q, p_1) = \dot{V}_2(-q, p_2) \;\; \text{for} \;\; p_1 = p_2. \tag{8}$$

The air mass flow rate $\dot{m}_i$ through the valve can be modeled by the standardized expression for air flow through an orifice (ISO 6385), whereby the orifice area varies with control input [85]. To provide a more exact model for servovalves, taking into consideration also the hard nonlinearities as the dead zones, an experimental procedure for obtaining of the real relationship

$$\dot{m}_i = f_m(u, p_{us}, p_{ds}) \tag{9}$$

between the air flow, the input voltage u to the valve, as well as the up- and downstream pressures (i.e. p_{us} and p_{ds} respectively) were applied [64, 137]. To achieve a model that describes the complete behavior of valves, both charging and discharging processes were explored for constant chamber value, inflow and exhaust flow maps were

calculated and then combined. In case of chamber charging, the upstream pressure is equivalent to the supply pressure (i.e. $p_{us} = p_s$), while the downstream pressure is equal

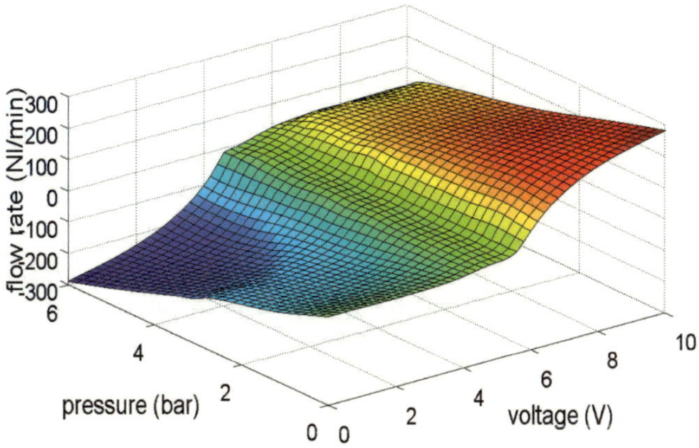

Figure 23. Experimentally determined flow map for the servo-valve Festo MPYE-5-1/8LF-010-B.

to the chamber pressure ($p_{ds} = p_i$). In case of chamber discharging, the upstream pressure is equivalent to chamber pressure ($p_{us} = p_i$) and the downstream pressure equals to atmospheric pressure ($p_{ds} = p_a$). Assuming that p_s and p_a are constant, expression 9 can be rewritten as

$$\dot{m}_i = f_m\left(u, p_i\right) \tag{10}$$

Fig. 23 shows an example of the experimentally determined characteristic of a servo valve.

6.2. Embedded control approach

In section 4 different kinds of FFAs and joint modules have been described. These actuators and joint modules were not developed with only one special application in mind. The main advantages of FFAs are their compactness and high torque to weight ratio as well as their inherent compliance. Thus it is wise to have a broad palette of solutions w.r.t. control of joint modules. Constituents for requirements for a joint module toolbox might include

- economic solutions,
- degree of compactness,
- quality of control and maximal joint rate, and
- modularity

to name the most important ones. Of course many other aspects like environmental ones may be considered (e.g. under water operation). Economic aspects might include overall costs, especially relative costs of joint module and components needed for control. Other aspects might address the lifespan of actuators, which is a compromise between maximum pressure to be applied and lifespan of actuator, type of actuator etc. Economic aspects may also include energy, consumption of compressed media, but also system integration aspects will play a role. The degree of compactness is dependent on the degree of integration of control components like sensors, valves and control logic. It also depends on the amount of tubing and cables needed for operation. High quality of control most often relies on stability, precision of position, tracking accuracy, torque precision, degree of overshooting, and bandwidth of control. In general, the more ambitious the requirements for control are, the

higher will be the costs especially for the valves. The higher the required bandwidth of control, the larger the valves will be. The pure size of the valves are in contradiction to compactness and weight requirements [**?**]. To some degree this also holds for sensor system and control and communication logic. High integration will enhance modularity when viewed from a higher standpoint where a joint module is one of many more other parts of a plant. With a closer view of a joint module, high integration will in general restrict the range of control components which can be used. Modularity should not only mean mechanical modularity but also different options for supply of compressed air, for electrical power supply, and for communication and system integration. A range of actuator solutions arranged by the degree of mechatronic integration is shown in Figure 24.

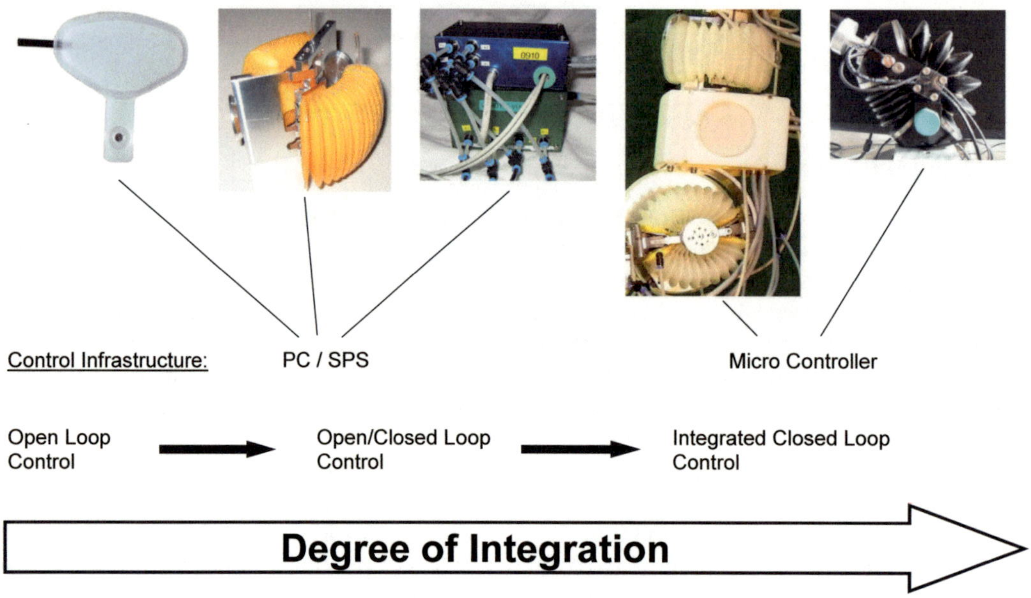

Figure 24. Different Degrees of Integration of the Control Components and Drive Systems

In this section our approach and solutions for embedded control as well as the achieved results will be described in more detail. The realization is around a 16-bit microcontroller (MC), a PIC24HJ128GP506 of Microchip with 40 MIPS performance and a broad variety of peripherals including communication via CAN. A miniature controller board CBR2 has been developed, and is able to control up to two joint modules reading joint values via 2 SPI-interfaces, pressure values via I2C or analog channels and control of switching valves by 8 output compare modules configured for PWM-mode. Control system integration is handled by an upper level control system on a PC running a user interface and coordinating different joint modules. Communication between MC and PC may be via RS232 and/or via CAN. The MC understands a simple 2-byte-command set with instructions for position set point, position ramp, torque command, pressure control, gravity compensation, parameter setting and for definition for periodical delivery of local information (to be specified by the PC) in a programmable rate with periods down to 21ms via CAN for one joint and 45 ms for 6 joints. General programming considerations relate efficiency of control calculation programming and means for utilizing parallelism. As no operating system kernel is used, mainly interrupt routines and the chipset inherent and in parallel running subunits are used. This works

very well for analog input of pressure signals and for CAN-message communication. Up to a certain level of analog signal information blocks as well as CAN message blocks are handled by DMA transfer autonomously without putting load on the CPU kernel. Message processing is implemented by a main loop, and the sampling task is a high priority interrupt task. The microcontroller does not offer floating point operations at the instruction level. Table 2 gives a hint of the relative efficiency programming int-, long int- and floating point operations [5]. Therefore mainly int16 was needed (e.g. to get more decimal places) long int (=int32) operations including shifting were used. This also means, that calculations for control algorithms were kept relatively simple. Trigonometric functions are implemented via tables:

Operator	+	-	$*$	$\div$	»const.	<
int	2	2	2	20	1	13
long int	6	6	11	469	6	16
float	124	148	118	380	-	77
double	140	152	113	383	-	77

Table 2. Instruction cycles for different operators and operands

For control of pneumatic actuators by an MC we use 2/2-way solenoid switching valves, (2 valves per actuator) one for inflation and one for deflation. Thus four valves per joint are needed, although for some cases one actuator might be replaced by a passive spring. This case will not be considered here. Presuming that a valve is sufficiently tight for higher pressure values, the maximum torque of the joint is not dependent on the maximal flow of the valves. Thus relatively small valves may be used if no special requirements for maximum speed/control bandwidth are given. The most compact and light weight valves are spider type valves. Additionally we mainly use modified double arranged switching valves of mass flow of $Q_n = 16$ SLPM (standard liters per minute) for EV08-type valves and 22 SLPM for EV09-type valves. Switching time is about $2\,ms$. These valves are light and compact enough to be integrable as demonstrated in Figure 24. For higher control bandwidth, higher mass flow is necessary. One may either operate two or more valves in parallel as one logical valve or use other models with higher mass flow. For torque and for position control pressure control is used. To measure pressure we use amplified pressure sensors with analog or I2C interface, 1ms response time and 10 bit resolution for a pressure range up to 7 bar. Using the MC's PWM features one can control mass flow by manipulating the length of the duty cycle. In our case a duty cycle has a maximum length of $7\,ms$, the sampling time of our control loop (142 Hz), which amounts to a PWM value of 1094. Unfortunately mass flow is also dependent on the type and the exemplar of valve, power supply voltage, and pressure difference. This means a model for each valve is needed. One must distinguish between the valve responsible for inflation and the one for deflation. In the first case, the pressure difference is given by the constant supply pressure and the actual pressure in the actuator. In the second case the pressure difference is between internal pressure and the $0\,bar$ environment. In Figure 25 the measured flow model is shown for inflating and deflating direction.

If one transforms flow values to the percentage scale, one gets a fairly linear relation, except for PWM values where inflation/deflation starts as can be seen in Figure 25(d). By means of interpolation using two characteristic curves for valve modelling, the MC can approximately set the percentage of mass flow given the actual pressure in the actuator. For control of mass flow a third characteristic curve will be needed.

[5] C30-Compiler, no optimization

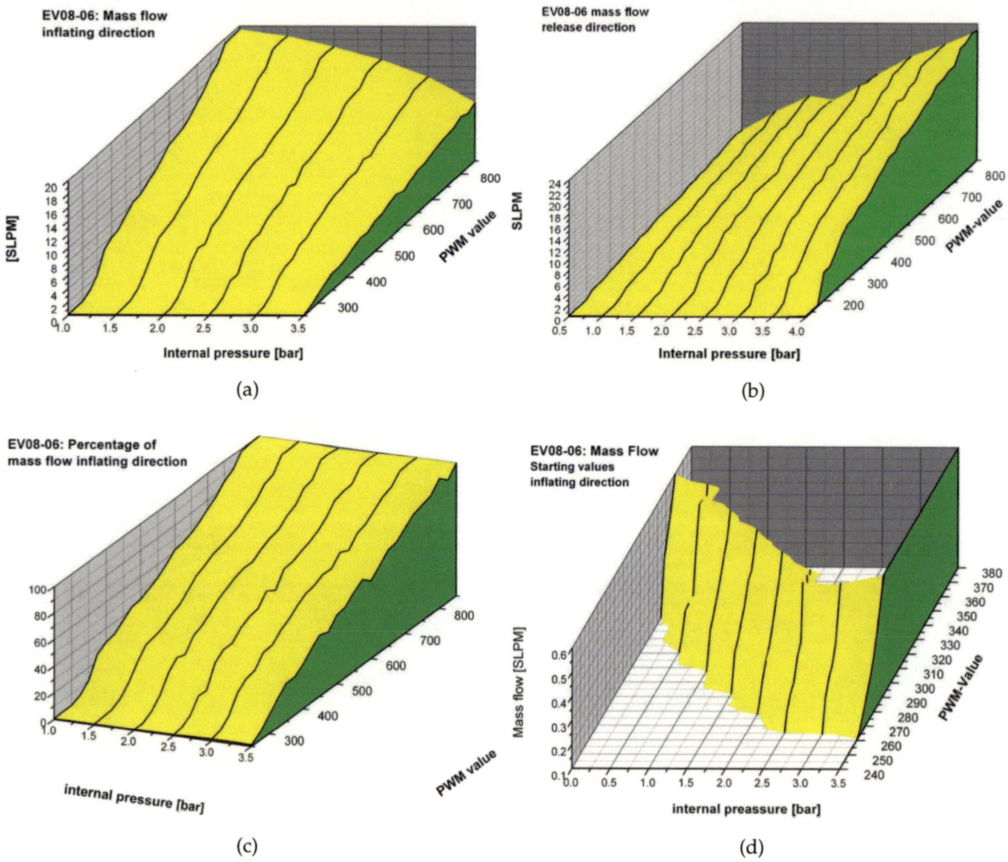

Figure 25. Flow Charcteristics of an EV08-Switching Valve

Based on such models, pressure control according to the scheme shown in Figure 27(a) give results shown in Figures 26(a) and 26(b).

In Figure 27(a), q denotes angular position, q_s position set point, p_i denotes internal pressure of actuator 1 and 2 respectively. Position control may be designed as a cascade on the base of pressure control and thus can build up (virtual) torque level using the linearized torque characteristic. Position control may also be implemented without pressure control (avoiding pressure sensors). Such schemes rely heavily upon the antagonistic actuator and use it as a kind of brake. Generally such schemes show decreased performance, may clip the angular range, have less stability and in many cases consume more compressed air. Figures 26(c) and 26(d) show position control results for different loads and different valve arrangements. The controller scheme is shown in figure 27(b).

It is difficult to achieve higher accuracy than for example $0.5 - 1^{\circ}$. This is due to the accuracy of valve control and to some degree of stochastic behavior in the flow starting part of the valves as well as limited position sensor accuracy giving poor velocity information. 14-bit sensors [AS5048 from AMS] are on the market and MC-technology is developing at a fast pace, so that there is a good chance tuning and quality of embedded control will soon improve.

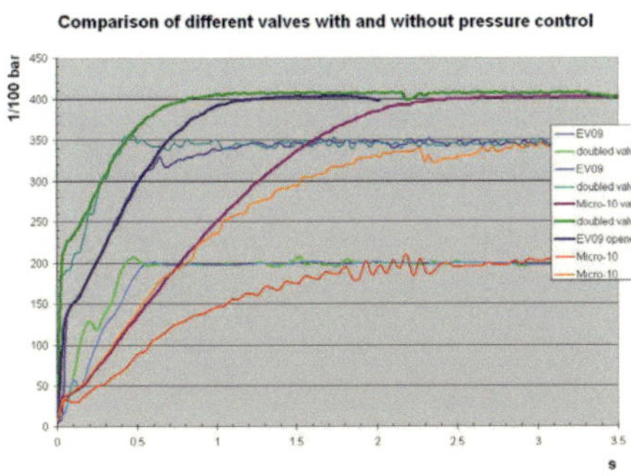

(a) Pressure Control - Different Valves

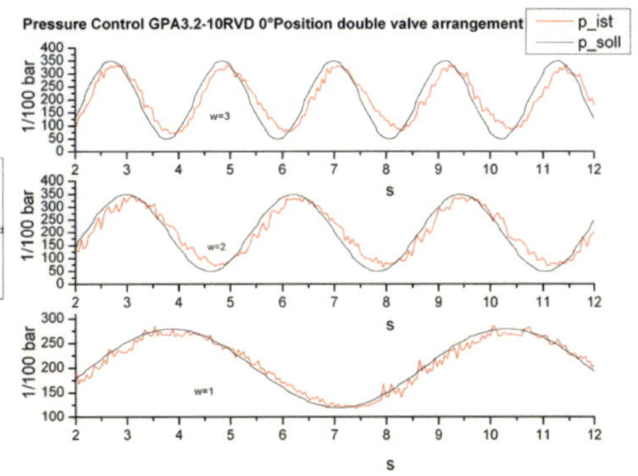

(b) Pressure Control - Double Valve Arrangement

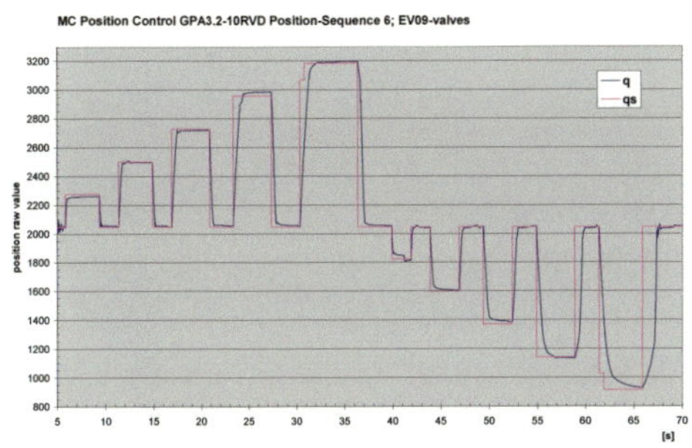

(c) Position Control - Single EV09-Valve

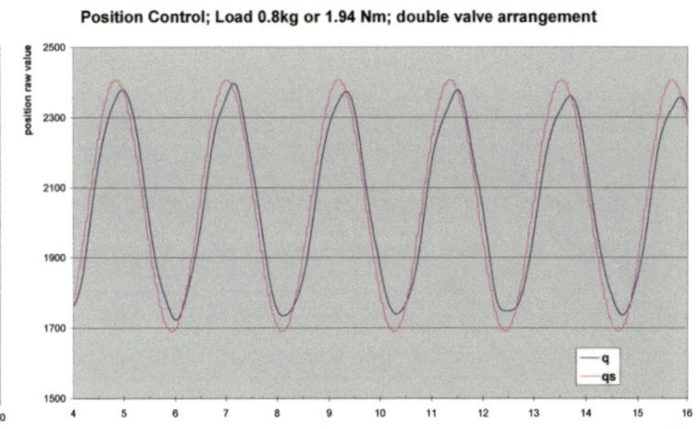

(d) Position Control - Double EV09-Valve

Figure 26. Results of Embedded Control using Switching Valves

6.3. Advanced control concepts

Implementation of advanced control concepts requires the use of a high-performance PC-based controller. The control concept has a cascade structure and is applied for both (pneumatic and hydraulic) FFA realizations [85]. Control algorithms for pressure, position and interaction control for single FFA as well as for compliant robotic arms, using different valve types, were developed and tested.

6.3.1. Pressure control

Fast and precise pressure control in the inner loop of the cascade control scheme is necessary for effective control of FFAs. To decouple the pressure subsystem from the mechanical subsystem, a control law was derived using the feedback linearization approach, which was modified with regards to specific traits of soft actuators [85]. In order to compensate non-linearities of FFAs pressure dynamics, model (6) - (7) of actuator chamber volume is used in control law as feedback and for the determined flow rate the control voltage is obtained by inverting of experimental flow map (figure 10). The effectiveness of designed pressure controller has been confirmed through several experiments (step response), in pneumatic case both for PWM-controlled on-off valves [86] as well as for different types of servo valves [64, 137].

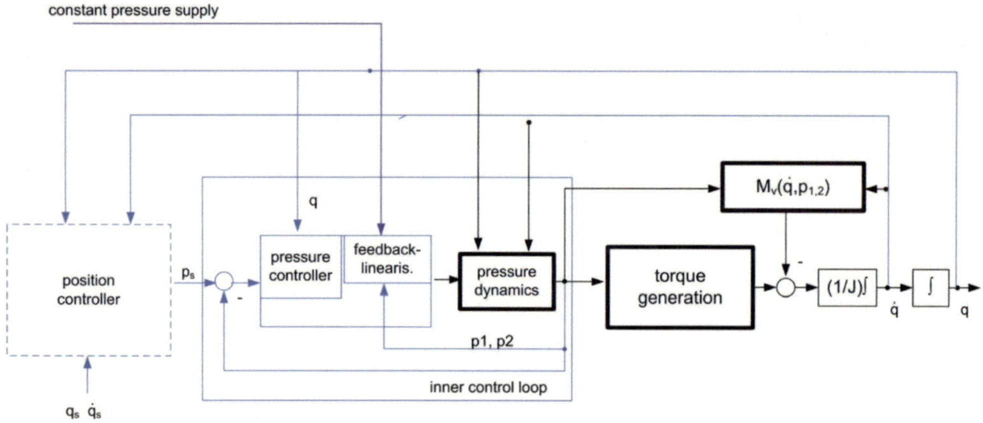

(a) Pressure Control Scheme

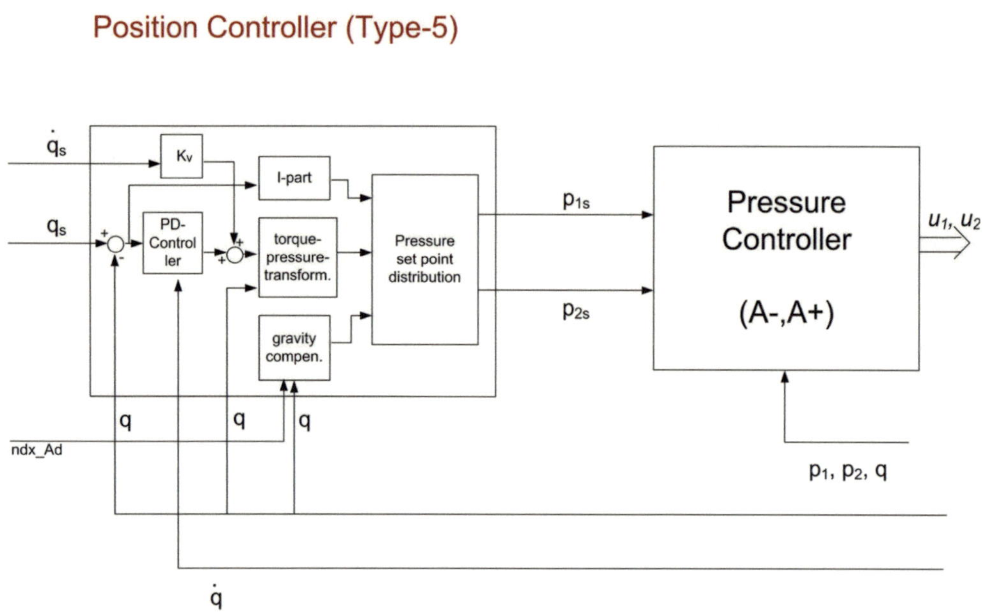

(b) Position Control Scheme

Figure 27. Different Control Schemes

6.3.2. Position control

For the position control in the outer loop, a sliding mode control with time delay estimation (SMCTE) was proposed and successfully tested for a single rotary joint. In the control design, the specific experimentally investigated dynamic model of pneumatic driven FFA was taken into account [85, 86]. In hydraulic case the iterative feedback tuning technique was applied [85]. For pneumatic driven planar robots with two soft fluidic actuators, the SMCTE approach was implemented for decentralized joint position control and shows better results than Fuzzy control, optimized using genetic algorithm [64]. Position control for pneumatic soft-robots

with spatial modular kinematics based on FFA modules is considered in [127]. In addition to decentralized SMCTE position controller the active gravity compensation-based on the quasi-static robot model is used in feed-forward loop to take the weight of robot mechanics into account. Experimental investigations, conducted with different loads for soft-robots with 4 and 6 degrees of freedom (DOF), show the behavior, the quality and the limits of the decentralized control concept with and without active gravity compensation.

6.3.3. Interaction control

The application of different control strategies for physical interaction of pneumatic soft-robots with the environment is studied in [147] by simulation and in [9] also by experiment. For control feedback the current measurements of pressure and joint angle position as well as a force/torque observer based on inverted experimental torque characteristics of FFA are used. Hereby the force sensor abilities of FFAs as of a kind of smart actuators are utilized. An adaptive admittance control with trajectory modification (ACTM) is compared by simulation to an adaptive admittance control with variable stiffness regulation (ACSR) using a model of a planar robot with two rotary joints. Both concepts enable desired force tracking in constraint direction and compliant position control in unconstrained direction. Furthermore the more promising ACSR approach was implemented and validated within an experimental set-up using a planar soft-robot with two FFA joint units by tracking even or lightly curved surfaces without knowledge of the environment stiffness.

7. Inflatable structural elements

Chapter 2.2 showed some examples of how inflatable structures can contribute to compliant robotics. This chapter shows how a modular design can help to integrate inflatable structures in robots independently from the drive concept. The main load cases of robotic structures are bending and torsion. In the shell of inflatable structures we have a state of plane stress. The shell cannot carry significant compressive force. However, when pressurized the shell is pretensioned. Compressive forces thus decrease this pretension in the shell. When the compressive forces overcome the pretension the whole structure deflects and yields the external loads.

The general fabrication process of structural elements is identical to the process described in chapter 3.2.1. The different load cases require different reinforcements in the shell. Two different layers are integrated in the shell in order to carry bending and torsion respectively. The fibers of the braided reinforcement follow the directions of the principal stresses for torsion on the surface of a cylinder as shown in figure 28(a). The reinforcement for pressure stability and bending are are shown in figure 28(b). This second reinforcement layer is a woven fabric tube with radial and axial fiber directions. The relationships between internal pressure and bending or torsional stiffness are presented in figure 28(c) and 28(d). These graphs show how the stiffness can be adjusted depending on the compliance requirements. Each front end of the structural element is equipped with a four screw flange, which allows for easy mounting and pressure sensor integration [36].

8. Flexure hinges in the field of robotics

The combination of flexible fluidic actuators and flexures leads to robotic structures with extraordinary characteristics in terms of weight, compliance, and degree of integration. This

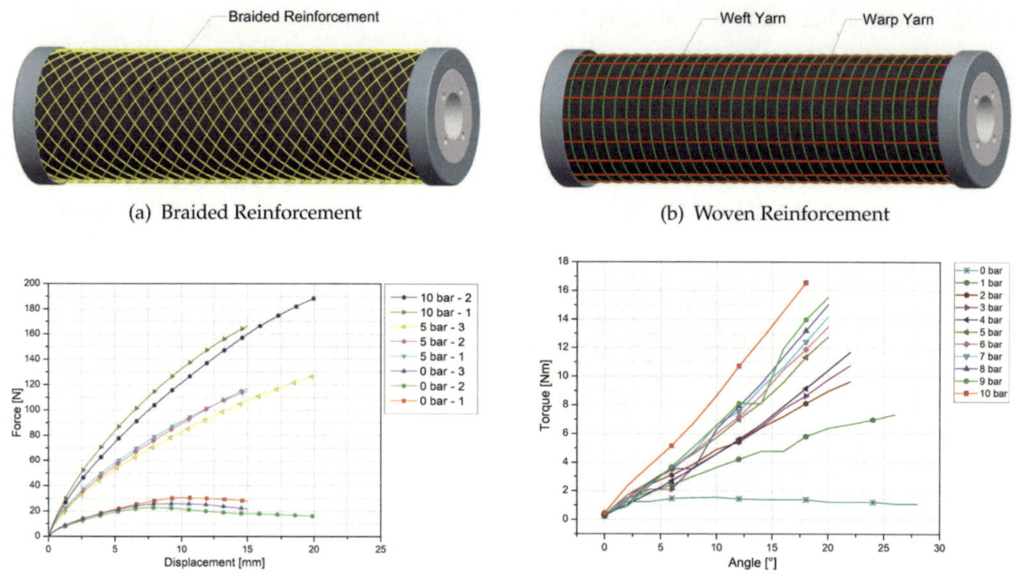

(a) Braided Reinforcement

(b) Woven Reinforcement

(c) Bending Stiffness of Compliant Structural Element

(d) Torsional Stiffness of Compliant Structural Element

Figure 28

section introduces fiber reinforced flexures fabricated in a VARTM-process (Vacuum Assisted Resin Transfer Molding).

A flexure hinge transforms an applied force into a rotary motion due to its different structural stiffnesses. The shorter the flexure length the more precise is the rotating motion. General consideration of beam theory show how the maximum bending stress σ_{max} limits the deflection Δx of the beam [51, 78].

$$\sigma_{max} = \frac{3\Delta x E \frac{h}{2}}{L^2} \tag{11}$$

Given that the desired deflection cannot be changed, the bending stress can only be influenced by the beam's height h and length L. By subdividing the bending beam of a flexure hinge the bending stress stays low without losing structural strength (figure 29).

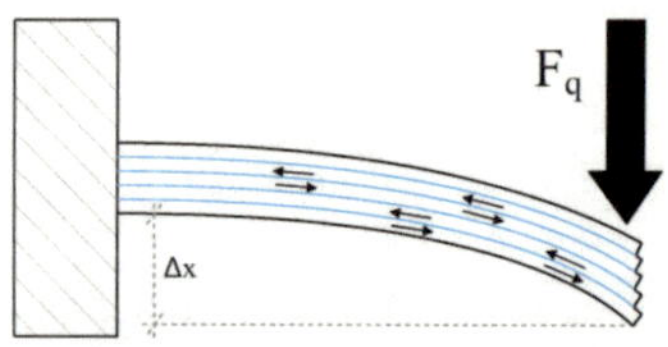

Figure 29. Simplified schematic view of a flexure hinge as a subdivided bending beam

Practically this approach is implemented by using woven fiber tapes to reinforce the bending section. The single fiber filaments represent the subdivided beams. After evaluation of technical fibers UHMWPE-fibers (also know under the brand name DYNEEMA®) have been found to meet the requirements best [36].

Dynamic testing has shown that after $100,000$ cycles, no visible damage is present. Static testing has also been described in earlier work [36]. The properties have found to be extraordinary with operating loads up to $100N$ and a maximum carrying capacity of over $3,000N$. The enhanced fabrication process for flexures allows for the production of single-acting (mass $9.1\,g$) and double-acting (mass $11.6\,g$) drives with full integration of the flexible fluidic actuator as well as the position sensor as described in [36]. Figure 30 shows the different joint-modules. These modules now can be freely combined as shown in figure 35.

(a) Single-Acting composite flexure drive

(b) Double-Acting composite flexure drive

Figure 30. Different Designs for Composite Flexure Drives in Robots

9. Applications

A large amount of robotic systems with flexible fluidic actuators have been developed over the last ten years. This is represented in many publications and patents [35, 65, 66, 108–111, 113–116]. These systems come from fields such as prosthetics, orthotics, medical devices as well as humanoid robotics and automation. Here some of the latest developments in the fields of vulcanized and HF-welded actuators are presented.

9.1. Robotic systems based on vulcanized FFAs

9.1.1. Modular flexible fluidic drive elements

These drives are based on the $\varnothing 36\,mm$ actuator. Integrated in the modules are the valves and one position sensor (AS5045-austriamicrosystems) as well as two pressure sensors (MS5803-14BA-measurement specialties) (figure 31).

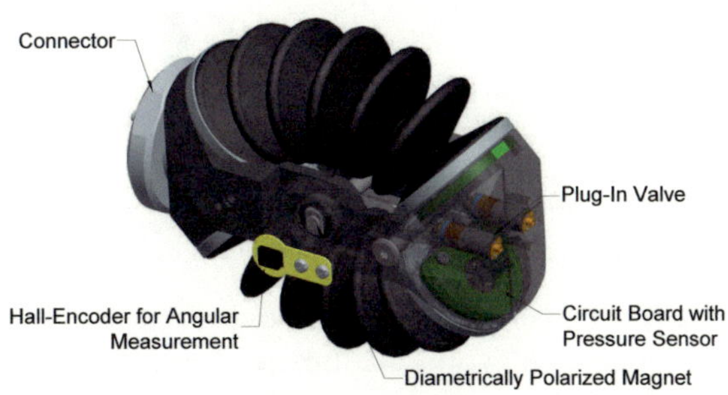

Figure 31. Highly Integrated Flexible Fluidic Drive Module

The modules can be combined in several parallel and serial configurations. Hence the torque-angle characteristics can be varied and adjusted to the case of operation, which is presented in figure 32-34.

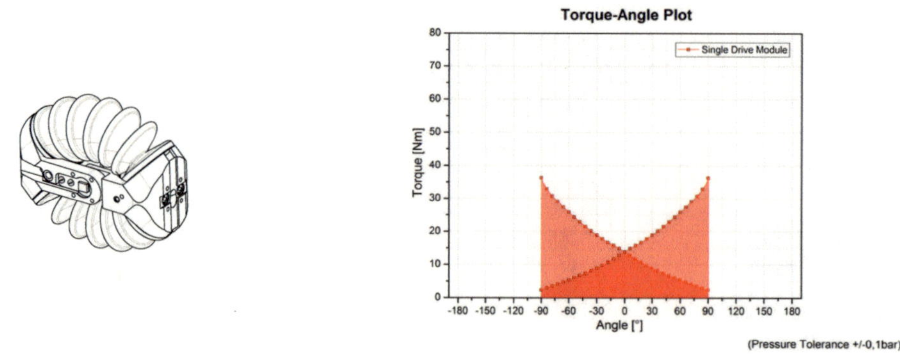

Figure 32. Single Drive Module

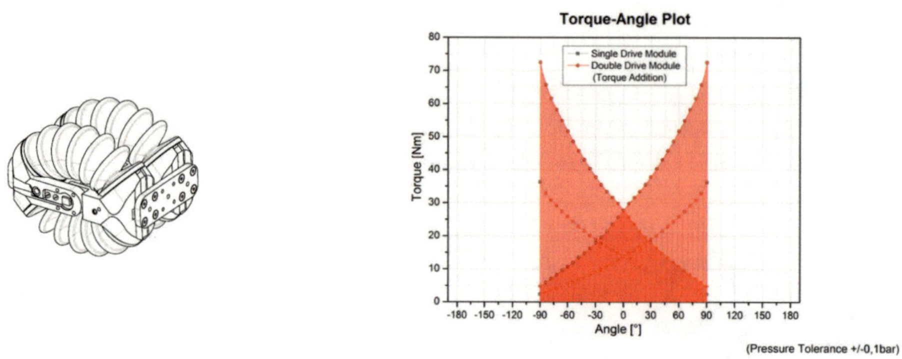

Figure 33. Double Module - Parallel Configuration

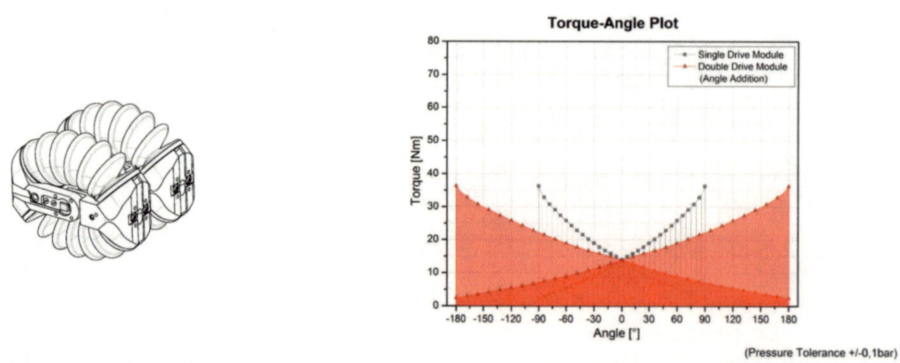

Figure 34. Double Module - Serial Configuration

9.1.2. Three-fingered composite gripper

The fiber reinforced flexure hinges are combined to form a three-fingered gripper as shown in figure 35. The base of this gripper is built of three double-acting flexure drives. Each finger is designed with two single acting flexure drives. The base of the gripper has three main positions. These positions are for spherical, cylindrical and precision grasping (figure 35). However, the compliance of the gripper allows a large variety of grasps. Figure 36 shows the taxonomy of the different grasps. The total weight of the gripper is 400 g. Each finger has a length of 140 mm.

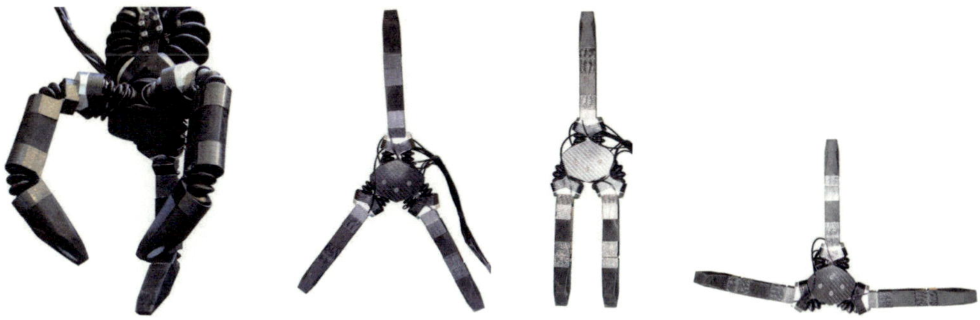

(a) Overview (b) Spherical (c) Parallel (d) Precision

Figure 35. Main Base Positions of the Composite Gripper

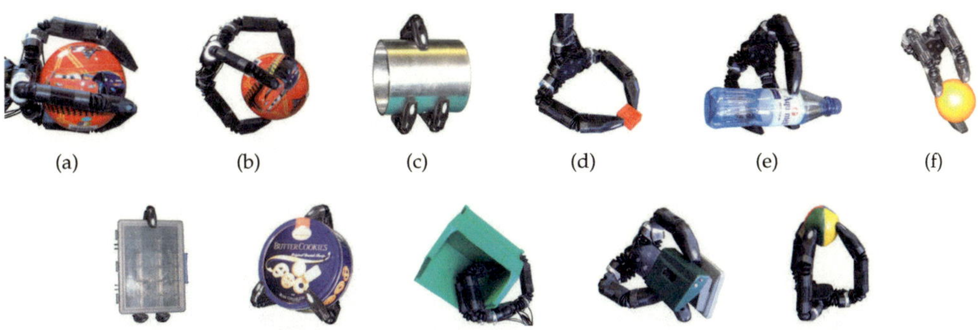

(a) (b) (c) (d) (e) (f) (g) (h) (i) (j) (k)

Figure 36. Grasp Taxonomy of the Composite Gripper

9.1.3. Lightweight Robotic Arm (LRA)

The drive modules, compliant structural elements, and the composite gripper have modular interfaces and can be combined freely. To evaluate the whole system a 6 DOF arm has been designed and built (figure 37). The proximal joint consists of two combined drive modules in parallel configuration. The weight of the whole arm is 3.45 kg and has a total length of 735 mm.

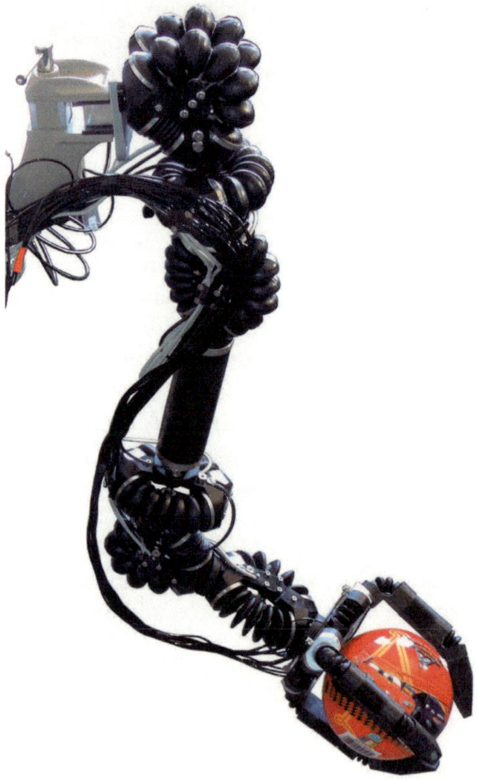

Figure 37. Lightweight Robotic Arm

9.2. Robotic systems based on HF-welded FFAs

The most recent achievements concerning HF-welded actuators are in the field of orthotics. The wide range of design possibilities of this manufacturing technology enable tailored geometries. Figure 38(a) shows an active elbow orthesis actuated with HF-welded FFAs.

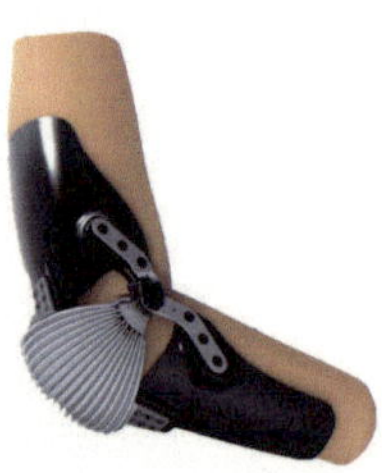

(a) Elbow Orthesis

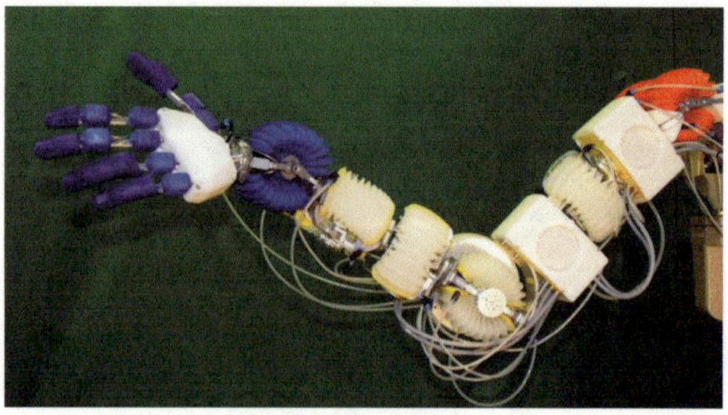

(b) 6-DOF GPA-Arm

Figure 38. Systems with HF-welded Actuators

The GPA modules also allow modular combination. Figure 38(b) shows a complete 6-DOF arm including two controller boxes and an anthropomorphic hand. The controller boxes provide valves, pressure sensors and microcontrollers to operate the whole arm independently. The whole arm-hand system has a weight of 3.2*kg* and a overall length of 620*mm*.

10. Conclusions

The future of robotics will require more and more inherent compliance. Inherent compliance can only be achieved by integrating elastic elements in the drive chain of a robotic system. Flexible Fluidic Actuators can be an appropriate solution for this problem. FFAs exhibit inherent compliance no matter if they are operated pneumatically or hydraulically. A big advantage is that FFAs do not require any transmission elements to create a rotary motion. The high torque-to-weight ratio make them suitable drives for robotics, prosthetics, orthotics, and general automation tasks. The modularity allows easy implementation of various kinematics, either for grippers and/or arms. The fluidic operation principle also make FFAs a very promising drive system for under water and deep sea operation. The challenges regarding fabrication and fatigue resistance of FFAs have been solved. The modeling and control approaches are very promising and the required control infrastructure will definitely shrink with future valve technologies and high-performance micro controllers.

Acknowledgments

The authors would like to thank all contributors and project partners. Especially we thank Maika Torge (Institute for Applied Materials/KIT) for her support regarding SEM analysis and tensile testing as well as Stefan Griebel (BMTI/TU Ilmenau) for his support regarding hyperelastic material modeling. This work was supported by the German Federal Ministry of Education and Research (BMBF) within the joint research project PortaSoR, Grants 16SV2290 - 16SV2294.

Author details

I. Gaiser, R. Wiegand, O. Ivlev, A. Andres, H. Breitwieser, S. Schulz and G. Bretthauer
Karlsruhe Institute of Technology (KIT), Institute for Applied Computer Science/Automation (AIA), Institute for Applied Computer Science (IAI); FWBI Friedrich-Wilhelm-Bessel-Institute Research Company and University of Bremen, Institute of Automation (IAT), Germany

11. References

[1] A. Kargov, T. Werner, C. P. S. S. [2008]. Development of a miniaturised hydraulic actuation system for artificial hands, *Sensors and Actuators A: Physical* 141: 548–557.

[2] Abele, G. . F. [1970]. *Arbeitsmappe System 70*, Zechner und Hüthig Verlag.

[3] Akhtar, S., De, P. P. & De, S. K. [1986]. Short fiber-reinforced thermoplastic elastomers from blends of natural rubber and polyethylene, *J. Appl. Polym. Sci.* 32(5): 5123–5146. URL: *http://dx.doi.org/10.1002/app.1986.070320530*

[4] Alami, R., Albu-Schaeffer, A., Bicchi, A., Bischoff, R., Chatila, R., Luca, A. D., Santis, A. D., Giralt, G., Guiochet, J., Hirzinger, G., Ingrand, F., Lippiello, V., Mattone, R.,

Powell, D., Sen, S., Siciliano, B., Tonietti, G. & Villani, L. [2006]. Safe and dependable physical human-robot interaction in anthropic domains: State of the art and challenges, *in* A. Bicchi & A. D. Luca (eds), *Procceedings IROS Workshop on pHRI - Physical Human-Robot Interaction in Anthropic Domains*, Beijing, China.

[5] Andorf, P., Franz, D., Lieb, A., Upper, G. & Guttropf, W. [1976]. Robot finger. URL: *http://www.freepatentsonline.com/3981528.html*

[6] Antonelli, K. & Immega, G. [1997]. An extensible robotic tentacle, *Industrial Robot: An International Journal* 24(6): 423 – 427.

[7] Baer, J. I. [1967]. Material handling apparatus and the like. URL: *http://www.freepatentsonline.com/3343864.html*

[8] Baer, J. I. [1975]. Fluid motor and material handling apparatus and the like utilizing same. URL: *http://www.freepatentsonline.com/RE28663.html*

[9] Baiden, D., Bartuszies, S. & Ivlev, O. [2012]. Position control of soft-robots with rotary-type pneumatic actuators, *7th German Conf. On Robotics (Robotik 2012)*, Munich, Germany, pp. 405–410.

[10] Bayerkohler, I. L. [1957]. Pneumatic bellows type jacks. URL: *http://www.freepatentsonline.com/2804118.html*

[11] Bögelsack, G., Karner, M. & Schilling, C. [2000]. On technomorphic modelling and classification of biological joints, *Theory in Biosciences* 119(2): 104–121. URL: *http://dx.doi.org/10.1007/s12064-000-0007-3*

[12] Bicchi, A. & Tonietti, G. [2004]. Fast and "soft-arm" tactics (robot arm design), *Robotics Automation Magazine, IEEE* 11(2): 22 – 33.

[13] Bousso, D. E. [1970]. Bellows devices. URL: *http://www.freepatentsonline.com/3495502.html*

[14] Brown, E., Rodenberg, N., Amend, J., Mozeika, A., Steltz, E., Zakin, M., Lipson, H. & Jaeger, H. [2010]. Universal robotic gripper based on the jamming of granular material, *Proceedings of the National Academy of Sciences (PNAS)*, Vol. 107, pp. 18809–18814.

[15] Bubic, F. R. [1992]. Flexible robotic links and manipulator trunks made thereform. URL: *http://www.freepatentsonline.com/5080000.html*

[16] CLAUS, H.-J. [2003]. Fluid-operated actuator has contraction hose fitting with at least one long contraction hose which can be pressurized by operating fluid.

[17] ContiTech AG [2011]. ContiTech Luftfedersysteme. URL: *http://213.164.133.30/lufecatalog/*

[18] Craig, P. S. & Fisher, J. A. [1989]. Grappling device. URL: *http://www.freepatentsonline.com/4815782.html*

[19] Daerden, F. [1999]. *Conception and Realization of Pleated Pneumatic Artificial Muscles and their Use as Compliant Actuation Elements*, PhD thesis, Vrije Universiteit Brussel.

[20] Daerden, F. & Lefeber, D. [2000]. Pneumatic artificial muscles: actuators for robotics and automation, *European journal of Mechanical and Environmental Engineering* 47: 10–21.

[21] Daerden, F. & Lefeber, D. [2001]. The concept and design of pleated pneumatic artificial muscles, *International Journal of Fluid Power* 2(3): 41–50.

[22] de Lavaud, D. S. [1929]. Vorrichtung zur Erzeugung eines Über- oder Unterdruckes in Gasen oder Flüssigkeiten.

[23] DeLepeleire, G. A. [1974]. Bag diaphragms and bag diaphragm operated air dampers. URL: *http://www.freepatentsonline.com/3804364.html*

[24] Edsinger, A. [2011]. Meka robotics. URL: *http://mekabot.com/MEKA_brochure02.pdf*

[25] Erickson, Joel R. (7636 Circle Hill Dr., O. C. . [2001]. Artificial muscle actuator assembly. URL: *http://www.freepatentsonline.com/6223648.html*

[26] Eurovinil S.P.A. [2011]. *Aufblasbares Zelt TPSE*, Eurovinil S.P.A.

[27] Ewing [1973]. Flexible actuator. URL: *http://www.freepatentsonline.com/3713685.html*

[28] Fang, H. & Lou, M. C. [2004]. Analytical characterization of space inflatable structures - an overview, *AIAA/ASME/ASC Structures, SDM Conference*, St. Louis, MO. URL: *http://hdl.handle.net/2014/16760*

[29] Festo AG & Co KG [1999]. Betätigungseinrichtung.

[30] Festo AG & Co. KG [2009]. *Fluidic Muscle DMSP/MAS.*

[31] Fleury, L. A. [1919]. Pneumatic jack.

[32] Frank, B. [1924]. Pneumatic incline. URL: *http://www.freepatentsonline.com/1493729.html*

[33] Freeland, R. E., Bilyeu, G. D., Veal, G. R., Steiner, M. D. & Carson, D. E. [1997]. Large inflatable deployable antenna flight experiment results, *Acta Astronautica* 41(4-10): 267 – 277. Developing Business.
URL: *http://www.sciencedirect.com/science/article/pii/S0094576598000575*

[34] Fuchs, G. [1994]. Faltenbalg.

[35] Gaiser, I., Pylatiuk, C., Schulz, S., Kargov, A., Oberle, R. & Werner, T. [2009]. The FLUIDHAND III: A Multifunctional Prosthetic Hand, *JPO Journal of Prosthetics & Orthotics* 21: 91–96.

[36] Gaiser, I., Schulz, S., Breitwieser, H. & Bretthauer, G. [2010]. Enhanced flexible fluidic actuators for biologically inspired lightweight robots with inherent compliance, *Robotics and Biomimetics (ROBIO), 2010 IEEE International Conference on*, pp. 1423 –1428.

[37] Giesler, J. V. [1932]. Attachment of heads to bellows.
URL: *http://www.freepatentsonline.com/1870904.html*

[38] Griebel, S., Fiedler, P., Streng, A., Haueisen, J. & Zentner, L. [2010]. Medical sensor placement with a screw motion, *Actuator 10 / International Conference on New Actuators*, Bremen, Germany, pp. 1047–1050.

[39] Gudo, M. [2002]. The development of the critical theory of evolution: The scientific career of wolfgang f. gutmann, *Theory in Biosciences* 121(1): 101–137.
URL: *http://dx.doi.org/10.1078/1431-7613-00052*

[40] Gudo, M. & Grasshoff, M. [2002]. The origin and early evolution of chordates: The hydroskelett-theorie and new insights towards a metameric ancestor, *Palaeobiodiversity and Palaeoenvironments* 82(1): 325–345.
URL: *http://dx.doi.org/10.1007/BF03043792*

[41] Gutmann, W. F. [1966]. *Zu Bau und Leistung von Tierkonstruktionen (4-6)*, Abhandlungen der Senckenbergischen Naturforschenden Gesellschaft ; 510Senckenberg am Meer ; 238, Frankfurt a. M.

[42] Gutmann, W. F. [1972]. *Die Hydroskelett-Theorie : AbriSS der Coelomaten-Herleitung von einer metameren Vorläufer-Konstruktion*, Aufsätze und Reden der Senckenbergischen Naturforschenden Gesellschaft ; 21, Kramer, Frankfurt a. M.

[43] Haddadin, S., Albu-Schaeffer, A. & Hirzinger, G. [2007]. Safety evaluation of physical human-robot interaction via crash-testing.
URL: *http://citeseerx.ist.psu.edu/viewdoc/summary?doi=10.1.1.149.1487*

[44] Haddadin, S., Albu-Schaffer, A., Frommberger, M., Rossmann, J. & Hirzinger, G. [2009a]. The dlr crash report : Towards a standard crash-testing protocol for robot safety - part i: Results, *Robotics and Automation, 2009. ICRA '09. IEEE International Conference on*, pp. 272 –279.

[45] Haddadin, S., Albu-Schaffer, A., Frommberger, M., Rossmann, J. & Hirzinger, G. [2009b]. The dlr crash report: Towards a standard crash-testing protocol for robot

safety - part ii: Discussions, *Robotics and Automation, 2009. ICRA '09. IEEE International Conference on*, pp. 280 –287.

[46] Haegele, M., Neugebauer, J. & Schraft, R. [2001]. From robots to robot assistants, *32nd ISR(International Symposium on Robotics)*.

[47] Hennequin, J. R. & Fluck, P. [1990]. Motorized joint.
URL: *http://www.freepatentsonline.com/4944755.html*

[48] Henriksen, T. K. [2008]. Trends in materials requirements in spacecraft structures and mechanisms, *Materials KTN 2nd Annual General Meeting*.

[49] Hertel, B. [1984]. Personnel qualification and work structuring in robot-automated manufacturing systems - exemplified by the application of industrial robots for body shell assembly at audi, ingolstadt, *in* N. Martensson (ed.), *Proceedings on 14th International Symposium on Industrial Robots*, Gothenburg, Sweden.

[50] Hirzinger, G., Sporer, N., Albu-Schäffer, A., Hahnle, M., Krenn, R., Pascucci, A. & Schedl, M. [2002]. DLR's torque-controlled light weight robot III. Are we reaching the technological limits now?, *Robotics and Automation, 2002. Proceedings. ICRA '02. IEEE International Conference on*, Vol. 2, pp. 1710 – 1716.

[51] Howell, L. [2002]. *Compliant Mechanisms*, John Wiley & Sons INC.

[52] IFR - International Federation of Robotics [2010a]. Executive summary of world robotics 2010, *Technical report*, IFR - International Federation of Robotics.

[53] IFR - International Federation of Robotics [2010b]. Serviceroboter etablieren sich erfolgreich, *Technical report*, IFR - International Federation of Robotics.

[54] IFR - International Federation of Robotics [2011a]. Ifr - charts for press, servicerobotik 2011, *Technical report*, IFR - International Federation of Robotics.

[55] IFR - International Federation of Robotics [2011b]. Ifr: Allzeithoch für die roboterindustrie - das kräftige wachstum der roboterverkäufe setzt sich fort, *Technical report*, IFR - International Federation of Robotics.

[56] IFR - International Federation of Robotics [2011c]. Professionelle service roboter - einsatz im militärischen bereich und der landwirtschaft überwiegen, *Technical report*, IFR - International Federation of Robotics.

[57] Ilievski, F., Mazzeo, A. D., Shepherd, R. F., Chen, X. & Whitesides, G. M. [2011]. Soft robotics for chemists, *Angewandte Chemie International Edition* 50(8): 1890–1895.
URL: *http://dx.doi.org/10.1002/anie.201006464*

[58] Immega, G. & Kukolj, M. [1990]. Axially contractable actuator.
URL: *http://www.freepatentsonline.com/4939982.html*

[59] *Industrieroboter- Sicherheitanforderungen - Teil 1, DIN EN ISO 10218-1* [2009].

[60] *Industrieroboter - Sicherheitanforderungen - Teil 2, DIN EN ISO 10218-2* [2008].

[61] Ivlev, O. [2009]. Soft fluidic actuators of rotary type for safe physical human-machine interaction, *Rehabilitation Robotics, 2009. ICORR 2009. IEEE International Conference on*, pp. 1 –5.

[62] Johnston, L. B. [1965]. Fluid actuated displacement and positioning system.
URL: *http://www.freepatentsonline.com/3202061.html*

[63] Jones, J. A. [2001]. Inflatable robotics for planetary applications, *Proceeding of the 6th International Symposium on Artificial Intelligence and Robotics & Automation in Space: i-SAIRAS*, Canadian Space Agency, St-Hubert, Quebec, Canada.

[64] Jordan, M., Pietrusky, D., Mihajlov, M. & Ivlev, O. [2009]. Precise position and trajectory control of pneumatic soft-actuators for assistance robots and motion therapy devices, *Rehabilitation Robotics, 2009. ICORR 2009. IEEE International Conference on*, pp. 663 –668.

[65] Kargov, A., Asfour, T., Pylatiuk, C., Oberle, R., Klosek, H., Schulz, S., Regenstein, K., Bretthauer, G. & Dillmann, R. [2005]. Development of an anthropomorphic hand for a mobile assistive robot, *Proceedings of the 9th International Conference on Rehabilitation Robotics*, ICORR, pp. 182–186.

[66] Kargov, A., Werner, T., Pylatiuk, C. & Schulz, S. [2008]. Development of a miniaturised hydraulic actuation system for artificial hands, *Sensors and Actuators A: Physical* 141(2): 548 – 557.
URL: *http://www.sciencedirect.com/science/article/B6THG-4PXDM21-4/2/56ea17517c8be38c771afc6d2c49b310*

[67] Karwowski, W., Rahimi, M., Parsaei, H., Amarnath, B. R. & Pongpatanasuegsa, N. [1991]. The effect of simulated accident on worker safety behavior around industrial robots, *International Journal of Industrial Ergonomics* 7(3): 229 – 239.
URL: *http://www.sciencedirect.com/science/article/B6V31-481D149-4H/2/f5d48f31d92b445ab7192e795b1d2ffd*

[68] Khatib, O., Yokoi, K., Brock, O., Chang, K. & Casal, A. [1999]. Robots in human environments, *Robot Motion and Control, 1999. RoMoCo '99. Proceedings of the First Workshop on*, pp. 213 –221.

[69] Kimura, H., Kajimura, F., Maruyama, D., Koseki, M. & Inou, N. [2006]. Flexible hermetically-sealed mobile robot for narrow spaces using hydrostatic skeleton driving mechanism, *Intelligent Robots and Systems, 2006 IEEE/RSJ International Conference on*, pp. 4006 –4011.

[70] Kleinknecht, H., Schuch, A. & Degen, I. [2000]. Process for the production of masterbatches containing short fibres or pulps.
URL: *http://www.freepatentsonline.com/6160039.html*

[71] Konishi, S., Kawai, F. & Cusin, P. [2001]. Thin flexible end-effector using pneumatic balloon actuator, *Sensors and Actuators A: Physical* 89(1-2): 28 – 35.
URL: *http://www.sciencedirect.com/science/article/B6THG-42DX434-5/2/0f8148b596cfdea63ca49836b9dde497*

[72] Koren, Yoram (Ann Arbor, M. & Weinstein, Yechiel (Misgav, I. [1991]. Inflatable structure. URL: *http://www.freepatentsonline.com/5065640.html*

[73] Kukolj, M. [1985]. Axially contractable actuator.

[74] Larsson, Ove (Gothenburg, S. [1988]. Flexible actuator.
URL: *http://www.freepatentsonline.com/4777868.html*

[75] Lens, T., Kunz, J., Trommer, C., Karguth, A. & von Stryk, O. [2010]. Biorob-arm: A quickly deployable and intrinsically safe, light- weight robot arm for service robotics applications, *41st International Symposium on Robotics (ISR 2010) / 6th German Conference on Robotics (ROBOTIK 2010)*, Munich, Germany, pp. 905–910.

[76] Lewis, D. J. [1974]. Fiber reinforced inflatable restraining band for vehicles. URL: *http://www.freepatentsonline.com/3830519.html*

[77] Lewis, J. [1847]. Car spring.
URL: *http://www.freepatentsonline.com/0004965.html*

[78] Lobontiu, N. [2003]. *Compliant mechanisms : design of flexure hinges*, CRC Press, Boca Raton, Fla. [u.a.]. Includes bibliographical references and index.
URL: *http://www.gbv.de/dms/ilmenau/toc/350071101.PDF*

[79] Lou, M. C. [2000]. Development and application of space inflatable structures, *22"d International Symposium on Space Technology and Science Jet Propulsion Laboratory, Californialn stitute of Technology*, Pasadena, CA, USA.

[80] Malloy, H. C. [1929]. Method of producing expansible collapsible elements. URL: *http://www.freepatentsonline.com/1698164.html*

[81] Marette, R. T. [1961]. Actuator. URL: *http://www.freepatentsonline.com/2991763.html*

[82] Marshall Space Flight Center, A. [2010]. Inflatable tubular structures rigidized with foams, *Technical report*, Marshall Space Flight Center, Alabama.

[83] Maruyama, D., Kimura, H., Koseki, M. & Inou, N. [2010]. Driving force and structural strength evaluation of a flexible mechanical system with a hydrostatic skeleton, *Journal of Zhejiang University - Science A* 11: 255–262. 10.1631/jzus.A000030. URL: *http://dx.doi.org/10.1631/jzus.A000030*

[84] McMahan, W., Chitrakaran, V., Csencsits, M., Dawson, D., Walker, I., Jones, B., Pritts, M., Dienno, D., Grissom, M. & Rahn, C. [2006]. Field trials and testing of the octarm continuum manipulator, *Robotics and Automation, 2006. ICRA 2006. Proceedings 2006 IEEE International Conference on*, pp. 2336 –2341.

[85] Mihajlov, M. [2008]. *Modelling and Control Strategies for Inherently Compliant Fluidic Mechatronic Actuators with Rotary Elastic Chambers*, PhD thesis, Institute of Automation, University of Bremen, Bremen, Germany. ISBN 978-3-8322-7275-3.

[86] Mihajlov, M., Hubner, M., Ivlev, O. & Graser, A. [2006]. Modeling and control of fluidic robotic joints with natural compliance, *Computer Aided Control System Design, 2006 IEEE International Conference on Control Applications, 2006 IEEE International Symposium on Intelligent Control, 2006 IEEE*, pp. 2498 –2503.

[87] Mihajlov, M., Ivlev, O. & Gräser, A. [2008]. Modelling and identification for control design of compliant fluidic actuators with rotary elastic chambers: Hydraulic case study, *17th IFAC World Congress*, Seoul Korea.

[88] Monroe, John W. (Warren, M. [1994]. Jointed assembly actuated by fluid pressure. URL: *http://www.freepatentsonline.com/5351602.html*

[89] Morin, A. H. [1953]. Elastic diaphragm. URL: *http://www.freepatentsonline.com/2642091.html*

[90] Morse, R. V. [1938]. Pneumatic jack. URL: *http://www.freepatentsonline.com/2140325.html*

[91] Mårtensson, L. [1987]. Interaction between man and robots, *Robotics* 3(1): 47 – 52. Special Issue: Industrial Robotics. URL: *http://www.sciencedirect.com/science/article/B756Y-480TX1T-N/2/e9dc7e14bd276f818 46f0ac477638069*

[92] Negishi, Koichi (Kodaira, J. [1991]. Double-acting flexible wall actuator. URL: *http://www.freepatentsonline.com/5067390.html*

[93] Nicolaisen, P. [1987]. Safety problems related to robots, *Robotics* 3(2): 205 – 211. Special Issue: Sensors. URL: *http://www.sciencedirect.com/science/article/B756Y-4808M3V-9/2/39a6988a29109bc9 bafa999095e31aa2*

[94] Oñate, E. & Kröplin, B. (eds) [2008]. *Textile Composites and Inflatable Structures II*, Springer.

[95] Orndorff, R. L. [1971]. Gripping device. URL: *http://www.freepatentsonline.com/3601442.html*

[96] Patterson, J. M. [1935]. Lifting jack. URL: *http://www.freepatentsonline.com/2001744.html*

[97] PAUL, L. J. [1998]. Inflatable wall for sealing manholes or pipes to test for water tightness, fr2751006, *Technical report*, PRONAL.

[98] Paynter, H. M. [1974]. Tension actuated pressurized gas driven rotary motors.
URL: *http://www.freepatentsonline.com/3854383.html*

[99] Paynter, Henry M. (35 Scotland Rd., R. M. . [1978]. Fluid-driven torsional operators for turning rotary valves and the like.
URL: *http://www.freepatentsonline.com/4108050.html*

[100] Paynter, Henry M. (35 Scotland Rd., R. M. . [1988a]. High pressure fluid-driven tension actuators and method for constructing them.
URL: *http://www.freepatentsonline.com/4751869.html*

[101] Paynter, Henry M. (35 Scotland Rd., R. M. . [1988b]. Hyperboloid of revolution fluid-driven tension actuators and method of making.
URL: *http://www.freepatentsonline.com/4721030.html*

[102] Paynter, Henry M. (Pittsford, V. [1992]. All-elastomer fluid-pressure-actuatable twistors and twistor drive assemblies.
URL: *http://www.freepatentsonline.com/5090297.html*

[103] Prior, S., Warner, P., White, A., Parsons, J. & Gill, R. [1993]. Actuators for rehabilitation robots, *Mechatronics* 3(3): 285 – 294. Special Issue Robot Actuators.
URL: *http://www.sciencedirect.com/science/article/B6V43-47YVPRC-3/2/0a2fb2c92d4ec09d6493c32cdf2ed09e*

[104] Rybski, M., Shoham, M. & Grossman, G. [1996]. Robotic manipulators based on inflatable structures, *Robotics and Computer-Integrated Manufacturing* 12(1): 111 – 120.
URL: *http://www.sciencedirect.com/science/article/B6V4P-3VV72YC-C/2/ad2533e8ae922da2adf9e131f35cf7f8*

[105] Salomonski, N., Shoham, M. & Grossman, G. [1995]. Light robot arm based on inflatable structure, *CIRP Annals - Manufacturing Technology* 44(1): 87–90.
URL: *http://www.sciencedirect.com/science/article/B8CXH-4P3DTXT-P/2/ce613ff549a49cb798f0d515bbc42c0f*

[106] Sanan, S., Moidel, J. & Atkeson, C. [2009]. Robots with inflatable links, *Intelligent Robots and Systems, 2009. IROS 2009. IEEE/RSJ International Conference on*, pp. 4331–4336.

[107] Schneider, E. [1993]. Air spring having an elastomeric air-spring flexible member.
URL: *http://www.freepatentsonline.com/5269496.html*

[108] Schulz, S. [2000]. Wurmförmiger Arbeitsmechanismus.

[109] Schulz, S. [2004a]. *Eine neue Adaptiv-Hand-Prothese auf der Basis flexibler Fluidaktoren*, PhD thesis, Universität Karlsruhe (TH).
URL: *http://books.google.de/books?id=A9VWAgAACAAJ*

[110] Schulz, S. [2004b]. Vorrichtung mit fluidischem schwenkantrieb.

[111] Schulz, S. [2005]. Fluidischer antrieb.

[112] Schulz, S., Pylatiuk, C. & Bretthauer, G. [1999a]. A new class of flexible fluidic actuators and their applications in medical engineering, *Automatisierungstechnik* 8: 390–395.

[113] Schulz, S., Pylatiuk, C. & Bretthauer, G. [1999b]. A new class of flexible fluidic actuators and their applications in medical engineering, *at-Automatisierungstechnik* 47(8): 390–395.

[114] Schulz, S., Pylatiuk, C., Kargov, A., Gaiser, I., Schill, O., Reischl, M., Eck, U. & Rupp, R. [2009]. Design of a hybrid powered upper limb orthosis, *Proc., World Congress Medical Physics and Biomedical Engineering*, München.

[115] Schulz, S., Pylatiuk, C., Kargov, A., Oberle, R., Klosek, H., Werner, T., RöSSler, W., Breitwieser, H. & Bretthauer, G. [2005]. Fluidically driven robots with biologically inspired actuators, *Proc. of the 8th International Conference on Climbing and Walking Robots (CLAWAR)*, London, UK, p. 39.

[116] Schulz, S., Pylatiuk, C., Reischl, M., Martin, J., Mikut, R. & Bretthauer, G. [2005]. A hydraulically driven multifunctional prosthetic hand, *Robotica* 23(3): 293–299.
[117] Shin, D., Khatib, O. & Cutkosky, M. [2009]. Design methodologies of a hybrid actuation approach for a human-friendly robot, *Robotics and Automation, 2009. ICRA '09. IEEE International Conference on*, pp. 4369 –4374.
[118] Shoval, S. & Shoham, M. [2001]. Sensory redundant parallel mobile mechanism, *Conference on Robotics and Automation, 2001. Proceedings 2001 ICRA. IEEE International*, Vol. 3, pp. 2273 – 2278 vol.3.
[119] Sigmon, J. W. [1976]. Rotary motion valve and actuator.
URL: *http://www.freepatentsonline.com/3977648.html*
[120] Staines, A. J. [1962]. Sheet lifting device.
URL: *http://www.freepatentsonline.com/3039767.html*
[121] Stienenw, A., Hekman, E., ter Braak, H., Aalsma, A., van der Helm, F. & van der Kooij, H. [2010]. Design of a rotational hydroelastic actuator for a powered exoskeleton for upper limb rehabilitation, *Biomedical Engineering, IEEE Transactions on* 57(3): 728 –735.
[122] Suzumori, K. [1992]. Gripping actuator with independently flexible cylinders.
URL: *http://www.freepatentsonline.com/5156081.html*
[123] Suzumori, K. [1995]. Actuator with flexible cylinders.
URL: *http://www.freepatentsonline.com/5385080.html*
[124] Suzumori, K. [1996]. Elastic materials producing compliant robots, *Robotics and Autonomous Systems* 18: 135–140(6).
[125] Suzumori, K., Endo, S., Kanda, T., Kato, N. & Suzuki, H. [2007]. A bending pneumatic rubber actuator realizing soft-bodied manta swimming robot, *IEEE International Conference on Robotics and Automation*, Roma, Italy, pp. 4975 –4980.
[126] Suzumori, K., Iikura, S. & Tanaka, H. [1992]. Applying a flexible microactuator to robotic mechanisms, *Control Systems Magazine, IEEE* 12(1): 21 –27.
[127] Taghia, J., Wilkening, A. & Ivlev, O. [2012]. Position control of soft-robots with rotary-type pneumatic actuators, *7th German Conf. On Robotics (Robotik 2012)*, Munich, Germany, pp. 399–404.
[128] Takagi, Takeo (Yokohama, J. & Sakaguchi, Yuji (Kawasaki, J. [1986]. Pneumatic actuator for manipulator.
URL: *http://www.freepatentsonline.com/4615260.html*
[129] Tillett, N., Vaughan, N. & Bowyer, A. [1994]. An improved flexible pneumatic joint for horticultural robots, *Mechatronics* 4(7): 653 – 671.
URL: *http://www.sciencedirect.com/science/article/B6V43-480TWR6-T/2/59208c93bf4b07b44 17f3c73d0a3efd1*
[130] Trivedi, D., Rahn, C. D., Kier, W. M. & Walker, I. D. [2008]. Soft robotics: Biological inspiration, state of the art, and future research, *Applied Bionics and Biomechanics* 5(3): 99–117.
URL: *http://www.informaworld.com/10.1080/11762320802557865*
[131] Tsimpris, C., Wartalski, J., Ferradino, A. & Vanderbilt, R. [2001]. Compounding with para-aramid fiber engineered elastomers., *Rubber World* 224(1): 35.
URL: *http://www.redi-bw.de/db/ebsco.php/search.ebscohost.com/login.aspx?direct=true&db= buh&AN=4385775&site=ehost-live*
[132] VanBlaricum, M. L., Reilley, J., Gilbert, M. A. & VanBlaricum Jr., G. F. [1999]. Quick feasibility demonstration for an inflatable antenna system in space, *Proceedings of the*

Ninth Annual DARPA Symposium on Photonic Systems for Antenna Applications, Monterey, CA.

[133] Vaughen, J. F. [1993]. Structure for an inflatable lift device.
URL: *http://www.freepatentsonline.com/5178367.html*

[134] Vetter GmbH [2011]. *VETTER Rettungstechnik*, Vetter GmbH.
URL: *www.vetter.de*

[135] Voisembert, S., Riwan, A., Mechbal, N. & Barraco, A. [2011]. A novel inflatable robot with constant and continuous volume, *Robotics and Automation (ICRA), 2011 IEEE International Conference on*, pp. 5843 –5848.

[136] Widmer, S. W. [1985]. Air bag jack.
URL: *http://www.freepatentsonline.com/4560145.html*

[137] Wilkening, A., Mihajlov, M. & Ivlev, O. [2010]. Model-based pressure and torque control for innovative pneumatic soft-actuators, *Proc. of 7th Int. Conf. on Fluid Power (7th IFK)*, Vol. 4, Aachen, Germany, pp. 291–302.

[138] Williamson, M. M. [1995]. Series elastic actuators, *Technical Report 1524*, MIT, Artificial Iintelligence Laboratory.

[139] Wilson, James F. (Durham, N. [1988]. Fluid actuated limb.
URL: *http://www.freepatentsonline.com/4792173.html*

[140] Wolf, S. & Hirzinger, G. [2008]. A new variable stiffness design: Matching requirements of the next robot generation, pp. 1741 –1746.

[141] Woods, J. E. [1957]. Mechanical transducer with expansible cavity.
URL: *http://www.freepatentsonline.com/2789580.html*

[142] Wypych, G. [2000]. *Handbook of Fillers - A Definitive User's Guide and Databook (2nd Edition)*, ChemTec Publishing.
URL: *http://www.knovel.com/web/portal/browse/display?_EXT_KNOVEL_DISPLAY_ bookid =1011*

[143] Yarlott, J. M. [1972]. Fluid actuator.
URL: *http://www.freepatentsonline.com/3645173.html*

[144] Yoon, C. S., Lee, E. M., Jang, M. S., Song, C. K., Kang, C. Y., Kim, K. H. & Kim, B. K. [2008]. High-frequency welding of thermoplastic lldpe/pa6/pe-g-mah ternary blends, *Journal of Applied Polymer Science* 109: 3355–3360.

[145] Zentner, L. & Böhm, V. [2007]. Nachgiebige monolithische fluidisch angetriebene aktuatoren mit neuartigem verformungsverhalten, *Technische Mechanik* 27(1): 18–27.

[146] Zentner, L., Böhm, V. & Minchenya, V. [2009]. On the new reversal effect in monolithic compliant bending mechanisms with fluid driven actuators, *Mechanism and Machine Theory* 44(5): 1009 – 1018.
URL: *http://www.sciencedirect.com/science/article/B6V46-4SVM0W6-1/2/0ece208ab278bd 053c985491f2e03d53*

[147] Zhang, X. & Ivlev, O. [2010]. Simulation of interaction tasks for pneumatic soft robots using simmechanics, *Robotics in Alpe-Adria-Danube Region (RAAD), 2010 IEEE 19th International Workshop on*, pp. 149 –154.

Section 5

Piezo-Based Systems

Chapter 23

A Tactile Sensing System Based on Arrays of Piezoelectric Polymer Transducers

Lucia Seminara, Luigi Pinna, Marco Capurro and Maurizio Valle

Additional information is available at the end of the chapter

http://dx.doi.org/10.5772/50112

1. Introduction

Tactile sensing enables robots to interact safely and effectively with unstructured environments and humans in case of both voluntary and reactive interaction tasks. Focusing on humanoid robots, there is increasing interest in avoiding negative human feelings towards the "entity" [1], enabling robots to interact with humans in more intuitive and meaningful ways [2-4]. This requires designing methods for extracting important information from tactile stimuli leading to classification of touch modalities [5-8]. It is still unclear, however, whether touch modality actually plays an important role in the communication of social messages. A very interesting research area consists in exploring social touch for robotics through an interpretation of emotions and other social messages [9].

Together with the social aspects of human-robot interactions, research in this field of robotics has focused largely on transduction principles and transduction technologies [10]. A survey of the state-of-the-art in robot tactile sensing is given by [11], with references to various sensor types. The functional requirements of a robotic skin remain debatable and are at least partially dependent on the specific application. However, some basic requirements can be identified. Sensors should be robust, stable, reproducible, showing a high sensitivity, capable of detecting a wide range of information and both static and dynamic touch (approximately in the 0 - 1 kHz frequency range). Moreover, this application demands mechanical flexibility and conformability, and the skin should be adaptable to various three-dimensional robotic platforms. Finally, the fabrication of the skin should be simple, low cost, replicable and scalable. All hardware should be portable and easily adaptable and electronics should be designed for minimum power consumption and minimum size.

Various technical issues have strongly limited the transition from small matrix prototypes to a large scale integrated solution. An example of truly scalable robot skin systems for humanoid robots has been proposed by Ohmura and Kunihoshi [12]. They approached the

problem at system level introducing a networked architecture featuring peripheral nodes (chips) scanning (locally) a limited number of taxels. All the electronics and the transducers are embedded on a PCB support which allows a simple mechanical integration of the sensor over curved surfaces. However, the spatial resolution is quite low, and the adopted infrared (IR) optical transducers have quite large power consumption. The issue of fault tolerance, for infrared based sensors was previously addressed by Um and Lumelsky [13], who tackled the problem via component redundancy for a system featuring over 1000 sensing elements. Another example of artificial skin system for a humanoid robot has been proposed by Tajika et al. [14]. The sensor has been designed with the aim of detecting stimuli coming from people trying to interact with the robot and features PVDF based transducers, but the skin has low spatial resolution (the transducer area is of about 25 cm^2).

At a high level, the goal of the ROBOSKIN European Project[1] is to develop and demonstrate a range of new robot capabilities based on the tactile feedback provided by a robotic skin from large areas of the robot body. To mimic the complex behavior of the human skin in a humanoid robot a multimodal system would be required, which employs different kinds of transducers. In particular, the present research is oriented to the development of distributed and modular components for general-purpose large-area tactile sensors based on piezoelectric polymer transducers. Piezoelectric sensors in the form of thin polymer films of Polyvinylidene Fluoride (PVDF) have been chosen [10,11,15-20], as they globally meet the given above sensor requirements except from perceiving static mechanical stimuli. Piezoelectric materials intrinsically convert the mechanical stimulus into an electrical signal on the basis of their electromechanical properties. The piezoelectric "functional" material must be integrated into structures which also include a substrate and a protective layer. How to integrate the PVDF transducer is not an easy task, because its response depends on several aspects including the properties of the whole mechanical chain, in particular material and thickness of the protective layer. Moreover, the design features also influence the requirements of the interface electronics and the data processing, to cite some of the most important aspects.

Following the system approach of the ROBOSKIN project, the research activities have been carried out by considering the skin as a system, which is as well a part of the overall robot architecture. Towards this perspective, a combined approach based on modelling and experimental testing is at the basis of the results achieved so far in the sensor manufacturing technology, the interface electronics, the tactile data processing, the embedded system architecture and the system integration. Main achievements in each of the cited fields will be outlined in this chapter.

The chapter will be organized as follows. To move towards an optimized design of the tactile sensing system, a preliminary experimental study has been carried out to classify the tactile stimuli in basic human-robot interactions. This aspect will be detailed in Section 2. The identified contact stress/force range has been used as reference for the design of the skin system.

---

[1] VII FP, http://www.roboskin.eu

In Section 3 attention will be focused on tactile sensing systems based on piezoelectric transducers. First, the electromechanical characterization of the *thickness-mode* behavior of piezoelectric polymer films will be presented. Therefore, first prototypes consisting in a single piezoelectric sensor covered by different protective elastomer layers will be described. Such prototypes have been employed both to validate the skin electromechanical model and to appropriately design the interface electronics., whose basic principles and circuits are reported in Section 4. A variable gain solution is also proposed to measure the wide range of tactile stimuli expected for the application.

The proposed skin system is made by conformable patches of triangular shape, interconnected in order to form a network structure [21]. Manufacturing and testing of triangular prototypes based on piezoelectric arrays are discussed in Section 5. Consequently, some system aspects related to prototype integration into the target robot platform, data transmission to the robot communication infrastructure and data processing requirements and algorithms are reported in Section 6. In relation to data processing, one of the goals of our research is the real-time implementation of tactile data processing by dedicated embedded digital circuits. In this view, an algorithm to provide the contact forces from sensor readings will be discussed.

An assessment of the achieved results and of the open issues paving the way for new research targets and novel design solutions will conclude the chapter.

2. From perception to tactile stimulus: human-robot interaction

A big challenge in robotics consists in designing a robot that is able to "feel," "understand" and respond to touch in accord with human expectations. The human-like robot response to external stimuli is far beyond the scopes of this chapter, but a first step in this direction is to design a tactile sensing system which enables the robot to "effectively" interact with humans. This requires recognizing and classifying different *typical* modalities of touch. At a lower level, a first assessment of the expected dynamics of the tactile stimulus is needed, as a starting point to adequately design the interface electronics and the whole sensing system. This is the scope of this section.

More concretely, we have tried to quantify the magnitude of the mechanical stimuli involved in basic human-robot interactions. As we deal with piezoelectric transducers which do not perceive static stimuli, only impact forces have been considered. Basically, we aim at designing a tactile system which is able to measure and distinguish between light/tender touches and strong impacts.

To enable this quantitative study, a "modally tuned impulse force hammer" (PCB Piezotronics, Model 086C03) which is equipped with an integrated piezoelectric load cell has been used. The load cell acquires the temporal signal of an impact force, i.e. it measures the *effect* of an external force while impacting on another surface. A contact force is thus measured, which depends both on the hammer indenter and on the mechanical properties of the touched surface.

The idea is to associate a *perception* to a *quantification* of the tactile stimulus. Therefore, we reproduced with the hammer the elementary interactions listed in Table 1 on different areas of some candidate skin. Essays either on biological tissues (different hardness / different parts of the body, as specified) or on clothes have been performed. Among the different choices for the hammer indenter, the one has been chosen which was the most similar to a human finger. Another indenter with a plate shape has been used to simulate light / strong interactions with an open hand (caresses / slaps). This allowed us to find (at least partial) quantitative information about the external stimuli which are relevant for our application, in order to design the skin prototypes and the interface electronics.

To set the skin input stress range, maximum and minimum applied stimuli have to be quantified. To define those limits, the question is how to play with the involved variables, i.e. the indenter size/material and the material of the touched surface, in order to decide whether "force" or "pressure" is the meaningful quantity. The idea is to use "perception" as a definition criterion.

Contrary to our intuitive understanding based on personal experience, the brain is not a camera that passively records the external world. Perception is a product of the brain's abstraction and elaboration of sensory input. The somatic sensory system transmits information about four major modalities: touch, proprioception, pain, and temperature sense. The four modalities are conveyed in separate ascending pathways to the thalamus and cerebral cortex. The perception of touch or pressure is consistent when touch-pressure receptors are electrically stimulated [22].

When touching the human skin, upper limits for the mechanical stimulus can be considered those associated with a "pain" feeling. Pain is, of course, a sub-modality of somatic sensation like touch, pressure, and position sense and serves as an important protective function. It is a complex perception. Its highly individual and subjective nature is one of the factors which makes it difficult to define and to treat clinically. More than any other sensory modality it is influenced by emotional state and environmental contingencies. Therefore, the same stimulus can produce different responses in different individuals under similar conditions. However, average behaviors are interesting in this context and the pain feeling can be used to approximately identify the stimulus upper limit. In particular, when testing different hardness indenters on different parts of the body, we observed that pain limits the maximum stress. The *maximum force* can be determined on the basis of the contact surface. The lower limit is on the contrary determined by the minimum perceived force. In this case *force* seems to be the relevant parameter and the *minimum stress* is thus calculated using the largest employed contact surface, approximately corresponding to a human hand.

Given the considerations above, the maximum stress corresponds to approximately 5.5 MPa, which corresponds to a force of 120-130 N (small indenter radius equal to 2.7 mm). On the other hand, to quantify the smallest force we used the PLATE-like indenter (radius equal to 35 mm), and the value we found is F = 0.2 N, which corresponds to approximately 50 Pa. Therefore, the application range goes from 50 Pa to 5 MPa (over 5 orders of magnitude). Moreover, the stiffness difference in various skin areas seems to influence the measured

contact force by a factor of approximately 3-4, with softer samples producing lower amplitudes. Results are summarized in Table 1.

TYPE of interaction	CONTACT FORCE (N)	PRESSURE (MPa)	Employed contacting tools	Contact DIAMETER (mm)	TESTED zones (different hardness, different sensitivity)	Impact type
Poke / finger mark	0.2 - 1.4	0.016 - 0.11	SOFT SPHERICAL INDENTER	4	Cheek, arm, belly (SOFT); shinbone, knee, back (HARD)	LIGHT
Poke / finger mark	20 - 130	0.86 - 5.6	SOFT SPHERICAL INDENTER	5,45	Thigh, Cheek, forearm, belly (SOFT); cheekbone, knee, back (HARD)	HARD
Light touch / caress	0.2 - 1.4	$5.2 \cdot 10^{-5} - 3.6 \cdot 10^{-4}$	BIG PLATE	70	Cheek, belly (SOFT); back (HARD)	SUPER SOFT
Smack / slap	70 - 170	0.018 - 0.044	BIG PLATE	70	Cheek, belly (SOFT); back (HARD)	HARD

Table 1. Basic human-robot interactions (impacts).

Our results are in accordance with literature, where typical human interactions are basically considered to range from approximately 1 N for a soft stroke up to approximately 100 N for a push or slap. In a recent paper published on Nature Materials [23], authors state that normal manipulation tasks involve stresses of the order of 10-100 kPa, while gentle touches correspond to stress values which are lower than 10 kPa.

We also compared our estimations with the specifications of commercial products which are today on the market, i.e. the Barrett hand[2] and BioTAC[3]. The Barrett Hand has a sensor resolution which is a bit lower (but of compatible order of magnitude on the palm) with respect to the smallest forces the application would require, and which correspond to 660Pa on fingertips, 330 Pa on fingers and 200 Pa on the palm. As regards the BioTAC Multimodal Biomimetic Tactile Sensor, a typical 0.03 N to 30 N force dynamic range is guaranteed, which is again compatible with the proposed dynamics which has been given in terms of the more appropriate *pressure* parameter (considering different contact areas). These are commercial products and specifications are related to *manipulation* tasks, however such data can be used as an indicative benchmark.

[2] http://www.barrett.com/robot/DS_BarretHand.pdf

[3] http://www.syntouchllc.com/TechSpecSheet.pdf

3. Assessment of the skin behavior

3.1. Electromechanical characterization of PVDF films

Commercial 100μm thick PVDF sheets from Measurement Specialties Inc.[4] have been purchased. PVDF samples have been cut from those sheets in a square geometry of 7x7 mm². Purchased sheets have been already stretched and poled. The process begins with the melt extrusion of the polymer resin pellets into sheet form, followed by a stretching step that reduces the sheet to about one-fifth its extruded thickness. Stretching at temperatures well below the melting point of the polymer causes chain packing of the molecules into parallel crystal planes (beta phase). The beta phase polymer is poled by application of very high electric fields (of the order of 100 V/μm) to align the crystallites to the poling field. In such conditions the piezoelectric film exhibits a material symmetry in the orthorhombic crystal system (C2V class), corresponding to that of the so-called orthotropic materials. The samples are oriented with axis 1 along the stretching direction – axis 2 in the in-plane orthogonal direction and axis 3 along the through-thickness direction (see Figure 1).

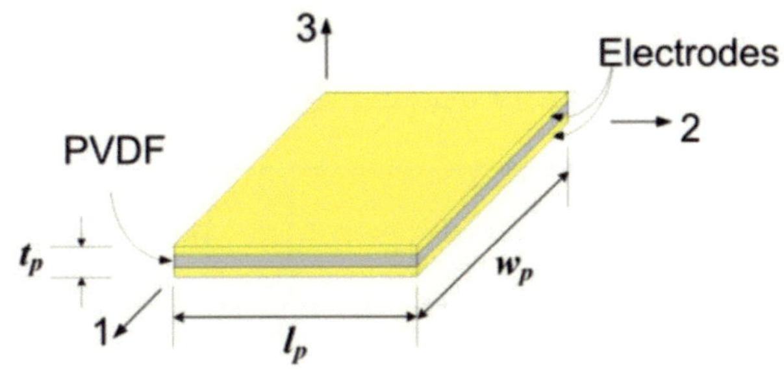

Figure 1. PVDF provided of electrodes to extract the charge signal. Reference axes are shown.

Linear electro-elastic constitutive equations are commonly used to describe the coupling of dielectric, elastic, and piezoelectric properties in piezoelectric materials [24].

In the frequency domain ($\hat{f}(\omega)$ means the Fourier Transform of any function ($f(t)$)) such equations are:

$$\begin{bmatrix} \hat{S} \\ \hat{D} \end{bmatrix} = \begin{bmatrix} \hat{s} & \hat{d}^T \\ \hat{d} & \hat{\varepsilon} \end{bmatrix} \begin{bmatrix} \hat{T} \\ \hat{E} \end{bmatrix} \tag{1}$$

where strain $\hat{S}$ and stress $\hat{T}$ are represented by 1x6 column vectors, while the electric displacement $\hat{D}$ and the electric field $\hat{E}$ are expressed by 1x3 column vectors. $\hat{s}$ is the 6x6 *compliance* matrix, $\hat{\varepsilon}$ the 3x3 *permittivity* matrix, both assumed to be symmetric, and $\hat{d}$ the *piezoelectric* 3x6 matrix.

In our application, the PVDF behavior is usefully described by the second row of constitutive equations:

---

[4] http://www.meas-spec.com/default.aspx

$$\begin{bmatrix} \hat{D}_1 \\ \hat{D}_2 \\ \hat{D}_3 \end{bmatrix} = \begin{bmatrix} 0 & 0 & 0 & 0 & \hat{d}_{15} & 0 \\ 0 & 0 & 0 & \hat{d}_{24} & 0 & 0 \\ \hat{d}_{31} & \hat{d}_{32} & \hat{d}_{33} & 0 & 0 & 0 \end{bmatrix} \begin{bmatrix} \hat{T}_{11} \\ \hat{T}_{22} \\ \hat{T}_{33} \\ \hat{T}_{23} \\ \hat{T}_{13} \\ \hat{T}_{12} \end{bmatrix} + \begin{bmatrix} \varepsilon_{11} & 0 & 0 \\ 0 & \varepsilon_{22} & 0 \\ 0 & 0 & \varepsilon_{33} \end{bmatrix} \begin{bmatrix} \hat{E}_1 \\ \hat{E}_2 \\ \hat{E}_3 \end{bmatrix} \tag{2}$$

The structures of the piezoelectric and the permittivity matrices are due to the reported material symmetry.

As the charge signal is measured with a charge amplifier, which converts the charge input into a voltage without supplying any electric field, the electric field can be set to null. Therefore, when the film is used in *thickness mode* the previous set of equations reduces to:

$$\hat{D}_3 = \hat{d}_{33}\hat{T}_{33} \tag{3}$$

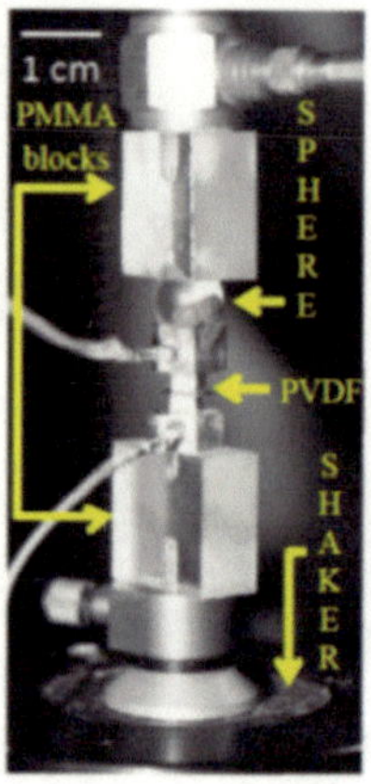

Figure 2. Compression test setup to measure the d_{33} coefficient.

For the present application, characterizing the electromechanical behavior of the piezoelectric polymer means to retrieve the frequency behavior of the d_{33} piezoelectric coefficient.

The employed experimental equipment has been thoroughly discussed in a previous publication [25], which reports more complete characterization results of the PVDF electromechanical behavior. Briefly, the experimental setup (see Figure 2) consists of a rigid frame with a lower fixed plate to which an electro-mechanical shaker is assembled. A piezoelectric force transducer is fixed to the upper head of the frame. Samples are mounted between the force transducer and the shaker. They are pressed between two metals heads of square cross-section with machined and polished contact surfaces. The lower head is assembled to the shaker head with an interposed PMMA block, and the upper head contrasts through a spherical joint with a similar block connected to the force transducer, for

self-alignment of the contact planes. Conductive glue creates a stiff connection between the sample and the heads thus excluding possible variations in the contact area during measurements.

The test is controlled by a computer in a completely automatic way. A swept sine signal is fed into the shaker. Frequency spacing and total duration of the test are determined by the settable frequency range and number of steps. The output charge signals (response) and force transducer (stimulus) are continuously acquired and processed in frequency to give the complex piezoelectric modulus. Normally the range between 10 and 1000 Hz can be explored without difficulties.

The frequency behavior of the d_{33} piezoelectric coefficient is shown in Figure 3. The most relevant result in this context is the almost flat behavior of both the real and imaginary parts of the modulus in the considered frequency range, in accordance with literature [26].

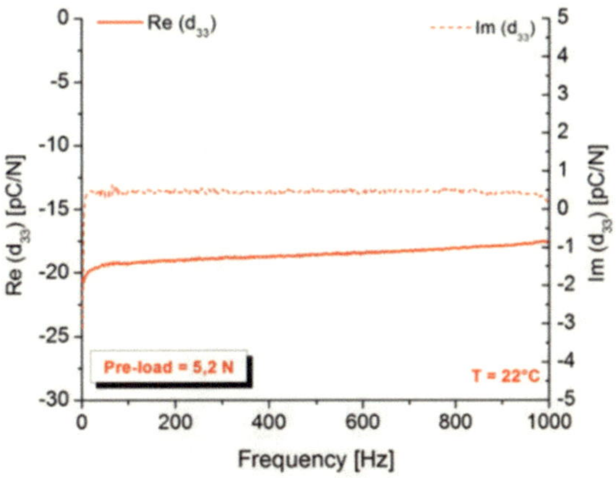

Figure 3. Frequency behavior of the real and imaginary parts of the d33 piezoelectric coefficient.

3.2. Electromechanical modeling of the skin structure

In order to associate the PVDF charge response to the effective load applied on the outer skin surface, a skin model which is based on the Boussinesq's equation has been considered.

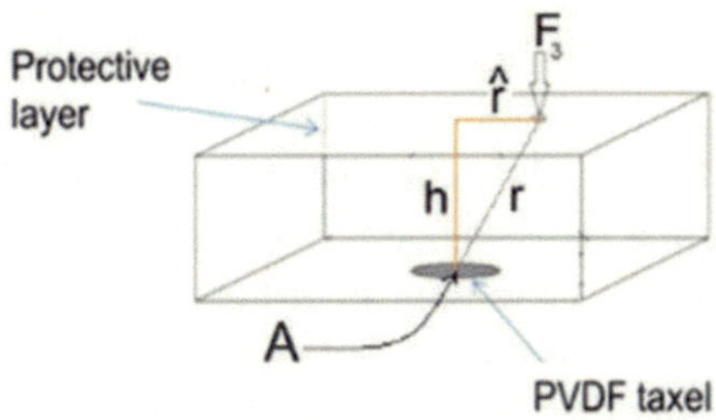

Figure 4. The PVDF sensor is located at the bottom of the protective layer of thickness h and a point force is applied on the outer surface.

Approximately, the relation between a point load **F** applied on the outer surface and the stress at a given point inside the cover layer is given by the Boussinesq's equation [27]:

$$T = \frac{3}{2\pi}\frac{F\cdot \mathbf{e}_r}{r^2}\mathbf{e}_r \otimes \mathrm{e}_r \tag{4}$$

where all bold faced symbols represent tensors or vectors, $\mathbf{e}_k$ is the unit vector in the *k*-direction and ⊗ is the symbol of the tensor product. The Boussinesq's problem considers a linearly elastic half-medium on which a point force is applied. Truly, the materials employed in the present application (typically elastomers or gels) are both non-linear elastic and visco-elastic. Nevertheless, Equation (4) does not depend on the elastic modulus and deviations from linearity are only expected as a consequence of geometry changes due to large deformations. For the applicability of the model, therefore, a sufficiently rigid protective layer should be chosen. In addition, Equation (4) holds whenever the Poisson ratio of the medium is close to 0.5. This holds for an elastomer ($\nu = 0.48$), but not for other materials (e.g. a foam). A complete solution valid for all ν is available in [28], but it is considerably more complex. The major approximation concerns the finite thickness of the layer, compared with the semi-infinite medium. The advantage of (4) is its simplicity, as the stress is uniaxial in the radial direction, and its independence of elastic parameters.

Equation (4) is applied to a sensor located on the bottom of the elastic cover of thickness *h*. The sensor works in the thickness mode, i.e. it can read the T_3 stress component ('3' is the direction normal to the bottom surface) via the d_{33} piezoelectric modulus. Letting $\hat{r}$ be the radial distance of the point where the force is applied from the sensor center projected on the outer surface, we have $r^2 = \hat{r}^2 + h^2$ and $\mathrm{e}_r = \sin\theta(\mathrm{e}_1\cos\varphi + \mathrm{e}_2\sin\varphi) - \mathrm{e}_3\cos\theta$, where $\sin\theta = \hat{\mathbf{r}}/r$.

Then we obtain:

$$\mathrm{T}_3 = \frac{3}{2\pi}\frac{\mathrm{h}^2}{\left(\hat{r}^2 + \mathrm{h}^2\right)^{5/2}}\left\{\left(\mathrm{F}_1\cos\varphi + \mathrm{F}_1\cos\varphi\right)\hat{r} - \mathrm{F}_3\mathrm{h}\right\} \tag{5}$$

For a vertical force ($F_1 = F_2 = 0$) Equation (5) reduces to:

$$\mathrm{T}_3 = -\frac{3}{2\pi}\frac{\mathrm{F}_3\mathrm{h}^3}{\left(\hat{r}^2 + \mathrm{h}^2\right)^{5/2}} \tag{6}$$

As T_3 is related to the charge density D_3 on the sensor surface by the piezoelectric constitutive equation:

$$D_3 = \mathrm{d}_{33}T_3 \tag{7}$$

under the hypotheses that the sensor size is sufficiently smaller than its distance from the point force, the total charge measured by the sensor can be approximated by:

$$q = -\frac{3}{2\pi} A \frac{d_{33} F_3 h^3}{\left(\hat{r}^2 + h^2\right)^{5/2}} \quad (8)$$

3.3. Experimental assessment of the single taxel behavior

In order to optimize the design of the skin prototypes we checked the pertinence/limits of the mechanical model, thus quantifying the PVDF electrical response to the external mechanical stimuli and testing the effect of the protective layer (material and thickness) on the PVDF response. We performed the tests on basic single taxel prototypes, where a single circular PVDF film is covered with an elastomeric cover to protect the sensor from physical damage by shocks or chemical contamination by oil and other materials. Attention has been focused on gel/rubber layers, in that they are more controllable and reproducible systems which also allowed for the required dynamics.

An experimental campaign has been carried out in order to quantify the sensor output charge in common tactile interaction tasks. The mechanical input stimuli range from 50 Pa to 5 MPa, as reported in section 2.

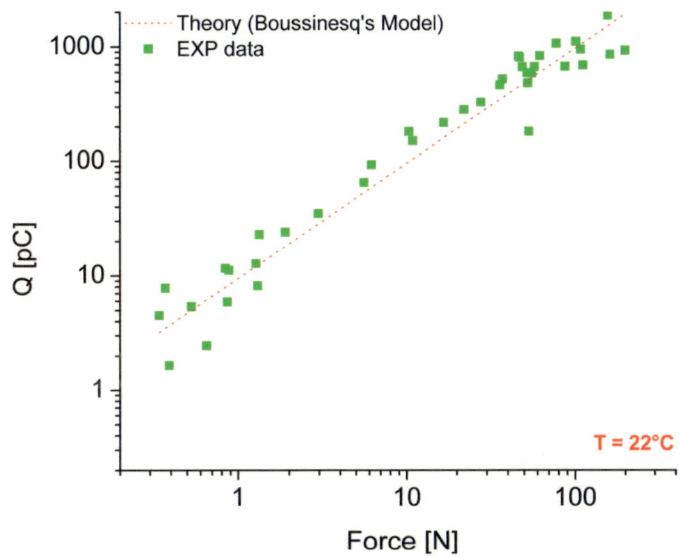

Figure 5. Single taxel skin prototype: comparison between model and experiments.

The "modally tuned impulse force hammer" (see Section 2) has been used to mechanically stimulate the single taxel sample. In this case, we recall that the load cell measures a contact force, which depends both on the indenter and on the protective layer materials. The measured charge over the 5 kPa - 5 MPa stress range has been compared with the model presented in the previous section and results are reported in Figure 5. In this case, on the horizontal axis the peak amplitude of the *time behavior* of the contact force is reported. Similarly, the peak amplitude of the charge response is reported on the y axis. It is

important to remark here that these measurements are not at all easy to perform, as it is not yet possible to accurately control the position of the hammer impact. A good statistics would be thus required to achieve reliable results. Results are however useful to understand the "order of magnitude" of the charge response and the limits of applicability of the models.

A good accordance between model and experimental data is recorded for high loads, which allows extending the model linear behavior for lower loads, thus covering the whole stress range which is of interest for the application (50Pa - 5MPa). Therefore, both charge and force ranges cover 5 orders of magnitude, the typical output charge ranging from 0.01 pC to 1 nC. This information has been a reference point for the design of the electronics.

4. Interface electronics

In Figure 6 the basic block diagram of the interface electronics is shown. The interface electronics converts the charge developed by the PVDF – as a result of the applied stress – to a voltage signal. It includes a Charge Amplifier (CA) cascaded with a Low Pass Filter (LPF) with high *cutoff frequency*, f_H. The CA has a high pass response with low *cutoff frequency*, f_L. The frequency band of interest at constant gain is, therefore, defined as BW = f_H - f_L.

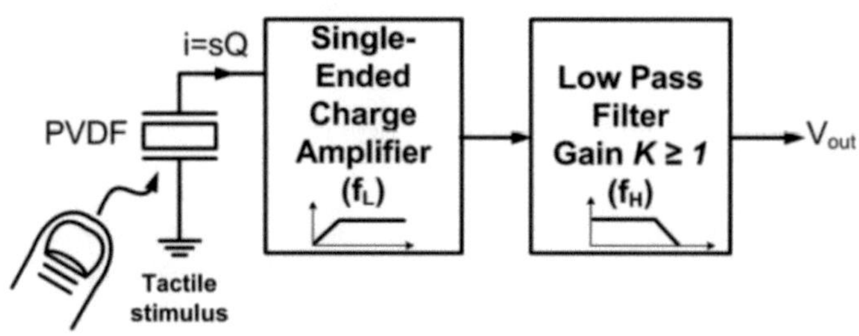

Figure 6. Basic block diagram of the interface electronics.

The mathematical expression of the charge generated by the PVDF sensor can be found from the piezoelectric constitutive equation (2) of Section 3.1. Considering the op amp as ideal, the electric field E_3 across the PVDF sensor is negligible because of the virtual ground at the op amp non inverting terminal. Therefore, under the assumption of the thickness mode operation, the expression for the electrical displacement reduces to (3). The charge generated by the PVDF sensor can be found by integrating the electrical displacement D_3 over the loading area A_c:

$$q = \iint_{A_C} D_3 dA_3 = d_{33} A_C T_3 = d_{33} F_C \tag{9}$$

where the stress $T_3 = F_c / A_c$ is assumed to be uniform over the loading area, and F_c is the applied contact force (see Figure 7).

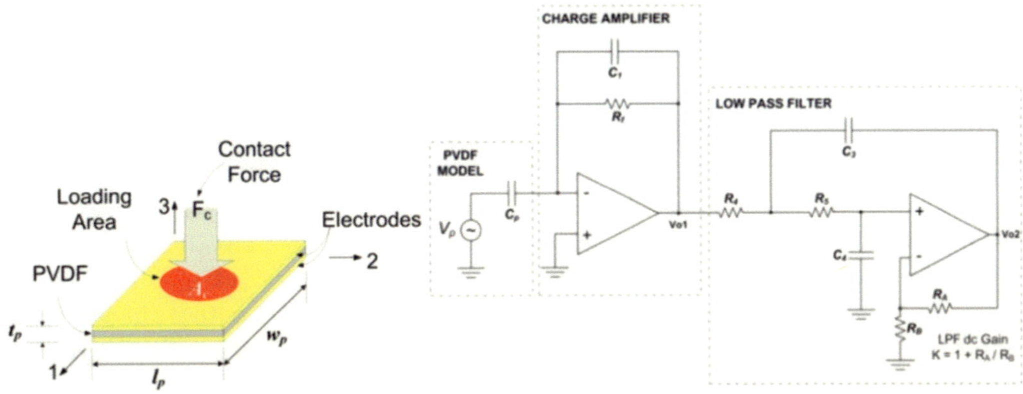

Figure 7. PVDF tactile sensor: The PVDF material (in the middle) is provided of two electrodes *(left)*. Charge amplifier connected to the PVDF equivalent electrical circuit model and the cascaded low pass filter *(right)*.

Figure 7 *(right)* shows the simplified equivalent circuit model of the PVDF sensor connected to the CA. In order to find the mathematical expression for the equivalent voltage source V_p Equation (2) can be used. In case of open circuit across the electrodes of the PVDF film, D_3 is zero. Therefore, expressing the electric field as the ratio of the open circuit voltage to the PVDF thickness (i.e. $E_3 = V_p/t_p$) and considering the loading area A_c equal to the PVDF area A_{PVDF}, we found:

$$V_p = -\left(\frac{t_p}{\varepsilon_{33} A_c}\right) d_{33} F_c = -\frac{q}{C_p} \tag{10}$$

where C_p (= $\varepsilon_{33} A_{PVDF}/t_p$) is the equivalent capacitance of the PVDF film and q (= $d_{33}F_c$) is the resulting charge. Using (10) and by analyzing the circuit of Figure 7 *(right)*, the transfer function of the CA – in terms of the input charge to the output voltage ratio – can be found:

$$H_{CA}(s) = \frac{V_{o1}}{q} = \frac{sR_f}{1 + sC_1 R_f} \tag{11}$$

The sensitivity of the CA is given by the ratio between the maximum output voltage (i.e. the supply voltage in the ideal case) to the maximum input charge. The sensitivity sets the value to be given to the feedback capacitance C_1 (i.e. $V_{o1,max} / q_{max} = 1/C_1$). Moreover, the CA low cutoff frequency defines the value to be given to the feedback resistance R_f (i.e. $f_L = 1/(2\pi R_f C_1)$).

4.1. Effects of the operational amplifier non idealities

The objective of the present section is to analyze the limitations which come into play when the op amp non idealities are taken into account. The non-idealities influence the behavior both of the op amp and of the PVDF tactile sensor.

4.1.1. Finite op amp open-loop gain

The open-loop gain of an op amp, a, is not infinite, therefore, the electric field, E_3 across the electrodes of the PVDF is not negligible. Hence, the dielectric contribution due to the electric field cannot be neglected in the piezoelectric constitutive Equation (2). Therefore, the total charge in input to the CA, is:

$$q_{tot} = \left(d_{33}F_3\right) + \left(-\frac{\varepsilon_{33}A_{PVDF}}{t_p}\frac{V_{o1}}{a}\right) = q_F + q_E \tag{12}$$

where the first term q_F is due to the applied force (i.e. direct piezoelectric effect), while, the second term q_E is due to the electric field caused by the op amp finite open-loop gain (i.e. inverse piezoelectric effect).

Component Name	Value	Unit
a [(b)]	10^6 (120dB)	V/V
C_1	100	pF
C_3	2.7	nF
C_4	1	nF
R_1	100	MΩ
R_2	9	kΩ
R_3	1	kΩ
R_4	3.6	kΩ
R_5	7.5	kΩ
R_A [(b)]	N.A.	Ω
R_B [(b)]	N.A.	Ω

[(a)] The LPF has a unity gain
[(b)] Open-loop gain OPA703

Table 2. Charge amplifier and low pass filter component values.

In order to quantify how the finite open-loop gain could contribute to the generated charge let us give the following example. Let the op amp be the OPA703[5] which is supplied at ±5 V. The CA parameters are the ones reported in Table 2; let the range of contact forces be 0.1 N ÷ 25 N (where the maximum force makes the CA output to saturate to 5 V); and let us assume $A_{PVDF} = A_C$. Finally, let the PVDF parameters be the ones reported in Table 3. Using Equation (12) the charge contribution due to the op amp finite open-loop gain is reported in Table 4.

[5] http://www.ti.com/lit/ds/symlink/opa703.pdf

As it can be seen from the values reported in Table 4, the charge, q_E due to the electric field contribution can be considered negligible, with respect to the dielectric contribution, the higher the op amp open-loop gain is.

Name	Parameter	Value	Unit
Piezoelectric constant [(a)]	d_{33}	-20	pC/N
Permittivity [(a)]	ε	106	pF/m
Relative permittivity [(a)]	ε_r	12	
PVDF thin film Length	l_p	5	mm
PVDF thin film Width	b_p	5	mm
PVDF thin film Thickness [(a)]	t_p	110	μm
PVDF electrode surface [(b)]	$A_{PVDF}=A_e= b_p \cdot l_p$	25e-6	m²
Static Capacitance [(b)]	$C_p = (\varepsilon_{33} A_c)/t_p$	24	pF
Input charge	$q = d_{33} F_c$	0.2 (F_c = 0.01N) ÷ 500 (F_c = 25N)	pC
Equivalent PVDF voltage source	$V_p = - q/[(\varepsilon_{33} A_c)/t_p]$	2e-3 (F_c = 0.01N) ÷ 5 (F_c = 25N)	V

(a) Piezo Film Sensor Technical Manual, Measurement Specialties, Inc.
(b) Refer to Figure 7.

Table 3. PVDF parameters.

F_c [N]	q_F [pC]	q_E [pC]
0.01	-2	-48.2e-9
25	-500	-120 e-9

Table 4. Contribution of the op amp finite open-loop gain to the input charge.

Moreover, if we consider the finite op amp open-loop gain, the CA frequency response becomes:

$$A_{CA,q}(j\omega) = \left(\frac{a}{1+a}\right)\left(\frac{t_p}{\varepsilon_{33} A_c}\right)\left[\frac{j\omega C_p R_f}{1 + j\omega\left(\frac{a}{1+a}\right) R_f \left(C_1 + \frac{C_1 + C_p}{a}\right)}\right] \quad (13)$$

The module of (13) at low frequency tends to zero, while for high frequencies, it tends to:

$$\left|A_{CA,q}(j\omega)\right| \cong \left(\frac{t_p}{\varepsilon_{33}A_c}\right)\frac{C_p}{\left(C_1+\dfrac{C_1+C_p}{a}\right)} \tag{14}$$

For high values of the op amp finite open-loop gain the term $(C_1+C_p)/a$ of Equation (14) becomes negligible.

The influence of the finite op amp open-loop gain can be considered negligible both as contribution to the generated charge and as contribution to the output frequency response of the CA. Therefore, for the analysis the op amp can be approximately considered as ideal.

4.1.2. Finite Gain-bandwidth product (GBP)

The gain-bandwidth product (GBP) defines the limit of the op amp amplification, introducing a *cutoff frequency* at high frequency. When choosing an op amp it is worth to check accurately if the op amp has a GBP that allows the CA to achieve the target pass-band frequency range. Table 5 reports the results of the comparison among the three op amps which have been chosen for their suitable features in terms of GBP, open-loop gain, supply voltage and package size for the design of the interface electronics.

Both the package size and supply voltage are two important specs to be considered because of the small space available on the robot. In fact, the use of single supply op amps allows reducing the number of components without degrading the circuit performance – negative voltage regulators are not necessary thus reducing the required number of components. Components with small packages should be preferred to save space. Therefore, according to the considered op amp specs summarized in Table 5 the OPA347 is the better tradeoff.

		OPA703	OPA347[*]	OPA348[†]
GBP		1 MHz	350 kHz	1 MHz
Open-loop gain		120 dB	115 dB	108 dB
Supply voltage		Double/Single ±2V to ±6V 4V to 12V	Single +2.3V to +5.5V	Single +2.1V to +5.5V
Package	Type	SOT23-8 SO-8	WCSP-8	SOT23-8 SO-8
	Size (mm)	W x L x H 2.8x2.9x1.45	W x L x H 2x1x0.625	W x L x H 2.8x2.9x1.45

[*] http://www.ti.com/lit/ds/symlink/opa2347.pdf
[†] http://www.ti.com/lit/ds/symlink/opa348.pdf

Table 5. Comparison among some off-the-shelf op amps characteristics in terms of GBP, supply voltage and package size.

Figure 8 shows that op amps in Table 5 satisfy the 1 Hz ÷ 1 kHz bandwidth specification for the *frequency band of interest*.

The experimental results validate the proposed circuit approach. The behaviour of the interface electronics is linear in the given frequency range. The input signal range of the interface electronics was able to measure up to 3 orders of input force magnitude range. In order to extend the input range of the detectable signal up to the 5 orders of magnitude expected for the application, a variable gain amplifier is currently under development.

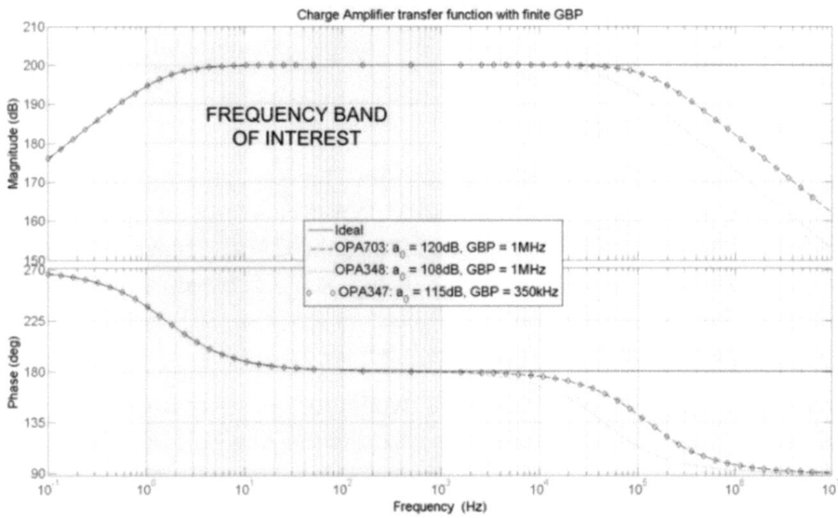

Figure 8. Comparison between the ideal and CA frequency responses using an ideal op amp, the OPA703 and the OPA348/7 with their corresponding finite GBPs.

5. Skin technology and experimental results

The basic structure of skin prototypes based on PVDF arrays is reported in Figure 9. Piezoelectric sensors have to be integrated on a rigid substrate and an elastomer layer is integrated on top as protective layer. This section briefly describes the technological solutions adopted to assemble the different building blocks. Details will be contained in a forthcoming publication.

The triangular geometry, taxel size and positions have been chosen to reproduce existing conformable skin patches manufactured at IIT (Italian Institute of Technology, Genoa) and based on capacitive transducers (*Roboskin project*).

Among technology solutions, how to provide PVDF films with metal contacts is a crucial point. Although metalized PVDF is commercially available, the brittleness of some of the available coatings (Copper over nickel, aluminum on chrome), the high roughness of the screen-printed silver ink solution and the impossibility of obtaining - as a prototyping service - custom patterning of the conducting film from the suppliers, convinced us that operating a "home-made" metallization was a better choice. We thus purchased *bare* piezoelectric polymer foils and made ad-hoc electrodes employing inkjet deposition technologies.

A whole metal layer (common ground) is deposited on top of the PVDF film, while patterned contacts with the same geometry of metal contacts on PCB (Figure 9) are deposited on the PVDF bottom side. Conductive glue has been used to fix the PVDF film on the PCB. The cover lay avoids short-circuits between taxels due to the employed glue. The protective layer is finally integrated on top. A Two-part silicone Sylgard® 184 Silicone Elastomer (*Dow Corning*) that cures to a flexible elastomer (PDMS) has been employed and it is directly polymerized on the PCB. Adhesion between PDMS and PVDF has been increased by the use of an *adhesion promoter*. A 2,5mm thick layer has been finally chosen as optimally meeting the application requirements.

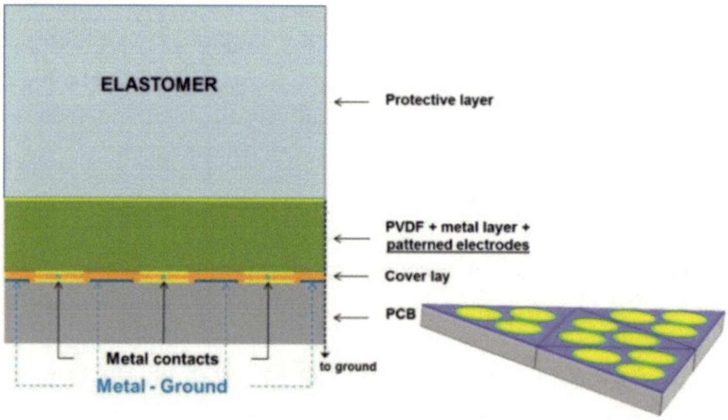

Figure 9. Structure of skin prototypes based on piezoelectric polymer arrays.

Figure 10 shows some results of experimental tests on triangular prototypes. A sinusoidal mechanical stimulus has been applied in correspondence of a certain taxel, by means of an indenter stiffly connected to an electromechanical shaker.

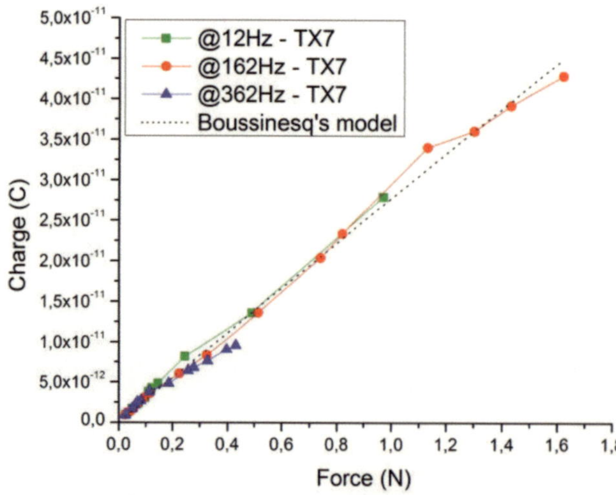

Figure 10. Comparison between experimental data at different frequencies and Boussinesq's model. Mean of peak values of sinusoidal waveforms at different frequencies are shown.

The tests have been conducted by varying the amplitude of the input force for different frequency values. Figure 10 shows the measured charge at 12 Hz, 162 Hz and 362 Hz compared with the value obtained using the Boussinesq's equation (see Section 3.2). As it can be seen a good correspondence can be found. A good linearity is achieved over the whole explored range, and the same voltage-to-force behavior for different frequencies is recorded (in accordance with d_{33} results reported in Section 3.1).

6. Towards system integration

6.1. Data processing

As for humans the cerebral cortex constructs an image of touched objects from the fragmented information provided by skin receptors, in a similar way tactile data processing is needed to enable the robot response to external stimuli. Starting from sensor data, the development of methods for touch interpretation is not an easy task.

Towards this scope, this chapter presents the basic ideas of a *force reconstruction* algorithm. The tactile stimuli applied on the outer surface of the elastomer and filtered by the soft layer, are assumed to be received by the sensors as stress components. Any tactile recognition task is haunted by the following problem: given the output of the sensor array, derive the force distribution on the outer surface. The problem is fatally an ill-posed one: in fact the set of the data is discrete (the sensor number is finite) and the set of the unknowns is a continuum (a force distribution is a vector field).

The proposed model exploits a discretization of the original continuous problem by constructing an array of concentrated forces at discrete points on the skin surface, starting from the information provided by the sensor array (normal stress measurements).

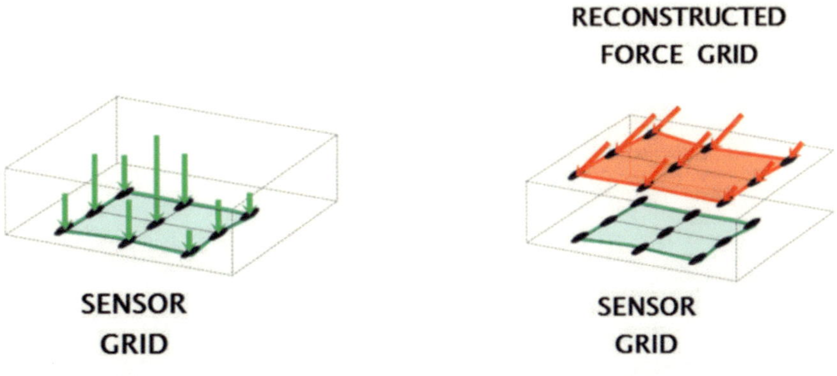

Figure 11. (a) A distribution of forces (not shown in the figure) acts on the cover layer. The pure normal stress component is recorded by the sensor array which is located beneath the elastomer layer. (b) From the sensor electrical outputs - through the developed algorithm – the force is reconstructed at discrete points over the outer layer.

If the force distribution is discretized into an array of point forces, a superposition of the Boussinesq equations (Section 3.1) for the single point forces can be used as a particularly simple and attractive approximation in the case of linear elasticity.

Mathematically it reduces to a linear vector equation $b = Cx$, where C is in general a rectangular matrix. The discretized problem consists in solving this equation for the force vector x.

The inverse Boussinesq's problem has a direct and unique solution only if C is a square full-rank matrix. This would imply that the number of sensor outputs is the same as the number of force components. If we have one output per sensor (typically, the normal stress) and want to explore 3 -component surface forces, the number of points in the force array should be one third of that in the sensor array, to meet the condition. Such a solution would be rather poor in definition.

We examined a solution where the number of point forces (applied to the nodes of a pre-determined grid) is equal to the number of sensors. Initial assumptions are made (some can be relaxed in further developments) in that the problem is considered to be static, the surface plane, deflections are small and the elastomer layer Poisson ratio close to 0.5. The method consists in creating a series of x solutions – written as a function of two scalar parameters - to the ill-posed problem. Those are built by using the Moore-Penrose pseudo inverse matrix [29] and an orthogonal projector acting on a not completely defined w vector which, however, fulfills the essential physical constraints (for instance, in a contact problem compressive normal forces and horizontal forces in the friction cone are expected). The eligible solution is the one which minimizes a cost functional corresponding to the norm of the difference between the generic x solution and the w vector.

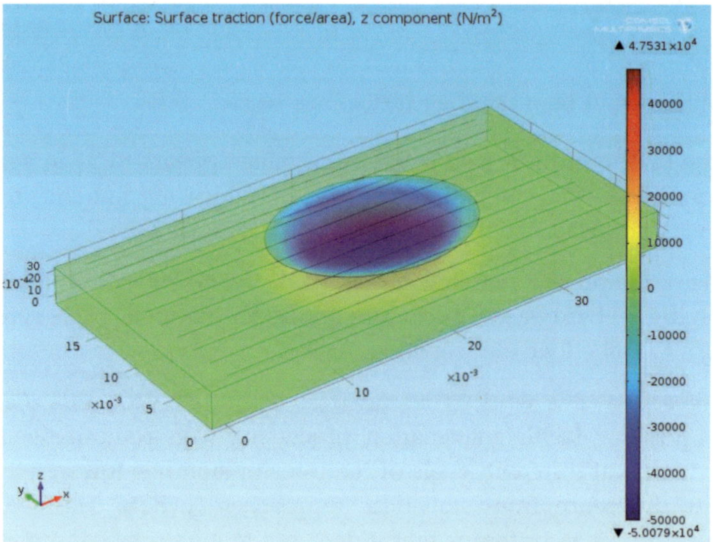

Figure 12. 3D representation of the test case, with only the normal traction field evidenced.

A check of the algorithm was made in MATLAB on a modeled skin structure consisting in a rectangular patch 40×20 mm^2 in size with a 10×10 point-like sensor grid at the bottom and 3mm thick elastomer layer (Figure 12). An hertzian force distribution with normal and tangential components (directed at 45° to the longitudinal axis of the patch) was applied to

an elliptical area on the outer surface. The model was previously evaluated by a FEM computation and the normal stress at the center of each sensor was calculated.

The algorithm described above has been applied to retrieve the 3 traction components on the loading area, having the sensor stress values as inputs (*b* vector). The result concerning the sole normal component is shown in Figure 13.

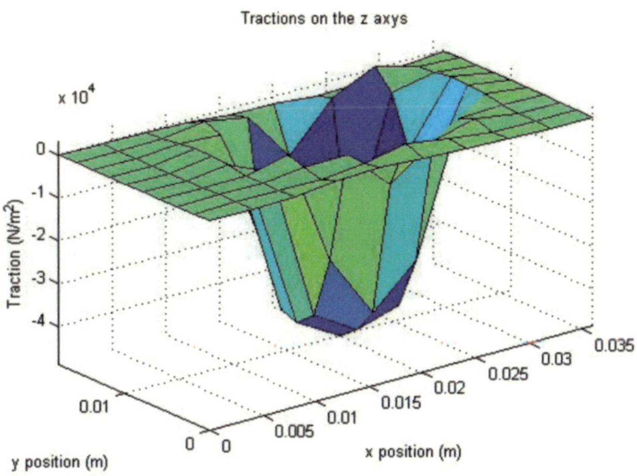

Figure 13. Algorithm output: normal traction on the outer surface.

The shape of the distribution as well as its peak value appears to be in very good accord with the assigned data. The tangential components, which are not shown in the present context, can be also retrieved within a certain approximation.

A crucial feature of the present research is that the proposed algorithm should be efficiently implemented on digital hardware. This in turn allows for real-time implementation of tactile data processing into embedded electronics systems.

In a wider perspective, this work is intended as a first step towards the integration of different techniques for tactile information processing (e.g. computational intelligence), possibly implemented at different levels of the transmission line towards the robot central processor. In its difference from statistical approaches such as Machine Learning, the advantage of the present algorithm is that it does not require time-consuming training and data set analysis.

6.2. System architecture

A robot skin can be seen as a complex system formed by a large number of spatially distributed sensing elements with embedded local processing electronics [30]. The robotic

skin should be designed in order to be independent from and compliant to different robotic platforms. Small/large scale-tailor-made robotic skin could require different transduction principles (capacitive, piezoelectric, etc.) for the multimodal transduction of contact features.

The robot skin design should focus on the following topics:

- Robot skin as a system. The Robot skin is intended to be a set of flexible or curved and compliant patches covering a large area of the robot body. Each patch is composed of a given set of tactile sensing arrays (i.e. 2-D arrays of taxels) and of embedded and dedicated electronics. Patches can be interconnected and networked to each other in order to achieve large sensing surfaces, where contacts or interactions are mostly expected to happen.
- Robot skin as a sensor. The tactile sensor is intended to be multimodal and it should include detection and measurement capabilities (i.e. contact event, force distributions, temperature, etc.).
- Robot skin electronics. Development of large parts of robot skin with a modular approach: i.e. configuring robot skin parts as the functional, electrical and mechanical aggregation of basic tactile sensing arrays with dedicated and embedded interface electronics, local data acquisition and processing. Each module will be networked with the others in order to configure scalable and modular arrangements.

Despite innovative designs, a large number of sensors have been rendered "bench top," as the emphasis has been on the sensors, and the system (in particular the underlying embedded electronic system) has largely been ignored [11]. Only few tactile sensing arrays with electronic circuitry on chip with sensors have been presented so far. Those having any possess circuitry with minimal complexity, e.g., a single MOS transistor associated with each transducer.

To overcome the limits of current implementations, i.e. to design and develop large area tactile sensing arrays to be flexible, conformable, and stretchable and at the same time to be intimately integrated with the embedded electronic system and with functional (i.e. PVDF in our case) and structural materials, researchers need to address the development of *the tactile embedded electronic system* which intimately copes and integrates with technology and devices on one side and with system features and constraints on the other one.

The functions to be implemented by the *tactile embedded electronic system* are various and demand covering the whole signal chain: sensors biasing, signal conditioning (e.g. low noise amplification, low pass filtering, etc.), matrix readout, signal acquisition (i.e. Analog to Digital conversion), local digital signal processing, communication bus interface, etc. For instance one of the tasks of the tactile embedded electronic system is interfacing heterogeneous sensors, with different read out circuit modes, to the robot skin electronic infrastructure.

Based on the assessment of the sensing devices performance and on the system requirements, researchers need to address a proper and cost effective partitioning of the

tactile embedded electronic system between dedicated and COTS implementation. The design methodology and the partitioning must take into account the high number and types of sensors to be read, the high sampling bandwidth for some of them, the high expected data throughput, the limited communication bus bandwidth, the need for a low complexity implementation i.e. a small number of devices and interconnections, etc. Due to the complexity and diversity of tasks to be implemented, the digital core and part of the signal conditioning/data acquisition blocks need to be implemented by dedicated silicon electronic circuits (i.e. Application Specific Integrated Circuit, ASIC).

Dedicated communication strategies are needed to transmit the large amount of data (due to the high sampling rate as in the case of the PVDF sensors, i.e. at least 2 kSamples/s per sensor, and due to the large number of sensors in each array) collected by the tactile sensors arrays distributed over the whole body [31]. In this context, the hierarchical architecture of the communication bus and the local data processing (for a number of tasks e.g. feature extraction, data compression, etc.) is explored. Going from lower levels (i.e. skin) to higher levels (i.e. central processing unit) protocols are different. The desired operation speed, noise and number of wires put a constraint on the type of communication channel used for interaction with higher levels. Serial communication buses are used (e.g. I^2C, Can bus, Flexray, Ethernet, etc.) to decrease wiring. The buses using CAN protocol are generally a preferred choice mostly due to the real-time capabilities, high reliability, and readily availability on most microcontrollers. But, the CAN bus suffers from a moderate transmission bandwidth (up to 1 Mbits/s) which will either slow down transmission of tactile data from a large number of sensors or put a cap on the number of touch sensors on the body. These issues can be solved either by using buses with higher transmission bandwidth (e.g. up to 10 Mbits/s can be achieved with FlexRay) or using more buses in parallel - which is anyway undesirable.

Due to the requirement of real time needed to use the tactile feedback in the control loop, deterministic protocols are mandatory. Going from periphery upwards, bandwidth of the bus increases in order to accommodate an increasing amount of data; protocol complexity increases as well. At lower levels high speed, lower connectivity and short distance wiring buses are preferred (e.g. I^2C). Moving up in the hierarchy, more complex protocols and longer wiring buses are preferred (e.g. CAN, Flexray, real-time Ethernet like Ethercat).

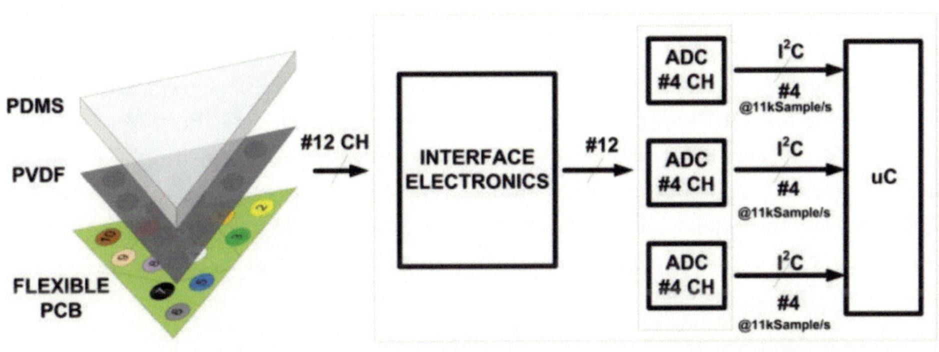

Figure 14. Block diagram of the system architecture with a triangle made of PVDF transducers.

The tactile sensing system based on PVDF transducer arrays proposed in this chapter [30] is based on a conformable mesh of sensor patches having a triangular shape [32]. On the bottom of the triangular substrate – which is in contact with the robot structure – the blocks of the interface and the local data acquisition/processing electronics are embedded (see Figure 14). The top surface hosts an array of 12 sensors/detectors covered by a PDMS protective layer. Each triangle is interconnected to each other to create a networked structure. Each patch is implemented on a flexible substrate allowing the system to conform to smooth curved surfaces of the robot body (i.e. upper limbs, torso, back, etc.). The PVDF film transducer is provided of two electrodes connected to wires which transfer the generated charge to the interface electronics. The PVDF sensor is used to detect tactile stimuli in the 1 Hz to 1 kHz range (corresponding to the human tactile sensing bandwidth). In the architecture shown in Figure 14, the 12 output signals from the triangle array are in input to the interface electronics (see Section 4). The interface electronics outputs the signals to three ADCs (MAX11613[6]) - each one manages 4 channels – and subsequently, through I²C buses, the signals are routed at 11kSample/s to the microcontroller (PIC24FJ64GB004 Family[7]). At present, skin patches (i.e. triangles) are interconnected through a CAN bus (the reference robotic platform is iCub[8]).

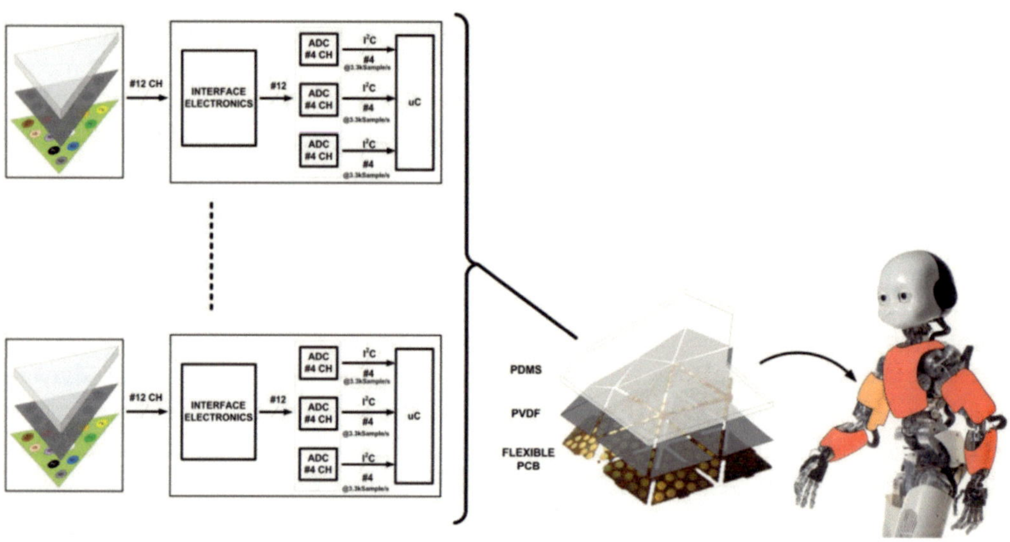

Figure 15. Tactile sensing system. iCub picture is printed by courtesy of IIT (Italian Institute of Technology).

[6] http://datasheets.maxim-ic.com/en/ds/MAX11612-MAX11617.pdf
[7] http://ww1.microchip.com/downloads/en/devicedoc/39940c.pdf
[8] www.icub.org

To be able to integrate the interface and the data acquisition/local processing electronics onto the bottom of each triangle, and due to the small space available, an application-specific integrated circuit (ASIC), see Figure 15, which embeds: a) the interface electronics; b) data acquisition; c) dedicated signal processing; d) communication bus interface should be pursued. The feasibility and integration of the tactile sensing system on the robot will mainly depend on the design of the architecture of the ASIC.

7. Conclusions

The main objective of the research activity in this area is the design and manufacturing of robotic skin systems based on arrays of PVDF film transducers. We addressed issues concerning the manufacturing technology, the interface electronics and the system integration. The main achievements are summarized below.

The electromechanical characterization of PVDF thin films working in thickness mode has been achieved. The experimental d_{33} characterization shows a flat behavior in the whole range of frequencies of interest for the present application. Future steps would include the characterization of the PVDF tangential piezoelectric moduli and of the polymer behavior when used in bending mode. Another field of interest is related to the interpretation of PVDF creep and recovery following a step in load or temperature.

A number of elementary qualitative tactile perceptions have been experimentally quantified i.e. the corresponding mechanical contact force/stress have been evaluated. The contact pressure spans on about 5 orders of magnitude ranging from 50Pa to 5MPa. Achieved results complete relevant information in the latest literature about skin and perceived contacts. The contact stress/force range has been used as reference for the development of the interface electronics.

The PVDF transducer charge response to contact stresses has been evaluated in the defined range. We measured the charge response of a set of single taxel prototypes (i.e. PVDF thin film + cover layer) and we focused on gel/rubber layers excluding foams, to get more controllable and reproducible systems which also allow for the required dynamics. Measured charge spans over a range from about 0.01pC up to 1-2nC. The charge response results have been compared with an electromechanical model of the skin structure and have been used as reference specifications to design the interface electronics.

The interface electronics has been designed, implemented and tested. Experimental results validate the proposed solution in the frequency band of interest. The output signal ranges over three orders of magnitude. The experimental tests highlighted the need of a variable gain electronics solution to be able to measure the wide range of tactile stimuli expected for the application (i.e. 5 orders of magnitude for the expected mechanical stress due to contact).

The manufacturing technology for PVDF sensors arrays has been identified. In particular, the thickness (i.e. approximately 3mm) and material (PDMS) of the protective layer for the

triangular prototypes have been assessed and tested. Some prototypes of triangular sensor arrays have been manufactured and experimental results are encouraging. In this context, the next development is related to assess the reproducibility of the array behavior. Moreover, durability and calibration tests will be necessary to demonstrate the effective viability and reliability of the proposed technology.

A data processing algorithm for the estimation of the force distribution at the top surface of the covering layer is being currently investigated. Some additional work is however required, which is related to the optimization of the cost functional. Next step is toward the algorithm implementation into a dedicated embedded electronic system interfacing the tactile skin with the upper levels of the robot processing system. Once the force distribution has been obtained, the *Theorem of Work* can be employed to estimate the displacements at the surface of the elastomer. Extensions to non-parallel forces (torques, pinching and so on) and to large deflections are considered as next-future developments.

The system architecture (i.e. from the sensors arrays to the local microcontroller) has been identified. It is based on the following structure: PVDF sensor array (i.e. triangle), interface electronics (one for each channel), ADC converter with I^2C communication interface, I^2C bus, microcontroller.

Author details

Lucia Seminara and Luigi Pinna
Department of Biophysical and Electronic Engineering - University of Genoa, Italy

Maurizio Valle
Department of Biophysical and Electronic Engineering - University of Genoa, Italy
Research Center on Materials Science and Technology - University of Genoa, Italy

Marco Capurro
Research Center on Materials Science and Technology - University of Genoa
Department of Civil, Environmental and Architectural Engineering - University of Genoa, Italy

Acknowledgement

This work is supported by the ROBOSKIN European Project about "Skin-Based Technologies and Capabilities for Safe, Autonomous and Interactive Robots", under grant agreement no. 231500.

8. References

[1] Shimada M., Minato T., Itakura S., and lshiguro H. (2007). Uncanny valley of androids and its lateral inhibition hypothesis. In Proc. IEEE International Symposium on Robot and Human interactive Communication, pp 374–379.

[2] Breazeal, C. (2003). Toward sociable robots. Robotics and Autonomous Systems, 42:167–175.

[3] Velonaki, M., Rye, D., Scheding, S., and Williams, S. (2005). Fish-Bird: Autonomous interactions in a new media arts setting. In Proc. Vital Signs: Creative Practice and New Media Now, volume 29.

[4] Robins, B, Dautenhahn K, Dubowski, J (2005) Robots as isolators or mediators for children with autism? A cautionary tale. Proceedings of the AISB'05 Symposium on Robot Companions Hard Problems and Open Challenges in Human-Robot Interaction, pp. 82-88.

[5] Naya, F., Yamato, J., and Shinozawa, K. (1999). Recognizing human touching behaviors using a haptic interface for a pet-robot. In Proc. IEEE International Conference on Systems, Man, and Cybernetics, volume 2, pp 1030–1034.

[6] Iwata, H. and Sugano, S. (2005). Human-robot-contact-state identification based on tactile recognition. IEEE Transactions on Industrial Electronics, 52(6):1468–1477.

[7] Stiehl, W D, Lieberman, J, Breazeal, C, Basel, L, Lalla, L, Wolf, M (2005) Design of a Therapeutic Robotic Companion for Relational, Affective Touch. Proceedings of the International Workshop on Robots and Human Interactive Communication. pp. 408-415.

[8] Koo, S., Lim, J. G., and Kwon, D. (2008). Online touch behavior recognition of hard-cover robot using temporal decision tree classifier. In Proc. IEEE International Symposium on Robot and Human Interactive Communication, pp 425–429.

[9] Silvera Tawil, D., Rye, D., and Velonaki, M. (2011). Touch modality interpretation for an EIT-based sensitive skin. In Proc. IEEE International Conference on Robotics and Automation, pp. 3770–3776.

[10] Lee, M, Nicholls, H (1999) Tactile sensing for mechatronics: A state of the art survey. Mechatronics. Vol. 9. No. 1: 1-31.

[11] Dahiya, R S, Metta, G, Valle, M, Sandini, G (2010) Tactile Sensing: From Humans to Humanoids. IEEE Transactions on Robotics. 26 (1): 1-20.

[12] Ohmura, Y., Kuniyoshi, Y., and Nagakubo, A. (2006). Conformable and scalable tactile sensor skin for curved surfaces. In Proc. IEEE International Conference on Robotics and Automation, pp. 1348–1353.

[13] Um, D., Lumelsky, V. (1999). Fault tolerance via component redundancy for a modularized sensitive skin. In Proceedings IEEE International Conference on Robotics and Automation, v 1: 722-727

[14] Tajika, T., Miyashita, T., Ishiguro, H. et al. (2006). Automatic Categorization of Haptic Interactions-What are the Typical Haptic Interactions Between a Human and a Robot. In Proceeding of IEEE-RAS International Conference on Humanoid Robots, Humanoids 2006, Genova, 4-6 December 2006.

[15] Reston, R. and Kolesar, E. (1989). Pressure-sensitive field-effect transistor sensor array fabricated from a piezoelectric polyvinylidene fluoride film. In Proc. Annual

International Conference of the IEEE Engineering in Medicine and Biology Society, 3: 918–919

[16] Hosoda, K. (2003). Robot finger design for developmental tactile interaction: Anthropomorphic robotic soft fingertip with randomly distributed receptors. In Proc. Embodied Artificial Intelligence, 2865: 219–230.

[17] Tanaka, Y., Tanaka, M., and Chonan, S. (2006). Development of a sensor system for measuring tactile sensation. In Proc. IEEE Conference on Sensors, pp. 554–557.

[18] Dahiya, R., Valle, M., Metta, G., and Lorenzelli, L. (2007). POSFET tactile sensor arrays. In Proc. IEEE International Conference on Electronics, Circuits and Systems, pp. 1075–1078.

[19] Ishiguro, H. and Nishio, S. (2007). Building artificial humans to understand humans. Journal of Artificial Organs, 10(3):133–142.

[20] Takamuku, S., Gómez, G., Hosoda, K., and Pfeifer, R. (2007). Haptic discrimination of material properties by a robotic hand. In Proc. IEEE International Conference on Development and Learning, pp. 1–6.

[21] Sandini, G, Maggiali, M, Cannata, G, Metta, G (2007) Tactile sensor arrangement and corresponding sensory system. United States Patent No. 20100234997.

[22] E. R. Kandel, J. H. Schwartz, T. M. Jessell. Principles of Neural Science. McGraw-Hill/Appleton & Lange; 4th edition (January 5, 2000)

[23] Mannsfeld, S.C.B., Tee B.C-K., Stoltenberg R.M., Chen C.V.H-H, Barman S., Muir B.V.O., Sokolov A.N., Reese C. and Bao Z. (2010). Highly sensitive flexible pressure sensors with microstructured rubber dielectric layers, Nature Materials, 9: 859-864

[24] Ikeda, T (1996) Fundamentals of Piezoelectricity. Oxford Science Publications.

[25] Seminara L, Capurro, M, Cirillo, P, Cannata, G, Valle, M (2011) Electromechanical characterization of piezoelectric PVDF polymer films for tactile sensors in robotics applications. Sens. Actuators A. 169 (1): 49-58.

[26] Vinogradov, A M, Holloway F (1999) Electro-mechanical properties of the piezoelectric polymer PVDF. Ferroelectrics, 226: 169-181.

[27] Johnson, K.L. (1985). Description of Contact Mechanics and Boussinesq equation in elastic surface, in Contact Mechanics, Cambridge University Press, New York, NY.

[28] Selvadurai A.P.S. (2001). On Boussinesq's problem. International Journal of Engineering Science, 39: 317-322

[29] Albert, A (1972) Regression and the Moore-Penrose pseudoinverse. Academic Press Ink. New York.

[30] Pinna, L, Carlini, G, Seminara, L, Valle, M (2011) Interface electronics for tactile sensing arrays. Electronics, Circuits and Systems (ICECS), 2011. 18th IEEE International Conference on. pp. 468-471.

[31] Ravinder, R S, Valle, M (2012) Robotic Tactile Sensing – The Silicon way to System. Elsevier Publishing.

[32] Cannata, G, Maggiali, M, Metta, G, Sandini, G (2008) An embedded artificial skin for humanoid robots. IEEE International Conference on Multisensor Fusion and Integration for Intelligent Systems, 2008. MFI 2008. pp. 434-438.

Chapter 24

Piezomechanics in PZT Stack Actuators for Cryogenic Fuel Injectors

Yasuhide Shindo and Fumio Narita

Additional information is available at the end of the chapter

http://dx.doi.org/10.5772/50495

1. Introduction

Lead zirconate titanate (PZT) ceramics in special electronic devices such as structural health monitoring systems of liquid rocket engines and microvalves for space applications are subjected to cryogenic temperatures. PZT ceramics are also used in active fuel injectors under severe environments (Senousy et al. 2009a; 2009b). In the application of the PZT stack actuators to hydrogen fuel injectors, the actuators are operated under electric fields at cryogenic temperatures. Hence, it is important to understand the cryogenic electromechanical response of the PZT actuators under electric fields.

In this chapter, we address the present state of piezomechanics in PZT stack actuators for fuel injectors at cryogenic temperatures. First, we discuss the cryogenic response of PZT stack actuators under direct current (DC) electric fields (Shindo et al. 2011). A thermodynamic model is used to predict a monoclinic phase around a morphotropic phase boundary (MPB). A shift in the boundary between the tetragonal and rhombohedral/monoclinic phases with decreasing temperature is determined, and the temperature dependent piezoelectric coefficients are evaluated. Temperature dependent coercive electric field is also predicted based on the domain wall energy. A finite element analysis (FEA) is then performed, considering the shift in the MPB and polarization switching, to calculate the electromechanical fields of the PZT stack actuators from room to cryogenic temperatures. In addition, experimental results on the DC electric field induced strain, which verify the model, are presented. Next, we discuss the dynamic response of PZT stack actuators under alternating current (AC) electric fields at cryogenic temperatures (Shindo et al. 2012). Dynamic electromechanical fields of the PZT stack actuators from room to cryogenic temperatures are simulated by the FEA with MPB shift and domain wall motion effects. Dynamic strain measurements of the PZT stack actuators under AC electric fields are also presented, and a comparison is made between calculations and measurements to validate the predictions. Moreover, a parametric study using FEA is performed to

investigate the factors affecting the cryogenic response of PZT stack actuators and to provide a basis for selecting desirable design details.

2. Analysis

2.1. Basic equations

Consider the orthogonal coordinate system with axes x, y, and z. The Newton's second law (the equations of motion) and Gauss' law for piezoelectric materials are given by

$$\begin{aligned}
\sigma_{xx,x} + \sigma_{yx,y} + \sigma_{zx,z} &= \rho u_{x,tt} \\
\sigma_{xy,x} + \sigma_{yy,y} + \sigma_{zy,z} &= \rho u_{y,tt} \\
\sigma_{xz,x} + \sigma_{yz,y} + \sigma_{zz,z} &= \rho u_{z,tt}
\end{aligned} \tag{1}$$

$$D_{x,x} + D_{y,y} + D_{z,z} = 0 \tag{2}$$

where (σ_{xx}, σ_{yy}, σ_{zz}, $\sigma_{xy} = \sigma_{yx}$, $\sigma_{yz} = \sigma_{zy}$, $\sigma_{zx} = \sigma_{xz}$) and ($D_x$, D_y, D_z) are the components of stress tensor and electric displacement vector, (u_x, u_y, u_z) are the components of displacement vectors, ρ is the mass density, and a comma denotes partial differentiation with respect to the coordinates or the time t. Constitutive relations for PZT ceramics poled in the z-direction can be written as

$$\begin{aligned}
\begin{Bmatrix} \varepsilon_{xx} \\ \varepsilon_{yy} \\ \varepsilon_{zz} \\ \varepsilon_{yz} \\ \varepsilon_{zx} \\ \varepsilon_{xy} \end{Bmatrix} &= \begin{bmatrix} s_{11} & s_{12} & s_{13} & 0 & 0 & 0 \\ s_{12} & s_{11} & s_{13} & 0 & 0 & 0 \\ s_{13} & s_{13} & s_{33} & 0 & 0 & 0 \\ 0 & 0 & 0 & s_{44}/2 & 0 & 0 \\ 0 & 0 & 0 & 0 & s_{44}/2 & 0 \\ 0 & 0 & 0 & 0 & 0 & s_{66}/2 \end{bmatrix} \begin{Bmatrix} \sigma_{xx} \\ \sigma_{yy} \\ \sigma_{zz} \\ \sigma_{yz} \\ \sigma_{zx} \\ \sigma_{xy} \end{Bmatrix} \\
&+ \begin{bmatrix} 0 & 0 & \bar{d}_{31} \\ 0 & 0 & \bar{d}_{31} \\ 0 & 0 & \bar{d}_{33} \\ 0 & \bar{d}_{15}/2 & 0 \\ \bar{d}_{15}/2 & 0 & 0 \\ 0 & 0 & 0 \end{bmatrix} \begin{Bmatrix} E_x \\ E_y \\ E_z \end{Bmatrix} + \begin{Bmatrix} \varepsilon^r_{xx} \\ \varepsilon^r_{yy} \\ \varepsilon^r_{zz} \\ \varepsilon^r_{yz} \\ \varepsilon^r_{zx} \\ \varepsilon^r_{xy} \end{Bmatrix}
\end{aligned} \tag{3}$$

$$\begin{Bmatrix} D_x \\ D_y \\ D_z \end{Bmatrix} = \begin{bmatrix} 0 & 0 & 0 & 0 & \bar{d}_{15} & 0 \\ 0 & 0 & 0 & \bar{d}_{15} & 0 & 0 \\ \bar{d}_{31} & \bar{d}_{31} & \bar{d}_{33} & 0 & 0 & 0 \end{bmatrix} \begin{Bmatrix} \sigma_{xx} \\ \sigma_{yy} \\ \sigma_{zz} \\ \sigma_{yz} \\ \sigma_{zx} \\ \sigma_{xy} \end{Bmatrix} + \begin{bmatrix} \epsilon^T_{11} & 0 & 0 \\ 0 & \epsilon^T_{11} & 0 \\ 0 & 0 & \epsilon^T_{33} \end{bmatrix} \begin{Bmatrix} E_x \\ E_y \\ E_z \end{Bmatrix} + \begin{Bmatrix} P^r_x \\ P^r_y \\ P^r_z \end{Bmatrix} \tag{4}$$

where (ε_{xx}, ε_{yy}, ε_{zz}, $\varepsilon_{xy} = \varepsilon_{yx}$, $\varepsilon_{yz} = \varepsilon_{zy}$, $\varepsilon_{zx} = \varepsilon_{xz}$) and ($E_x$, E_y, E_z) are the components of strain tensor and electric field intensity vector, ($\varepsilon^r_{xx}, \varepsilon^r_{yy}, \varepsilon^r_{zz}, \varepsilon^r_{xy}, \varepsilon^r_{yz}, \varepsilon^r_{zx}$) and (P^r_x, P^r_y, P^r_z) are the remanent strain and polarization components, (s_{11}, s_{12}, s_{13}, s_{33}, s_{44}, s_{66} = $2(s_{11}- s_{12})$) are the elastic compliances, ($\bar{d}_{31}, \bar{d}_{33}, \bar{d}_{15}$) are the temperature dependent piezoelectric coefficients, and ($\in^{\mathrm{T}}_{11}, \in^{\mathrm{T}}_{33}$) are the dielectric constants. The quantities ε^r_{ij} and P^r_i are taken to be due entirely to polarization switching. The strain components are

$$\begin{aligned} &\varepsilon_{xx} = u_{x,x}, \quad \varepsilon_{yy} = u_{y,y}, \quad \varepsilon_{zz} = u_{z,z}, \\ &\varepsilon_{xy} = \frac{1}{2}\left(u_{x,y} + u_{y,x}\right), \quad \varepsilon_{yz} = \frac{1}{2}\left(u_{y,z} + u_{z,y}\right), \quad \varepsilon_{zx} = \frac{1}{2}\left(u_{z,x} + u_{x,z}\right) \end{aligned} \tag{5}$$

The electric field components are related to the electric potential ϕ by

$$E_x = -\phi_{,x}, \quad E_y = -\phi_{,y}, \quad E_z = -\phi_{,z} \tag{6}$$

2.2. Temperature dependent piezoelectric coefficient

Temperature dependent piezoelectric coefficient is outlined here. Figure 1 shows the phase diagram of PZT established in Jaffe et al. (1971) and Noheda et al. (2000). As the temperature T is lowered, PZT undergoes a paraelectric-to-ferroelectric phase transition, and the cubic unit cell is distorted depending on the mole fraction X of $PbTiO_3$. In the Zr-rich region, the paraelectric phase changes to orthorhombic phase. An intermediate monoclinic phase exists between the Zr-rich rhombohedral perovskite and Ti-rich tetragonal perovskite phases. Compositions between Zr/Ti ratios 90/10 and 65/35 reveals a ferroelectric-to-ferroelectric transition between rhombohedral space groups. This transition involves the oxygen octahedral tilt.

The MPB between the tetragonal and rhombohedral/monoclinic phases is the origin of the unusually high piezoelectric response of PZT, and this MPB is numerically predicted. For simplicity here, we ignore the octahedral tilt transition which differentiates the high temperature (HT) and low temperature (LT) rhombohedral phases, and the orthorhombic phase.

An energy function for the solid solution between the two end-members $PbTiO_3$ and $PbZrO_3$ is given by (Bell & Furman 2003)

$$\Delta G_{PZT} = XG_{PT}(p_i) + (1-X)G_{PZ}(q_i) + G_C(p_i, q_i) \tag{7}$$

where p_i and q_i ($i = 1,2,3$) denote the polarizations of the $PbTiO_3$ and $PbZrO_3$, respectively, and

$$\begin{aligned} G_{PT}(p_i) &= 3.74\times10^5(T-T_{PT})\left(p_1^2+p_2^2+p_3^2\right) - 7.9\times10^7\left(p_1^4+p_2^4+p_3^4\right) \\ &+7.5\times10^8\left(p_1^2p_2^2+p_2^2p_3^2+p_3^2p_1^2\right) + 2.61\times10^8\left(p_1^6+p_2^6+p_3^6\right) \\ &+6.3\times10^8\left\{p_1^4\left(p_2^2+p_3^2\right)+p_2^4\left(p_3^2+p_1^2\right)+p_3^4\left(p_1^2+p_2^2\right)\right\} - 3.66\times10^9 p_1^2p_2^2p_3^2 \end{aligned} \tag{8}$$

$$G_{PZ}(q_i) = 2.82\times10^5 (T - T_{PZ})(q_1^2 + q_2^2 + q_3^2) + 5.12\times10^8 (q_1^4 + q_2^4 + q_3^4)$$
$$-6.5\times10^8 (q_1^2 q_2^2 + q_2^2 q_3^2 + q_3^2 q_1^2) + 5.93\times10^8 (q_1^6 + q_2^6 + q_3^6) \tag{9}$$
$$+2\times10^9 \{q_1^4 (q_2^2 + q_3^2) + q_2^4 (q_3^2 + q_1^2) + q_3^4 (q_1^2 + q_2^2)\} - 9.5\times10^9 q_1^2 q_2^2 q_3^2$$

$$G_C(p_i, q_i) = \gamma_{200} (p_1^2 q_1^2 + p_2^2 q_2^2 + p_3^2 q_3^2) + \gamma_{220} \{p_1^2 (q_2^2 + q_3^2) + p_2^2 (q_3^2 + q_1^2) + p_3^2 (q_1^2 + q_2^2)\} \tag{10}$$

In Eqs. (8) - (10), G_{PT} and G_{PZ} are identical to a Landau-Devonshire potential (free energy of a ferroelectric crystal) up to sixth order, G_C represents the coupling energy, T_{PT} = 766 K and T_{PZ} = 503 K are the Curie temperatures of $PbTiO_3$ and $PbZrO_3$, respectively, and γ_{200} and γ_{220} are unknown coefficients.

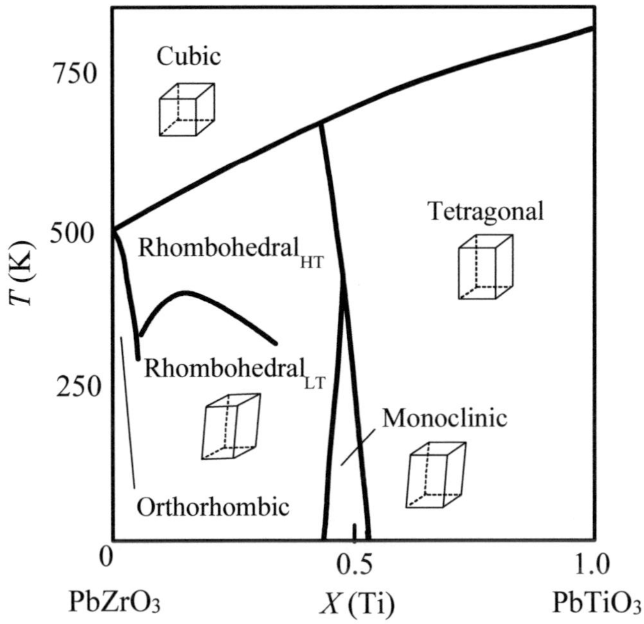

Figure 1. PZT phase diagram

The thermodynamic equilibrium state can be determined via minimization of ΔG_{PZT} with respect to p_i and q_i. For simplicity, only positive values of p_i and q_i are considered. For each temperature T and mole fraction X, the local minima in ΔG_{PZT} are systematically obtained for the following phases:

Cubic (C)

$$p_1 = p_2 = p_3 = 0, \quad q_1 = q_2 = q_3 = 0 \tag{11}$$

Tetragonal (T)

$$p_1 = p_2 = 0, p_3 \neq 0, \quad q_1 = q_2 = 0, q_3 \neq 0 \tag{12}$$

Rhombohedral (R)

$$p_1 = p_2 = p_3 \neq 0, \quad q_1 = q_2 = q_3 \neq 0 \tag{13}$$

Monoclinic (M)

$$p_1 = p_2 \neq 0, p_3 \neq 0, \quad q_1 = q_2 \neq 0, q_3 \neq 0 \tag{14}$$

The energies of the minima are then compared to define the stable state. In the simulation, $\gamma_{200} = 6 \times 10^8$ and $\gamma_{220} = 1.2 \times 10^8$ are assumed.

We now show a numerical example and comparison with experiments. Fig. 2 shows the

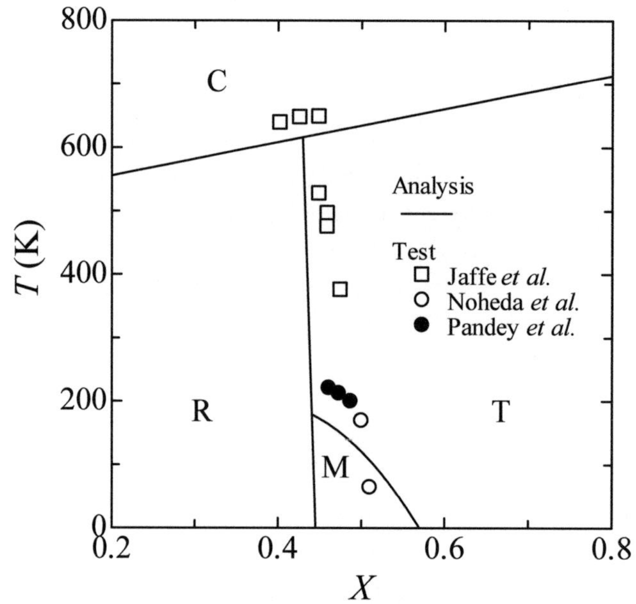

Figure 2. Calculated and experimental phase diagram of PZT

calculated PZT phase diagram. We also plot the experimental phase diagrams of PZT. The open square, open circle and solid circle are the results from Jeffe et al. (1971), Noheda et al. (2000) and Pandey et al. (2008). The simulation result reasonably agrees with the experimental data. It is noted that at room temperature, the MPB is located at about $X = 0.44$, and that there is an apparent shift of about 13 mol% in the MPB location on cooling to 0 K. In the work of Boucher et al. (2006), the piezoelectric coefficient d_{333} of PZT (Mn-doped) presented a significant decrease (about 85 %) due to a shift of 12 mol% of Zr at room temperature. Thus, from the consideration of Fig. 2, we propose the temperature dependent piezoelectric coefficient $\bar{d}_{ijk}$ ($\bar{d}_{311} = d_{31}$, $\bar{d}_{333} = d_{33}$, $\bar{d}_{131} = d_{15} / 2$ here) for $X = 0.44$, i.e.,

$$\bar{d}_{ijk} = \begin{cases} (1.7 \times 10^{-4} T + 0.95) d_{ijk} & 192 \leq T \\ (1.0 \times 10^{-5} T^2 + 2.4 \times 10^{-3} T + 0.15) d_{ijk} & 0 < T \leq 192 \end{cases} \tag{15}$$

where d_{ijk} ($d_{311}=d_{31}$, $d_{333}=d_{33}$, $d_{131}=d_{15}/2$ here) is the piezoelectric coefficient at 298 K. In Eqs. (3) and (4), reduced indices for the full notations of elastic compliances s_{ijkl} and temperature dependent piezoelectric coefficients $\bar{d}_{ijk}$ are used, with the following correspondence between the one and two indices: 1 = 11, 2 =22, 3 = 33, 4 = 23, 5 = 31, 6 = 12. We can also predict the piezoelectric coefficient for other mole fractions. For example, the temperature dependent piezoelectric coefficient for $X = 0.56$ can be expressed as

$$\bar{d}_{ijk} = \begin{cases} (-1.7\times10^{-4}T + 0.25)d_{ijk} & 192 \leq T \\ (-1.0\times10^{-5}T^2 - 2.4\times10^{-3}T + 1.1)d_{ijk} & 20 < T \leq 192 \\ (1.0\times10^{-5}T^2 + 2.4\times10^{-3}T + 0.95)d_{ijk} & 0 < T \leq 20 \end{cases} \tag{16}$$

2.3. Polarization switching

High electromechanical fields lead to the polarization switching. We assume that the direction of a spontaneous polarization P^s of each grain can change by 180° or 90° for ferroelectric switching induced by a sufficiently large electric field opposite to the poling direction. The 90° ferroelastic domain switching is also induced by a sufficiently large stress field. The criterion states that a polarization switches when the electrical and mechanical work exceeds a critical value (Hwang et al. 1995)

$$\begin{aligned} &\sigma_{xx}\Delta\varepsilon_{xx} + \sigma_{yy}\Delta\varepsilon_{yy} + \sigma_{zz}\Delta\varepsilon_{zz} + 2\sigma_{xy}\Delta\varepsilon_{xy} + 2\sigma_{yz}\Delta\varepsilon_{yz} + 2\sigma_{zx}\Delta\varepsilon_{zx} \\ &+ E_x\Delta P_x + E_y\Delta P_y + E_z\Delta P_z \geq 2P^s\bar{E}_c \end{aligned} \tag{17}$$

where $\bar{E}_c$ is a temperature dependent coercive electric field, and $\Delta\varepsilon_{ij}$ and ΔP_i are the changes in the spontaneous strain and polarization during switching, respectively. The values of $\Delta\varepsilon_{ij} = \varepsilon^r_{ij}$ and $\Delta P_i = P^r_i$ for 180° switching can be expressed as

$$\begin{aligned} &\Delta\varepsilon_{xx} = 0, \quad \Delta\varepsilon_{yy} = 0, \quad \Delta\varepsilon_{zz} = 0, \quad \Delta\varepsilon_{xy} = 0, \quad \Delta\varepsilon_{yz} = 0, \quad \Delta\varepsilon_{zx} = 0, \\ &\Delta P_x = 0, \quad \Delta P_y = 0, \quad \Delta P_z = -2P^s \end{aligned} \tag{18}$$

For 90° switching in the zx plane, the changes are

$$\begin{aligned} &\Delta\varepsilon_{xx} = \gamma^s, \quad \Delta\varepsilon_{yy} = 0, \quad \Delta\varepsilon_{zz} = -\gamma^s, \quad \Delta\varepsilon_{xy} = 0, \quad \Delta\varepsilon_{yz} = 0, \quad \Delta\varepsilon_{zx} = 0, \\ &\Delta P_x = \pm P^s, \quad \Delta P_y = 0, \quad \Delta P_z = -P^s \end{aligned} \tag{19}$$

where γ^s is a spontaneous strain. For 90° switching in the yz plane, we have

$$\begin{aligned} &\Delta\varepsilon_{xx} = 0, \quad \Delta\varepsilon_{yy} = \gamma^s, \quad \Delta\varepsilon_{zz} = -\gamma^s, \quad \Delta\varepsilon_{xy} = 0, \quad \Delta\varepsilon_{yz} = 0, \quad \Delta\varepsilon_{zx} = 0, \\ &\Delta P_x = 0, \quad \Delta P_y = \pm P^s, \quad \Delta P_z = -P^s \end{aligned} \tag{20}$$

The constitutive equations after polarization switching are

$$\begin{Bmatrix} \varepsilon_{xx} \\ \varepsilon_{yy} \\ \varepsilon_{zz} \\ \varepsilon_{yz} \\ \varepsilon_{zx} \\ \varepsilon_{xy} \end{Bmatrix} = \begin{bmatrix} s_{11} & s_{12} & s_{13} & 0 & 0 & 0 \\ s_{12} & s_{11} & s_{13} & 0 & 0 & 0 \\ s_{13} & s_{13} & s_{33} & 0 & 0 & 0 \\ 0 & 0 & 0 & s_{44}/2 & 0 & 0 \\ 0 & 0 & 0 & 0 & s_{44}/2 & 0 \\ 0 & 0 & 0 & 0 & 0 & s_{66}/2 \end{bmatrix} \begin{Bmatrix} \sigma_{xx} \\ \sigma_{yy} \\ \sigma_{zz} \\ \sigma_{yz} \\ \sigma_{zx} \\ \sigma_{xy} \end{Bmatrix}$$
$$+ \begin{bmatrix} \bar{d}^*_{111} & \bar{d}^*_{211} & \bar{d}^*_{311} \\ \bar{d}^*_{122} & \bar{d}^*_{222} & \bar{d}^*_{322} \\ \bar{d}^*_{133} & \bar{d}^*_{233} & \bar{d}^*_{333} \\ \bar{d}^*_{123} & \bar{d}^*_{223} & \bar{d}^*_{323} \\ \bar{d}^*_{131} & \bar{d}^*_{231} & \bar{d}^*_{331} \\ \bar{d}^*_{112} & \bar{d}^*_{212} & \bar{d}^*_{312} \end{bmatrix} \begin{Bmatrix} E_x \\ E_y \\ E_z \end{Bmatrix} + \begin{Bmatrix} \varepsilon^r_{xx} \\ \varepsilon^r_{yy} \\ \varepsilon^r_{zz} \\ \varepsilon^r_{yz} \\ \varepsilon^r_{zx} \\ \varepsilon^r_{xy} \end{Bmatrix} \tag{21}$$

$$\begin{Bmatrix} D_x \\ D_y \\ D_z \end{Bmatrix} = \begin{bmatrix} \bar{d}^*_{111} & \bar{d}^*_{122} & \bar{d}^*_{133} & \bar{d}^*_{123} & \bar{d}^*_{131} & \bar{d}^*_{112} \\ \bar{d}^*_{211} & \bar{d}^*_{222} & \bar{d}^*_{233} & \bar{d}^*_{223} & \bar{d}^*_{231} & \bar{d}^*_{212} \\ \bar{d}^*_{311} & \bar{d}^*_{322} & \bar{d}^*_{333} & \bar{d}^*_{323} & \bar{d}^*_{331} & \bar{d}^*_{312} \end{bmatrix} \begin{Bmatrix} \sigma_{xx} \\ \sigma_{yy} \\ \sigma_{zz} \\ \sigma_{yz} \\ \sigma_{zx} \\ \sigma_{xy} \end{Bmatrix} + \begin{bmatrix} \epsilon^T_{11} & 0 & 0 \\ 0 & \epsilon^T_{11} & 0 \\ 0 & 0 & \epsilon^T_{33} \end{bmatrix} \begin{Bmatrix} E_x \\ E_y \\ E_z \end{Bmatrix} + \begin{Bmatrix} P^r_x \\ P^r_y \\ P^r_z \end{Bmatrix} \tag{22}$$

where

$$\bar{d}^*_{ijk} = \{\bar{d}_{33} n_i n_k n_l + \bar{d}_{31}(n_i \delta_{kl} - n_i n_k n_l) + \frac{1}{2}\bar{d}_{15}(\delta_{ik} n_l - 2 n_i n_k n_l + \delta_{il} n_k)\} \tag{23}$$

In Eq. (23), n_i is the unit vector in the poling direction and δ_{ij} is the Kronecker delta.

2.4. Domain wall motion

A domain wall displacement causes changes of strain and polarization (Cao et al. 1999). For simplicity, here the applied AC electric field $E_z = E_0 \exp(i\omega t)$ is parallel to the direction of spontaneous polarization P^s in one of the domains (see Fig. 3); E_0 is the AC electric field amplitude and ω is angular frequency($=2\pi f$ where f is frequency in Hertz). The changes of the strains and polarization due to the domain wall displacement Δl (Arlt and Dederichs, 1980) can be written as

$$\begin{aligned} \Delta\varepsilon^*_{xx} &= \Delta s_{11}\sigma_{xx} + \Delta s_{13}\sigma_{zz} + \Delta d_{311}E_z \\ \Delta\varepsilon^*_{zz} &= \Delta s_{13}\sigma_{xx} + \Delta s_{33}\sigma_{zz} + \Delta d_{333}E_z \\ \Delta P^*_z &= \Delta d_{311}\sigma_{xx} + \Delta d_{333}\sigma_{zz} + \Delta\epsilon^{T}_{33} E_z \end{aligned} \quad (24)$$

where all terms with Δ are the contributions from the domain wall motion, and

$$\begin{aligned} \Delta s_{11} &= \frac{(\gamma^s)^2}{2lf_D}, \quad \Delta s_{13} = -\frac{(\gamma^s)^2}{2lf_D}, \quad \Delta s_{33} = \frac{(\gamma^s)^2}{2lf_D}, \\ \Delta d_{311} &= -\frac{\gamma^s P^s}{2lf_D}, \quad \Delta d_{333} = \frac{\gamma^s P^s}{2lf_D}, \quad \Delta\epsilon^{T}_{33} = \frac{(P^s)^2}{2lf_D} \end{aligned} \quad (25)$$

In Eq. (25), l is the domain width and f_D is the force constant for the domain wall motion process. The strains ε_{xx}, ε_{zz} and electric displacement D_z are given by

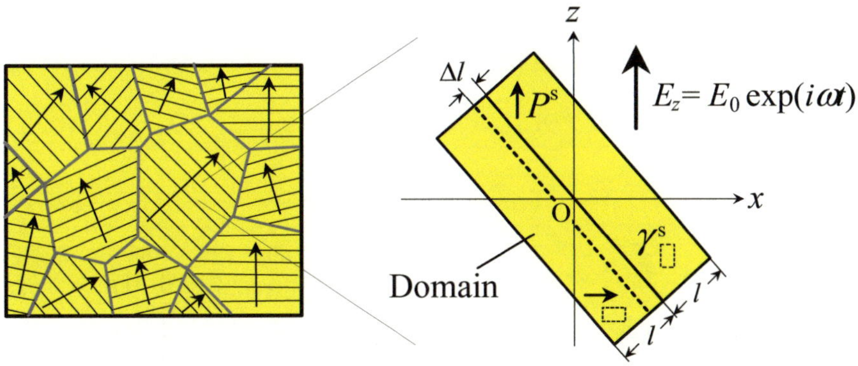

Figure 3. Schematic drawing of many grains which in turn consist of domains and basic unit of a piezoelectric crystallite with a domain wall.

$$\begin{aligned} \varepsilon_{xx} &= s^*_{11}\sigma_{xx} + s_{12}\sigma_{yy} + s^*_{13}\sigma_{zz} + d^*_{31}E_z \\ \varepsilon_{zz} &= s^*_{13}\sigma_{xx} + s^*_{13}\sigma_{yy} + s^*_{33}\sigma_{zz} + d^*_{31}E_z \\ D_z &= d^*_{31}\sigma_{xx} + d^*_{31}\sigma_{yy} + d^*_{33}\sigma_{zz} + \epsilon^{T*}_{33} E_z \end{aligned} \quad (26)$$

where

$$\begin{aligned} s^*_{11} &= s_{11} + \Delta s_{11}, \quad s^*_{13} = s_{13} + \Delta s_{13}, \quad s^*_{33} = s_{33} + \Delta s_{33} \\ d^*_{31} &= \bar{d}_{31} + \Delta d_{311}, \quad d^*_{33} = \bar{d}_{33} + \Delta d_{333}, \\ \epsilon^{T*}_{33} &= \epsilon^{T}_{33} + \Delta\epsilon^{T}_{33} \end{aligned} \quad (27)$$

Experimental studies on PZT ceramics have shown that 45-70% of dielectric and piezoelectric moduli values may originate from the extrinsic contributions (Luchaninov et al. 1989, Cao et al. 1991). The extrinsic dielectric constant $\Delta\epsilon^{T}_{33}$ is approximately estimated as the two thirds of the bulk properties (Li et al. 1993). Here, the following equation (Narita

et al. 2005) is utilized to describe $\Delta\epsilon_{33}^{T}$ in terms of AC electric field amplitude E_0 and temperature dependent coercive electric field $\overline{E}_c$:

$$\Delta\epsilon_{33}^{T}=\epsilon_{33}^{T}\frac{2E_0}{3\overline{E}_c} \tag{28}$$

By substituting Eq. (28) into the sixth of Eq. (25), $l_{fD} = 3(P^s)^2 \overline{E}_c /(4\epsilon_{33}^{T} E_0)$ is obtained. By eliminating l_{fD}, the changes in the elastic compliances and piezoelectric coefficients in Eq. (25) can be rewritten in terms of AC electric field amplitude etc.

2.5. Finite element model

Consider a PZT stack actuator with 300 PZT layers of width W_p = 5.2 mm and thickness h_p = 0.1 mm, thin electrodes, and elastic coating layer of thickness h_e = 0.5 mm as shown in Fig. 4. A rectangular Cartesian coordinate system O-xyz is used and the origin of the coordinate system coincides with the center of the stack actuator. Each PZT layer is sandwiched between thin electrodes. An external electrode is attached on both sides of the actuator to address the internal electrodes.

In order to discuss the electromechanical fields near the internal electrode, the problem of the stack actuator is solved using the unit cell model (two layer piezoelectric composite with $|x| \leq W_p/2 + h_e$, $|y| \leq W_p/2 + h_e$, $0 \leq z \leq 2h_p$) shown in Fig. 5. Electrodes of length a and width W_p = 5.2 mm are attached to the PZT layer of thickness h_p = 0.1 mm, and a W_p - a tab region exists on both sides of the layer. Because of the geometric and loading symmetry, only the half needs to be analyzed. The electrode layers are not incorporated into the model.

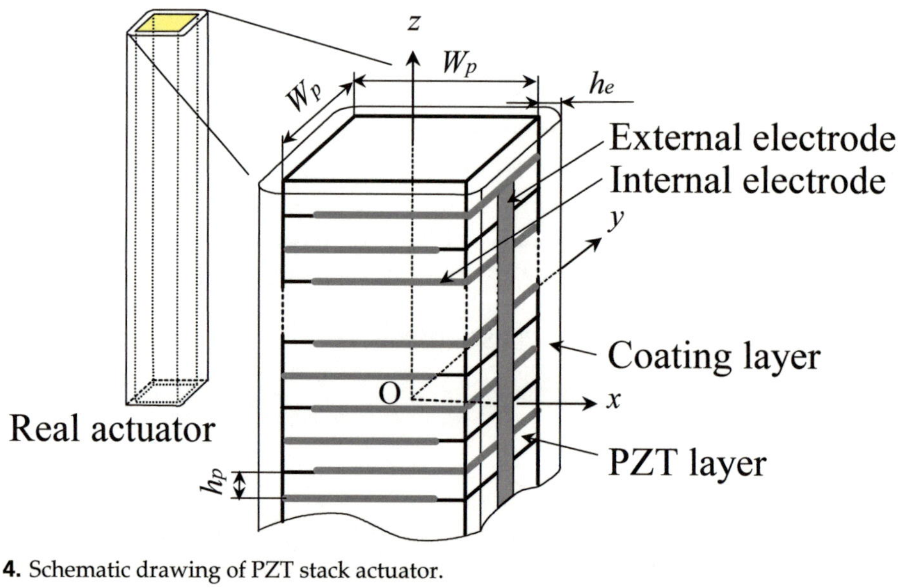

Figure 4. Schematic drawing of PZT stack actuator.

First, we consider the PZT stack actuator under DC electric fields. The electric potential on two electrode surfaces ($-W_p/2 \le x \le -W_p/2+a$, $0 \le y \le W_p/2$, $z = 0, 2h_p$) equals the applied voltage, $\phi = V_0$. The electrode surface ($W_p/2-a \le x \le W_p/2$, $0 \le y \le W_p/2$, $z = h_p$) is connected to the ground, so that $\phi = 0$. The mechanical boundary conditions include the traction-free conditions on the coating layer surfaces at $x = \pm(W_p/2 + h_e)$, $y = W_p/2 + h_e$ and the symmetry conditions on the xz plane at $y = 0$ and xy planes at $z = 0, 2h_p$. In addition, the origin is constrained against the displacement in the x-direction, to avoid rigid body motion. We next consider the PZT stack actuator under AC electric fields. The electric potential on two electrode surfaces ($-W_p/2 \le x \le -W_p/2+a$, $0 \le y \le W_p/2$, $z = 0, 2h_p$) equals the applied voltage, $\phi = V_0\exp(i\omega t)$, and the electrode surface ($W_p/2-a \le x \le W_p/2$, $0 \le y \le W_p/2$, $z = h_p$) is connected to the ground.

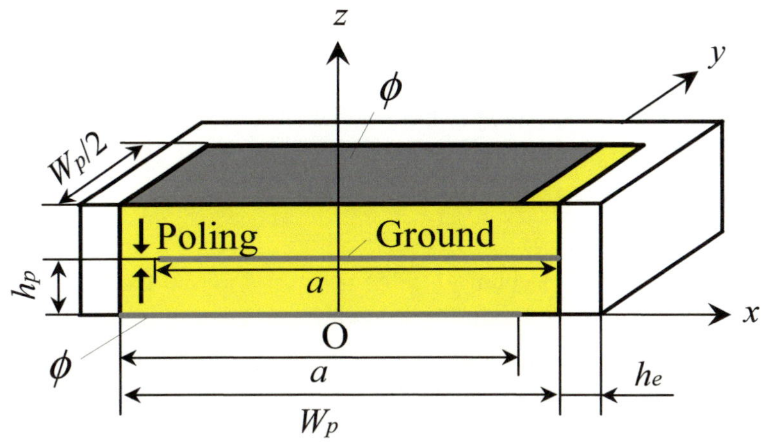

Figure 5. Unit cell of the PZT stack actuator.

Each element in ANSYS is defined by eight-node 3-D coupled field solid for the PZT layers and eight-node 3-D structural solid for the coating layer. $P^s = 0.3$ C/m^2 and $\gamma^s = 0.004$ are used. In order to calculate the electromechanical fields, we need the temperature dependent coercive electric field $\bar{E}_c$. First-principles free energy calculations for ferroelectric perovskites (Kumar and Waghmare 2010) have shown that the domain wall energy increases linearly about 50% as the temperature T decreases from room temperature to 260 K. Since higher domain wall energy leads to higher coercive electric field, the temperature dependent coercive electric field is assumed to be the following equation:

$$\bar{E}_c = (4.84 - 0.0129T)E_c \tag{29}$$

where E_c is a coercive electric field at 298 K.

3. Experiment

The stack actuator is fabricated using 300 soft PZT N-10 layers (NEC/Tokin Co. Ltd., Japan) of width $W_p = 5.2$ mm and thickness $h_p = 0.1$ mm (see Fig. 6). A rectangular Cartesian

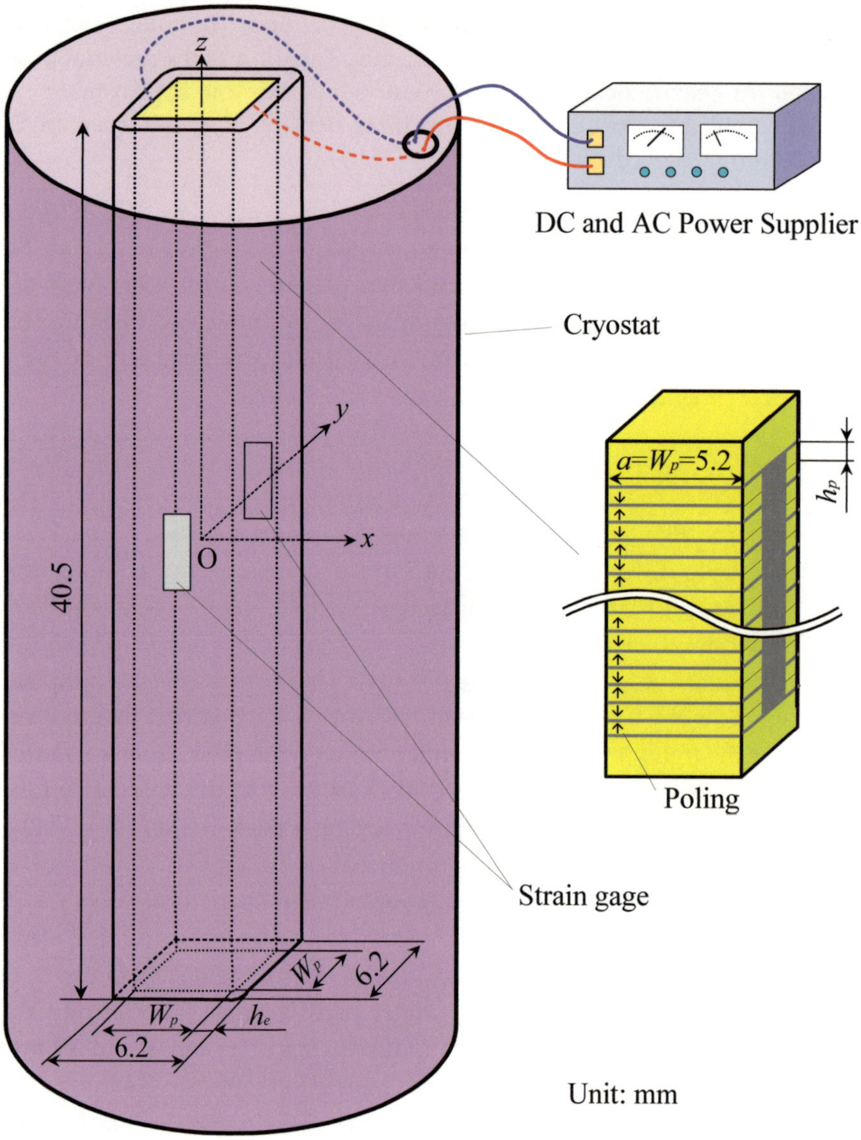

Figure 6. Experimental setup.

Elastic compliance ($\times10^{-12}$ m²/N)					Piezoelectric coefficient ($\times10^{-12}$ m/V)			Dielectric constant ($\times10^{-10}$ C/Vm)		Density (kg/m³)
s_{11}	s_{33}	s_{44}	s_{12}	s_{13}	d_{31}	d_{33}	d_{15}	ϵ_{11}^{T}	ϵ_{33}^{T}	ρ
14.8	18.1	44.9	-5.1	-5.8	-287	635	930	443	481	8000

Table 1. Material properties of N-10.

coordinate system O-xyz is used and the origin of the coordinate system coincides with the center of the actuator. The electrode length is $a = W_p = 5.2$ mm (full electrodes), and the

actuator is coated with epoxy layer of thickness h_e = 0.5 mm. The total dimensions of the specimen are width of 6.2 mm and length of 40.5 mm. Table 1 lists the material properties of N-10. The coercive electric field of N-10 at room temperature is approximately E_c = 0.36 MV/m. Young's modulus, Poisson's ratio and mass density of epoxy layer are 3.35 GPa, 0.214 and 1100 kg/m^3, respectively.

The actuator is bonded to the test rig of a SUS304 stainless steel plate using epoxy bond, and DC voltage (0 Hz) and AC voltage (400 Hz) are applied using a power supply. Two strain gages are attached at the center of the $y = \pm 3.1$ mm planes, and the magnitude of strain is measured. To control the temperature of the actuator, an automated helium refill system (TRG-300, Taiyo Toyo Sanso Co. Ltd., Japan) is used. For the reliability of the test, two specimens are experimented, and four strain values are obtained.

4. Results and discussion

We first consider the PZT stack actuators under DC electric fields. Figure 7 shows the predicted normal strain ε_{zz} versus temperature T at x = 0 mm, $y = W_p/2 + h_e$ = 3.1 mm and z = 0 mm of the PZT stack actuators for mole fraction X = 0.44 and 0.56 with $a = W_p$ = 5.2 mm (full electrodes) under DC electric field $E_0 = V_0/h_p$ = 0.1 MV/m (f = 0 Hz). Also shown are the measured data (average of four values) for X = 0.44. The electric field induced stain of the PZT stack actuator for X = 0.44 decreases with decreasing temperature due to a shift in the MPB. We see that the trend is sufficiently similar between the prediction and measurement. On the other hand, the electric field induced stain of the PZT stack actuator for X = 0.56 tends to increase with decreasing temperature reaching a peak at about T = 20 K and then decrease in magnitude. Figure 8 shows the predicted normal strain ε_{zz} versus DC electric field E_0 at x = 0 mm, y = 3.1 mm and z = 0 mm (strain measurement location) of the PZT stack actuators for X = 0.56 with a = 5.0 mm (partial electrodes) and $a = W_p$ = 5.2 (full electrodes) at liquid hydrogen temperature (20 K). Strain versus electric field curves show nonlinear behavior due to the polarization switching under high negative DC electric fields. Small difference is observed in the DC electric field induced strains for a = 5.0 and 5.2 mm. Figure 9 shows the polarization switching zones near the electrode tip (x = 2.4 mm) at y = 0 mm plane of the PZT layer for X = 0.56 with a = 5.0 mm (partial electrodes) at 20 K. The coercive electric field at 20 K is about 1.65 MV/m. As the negative DC electric field increases, the area of the switched region grows.

We next consider the PZT stack actuators under AC electric fields. Figure 10 shows the normal strain ε_{zz} versus electric field amplitude E_0 at x = 0 mm, y = 3.1 mm and z = 0 mm of the PZT stack actuators for X = 0.44 and 0.56 at frequency f = 400 Hz and T = 20 K. The dashed line represents the strain computed by the FEA without domain wall motion effect, and the solid line represents the strain after the domain wall motion effect has been applied. The open circle denotes the experimental data. As AC electric field increases, the strain increases gradually away from the linear curve. This is due to the domain wall motion under the influence of AC electric fields. It can be seen that agreement between analysis

with domain wall motion effect and experiment is fair. Figure 11 shows the distribution of the normal component of stress σ_{zz} as a function of x at y = 0 mm and z = 0.025 mm of the PZT stack actuators for X = 0.44 and 0.56 with a = 5.0 mm (partial electrodes) and $a = W_p$ = 5.2 mm (full electrodes) under E_0 = 1.65 MV/m at f = 400 Hz and 20 K. The coercive electric field at 20 K is about 1.65 MV/m. In the case of the PZT stack actuator with full electrodes, small difference is observed for X = 0.44 and 0.56. For the PZT stack actuators with partial electrodes, a high normal stress occurs near the electrode tip (x = 2.4 mm) as is expected. Figure 12 shows the electric field distribution E_z as a function of x at y = 0 mm and z = 0.025 mm for the PZT stack actuators with a = 5.0 mm (partial electrodes) and a = 5.2 mm (full electrodes) under E_0 = 1.65 MV/m at f = 400 Hz and 20 K. For the PZT stack actuators with partial electrodes, a high electric field is observed near the electrode tip.

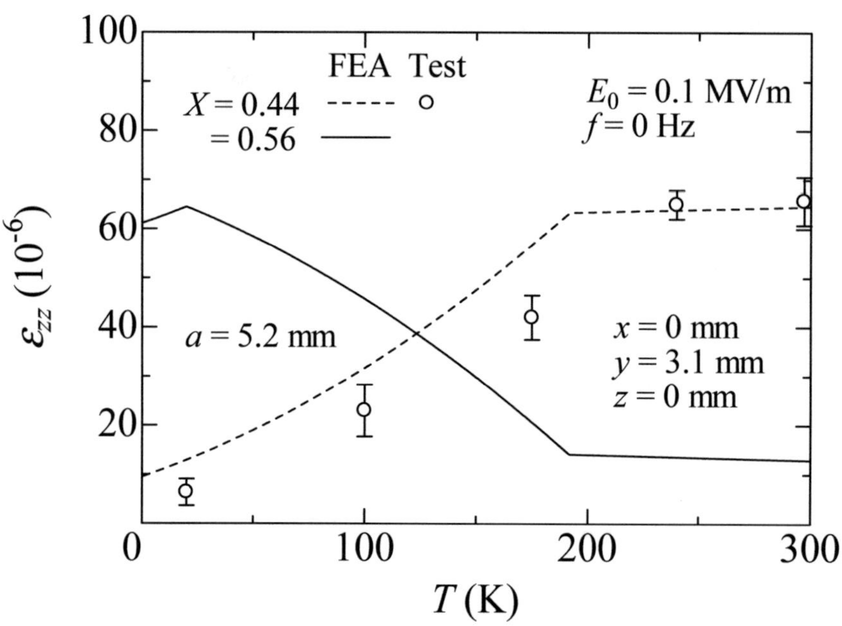

Figure 7. Strain vs temperature of PZT stack actuators for X = 0.44 and 0.56 under DC electric field E_0 = 0.1 MV/m.

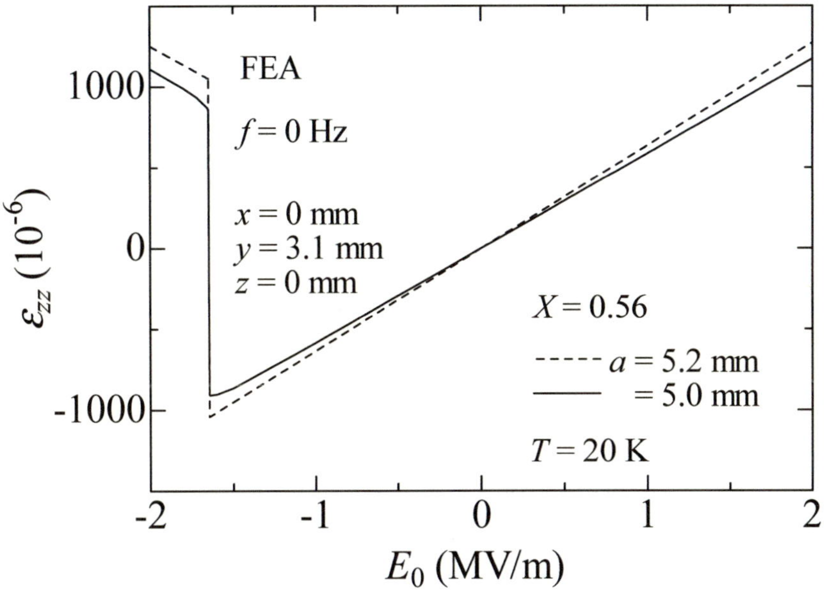

Figure 8. Strain vs DC electric field of PZT stack actuators for X = 0.56 at 20 K.

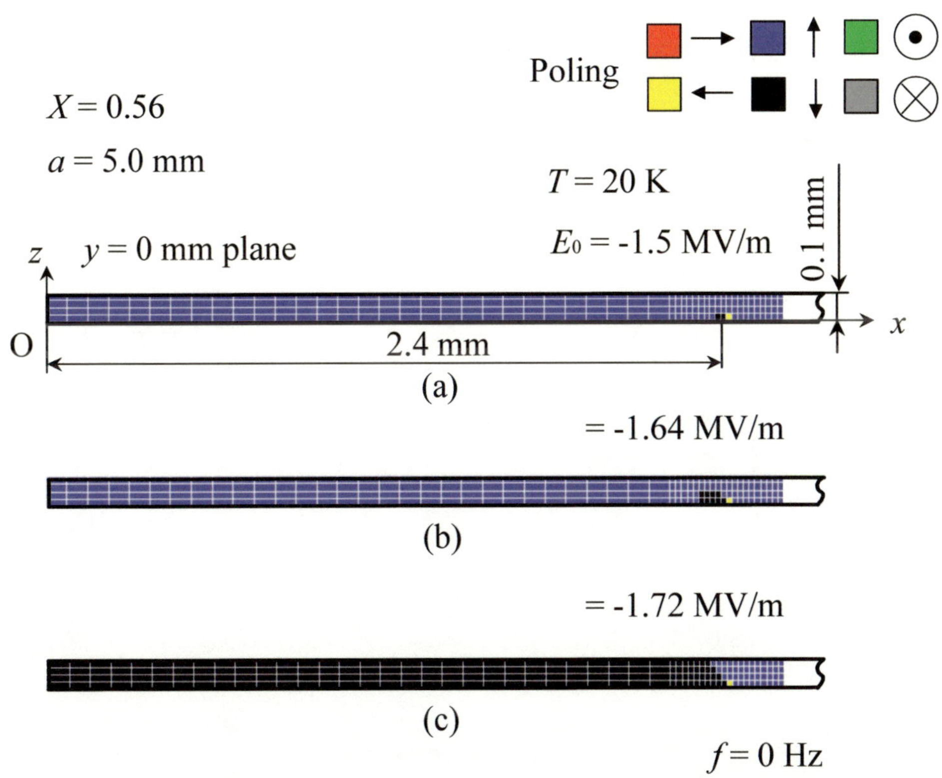

Figure 9. Polarization switching zones at $y = 0$ mm plane of PZT layer for $X = 0.56$ at 20 K, induced by DC electric field (a) $E_0 = -1.5$ MV/m, (b) $E_0 = -1.64$ MV/m and (c) $E_0 = -1.72$ MV/m.

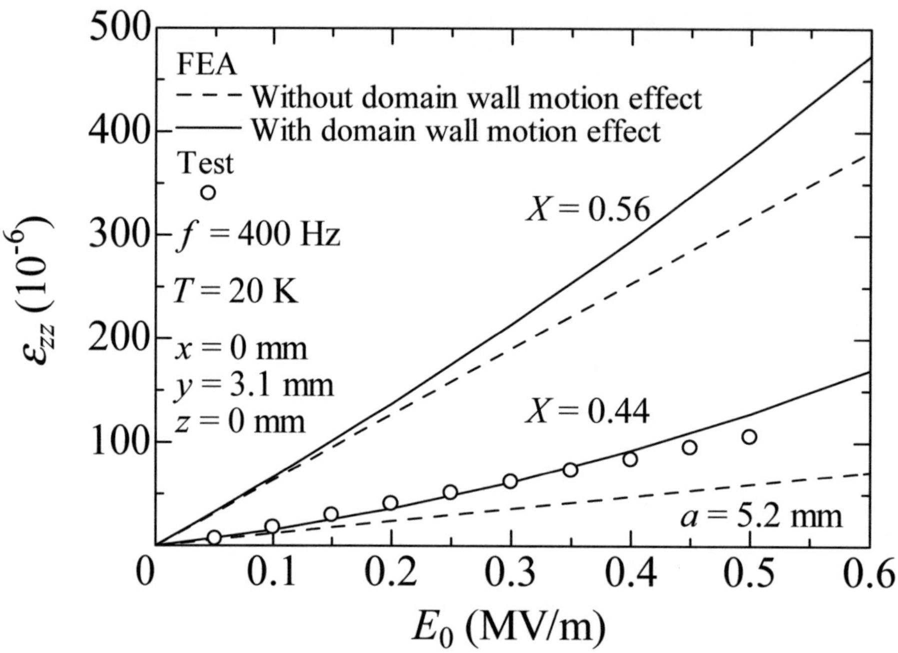

Figure 10. Strain vs AC electric field of PZT stack actuators for X = 0.44 and 0.56 at 20 K.

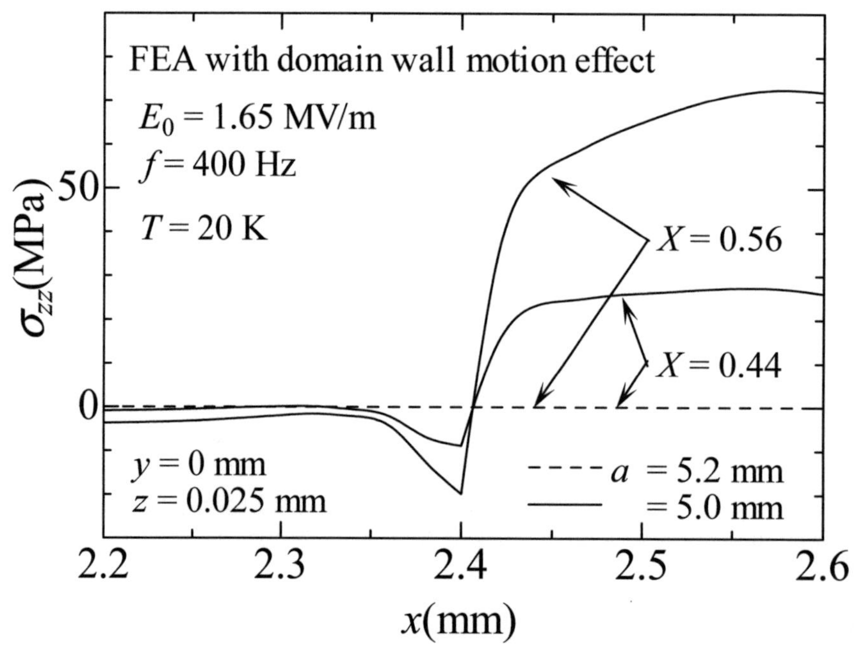

Figure 11. Variation of normal stress σ_{zz} as a function of x at $y = 0$ mm and $z = 0.025$ mm for PZT stack actuators under AC electric field $E_0 = 1.65$ MV/m at $T = 20$ K.

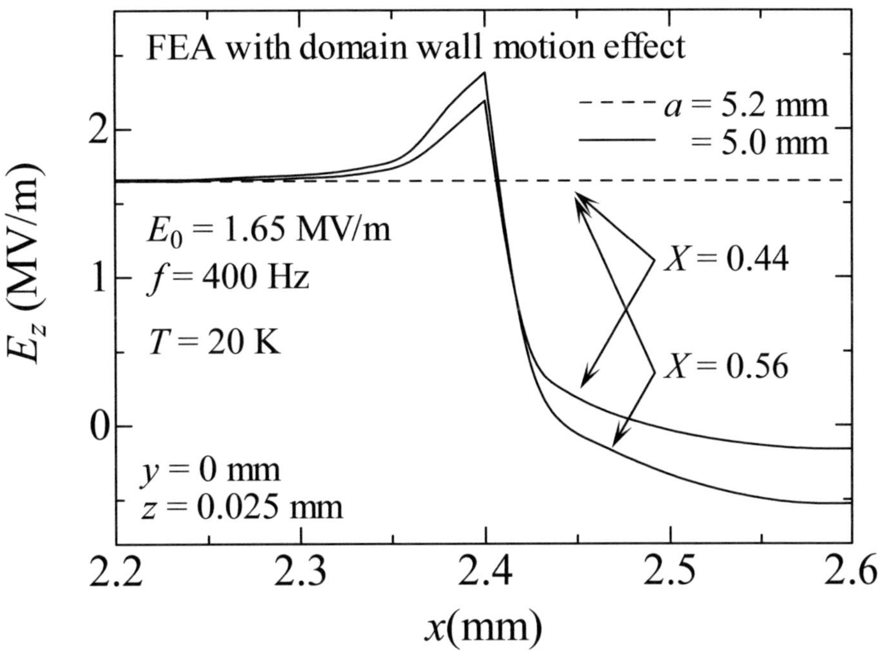

Figure 12. Variation of electric field E_z as a function of x at $y = 0$ mm and $z = 0.025$ mm for PZT stack actuators under $E_0 = 1.65$ MV/m at $T = 20$ K.

5. Conclusions

Numerical and experimental examination on the electromechanical response of PZT stack actuators at cryogenic temperatures is reported. It is found that the electric field induced strain decreases or increases with decreasing temperature depending on the mole fraction. That is, in the case of high performance PZTs for X = 0.44, the electric field induced strain is very high at room temperature, whereas in the case of PZTs for X = 0.56, the electric field induced strain at cryogenic temperatures will seem to be higher than at room temperature. It is also shown that the stress and electric field in the PZT layers are very high near the electrode tip for the PZT stack actuators with partial electrodes, although the electric field induced strains at the center of the surface for the partially and fully electroded PZT stack actuators have the same level. This study may be useful in designing high performance hydrogen fuel injectors.

Author details

Yasuhide Shindo and Fumio Narita
Tohoku University, Japan

6. References

Arlt, G. & Dederichs, H. (1980). Complex elastic, dielectric and piezoelectric constants by domain wall damping in ferroelectric ceramics, *Ferroelectrics*, Vol. 19: 47-50.

Bell, A. J. & Furman, E. (2003). A two-parameter thermodynamic model for PZT, *Ferroelectrics* Vol. 293: 19-31.

Boucher, E., Guiffard, B., Lebrum, L. & Guyomar, D. (2006). Effects of Zr/Ti ratio on structural, dielectric and piezoelectric properties of Mn- and (Mn, F)-doped lead zirconate titanate ceramics, *Ceramics International* Vol. 32 (No. 5): 479-485.

Cao, S., Li, W. and Cross, L. E. (1991). The extrinsic nature of nonlinear behavior observed in lead zirconate titanate ferroelectric ceramic, *Journal of Applied Physics*, Vol. 69: 7219-7224.

Hwang, S.C., Lynch, C.S. & McMeeking, R.M. (1995). Ferroelectric/ferroelastic interactions and a polarization switching model, *Acta Metallurgica et Materialia*, Vol. 43 (No. 5): 2073-2084.

Jaffe, B., Cook, W. R. & Jahhe, H. (1971). *Piezoelectric Ceramics*, Academic, New York.

Kumar, A. & Waghmare, U. V. (2010). First-principles free energies and Ginzburg-Landau theory of domains and ferroelectric phase transitions in $BaTiO_3$, *Physical Review B*, Vol. 82: 054117.

Li, S., Bhalla, A. S., Newnham, R. E. & Cross, L. E. (1993). Quantitative evaluation of extrinsic contribution to piezoelectric coefficient d_{33} in ferroelectric PZT ceramics, *Materials Letter*, Vol. 17: 21-26.

Luchaninov, A. G., Shil'nikov, A. V., Shuvalov, L. A. & Shipkova, I. J. U. (1989). The domain processes and piezoeffect in polycrystalline ferroelectrics, *Ferroelectrics*, Vol. 98: 123-126.

Narita, F., Shindo, Y. & Mikami, M. (2005). Analytical and experimental study of nonlinear bending response and domain wall motion in piezoelectric laminated actuators under AC electric fields, *Acta Materialia*, Vol. 53 (No. 17): 4523-4529.

Noheda, B., Cox, D. E., Shirane, G., Guo, R., Jones, B. & Cross, L. E. (2000). Stability of the monoclinic phase in the ferroelectric perovskite $PbZr_{1-x}Ti_xO_3$, *Physical Review B* Vol. 63 (No. 1): 014103.

Pandey, D., Singh, A. K. & Baik, S. (2008). Stability of ferroic phases in the highly piezoelectric $Pb(Zr_xTi_{1-x})O_3$ ceramics, *Acta Crystallographica Section A* Vol. 64: 192-203.

Senousy, M. S., Li, F. X., Mumford, D., Gadala, M. & Rajapakse, R. K. N. D. (2009a). Thermo-electro-mechanical performance of piezoelectric stack actuators for fuel injector applications, *Journal of Intelligent Material Systems and Structures* Vol. 20: 387-399.

Senousy, M. S., Rajapakse, R. K. N. D., Mumford D. & Gadala, M. S. (2009b). Self-heat generation in piezoelectric stack actuators used in fuel injectors, *Smart Materials and Structures* Vol. 18: 045008.

Shindo, Y., Narita, F. & Sasakura, T. (2011). Cryogenic electromechanical behavior of multilayer piezo-actuators for fuel injector applications, *Journal of Applied Physics* Vol. 110 (No. 8): 084510.

Shindo, Y., Sasakura, T. & Narita, F. (2012). Dynamic electromechanical response of multilayered piezoelectric composites from room to cryogenic temperatures for fuel injector applications, *ASME Journal of Engineering Materials and Technology*, in press.

Chapter 25

Smart Actuation for Helicopter Rotorblades

A. Paternoster, R. Loendersloot, A. de Boer and R. Akkerman

Additional information is available at the end of the chapter

http://dx.doi.org/10.5772/51438

1. Introduction

The first helicopter to successfully achieve stable hovering flight and decent forward flight performances was demonstrated in 1935 and is attributed to Louis Breguet and René Dorand [22] as shown in Figure 1. Their patent details the two coaxial counter-rotating blades which resulted in an unprecedented level of performance, stability and safety for a rotorcraft [7]. This success was soon matched by other helicopter pioneers. Henrich Focke at the Focke Wulf Company and Juan de la Cierva within the Weir company demonstrated the hovering capabilities of a side by side rotor configuration in 1936 and 1938 respectivelly [3, 22, 47]. In 1940 Sikorsky flew a single rotor helicopter configuration with three auxiliary tail rotors to negate the counter-torque effect [22, 51] as shown in Figure 2. Sikorsky refined his design and produced a significant number of helicopters during the war, some of which were used during World War II in the Pacific [22]. After the war, this configuration was widely adopted by the emerging industry. Today almost every helicopter uses this single-rotor configuration.

Figure 1. Picture of the Breguet-Dorand helicopter.

These successes and the birth of the modern helicopter are the results of the convergence of technology, knowledge and experience. Before Breguet and Sirkorsky flights, many aircraft enthusiasts and pioneers built contraptions that merely hopped a few meters. Successful machines came when mature engine and mechanical technologies met scientific

Figure 2. Picture of Sikorsky VS300 prototype.

study and a good understanding of the specificity of helicopter aerodynamics. In the 1930s, engines were refined by the booming aircraft industry. They were delivering unprecedented power-to-weight ratios [47], enabling helicopters to sustain more efficiently hovering flights. The counter-torque effect was tackled in many ingenious ways. Breguet used two counter-rotating shafts on the same axis to balance the torques. Other concepts used two and even quad-rotor configurations to balance this effect [3, 22, 34]. Patents show the level of engineering that was achieved to overcome the complexity of the various designs. Nevertheless, compelling forward flight performance came when the airflow asymmetry on the rotorblades was balanced.

When hovering, each blade experiences the same distribution of incident airflow velocity. This distribution is linear and proportional to the blade radius and the blade rotation. The lift generated by each blade can be estimated using the lift formula

$$L = \frac{1}{2}\rho v^2 A C_l \tag{1}$$

where L is the lift force on the rotorblade profile, ρ is the density of the air, v is the velocity of the airflow on the profile, A is the surface of the profile considered and C_l is the lift coefficient. The lift coefficient is function of the pitch angle of the blade. Assuming the helicopter is hovering, the pitch angle and the airspeed distribution are the same regardless of the position of the blade relative to the helicopter. The lift force of each blade is obtained from the integration of the lift formula along the length of a helicopter blade R

$$L = \int_0^R \frac{1}{2}\rho C_l (\omega r)^2 c dr \tag{2}$$

where c is the chord length of the blade profile and ω, the rotational velocity of the blade. After integration, we obtain

$$L = \frac{1}{6}\rho c C_l \omega^2 R^3 \tag{3}$$

As soon as the helicopter goes forward an extra velocity component is added to the velocity profile [6, 48]. We can distinguish the retreating side where the blade motion points in the opposite direction of the helicopter motion and the advancing side where the blade motion is in the same direction as the helicopter motion alike shown in Figure 3. Therefore, the incident airflow speed is increased on the advancing side and reduced in the retreating side. This

asymmetry causes a difference in the lifting capabilities of the two helicopter sides. The lift for a blade in the retreating side becomes

$$L = \int_0^R \frac{1}{2}\rho C_l(\omega r - v_n)^2 c dr \tag{4}$$

where v_n is the component of the helicopter velocity normal to the blade and ω is the rotational velocity of the blade. After integration, we obtain

$$L = \frac{1}{2}\rho c C_l \left(\frac{1}{3}\omega^2 R^3 - \omega v_n R^2 + v_n^2 R\right) \tag{5}$$

The difference between equations 3 and 5 gives the loss of lift Δ on the reatreating side due to the helicopter overall motion

$$\Delta L = \frac{1}{2}\rho c C_l \left(v_n^2 R - \omega v_n R^2\right) \tag{6}$$

The quadratic relation between the loss of lift on the retreating side and the helicopter forward motion velocity shows the importance of this phenomenon. Reverse flow is another important consequence of the forward motion of the helicopter. It happens where the helicopter speed is larger than the velocity of the blade due to its rotation. At high speeds, this region can cover a significant portion of the blade, meaning most of the lift is generated by the outer part of the blade. A cyclic control input was the key to balance the lift. Breguet-Dorand aircraft as well as Cierva and Sikorsky helicopters used a swashplate to vary the pitch of each blade during its revolution [8, 22, 51, 52]. Modifying the pitch of the blade changes the angle of attack and thus the lift for various positions of the blade around the helicopter. The angle of attack is increased on the retreating side and decreased on the advancing side. The lift is therefore evened on the two sides of the helicopter. Other early rotorcrafts pioneer considered a change of the twist of the blade or the deployment of flaps at the trailing edge of the blade to control on the lift [22, 43].

Today, all helicopters use cyclic pitch control for tuning the lift as the blades rotate. But lift can only be maintained by improving the angle of attack up to the stalling point of the blade profile. The maximal speed of a rotorcraft is therefore limited to the amount of lift the rotorblade can develop on the retreating side. In the case of rotorblade, the stall is dynamic, due to the unsteady nature of the flow. The vertical motion of the blade along with time-dependent pitching moments allows the angle of attack of the blade to exceed the quasi-static stalling angle of the profile. This favourable effect is followed by the development of vortexes close to the leading edge which can move towards the trailing edge causing large downward pitching moments [6, 22, 48]. Consequently, the rotor performance and the stability of the aircraft are reduced.

To further improve the helicopter blade performance, adaptive blade concepts are studied. The aim is to adapt the aerodynamic characteristics of the blade to maximise performance on both the advancing and the retreating side of the blade and improve the stall performance for large angles of attack. These systems range from morphing and changing the shape of a full blade profile to smaller devices acting on the boundary layer of the airflow to control its separation.

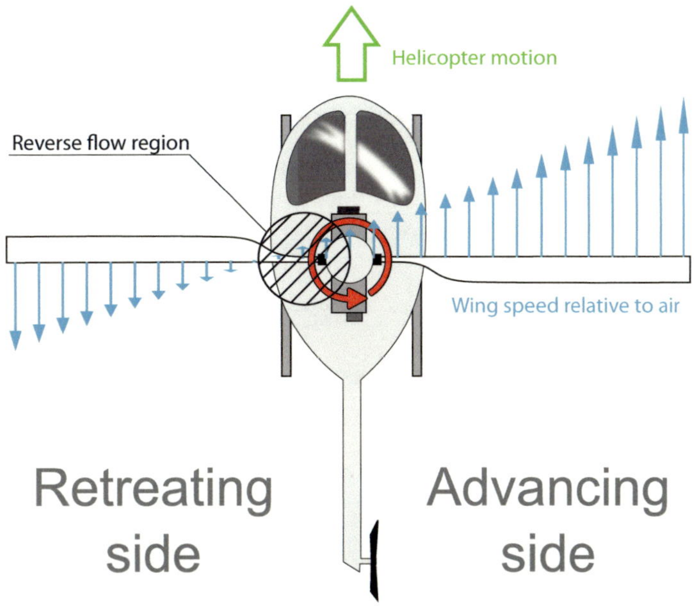

Figure 3. Helicopter in forward flight.

2. Capabilities of smart systems in rotor blades

Smart-blades can greatly enhance the performances of modern helicopters. Local modifications of the aerodynamic characteristics of a blade profile provide an optimised performance across the full blade revolution. The control of such systems conditions its capabilities and usage. In the introduction, focus was made on the lift characteristics of the retreating side. Most concepts improve directly the lift of the profile at fixed angle of attack. Other systems increase the efficiency of the helicopter improving the stall behavior of the profile or by reducting the vibration on the rotor. Vibrations decrease an helicopter blade efficiency, influence the dynamic stall and generate noise. The latter is a great concern for helicopter operating in a urban environment.

2.1. Flaps

Flaps on helicopter blades are not designed as a primary control surface like in airplanes. They act as a secondary control to improve the efficiency of the rotorblade by modifying the lift of the profile and by reducing vibrations on the rotor.

2.1.1. Active trailing edge flaps

Active trailing edge flaps are flaps situated at the trailing edge that actively modify the rotorblade performance. A schematic of a trailing edge flap for a helicopter blade taken from Koratkar [28] is shown in Figure 4. Although research has been conducted to study the possibility to use them for control in a swashplateless configuration [49], most of the studies focus on their ability to reduce the vibrations of helicopter blades [1, 15, 18, 27, 28, 35, 62]. The angle of the flap directly relates to a change of the bending of the blade during rotation

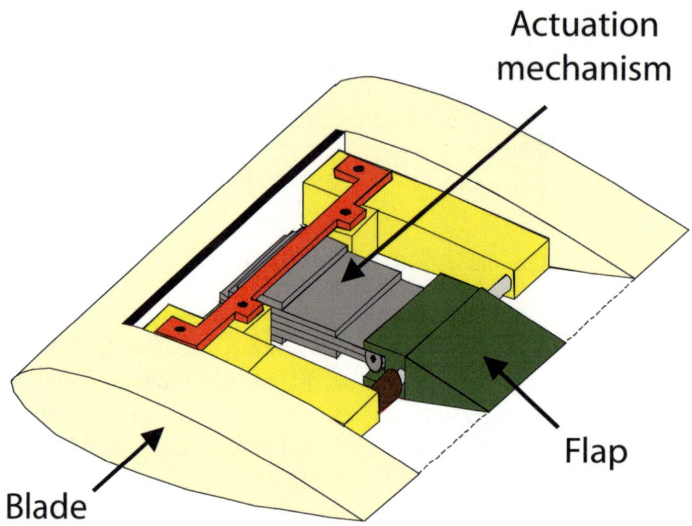

Figure 4. Schematic of a trailing-edge flap and its actuation system (adapted from Koratkar paper [28]

[18, 28]. The position of the flaps along the rotor has a great influence on the final performances of the mechanism. Although the optimum positioning of the flap depends on the application objective, studies show that multiple flaps achieve a better vibration reduction that a single flap [18, 27, 62]. These flaps only need a few degrees of deflection to affect the system dynamics [15, 28]. Loads on the helicopter rotor are function of the rotational frequency of the blade. The largest loads happens at 1, 2, 3 and 4 times the rotational frequency of the blade [6]. Therefore the flaps need to be actuated at similar frequencies to cancel undesired vibrations. With multiple flaps, the phase between the flaps is another key element. Under a suitable control authority, literature shows that the vibratory loads on the rotor can be reduced up to 80% [15, 18].

The amount of noise generated by helicopters is another important issue, especially because many helicopter missions involve flying over dense populated area. Noise generated by the helicopter blade comes mainly from the interaction between one blade and the vortexes generated by the previous blade [69]. This phenomenon is called blade-vortex interaction (BVI). Decreasing the effects of blade-vortex interaction can not only lead to a reduction in the noise emitted but also to a decrease in the power requirement. Active trailing edge flaps are actively studied to limit this effect by an individual control on each blade [1, 18, 69]. Controlling the trailing edge flap at 2 cycles per revolution (hereafter indicated as 2/rev) shows potential for consequent noise reduction [1].

2.1.2. *The Gurney flap*

The Gurney flap is a small flap deployed at 90 degrees at the edge of the trailing edge of the rotorblade, as shown in Figure 5. Typically its length is 2% of the chord length of the blade profile. The Gurney flap modifies the flow at the blade trailing edge and induces a low pressure zone which brings the separation point closer to the trailing edge [53]. The result is an increase of the lift over a large range of angles of attack with a small drag penalty [38, 53, 59,

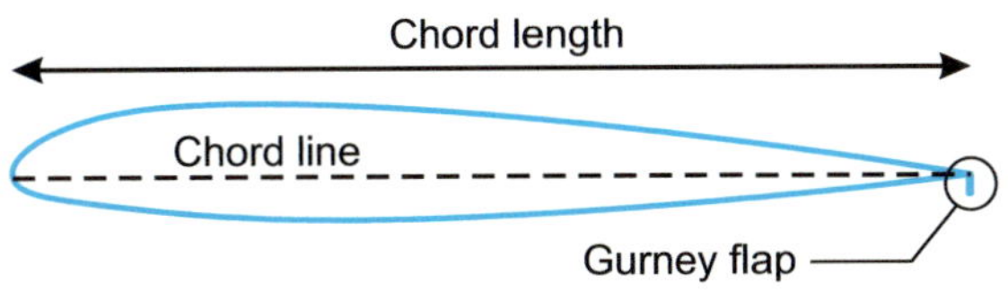

Figure 5. Sketch of a Naca 23012 profile with a 2% Gurney flap.

68]. Although the Gurney flap induces pitching moment, it provides a beneficial improvement of the efficiency of the rotorblade profile for the hovering situation [68]. In forward flight, the Gurney flap provides the blade with additional lift on the retreating side to balance the lift distribution [59]. For large forward velocity, the Gurney flap improves the airfoil behaviour in light stall conditions, which increases directly the flight envelope of a helicopter [68]. The behaviour of the Gurney flap is related to its length and placement. Studies about the length of the Gurney flap show an increase in drag and pitching moments with increasing lengths [38, 53, 63, 68]. Depending on the application, the Gurney flap length is limited to the point where these disadvantages outweigh its benefits in lift and stall characteristics.

Furthermore, the Gurney flap can have a positive effect on blade-vortex interaction. Similarly to a trailing edge flap, the Gurney flap acts on the blade mechanical behaviour [68]. Actuating the Gurney flap at 2/rev with suitable control would lead to a decrease in vibration and noise in a similar way than active trailing edge flaps [69].

2.2. Morphing blades

The idea behind morphing blades is to change the aerodynamic characteristics of the blade by a continuous change in its shape. This approach is inspired by the way birds and flying animals are modifying their wings to adapt to the various situations they encounter while flying. Most of these solutions involve a stiff structure that supports the loads and a flexible skin to keep the outer surface of the rotorblade without discontinuities.

2.2.1. Variable droop leading edge

The concept behind the nose drop is to advance the front part of the profile at an angle. It increases the profile as well as the curvature, as shown in Figure 6. The variable droop leading edge is used to alleviate the dynamic stall by ensuring that the flow passes smoothly over the leading edge for high angles of attack [32]. Although the lift is increased during the downward motion of the leading edge [32], the maximum lift is reduced by 10% [13, 21, 37]. More significantly, the drag and pitching moments are reduced by 50% [13]. The variable droop leading edge concept provides a decrease in helicopter vibrations and loads due to the suppression of dynamic stall within the retreating blade region. However, the helicopter maximum speed is reduced due to a decrease in lift when the droop nose is deployed. Therefore, the variable droop leading edge is studied in combination with the Gurney flap to negate the lift reduction [14]. This concept can also be applied to reduce the noise generated by a helicopter [9].

Figure 6. Sketch of the VR-12 profile used for wind tunnel testing at NASA Research Center from Lee paper [32].

2.2.2. Camber change

Changing the camber of a profile increases its lift for the same chord length [54]. The benefit is a larger flight envelope of the helicopter by improving the lift on the retreating side of the rotorcraft in the same way than the Gurney flap concept. Once again harmonic actuation at 2/rev could reduce the noise and the vibratory loads on the rotor, improving the rotor performance [18, 20]. Most studies on this concept consider the plane as the main application, envisioning morphing flaps as a main control surface [39].

2.2.3. Active-twist

Among early helicopter prototypes, some developer considered cyclic twist control [43] for changing the lift of the rotating blades. The idea behind active twist is to modify the twist and the torsional stiffness of rotating blade not only to improve the lift and the global helicopter performance but also to actively damp vibrations. Early experiments on active twist involved changing the twist of the helicopter blade at the root of the blade [2]. Later experiments used a distributed actuation actuation system to modify the blade twist [46, 50, 58, 66]. In a similar manner to the active trailing edge, the placement and the number of actuators modify the amount of vibration that are reduced. Thakkar study shows that up to 69% of reduction in vibrations can be achieved with the actuation of two sections [58]. Wind tunnel tests on a helicopter model demonstrated a 95% reduction in vibrations [66]. In the tests, each of the four blades mounted on the helicopter model was equipped with 24 actuators bonded onto the skin of the blade. Although only up to 1.4 degree of change in the pitching angle of the blade was achieved, torsional vibrations at 3/rev and 5/rev were successfully damp. In addition, the noise generated by the blade-vortex interaction can be reduced by up to 90% using an appropriate control of the blade twist [5].

2.2.4. Extended trailing edge

The amount of lift a profile can deliver depends on its chord length. For the same geometry, the lift is proportional to the blade surface area as shown in equation 1. Therefore, extending the chord length of a profile increases the lift generated. Studies on an extended trailing edge

active blade have shown an increase in the lift without significant increase in the lift-to-drag ratio [36].

2.3. Active flow control

Active flow control devices take another approach to improve the lift on a profile. Instead of modifying the airfoil geometry to act on the flow, they directly modify the air flow by re-energising the boundary layer on the top of the profile with a high speed jet. Such a flow is called a synthetic jet. The objective is to bring the separation point closer to the trailing-edge and therefore improve laminar flow over a larger portion of the airfoil [23]. Actuators for this application are placed inside a cavity which has a tiny opening [26] or a full slot perpendicular to the flow direction on the top part of the profile [67]. Figures 7 and 8 show these two types of synthetic jet actuator.

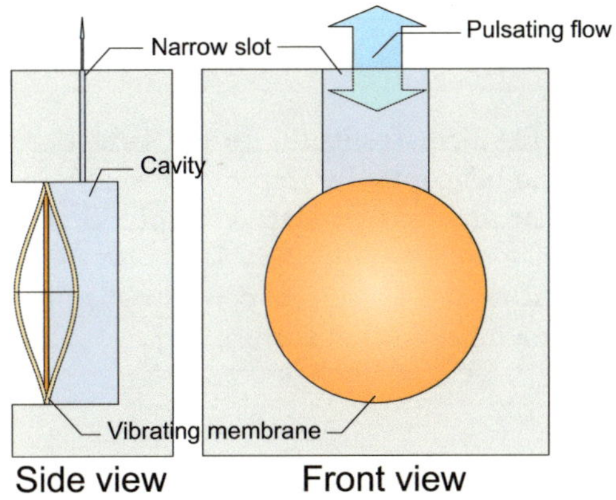

Figure 7. Sketch of a slot synthetic jet system.

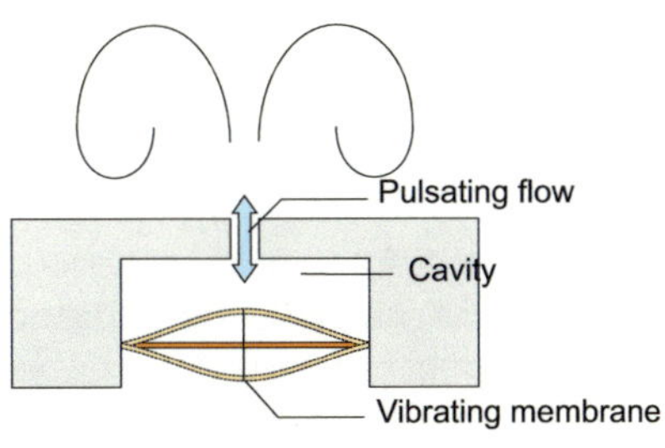

Figure 8. Sketch of a synthetic jet system with a circular orifice.

Wind tunnel experimentations have shown that synthetic jets improve the aerodynamic performance when driven at a specific frequency [23]. Much better performance is obtained when the actuation mechanism is combined with sensors arrays before and after the position of the synthetic jet system [57]. The sensors monitor the instabilities that will trigger the flow transition and actuate the synthetic jet system so that it damped the instabilities delaying further the transition. The actuation frequencies are in the kHz range and are related to the

airflow speed [10]. Most of the literature focuses on fixed wind tunnel test [10] but simulations show a potential increase in the maximum lift of an airfoil by 34% with an increase in the maximum stall angle of a profile [17]. These characteristics make synthetic jets system very promising for improving the characteristics of a profile for helicopter applications.

3. Challenges for smart-systems inside helicopter blades

Smart-systems need to answer challenges specific to the integration in helicopter blades. The combination of these challenges make smart-blade concepts very difficult to design.

3.1. Weight and space constraints

The weight and space are the main constrains in helicopter blades. Helicopter blades are designed to handle large centrifugal loads. The structural material takes most of the section of a rotorblade. Carbon fibre composites provide strength in the direction of the blade and reinforced layers give the blade impact resistance. The only space available is in the tail of the profile. Therefore, it is very difficult to integrate a system directly in the rotorblade skin for structural reasons. Concerning the weight, not only it cannot increase much, but its distribution in the profile should not affect the chordwise balance of the blade. Therefore, any weight added behind the aerodynamic centre needs to be compensated by an extra mass in the leading edge. For the whirl tower test of the SMART active flap rotor, weight was added in the leading edge to maintain the blade balance [55]. This constraint makes distributed and light systems like the active twist very relevant to maintain the distribution of mass along the profile chord. In comparison, the variable droop leading edge requires a very heavy mechanism to deform the leading edge of the profile that would change completely the weight distribution around the aerodynamic centre [29].

3.2. Mechanical constraints

Mechanical constraints are significant in a helicopter blade. The centrifugal loads are by far the main issue. The centrifugal loads come from the large rotational speeds of the blade. The centrifugal acceleration a is calculated with the following formula

$$a = \omega^2 r \tag{7}$$

where ω is the rotational speed of the blade in rad/sec and r is the position along the length of the blade. An 8m rotorblade rotating at 250rpm will generate close to 560g of acceleration at the tip. Because of the aerodynamics of an helicopter blade, the active system needs to be integrated near the tip of the blade where the centrifugal acceleration is the largest. The centrifugal loads resulting from the acceleration depend on the mass of the actuation system. Thus, a very light system does not lead to large loads. Some small actuation systems are very small and robust. The "Squiggle" linear drive motor, developed by NewScale technology, features a shock resistance of 2500g [40, 64]. For larger mechanisms most of the designs limit the load transfer along the blade [42, 55, 60]. Hence the design can be approximated to a bi-dimensional structure that is extended along the blade axis. For distributed systems that use patch actuators bonded onto the structure, like the active-twist technology developed by DLR, the actuators are being supported by the blade structure [46]. The main concern with these actuators is related to the deformation of the blade during its rotation. The peak strain

Active concept	Actuation frequency	Lifetime
Retreating side actuation	1/rev	66,460 hours
BVI noise	2/rev	33,230 hours
Vibrations	4/rev	16,615 hours
Flow control	2kHz	278 hours

Table 1. Comparison of the life time of an 10^9 cycles actuation mechanism in a 250rpm rotor blade system for various active control concepts. For the Active flow control system, the system is in operation only on the retreating side of the helicopter.

at the surface of the blade must not exceed the maximum strain of the actuator that will lead to breaking or debonding.

In addition to centrifugal loads, helicopter blades are subjected to large vibrations. In the most common configuration, helicopter blades are attached to the rotor by a joint that allows rotations with three degrees of freedom. The motions of the blade relative to the joint are defined as flapping, leading-lagging and feathering. Each of these motions are associated with one degree of freedom. While the blade is rotating the cyclic loads excite each degree of freedom causing vibrations at frequencies that are multiples of the blade rotational frequency [6, 48]. As a consequence, the design of a mechanism must address the resulting loads. Moreover, these loads constrain the use of mechanical elements like hinges or friction surfaces that will tend to jam and fail.

3.3. Reliability and environmental constraints

Any mechanism built in a commercial aircraft must comply to a set of rules to ensure the reliability of the system after a large number of actuation cycles, as well as the safety and the integrity of the aircraft in the event of a failure. Helicopter blades in a general purpose helicopter are designed for 10,000 flight hours [30]. Although composite blade design can handle even more loading cycles, manufacturers specify helicopter blades to be maintained and replaced on a much shorter basis [25, 30]. Actuation mechanisms for the active blade need to have a design life superior to the blade design life and have to maintain performances through their operational life. In aircrafts, hydraulic and pneumatic mechanisms are widely used due to these concerns. They are especially utilized for moving control surfaces that must satisfy a reliability requirement of 10^{-9} failure per hour and their performance is hardly affected even after a large number of cycles. It is only recently that electrical mechanisms have reached equivalent levels of safety and have been used to control control-surfaces in aircraft [4].

The reliability of the actuation system is also depending a lot on the application type. For alleviating the lift asymmetry on the retreating side, the mechanism performs a cycle once per revolution. This figure is to be compared with an actuation system for actively cancelling vibrations that need to operate at 2/rev and 4/rev or even at 5/rev in the case of torsional frequencies [18, 66]. Table 1 shows the various expected life time for a mechanism that has been design for 10^9 cycles in the case of various active blade concepts for a 250 rpm rotor system.

Although all the cases satisfy the 10,000 hours of operating life, the vibration damping case shows that a high quality actuation system certified for 10^9 has an operating life close to that

of the blade. Therefore, it would be very difficult to qualify an actuation system for damping vibration at 6/rev. For flow control devices, very much higher quality actuators are required to be certified for active flow control.

Finally, helicopters need to operate under a large range of environmental conditions. Blades are certified to perform over a large temperature range: from high altitude freezing conditions to high temperatures and with very high moisture content. It is therefore very difficult to design a fail-safe mechanism in these conditions, especially on small helicopters which do not have a de-icing system. Furthermore, some specific environments subject helicopter components to very difficult conditions such as sea and desert environments where corrosion and erosion are important matters.

3.4. Failure

In addition to being designed to exceed the life time of the blade, the active blade actuation system must also be developed not to influence the performance of the helicopter in the event of a failure. For distributed systems like the active-twist, if the patch actuators are not working, they will not reduce the performance of the blade profile. On the contrary, for the Gurney flap, the variable droop leading edge and the trailing edge active blade concept, in the event of a jamming, the blade profile will be modified during the full rotation of the rotor blade. Hence, care must be taken to make sure the helicopter is stable and able to be controlled with a modified profile. Furthermore, in the event of a power failure, the mechanism must go back to its initial state. This can be done by prestressing the mechanism or making sure that the aerodynamic loads are sufficient to bring the mechanism back to its inactive state.

3.5. Power requirement

To operate an actuator in a rotorblade, power needs to be transfered from the helicopter to the blades. Electrical, pneumatic and hydraulic power can be provided to a rotating blade by the use of specialized rotor mounts which add to the complexity of the rotor hub [24]. The type and the amount of power that can be drawn for an actuation system is a serious limitation to some active system. Large helicopter blades include a de-icing system for high altitude flight. Such a system draws up to 1kW of electrical power that is transfer to each blade. This gives a good estimation of the power available for an electrical actuation system.

3.6. Complexity of the system

Developing smart systems for helicopters is tremendously complex due to the number and the variety of the domains involved. For designing an active helicopter rotorblade, knowledge is required not only in aerodynamics and mechanics, but also in control, material science and electronics. Simulating for validating a concept, selecting of a suitable actuation technology and defining an application demand skills in all these domains. In order to move from research and laboratory experiments to flying prototypes and commercial products, the European union has created a Consortium within the Clean Sky Joint Technology Initiative to bring various research partners together. The Green Rotorcraft Consortium, among its lines of research, manages the evaluation of the Gurney flap technology to improve helicopter performance and noise reduction with both academic and industrial partners [16].

4. Piezoelectric actuators for smart-rotor blade systems

Many actuation technologies are available to actuate every smart blade concepts. Among them, piezoelectric actuators have a tremendous potential to meet and exceed the various requirements of these specific applications. This section focuses on piezoelectric actuators and their potential for the actuation of active systems for helicopter blades.

4.1. Piezoelectricity

Piezoelectric materials are materials that have the property to convert mechanical energy into electrical energy. When such a material is subjected to a strain, an electrical charge is created inside the material. This property is called the direct piezoelectric effect. Additionally, when the material is subjected to an electrical field, it deforms according to the electrical field magnitude. This is called the converse piezoelectric effect. A piezoelectric material is characterized by the piezoelectric strain constant d_{ij} which relates the strain to the electrical field. The subscript i indicates the direction of the applied electrical field and the subscript j indicates the direction of the deformation. Prior to be used, the piezoelectric material is poled. Conventionally the poling direction is along the vertical axis (3-axis) as shown in Figure 9. When an electrical field is applied in the poling direction, the material is contracting in that direction and extending in other directions (1- and 2-axis). Changing the direction of the electrical field will result in a contraction along the 1- and 2-axis and extension along the vertical axis. To quantify these piezoelectric effects, the direct and shear strains are related to the electrical field by the following constants: $d_{31} = d_{32}$, d_{33}, $d_{25} = d_{15}$. Equation 8 is the equilibrium equation that relates the electrical field E to the strain and shear components of the material (ε and γ) when no mechanical constraint is applied on the material.

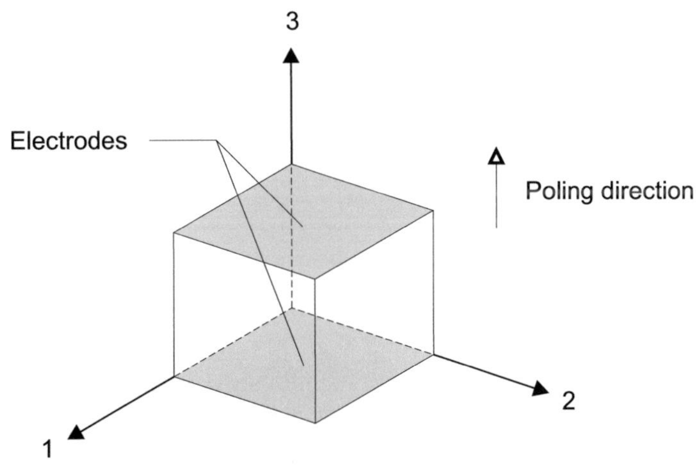

Figure 9. Axis reference system for piezoceramic components.

$$\begin{pmatrix} \varepsilon_1 \\ \varepsilon_2 \\ \varepsilon_3 \\ \gamma_{23} \\ \gamma_{31} \\ \gamma_{12} \end{pmatrix} = \begin{bmatrix} 0 & 0 & d_{31} \\ 0 & 0 & d_{32} \\ 0 & 0 & d_{33} \\ 0 & d_{25} & 0 \\ d_{15} & 0 & 0 \\ 0 & 0 & 0 \end{bmatrix} \times \begin{pmatrix} E_1 \\ E_2 \\ E_3 \end{pmatrix} \quad (8)$$

Name	d_{31}(m/V)	d_{33}(m/V)	e_{33}	CT(°C)	ρ(kg/m^3)	S_{33}(m^2/N)
PZT-SP4	-1.23e-10	3.1e-10	1300	325	7500	1.81e-11
PZT-5A1	-1.85e-10	4.4e-10	1850	335	7500	2.07e-11
PZT-5H	-2.74e-10	5.93e-10	3400	193	7500	2.083e-11
PZT-PSt-HD	-1.9e-10	4.5e-10	1900	345	7500	2.1e-11
PZT-PSt-HPSt	-2.9e-10	6.4e-10	5400	155	8000	1.8e-11
PZT-PIC-255	-1.8e-10	4.0e-10	1750	350	7800	2.07e-11
PZT-PIC-151	-2.1e-10	5.0e-10	2400	250	7800	1.9e-11

Table 2. Table of some piezoceramic materials. PZT-SP4 and PZT-5A1 are from Smart-Material Company. PZT-5H is taken from Chopra review on actuators [15]. PZT-PSt-HD and PZT-PSt-HPSt are from Piezomechanik Company. Finally, PZT-PIC-255 and PZT-PIC -151 are from Physic Instrumente Company [45].

Consequently, the knowledge of the 3 constants d_{31}, d_{33} and d_{15} is sufficient to fully characterize the electromechanical properties of a piezoelectric material. A deformation along the 1- and 2-axis which is characterized by the d_{31} coefficient is called the d_{31} effect and a deformation along the 3-axis is called the d_{33} effect as the d_{33} coefficient characterizes this deformation as shown in Figure 10.

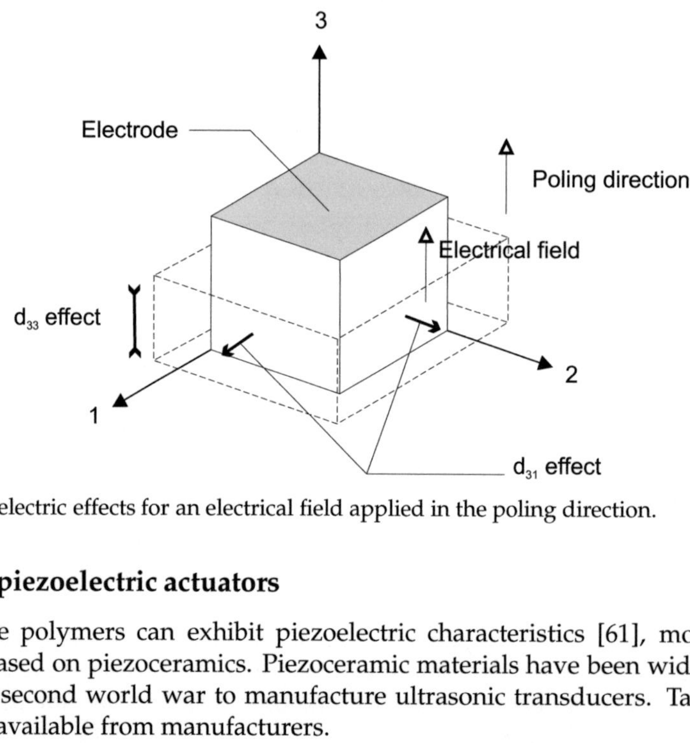

Figure 10. Piezoelectric effects for an electrical field applied in the poling direction.

4.2. Types of piezoelectric actuators

Although some polymers can exhibit piezoelectric characteristics [61], most piezoelectric actuators are based on piezoceramics. Piezoceramic materials have been widely studied and used since the second world war to manufacture ultrasonic transducers. Table 2 lists some piezoceramics available from manufacturers.

4.2.1. d_{31} *effect actuators*

Piezoceramic actuators using the d_{13} effect are based on the fact that a through-thickness electrical field will contract the material's width and length. The components manufactured

using this principle are consequently laminates of fine piezoceramic sheets. Electrodes are bonded on the upper and lower faces of the piezoceramic patch. Applying a voltage through the patch's thickness causes a contraction in the plane of the patch as shown in Figure 11. These can be easily bonded or embedded inside a structure, thanks to their low weight and volume. Thus, they are mainly used to manufacture unimorph and bimorph structures.

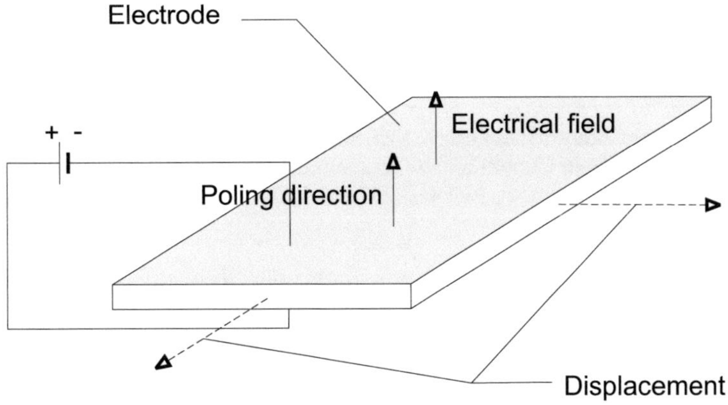

Figure 11. Principle of a piezoelectric laminar actuator

4.2.2. d_{33} effect actuators

Piezoceramic actuators using the d_{33} effect are based on the fact that a through-thickness electrical field will modify the material's thickness. Within piezoceramics, the d_{33} coefficient is always more important than the d_{31} coefficient, therefore using the d_{33} effect is preferable. Consequently, there are many types of actuators that try to take advantage of the larger d_{33} coefficient using various geometries.

- Stack actuator
  Stack actuators use the d_{33} effect to achieve deflection. They consist of multiple layers of piezoceramic plates separated by electrodes as shown in Figure 12. This configuration allows long elements to be made for higher displacement capabilities while high voltage is not needed to obtain high electrical fields if the piezoceramic layers are small enough between two electrodes. Stack actuators are capable of delivering higher forces than laminar actuators as they are fully using the highest strain coefficient available.
- Macro Fibre Composites (MFC) & Active Fibre Composite (AFC)
  Long piezoceramic components have better displacement capabilities. Active Fibre Composite (AFC) consists of piezo ceramic fibres embedded into a protective polymer substrate and poled into the fibre direction to use the highest strain coefficient. Interdigitated electrodes bonded onto the fibres ensure high electrical fields. The voltage required by these components depends on the fibre diameter and the distance between electrodes. Compared with laminar actuators, these components require less voltage to achieve the same force and displacement. Macro Fibre Composite exploit the same principles as AFC, except that they are made of fibres having an improved contact with electrodes, which plays an important role in the electrical field magnitude inside the material. Furthermore, those actuators are much more flexible than a strip made of the same material.

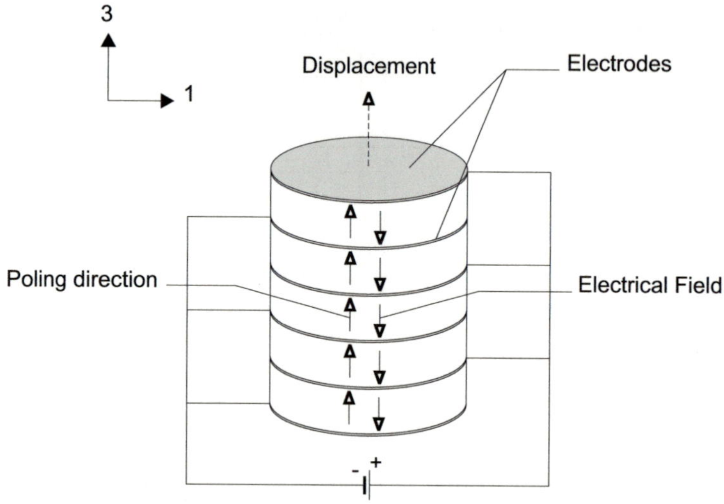

Figure 12. Principle of a piezoceramic stack actuator.

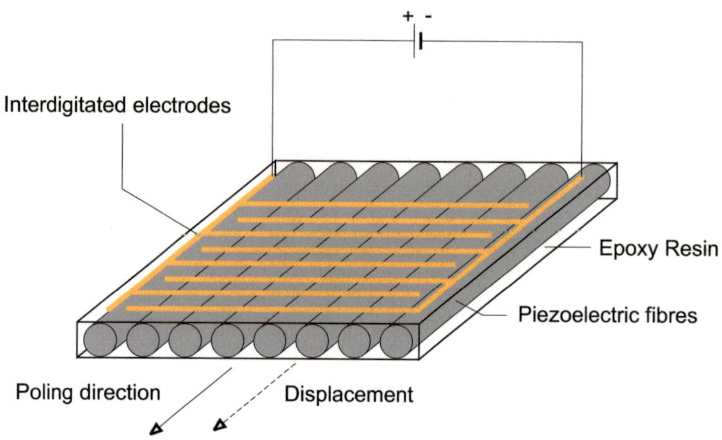

Figure 13. Sketch of an Active Fibre Composite (AFC) actuator.

4.2.3. d_{15} *effect actuators*

The d_{15} effect is a shearing effect. The material shears in the 3-1 or 3-2 plane when an electrical field is applied in the 1-axis or 2-axis respectively, perpendicular to the poling direction as shown in Figure 14.

The d_{15} shearing effect can be used to manufacture a tube actuator that twists when actuated as shown in Figure 15. Centolanza has discussed the manufacturing of an induced shear actuator and its testing [12]. Kim has tested the same component with feedforward control and pointed out fatigue and heating issues [27]. Their conclusions are that such an actuator is a promising option and more studies of the piezoceramic material would improve their models accuracy.

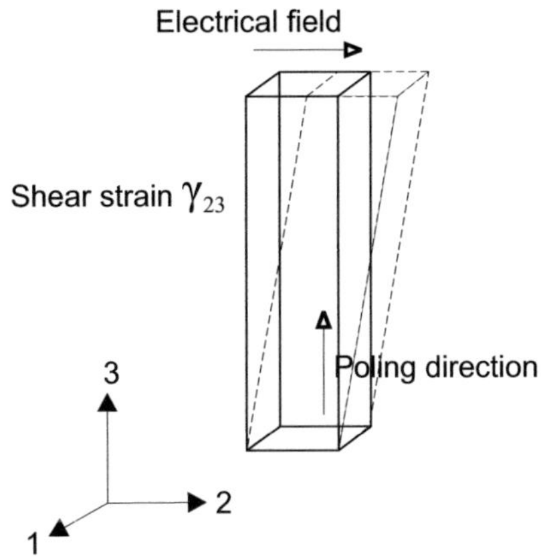

Figure 14. d_{15} effect in piezoceramics.

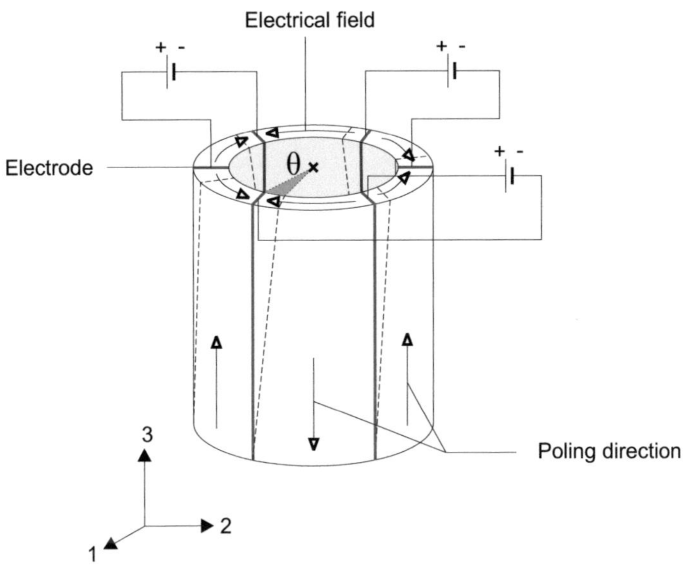

Figure 15. Twisting motion for the d_{15} effect in a shear tube actuator.

4.3. Performances

4.3.1. Block force and free displacement

The performance of piezoelectric actuators is evaluated in block force and free displacement. The block force is the maximum force the piezoelectric actuator can deliver when clamped. The free displacement is the displacement achieved by the actuator when no force is applied on it. These two parameters depend on the values of the piezoelectric strain coefficient, the electrical field inside the material and the geometry of the actuator. For a d_{31} patch actuator the free displacement Δ and the block force F_{block} can be directly derived from the Equation 8.

$$\Delta = d_{31} L \frac{V}{t} \tag{9}$$

where d_{31} is the piezoelectric strain coefficient, L the length of the actuator, V the applied voltage and t the distance between the two electrodes.

$$F_{block} = \frac{d_{31} AV}{S_{11} t} \tag{10}$$

where A is the cross-section area of the actuator, t the distance between two electrodes and S_{11} the compliance in the plane of the actuator.

Piezoelectric actuators provide large blocking forces but only very low displacements. Therefore, typical mechanisms that involve piezoelectric components contain systems to amplify the displacement of the actuator. The amplification can be achieved by the use of level arms or by the integration of the piezoelectric component in a structure providing that amplification. Cedrat Technologies develops these systems based on stack actuators [11, 44].

4.4. Bandwidth

The first widespread use of piezoelectric actuators was for manufacturing acoustic sources for sonar because of their large actuation frequency bandwidth. Overall, piezoelectric actuators have a very small actuation time and can achieve large motion speed and acceleration. Cedrat Technologies manufactures actuators with a response time below 1ms [11]. Moreover, the typical resonance frequency for piezoelectric actuators is in the kHz range, leaving a very comforting margin for structural applications [11, 45].

4.5. Power consumption and voltage

The power consumption of piezoelectric actuators depends on the type of actuation. Integrated inside an electrical circuit, a piezoelectric actuator behaves like a capacitor. Under harmonic actuation, the energy required to charge the piezoelectric actuator can be recuperated in the system for the next charge. When fast positioning is required, the electrical components which drive the piezoelectric actuator must be able to provide large currents. For the discharge, the circuit and its components must be able to dissipate quickly the energy stored in the piezoelectric. Therefore, active systems using harmonic actuation only need around 100W in operation, while a fast actuation system requires close to 1000W of power depending on the actuation profile. However, the main problem is not the amount of electrical power required but the time during which the power is required.

4.6. Reliability and operational environment

Piezoelectric actuators are very reliable and can perform a large number of cycles under good operating conditions. For instance piezoelectric stack actuators manufactured by Physik Instrumente feature 10^9 cycles [45]. Cedrat technologies is developing high end actuators with 10^{10} cycles before failure [11]. Furthermore, piezoelectric actuators can operate in very harsh environments. Physik Instrumente provides piezoelectric stack actuators capable of operating in cryogenic environments [45]. Moreover, piezoelectric actuators can operate at high temperatures as long as the Curie temperature is not reached. The Curie temperature is the temperature at which the piezoelectric material looses its electro-mechanical coupling and this temperature is usually higher than 200°C as shown in Table 2.

4.7. Applications

The relevance of piezoelectric actuators for active blade systems comes from the large specific work they can output [41] while being small and easily integrable. Moreover their reliability makes them suitable for safely powering mechanisms in smart blade concepts.

Many actuation systems for the active trailing-egde rotorblade are actuated using amplified stack actuators [33, 44, 56]. Other designs use piezoelectric patch actuators bonded onto a beam [28] or piezoelectric shear actuators built as a torsional actuator as shown in Figure 15 [12].

Some studies on the design of airfoils with controllable camber involve stack actuators inside a structure that convert and amplify the motion into a change of curvature of the airfoil [19, 20].

For the deployment of the Gurney flap at the trailing edge, bimorph piezoelectric actuation mechanisms are being studied [59], as well as more complex structures to amplify the displacement of piezoelectric patch actuators and MFCs [42]. AFC and MFC actuators have also been tested successfully for the active-twist application [65]. They provide distributed strain over all the rotorblade surface to successfully twist the blade under operating conditions [46, 50].

Research on flow control systems has considered piezoelectric diaphragms to deliver enough airflow speed for synthetic jets and achieve proper flow control [31, 67].

5. Conclusion

Today's helicopters are the results of some tremendous work and collaborations in mechanical engineering and aeronautics. The first successes came from inventors that could understand the complexity of a rotating lift surface while designing advanced mechanical mechanisms. To further improve today's helicopters, research is focussing on active blade systems to adapt the aerodynamic properties of the blade to the local aerodynamic conditions. Two aspects are especially studied: enhancing the lift on the retreating side and alleviating the large vibrations in the rotor. Both these aspects will provide improvements on the helicopter performances. Besides the efficiency of the rotor system, the objective is to push the flight envelope of these aircrafts and to make them faster, smoother and quieter.

Many active concepts are being studied but they all face a large number of challenges to be successfully integrated within a helicopter blade. The rotation speed generates critical loads on the blade and any system within it. With the helicopter blade being the component providing both lift and control in a helicopter, any mechanism influencing its behaviour is required to be durable, reliable and safe. Actuation of the active system is the most critical aspect of a smart adaptive blade. Piezoelectric actuators have the potential to provide compelling actuation for these systems. They are actively tested for many of these concepts. Their toughness, size and reliability make them especial candidates for delivering the required mechanical power.

The key aspect of helicopter progress remains in the collaboration between partners from various domains, combinig different skills and expertise, to answer these challenges and develop tomorrow's aircrafts.

Acknowledgements

This project is funded by the European Union in the framework of the Clean Sky program - Green Rotorcraft.

Author details

Paternoster A.
Chair of Applied Mechanics, Faculty of Engineering Technology, University of Twente, Enschede, The Netherlands
Chair of Production Technology, Faculty of Engineering Technology, University of Twente, Enschede, The Netherlands

Loendersloot R. and de Boer A.
Chair of Applied Mechanics, Faculty of Engineering Technology, University of Twente, Enschede, The Netherlands

Akkerman R.
Chair of Production Technology, Faculty of Engineering Technology, University of Twente, Enschede, The Netherlands

6. References

[1] Altmikus, A., Dummel, A., Heger, R. & Schimke, D. [2008]. Actively controlled rotor: aerodynamic and acoustic benefit for the helicopter today and tomorrow, *34th European Rotorcraft Forum*, Liverpool.
[2] Barrett, R. & Stutts, J. [1997]. Design and testing of a 1/12th-scale solid state adaptive rotor, 491.
[3] Bennett, J. [1944]. Helicopter and Gyroplane, *US patent 2,344,967* .
[4] Bennett, J., Mecrow, B., a.G. Jack, Atkinson, D., Sheldon, S., Cooper, B., Mason, G., Sewell, C. & Cudley, D. [2005]. A prototype electrical actuator for aircraft flaps and slats, *IEEE International Conference on Electric Machines and Drives, 2005.* pp. 41–47.
[5] Booth, E. & Wilbur, M. [n.d.]. Acoustic Aspects of Active-Twist Rotor Control, *Journal of the American Helicopter Society* 49(1): 8.
[6] Bramwell, A., Done, G. & Balmford, D. [2000]. *Bramwell's Helicopter Dynamics*, Elsevier.
[7] Breguet, L. [1933]. Revolving Supporting Surfaces, *US Patent 1,919,089* .
[8] Breguet, L. [1935]. Flying machine having revolving supporting surfaces, *US Patent 1,986,709* .
[9] Breitbach, E. & Büter, A. [1996]. The main sources of helicopter vibration and noise emissions and adaptive concepts to reduce them, *Journal of Structural Control* 3(1-2): 21–32.
[10] Cattafesta, L., Song, Q., Williams, D., Rowley, C. & Alvi, F. [2008]. Progress in Aerospace Sciences Active control of flow-induced cavity oscillations, *Progress in Aerospace Sciences* 44: 479–502.
[11] CedratTechnologies [2012]. CEDRAT TECHNOLOGIES: Innovation in Mechatronics, *www.cedrat-technologies.com* .
[12] Centolanza, L., Smith, E. & Munksy, B. [2002]. Induced-shear piezoelectric actuators for rotor blade trailing edge flaps, *Smart Materials and Structures , February* .

[13] Chandrasekhara, M. [2007]. Compressible dynamic stall vorticity flux control using a dynamic camber airfoil, *Sadhana* 32(1-2): 93–102.

[14] Chandrasekhara, M. [2010]. Optimum Gurney flap height determination for lost-lift recovery in compressible dynamic stall control, *Aerospace Science and Technology* 14(8): 551–556.

[15] Chopra, I. [2002]. Review of State of Art of Smart Structures and Integrated Systems, *AIAA/ASME/ASCE/AHS/ASC Structures, Structural Dynamics, and Materials Conference, Seattle, WA, AIAA Journal* 40(11).

[16] CleanSky [2012]. Clean Sky Joint Technology Initiative, *http://www.cleansky.eu* .

[17] Duvigneau, R. & Visonneau, M. [2006]. Simulation and optimization of stall control for an airfoil with a synthetic jet, *Aerospace Science and Technology* 10(4): 279–287.

[18] Friedmann, P., de Terlizzi, M. & Myrtle, T. [2001]. New developments in vibration reduction with actively controlled trailing edge flaps, *Mathematical and Computer Modelling* 33(10-11): 1055–1083.

[19] Gandhi, F. & Anusonti-Inthra, P. [2008]. Skin design studies for variable camber morphing airfoils, *Smart Materials and Structures* 17(1): 015025.

[20] Gandhi, F., Frecker, M. & Nissly, A. [2008]. Design Optimization of a Controllable Camber Rotor Airfoil, *AIAA Journal* 46(1): 142–153.

[21] Geissler, W. & Trenker, M. [2002]. Numerical investigation of dynamic stall control by a nose-drooping device, *Technical Report C*.

[22] Gordon Leishman, J. [2006]. *Principles of Helicopter Aerodynamics*.

[23] Greenblatt, D. & Wygnanski, I. [2000]. The control of flow separation by periodic excitation, *Progress in Aerospace Science* 36.

[24] Haber, A. & Jacklin, S. [2002]. Development, manufacturing, and component testing of an individual blade control system for a UH-60 Helicopter Rotor, *American Helicopter* .

[25] Head, E. [2012]. Blade Trouble | Vertical - Helicopter News.

[26] Hong, G. [2006]. Effectiveness of micro synthetic jet actuator enhanced by flow instability in controlling laminar separation caused by adverse pressure gradient, *Sensors And Actuators* 132(2006): 607–615.

[27] Kim, J.-S., Wang, K. & Smith, E. [2007]. Development of a resonant trailing-edge flap actuation system for helicopter rotor vibration control, *Smart Materials and Structures* 16(6): 2275–2285.

[28] Koratkar, N. & Chopra, I. [n.d.]. Wind Tunnel Testing of a Mach-Scaled Rotor Model with Trailing-Edge Flaps, *Journal of the American Helicopter Society* 47(4): 263–272.

[29] Kota, S., Ervin, G. & Osborn, R. [2008]. Design and Fabrication of an Adaptive Leading Edge Rotor Blade, *American Helicopter Society Annual Forum* .

[30] Kwon, J.-H., Hwang, K.-J., Kim, S.-S., Kim, P.-J. & Kim, C.-S. [n.d.]. Fatigue life evaluation in composite rotor blade of multipurpose helicopter, *Proceedings 6th Russian-Korean International Symposium on Science and Technology. KORUS-2002 (Cat. No.02EX565)*, IEEE, pp. 15–20.

[31] Lee, C., Hong, G., Ha, Q. & Mallinson, S. [2003]. A piezoelectrically actuated micro synthetic jet for active flow control, *Sensors And Actuators* 108(April 2003): 168–174.

[32] Lee, S. & McAlister, K. [1993]. Characteristics of deformable leading edge for high performance rotor, *AIAA Journal* 35.

[33] Lee, T. & Chopra, I. [2001]. Design of a piezostack driven trailing-edge flap actuator for helicopter rotors, *Smart Mater. Struct.* 10: 15–24.

[34] LePage, L. [1936]. Flight on rotating wings, *Journal of the Franklin Institute* 222.

[35] Lim, I.-G. & Lee, I. [2009]. Aeroelastic analysis of rotor systems using trailing edge flaps, *Journal of Sound and Vibration* 321(3-5): 525–536.

[36] Liu, T., Montefort, J., Liou, W., Pantula, S. R. & Shams, Q. a. [2007]. Lift Enhancement by Static Extended Trailing Edge, *Journal of Aircraft* 44(6): 1939–1947.

[37] Martin, P., McAlister, K. & Chandrasekhara, M. [2003]. Dynamic Stall Measurements and Computations for a VR-12 Airfoil with a Variable Droop Leading Edge, *Technical report*.

[38] Maughmer, M. & Bramesfeld, G. [2008]. Experimental Investigation of Gurney Flaps, *Journal of Aircraft* 45(6): 2062–2067.

[39] Monner, H. [2001]. Realization of an optimized wing camber by using formvariable flap structures, *Aerospace Science and Technology* 5: 445–455.

[40] NewScaleTechnologies [2012]. New Scale Technologies - precision motion for micro imaging and vision systems, *http://www.newscaletech.com/* .

[41] Paternoster, A., DeBoer, A., Richard, L. & Remko, A. [2010]. Actuators for Smart Applications, *Proceedings of the ASME 2010 Conference on Smart Materials, Adaptive Structures and Intelligent Systems*, Asme.

[42] Paternoster, A., Loendersloot, R., de Boer, A. & Akkerman, R. [2011]. Geometric Optimisation of Hinge-less Deployment System for an Active Rotorblade, *Proceedings of the ASME 2011 Conference on Smart Materials, Adaptive Structures and Intelligent Systems*, ASME.

[43] Pescara, R. [1923]. Screw propeller of helicopter flying machines, *US Patent 1,449,129* .

[44] Petitniot, J., Rochettes, H.-M. & Leconte, P. [2002]. Experimental assessment and further development of amplified piezo actuators for active flap devices, *Actuator*, number June, pp. 10 – 12.

[45] PhysikInstrumente [2012]. PI Leader in: Precision Nano-Positioning & Piezo Engineering, NanoAutomation, Piezo Stage, Hexapod, PZT, Piezo Actuator, Transducer: Sub-Nanometer Resolution, Metrology, Photonic Packaging Automation, Piezo Linear Motor, Steering Mirror, Translation, *http://www.physikinstrumente.com/* .

[46] Riemenschneider, J., Opitz, S., Schulz, M. & Plaß meier, V. [2010]. Active Twist Rotor for Wind Tunnel Investigations, *Proceedings of the ASME 2010 Conference on Smart Materials, Adaptive Structures and Intelligent Systems*, Vol. 2010, ASME, pp. 371–378.

[47] Rosen, K. [n.d.]. A Prospective: The Importance of Propulsion Technology to the Development of Helicopter Systems with a Vision for the Future. The 27th Alexander A. Nikolsky Lecture, *Journal of the American Helicopter Society* 53(4): 31.

[48] Saunders, G. [1975]. *Dynamics of helicopter flight*, Wiley-Interscience.

[49] Shen, J. & Chopra, I. [2004]. A Parametric Design Study for a Swashplateless Helicopter Rotor with Trailing-Edge Flaps, *Journal of the American Helicopter Society* 49(1): 43.

[50] Shin, S., Cesnik, C., Wilkie, W. & Wilbur, M. [2008]. Design and Manufacturing of a Model-scale Active Twist Rotor Prototype Blade, *Journal of Intelligent Material Systems and Structures* 19(12): 1443–1456.

[51] Sikorsky, I. [1943]. Helicopter and Controls therefor, *US Patent 2,318,260* .

[52] Sikorsky, I. [1947]. Helicopter rotor, *US Patent 2,627,929* .

[53] Singh, M., Dhanalakshmi, K. & Chakrabartty, S. [2007]. Navier-Stokes Analysis of Airfoils with Gurney Flap, *Journal of Aircraft* 44(5): 1487–1493.

[54] Stanewsky, E. [2001]. Adaptive wing and flow control technology, *Progress in Aerospace Sciences* 37(7): 583–667.

[55] Straub, F., Kennedy, D. & Stemple, A. [2004]. Development and whirl tower test of the SMART active flap rotor, *SPIE Conf. on Smart* .

[56] Straub, F., Ngo, H. T., Anand, V. & Domzalski, D. [2001]. Development of a Piezoelectric Actuator for Trailing Edge Flap Control of Full Scale Rotor Blades, *Smart Materials and Structures, Vol. 10, No.* 1: 101088/0964–1726/10/1/303.

[57] Sturzebecher, D. & Nitsche, W. [2003]. Active cancellation of TollmienâĂŞSchlichting instabilities on a wing using multi-channel sensor actuator systems, *International Journal of Heat and Fluid Flow* 24(4): 572–583.

[58] Thakkar, D. & Ganguli, R. [2007]. Induced shear actuation of helicopter rotor blade for active twist control, *Thin-Walled Structures* 45(1): 111–121.

[59] Thiel, M. [2006]. *Actuation of an active Gurney flap for rotorcraft applications*, PhD thesis, The Pennsylvania State University.

[60] Thiel, M. & Lesieutre, G. [2009]. New Actuation Methods for Miniature Trailing-Edge Effectors for Rotorcraft, *AIAA/ASME/ASCE/AHS/ASC Structures, Structural Dynamics, and Materials Conference*, number May.

[61] Vinogradov, A., Schmidt, V., Tuthill, G. & Bohannan, G. [2004]. Damping and electromechanical energy losses in the piezoelectric polymer PVDF, *Mechanics of Materials* 36(10): 1007–1016.

[62] Viswamurthy, S. & Ganguli, R. [2004]. An optimization approach to vibration reduction in helicopter rotors with multiple active trailing edge flaps, *Aerospace Science and Technology* 8(3): 185–194.

[63] Wang, J., Li, Y. & Choi, K.-S. [2008]. Gurney flap-Lift enhancement, mechanisms and applications, *Progress in Aerospace Sciences* 44(1): 22–47.

[64] Watson, B., Friend, J. & Yeo, L. [2009]. Piezoelectric ultrasonic micro/milli-scale actuators, *Sensors And Actuators* 152: 219–233.

[65] Wickramasinghe, V. & Hagood, N. [2004]. Material characterization of active fiber composites for integral twist-actuated rotor blade application, *Smart Materials and Structures* 13(5): 1155–1165.

[66] Wilbur, M., Mirick, P., Yeager, W., Langston, C., Cesnik, C. & Shin, S. [n.d.]. Vibratory Loads Reduction Testing of the NASA/Army/MIT Active Twist Rotor, *Journal of the American Helicopter Society* 47(2): 11.

[67] Yang, A., Ro, J., Yang, M. & Chang, W. [2009]. Investigation of piezoelectrically generated synthetic jet flow, *Journal of Visualization* 12(1): 9–16.

[68] Yee, K., Joo, W. & Lee, D.-H. [2007]. Aerodynamic Performance Analysis of a Gurney Flap for Rotorcraft Application, *Journal of Aircraft* 44(3): 1003–1014.

[69] Yu, Y., Gmelin, B. & Splettstoesser, W. [1997]. Reduction of helicopter blade-vortex interaction noise by active rotor control technology, *Progress in Aerospace* 33(97): 647–687.

Chapter 26

Active Control of Plate-Like Structures for Vibration and Sound Suppression

Ipek Basdogan, Utku Boz, Serkan Kulah and Mustafa Ugur Aridogan

Additional information is available at the end of the chapter

http://dx.doi.org/10.5772/51755

1. Introduction

Mechanical vibrations cause undesirable effects such as noise and acoustic signature and consequences of severe vibrations may lead to a decrease in service life of machines, or even cause damages on the components. In order to reduce vibration levels in ground, marine and aerospace applications as well as nano and micro devices, active control methods for vibration and noise reduction has been proposed in the last decade.

In 1980s, the usages of large structures as the components of the space vehicles were resulted in new challenges in vibration control since these structures are lightly damped and have long decay times. Researchers were motivated to design and implement active vibration control to these lightly damped structures with piezoelectric materials. In one of the first research studies relevant to this area, Bailey and Hubbard [1] designed an active damper system by using piezoelectric material as distributed actuator for the vibration suppression of a cantilever beam. Subsequently, Crawley and Luis [2] developed and experimentally verified the analytical model of piezoelectric materials. The main contribution of Crawley and Luis was the accurate prediction of performance and effectiveness of piezoelectric materials when they are used as actuators. While passive isolation techniques are not generally feasible and efficient for low frequency vibration suppression; aforementioned initiative studies showed that piezoelectric materials can be used to suppress the low frequency vibration. Indeed, Ro and Baz [3] investigated and compared the overall effectiveness of the active and passive treatments for vibration and noise reduction. Their results revealed that active treatments such as bonded piezoelectric patch actuators can significantly reduce the vibration and noise radiation better than the passive treatments. Currently, different types of piezoelectric elements are available in market which can be employed widely as actuators and sensors for active vibration reduction of structures.

In the literature, there are published review articles discussing fundamental aspects of active vibration control. Rao and Sunar [4] published a review paper on the use of piezoelectric materials as sensors and actuators for control of flexible structures. In this review paper, they presented the general framework of structural control strategies and presented the applications in various fields. One of the most common application areas is aerospace industry, since the aerospace structures are flexible and have limited tolerance for vibration. In a different review study, Loewy [5] presented the key applications of smart structures in aeronautical applications with potential usages. In this work, smart materials are categorized in terms of their energy-exchange capabilities. The benefits of using smart structures in aeronautical applications are also presented.

The active vibration control systems should be feasible and reliable to be used in marine, aerospace and automotive applications. The reliability and feasibility of the control systems are related to dynamics of piezoelectric actuators and implementation of controller algorithms. Thus, it is needed to predict performance of the piezoelectric sensor and actuators embedded in control systems before implementation and production. For this purpose, Chee et al. [6] presented the modeling approaches for performance investigation of piezoelectric materials. As discussed in their article, analytical and finite element modeling of piezoelectric materials with host structures can be built using the linear constitutive piezoelectric equations only for low actuation authority. In case of high actuation authority; the nonlinearities occurs and different methodologies should be considered.

The choice of the controller type and optimal positioning of sensors and actuators are other important aspects of design and implementation of active vibration control systems. There are also published technical review papers discussing the controller algorithms and optimal placement of actuators and sensors. For instance, Alkhatib and Golnaraghi [7] reviewed the controller architectures and presented the general design procedures for the active vibration control systems. In this review paper, the advantages and disadvantages of different controller architectures with various application examples are presented. Specifically that review paper states that the active damping system is advantageous when the vibration suppression is aimed only around resonance frequencies. It is also noted that the active damping system ensures stability when collocated sensor and actuator pairs are used. In another review paper, Gupta et al. [8] concentrated on the optimal positioning of piezoelectric sensors and actuators. In this technical review, the optimization techniques are presented based upon six optimizing criteria, namely (i) maximizing modal forces/moments applied by piezoelectric actuators, (ii) maximizing deflection of the host structure, (iii) minimizing control effort/maximizing energy dissipated, (iv) maximizing degree of controllability, (v) maximizing degree of observability, and (vi) minimizing spill-over effects. Gupta et al. presented the optimal positioning results in the literature in a tabular form for beam and plate like structures.

For robust and adaptive active vibration control systems, the structural modeling techniques and estimation of uncertainties are important. A comprehensive methodology for the design

and validation of a robust controller is presented by Iorga et al. [9]. In their paper, structural modeling techniques with uncertainty analysis are explained and H_∞ controller design is pursued by placing emphasis on robust control concepts.

The piezoelectric materials also attracted attention of the researchers working on noise reduction. In a review paper on smart structures and integrated systems, Chopra [10] stated that most of the applications for aerospace vehicles are focused on minimization of vibration levels; however, interior and exterior noise reduction, using piezoelectric materials is also a potential research area for transport vehicles. Hanselka [11] named the use of smart materials in active noise and vibration control as an innovative technology and gave an application example of active noise reduction. In that application, piezoelectric materials served as sensors and actuators to reduce structural-borne noise and the results showed that the application of current developments is possible and advantageous.

This chapter is organized as follows. In section 2, the design procedure and algorithms for active vibration and noise control systems are explained briefly. In section 3, a proportional velocity feedback implementation for active vibration control system developed in our laboratory is presented. Finally, the vibration and noise reduction performance of the controller is discussed.

2. Active vibration and noise control system design

2.1. Design procedure

Undesired mechanical vibrations and noise can be avoided or minimized by active vibration and noise control systems. The design of such systems starts with the dynamic characterization of the target structure that is called host structure for embedding or bonding the smart materials. The dynamic characterization procedure can be carried out numerically and/or experimentally by investigating the vibration characteristics of the structure. By utilizing sensors and actuators, system identification methods can be applied for modeling the whole system. The next step is the reduction of the developed model such that the computations can be performed in a reasonable time. In fact, the model reduction techniques shall be applied while considering the controllability and observability properties of the system. The steps between the characterization of the host structure and the reduced order model are referred as "modeling stage". Having determined the reduced-order model, the controller is designed and configured until the performance objectives are met. This stage is called the controller stage where the aim is to design and implement the controller for the performance requirements. The design steps of modeling and controller stages are presented in flow chart by Preumont [12] and depicted in Fig. 1. In the next section, the controller architectures will be discussed in detail.

2.2. Selection of controller architecture

This section presents an overview of controller algorithms for active vibration and noise control systems. In general, the controllers are designed for two different tasks: tracking a

trajectory or rejecting disturbances. For active vibration suppression systems, the task is the rejection of external disturbances and reduction of vibration levels. These can be achieved via feedback and feedforward controllers. The generic block diagrams [7] of these controller algorithms are presented in Fig. 2a and Fig. 2b. Basically, the feedback controller generates controller output signal based on summation of plant response and external disturbance. On the other hand, feedforward controller generates controller output by measuring external disturbance and predicting the plant response.

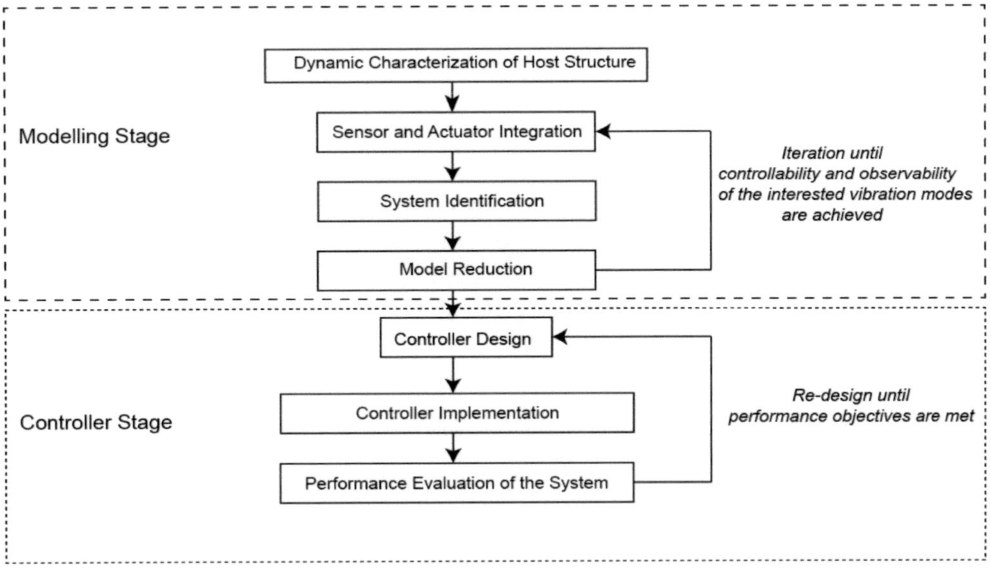

Figure 1. Control System Design Steps

The feedback controller algorithms can be divided into two categories [13] : active damping systems and model-based controllers. In an active damping system, the vibration of the host structure (acceleration, velocity or displacement) can be suppressed around the resonance frequencies. The closed-loop transfer function of the active damping system can be derived by using the block diagram shown in Fig. 2a as it follows:

$$y = d + uG(s) \tag{1}$$

Here, the sensor output signal is denoted by y, external disturbance is shown with d and controller output is u. The open-loop system is presented with G(s) in Laplace domain. After algebraic manipulations, one can obtain the following relation for the closed-loop system:

$$y = \frac{1}{1+C(s)G(s)} d \tag{2}$$

In Equation 2, the closed-loop transfer function shows that the effect of external disturbance on the sensor output can be minimized by increasing the magnitude of the *C(s)G(s)* as the phase and gain margin of the open-loop system *G(s)* allows. Since the collocated sensor and actuator pair provides infinite gain and phase margins, the active damping system works very efficiently when the collocated sensor and actuator pair is used.

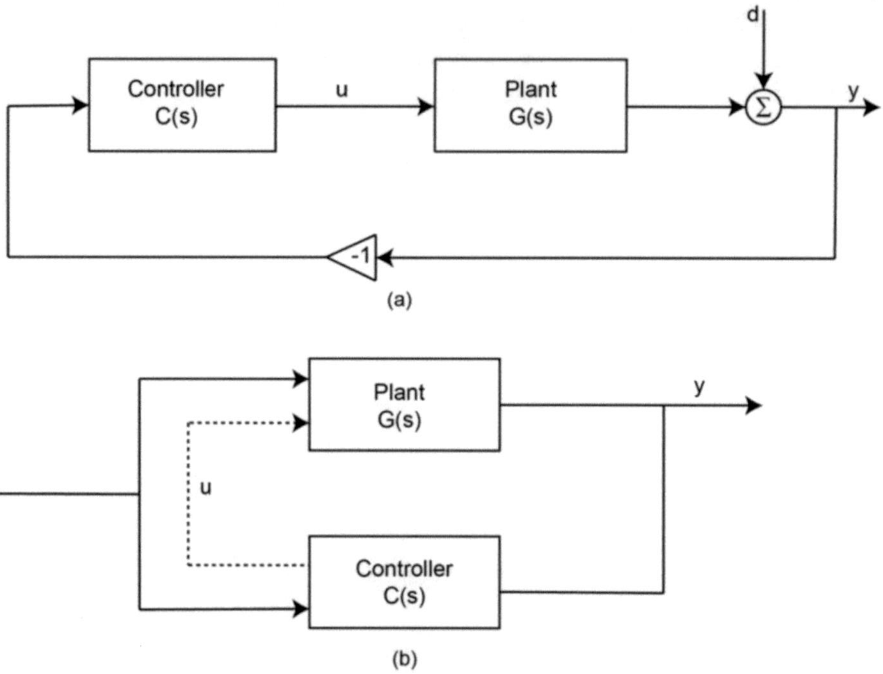

Figure 2. a) Feedback control block diagram b) Feedforward control block diagram

In the case of model-based controller, the open-loop system dynamics is represented with the state-space model as in Equation 3:

$$\begin{aligned} \dot{\mathbf{x}} &= \mathbf{Ax} + \mathbf{Bu} + \mathbf{B_d d} \qquad \text{(a)} \\ \mathbf{y} &= \mathbf{Cx} + \mathbf{Dy} \qquad \text{(b)} \end{aligned} \tag{3}$$

In this state-space form, **x** is the state variable, **y** is the measured output, **u** is the input signal and **d** is the external disturbance signal. In addition to these, **A** presents the state matrix, **B** is the controller input matrix, $\mathbf{B_d}$ is the disturbance matrix, **C** is the output matrix and **D** is the feed-through matrix. For full-state feedback controller ($u = -\mathbf{K}x$), the closed-loop system can be obtained as

$$\begin{aligned} \dot{x} &= (\mathbf{A} - \mathbf{BK})\mathbf{x} + \mathbf{B_d d} \qquad \text{(a)} \\ \mathbf{y} &= (\mathbf{C} + \mathbf{DK})\mathbf{x} \qquad \text{(b)} \end{aligned} \tag{4}$$

The aim of this state-space closed-loop system is to eliminate the effect of disturbance on output signal similar to the active damping. The gain matrix **K** can be determined by applying pole-placement or optimal control design methodologies.

2.3. Design of controller architecture

For control applications, it is necessary to understand the system dynamics appropriately since the controller design parameters are determined based on the system dynamics. As

stated in section 2.2 there are different types of the control algorithms and for each type of the controller, system identification is essential. This system identification can be carried out via a vibration testing & analysis methods. Aim of such method is to extract Frequency Response Functions (FRF) where these functions presents the system response to a specific input in frequency range of interest by means of magnitude and phase. A representative FRF is shown in Fig. 3 for a structure where piezoelectric materials are used as bonded patch actuators (to excite the structure) and LDV as velocity sensor (to measure the vibration response) .

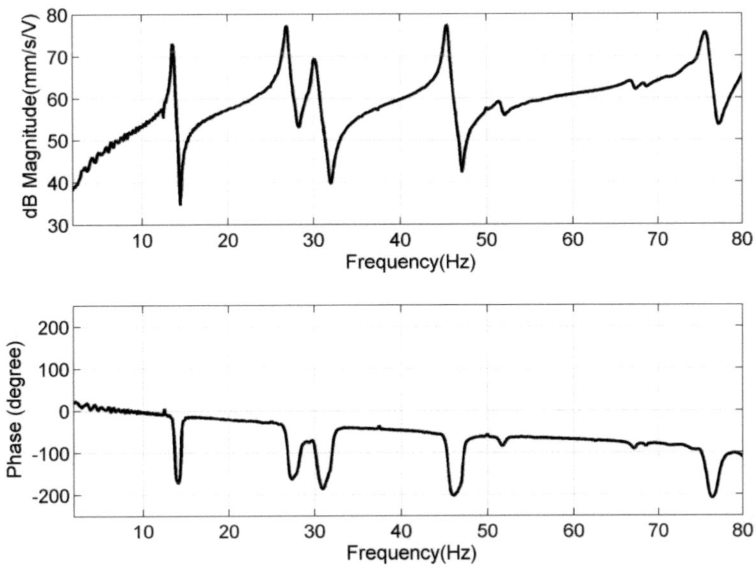

Figure 3. Example FRF of a system

The modal (resonance) frequencies of the systems are indicated in magnitude plot of FRFs as sharp peaks where the active damping systems are most effective. In active damping systems, the higher frequency modes may deteriorate the stability due to phase lag. Such effects of the higher frequency modes can be suppressed by including the low-pass filters. On the other hand, model based controllers use mathematical relations which in turn represents the system response based on FRF. Since the scope of this chapter is devoted only to proportional analog velocity feedback controller, more detailed information on model-based controllers and mathematical system representations can be found in references [6, 7, 9, and 12].

3. Application of analog velocity feedback controller

In this section an active vibration control system using an analog velocity feedback algorithm is described. At first, the system architecture is explained. Later, the controller design and implementation steps are discussed. In the final step, the evaluation of the controller performance is presented.

3.1. System architecture

The structure used in this study is a rectangular box (1 x 1 x 2 m) and shown in Fig. 4. The surrounding cavity walls are constructed using wood filled concrete with a thickness of 10 mm. This box is separated inside with a thin plate to obtain two enclosed compartments. The two compartments are used for investigation of vibration and noise transmission. The thin steel separation plate (1 x 1 m) in the rectangular box has a thickness 1.8 mm and clamped along on all four edges.

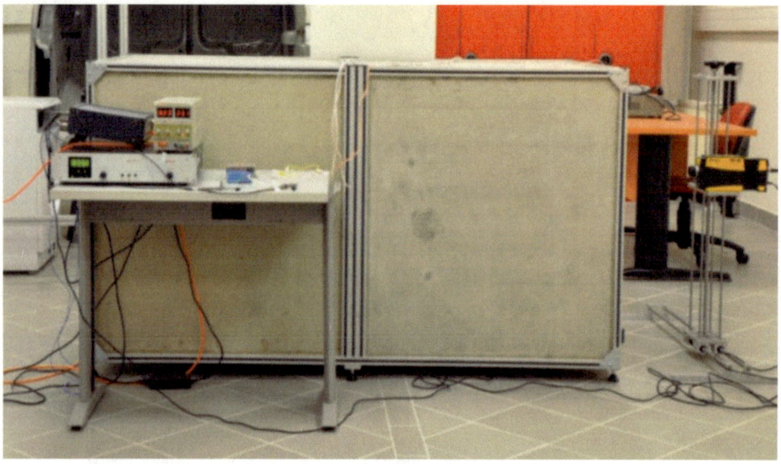

Figure 4. Rectangular box for vibration and noise reduction experiments

In the presence of mechanical and acoustical vibrations, the thin separation plate transmits the vibration and noise to the compartments. By attenuating vibrations of the separation plate, the noise radiation can be also reduced. In this study, the piezoelectric patch actuators bonded on the separation plate are employed as actuators to control and suppress vibration. The piezoelectric actuator is a PI Dura-act 875 bender type piezoelectric patch. This type of patch is compact, lightweight and insulated so it can be easily attached to the host structure and it does not add additional weight. The mechanical properties of the steel and piezoelectric patches are given in Table 1.

Since the piezoelectric materials generate high strain on the host structure, the most effective position for the piezoelectric patch is around the high strain regions on the host structure. In fact, for a clamped-clamped plate, the maximum curvature (highest strain region) changes over the surface for each vibration mode. The maximum curvature of the first mode of a clamped-clamped plate occurs at the center and along the edges; whereas in the second vibration mode, maximum curvature is in the middle of the upper and lower halves of the plate. After these high strain regions are identified, the piezoelectric patches are attached to the center and upper-left (-35 cm, 35 cm from the left upper corner) of the plate in order to suppress first two modes. For suppression of vibration and noise, two piezoelectric patches are used in independent and pair configuration. Fig. 5 shows the thin-separation plate and piezoelectric patches.

Property	Steel	Piezoelectric Patch
Length (mm)	1000	65
Width (mm)	1000	31
Depth (mm)	1.5	0.5
Elastic modulus (GPA)	190	70.2
Poisson's ratio	0.33	0.36
Density (kg/m3)	6305	7800
Piezoelectric Strain Constant d_{31} (10^{-10} m/V)	-	1.74
Piezoelectric Strain Constant d_{33} (10^{-10} m/V)	-	3.94
Piezoelectric Strain Constant d_{51} (10^{-10} m/V)	-	5.35

Table 1. Mechanical properties of host structure and piezoelectric patch

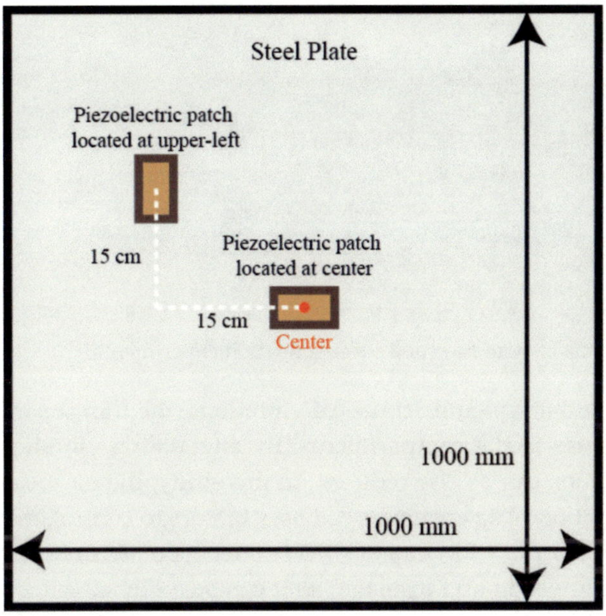

Figure 5. Thin steel plate with piezoelectric patches

3.2. Controller architecture

This section presents design and implementation of the active vibration controller via piezoelectric actuators. At first, proportional feedback controller is presented and then analog implementation of the controller is explained.

3.2.1. Design of proportional feedback controller

Proportional feedback control is a very simple and easily implementable controller methodology. Since only a sensor signal is passed through a negative amplifier and fed back to the system, the total computational and implementation cost is very low. Thus, it is possible to apply this control scheme using analog circuits since it does not need online

calculation of parameters such as controller gains and current states' values. In addition to these, by using collocated sensor and actuator pair, proportional feedback controller provides active damping at the resonance frequencies of the structure. In fact, this active damping can be employed effectively in the lower-frequency modes of the structure because of stability issues. At the high-frequencies, the phase lag may deteriorate the stability of the collocation and the closed-loop system can be unstable. This high-frequency stability problem is eliminated usually by decreasing the proportional gain or including a low-pass filter in the closed-loop. In this paper, we consider a proportional velocity feedback controller along with a low-pass filter to eliminate high-frequency dynamics.

The undesired vibrations of the thin separation plate are originated by external disturbances and the controller is designed for the expected disturbances including frequencies up to the third resonance mode. By only aiming low-frequency vibration suppression of the plate, high frequency performance of the controller is not considered. For this purpose, a low pass filter which has a cut-off frequency around third resonance mode is considered. In order to have decay at the frequencies higher than the cut-off frequency, the filter is included as an input low-pass filter and is placed between the sensor signal and the controller.

Then, the amplifier gain is adjusted by utilizing the tools of Ziegler-Nichols method. In Ziegler-Nichols method, the important step is the determination of critical gain (gain margin of the closed-loop system). This critical gain is defined as the amplifier ratio at which response of the controlled plant has sustained oscillations and closed-loop system is at the stability limit. Proportional feedback criterion of Ziegler-Nichols method claims that 0.8 times the critical gain yields the best proportional controller performance.

Having determined the cut-off frequency of the filter and amplifier ratio, the proportional velocity feedback controller is finalized for the implementation and experiments.

3.2.2. Implementation of velocity feedback controller

The designed proportional velocity feedback controller is implemented on a breadboard by using the circuit shown in Fig. 6. This analog circuit includes one inverting amplifier and low-pass filter.

The measured sensor voltage signal (V_{in}) is passed through a low pass filter which is in the form of resistor-capacitor (RC) filter. The cut-off frequency of this filter equals to one divided by multiplication of resistor and capacitance value. The aim of this low-pass filter is the reduction of high-frequency components of the sensor signal. After passing through this filter, the filtered signal is fed to the inverting amplifier. This inverting amplifier includes an operational amplifier (op-amp) and two resistors as indicated in Fig. 6. The op-amp is powered by dual polarity. The filtered sensor voltage is connected through input resistor R_c to the positive input channel of the op-amp. Then, a jumper resistor R_f is included between positive input channel and the output channel, whereas negative input channel is connected to the ground. By passing through this inverting amplifier, the sign of filtered sensor signal is inverted and its voltage value is amplified. The amplifier ratio is equivalent the ratio of jumper resistor R_f and input resistor R_c. Finally, this amplified and inverted signal serves as the proportional controller output signal.

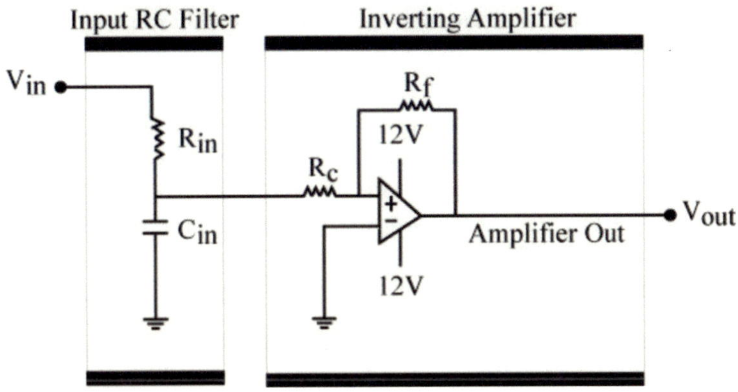

Figure 6. Implementation diagram for analog proportional velocity feedback controller

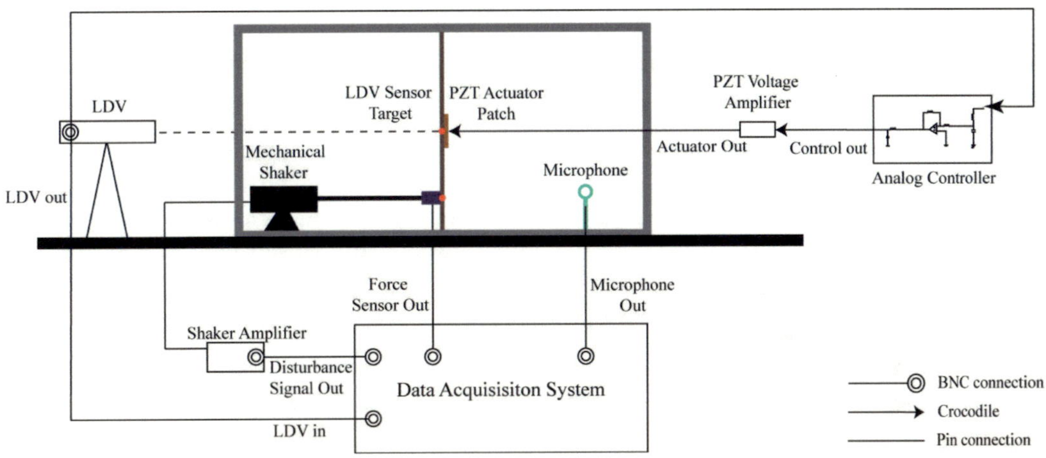

Figure 7. Experimental setup for active vibration and noise reduction

3.3. Experimental setup

The experimental setup for active vibration control via analog velocity feedback controller is presented in Fig. 7 with the data acquisition system. The data acquisition system has the ability of generation and recording analog voltage signal while communicating with a PC through Ethernet connection. The system provides high frequency sampling with high measurement accuracy and also generates the disturbance signal to drive the shaker amplifier. The mechanical shaker is located in the left compartment in Fig.4, and attached to the thin-separation plate via a connecting rod and a force sensor. The vibration of this plate is measured by Polytech PDV100 laser Doppler vibrometer (LDV). The target location of the vibrometer is determined as the same point of the piezoelectric patch to obtain a collocated sensor and actuator pair. The voltage output of the LDV is connected to the analog controller as a sensor signal and sent to the data acquisition system for recording. By connecting the sensor signal to the analog circuit, the controller output signal is acquired. This controller signal is amplified by E-413 Dura-act Piezo Driver and sent to the PZT patches (Dura-act P876.A12). Control input voltage between -2V and +8V are accepted by

the voltage amplifier and the input signal is amplified by factor of 50. By closing-the-controller loop, an active damping system for vibration suppression is created. While the vibration suppression experiments are conducted, the output signal of a microphone properly located in one of the compartments is also recorded at the same time to demonstrate the suppression of noise.

3.4. Results

In this section, active vibration suppression of the thin separation plate and noise reduction inside the compartment via analog velocity feedback controller is presented. The experimental results are given in frequency and time domains for three different piezoelectric patch configurations. At first, the piezoelectric patch located at center and upper-left is used as an actuator independently, and then they are employed together as an actuator.

3.4.1. Suppression of vibration in time domain

The performance of the analog velocity feedback control for the vibration suppression is evaluated in the time domain by disturbing the thin separation plate with the mechanical shaker. For this purpose, disturbance signal is applied to the mechanical shaker via data acquisition unit as a sine-wave at the first resonance frequency (approximately 13 Hz) and at the second-resonance frequency (approximately 26 Hz) of the plate. The vibration of the plate is monitored by targeting laser displacement to the center of the piezoelectric actuator patch. The experiments are conducted for each piezoelectric actuator configuration. Fig. 8 presents the results for forced vibration at the first resonance frequency. Fig. 8 (a) presents the open-loop vibration at the center of the plate. The open-loop response corresponds to the structural vibrations of the thin plate when the analog feedback controller is inactive. Fig. 8(b) shows the closed-loop vibration response of the plate when the active vibration suppression is obtained via piezoelectric patch located at center. In Fig. 8(c), the piezoelectric patch located at upper-left and is employed as actuator independently. The Fig. 8(d) shows the closed-response for the piezoelectric actuator in pair configuration where the center and upper-left patch are used as actuator jointly and laser displacement sensor is targeted to the center of the upper-left patch.

As can be seen from the subfigures of Fig. 8, the vibration of the plate at the first resonance frequency is attenuated for each piezoelectric actuator configuration. In fact, the best performance is acquired when the piezoelectric center patch is used as actuator independently. The performance of this configuration is better than the pair configuration because of phase mismatch of the piezoelectric patches in pair. Fig. 8 presents the open and closed vibration of the plate for the second resonance forced vibration. To measure the open-loop vibrations of the plate, the LDV is targeted to the center of the upper-left patch. Fig. 9(a) presents the open-loop vibration measured at this location. Fig. 9(b) shows the performance of the active vibration control via piezoelectric actuator patch located at the upper-left of the plate. Besides, Fig. 9(c) presents the effectiveness of piezoelectric pair configuration for the vibration suppression of the plate.

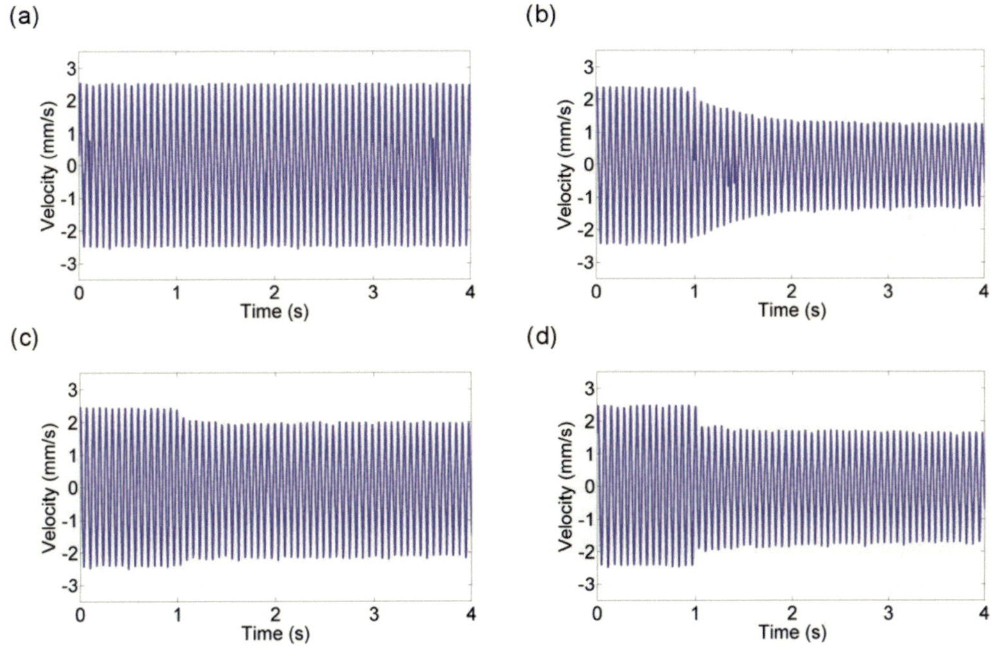

Figure 8. Vibration response of the plate at the first resonance frequency (a) Open-loop (b) Closed-loop with piezoelectric patch located at center (c) Closed-loop with piezoelectric patch located at upper-left (d) Closed-loop with piezoelectric actuator patches in pair

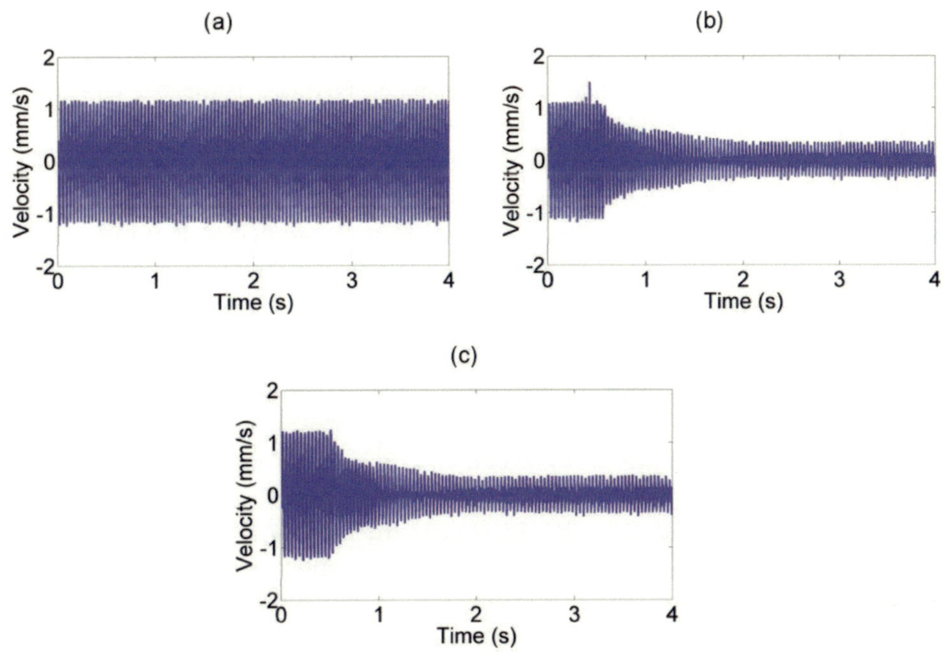

Figure 9. Vibration response of the plate at the second resonance frequency (a) Open-loop (b) Closed-loop with piezoelectric patch located at upper-left (c) Closed-loop with piezoelectric actuator patches in pair

Piezoelectric Patch Configuration	LDV Target Location	First Mode	Second Mode
Center	Center	44.6% (5.1dB)	-
Upper-left	Center	20.4% (1.9 dB)	73.8% (11.6 dB)
Center and Upper-left (Pair Configuration)	Upper-left	32.3% (3.4 dB)	74.9% (11.9 dB)

Table 2. Vibration suppression levels in time domain for different piezoelectric actuator configurations

To employ piezoelectric patches in pair for the second resonance mode of the plate, the LDV is targeted again to the center of the upper-left piezoelectric patch. It is obvious that the independent configuration of piezoelectric patch located at upper-left provides better vibration suppression when it is compared with the pair configuration. The reduction levels for each configuration are presented in Table 2.

3.4.2. Supression of vibration in frequency domain

The aim of this section is to present the performance of the active vibration suppression in the frequency domain. The experiments are conducted for the same piezoelectric actuator configurations. In contrast to time-domain results, the disturbance signal applied to the mechanical shaker is changed to sine-sweep in the bandwidth of 2 Hz to 100 Hz. By monitoring the open-loop and closed-loop response of the plate, frequency responses are gathered using the data acquisition system. Fig. 10 presents the frequency responses for different piezoelectric configurations between vibrometer and the force transducer at the tip of the mechanical shaker. In Fig. 10(a), the performance of the active vibration suppression system via piezoelectric patch located at center is presented. It is noticeable that the vibration is attenuated very well in the first resonance mode of the plate; however the other resonance frequencies of the plate are not affected. This is due to the fact that the position of piezoelectric patch located at center corresponds to the anti-mode of the higher order modes. So, the higher structural modes are not affected. Fig. 10(b) presents the closed-loop performance of the piezoelectric patch located at upper-left. As can be seen this figure, the closed-loop system reduces the vibration of the plate in a broad frequency range for this configuration. Fig. 10(c) shows the effectiveness of the closed-loop system via piezoelectric actuator pair. It is noticeable that the performance of the closed-loop is improved considerably when piezoelectric pair configuration is adopted instead of using upper-left patch independently. Indeed, the closed-loop system reduces the vibration levels as compared to open-loop in all piezoelectric actuator configurations. Table 3 lists the vibration reduction levels in the frequency domain.

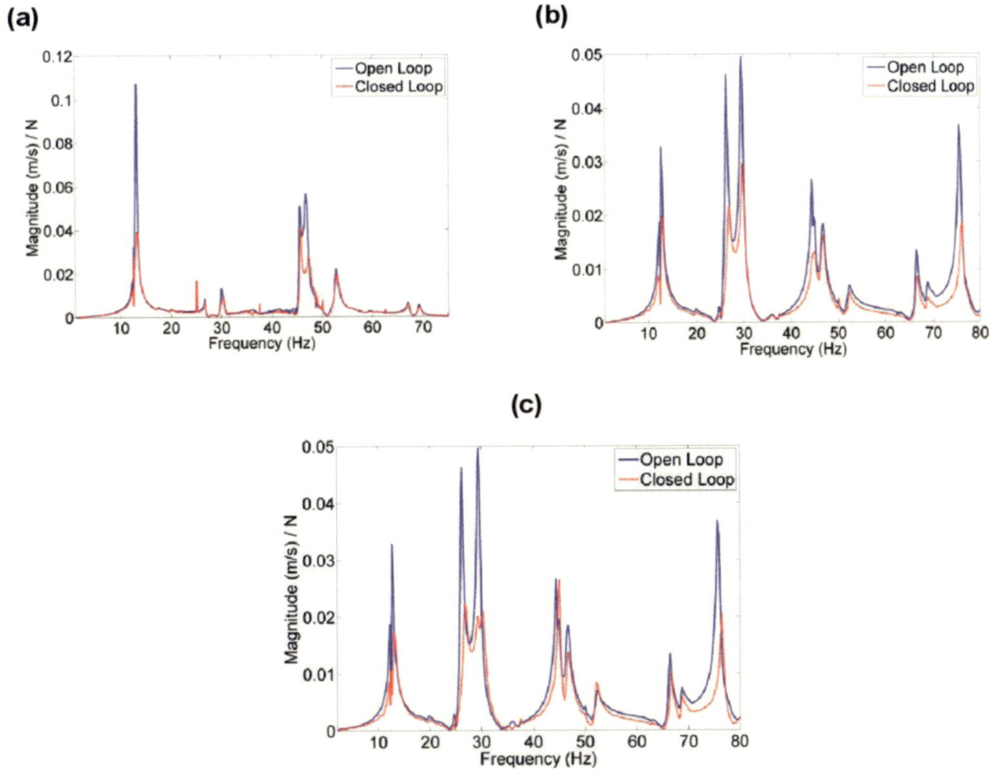

Figure 10. Frequency responses for different piezoelectric actuator configurations (a) Closed-loop with piezoelectric patch located at center(LDV at center) (b) Closed-loop with piezoelectric patch located at upper-left(LDV at upper-left) (c) Closed-loop with piezoelectric actuator patches in pair(LDV at upper-left)

Piezoelectric Patch Configuration	LDV Target Location	First Mode	Second Mode
Center	Center	63.7 % (8.8 dB)	-
Upper-left	Center	34.1 % (4.4 dB)	52.8 % (6.5 dB)
Center and Upper-left (Pair Configuration)	Upper-left	50.6 % (6.1 dB)	54.0 % (6.7 dB)

Table 3. Vibration suppression levels in time domain for different piezoelectric actuator configurations

3.4.3. Suppression of noise in frequency domain

This section presents the noise reduction in one of the compartments for controller-inactive and active cases while disturbing the thin plate via shaker. The noise inside the compartment is measured with a microphone. The frequency responses between the microphone and the force transducer are gathered using the data acquisition system. Fig. 11

presents the results for noise in the frequency domain. As in the same manner with vibration suppression experiments, open-loop response shows the noise radiation in the compartment when the controller is inactive, whereas closed-loop response is monitored when the controller is active. In Fig. 11(a), the performance of noise reduction via piezoelectric patch located at center is given. It is noticeable that the low-frequency acoustic mode in the compartment coincides with the vibration mode of the plate. Since the vibration of plate at the first resonance mode is considerably attenuated, the noise radiation is also reduced for this mode. At the higher-acoustics modes, the closed-loop system does not improve and deteriorate the noise radiation in the compartment. Fig. 11(b) presents the effectiveness of the piezoelectric patch located at upper-left. Similar to the previous results, noise radiation is reduced by suppressing the vibration of the plate. However, it is obvious that the performance of the piezoelectric patch located at center is fairly better than the performance of the piezoelectric patch located at upper-left.

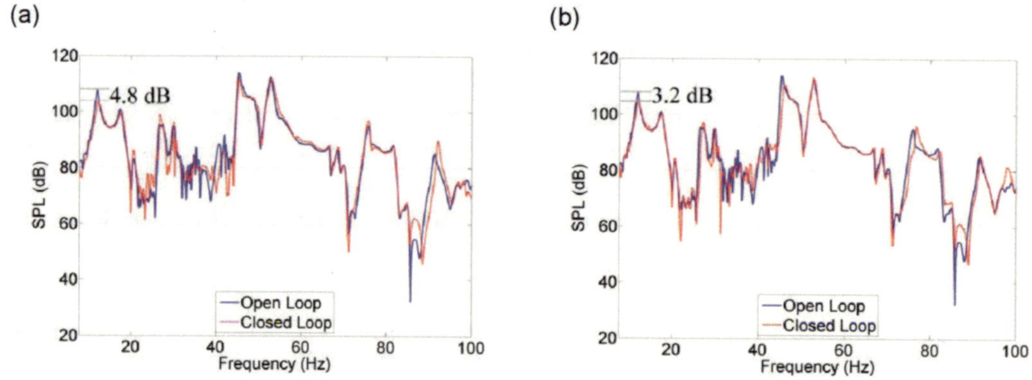

Figure 11. Noise levels in the one of the compartments for different piezoelectric patch configurations (a) Open and closed-response for the piezoelectric patch located at center (b) Open and closed-loop response for the piezoelectric patch located at upper-left

4. Conclusion

In the first part of this chapter, a short review of literature about active vibration control is presented. The general controller design approach and the differences between various controller algorithms are discussed.

In the second part of this chapter, an application using an analog velocity feedback controller for vibration suppression of a flexible plate is presented. Experiments for vibration suppression of the flexible plate are conducted and by measured the vibration of the plate; the time-responses and frequency-responses are presented. Finally, the acoustic response in one of the compartments is also obtained for different piezoelectric actuator configurations. The results revealed that vibration reduction of the separation plate improves the noise radiation and the sound pressure level is decreased due to the reduction of the vibration. However, for an effective and powerful noise reduction, robust controller algorithms with multiple piezoelectric actuator patches can be preferred.

Author details

Ipek Basdogan*, Utku Boz, Serkan Kulah and Mustafa Ugur Aridogan
Koç University, Turkey

5. References

[1] Bailey T, Hubbard J.E. Distributed Piezoelectric Polymer Active Vibration Control of a Cantilever Beam. Journal of Guidance Control and Dynamics 1985;8(5): 605-611.

[2] Crawley E.F, De Luis J. Use of Piezoelectric Actuators as Elements of Intelligent Structures. AIAA Journal 1987; 25: () 1373-1385.

[3] Ro J, Baz A, Control of Sound Radiation from a Plate into an Acoustic Cavity Using Active Constrained Layer Damping. Smart Materials and Structures 1999; 8(3): () 292.

[4] Rao S.S, Sunar M. Piezoelectricity and Its Use in Disturbance Sensing and Control of Flexible Structures: A Survey. Applied Mechanics Reviews 1994; 47(4):113-123.

[5] Loewy R.G, Recent Developments in Smart Structures with Aeronautical Applications. Smart Materials and Structures 1997; 6(5): R11.

[6] Chee C.Y.K, Tong L, Steven G.P. A Review on the Modelling of Piezoelectric Sensors and Actuators Incorporated in Intelligent Structures. Journal of Intelligent Material Systems and Structures 1998; 9(1): 3-19.

[7] Alkhatib R, Golnaraghi M.F. Active Structural Vibration control: A review, Shock and Vibration Digest 2003; 35(5):367-383.

[8] Gupta V, Sharma M, Thakur N. Optimization Criteria for Optimal Placement of Piezoelectric Sensors and Actuators on a Smart Structure: A Technical Review. Journal of Intelligent Material Systems and Structures 2010; 21(12):1227-1243.

[9] Iorga L, Baruh H, Ursu I. A Review of H_∞ Robust Control of Piezoelectric Smart Structures. Applied Mechanics Reviews 2008; 61(4): 040802-040815.

[10] Chopra I. Review of State of Art of Smart Structures and Integrated Systems. AIAA Journal 2002; 40(11):2145-2187.

[11] Hanselka H. Adaptronics as a Key Technology for Intelligent Lightweight Structures. Advanced Engineering Materials 2001;, 3: 205-215.

[12] Preumont A. Vibration Control of Active Structures : An Introduction, 2 ed. Kluwer Academic: New York; 2002.

[13] Hurlebaus S, Gaul L. Smart Structure Dynamics. Mechanical Systems and Signal Processing 2006; 20(2):255-281.

* Corresponding Author

Chapter 27

Shunted Piezoceramics for Vibration Damping – Modeling, Applications and New Trends

Marcus Neubauer, Sebastian M. Schwarzendahl and Xu Han

Additional information is available at the end of the chapter

http://dx.doi.org/10.5772/50609

1. Introduction

Piezoelectric shunt damping is a well known technique to damp the vibrations of mechanical structures. This technique relies on the piezoelectric effect that converts mechanical energy into electrical energy. A damping effect on the host structure is observed when the electrical energy is dissipated. In order to optimize the damping performance, the transferred energy as well as the dissipated energy must be maximized. The transferred energy depends on the piezoelectric constants as well as the vibration mode and the location of the piezoelements within the structure. While higher piezoelectric constants generally increase the amount of transferred energy, the location can typically only be optimized for one eigenform of the structure. As a consequence, the piezoelectric transducer can be placed in such a way that it only affects one eigenform of the system. One measure for the coupling of the piezoceramics is the generalized coupling coefficient, which is defined for every vibration mode and generally takes different values for the individual modes. The design of the electrical shunt aims at maximizing the energy dissipation. Different networks have been developed, which can be classified into the categories passive, active, linear or non-linear. The best choice of network depends on the performance target, the availability of electrical power supply and the vibration behavior and excitation type of the mechanical structure.

Passive resonant circuits have been among the first networks for piezoelectric shunt damping. Forward studied inductance-resistance networks for the damping of optical systems [4] which were tuned to the resonant frequency of the mechanical system. Hagood and von Flotow then studied the performance and tuning of these *LR*-networks in more detail [7]. They described the shunted piezoceramics as a frequency depending stiffness and damping element and they showed the analogies of *LR*-shunted piezoceramics and tuned mass dampers. They obtained from calculations that the damping effect grows with the piezoelectric coupling coefficient. The standard *LR*-network can only be tuned to one frequency, therefore in the subsequent years new circuits were proposed that are capable to damp several frequencies at the same time [2, 9]. These networks basically consists of multiple *LR* branches that are tuned to the individual frequencies to be damped.

In order to enhance the limited damping performance of passive shuntings, active elements have been proposed. The mostly studied element is a negative capacitance, which can be realized by a negative impedance converter circuit [5]. Initially considered by Forward [3], a negative capacitance proves to be able to increase the effective piezoelectric coupling factor. Especially the combination of passive networks with active elements is a promising approach. This class of networks is called 'active-passive hybrid piezoelectric network' (APPN) by Tang [21]. Most prominent APPN networks are a negative capacitance with a resistor and a negative capacitance with an inductor and resistor.

The drawback of these linear resonant networks is that they all must be tuned to a certain frequency, which has to be known in advance and which should not change during operation. For many applications they are therefore not suitable. In these cases adaptive, non-linear networks are a better choice. The most common one is the 'synchronized switch damping on inductor' (SSDI) technique, which consists of an *LR*-branch and a switch that can connect and disconnect the network to the electrodes of the piezoceramics [10]. For the case of monoharmonic excitation the switch is closed at the moments of maximum deformation of the piezoceramics. In this moment, the electrical charge is inverted via the inductance. The inductance value is very small in order to realize a fast inversion. When fully inverted, the switch is closed so that the charge cannot flow anymore. During the following half period of excitation the charge stays nearly constant, so that the piezoceramics generates a force acting against the deformation velocity. The resulting force signal is nearly rectangular shaped. Like for the passive *LR* shunting the damping strongly depends on the electrical damping ratio, which can be set by the resistance value. A small damping results in a good inversion of the charge, which amplifies the stationary charge amplitudes and the dissipated energy. The adaptive capability of the SSDI technique comes from the triggering of the switching times. Therefore, typically one additional sensor is used which monitors the vibration of the mechanical structure. Due to this triggering, the force signal from the piezoceramics is always in phase with the structure vibration and the performance is only minimally dependend on the excitation frequency.

2. Modeling of mechanical structures with a shunted piezoceramics

In order to obtain general results which can be transferred to various mechanical structures, the structure is reduced to a one degree of freedom oscillator. The piezoceramics is shunted to an arbitrary electrical impedance Z, as shown in Figure 1. The equations of motion for this general case read

$$\begin{bmatrix} m & 0 \\ 0 & 0 \end{bmatrix} \begin{bmatrix} \ddot{q} \\ \ddot{Q} \end{bmatrix} + \begin{bmatrix} d & 0 \\ 0 & 0 \end{bmatrix} \begin{bmatrix} \dot{q} \\ \dot{Q} \end{bmatrix} + \begin{bmatrix} c + \frac{\alpha^2}{C_\mathrm{p}} & \frac{\alpha}{C_\mathrm{p}} \\ \frac{\alpha}{C_\mathrm{p}} & \frac{1}{C_\mathrm{p}} \end{bmatrix} \begin{bmatrix} q \\ Q \end{bmatrix} = \begin{bmatrix} F(t) \\ -u_\mathrm{p} \end{bmatrix}, \tag{1}$$

where the parameters m, d and c denote respectively the modal mass, damping and stiffness. Stiffness c is herein the sum of the mechanical stiffness c_mech and the stiffness of the piezoelement c_p. $F(t)$ represents the external force, u_p is the voltage at the electrodes of piezoceramics, C_p the capacitance of the piezoceramics and α the force factor which can be deduced from geometry and characteristics of the piezoceramics and the mechanical structure. The variables q and Q are respective the modal displacement and electrical charge.

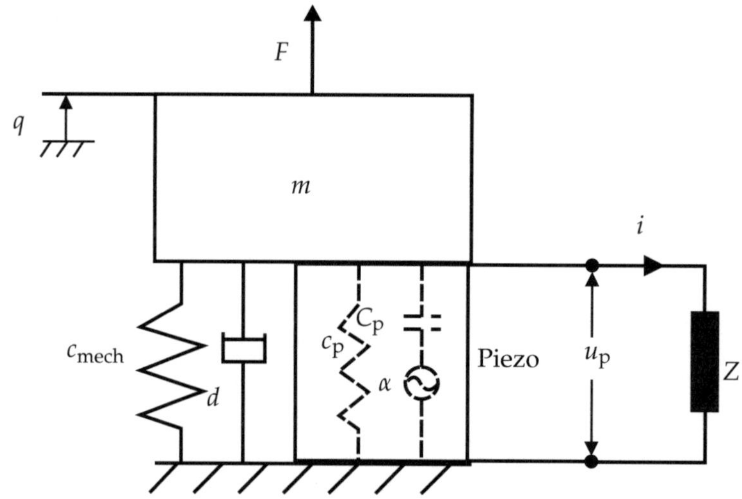

Figure 1. Single Degree-of-Freedom oscillator with piezoceramics and shunt circuit.

When the piezoceramics is shunted to an electrical circuit, the voltage u_p depends on the charge Q as well as the impedance Z of the shunt,

$$u_p = ZQ. \tag{2}$$

In this context, Z describes the relationship between voltage and charge rather than between voltage and current. Inserting (2) into (1), the generalized equation of a shunt damping system reads as

$$\begin{bmatrix} m & 0 \\ 0 & 0 \end{bmatrix} \begin{bmatrix} \ddot{q} \\ \ddot{Q} \end{bmatrix} + \begin{bmatrix} d & 0 \\ 0 & 0 \end{bmatrix} \begin{bmatrix} \dot{q} \\ \dot{Q} \end{bmatrix} + \begin{bmatrix} c + \frac{\alpha^2}{C_p} & \frac{\alpha}{C_p} \\ \frac{\alpha}{C_p} & \frac{1}{C_p} + Z \end{bmatrix} \begin{bmatrix} q \\ Q \end{bmatrix} = \begin{bmatrix} F(t) \\ 0 \end{bmatrix}. \tag{3}$$

This equation is the basis for all further calculations.

3. Optimization of resonant LR-shunting for damped mechanical systems

Let us first consider a resonant LR-shunt with impedance $Z = Ls^2 + Rs$. Substituting this term into (3) leads us to

$$\begin{bmatrix} m & 0 \\ 0 & L \end{bmatrix} \begin{bmatrix} \ddot{q} \\ \ddot{Q} \end{bmatrix} + \begin{bmatrix} d & 0 \\ 0 & R \end{bmatrix} \begin{bmatrix} \dot{q} \\ \dot{Q} \end{bmatrix} + \begin{bmatrix} c + \frac{\alpha^2}{C_p} & \frac{\alpha}{C_p} \\ \frac{\alpha}{C_p} & \frac{1}{C_p} \end{bmatrix} \begin{bmatrix} q \\ Q \end{bmatrix} = \begin{bmatrix} F(t) \\ 0 \end{bmatrix}. \tag{4}$$

For maximum damping performance the circuit parameters L and R have to be tuned in such a way that the system has double eigenvalues. For the undamped case ($d = 0$) the optimal parameters are well known.

With the help of normalized, non-dimensional parameters, the equations can be written in a more compact form. In detail, the generalized piezoelectric coupling coefficient K, the eigenfrequency of the system with isolated electrodes ω_{iso}, the electrical eigenfrequency ω_{el},

the electrical damping ratio ζ_0 of the LR-branch and the frequency ratio η_{el} are introduced,

$$K^2 = \frac{\alpha^2}{cC_p + \alpha^2}, \quad \omega_{\text{el}} = \sqrt{\frac{1}{C_p L}}, \quad \omega_{\text{iso}}^2 = \frac{c + \alpha^2/C_p}{m},$$

$$\zeta_0 = \frac{R}{2}\sqrt{\frac{C_p}{L}}, \quad \eta_{\text{el}} = \frac{\omega_{\text{el}}}{\omega_{\text{iso}}} = \sqrt{\frac{m}{\alpha^2 L}} K. \tag{5}$$

The generalized piezoelectric coupling coefficient K is a measure of the effectiveness of the piezoceramics. It depends on the piezoceramics characteristics as well as on the structure vibration form. The shunt parameters L, R are substituted by non-dimensional parameters $\eta_{\text{el}}, \zeta_0$. The matrix $\mathbf{A}$ of the corresponding state-space system then reads as

$$\mathbf{A} = \begin{bmatrix} 0 & 0 & 1 & 0 \\ 0 & 0 & 0 & 1 \\ -1 & -K^2 & 0 & 0 \\ -\eta_{\text{el}}^2 & -\eta_{\text{el}}^2 & 0 & -2\eta_{\text{el}}\zeta_0 \end{bmatrix}, \tag{6}$$

and the characteristic equations become

$$\det(\mathbf{A} - \lambda \mathbf{I}) = \lambda^4 + 2\zeta_0\eta_{\text{el}}\lambda^3 + \left(1 + \eta_{\text{el}}^2\right)\lambda^2 + 2\zeta_0\eta_{\text{el}}\lambda + \left(1 - K^2\right)\eta_{\text{el}}^2 = 0. \tag{7}$$

Double eigenvalues are obtained when the shunt parameters are tuned according to

$$\eta_{\text{el,opt}} = \frac{1}{\sqrt{1 - K^2}} \approx 1, \quad \zeta_{0,\text{opt}} = K. \tag{8}$$

These equations basically state that the inductance has to be tuned in such a way that the electrical resonant frequency nearly equals the mechanical resonant frequency. The damping ratio of the circuit must match the generalized piezoelectric coupling coefficient K. The resulting damping ratio from such a shunting then becomes

$$D = \frac{K}{2\sqrt{1 - K^2}} \approx \frac{K}{2}. \tag{9}$$

This equation proves the importance of the generalized piezoelectric coupling coefficient K, as the damping performance grows with K. Figure 2 shows the influence of the network parameters upon the location of the complex eigenvalues.

For the damped mechanical oscillator, the same strategy can be followed to optimize the system. Normalizing the mechanical damping by $D_m = \frac{d}{2m\omega_{\text{iso}}}$ the characteristic equation of the system becomes

$$\lambda^4 + (2\zeta_0\eta_{\text{el}} + 2D_m)\lambda^3 + \left(1 + \eta_{\text{el}}^2 + 4\zeta_0\eta_{\text{el}}D_m\right)\lambda^2 + \left(2\zeta_0\eta_{\text{el}} + 2\eta_{\text{el}}^2 D_m\right)\lambda + \left(1 - K^2\right)\eta_{\text{el}}^2 = 0. \tag{10}$$

The results for the optimal network parameters $\eta_{\text{el,opt}}$ and $\zeta_{0,\text{opt}}$ as well as the resulting damping performance are very lengthy terms. It is useful to express them in a Taylor series,

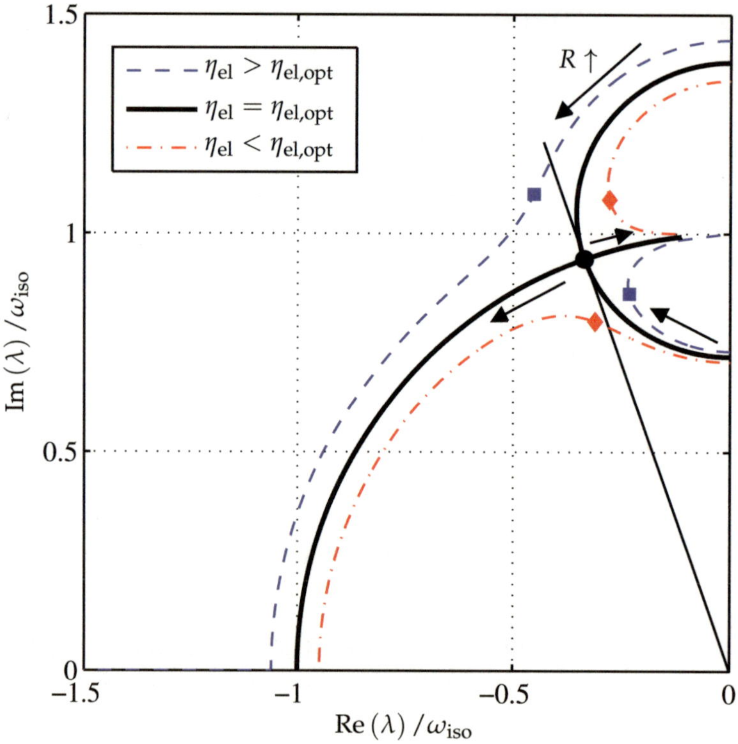

Figure 2. Eigenvalue of mechanical 1 DOF oscillator with *LR*-shunted piezoceramics.

and only consider the first elements of the series. This gives

$$\eta_{el,opt} \approx \frac{1}{\sqrt{1-K^2}} + \frac{K}{1-K^2} D_m,$$

$$\zeta_{0,opt} \approx K + \sqrt{1-K^2} D_m,$$

$$D \approx \frac{1}{2}\frac{K}{\sqrt{1-K^2}} + \frac{4-3K^2}{4-4K^2} D_m. \tag{11}$$

These approximations are valid for small mechanical damping ratios D_m, which is practically fulfilled in most cases. The equations clearly show the trend when mechanical damping is included. For the undamped case, $D_m = 0$, the results are per definition equal to the values obtained in (8) and (9). But additional mechanical damping leads to a slight increase of the optimum electrical resonant frequency η_{el} and damping ratio $\zeta_{0,opt}$. Naturally also the resulting damping performance grows with additional mechanical damping. One can realize that for high mechanical damping D_m (compared to the coupling coefficient K), the overall damping converges the mechanical damping, $D \approx D_m$, and the additional damping caused by the shunted piezoceramics is negligible.

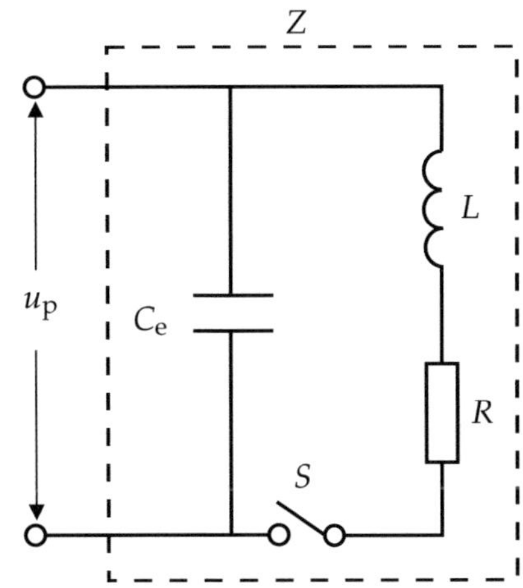

Figure 3. SSDNCI network.

4. Synchronized switch damping on negative capacitance and inductance (SSDNCI)

This section describes a novel combination of a negative capacitance together with the well-known SSDI-technique schematized in Figure 3, which has recently also been studied by other research groups [11]. It is intended to combine the adaptive ability of the SSDI with the enhanced performance and coupling of a negative capacitor. Again the calculations will be performed using nondimensional parameters,

$$\delta = \frac{C_e}{C_p}, \quad \omega_{el} = \frac{1}{\sqrt{(1+\delta)LC_p}}, \quad \zeta = \sqrt{1+\delta}\frac{R}{2}\sqrt{\frac{C_p}{L}} = \sqrt{1+\delta}\zeta_0, \quad \tau = \omega_{el}t. \tag{12}$$

The electrical damping ratio of the LR-branch of the circuit is again termen ζ_0, while the overall electrical damping ratio with negative capacitance is ζ. The capacitance ratio δ can be set by choosing appropriate values for the negative capacitance. For a positive external capacitance, the parameter δ is positive, for a negative capacitance it is negative. The electrical resonance frequency as well as the electrical damping ratio both depend on the capacitance ratio ζ. Setting $\delta = 0$ results in the standard SSDI technique without negative capacitance, with the corresponding electrical damping ratio ζ_0. Obviously, the negative capacitance influences the electrical resonance frequency as well as the damping ratio. Especially the latter one is important, as the damping ratio should be as small as possible. Using a negative capacitance reduces the damping ratio.

The switching network is a nonlinear system, but it can be considered as linear during the periods with open and closed switch. When the switch is closed, the shunt impedance Z_{cl} reads

$$Z_{cl} = \frac{1}{\frac{1}{Ls^2+Rs} + \frac{1}{1/C_e}} = \frac{Ls^2+Rs}{C_eLs^2 + C_eRs + 1}. \tag{13}$$

Substitution of Z_{cl} into (3) and representation in terms of the non-dimensional parameters yields

$$\begin{bmatrix} m & 0 \\ \alpha\delta L & (1+\delta)L \end{bmatrix} \begin{bmatrix} \ddot{q} \\ \ddot{Q} \end{bmatrix} + \begin{bmatrix} 0 & 0 \\ \alpha\delta R & (1+\delta)R \end{bmatrix} \begin{bmatrix} \dot{q} \\ \dot{Q} \end{bmatrix} + \begin{bmatrix} c+\frac{\alpha^2}{C_p} & \frac{\alpha}{C_p} \\ \frac{\alpha}{C_p} & \frac{1}{C_p} \end{bmatrix} \begin{bmatrix} q \\ Q \end{bmatrix} = \begin{bmatrix} F(t) \\ 0 \end{bmatrix}. \tag{14}$$

The standard SSDI network without negative capacitance is the special case with $\delta = 0$.

In this study, a harmonic mechanical vibration q is assumed, and the influence of the shunted piezoceramics upon the vibration is neglected. This is fulfilled in good appromixation for systems where the piezoelectric coupling is not excessively large. Therefore, the second equation from (14) can be rewritten, summarizing all terms with q as excitation on the right side,

$$(1+\delta)L\ddot{Q} + (1+\delta)R\dot{Q} + \frac{1}{C_p}Q = -(\alpha\delta L\ddot{q} + \alpha\delta R\dot{q} + \frac{\alpha}{C_p}q). \tag{15}$$

When the switch is open, the piezoceramics is connected to the negative capacitance, $Z_{iso} = \frac{1}{C_e}$. Inserting this impedance into (3) yields

$$\begin{bmatrix} m & 0 \\ 0 & 0 \end{bmatrix} \begin{bmatrix} \ddot{q} \\ \ddot{Q} \end{bmatrix} + \begin{bmatrix} c+\frac{\alpha^2}{C_p} & \frac{\alpha}{C_p} \\ \frac{\alpha}{C_p} & \frac{1}{C_p}+\frac{1}{C_e} \end{bmatrix} \begin{bmatrix} q \\ Q \end{bmatrix} = \begin{bmatrix} F(t) \\ 0 \end{bmatrix}. \tag{16}$$

The charge Q is directly coupled with mechanical displacement q. During the switch open period, the charge Q changes according to

$$Q(t) = -\frac{\alpha}{C_p}\frac{1}{\frac{1}{C_p}+\frac{1}{C_e}}q(t) + C = -\alpha\frac{\delta}{1+\delta}q(t) + C, \tag{17}$$

where C is the offset of charge signal, which still has to be determined. For a harmonically excited system, it can be calculated by assuming that the voltage signal is periodic with the same period time. It is therefore sufficient to consider one half period time of excitation only. Every half period consists of a period of open switch and of closed switch. Here the magnitude change of switch open circuit and switch closed circuit are defined respectively as ΔQ_{open} and ΔQ_{close}. The steady state is then characterized by

$$\Delta Q_{open} + \Delta Q_{close} = 0. \tag{18}$$

This equation basically means that the voltage is the same after each period time. In the case of harmonic excitation with an amplitude of $\hat{q}$ and excitation frequency of Ω, the mechanical displacement, velocity and acceleration can be expressed as

$$q(t) = \hat{q}\cos(\Omega t), \quad \dot{q} = -\hat{q}\Omega\sin\Omega t, \quad \ddot{q} = -\hat{q}\Omega^2\sin\Omega t. \tag{19}$$

The change of charge ΔQ_{open} is proportional to the change in displacement during switch open period. With the absolute charge value after inversion (which is the initial condition of the closed switch period) termed Q_0, and the absolute value before inversion Q^*, we can write

$$\Delta Q_{\text{open}} = Q_0 - Q^* = -2\alpha\frac{\delta}{1+\delta}\hat{q},$$
$$Q_0 = \frac{1}{2}\Delta Q_{\text{open}} + C = -\alpha\frac{\delta}{1+\delta}\hat{q} + C. \tag{20}$$

Compared to the mechanical periodic time, the electric periodic time is normally very short. Additionally, the switching occurs at the times when the deformation q is maximized, which means that the velocity is zero. It is demonstrated in [14], that it is therefore feasible to neglect the change of mechanical signals during the time the switch is closed. Thus we can approximate the right side of (15) with the following terms: $q(t) = \hat{q}\cos(\Omega t) \approx \hat{q}$, $\dot{q} = -\hat{q}\Omega\sin(\Omega t) \approx 0$, $\ddot{q} = -\hat{q}\Omega^2\cos\Omega t \approx -\hat{q}\Omega^2$. As a result, the right side of the differential equation becomes a constant,

$$(1+\delta)L\ddot{Q} + (1+\delta)R\dot{Q} + \frac{1}{C_p}Q = \alpha\hat{q}(\delta L\Omega^2 - \frac{1}{C_p}). \tag{21}$$

The solution of (21) is the superposition of the general solution and the particular solution. The particular solution can be obtained with the Duhamel integral. After some mathematical calculations, the value of charge at τ^*, which is the moment of opening the switch, is obtained as

$$Q(\tau^*) = -e^{-\pi\zeta}Q_0 - \alpha\hat{q}(1 + e^{-\pi\zeta}), \quad Q^* = |Q(\tau^*)|. \tag{22}$$

The difference between Q^* and Q_0 is the magnitude change of charge for closed switch ΔQ_{close}. Combining all results, the stationary value of charge Q_0 and the constant component C are obtained as

$$Q_0 = \alpha\hat{q}(\frac{1+e^{-\pi\zeta}}{1-e^{-\pi\zeta}} - \frac{2\delta}{1+\delta}\frac{1}{1-e^{-\pi\zeta}}), \quad C = \frac{1}{1+\delta}\alpha\hat{q}\frac{1+e^{-\pi\zeta}}{1-e^{-\pi\zeta}}. \tag{23}$$

The results for C and Q_0 are the absolute values, their signs periodically change so that they are always in antiphase with the velocity $\dot{q}$. Further on, this result can be approximated for low damping $\zeta \ll 1$,

$$C \approx \frac{1}{1+\delta}\alpha\hat{q}\frac{2}{\pi\zeta} = (1+\delta)^{-\frac{3}{2}}\alpha\hat{q}\frac{2}{\pi\zeta_0}. \tag{24}$$

Equation (24) demonstrates that the stationary charge is increased for $\delta < 0$, which means that only a negative capacitance increases the charge buildup. Especially when δ approaches -1, the constant C is theoretically infinity. The negative capacitance is an active analog circuit, so in practice the stationary charge cannot be infinitely high due to the limited maximal output of the operational amplifier. Additionally, the overall capacitance has to be positive in order to keep the electrical network stable. Therefore, the theoretical available range of the negative capacitance is the same as for the *LRC* shunt circuit

$$-C_p < C_e < 0 \quad \text{or} \quad -1 < \delta < 0. \tag{25}$$

The time signals of the SSDI and the SSDNCI with different capacitance ratios are given in Figure 4. For a clear illustration of the switching times t_{cl} and t_{op}, the inversion of charge does not occur instantaneously, as it is assumed in the calculations. Obviously, a larger negative capacitance increases the charge amplitudes as compared to the SSDI technique ($\delta = 0$).

Finally, the dissipated energy E_{diss} per vibration period, which is a measure of the damping

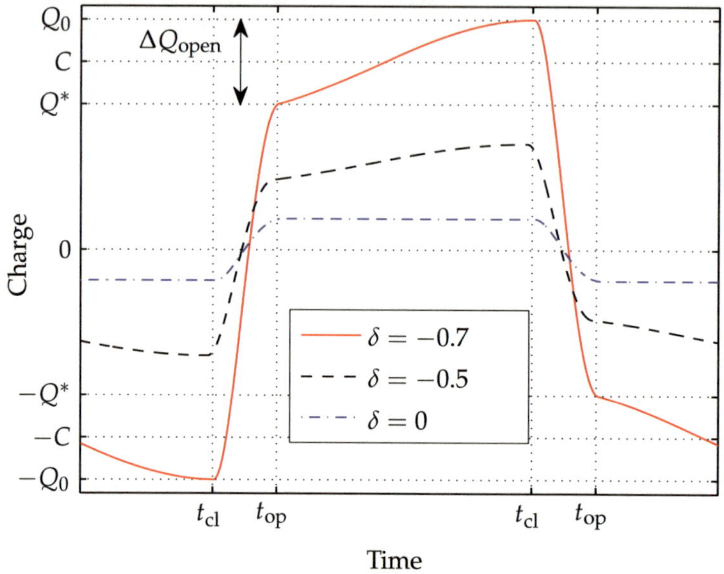

Figure 4. Time signals of the electrical charge for different capacitance values δ.

performance, can be obtained by integrating the product of piezoelectric force and mechanical velocity over a mechanical period time T_{mech},

$$E_{\text{diss}} = -\alpha \int_{t^*}^{t^*+T_{\text{mech}}} u_{\text{p}}(t)\dot{q}(t)\,\mathrm{d}t. \tag{26}$$

When the charge inversion occurs nearly instantaneously, it is sufficient to consider the time with open switch only. With above results the piezovoltage can be obtained as

$$u_{\text{p}}(t) = \frac{\alpha}{C_{\text{p}}} q(t) + \frac{Q(t)}{C_{\text{p}}} = \frac{\alpha}{C_{\text{p}}} \frac{1}{1+\delta} q(t) + \frac{C}{C_{\text{p}}}. \tag{27}$$

Inserting (27) into (26), the expression of dissipated energy is rewritten as

$$E_{\text{diss}} = -\frac{\alpha}{C_{\text{p}}} \int_{t^*}^{t^*+T_{\text{mech}}} \left(\alpha \frac{1}{1+\delta} q(t)\dot{q}(t) + C\dot{q} \right) \mathrm{d}t = -\frac{\alpha}{C_{\text{p}}} \int_{t^*}^{t^*+T_{\text{mech}}} C\dot{q}\,\mathrm{d}t. \tag{28}$$

As it is shown in (28), the amount of dissipated energy only depends on the charge offset C. Therefore the aim in the design of the nonlinear shunt network is to maximize the offset of the charge. Another way to illustrate the damping performance is the hystereisis cycle, in which the piezoelectric voltage or force is drawn versus the deformation. Periodic vibrations are characterized by closed loops, and the energy dissipation is proportional to the enclosed area. Fig. 5 depicts the hysteresis loops for the standard SSDI ($\delta = 0$) and the SSDNCI with two different capacitance ratios. The voltage amplitude immediately before inversion is maximal, $\pm\hat{u}_{\text{p}}$, and after inversion, $\mp\hat{u}_{\text{p}}\mathrm{e}^{-\pi\zeta}$. For the case of an instantaneous voltage inversion, the hysteresis cycles are parallelograms. The slope of these lines is proportional to the force factor

α. However, for the extension of the area, only the voltage amplitude, i.e. the charge offset, is relevant. Clearly, a negative capacitance has a positive effect in both states, therefore resulting in a higher charge offset. Inserting (24) into (28) we can get the expression of the dissipated

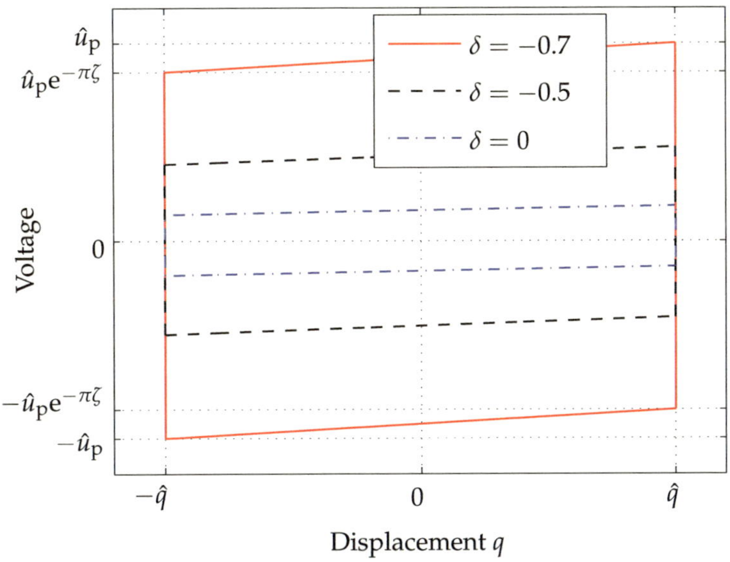

Figure 5. Hysteresis cycles for different capacitance values δ.

energy per period,

$$E_\mathrm{diss} = 4\frac{\alpha^2}{C_\mathrm{p}}\hat{q}^2\frac{1+\mathrm{e}^{-\pi\zeta}}{1-\mathrm{e}^{-\pi\zeta}}\frac{1}{1+\delta}. \tag{29}$$

The increase in dissipated energy has the same trend as for the charge offset. Comparing with SSDI shunt, the dissipated energy is scaled by $(1+\delta)^{-\frac{3}{2}}$. For a linear LRC shunt, the dissipated energy is scaled by $1/(1+\delta)$, see also in [15].

5. Optimized switching law for bimodal excitation

The assuption of a harmonic excitation is not valid for all situations. In many cases, the signal also contains additional frequencies. In order to discuss the influence of more general excitations, in the following a bimodal excitation is considered, which contains two frequencies Ω_1 and Ω_2 with $\Omega_2 > \Omega_1$,

$$q(t) = \hat{q}_1\cos(\Omega_1 t) + \hat{q}_2\cos(\Omega_2 t + \varphi). \tag{30}$$

Both signals have in general different amplitudes and a phase shift between them.

It is obvious that the standard switching law, which means switching at the maxima of the first mode, does not yield optimal results anymore. One can show that - using the standard switching law - the dissipated energy per vibration period is exactly the same as

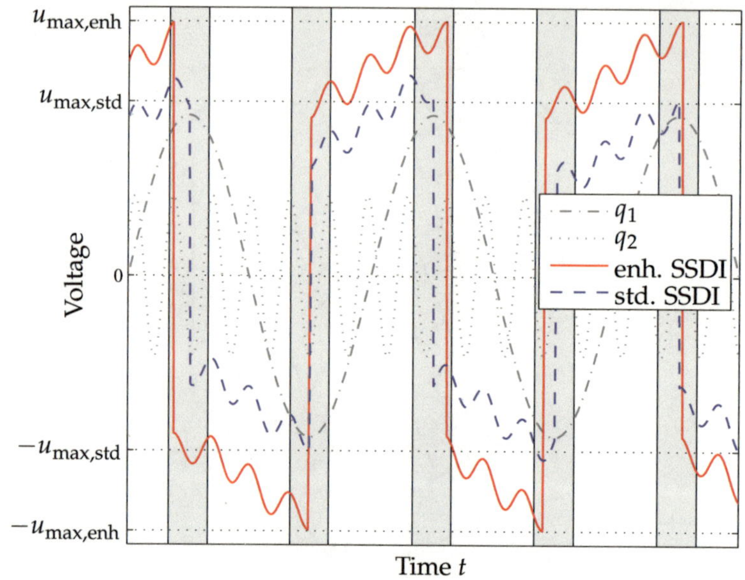

Figure 6. Time signals for standard and enhanced SSDI with bimodal excitation.

for a monoharmonic excitation with frequency Ω_1 only,

$$E_{\text{diss}} = 4\frac{\alpha^2}{C_\text{p}}\hat{q_1}^2\frac{1+\text{e}^{-\pi\zeta}}{1-\text{e}^{-\pi\zeta}}, \tag{31}$$

which is the result for the SSDNCI circuit with $\delta = 0$. Therefore more sophisticated switching laws have been developed, which target to extract energy from the higher frequency oscillations and use it to increase the damping of the main mode [17].

The new switching law described in the following is defined according to these positions:

- A modal observer reconstructs both vibrations in the first and second frequency of the excitation.
- A timeframe $-T_2/2 < t < T_2/2$ around each first mode extremum is defined, where T_2 is the period time of the second vibration mode. This assures that exactly one maximum and one minimum of the second mode is located within this timeframe.
- The switching is triggered at the moments of the second mode extremum. It the timeframe is defined around a maximum of the first mode, then it is triggered at the second mode maximum, if it is defined around a minimum of the first mode, then it is triggered at the second mode minimum within this timeframe.

For such a switching law it is assured that the voltage induced by the second mode is added to the value caused by the first mode. Figure 6 shows a comparison of the standard and the enhanced switching law for a biharmonic excitation. The higher frequency is recognizable

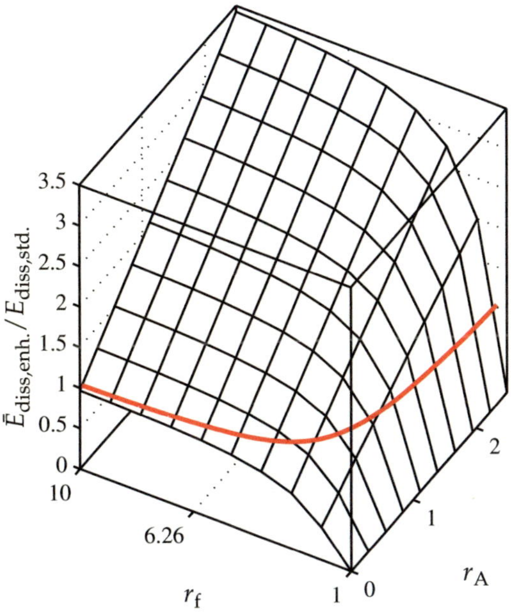

Figure 7. Amplification of dissipated energy with enhanced SSDI technique versus amplitude ratio r_A and frequency ratio r_f.

in the high frequency oscillations during the open switch phases. One can realize that in the standard switching law the switching always occurs exactly during the first mode extrema. At these moments, the voltage at the piezoceramics might be increased or decreased by the influence of the higher frequency, so that in mean this effect cancels out. With the enhanced switching law, the switch is always triggered when the second mode is maximum and augments therefore the voltage buildup. However, the switching is no longer occuring in phase with the first mode velocity, which reduces slightly the energy dissipation. For more details the reader is referred to [12].

Obviously the increase in energy dissipation grows with the second mode amplitude. But also the frequency ratio $r_f = \Omega_2/\Omega_1$ between the first and second mode has an influence. The higher the second frequency, the smaller is the period time T_2 and therefore the timeframe. This means that the second mode maximum is in average closer to the first mode maximum, which is ideal for the energy dissipation. Figure 7 shows the amplification of energy dissipation versus the frequency ratio r_f and the amplitude ratio $r_A = \hat{q}_2/\hat{q}_1$. It can be concluded that for a given frequency ratio r_f (this ratio is approximately $2\pi = 6.26$ for the clamped beam), the energy dissipation grows linearly with the second mode amplitude. Additionally, the energy dissipation grows with a higher second frequency. Theoretically, for very low second mode amplitude, this enhanced switching law actually might give less damping than the standard law (the borderline is marked by a red line). This is due to the non-optimal phase shift of the switching signal, which is not in exact antiphase with the first mode velocity anymore. But these regions are practically not very relevant.

6. Technical applications

After discussing the performance of various shunt damping techniques, in the following section two technical systems, namely a squealing disc brake and a bladed disc, are investigated as potential applications for piezoelectric shunt damping.

6.1 Brake squeal

Brake noise that is dominated by frequencies above 1 kHz is usually called 'brake squeal'. It is widely accepted that brake squeal is caused by friction induced vibrations. A friction characteristic that is decreasing with relative velocity results in an energy input and can excite vibrations. Other works explain the instability with nonconservative restoring forces [6, 18]. This mechanism does not need the assumption of a decreasing friction characteristic, and it is not depending on certain damping properties. Although the brake function itself is not affected by these vibrations, the generated noise marks a significant comfort problem. Brake squeal remains unpredictable, even state-of-the-art FE analyses cannot cope with the complexity of the problem. Therefore, brake manufacturers typically reduce the tendency to squeal in a time consuming process of designing, building and testing of prototypes in a mostly empirical way.

Recently, the use of piezoceramics has been investigated for the suppression of brake squeal [22] in an active feedback control. The authors succeeded in controlling the squaling, however this method requires sensing electronics, complex amplifiers and a power supply. Therefore, this technology is expensive and unsuitable for many applications like automotive brakes. Piezoelectric shunt damping for brake squeal control might be a cheaper alternative.

6.1.1 Brake prototype and stability analysis

Before designing the shunt damping network, the stability of the brake is studied using a multibody system, as shown in Figure 8. This model has been introduced in [13] to simulate

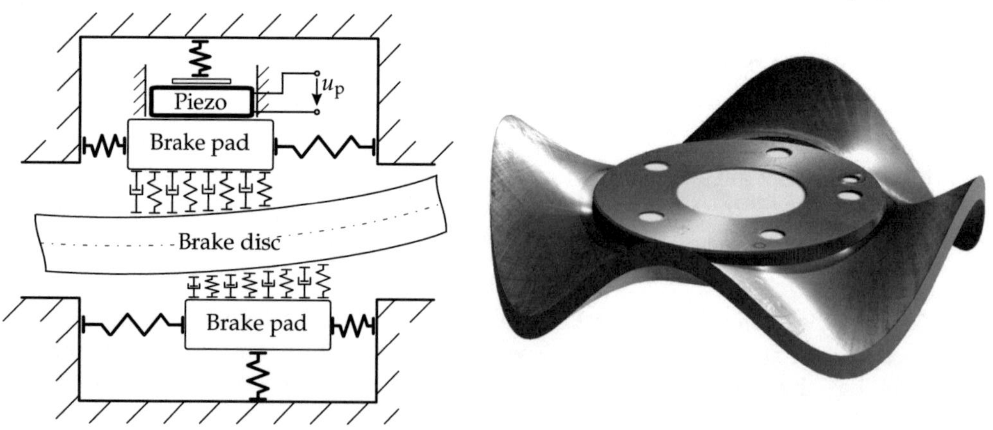

Figure 8. Brake model and disc eigenform.

the efficiency of linear *LR* and *LRC* shunts as well as a feedback control for brake squeal suppression. The two brake pads are modelled as rigid bodies and the contact area is

represented as a layer with distributed stiffness and damping properties. Both pads have two translational degrees of freedom (out-of-plane and in-plane direction) and stay in contact with the brake disc. The coefficient of friction μ between disc and pads is assumed to be constant. The brake disc is described as an annular disc according to the Kirchhoff plate theory. Only the mode with four nodal diameter and one nodal circle is considered, this mode is depicted in Figure 8, as the corresponding frequency agrees best with the squealing frequency. The rotation of the disc introduces gyroscopic terms. Further more, the brake model contains nonconservative restoring forces as a result of the friction forces in the contact area between the pads and the disc. These forces can be identified in the unsymmetric stiffness matrix. Because of these forces, the mechanical model is possibly unstable. This can be shown by a complex eigenvalue analysis, as reported in Figure 9. The stability of the brake system

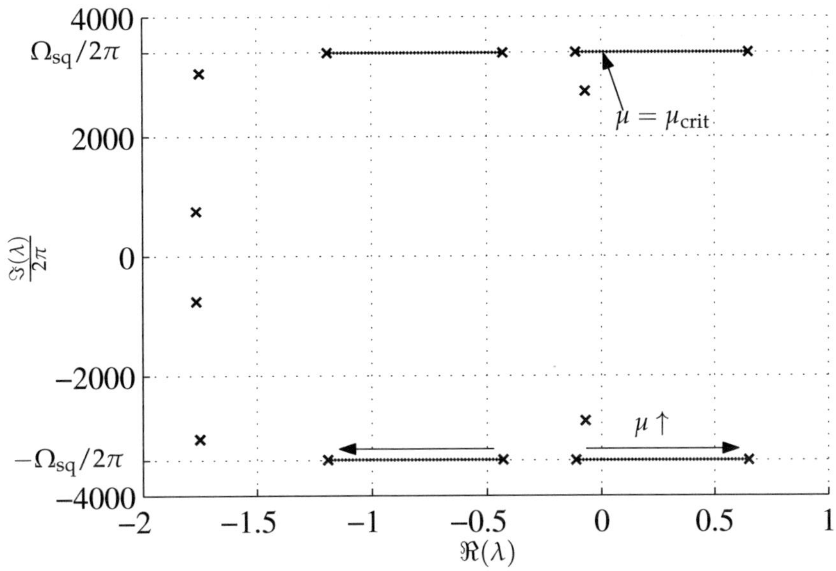

Figure 9. Imaginary part versus real part of the eigenvalues of the uncontrolled brake.

is determined by the largest real part of the eigenvalues, termed λ_{max}. A variation of the coefficient of friction μ shows the influence of the nonconservative restoring forces. Without friction forces, $\mu = 0$, the brake is asymptotically stable, as λ_{max} is negative. With increasing friction, two pairs of eigenvalues move in opposite direction. The system becomes unstable above a critical friction force μ_{crit} with $\lambda_{max}(\mu = \mu_{crit}) = 0$. The imaginary part corresponds to the squealing frequency, and is termed Ω_{sq}.

Figure 10 shows the prototype disc brake at the Institute of Dynamics and Vibration Research with three piezoelectric stack actuators. Their forces act in the same direction as the brake pressure so that the out-of-plane vibrations of the brake disc can be influenced. The piezoceramics are placed between the inboard brake pad and the brake piston and protected by a cap construction against shear forces and debris. Other publications propose a similar placement of the actuators, for example the 'smart pads' [23] which include the piezoceramics directly into the back side of the brake pads. Another possibility is to place the actuators within the brake piston [1]. Three piezoelectric stack actuators with circular cross section

Figure 10. Prototype disc brake with embedded piezoceramics.

and material FPM 231 from company MARCO are used. They are designed to withstand brake pressures exceeding 30 bar and temperatures up to 200°C. This is certainly not enough for typical temperatures during strong brakings, but enough for principal feasibility studies in the lab. It is possible to connect all piezoceramics with *LR*- or *LRC*-shunts. When the SSDI-technique is used, one of the ceramics (typically the middle one) is used as a sensor and the remaining two are shunted.

6.1.2 Modeling of the combined system and control of brake squeal

The tuning of the resonant *LR*- and *LRC*-shunts is done like it is described in [16] for an assumed squealing frequency of $f_{sq} \approx 3400$Hz. The results for a passive *LR* and two negative capacitance shunts with different capacitance ratios δ are shown in Figure 11. The maximum real part λ_{max} is given versus the squealing frequency f_{sq}. The squealing frequency of the brake model is artificially changed by multiplicating the stiffness matrix by a constant term, which results in a change of all eigenfrequencies of the system.

All three networks are capable to stabilize the brake when tuned precisely, as λ_{max} is negative. However, the frequency bandwidth in which the brake is stable is very narrow for the passive *LR*-shunt. Practically this frequency range is not enough for a robust suppression of the brake squealing, as it might occur in a broad range due to the many possible eigenfrequencies of the brake. As expected, the negative capacitance networks perform better. The maximum reduction of λ_{max} is equal to that achievable with *LR*-networks, but this occurs in a broader frequency range. The closer the capacitance value is tuned to -1, the better the performance results.

6.1.3 Measurements on the brake test rig

Measurements are conducted on the brake test rig with the modified brake using the following procedure to experimentally determine the frequency bandwidth of the damping effect: The passive *LR* or active *LRC* shunt is disconnected from the piezoceramics, and the brake

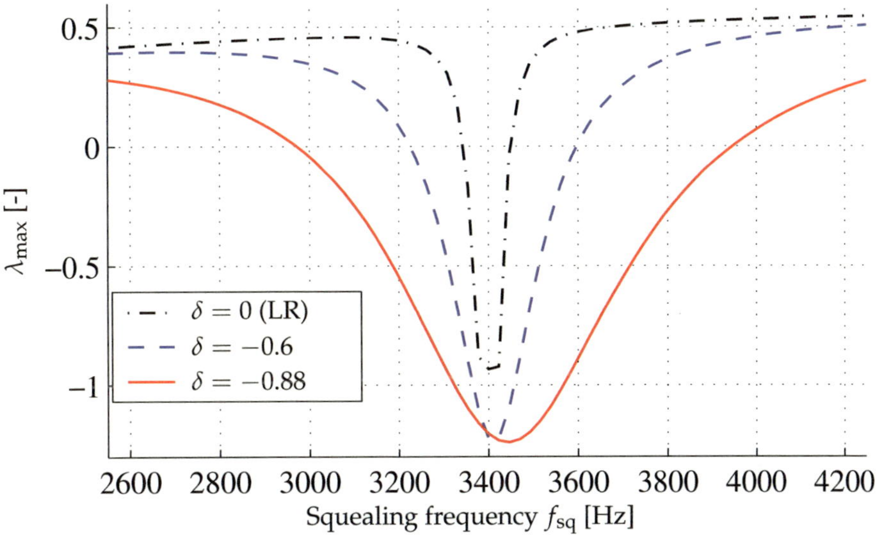

Figure 11. Stability of the brake model for LR-shunt ($\delta = 0$) and LRC-shunts with $\delta_1 = -0.6$ and $\delta_2 = -0.88$.

pressure and disc speed is varied until a proper and steady squealing arises. During the tests, this usually happens for pressures between 8 and 15 bar and velocities of 23 rpm of the beake disc. The squealing frequency could be located at approx. 3400Hz. After this, the shunt is connected to the electrodes, and the inductance and resistance are set to the calculated optimum values. Afterwards, the inductance value is reduced until the damping effect is vanishes, as the network is too strongly mistuned. This is the initial value of the inductance at the beginning of each measurement.

During the measurements, the shunt is periodically connected and disconnected for 10 seconds. After each cycle, the inductance is increased so that in the following 10 seconds of connection the shunt is tuned to a constant, new frequency. In the first half of each measurement, the electrical resonance frequency is successively tuned closer to the squealing frequency and the damping effect grows. In the middle of the measurement, the shunt is tuned nearly perfectly, and the effect is maximized. In the second half, the mistuning grows again as the inductance value is further increased, and the damping effect is diminished. The measurement is stopped when no squealing reduction is noticeable anymore. This procedure is repeated for different LR and LRC-shunts.

During the measurements, the sound pressure is recorded with a microphone, which is located in a distance of 50 cm from the brake. In the upper plot of Figure 12 the sound pressure is given versus the time for one exemplary measurement. In the lower plot, the corresponding sound pressure level (SPL) and the inductance values are shown. As shown, during the measurement time of more than 3 minutes, the SPL of the squealing brake remained nearly constant within 95-100 dB. In the very first and last switchings between connection and disconnection of the shunt, nearly no reduction in the SPL is noticed, as the mistuning is too strong. In the middle of the measurement the squealing stops immediately after connecting the shunt and starts again after disconnecting. The remaining sound without the squealing is environmental noise,

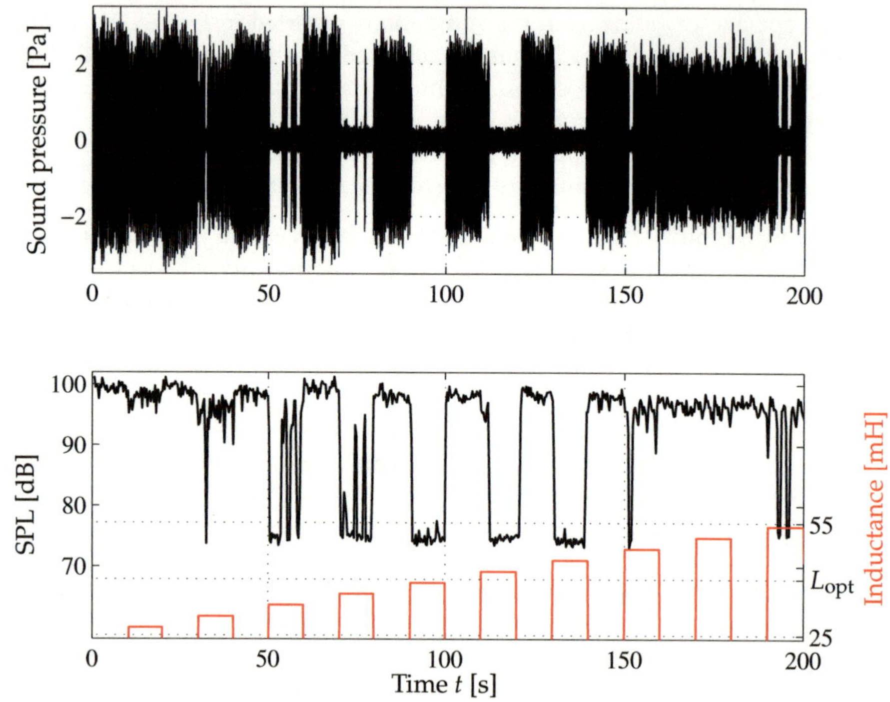

Figure 12. Sound pressure and SPL during one measurement with stepwise varied inductance.

which has been measured as high as 75 dB, and is dominated by the sound of the electric motor that drives the brake disc.

The performance of the shunted piezoceramics is evaluated by the reduction of the mean SPL during each 10 seconds of connection and disconnection for every inductance value. In Figure 13 this reduction is given versus the indunctance (normalized to the optimal value). The figure shows the results for the passive LR shunt ($\delta = 0$) and two different LRC shunts

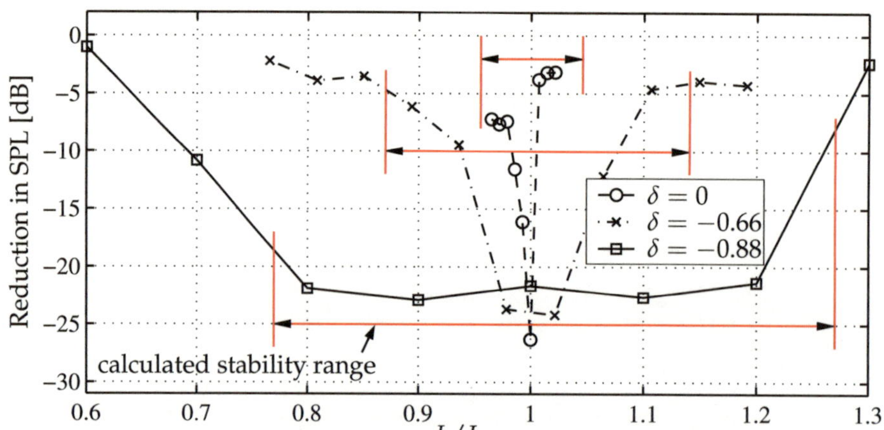

Figure 13. Reduction in SPL versus inductance tuning for LR- and LRC-shunts.

with the same capacitance ratios as in the simulations reported in Figure 11. It can be seen that the maximum reduction for each shunt is achieved for the perfectly tuned shunts ($L \approx L_{\text{opt}}$ respectively $\eta \approx 1$). In these cases, all shunts - including the passive LR shunt - are capable to suppress the squealing totally, as predicted by the simulations. The differences in the maximum reduction can be explained by different strength of the squealing. Naturally, a weak squealing delimits the maximum possible reduction compared to a strong squealing.

From the inductance ratio L/L_{opt}, the frequency ratio between the electrical eigenfrequency and the squealing frequency can be re-calculated. Defining the state 'silent' and 'squealing' by an arbitrary threshold of 12 dB SPL-reduction, the brake is stabilized in a range of $\Delta f = 40$Hz for the passive LR shunt. With actice LRC-shunts, the stabilized range covers $\Delta f = 212$Hz with $\delta = -0.66$ and $\Delta f = 950$Hz with $\delta = -0.88$. These results show a good accordance with the simulation results in Figure 11. However, some influences like the heating up of the piezoceramics lead to a reduction of the piezoelectric effect so that the performance at the end of each measurement is slightly lower than in the beginning.

6.2 Damping of turbine blades

Another application is the vibration damping of turbine blades. Here the excitation comes from high static and dynamic loads. Static loads are due to centrifugal forces and thermal strains while fluctuating gas forces are the cause of dynamic excitation which can lead to High Cycle Fatigue (HCF) failures. As the material damping is extremely low, any further damping provided to the structure is desireable. Coupling devices like underplatform dampers, lacing wires and tip shrouds are common in turbomachinery applications [19, 20]. The effectiveness of these damping concepts is limited to the relative vibrations of neighbouring blades and therefore they are often only efficient for specific engine speeds and mode shapes. Furthermore, the aerodynamics of the blades is influenced by these coupling devices.

In the following, the damping of turbine blades by shunted piezoceramics is studied with a bladed disc model (BLISC), depicted in Figure 14, which has been introduced by Hohl [8]. Each

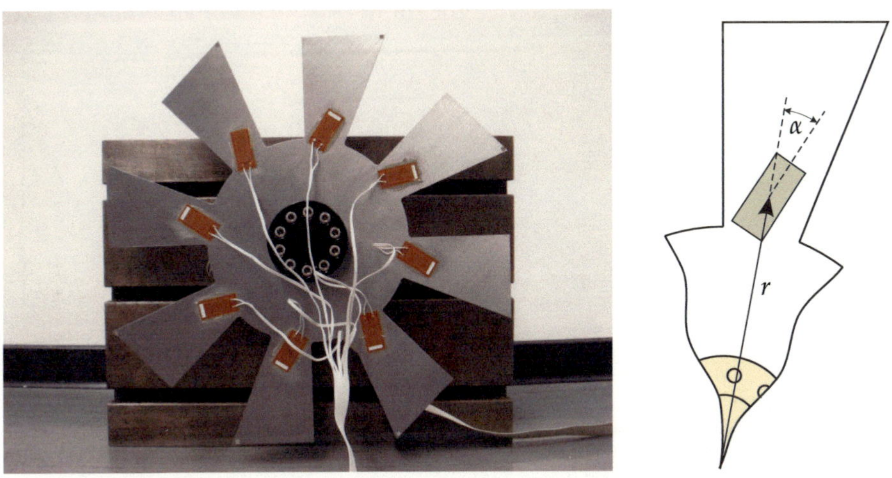

Figure 14. Photography and sketch of the BLISC test rig with attached piezoceramics.

blade is equipped with a MACRO FIBER COMPOSITE piezoceramics M2814 P1 from MARCO company for vibration damping.

6.2.1 Optimizing the location of the piezoceramics

The intention of this study was to optimize the placement of the piezoceramics within the structure. As the geometry is too complex for an analytical description, it is modeled by Finite Elements in Ansys using 3-D 20-Node structural solid elements (solid186) and a 3-D 20-node coupled-flied solid (solid226) for the piezoelectric material. Subsequently a modal reduction is performed. The location of every piezoceramics is described by the radius r and the orientation α, which have to be optimized, with the generalized coupling coefficient K taken as a measure of the coupling. This factor can be calculated by the system' eigenfrequencies with isolated and short circuit electrodes of the piezoceramics, which are both determined within the FE program. Generally, the coupling with the individual eigenforms of the system differ from each other. In Figure 15 the coupling coefficients for the first bending and first torsion mode of the blades are given versus α and r. For the bending mode, the piezoceramics should

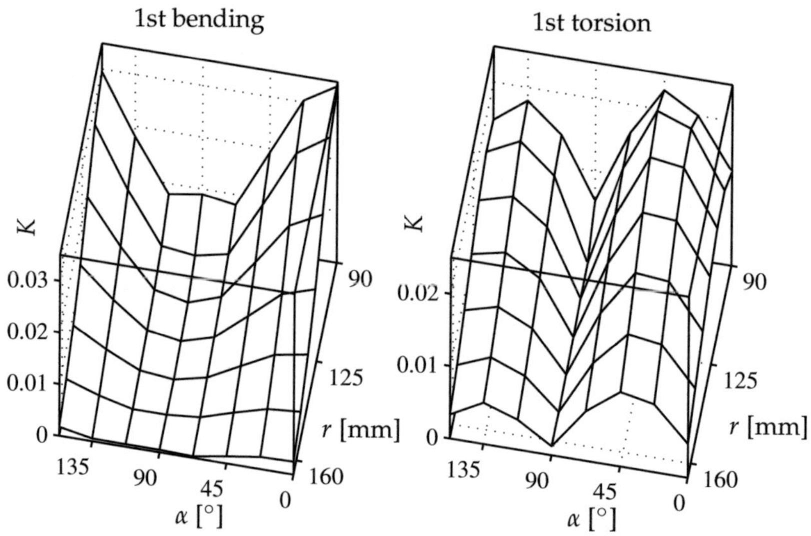

Figure 15. Generalized coupling coefficient K for the first bending and torsion modes versus the location of the piezoceramics.

be placed close to the clamped ending of the blade at $r = 90$mm, which is approximately the radius of the disc. This can be explained by the bending moment, which is maximized at this position. The bending moment reduces to zero at the free end of the blade, therefore also the coupling reduces in that direction. The dependency with the orientation α is nearly symmetric: the coupling is maximal when the piezoceramics is facing in radial direction ($\alpha = 0°$ or $\alpha = 180°$) and minimal for $\alpha = 90°$. The resulting maximum coupling is $K \approx 3.5\%$.

For the torsion mode, the optimal radius is similar, yet slightly larger than for the bending mode. However, the orientation is oppositional to the bending case: the best coupling results for $\alpha = 45°$, while it is nearly zero for $\alpha = 0°$ and $\alpha = 90°$. The maximum coupling with the torsion mode is $K \approx 2.25\%$ and thus smaller than for the bending.

Therefore, for the overall optimal location a trade-off is necessary, and the piezoceramics is placed with $r = 97.5$mm and $\alpha = 22.5°$. In this case the coupling with both the bending and torsion mode is about $K = 2\%$.

6.2.2 Measurements

Finally, measurements are conducted with the BLISC test rig. The system is excited harmonically by additional piezoceramics placed at the back side at identical positions as the shunted ones at the front side. One single passive *LR* network is connected to all piezoceramics simultaneously, and the electrical eigenfrequency and the damping ratio are set to the optimal values according to the previous sections. Figure 16 shows the measurement as well as the simulation results for isolated electrodes and optimal *LR*-shunting. Generally, the

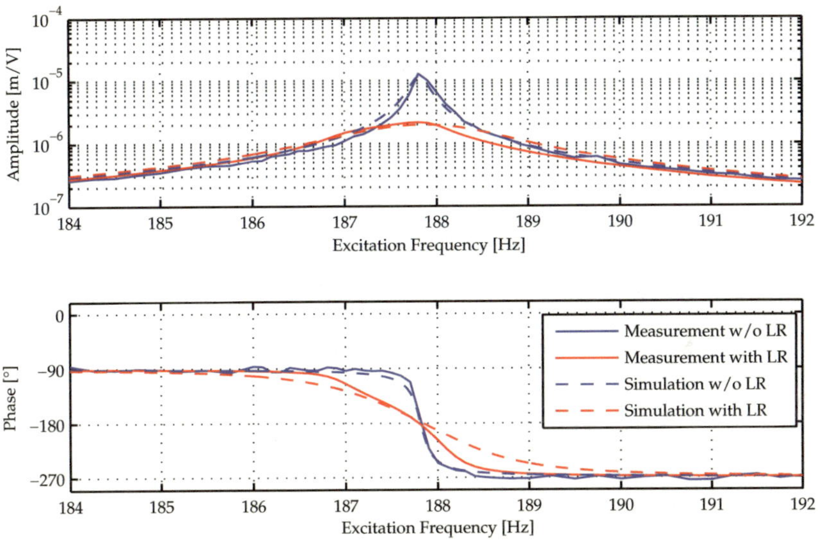

Figure 16. Simulated and measured frequency response of the BLISC model for isolated electrodes and *LR*-shunting.

simulation results are in very good agreement with the measured ones. The damping effect of the shunted piezoceramics is clearly visible.

7. Conclusions

This chapter deals with shunted piezoceramics for vibration damping. A small overview of typical shuntings is presented. Further on, a general model of a one degree of freedom mechanical oscillator with embedded piezoceramics and external electrical circuit is derived.

Based on this system, the optimal tuning of a resonant *LR*-shunt is performed for a damped mechanical system. The influence of the mechanical damping upon the optimal parameters and the resulting damping performance is studied. Further on, a novel combination of a 'SSDI' switching circuit and a negative capacitance is discussed. It is shown that this network

inherits the adaptive structure of the SSDI technique and combines it with the enhanced performance of a negative capacitance. An enhanced switching law for bimodal excited systems is presented as well. With this technique, the damping of the main mode can be maximized using the vibration energy stored in the higher mode.

Finally, a squealing disc brake and a bladed disc are introduced as two technical applications for piezoelectric shunt damping. For both cases the vibration behavior is studied by mechanical replacement models, and the location of the piezoceramics and the electrical shuntings are chosen based on these models. In both cases it is possible to control the vibrations and increase significanly the damping of the structure. Measurements are conducted which validate the theoretical models.

Author details

Marcus Neubauer, Sebastian M. Schwarzendahl and Xu Han
Institute of Dynamics and Vibration Research, Leibniz University Hannover, Germany

8. References

[1] Cunefare, K. A. & Graf, A. J. [2002]. Experimental active control of automotive disc brake rotor squeal using dither, *Journal of Sound Vibration* 250: 579–590.

[2] Fleming, A. J., Behrens, S. & Moheimani, S. O. R. [2003]. Reducing the inductance requirements of piezoelectric shunt damping systems, *Smart Material Structures* 12: 57–64.

[3] Forward, R. L. [1979a]. Electromechanical transducer-coupled mechanical structure with negative capacitance compensation circuit.
URL: *http://www.freepatentsonline.com/4158787.html*

[4] Forward, R. L. [1979b]. Electronic damping of vibrations in optical structures, *Applied Optics* 18: 690–697.

[5] Fukada, E., Date, M., Kimura, K., Okubo, T., Kodama, H., Mokry, P. & Yamamoto, K. [Apr 2004]. Sound isolation by piezoelectric polymer films connected to negative capacitance circuits, *Dielectrics and Electrical Insulation, IEEE Transactions on [see also Electrical Insulation, IEEE Transactions on]* 11(2): 328–333.

[6] Hagedorn, P. [2003]. Modeling disk brakes with respect to squeal, *COBEM*.

[7] Hagood, N. W. & von Flotow, A. [1991]. Damping of structural vibrations with piezoelectric materials and passive electrical networks, *Journal of Sound Vibration* 146: 243–268.

[8] Hohl, A., Neubauer, M., Schwarzendahl, S. M., Panning, L. & Wallaschek, J. [2009]. Active and semiactive vibration damping of turbine blades with piezoceramics, Vol. 7288, SPIE, p. 72881H.
URL: *http://link.aip.org/link/?PSI/7288/72881H/1*

[9] Hollkamp, J. J. [1993]. Multimodal passive vibration suppression with piezoelectrics, *in* R. B. Malla & B. B. Nalluri (eds), *AIAA/ASME/ASCE/AHS/ASC Structures, Structural Dynamics, and Materials Conference, 34th and AIAA/ASME Adaptive Structures Forum, La Jolla, CA, Apr. 19-22, 1993, Technical Papers. Pt. 6 (A93-33876 13-39), p. 3227-3237.*, pp. 3227–3237.

[10] Lefeuvre, E., Badel, A., Petit, L., Richard, C. & Guyomar, D. [2006]. Semi-passive Piezoelectric Structural Damping by Synchronized Switching on Voltage Sources, *Journal of Intelligent Material Systems and Structures* 17(8-9): 653–660.
URL: *http://jim.sagepub.com/cgi/content/abstract/17/8-9/653*

[11] Mokrani, B., Rodrigues, G., Ioan, B., Bastaits, R. & Preumont, A. [2012]. Synchronized switch damping on inductor and negative capacitance, *Journal of Intelligent Material Systems and Structures* .

[12] Neubauer, M., Han, X. & Schwarzendahl, S. M. [2011]. Enhanced switching law for synchronized switch damping on inductor with bimodal excitation, *Journal of Sound and Vibration* 330(12): 2707 – 2720.
URL: *http://www.sciencedirect.com/science/article/B6WM3-520V42F-3/2/9fea46dd56c528bf824c8190fb0d1b95*

[13] Neubauer, M. & Kroeger, M. [2006]. Suppression of brake squeal using piezoceramics, *Braking 2006: International Conference on Vehicle Braking Technology*, pp. 254–263.

[14] Neubauer, M. & Wallaschek, J. [2008]. Analytical and experimental investigation of the frequency ratio and switching law for piezoelectric switching techniques, *Smart Materials and Structures* 17(3): 035003 (9pp).
URL: *http://stacks.iop.org/0964-1726/17/035003*

[15] Neubauer, M. & Wallaschek, J. [2009]. Vibration damping with piezoceramics shunted to negative capacitance networks, *Proceedings of International Conference on Advanced Intelligent Mechatronics, Singapore*, pp. 1100–1105.

[16] Neubauer, M. & Wallaschek, J. [2010]. Vibration Damping with Shunted Piezoceramics: Fundamendals and Technical Applications, Presented at the 6th Int. Conference Mechatronic Systems and Materials (MSM).

[17] Richard, C., Harari, S. & Gaudiller, L. [2009]. Enhanced piezoelectric voltage build-up for semi-active control of smart structures, Vol. 7288, SPIE, p. 72881Y.
URL: *http://link.aip.org/link/?PSI/7288/72881Y/1*

[18] Rudolph, M. & Popp, K. [2001]. Brake squeal, *in* K. Popp (ed.), *Detection, Utilization and Avoidence of Nonlinear Dynamical Effects in Engineering Applications*, Shaker Verlag, pp. 197–225.

[19] Sextro, W. [2000]. The calculation of the forced response of shrouded blades with friction contacts and its experimental verification, ASME Paper 2000-GT-540, Int. Gas Turbine & Aeroeng. Congress & Exh., Munich.

[20] Szwedowicz, J. [1999]. Cyclic Finite Element Modeling of Shrouded Turbine Blades Including Frictional Contacts, ASME Paper 99-GT-92, Int. Gas Turbine & Aeroeng. Congress & Exh., Indianapolis.

[21] Tang, J. & Wang, K. W. [2001]. Active-passive hybrid piezoelectric networks for vibration control: comparisons and improvement, *Smart Material Structures* 10: 794–806.

[22] von Wagner, U., Hochlenert, D., Jearsiripongkul, T. & Hagedorn, P. [2004a]. Active control of brake squeal via 'smart pads'.

[23] von Wagner, U., Hochlenert, D., Jearsiripongkul, T. & Hagedorn, P. [2004b]. Active control of brake squeal via smart pads, *SAE 2004 Transactions Journal of Passenger Cars - Mechanical Systems* pp. 1186–1192.